NONVOLATILE MEMORY TECHNOLOGIES WITH EMPHASIS ON FLASH

IEEE Press
445 Hoes Lane
Piscataway, NJ 08854

IEEE Press Editorial Board
Mohamed E. El-Hawary, *Editor in Chief*

R. Abari	T. Chen	R. J. Herrick
S. Basu	T. G. Croda	S. V. Kartalopoulos
A. Chatterjee	S. Farshchi	M. S. Newman
	B. M. Hammerli	

Kenneth Moore, *Director of IEEE Book and Information Services (BIS)*
Catherine Faduska, *Senior Acquisitions Editor*
Jeanne Audino, *Project Editor*

IEEE Components, Packaging, and Manufacturing Technology Society, Sponsor
CPMT Liaison to IEEE Press, Joe E. Brewer

IEEE Electron Devices Society, Sponsor
EDS Liaison to IEEE Press, Joe E. Brewer

Technical Reviewers
R. Jacob Baker, *Boise State University*
Ashok K. Sharma, *NASA/Goddard Space Flight Center*

NONVOLATILE MEMORY TECHNOLOGIES WITH EMPHASIS ON FLASH

A Comprehensive Guide to Understanding and Using NVM Devices

Edited by

Joe E. Brewer

Manzur Gill

IEEE Press Series on Microelectronic Systems
Stuart K. Tewksbury and Joe E. Brewer, *Series Editors*
IEEE Components, Packaging, and Manufacturing Technology Society, *Sponsor*
IEEE Electron Devices Society, *Sponsor*

IEEE PRESS

WILEY-INTERSCIENCE
A JOHN WILEY & SONS, INC., PUBLICATION

Copyright © 2008 by the Institute of Electrical and Electronics Engineers, Inc.

Published by John Wiley & Sons, Inc., Hoboken, New Jersey. All rights reserved.
Published simultaneously in Canada

No part of this publication may be reproduced, stored in a retrieval system, or transmitted in any form or by any means, electronic, mechanical, photocopying, recording, scanning, or otherwise, except as permitted under Section 107 or 108 of the 1976 United States Copyright Act, without either the prior written permission of the Publisher, or authorization through payment of the appropriate per-copy fee to the Copyright Clearance Center, Inc., 222 Rosewood Drive, Danvers, MA 01923, (978) 750-8400, fax (978)750-4470, or on the web at www.copyright.com. Requests to the Publisher for permission should be addressed to the Permissions Department, John Wiley & Sons, Inc., 111 River Street, Hoboken, NJ 07030, (201) 748-6011, fax (201) 748-6008, or online at http://www.wiley.com/go/permission.

Limit of Liability/Disclaimer of Warranty: While the publisher and author have used their best efforts in preparing this book, they make no representations or warranties with respect to the accuracy or completeness of the contents of this book and specifically disclaim any implied warranties of merchantability of fitness for a particular purpose. No warranty may be created or extended by sales representatives or written sales materials. The advice and strategies contained herein may not be suitable for your situation. You should consult with a professional where appropriate. Neither the publisher nor author shall be liable for any loss of profit or any other commercial damages, including but not limited to special, incidental, consequential, or other damages.

For general information on our other products and services or for technical support, please contact our Customer Care Department within the United States at (800) 762-2974, outside the United States at (317) 572-3993 or fax (317) 572-4002.

Wiley also publishes its books in a variety of electronic formats. Some content that appears in print may not be available in electronic formats. For more information about Wiley products, visit our web site at www.wiley.com

Library of Congress Cataloging-in-Publication Data is available.

ISBN 978-0471-77002-2

Printed in the United States of America

10 9 8 7 6 5 4 3 2 1

CONTENTS

Foreword	xvii
Preface	xxi
Contributors	xxiii

1 INTRODUCTION TO NONVOLATILE MEMORY   1
Joe E. Brewer

1.1	Introduction		1
1.2	Elementary Memory Concepts		2
1.3	Unique Aspects of Nonvolatile Memory		9
	1.3.1	Storage	9
	1.3.2	Storage Mechanisms	12
	1.3.3	Retention	12
	1.3.4	Endurance	13
1.4	Flash Memory and Flash Cell Variations		13
1.5	Semiconductor Device Technology Generations		16
References			18

2 FLASH MEMORY APPLICATIONS   19
Gary Forni, Collin Ong, Christine Rice, Ken McKee, and Ronald J. Bauer

2.1	Introduction		19
	2.1.1	Spectrum of Memory Devices	20
	2.1.2	Evolving from EPROMs	21
	2.1.3	NOR and NAND	22
	2.1.4	Evolution of Flash Usage Models	23
	2.1.5	Understanding Flash Attributes	25
2.2	Code Storage		38
	2.2.1	Execute-in-Place	39
	2.2.2	Store and Download	43
	2.2.3	Contrasting Execute-in-Place Versus Store and Download	45
	2.2.4	Future Code Storage Applications	45
2.3	Data Storage		46
	2.3.1	Why Use Flash to Store Data?	46

	2.3.2	Architectural Decisions	46
	2.3.3	Embedded Flash Storage	49
	2.3.4	Removable Media	50
2.4	Code+Data Storage		54
	2.4.1	Relevant Attributes for Code+Data	55
	2.4.2	Fitting the Pieces Together for Code+Data	58
	2.4.3	Benefits of Code+Data	61
2.5	Conclusion		62

3 MEMORY CIRCUIT TECHNOLOGIES — 63
Giulio G. Marotta, Giovanni Naso, and Giuseppe Savarese

3.1	Introduction		63
3.2	Flash Cell Basic Operation		63
	3.2.1	Cell Programming	64
	3.2.2	Cell Erase	64
	3.2.3	Compaction	65
	3.2.4	Read	65
3.3	Flash Memory Architecture		66
	3.3.1	Memory Cell Array	69
	3.3.2	Analog Blocks	71
	3.3.3	Control Logic	73
3.4	Redundancy		75
	3.4.1	Defectivity and Process Variations	75
	3.4.2	Yield Improvement	75
	3.4.3	Yield Simulator	77
	3.4.4	Redundancy Fuses Design	78
	3.4.5	Row Redundancy Design	79
	3.4.6	Column Redundancy Design	80
	3.4.7	Advanced Redundancy Design	81
3.5	Error Correction Coding (ECC)		87
	3.5.1	On-Chip ECC and Endurance/Retention in Flash Memories	88
	3.5.2	On-Chip ECC and Multilevel Flash Memories	89
3.6	Design for Testability (DFT)		89
	3.6.1	Test Entry and Organization	91
	3.6.2	Fuse Cell	92
	3.6.3	Sense Amplifier Reference Trimming and Monitor	93
	3.6.4	High Voltages Trimming	94
	3.6.5	Timings Trimming and External Control	96
	3.6.6	Internal State Machine Algorithm Skips and Monitor	97
	3.6.7	Address Path Configuration	98
	3.6.8	Data Path Configuration and Trimming	99

		3.6.9	High Voltages External Forcing and Monitor	101
		3.6.10	Array Direct Access and Stresses	103
		3.6.11	Internal Pattern Write and Verify	105
		3.6.12	Data Compression	106
	3.7	Flash-Specific Circuit Techniques		108
		3.7.1	Voltage Level Shifting	109
		3.7.2	Sensing	112
		3.7.3	Voltage Multiplication	114
		3.7.4	Reference Voltage Generation	118
		3.7.5	Voltage Regulation	119
		3.7.6	I/O Signal Buffering	122
	References			123

4 PHYSICS OF FLASH MEMORIES — 129

J. Van Houdt, R. Degraeve, G. Groeseneken, and H. E. Maes

4.1	Introduction		129
4.2	Basic Operating Principles and Memory Characteristics		130
	4.2.1	Floating-Gate Principle	130
	4.2.2	Basic Definitions with Examples	131
	4.2.3	Basic Equations and Models	140
4.3	Physics of Programming and Erase Mechanisms		143
	4.3.1	Fowler–Nordheim Tunneling	145
	4.3.2	Polyoxide Conduction	148
	4.3.3	Channel Hot-Electron Injection (CHEI)	150
	4.3.4	Substrate Hot-Electron Injection (SHEI)	153
	4.3.5	Source-Side Injection (SSI)	155
	4.3.6	Secondary Impact Ionization Initiated Channel Hot-Electron Injection	156
4.4	Physics of Degradation and Disturb Mechanisms		158
	4.4.1	Band-to-Band Tunneling	158
	4.4.2	Oxide Degradation	159
	4.4.3	Oxide Breakdown	168
4.5	Conclusion		171
References			172

5 NOR FLASH STACKED AND SPLIT-GATE MEMORY TECHNOLOGY — 179

Stephen N. Keeney, Manzur Gill, and David Sweetman

5.1	Introduction		179
5.2	ETOX Flash Cell Technology		180
	5.2.1	Introduction	180
	5.2.2	Cell Structure	180

		5.2.3	Read (Sensing)	182
		5.2.4	Programming	183
		5.2.5	Erasing	183
		5.2.6	Array Operation	186
		5.2.7	Erase Threshold Control	187
		5.2.8	Process and Scaling Issues	190
		5.2.9	Key Circuits and Circuit/Technology Interactions	200
		5.2.10	Multilevel Cell Technology Circuits	206
	5.3	SST SuperFlash EEPROM Cell Technology		206
		5.3.1	Introduction	206
		5.3.2	Cell Cross Sections and Layout	207
		5.3.3	Charge Transfer Mechanisms	208
		5.3.4	Erase	209
		5.3.5	Programming	210
		5.3.6	Cell Array Architecture and Operation	212
		5.3.7	Erase Threshold Control and Distribution	214
		5.3.8	Process Scaling Issues	214
		5.3.9	Key Circuit Interactions	215
		5.3.10	Multilevel Cell Implementation	216
	5.4	Reliability Issues and Solutions		216
		5.4.1	Oxide Integrity	216
		5.4.2	Contact Integrity	217
		5.4.3	Data Retention	217
		5.4.4	Endurance	218
		5.4.5	Disturbs	219
		5.4.6	Life Test (Dynamic Burn-in)	220
	5.5	Applications		220
	References			220

6 NAND FLASH MEMORY TECHNOLOGY 223
Koji Sakui and Kang-Deog Suh

	6.1	Overview of NAND EEPROM		223
	6.2	NAND Cell Operation		227
		6.2.1	Cell Structure	227
		6.2.2	Erase Operation	227
		6.2.3	Program Operation	228
		6.2.4	Program Disturb	229
		6.2.5	Read Operation	230
	6.3	NAND Array Architecture and Operation		231
		6.3.1	Staggered Row Decoder	231
		6.3.2	Self-Boosted Erase Inhibit Scheme	233

		6.3.3	Self-Boosted Program Inhibit Scheme	235
		6.3.4	Read Operation	237
	6.4	Program Threshold Control and Program V_t Spread Reduction		237
		6.4.1	Bit-by-Bit Verify Circuit	237
		6.4.2	Sophisticated Bit-by-Bit Verify Circuit	242
		6.4.3	Overprogram Elimination Scheme	247
	6.5	Process and Scaling Issues		252
		6.5.1	Shallow Trench Isolation NAND Technology (256-Mbit NAND)	252
		6.5.2	Booster Plate Technology	256
		6.5.3	Channel Boost Capacitance Cell	258
		6.5.4	Negative V_{th} Cell	263
		6.5.5	Free Wordline Spacing Cell	268
	6.6	Key Circuits and Circuit/Technology Interactions		270
		6.6.1	Shielded Bitline Sensing Method	270
		6.6.2	Full Chip Burst Read Operation	272
		6.6.3	Symmetric Sense Amplifier with Page Copy Function	273
		6.6.4	Source Line Programming Scheme	278
	6.7	Multilevel NAND		283
		6.7.1	Multilevel Circuit Technology	283
		6.7.2	Array Noise Suppression Technology	286
		6.7.3	Side-Wall Transfer Transistor Cell	293
		6.7.4	Three-Level NAND	297
		6.7.5	High-Speed Programming	301
	References			307
	Bibliography			310
7	**DINOR FLASH MEMORY TECHNOLOGY**			**313**
	Moriyoshi Nakashima and Natsuo Ajika			
	7.1	Introduction		313
	7.2	DINOR Operation and Array Architecture		313
		7.2.1	DINOR Operation	313
		7.2.2	DINOR Cell Characteristics	314
		7.2.3	DINOR Array Architecture	316
		7.2.4	DINOR Advanced Array Architecture	316
		7.2.5	VGA-DINOR Device Structure and Fabrication	317
		7.2.6	Characteristics of the Cell with Asymmetrical Offset Source/Drain Structure	318
	7.3	DINOR Technology Features		320
		7.3.1	Low-Voltage Read	320
		7.3.2	Fast Read Access	321

7.4	DINOR Circuit for Low-Voltage Operation		321
	7.4.1	High-Voltage Generation [7]	321
	7.4.2	Wordline Boost Scheme	326
7.5	Background Operation Function		327
	7.5.1	Background Operation and DINOR	327
	7.5.2	Emulating Electrically Erasable Programmable Read-Only Memory (EEPROM) and Static Random-Access Memory (SRAM)	327
	7.5.3	Background Operation Fast Erase	328
7.6	P-Channel DINOR Architecture		328
	7.6.1	Introduction	328
	7.6.2	Band-to-Band Hot-Electron Injection Cell Operation	329
	7.6.3	DINOR BBHE Programmed Cell	332
	7.6.4	P-Channel DINOR Summary	334
References			334
Bibliography			335

8 P-CHANNEL FLASH MEMORY TECHNOLOGY   337

Frank Ruei-Ling Lin and Charles Ching-Hsiang Hsu

8.1	Introduction		337
8.2	Device Structure		338
8.3	Operations of P-Channel Flash		338
8.4	Array Architecture of P-Channel Flash		343
	8.4.1	NOR-Type Array Architecture	343
	8.4.2	NAND-Type Array Architecture	344
8.5	Evolution of P-Channel Flash		345
	8.5.1	Hsu et al. [1]	345
	8.5.2	Ohnakado et al. [4]	349
	8.5.3	Ohnakado et al. [5]	350
	8.5.4	Shen et al. [6]	353
	8.5.5	Chung et al. [7]	353
	8.5.6	Sarin et al. [8]	354
	8.5.7	Wang et al. [9]	357
	8.5.8	Ohnakado et al. [2]	359
	8.5.9	For Further Study	362
8.6	Processing Technology for P-Channel Flash		366
	8.6.1	NOR-Type Array Architecture	367
	8.6.2	NAND-Type Array Architecture	368
References			370
Bibliography			371

9 EMBEDDED FLASH MEMORY — 373
Chang-Kiang (Clinton) Kuo and Ko-Min Chang

9.1	Introduction	373
9.2	Embedded Flash Versus Stand-Alone Flash Memory	375
	9.2.1 Advantages of Embedded over Stand-Alone Flash Memory	375
	9.2.2 Disadvantages of Embedded over Stand-Alone Flash Memory	376
9.3	Embedded Flash Memory Applications	377
	9.3.1 Applications by Device Type	377
	9.3.2 Applications by Function	379
	9.3.3 Applications by End Product	380
	9.3.4 Applications by Usage	382
9.4	Embedded Flash Memory Cells	383
	9.4.1 Special Requirements and Considerations	383
	9.4.2 Cell Selection for Embedded Applications	385
9.5	Embedded Flash Memory Design	394
	9.5.1 Special Requirements and Consideration	394
	9.5.2 Flash Module Design for Embedded Applications	396
	9.5.3 Design Techniques for Embedded Flash Module	398
References		403

10 TUNNEL DIELECTRICS FOR SCALED FLASH MEMORY CELLS — 407
T. P. Ma

10.1	Introduction	407
10.2	SiO_2 as Tunnel Dielectric—Historical Perspective	408
10.3	Early Work on Silicon Nitride as a Tunnel Dielectric	409
10.4	Jet-Vapor Deposition Silicon Nitride Deposition	410
10.5	Properties of Gate-Quality JVD Silicon Nitride Films	411
10.6	Deposited Silicon Nitride as Tunnel Dielectric	417
10.7	N-Channel Floating-Gate Device with Deposited Silicon Nitride Tunnel Dielectric	425
10.8	P-Channel Floating-Gate Device with Deposited Silicon Nitride Tunnel Dielectric	429
10.9	Reliability Concerns Associated with Hot-Hole Injection	432
10.10	Tunnel Dielectric for SONOS Cell	432
10.11	Prospects for High-K Dielectrics	434

10.12	Tunnel Barrier Engineering with Multiple Barriers	437
	10.12.1 Crested Barrier	437
	10.12.2 U-Shaped Barrier	439
10.13	Summary	440
	References	440

11 FLASH MEMORY RELIABILITY 445
Jian Justin Chen, Neal R. Mielke, and Chenming Calvin Hu

11.1	Introduction	445
11.2	Cycling-Induced Degradations in Flash Memories	447
	11.2.1 Overview of Cycling-Induced Degradations	447
	11.2.2 Channel Hot-Electron Programming-Induced Oxide Degradation	449
	11.2.3 Tunnel-Erase-Induced Oxide Degradation	456
	11.2.4 Erratic Erase	462
11.3	Flash Memory Data Retention	466
	11.3.1 Activation Energy and Accelerated Data Retention Bake Tests	467
	11.3.2 Charge-Loss and Gain Mechanisms in EPROMs and Flash EPROMs	473
	11.3.3 Flash EEPROM Cycling-Induced Data Retention Issues	477
	11.3.4 Data Retention Characteristics Related to Tunnel Oxide and Floating-Gate Poly Texture	481
	11.3.5 Soft Errors	484
11.4	Flash Memory Disturbs	487
	11.4.1 Read Disturb and the Effects of Cycling	487
	11.4.2 Program Disturb	491
	11.4.3 Erase Disturb	495
	11.4.4 Block-to-Block Disturbs	495
11.5	Stress-Induced Tunnel Oxide Leakage Current	496
	11.5.1 Uniform SILC in Thin Oxide	497
	11.5.2 SILC in Thin Oxide after Bipolarity Stress	502
	11.5.3 Microscopic Characteristics of Stress-Induced Leakage Current (mSILC)	508
	11.5.4 Stress-Induced Leakage Current in Oxynitride	510
	11.5.5 Stress-Induced Leakage Current as the Limiting Factor for Tunnel Oxide Scaling	511
11.6	Special Reliability Issues for Poly-to-Poly Erase and Source-Side Injection Program	512
	11.6.1 Poly-to-Poly Erase and Its Reliability Issues	512
	11.6.2 Source-Side Injection and Its Reliability Issues	517
11.7	Process Impacts on Flash Memory Reliability	525

	11.7.1	Tunnel Oxide Process and Nitrogen Incorporation	526
	11.7.2	Effects of Floating-Gate Process and Morphology	526
	11.7.3	Stacked Gate SAS (Self-Aligned Source) Etch Process and Erase Distribution	528
	11.7.4	In-Line Plasma Charging Damage	530
	11.7.5	Impacts of Intermetal Dielectric and Passivation Films on Flash Memory Reliability	533
11.8	High-Voltage Periphery Transistor Reliability		536
	11.8.1	High-Voltage Transistor Technology	536
	11.8.2	Reliability of HV Transistors in Flash Memory Products	537
	11.8.3	Process Defects: The Role of Cycling and Burn-in	539
11.9	Design and System Impacts on Flash Memory Reliability		543
	11.9.1	Embedded Erase and Program Algorithm	544
	11.9.2	Redundancy and Defect Mapping	547
	11.9.3	Error Correction Concepts and Techniques	548
	11.9.4	Wear Leveling	552
11.10	Flash Memory Reliability Screening and Qualification		552
	11.10.1	Introduction to Reliability Testing and Screening	552
	11.10.2	Classification of Flash Memory Reliability Tests	554
	11.10.3	Acceleration Models of the Reliability Tests	557
	11.10.4	Flash Memory Sort and Reliability Test Flow	559
	11.10.5	Flash Memory Product Qualification Flow	561
	11.10.6	Burn-In and Reliability Monitoring Program	564
	11.10.7	Failure Rate Calculations	565
11.11	For Further Study		570
	11.11.1	Introduction	570
	11.11.2	Erratic Erase	570
	11.11.3	Stress-Induced-Leakage-Current Related Retention Effects	571
	11.11.4	Detrapping-Related Retention Effects	572
	11.11.5	Qualification Methods	573
	11.11.6	Flash Memory Floating-Gate to Floating-Gate Coupling	574
	11.11.7	New Program Disturb Phenomenon in NAND Flash Memory	575
	11.11.8	Impacts of Random Telegraph Signals and Few-Electron Phenomena on the Scaling of Flash Memories	576
References			579

12 MULTILEVEL CELL DIGITAL MEMORIES — 591
Albert Fazio and Mark Bauer

- 12.1 Introduction — 591
- 12.2 Pursuit of Low-Cost Memory — 592
- 12.3 Multibit Storage Breakthrough — 594
  - 12.3.1 Intel StrataFlash Technology — 594
  - 12.3.2 Evolution of MLC Memory Technology Development — 596
  - 12.3.3 Multilevel Cell Concept — 596
- 12.4 View of MLC Today — 599
  - 12.4.1 Multilevel Cell Key Features — 599
  - 12.4.2 Flash Cell Structure and Operation — 599
  - 12.4.3 Multilevel Cell Operation — 603
  - 12.4.4 Mixed Signal Design Implementation — 608
- 12.5 Low-Cost Design Implementation — 611
- 12.6 Low-Cost Process Manufacturing — 612
- 12.7 Standard Product Feature Set — 612
  - 12.7.1 Programming Speed — 613
  - 12.7.2 Read Speed — 613
  - 12.7.3 Power Supply — 613
  - 12.7.4 Reliability — 613
- 12.8 Further Reading: Multilevel Flash Memory and Technology Scaling — 614
- 12.9 Conclusion — 614
- References — 614

13 ALTERNATIVE MEMORY TECHNOLOGIES — 617
Gary F. Derbenwick and Joe E. Brewer

- 13.1 Introduction — 617
- 13.2 Limitations of Flash Memory — 619
  - 13.2.1 Introduction — 619
  - 13.2.2 Programming Voltage — 619
  - 13.2.3 Programming Speed — 623
  - 13.2.4 Endurance — 623
  - 13.2.5 Scaling — 623
- 13.3 NROM Memories — 624
  - 13.3.1 Introduction — 625
  - 13.3.2 Memory Cell and Array; Structure and Operation — 625
  - 13.3.3 Storage Mechanism — 632
  - 13.3.4 Reliability — 638
  - 13.3.5 Quad NROM Technology — 645
  - 13.3.6 Fabrication — 650

	13.3.7	Scaling	652
	13.3.8	Products	655
	13.3.9	Summary	658
13.4	Ferroelectric Memories		658
	13.4.1	Introduction	658
	13.4.2	Storage Mechanism	660
	13.4.3	Memory Cells and Arrays	664
	13.4.4	Fabrication	670
	13.4.5	Nonvolatile Characteristics	671
	13.4.6	Scaling	673
	13.4.7	Reliability	674
	13.4.8	Die and Test Cost	675
	13.4.9	Ferroelectric Products	676
	13.4.10	Ferroelectric Memory Summary	677
13.5	Magnetic Memories		678
	13.5.1	Introduction	678
	13.5.2	Magnetic Random-Access Memory with Giant Magnetoresistive Devices	679
	13.5.3	Magnetic Random-Access Memory with Magnetic Tunnel Junction Devices	684
	13.5.4	Programming Characteristics	685
	13.5.5	Fabrication	686
	13.5.6	Nonvolatile Characteristics	687
	13.5.7	Scaling	687
	13.5.8	Reliability	688
	13.5.9	Die and Test Cost	688
	13.5.10	Magnetic Memory Summary	689
13.6	Single-Electron and Few-Electron Memories		689
	13.6.1	Introduction	689
	13.6.2	Electric Charge Quantization in Solids	689
	13.6.3	Single-Electron Effects in Memory Cells	691
	13.6.4	Single-Electron Memories	693
	13.6.5	Few-Electron Memories	693
13.7	Resistive and Hybrid CMOS/Nanodevice Memories		696
	13.7.1	Introduction	696
	13.7.2	Programmable Diode Technologies	698
	13.7.3	Hybrid CMOS/Nanodevice Resistive Memories	700
	13.7.4	Expected Performance	701
	13.7.5	Resistive Memory Summary	703
13.8	NOVORAM/FRAM Cell and Architecture		703
	13.8.1	Introduction	703
	13.8.2	Crested Tunnel Barriers	703

		13.8.3 NOVORAM/FGRAM Cell and Architecture	706
		13.8.4 NOVORAM/FGRAM Summary	707
	13.9	Phase Change Memories	707
		13.9.1 Introduction	707
		13.9.2 Storage Mechanism	709
		13.9.3 GST Phase Change Material	709
		13.9.4 Memory Cell	712
		13.9.5 Memory Array and Support Circuitry	720
		13.9.6 Fabrication	721
		13.9.7 Scaling	722
		13.9.8 Reliability	725
		13.9.9 Products	727
		13.9.10 Summary	728
	References		728

Index — **741**

About the Editors — **759**

FOREWORD

The story of Flash memory is one of a unique technology that was almost a failure. Only after three unsuccessful attempts did Flash succeed in the marketplace. To succeed, Flash had to solve several technical problems and had to create a market for itself. Once these issues were addressed, Flash quickly became the highest volume nonvolatile memory displacing the EPROM only 5 years after its successful entry in the market. Today, Flash is challenging DRAM for the highest volume semiconductor memory used in the world. While the market forecast for Flash is bright, the technology is approaching fundamental limits in its scalability.

Flash memory was born in a time of turmoil during a semiconductor recession in the mid-1980s. The highest volume writeable nonvolatile memory, the EPROM, was under extreme price pressure due to a maturing market with over 20 competitors. In addition, customers were growing weary of the UV erase required to reprogram an EPROM. They wanted the electrically erase capability of the new high-priced EEPROM technology at EPROM prices. This spurred Toshiba, SEEQ, and Intel to search for the "holy grail of nonvolatile memory," the single-transistor electrically erasable memory. In 1985, Toshiba was the first to announce a single-transistor electrically erasable memory and coined the name "Flash memory" as the new device erased in a "flash." Unfortunately, Toshiba's initial Flash product was difficult to use and, as a result, a market failure. SEEQ followed a year later with another complex Flash device that did not succeed in the market. Meanwhile, Intel took a diversion in attempting to develop a single-transistor EEPROM technology by partnering up with Xicor, one of the early EEPROM memory pioneers. While the single-transistor EEPROM looked good on paper, the reality was the cell operating window was nonexistent and the partnership was dissolved. Fortunately for Intel, a parallel internal development on an EPROM-based Flash memory technology was started as a "skunkworks project." By 1985, Intel had a working 64-kb Flash memory in the lab. But to everyone's surprise, the Intel EPROM business unit was not interested in commercializing Flash, claiming that the market would not accept it based on Toshiba's lack of success and a fear of Flash cannibalizing Intel's own EPROM business. If it were not for Gordon Moore and a band of very dedicated pioneers, Intel would never have entered the Flash market. In 1986, Intel formed a separate Flash business unit and introduced a 256-kb Flash memory 2 years later. With over 95% of its manufacturing steps the same as an EPROM, the Intel Flash technology was able to quickly ramp up in volume using existing EPROM fabs. To ensure market success, Intel designed its 256-kb Flash as an EPROM replacement by having the same package pinout and control signals as EPROM devices.

Flash was able to easily cannibalize EPROM embedded applications where cost-effective reprogramming was required. One of the very first commercial Flash memory applications was on an adjustable oil well drill bit. Clearly, the oil drillers did not want to pull up the drill bit every time they needed to adjust it. Other embedded applications followed the conversion to Flash... including automotive engine control, laser printers, robotics, medical, factory automation, disk drive controllers, telecommunication, and military applications. With Flash commercially successful, the race was on and several other competitors entered the Flash market.

New customer-desired features were added to Flash that enabled Flash to penetrate markets beyond the simple EPROM replacement business. The memory array was segmented into independently erasable blocks and all the control circuitry was placed on-chip for automated programming and erase functions. Charge pumps were added to eliminate the need for a high-voltage program/erase supply voltage. New innovative packaging such as TSOP, "bumped" bare die, and memory cards were utilized to enter the portable equipment market.

The real high-volume Flash applications were developed based on these new features. One of these high-volume Flash applications was the digital camera. The first real-time demonstration of a digital camera using a Flash "film card" was in April of 1988 during the Intel 256-kb Flash product introduction at the Eiffel Tower in Paris. A picture was taken of a journalist in the audience; and then, electronically, the journalist was placed on a beach in Tahiti. While a novelty through much of the 1990s, the Flash "electronic film" business took off as the sales of digital cameras crossed over the sales of film-based cameras in 2003. The steep ramp in digital camera sales combined with the rapidly increasing digital picture resolution will drive an insatiable demand for Flash for many years to come. Another high-volume Flash application similar to the digital camera is the portable music player. The portability and the ability to store thousands of songs have made Flash the ideal storage media for these MP3 music players. One of the earliest high-volume Flash applications was the cellular phone. In the early 1990s, the Flash cell phone market took off. Every cell phone needed to have Flash to store the frequently changing cellular digital protocols. Today, cell phones utilize Flash chips up to 256 Mb in density to store games and pictures as well as the cellular protocols.

As Flash prices continue to fall, more high-volume applications will emerge for Flash. In the future, expect to see digital video recorders and portable DVD players that utilize Flash to store videos. In your PC, the portion of the magnetic disk drive that stores your program code (Windows, Word, Photoshop, etc.) will be replaced by "instantly on" Flash memory. These are a few of the new and exciting applications of Flash that are on the horizon.

Flash memory technology has evolved in two major directions. In the early years of Flash, there was only a single type of Flash, NOR Flash. NOR Flash has a random-access memory cell optimized for high-speed applications such as the cell phone and other code storage applications. A serial-based NAND Flash technology was created to meet the emerging needs of the low-cost file storage market that can live with a slow serial read access time. The primary NAND Flash applications are the digital camera and the portable MP3 music players. Due to technical problems, NAND Flash was not commercially successful until the mid-1990s. However, with the skyrocketing customer demand for digital cameras and MP3 players, NAND Flash volume crossed over NOR Flash volume in 2005.

Flash memory has scaled very nicely following "Moore's law." In 1988, the 256-kb Flash memory was fabricated on a 1.5-μm process and had a 36-μm^2 memory cell size. By 2003, Flash evolved to a 256-Mb density on a 0.13-μm process with a 0.154-μm^2 memory cell size. This scaling has driven the pricing of Flash memory from $1000/MB in 1988 to about 10 cents/MB in 2004. That is an astounding 10,000 times cost reduction in only 16 years! No wonder Flash has become the fastest growing memory market in history. Starting at $1 million of revenue in 1988, the Flash total available market has grown to over $10 billion in 2003. Flash is projected to cross over DRAM in sales in 2007 with over a $40 billion market size.

The future of Flash technology is not as rosy as the business projections. Flash scaling is running out of gas. At the 45-nm level, the physics of the Flash device begins to break down. The electric fields required for programming and erase are so high that materials begin to fall apart. Flash will continue to scale, but it will be a game of diminishing returns. As a result, almost all the major Flash suppliers have begun work on a Flash replacement technology. Intel is experimenting with Ovonics Unified Memory. Meanwhile, Motorola is looking at silicon nanocrystals. Other alternatives being pursued include magnetic RAM (MRAM), ferroelectric RAM (FeRAM), and polymer memory. It will be interesting to see if one of these innovative technologies can become a reliable high-volume memory replacement for Flash.

RICHARD D. PASHLEY
University of California at Davis
Intel Corporation (retired)

BIOGRAPHICAL NOTE

Dr. Pashley retired from Intel in 1998 after 25 years of service. His most significant accomplishment at Intel was the development of Intel's Flash memory business from its technical inception in 1983 to a $1 billion business in 1998. One of Dr. Pashley's technical teams invented the world's first NOR ETOX Flash memory cell and demonstrated product viability in 1985. In 1986, Gordon Moore gave Dr. Pashley the responsibility to start up a Flash memory business inside Intel. As general manager of the Flash Memory Business, Dr. Pashley directed all aspects of the business including engineering, manufacturing, marketing, and financial. Early in 1988, his Flash business team began selling its first Flash memory product. By 1990, Intel Flash Memory Business was so successful that it was merged with the Intel EPROM Business under Dr. Pashley. In 1993, he was appointed Vice President of Intel.

Dr. Pashley joined Intel in 1973 as a SRAM memory design engineer. One year later, he was promoted to HMOS program manager where he was responsible for the development of Intel's next-generation microprocessor and SRAM silicon technology. From 1977 to 1986, he held positions as director of microprocessor/SRAM technology development and director of nonvolatile memory technology development.

Dr. Pashley is currently an adjunct professor in the Graduate School of Engineering at the University of California Davis where he is teaching a class on technology management.

PREFACE

Nonvolatile Memory Technologies with Emphasis on Flash (NMT) was written to provide a detailed view of the integrated circuit main-line technologies that are currently in mass production. It also gives an introduction to technologies that have less manufacturing exposure, and it presents a description of various possible alternative technologies that may emerge in the future. All variations of Flash technology are treated. This includes Flash memory chips, Flash embedded in logic, binary cell Flash, and multilevel cell Flash. The scope of treatment includes basic device structures and physics, principles of operation, related process technologies, circuit design, overall design trade-offs, reliability, and applications.

The ambitious goal for NMT was to serve as *the* authoritative reference guide for nonvolatile memory users, engineers, and technical managers. Individual chapters were written by leading practitioners in the nonvolatile memory field who have participated in the pioneering research, development, design, and manufacture of technologies and devices. We wanted to provide authoritative background information to practicing engineers in the areas of circuit design, applications, marketing, device design, and reliability. We expected that the emergence of Flash embedded in logic chips would increase the audience and demand for a reference book of this nature.

New engineers and graduate students should find the book particularly useful. It may serve as a text or supporting reference for graduate courses. Persons unfamiliar with the nonvolatile memory field will find the inclusion of an elementary tutorial introductory chapter helpful as an orientation to the field. The objective of that chapter was to help the nonspecialist become familiar with the basic concepts of nonvolatile semiconductor memory.

Chapter 2, on applications, explains in detail the nature and utility of nonvolatile memory and delves into the complexities of memory organization, software/hardware internal control, and memory packaging. This is perhaps the most complete treatment of these matters provided in any existing book on nonvolatile memory.

Chapter 3 is devoted to an exposition of the circuit aspects of a modern memory device. Required circuit functions are described, dominant circuit architectures are presented, and key circuit blocks are explained. Techniques for management of defects and "design for test" approaches are provided.

Chapter 4 treats the physics of Flash memories at a mature tutorial engineering level. It begins with an examination of the basic operations of the memory transistor and explains the important characteristics from a physics perspective including circuit models and equations. Physical phenomena associated with programming and erase including conduction modes and key reliability issues are presented. The critical question of oxide degradation is examined in some detail.

The successive chapters are devoted to the mainstream classifications of Flash memory including stacked-gate standard ground Flash (Chapter 5), NAND Flash

(Chapter 6), DINOR Flash (Chapter 7), and P-channel Flash (Chapter 8). Both the history and design implementation of these specific Flash variants are treated in great detail. The important topic of embedded Flash is addressed in Chapter 9. It is expected that embedded Flash will become an almost universal feature of emerging "system on a chip" devices. The uses of embedded Flash are explained and supporting circuit and design technology is presented.

In Chapter 10, tunnel dielectrics are briefly examined with an eye toward implementation of dielectric processing. Chapter 11 is a comprehensive examination of Flash reliability that delves into all aspects of the complex mechanisms that impact device life and performance. Chapter 12 presents the important subject of multilevel cell memory. This is a basic tutorial introduction to the multilevel cells that includes perspective on the value and key issues associated with the approach.

Chapter 13, the final chapter, is devoted to memory technologies that may become successful alternatives to the conventional floating-gate Flash in various applications. Authored by leading developers and advocates of the respective technologies, this chapter provides an excellent engineering level introduction to NROM, ferroelectric memory, solid-state magnetic memory, single- and few-electron memories, resistive (crossbar) memory, and NOVORAM. NROM, of course, is already well on its way into high-volume production as a "non-floating-gate" version of Flash memory.

Credit for the original concept for NMT belongs to my friend and co-editor Manzur Gill. Manzur was the author of the Flash chapter in the 1998 book *Nonvolatile Semiconductor Memory Technology* that William Brown and I co-edited. It is a bit of trivia that a role reversal occurred with NMT. The idea for the early book was mine and Bill Brown bore the bulk of the editorial work in pulling it together. With NMT, my co-editor had the idea and I did the bulk of the editorial work. Of course, that means I am to blame for any errors that survived our many reviews.

The book chapters can be read sequentially, but most chapters provide sufficient background material that they can stand alone. We have tried to reflect the state of the art in 2006–2007, but we are well aware that the passage of time and the pace of the technology will rapidly make some of our pronouncements seem quaint. That is the risk and fun of being in this business.

Manzur and I feel privileged to have had the opportunity to work with the many talented contributors to this work. We have listed their names and affiliations in the list of contributors. Our chapter authors come from countries around the world, and most major NVM device manufacturers are represented in this group. We are also indebted to Richard Pashley for providing a historical view of the technologies in the Foreword.

We hope that you will find our labors useful, and invite your comments and corrections. It is only through feedback that we can improve.

JOE E. BREWER

Gainesville, Florida
October 2007

CONTRIBUTORS

Natsuo Ajika
Genusion, Inc., Hyogo, Japan

Greg Atwood
Intel Corporation, Santa Clara, California

Mark Bauer
Intel Corporation, Folsom, California

Ronald J. Bauer
Intel Corporation, Folsom, California

Yoram Betser
Saifun Semiconductors, Ltd., Netanya, Israel

Roberto Bez
STMicroelectronics, Milan, Italy

Ilan Bloom
Saifun Semiconductors, Ltd., Netanya, Israel

Joe E. Brewer
University of Florida, Gainesville, Florida

Ko-Min Chang
Freescale Semiconductor, Inc., Austin, Texas

Jian "Justin" Chen
SanDisk Corporation, San Jose, California

Guy Cohen
Saifun Semiconductors, Ltd., Netanya, Israel

Oleg Dadashev
Saifun Semiconductors, Ltd., Netanya, Israel

R. Degraeve
IMEC, Leuven, Belgium

Gary F. Derbenwick
Celis Semiconductor, Inc., Colorado Springs, Colorado

Alan D. DeVilbiss
Celis Semiconductor, Inc., Colorado Springs, Colorado

Shai Eisen
Saifun Semiconductors, Ltd., Netanya, Israel

Boaz Eitan
Saifun Semiconductors, Ltd., Netanya, Israel

Albert Fazio
Intel Corporation, Santa Clara, California

Gary Forni
Marvell Semiconductor, Inc., Santa Clara, California

Manzur Gill
Forman Christian College, Lahore, Pakistan
retired from Intel Corporation

G. Groeseneken
IMEC, Leuven, Belgium

Chenming Calvin Hu
University of California, Berkeley, California

Charles Ching-Hsiang Hsu
eMemory Technology, Inc., Hsinchu, Taiwan

Meir Janai
Saifun Semiconductors, Ltd., Netanya, Israel

Romney R. Katti
Honeywell International, Inc., Plymouth, Minnesota

Stephen N. Keeney
Intel Ireland, Ltd., Kildare, Ireland

Chang-Kiang Clinton Kuo (deceased)
Motorola Semiconductor

Stefan Lai
Intel Corporation, Santa Clara, California

Konstantin K. Likharev
SUNY-Stony Brook, New York

Frank Ruei-Ling Lin
Power Flash Inc., Hsinchu, Taiwan

Eli Lusky
Saifun Semiconductors, Ltd., Netanya, Israel

T. P. Ma
Yale University, New Haven, Conneticut

Eduardo Maayan
Saifun Semiconductors, Ltd., Netanya, Israel

H. E. Maes
IMEC, Leuven, Belgium

Giulio G. Marotta
Micron Technology Italia, Avezzano, Italy

Ken McKee
Intel Corporation, Folsom, California

Neal Mielke
Intel Corporation, Santa Clara, California

Moriyoshi Nakashima
Genusion, Inc., Hyogo, Japan

Giovanni Naso
Micron Technology Italia, Avezzano, Italy

Collin Ong
formerly of Intel Corporation, Santa Clara, California

Richard D. Pashley
University of California, Davis, California

Yan Polansky
Saifun Semiconductors, Ltd., Netanya, Israel

Christine M. Rice
Intel Corporation, Chandler, Arizona

Koji Sakui
Intel Corporation, Folsom, California
formerly of Toshiba Corporation

Giuseppe Savarese
Consultant, Napoli, Italy

Assaf Shappir
Saifun Semiconductors, Ltd., Netanya, Israel

Yair Sofer
Saifun Semiconductors, Ltd., Netanya, Israel

Kang-Deog Suh
Samsung Electronics Co., Seoul, Korea

David Sweetman
retired from Silicon Storage Technology, Dyer, Nevada

Jan Van Houdt
IMEC, Leuven, Belgium

1

INTRODUCTION TO NONVOLATILE MEMORY

Joe E. Brewer

1.1 INTRODUCTION

In this introductory chapter the ABCs of nonvolatile memory are reviewed. The purpose of this elementary discussion is to provide the perspective necessary for understanding the much more detailed chapters that follow. The emphasis is on communication of an overview, rather than on specifics of implementation. Simple memory concepts and terminology are presented, the parameters and features unique to nonvolatile memory (NVM for short) are examined, generic Flash memory variants are described, and finally the treatment of NVM in the International Technology Roadmap for Semiconductors (ITRS) is described.

Semiconductor memory is an essential part of modern information processors, and like all silicon technology it has been more or less growing in density and performance in accordance with Moore's law. Semiconductor NVM technology is a major subset of solid-state memory. Nonvolatility, of course, means that the contents of the memory are retained when power is removed. This book provides an in-depth description of semiconductor-based nonvolatile memory including basic physics, design, manufacture, reliability, and application. Flash memory is emphasized because for a long period of time it has been the dominate form of NVM both in terms of production volume and magnitude of sales dollars. Flash, however, is not the only alternative, and this book also attempts to describe some of the many NVM technologies that have some hope of achieving success in the marketplace.

Nonvolatile Memory Technologies with Emphasis on Flash. Edited by J. E. Brewer and M. Gill
Copyright © 2008 the Institute of Electrical and Electronics Engineers, Inc.

1.2 ELEMENTARY MEMORY CONCEPTS

All information processing can be viewed as consisting of the sequential actions of sensing, interpreting/processing, and acting. These actions cannot be accomplished without somehow remembering the item of interest at least long enough to allow the operations to take place, and most likely much longer to allow convenient use of the raw data and/or the end results.

The length of time that the memory can retain the data is the property called *retention*, and the *unpowered retention* time parameter is the measure of *nonvolatility*. A volatile memory will typically have a worst-case retention time of less than a second. A nonvolatile memory is usually specified as meeting a worst-case unpowered retention time of 10 years, but this parameter can vary from days to years depending on the specific memory technology and application.

Integrated circuit nonvolatile memories are frequently classified in terms of the degree of functional flexibility available for modification of stored contents. Table 1.1 summarizes the categories currently in frequent use [1]. This class of memory was evolved from ultraviolet (UV) erasable read-only memory (ROM) devices, and thus the category labels contain "ROM" as a somewhat awkward reminder of that heritage.

Flash memory [2] is an EEPROM where the entire chip or a subarray within the chip may be erased at one time. There are many variants of Flash, but present-day production is dominated by two types: NAND Flash, which is oriented toward data-block storage applications, and common ground NOR Flash, which is suited for code and word addressable data storage.

In general, information processing requires memory, but it is not at all clear that any constraints are placed on the structure or location of the storage relative to the processing elements. That is, the memory may be a separate entity and entirely different technology than the logic, or it may be that the logic is embedded in the memory and be a technology compatible with the logic, or any combination thereof.

At the heart of every memory is some measurable attribute that can assume at least two relatively stable states. Many common memory devices are *charge based* where charge can be injected into or removed from a critical region of a device, and

TABLE 1.1. Nonvolatile Memory Functional Capability Classifications

Acronym	Definition	Description
ROM	Read-only memory	Memory contents defined during manufacture and not modifiable.
EPROM	Erasable programmable ROM	Memory is erased by exposure to UV light and programmed electrically.
EEPROM	Electrically erasable programmable ROM	Memory can be both erased and programmed electrically. The use of "EE" implies block erasure rather than byte erasable.
E^2PROM	Electrically erasable programmable ROM	Memory can be both erased and programmed electrically as for EEPROM, but the use of "E^2" implies byte alterability rather than block erasable.

ELEMENTARY MEMORY CONCEPTS

the presence or absence of the charge can be sensed. The process of setting the charge level is called *writing*, and the process of sensing the charge level is called *reading*. Alternatively, the write operation may be referred to as the *store* operation, and the read operation may be called the *recall* operation.

Dynamic random-access memory (DRAM), a volatile technology, uses charge stored on a capacitor to represent information. Charge levels greater than a certain threshold amount can represent a logic ONE, and charge levels less than another threshold amount can represent a logic ZERO. The two critical levels are chosen to assure unambiguous interpretation of a ZERO or ONE in the presence of normal noise levels. (Here the higher charge level has been called a ONE, but it is arbitrary which level is defined to be the ONE or ZERO.)

Leakage currents and various disturb effects limit the length of time that the capacitor can hold charge, and thus limits "powered" retention to short periods. The word "dynamic" in the name "DRAM" indicates this lack of ability to hold data continuously even while the circuit is connected to power. Each time the data is read, it must be rewritten in order to assure retention, and regular data refresh operations must be performed when the cell is idle. Worst-case retention time (i.e., the shortest retention time for any cell within the chip) is typically specified as about 60 ms. DRAM is a volatile memory in terms of "unpowered" retention because the charge is not maintained when the circuit power supply is turned off.

Flash memory makes use of charge stored on a floating gate to accomplish nonvolatile data storage. Figure 1.1 provides a cartoon cross-section sketch of a floating-gate transistor and its circuit symbol representation. The floating-gate electrode usually consists of a polysilicon layer formed within the gate insulator of a field-effect transistor between the normal gate electrode (the control gate) and the channel. The amount of charge on the floating gate will determine whether the transistor will conduct when a fixed set of "read" bias conditions are applied. The fact that the floating gate is completely surrounded by insulators allows it to retain charge for a long period of time independent of whether the circuit power supply voltage is present. The act of reading the data can be performed without loss of the information.

Figure 1.2 compares an imaginary idealized transistor with no charge layer in the gate insulator with a transistor that has a charge per unit area, Q, at distance d from the silicon channel surface. The impact of the charge on the threshold voltage depends on the amount of charge per unit area, its distance from the silicon surface, and the permittivity of the insulator between the charge and the silicon. In a Flash device the floating gate provides a convenient site for the charge, but other means may serve the same purpose. For example, in silicon oxide nitride oxide silicon (SONOS) transistors the charge will reside in traps within the nitride layer.

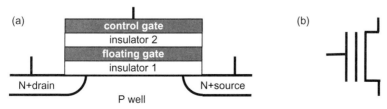

Figure 1.1. Floating-gate transistor: (a) elements of the transistor structure and (b) circuit symbol.

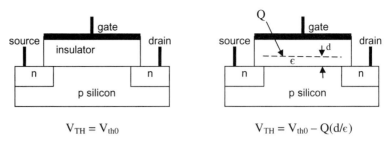

Figure 1.2. V_{TH} shift due to charge in gate insulator.

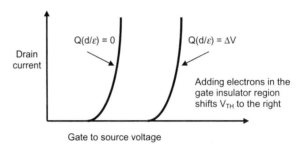

Figure 1.3. Shift of current–voltage characteristics because of inserted charge.

The threshold voltage, of course, impacts how the source-to-drain current of the transistor will change as a function of change in the gate-to-source voltage. Figure 1.3 shows how the characteristic curves can be made to shift as a function of the stored charge. As electrons are added to the charge within the gate insulator region, the curve will move in a positive direction.

For some memory technologies the process of reading destroys the data. This is referred to as *destructive readout* (DRO). For other technologies, Flash, for example, readout can be accomplished without significantly disturbing the data. This is referred to as *nondestructive readout* (NDRO). DRO memory has the disadvantage of requiring that every read operation must be followed by a write operation to restore the data.

Over time some forms of memory organization have been so firmly established that most engineers immediately assume those structures and the parameters that characterize those structures as being the norm. Probably the most pervasive assumption is that the information to be stored and recalled is in a digital binary format.

There are several schemes where a single transistor may be used to store more than just one bit. One approach is to store charge in physically separated parts of the device that can be sensed separately. Currently, the most common example of this concept is the NROM cell discussed later in this chapter. Another approach is to interpret the amount of charge stored in one physical location in the device as a representation of a multibit binary number. In this case the sensing process must reliably distinguish between different quantities of stored charge and the readout process must generate the corresponding binary number.

Consider the "one physical location" approach. A 1-bit-per-cell arrangement is a robust form of storage that allows relaxed margins and comparatively simple sensing circuitry. The read current needs only to be unambiguously above or below a preset value in order to establish whether a ONE or a ZERO was stored. For a 2-bit-per-cell memory, the recall process must reliably distinguish between four preset levels of charge, and the readout circuitry must translate the detected level into a 2-bit digital format. The storage process and the protection of the cell from disturb conditions must accurately set and maintain those four levels under all operating and nonoperating storage conditions. Considering that the usual requirement is for nonvolatile data retention for 10 years, assuring stability of charge levels and circuit characteristics is quite a challenge. Of course, the complexity rapidly increases as a cell is required to store larger numbers of bits. For example, a 4-bit-per-cell memory must reliably cope with 16 levels and still meet all specifications.

While the heart of a semiconductor memory is the cell, the surrounding circuitry is the mechanism that makes it usable. For economic reasons, cells are packed as close together in a rectangular planar array as available integrated circuit technology and noise management concerns will allow. This X, Y array arrangement contains the cells and conductive lines that allow access to each individual cell.

The lines that run in the X direction (rows) are called *wordlines*, and they are used to select a row of cells during the write or read operations. Wordlines tie to the control gates of the cells in a row. The lines that run in the Y direction (columns) are called *bitlines*. As shown in Figure 1.4, bitline connections for the NOR array architecture are tied to drain terminals of devices in a column. One end of a bitline connects to power and then goes through the array to sensing and writing circuitry. Thus the wordlines activate a specific group of cells in a row, and the bitlines for each intersecting column connect those cells to read and write circuits.

In the example of a NOR architecture, the cells in a column are connected in parallel where all the drain terminals tie to a bitline and all the source terminals tie to a common source line (ground). In this configuration a positive read mode voltage on one wordline while all other wordlines are at zero volts will result in a bitline current that is a function only of the selected row of cells.

This, of course, assumes that a zero control gate-to-source voltage actually turns off the unselected rows. For the NOR organization it is important that the process of initializing the array, called *erase*, not proceed to the point of overerasing transistors to the extent that the threshold voltage becomes negative and the transistors change from operation in the enhancement mode to operation in the depletion mode.

The NAND array architecture, shown in Figure 1.5, achieves higher packing density. Here the bitlines are formed using series-connected strings of cells that do not require contact holes. A string is typically 8, 16, or 32 cells long. If other strings

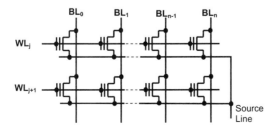

Figure 1.4. NOR array architecture.

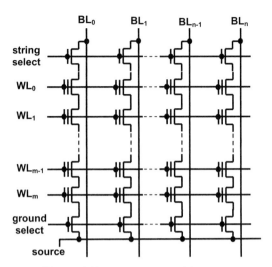

Figure 1.5. NAND array architecture.

were to be tied to the same bitlines, they would be connected in parallel between the bitline and the common source line in a manner similar to individual cells in the NOR architecture.

Reading and writing a NAND device is more complicated than dealing with a NOR device. The wordlines are used to select the transistor of interest in the string. To access data from a string, the reading process requires that all nonselected transistors be turned on while only the selected transistor is allowed to influence the current flow through the string. In contrast to the NOR architecture, it is not objectionable to allow the transistors to be shifted into depletion mode. There are, however, problems if transistors are shifted too far in the enhancement direction. It is important that the distribution of threshold voltages for the programmed state be limited to a specified design range in order to assure proper circuit operation.

There are a number of basic principles of operation that apply to both NOR and NAND organized devices. For a single read operation the individual bits that form a word are made to appear in an input/output register. For a single write operation the word present in an input/output register is used to determine the data inserted into the cells for that word. Reading or writing processes must be designed such that unselected cells are not disturbed while the selected cells are operated on. The design of the array must contend with basic circuit design issues associated with driving heavily loaded transmission lines as well as assuring proper operation of each individual cell.

In order to select a given row and column an integrated memory device is usually provided with a binary address word from external circuitry. The address word is routed to address decoder circuitry that is tightly tied to the sides of the array, and designed to drive the word and bitlines. For reasons of management of loading and data grouping considerations, large memory chips are usually partitioned into many arrays.

A modern integrated memory device incorporates control circuitry that accepts relatively simple commands as inputs and generates timed sequences of signals to accomplish writing, reading, and various other modes of operation. Also, the writing operations may require voltages that differ from the readily available normal logic supply voltages. In order to simplify the use of the device these voltages are typically generated on-chip. This feature also has the advantage of allowing the semiconductor manufacturer to assure that proper voltages and pulse sequences are applied to the memory cells during all operating conditions without having to depend on decisions made by the end user.

The data is arranged in binary *words* of some prescribed length and structure, and the words may be handled in groups usually called *blocks* or *sectors*. Figure 1.6 summarizes in a black box format the functional interfaces of present-day semiconductor memories. This diagram can be applied to a memory chip, a memory module, or a complete memory system.

While the diagram hides the internal complexity of the memory, the performance parameters for all memories can be understood in the context of this simple block diagram. Memory inputs consist of data, address, and control signals. Memory outputs consist of data and status signals.

The major dynamic performance parameter of a memory is the time between stable address input and stable output data. This latency is the key parameter that determines information processor performance. In a memory that is completely word oriented this delay is called the *access time*. By definition, in a *random-access memory* (RAM) the access times have the same value independent of the address. For memories that are block oriented, the memory performance is characterized by the latency parameter and a data flow parameter. That is, the memory has a delay (latency) before the first word or byte appears at the data out signal terminal, and then successive words or bytes can be clocked out at some byte/second rate.

For writing, the important dynamic characteristic is the time between stable address and input data and the time when the chip can accept a new set of input data. A status output signal (e.g., ready/busy signal) is usually provided to signal when a new operation can be initiated.

Data may be organized within the memory system in various ways. It has been a frequent practice to use chips that have a one-bit-wide output, and use separate chips to provide the bits that make up a word. This is convenient for reliability purposes where spare bits (chips) can be added to form a coded word such that all single-bit errors can be detected and corrected. In this manner single-chip failures can be tolerated, and system reliability is enhanced. For large bit capacity memory chips this approach can make it difficult to efficiently implement small memory systems. Chips 4 bits, 8 bits, or 16 bits wide are more appropriate in such cases.

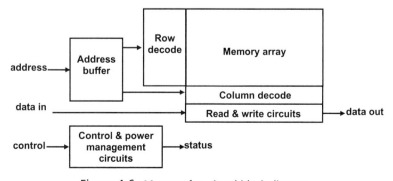

Figure 1.6. Memory functional block diagram.

For block- or sector-oriented chips it is common to serially input and output data in 8-bit-wide bytes. Internal to a chip the bytes are assembled into words and assigned to sectors. Reading and writing are accomplished by loading command and address information and then initiating an automatic sequence that accomplishes the commanded functions and moves and formats the data. Typically these on-chip automatic control sequence signal generators are implemented as state machines.

Historically, the design of information processors has been an art form where engineers made use of whatever logic and memory technologies were currently available and married the technologies to achieve viable systems. The Von Neumann model for a stored program computer was an abstraction that helped engineers to visualize specific forms of computing elements and begin to build practical systems. An accompanying concept, the memory hierarchy, provided a way to work around the limitations of whatever technology was currently available and achieve improved performance–cost trade-offs.

Figure 1.7 shows the 2005 typical hierarchy for a personal computer. The notion of a hierarchy is that a central processing unit (CPU) within a processor is serviced by a series of memory technologies where the memory closest to the CPU, the cache, provides very fast access to instructions and data. The information stored in cache is an image of a portion of the data stored in the next highest level of memory, that is, the information immediately needed by the program code being executed. The much larger, slower, less expensive main, or primary, memory stores the bulk of the programs and data to be processed at the present time. This relationship also occurs between the primary and secondary memory. The primary memory retrieves program and code from the secondary store and writes information to be retained in the form of files back to the secondary store. The cache memory is word oriented (addressable to the word level). The primary memory is both word and block oriented. It can be addressed at the word level and then roll out a page (block) of data in a burst mode. The secondary memory is nonvolatile and block oriented.

Variants of the diagram in Figure 1.7 appear in most texts on computer architecture. Note that nonvolatile semiconductor memory does not explicitly appear in Figure 1.7. For personal computers the nonvolatile magnetic disk is currently a much lower cost option for secondary memory as long as the use environment does not impose high mechanical stresses. NVM semiconductor drives can and do perform the same functions as the magnetic secondary store where requirements other than cost allow them to compete. For personal computers the semiconductor option is firmly established as relatively small manually transportable storage that can be accessed via a universal serial bus (USB) port, and for physically small mobile

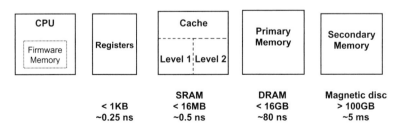

Figure 1.7. Memory hierarchy showing approximate latency and byte capacity bounds.

devices such as digital cameras the semiconductor option is dominate. As the integrated circuit technology progresses to densities where multiprocessors per chip dominate the market, it is likely that a mix of memory technologies (including nonvolatile memory) will be included on the devices, and most of the chip area will be devoted to memory circuitry.

Information processing systems dedicated to specific functions take different approaches to the mix of logic and memory. Engineering of processing systems is an ad hoc art focused on the specific needs of the product in question. Design for items intended to be produced in volume at low cost may attempt to meet requirements using a single processor chip that includes embedded solid-state memory of various kinds. High-performance systems may use multiple processors and a finely tuned mix of supporting memory in order to meet processing throughput requirements. Because the capabilities and options for both logic and memory grow rapidly, the system designer is always faced with a moving baseline. The optimum design of today most probably will not be optimum tomorrow.

1.3 UNIQUE ASPECTS OF NONVOLATILE MEMORY

Nonvolatile memory devices, like all semiconductor memory technology, have electrical alternating current (ac) and direct current (dc) characteristics that can be specified and measured, thermal and mechanical ratings, and reliability characteristics. However, the requirement for nonvolatile retention of data is unique to NVM. Each NVM technology makes use of some physical attribute to achieve nonvolatility, and the particular mechanism employed brings with it a number of technology-specific characteristics.

Device data sheets need to specify retention, and either the data sheet or supporting application information should explain the basis for retention claims. Stresses associated with the processes of writing and reading an NVM device may degrade retention capability and/or other properties of a device. This stress is generally related to the number of writes and/or reads and is summarized in an "endurance" specification stated in terms of the number of write and/or reads that can be tolerated. These matters may be different in nature for different NVM technologies. The particular features of Flash are examined here to illustrate the major issues involved.

1.3.1 Storage

Perhaps the easiest way to understand the nature of the floating-gate memory device is to consider the energy levels involved. Figure 1.8 shows the band structure for a simple floating-gate device where the silicon substrate is shown on the left. The N-type control gate is on the right, and an N-type polysilicon floating gate is in the middle, sandwiched between two silicon dioxide layers. The floating gate, embedded within insulators, is isolated from the exit or entry of charge by the high-energy barrier between the conduction band in the polysilicon and the conduction bands in the top and bottom SiO_2 layers. These barriers, much greater than the thermal energy, provide nonvolatile retention of the charge. In order to change the amount of charge stored on the floating gate it is necessary to change the potential of the floating gate relative to the potential on the opposite side of either SiO_2 layer until

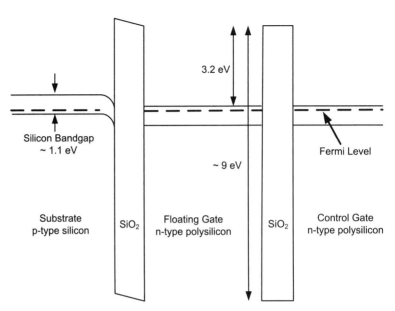

Figure 1.8. Energy band diagram for a typical floating-gate structure.

some conduction mechanism is invoked that can overcome or tunnel through the barrier.

Different strategies may be selected to overcome the energy barriers. Two conduction mechanisms in common use are channel hot-electron (CHE) injection and Fowler–Nordheim (FN) tunneling. Hot carrier injection may be used to add electrons to a floating gate (i.e., programming). FN tunneling may be used to remove or add electrons to a floating gate (i.e., erase or program). Present-day NOR devices typically use CHE to program floating gates and FN to erase floating gates. NAND devices employ FN for both program and erase.

To invoke CHE injection a lateral channel electric field on the order of 10^5 V/cm is required to accelerate electrons to energy levels above the barrier height. Some of those electrons will be scattered by the lattice and be directed toward the floating gate. To actually reach the floating gate, the scattered electrons must retain sufficient energy to surmount the silicon to insulator barrier and cross the insulator to the floating gate. CHE is an inefficient method in that less than about 0.001% of the channel current will be directed to the floating gate.

Fowler–Nordheim tunneling of cold electrons will occur when a field on the order of 8 to 10 MV/cm is established across the insulator next to the floating gate. The process is slower than CHE injection, but it is better controlled and more efficient.

The floating-gate memory transistor can be viewed as a capacitively coupled network as illustrated in Figure 1.9. A pulse applied to the control gate (or any other terminal) will be capacitively coupled to the floating gate, and the resulting potential on the floating gate relative to the other terminals can cause the movement of electrons to or from the floating gate.

Using the terminology of Figure 1.9, the total capacitance on the floating gate, C_t, is the sum $C_g + C_d + C_b + C_s$. If the network is driven from the control gate, the coupling ratio, k_g, would be C_g/C_t. If the network is driven from the drain, the

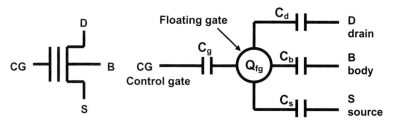

Figure 1.9. Capacitor model for a floating-gate transistor.

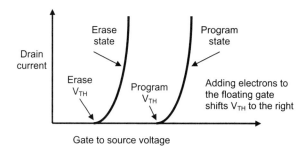

Figure 1.10. Erase state and program state current–voltage characteristics.

coupling ratio, k_d, would be C_d/C_t. In a similar fashion the coupling coefficients k_b and k_s would be C_b/C_t and C_s/C_t. The voltage on the floating gate resulting from pulses applied to the four terminals would be

$$V_{fg} = k_g V_{CG} + k_d V_D + k_s V_S + k_b V_B$$

The usual process of writing (storing) data into a Flash memory requires two operations. First, all of the cells in a common tub (i.e., a sector) are "erased" to initialize the state of the cells. Erasing refers to the removal of all charge from the floating gates. By convention this is usually taken to mean that all of the cells have been cleared to a ONE state. Second, the cells within the tub that are selected by a row address are "programmed" to ZERO or left at ONE in accord with the input data signals. The programming operation may continue over row addresses until all of the data sites in the sector have been programmed.

Erasing moves the threshold voltage in a negative direction while writing moves the threshold voltage in a positive direction. Figure 1.10 shows the relative relationship of the erase and program states in terms of the resulting current–voltage characteristics. Both the magnitude of shift and the statistical distribution of threshold voltages after an erase or program is a major design issue.

The erase procedure is complicated by several concerns. First, erasing (removal of electrons) shifts the threshold voltage negatively from a positive value toward a value nearer to zero. If continued too far, the threshold voltage can go through zero and become negative. This is referred to as *overerase*, and the transistor changes from enhancement mode to depletion mode. In NOR arrays depletion must be avoided in order for the array to work properly.

Second, the erase process is sensitive to the initial state of the transistor. Erase voltages are applied to transistors located in the same tub simultaneously. If some of those transistors are in a programmed state while others are in the erased state, the resulting final threshold voltages will have quite a spread. It is likely that for some devices overerase will occur. To manage this situation, it is common practice to program all the transistors before applying an erase voltage.

Third, caution should be exercised to avoid overstressing the insulator between the floating gate and the channel. A transistor in the programmed state has a static charge on the floating gate. A fast transient erase pulse to the device will be coupled through the capacitive divider circuit of Figure 1.9 and add directly to the potential across the insulator. This can result in very high electric fields, and repeated application of this stress can rapidly degrade the insulator. Slowing the leading edge of the pulse will allow the floating gate to begin to discharge as the applied voltage increases and avoid the peak stress condition. Because modern Flash chips generate erase pulses on-chip, this risk is easily managed by the chip designer.

1.3.2 Storage Mechanisms

Each nonvolatile memory technology exploits some physical or chemical approach to capture and retains a representation of information independent of the presence of a power source. Many technology options are charge based. Flash memory, as pointed out in the previous examples in this chapter, makes use of charge held on a conductive floating gate surrounded by insulators as its storage mechanism. Variants of the floating-gate approach such as nanocrystal memory operate similarly. Another common technology, SONOS (silicon oxide nitride oxide silicon) uses charge retained in nitride traps as its storage mechanism. Single and few electron transistor approaches confine charge on small islands.

There are several non-charge-based approaches being actively pursued. Ferroelectric memories make use of the switchable polarization of certain materials to store information and detect the resultant change in capacitance. Solid-state magnetic memory approaches currently being explored in integrated circuit form are based on either the giant magneto-resistance effect or magnetic tunnel junctions. In both cases readout is accomplished by detecting a resistance change. Phase change memories based on chalcogenide materials are reversibly switched between a low-resistance crystalline state and an amorphous high-resistance state through the application of heat.

This incomplete list of mechanisms is intended to convey a notion of the broad scope of approaches. It has been noted by many workers in the field that whenever it is observed that a material exhibits two or more relatively stable switchable states, that material has been considered a candidate for forming a memory technology.

1.3.3 Retention

The elapsed time between data storage and the first erroneous readout of the data is the *retention time*. Each nonvolatile storage technology employs a particular storage mechanism, and properties associated with that mechanism and its implementation format will determine the retention characteristics of the device. For Flash the storage mechanism is to represent data by quantities of charge held on a floating gate.

Each technology can be expected to have some natural processes where the data representation changes with time. Flash has some intrinsic charge decay characteristics that define the ultimate retention potential of the approach. Natural decay tracking points to very large retention on the order of thousands or millions of years. Natural decay is so slow as to not be a factor in determining practical retention specifications. At the present time a typical unpowered retention specification is 10 years.

Defects associated with materials, details of device geometry, or aspects of circuit design can impact retention. Each of these three potential problem areas can result in the addition or removal of charge to/from the floating gate. Gate insulator or interpoly insulator defects are typical causes of retention degradation. Phenomena associated with ionic contamination or traps can also be contributing factors. Management of these issues is a function of the general state of the integrated circuit reliability arts and of the specific practices and equipment of given manufacturers.

1.3.4 Endurance

Achievement of nonvolatility depends on exploitation of some natural phenomena peculiar to a given technology approach. In most nonvolatile technologies the normal processes employed to write and/or read cells will result in stresses that eventually degrade the properties of the memory or disturb the contents of the memory. *Endurance* is the term used to describe the ability of a device to withstand these stresses, and it is quantified as a minimum number of erase–write cycles or write–read cycles that the chip can be expected to survive. For quite a number of years the industry has used 100,000 cycles as the minimum competitive endurance requirement.

Knowledge of the reliability physics of the memory technology is essential in order to develop meaningful endurance specifications. The endurance capabilities of a device are a function of both the intrinsic properties of the technology and the quality control of the production line. The impact of cycling stress on retention is a key aspect of endurance, and window closure or shift is a concern. The *window* for a Flash memory is the difference between the erased state threshold voltage and the programmed state threshold voltage. This is the difference that must be reliably detected by a readout sense amplifier.

Both retention and endurance pose difficult test and verification quandaries. Economics and the limits of the human life span make direct observation of retention periods of decades impractical. The process of testing endurance by cycling each part implies that the testing would consume much of the useful life time of the product.

1.4 FLASH MEMORY AND FLASH CELL VARIATIONS

Flash has been used as the example technology throughout the discussions above where a simple stacked gate transistor was used to illustrate several points. In practice the design of Flash cells has several flavors each of which has advantages and disadvantages. Figure 1.11 groups some of the existing Flash varieties by addressing method and by program and erase technique (see Figure 5 in Bez et al. [2]). As mentioned earlier, the common ground NOR used for both code and data storage, and the NAND used for mass storage, presently dominate the market.

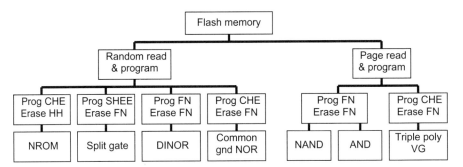

Figure 1.11. Flash cell architecture family tree suggested by [2].

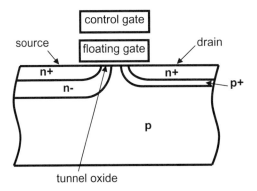

Figure 1.12. Source-erase stacked gate Flash cell [7].

The first modern Flash EEPROM was proposed by Masuoka et al. of Toshiba [3] at the 1984 IEEE International Electron Devices Meeting. The term *Flash* was coined to indicate that the complete memory array is erased quickly. This label is generally applied to all EEPROM devices where a block or page of data is erased at the same time. As indicated by the Figure 1.11 family tree, a large (and growing) number of variations of Flash have emerged that differ in the combination of cell structures, erase and program schemes, and array circuits. (These circuits are described later in this book.) The history of Flash [4] has been characterized by constant and aggressive introduction of modifications that seek to improve density, cost, performance, and reliability.

The Toshiba Flash cell [3, 5, 6] accomplished erase by having a special erase gate. Electrons were removed from the polysilicon floating gate by field emission directly to the polysilicon erase gate in response to a positive voltage pulse. A source-erase stacked gate Flash was proposed in 1985 by Mukherjee et al. [7]. As shown in Figure 1.12, this device looked like a conventional stacked gate UV-EEPROM with two modifications. The source junction was graded to support high voltages, and the oxide under the floating gate was thinned to allow FN tunneling.

It was not until 1988 when the reliability of the emerging devices was proven that volume production really began [8], and that structure, named ETOX (Intel trademark) for EPROM tunnel oxide [9], became an industry standard Flash cell. The ETOX device had a tunnel oxide thickness of about 10 nm, a graded source junction, and an abrupt drain junction. Erase was accomplished by FN tunneling of electrons from the floating gate to the source diffusion using a typical voltage of

12.5 V. Programming was done by CHE. The ETOX type of cell has been employed dominantly for NOR arrays. Historically, NAND arrays have employed a similar transistor structure, but the simple daisy chain interconnection of individual transistors to form a string allows the sharing of interconnection overhead over 8, 16, or 32 cells resulting in a much smaller area per bit.

Split-gate cells are distinctly different in structure because the floating-gate transistor incorporates a series nonmemory device that provides a select and over-erase mitigation function. As illustrated in Figure 1.13 a portion of the channel is controlled by the floating gate while a second portion is under direct control of the control gate. The nonmemory transistor assures that the overall transistor structure remains in the enhancement mode independent of the status of the memory transistor portion. The isolation afforded by the nonmemory structure is also beneficial in reducing disturb conditions.

In 2000 Eitan et al. [10] introduced the NROM (Saifun trademark) concept, which established a marked change from the conventional floating-gate structure. This cell makes use of localized charge trapping at each end of an N-channel memory transistor to physically store 2 bits. Figure 1.14 shows a cross section of a cell along a wordline. In this structure the gate left junction is at 0 V, and the wordline is at 9 V. The accelerated electrons will be injected into the oxide nitride oxide (ONO) near the right junction. To read bit 1 the right junction is held at 0 V, left junction at 1.5 V, and the wordline at 3 V. This bias condition maximizes the effect of the bit 1 charge on the threshold voltage. Reversing the bias conditions for the left and right junctions enables the programming and reading of bit 2. Erase is accomplished by establishing positive voltage on the bitlines relative to the wordlines. This causes band-to-band tunneling of hot holes in the depletion layer under the ONO above the n+ junctions. Chapter 13 treats the NROM cell in detail.

The bitlines (BLs) are buried diffusions and the wordlines (WLs) are polystripes. The array does not require contacts and is very flat because there is no field

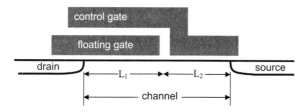

Figure 1.13. Typical split-gate cell structure cartoon.

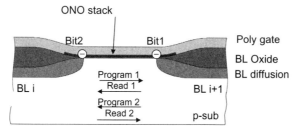

Figure 1.14. NROM cell showing localized electron storage regions.

isolation. NROM arrays are very efficient in utilization of area. A virtual ground array can be laid out using an ideal $2F$ wordline pitch and a $2.5F$ bitline pitch. This yields a $5F^2$ cell area that, since the cell contains 2 bits, is $2.5F^2$ per bit. In 2005 Eitan et al. [11] described the extension of the NROM concept to achieve 4 bits per cell. Here four threshold voltage levels are provided on each side of the cell to achieve an area per bit of $1.25F^2$.

1.5 SEMICONDUCTOR DEVICE TECHNOLOGY GENERATIONS

Integrated circuit technology has historically improved functional density at an exponential rate. This exponential trend was first pointed of by Gordon Moore in a 1965 article [12] where he observed that the number of transistors per chip was doubling every year. This trend has since slowed, but an exponential rate has been maintained. This so-called Moore's law has been the guiding rule for the International Technology Roadmap for Semiconductors (ITRS). The industry progresses by means of development of technology such that a new generation, referred to as a technology *node*, with minimum feature (F) sizes $0.7x$ of the previous generation emerges at 3-year intervals. The 2004 ITRS includes data for five nodes: 2004, 90 nm; 2007, 65 nm; 2010, 45 nm; 2013, 32 nm, and 2016, 22 nm. The most recent version of the ITRS is available at http://public.itrs.net.

Nonvolatile memory technology is treated in the ITRS in several places. The chapter developed by the Process Integration, Devices and Structures (PIDS) technology working group follows memory technologies that have matured to the point of being in production and addresses requirements for each technology to be realized in future technology nodes. Table 1.2 lists the cell size requirements given for DRAM and five NVM technologies.

Figure 1.15 provides a graphical view of this same data. The expected density advantage of Flash NAND is evident.

The ITRS plays an important part in the evolution of NVM. It is a document formulated and read by integrated circuit manufacturers worldwide. Each manufac-

TABLE 1.2. Memory Cell Sizes (μm^2) as Tabulated in the 2004 ITRS Update

Year	DRAM	Flash NAND	Flash NOR	FeRAM	SONOS	MRAM
2004	0.0650	0.0446	0.1010	0.4860	0.0660	0.4000
2005	0.0480	0.0352	0.0800	0.2030	0.0550	0.2000
2006	0.0360	0.0270	0.0610	0.2030	0.0450	0.1800
2007	0.0280	0.0190	0.0530	0.1730	0.0290	0.0900
2008	0.0190	0.0146	0.0420	0.1210	0.0250	0.0700
2009	0.0150	0.0113	0.0340	0.1210	0.0180	0.0600
2010	0.0122	0.0091	0.0270	0.0800	0.0150	0.0450
2011	0.0100	0.0072	0.0220	0.0650	0.0120	0.0360
2012	0.0077	0.0055	0.0170	0.0510	0.0100	0.0270
2013	0.0061	0.0046	0.0170	0.0390	0.0080	0.0230
2014	0.0043	0.0035	0.0110	0.0330	0.0070	0.0170
2015	0.0038	0.0028	0.0100	0.0260	0.0050	0.0140
2016	0.0025	0.0022	0.0080	0.0200	0.0040	0.0100
2017	0.0021	0.0018	0.0070	0.0180	0.0037	0.0090
2018	0.0016	0.0015	0.0050	0.0160	0.0030	0.0070

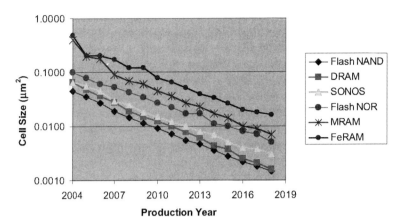

Figure 1.15. Plot of 2004 ITRS memory cell size requirements.

turer recognizes that the roadmap indicates approximately where the competition's capability will be at any given time, and each company strives to beat the roadmap goals as a matter of survival.

Up to this point in time the industry has been successful in maintaining the frantic pace of development set out in the ITRS. Every year at every major semiconductor meeting and in private discussions among technologists there has been someone who argues that the problems on the horizon are insurmountable and scaling will end. Usually the proponent of doom has one or two specific issues that he or she feels are going to be fatal. As the minimum feature size progresses closer to atomic dimensions, no doubt the nature of density improvement will shift from simple scaling to alternative ways of advancing functional integration. The challenge for nonvolatile memory is perhaps even more difficult than for CMOS logic because of the need to manage both dielectric stresses and maintain retention characteristics.

The history of Flash is an excellent demonstration of the innovative ways that engineers have met the scaling challenge. Intel, for example, has a publication track record for ETOX that shows how the desirable characteristics of a proven cell structure can be maintained while continuously improving the utilization of array area [8, 9 13–17]. For each technology generation an improved lithographic capability has become available, but the core portion of a Flash cell (i.e., that portion that actually stores charge) is difficult to scale. Engineering attention has focused on identifying and minimizing the area devoted to nonmemory portions of the array such as contacts, isolation, and alignment tolerances. Of course, the multibit cell was a major tactic in the fight to continue to reduce area per bit.

Another portion of the ITRS also includes consideration of memory: the Emerging Research Devices (ERD) section. ERD recognizes that simple scaling will eventually reach limits, and it attempts to monitor and evaluate published data for devices in the early research stage in hopes of identifying candidate technologies capable of maintaining desired density and performance growth [18]. The ERD chapter of the ITRS is necessarily a snapshot of a rapidly changing picture that must be revised every year. New technologies are almost always described overoptimistically by initial advocates, and very few approaches will survive to the development stage. The ERD working group strives to provide a balanced assessment of comparative technology potential that will be useful to the semiconductor industry for making research investment decisions.

REFERENCES

1. IEEE Std. 1005-1998, IEEE Standard Definitions and Characterization of Floating Gate Semiconductor Arrays, Feb. 1999.
2. R. Bez, E. Camerlenghi, A. Modelli, and A. Visconi, "Introduction to Flash Memory," *Proc IEEE*, Vol. 91, pp. 489–502, Apr. 2003.
3. F. Masuoka, M. Assano, H. Iwahashi, T. Komuro, and S. Tanaka, "A New Flash EEPROM Cell Using Triple Polysilicon Technology," *IEEE IEDM Tech. Dig.*, pp. 464–467, 1984.
4. M. Gill, and S. Lai, "Floating Gate Flash Memories," in W. D. Brown and J. E. Brewer (Eds.), *Nonvolatile Semiconductor Memory Technology*, IEEE Press, Piscataway, NJ, 1998.
5. F. Masuoka, M. Assano, H. Iwahashi, T. Komuro, and S. Tanaka, "A 256 K Flash EEPROM Using Triple Polysilicon Technology," *IEEE ISSCC Tech. Dig.*, pp. 168–169, 1985.
6. F. Masuoka, M. Assano, H. Iwahashi, T. Komuro, N. Tozawa, and S. Tanaka, "A 258 K Flash EEPROM Using Triple Polysilicon Technology," *IEEE J. Solid-State Circuits*, Vol. SC-22, pp. 548–552, 1987.
7. S. Mukherjee, T. Chang, R. Pan, M. Knecht, and D. Hu, "A Single Transistor EEPROM Cell and Its Implementation in a 512 K CMOS EEPROM," *IEEE IEDM Tech. Dig.*, pp. 616–619, 1985.
8. G. Verma and N. Mielke, "Reliability Performance of ETOX Based Flash Memories," *Proc. IRPS*, pp. 158–166, 1988.
9. V. N. Kynett, A. Baker, M. Fandrich, G. Hoekstra, O. Jungroth, J. Kreifels, S. Wells, and M. D. Winston, "An In-System Reprogrammable 256 K CMOS Flash Memory," *IEEE J. Solid-State Circuits*, Vol. 23, No. 5, pp. 1157–1163, Oct. 1988.
10. B. Eitan, P. Pavan, I. Bloom, E. Aloni, A. Frommer, and D. Finzi, "NROM: A Novel Localized Trapping, 2-Bit Nonvolatile Memory Cell," *IEEE Electron Device Lett.*, Vol. 21, No. 11, pp. 543–545, Nov. 2000.
11. B. Eitan, G. Cohen, A. Shappir, E. Lusky, A. Givant, M. Janai, I. Bloom, Y. Polansky, O. Dadashev, A. Lavan, R. Sahar, and E. Maayan, "4-Bit per Cell NROM Reliability," *IEEE IEDM Tech. Dig.*, pp. 539–542, Dec. 5, 2005.
12. G. E. Moore, "Cramming More Components onto Integrated Circuits," *Electronics*, Vol. 38, No. 8, Apr. 19, 1965.
13. V. Kynett, M. Fandrich, J. Anderson, P. Dix, O. Jungroth, J. Kreifels, R. Lodenquai, B. Vajdic, S. Wells, M. Winston, and L. Yang, "A 90 ns One Million Erase/Program Cycle 1-Mbit Flash Memory," *IEEE J. Solid-State Circuits*, Vol. 24, No. 5, pp. 1259–1264, Oct. 1989.
14. A. Fazio, "A High Density High Performance 180 nm Generation ETOX™ Flash Memory Technology," *IEEE IEDM Tech. Dig.*, pp. 267–270, Dec. 5–8, 1999.
15. S. N. Keeney, "A 130 nm Generation High Density ETOX™ Flash Memory Technology," *IEEE IEDM Tech. Dig.*, pp. 2.5-1–2.5-4, Dec. 2–5, 2001.
16. G. Atwood, "Future Directions and Challenges for ETOX Flash Memory Scaling," *IEEE Trans. Devices Mater. Reliabil.*, Vol. 4, No. 3, pp. 301–305, Sept. 2004.
17. R. Koval, V. Bhachawat, C. Chang, M. Hajra, D. Kencke, Y. Kim, C. Kuo, T. Parent, M. Wei, B-J. Woo, and A. Fazio, "Flash ETOXTM Virtual Ground Architecture: A Future Scaling Direction," paper presented at the 2005 Symposium on VLSI Technology Digest of Technical Papers, June 14–16, 2005, pp. 204–205.
18. J. E. Brewer, V. V. Zhirnov, and J. A. Hutchby, "Memory Technology for the Post CMOS Era," *IEEE Circuits Devices Mag.*, Vol. 21, No. 2, pp. 13–20, Mar.–Apr. 2005.

2

FLASH MEMORY APPLICATIONS

Gary Forni, Collin Ong, Christine Rice, Ken McKee, and Ronald J. Bauer

2.1 INTRODUCTION

The consumer electronics industry is rapidly transitioning from dominance by the mature home electronics and appliances market to the rapidly rising mobile electronics market. Cellular telephones, handheld computers, and digital music players are some of the most obvious examples of the electronic gadgetry that are filling consumer's pockets. Flash memory has enabled the creation of these devices by providing storage for the program code that defines their functionality as well as for the increasing amount of user data that must be carried. Flash memory provides the ideal combination of attributes for mobile applications: low power, nonvolatile, speed, and small size. Flash is a key enabling technology because the alternatives have trade-offs that would violate the primary design criteria for mobile devices. For example, using disk-based storage precludes miniaturization and ruggedness.

Cellular handsets provide an illuminating case study for the enabling role of Flash in mobile applications (Fig. 2.1). Early cell phones used read-only memory (ROM) devices to store their program code, which was an adequate option for their limited functionality and hard-wired analog protocols. However, designers using erasable programmable ROM (EPROM) for development quickly moved to Flash devices for faster prototyping cycles. Production engineers could also see benefits in code version inventory control by using Flash. But the driving impetus to switch to using Flash in production phones came with digital cellular protocols. The in-system reprogrammability of Flash was needed to support the variety of digital protocols implemented around the world and frequent updates to specifications. As standards stabilized, manufacturers looked to consolidate their memory subsystems

---

Nonvolatile Memory Technologies with Emphasis on Flash. Edited by J. E. Brewer and M. Gill
Copyright © 2008 the Institute of Electrical and Electronics Engineers, Inc.

2. FLASH MEMORY APPLICATIONS

Figure 2.1. Flash makes mobile devices an essential piece of our pocket's contents. (*Courtesy of Motorola, Inc.*)

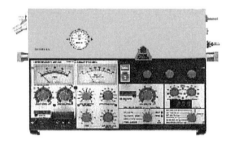

Figure 2.2. Flash is also an essential part of industrial and embedded equipment.

by moving their parameter data store from an electrically erasable programmable ROM (EEPROM) to the Flash and minimizing static random-access memory (SRAM) requirements by executing directly from Flash. To accomplish this, Flash needed to handle both code and data, functionality provided by upgraded Flash software media managers. The ability of Flash to handle both code and data provided the infrastructure needed to implement new functions, such as voice recording, messaging, and email. As the capacity of Flash devices grows even more sophisticated applications are appearing on handsets, like Internet browsing, voice recognition, and video. So, from its beginning as mere code storage, Flash has become a driver of new cell phone features and applications.

However, portable applications are not the only ones that have adopted Flash memory. Flash has become the standard system memory for embedded designs because it provides benefits for both engineering, in quick code updates during development, and for production and support, through ROM version inventory management and field updates (Fig. 2.2). Accordingly, Flash is now used in applications as diverse as storing address tables in network routers to capturing biometric data in medical monitoring equipment.

2.1.1 Spectrum of Memory Devices

System designers face two familiar decisions at the start of a new project. They must choose a processor and a memory, driven by the needs and requirements of the application. Together the processor and the memory define the core of the system, and the type of memory selected further dictates the system architecture.

TABLE 2.1. Strengths and Weaknesses of Common Memory Technologies

Memory Type	Strengths	Weaknesses
ROM	Cost, nonvolatile	Nonwritable, not erasable
EPROM	Erasable, nonvolatile	Not in-system writable
RAM	Fast reads/writes, byte alterable	Cost, volatile
EEPROM	Nonvolatile, byte alterable	Low density, cost
Disk	High capacity, cost, nonvolatile	Access time, reliability, floor cost, store-and-download execution model, not byte alterable
Flash	Cost, nonvolatile, high density, erasable, fast reads	Not byte alterable, writes not single cycle

With the profusion of memory technology types from which to choose—random-access memory (RAM), dynamic RAM (DRAM), SRAM, ferroelectric RAM (FRAM), hard disk drive (HDD), ROM, EPROM and, of course, Flash—how does one choose? All memories are a compromise of strengths and deficiencies, and the requirements of the specific application at hand will determine the trade-offs necessary. Table 2.1 summarizes the strengths and weaknesses of the memory types in common usage today.

2.1.2 Evolving from EPROMs

In the 1970s and first half of the 1980s, EPROMs ruled the electronics design world as the predominant means of storing instructions to control processors. They were nonvolatile, meaning they didn't lose their contents when their power source was shut off, and were relatively high density, meaning they could hold a lot. You could even erase them by sticking them into an ultraviolet (UV) lightbox for 20 minutes, or, in a pinch, stick them in the sun for a few days. Then you could reprogram them in a few minutes, put them back in their sockets, and be running with your new code. Senior system designers likely recalled the days when loading memory was done a nibble at a time by flipping toggle switches, or perhaps they were lucky enough to have loaded stacks of punch cards instead. Either way, they lost their program code and data at the end of the day, if they were fortunate enough not to lose everything in a power loss. To them, designing a system using a 16-kB (kilobyte) nonvolatile, erasable memory seemed like the good life.

EPROMs worked by trapping electrons on the floating, or insulated, gate of a transistor, thus storing information. One of the failure modes in manufacturing EPROMs was leakage of electrons from the floating gate. If those electrons leaked off, even slowly, that compromised the information being stored. To the engineers working on Flash, this was clearly a problem that had to be solved.

However, some engineers working on the problem thought they could use the leakage as an advantage instead. You were supposed to erase an EPROM with UV light, which kicked the electrons off the floating gate with a few extra electron volts of energy, but if the leakage also took the electrons off, then why not harness the bug to erase the chip? Some other engineers at Toshiba had realized the same thing, but the Intel engineers were the first to put the idea into practice and "Flash EPROMs," so-called because they could quickly be erased "in a flash" electrically and without UV light, went into production in 1988.

The change from UV-erased EPROMs to electrically erased Flash memories had an immediately apparent consequence: the window on top of the chip's package that let in the UV light disappeared. But other than a quicker and more convenient erase, Flash didn't seem all that different from regular EPROMs. In fact, the full ramifications of the change would take 12 years to evolve and, today, the Flash industry still hasn't run out of features to add.

Because of its roots as an extension of EPROMs, Flash was easily adopted as easier-to-use EPROMs, but at the same time, those roots brought a legacy of system architecture, design, and manufacturing techniques that slowed evolution of system memory usage models more specifically appropriate to Flash memory. The usage models that have become most prevalent in embedded design can be divided into three groups: code storage, data storage, and code+data storage.

This chapter will discuss the three usage models, how they relate to Flash applications, and the relevant attributes of Flash memory devices and system design.

2.1.3 NOR and NAND

The two main types of Flash memory technologies are NOR and NAND, and each has strengths and weaknesses. NOR-type Flash came to market first as an EPROM replacement and remains dominant for code and code+data storage usages for a broad variety of applications across nearly all computer and embedded designs. NAND-type Flash came later and has remained primarily a data storage memory in the removable-cards arena.

Intel, ST Micro, and most Spansion Flash memories are based on NOR technologies, while Toshiba and Samsung are the major suppliers for NAND Flash memory.

2.1.3.1 NOR Flash Memory. The strengths of NOR-type Flash memories are fast random-access reads and reliable storage.

Code execution generally requires fast random-access reads because code jumps from one memory location to another when branching. For this reason, the high-speed random-access capability of NOR Flash architecture is ideal for code storage applications.

Code stored in Flash for direct execution must be stored with total reliability, since any errors will produce a system fault. NOR Flash guarantees 100% good bits, removing any need for error correcting code or controllers. Guaranteed "good bits" simplifies implementation by removing the need for error correction in the system.

Both attributes make NOR Flash ideally suited for combined code and data applications (code+data). The code can be randomly accessed as necessary, while data can be reliably stored and retrieved from the same device. This memory architecture provides maximum functionality with the lowest component count.

The most common use of NOR Flash memory is for code and data storage in embedded applications and cellular phones. However, NOR Flash works well in any application where fast, reliable random-access reads are needed, like for code or combined code+data storage. Additionally, a 16- or 8-bit-wide parallel data bus allows for fast code and/or data transfers into or out of NOR Flash.

2.1.3.2 NAND Flash Memory. The strengths of NAND-type Flash memories are fast writes and high-density arrays, which lower die costs at the expense of random-access read capability.

NAND is strongly suited to data-only applications such as audio storage, video storage, or data logging where fast data writes are required and the data will be accessed sequentially. It works best with these types of applications because of its fast write performance.

NAND Flash memory has poor random-access read performance, but sequential read performance is good. NAND data is accessed sequentially because of the way its transistors are arranged in its array. The first access in a page is slow because the system must wait for the read current to propagate through a column of Flash cells. In applications that read data sequentially, like audio and video playback or data logging, this is a benefit. However, if random accesses are required, like they are for code execution, the performance will be poor.

NAND-based memory has a smaller cell size for a given lithography but requires more complex circuitry to implement. NAND Flash typically does not guarantee 100% good bits, but fixes this with error correction and cell redundancies [error correcting code (ECC) is used to make sure errors from bad blocks are removed from the system]. The redundant memory cells and required error correction add to the complexity and cost of the media or system.

2.1.4 Evolution of Flash Usage Models

Two Flash usage models, code and data storage, were evident from the beginning, and both evolved from the way prior memory technologies were used. Only recently has a third usage model, code+data developed specifically around the strengths of Flash memory.

2.1.4.1 Code Usage Model. Historically, the most basic system architecture model was to store central processing unit (CPU) instructions in a nonvolatile memory such as a ROM or an EPROM, which was directly connected to the CPU memory bus for instruction fetches. System designers evolved this model in various directions as they tried to meet other memory subsystem requirements, such as performance and nonvolatile storage.

Random-access memory added higher-speed code access to increase performance and provided scratchpad space. The various types of RAM devices excel at code execution because of their fast access time, but since they lose their contents at power down, another duplicate memory must be in the system to store code and data during downtime. This problem can be addressed by directly executing code stored in ROM with a smaller amount of RAM for temporary working copies of data, but slower ROM access times slow system performance and the unchangeable contents cause inconvenience during engineering, inventory problems during manufacturing, and upgradeability issues in the field.

Replacing the ROM with an EPROM allows the code to be updated during engineering and after the product has shipped, but still requires physically pulling the device to be erased in a UV lightbox. Neither the ROM nor EPROM have any provision for persistent storage of data when the system is powered down, so yet another memory, an EEPROM, may be added for nonvolatile parameter storage.

The EEPROMs were added to support the nonvolatile storage of small system parameters and settings. Disks were added to support growing software code and data sizes. This proliferation of memory technologies complicated system and added memory subsystem redundancy. Now the system must include many memory devices: disk for persistent code/data storage, ROM/EPROM for CPU boot code, RAM for execution of downloaded code and working data, and perhaps an EEPROM for configuration parameter storage.

An example of this model taken to the extreme is the standard personal computer system architecture, which utilizes a ROM, EPROM, or Flash device for the basic input/output system (BIOS), DRAM for main code execution and temporary data storage, EEPROM/CMOS (complementary metal–oxide–semiconductor) for configuration parameters, and disk for nonvolatile code and data storage. This architecture is nonoptimal in that it duplicates storage space when it copies code and data between disk and RAM and uses a separate memory for BIOS that is used almost only at startup.

This tangle of duplicated memories may be viable in systems of the scale of a personal computer or larger, but many embedded or portable applications cannot support the size, cost, and power consumption of all these redundant devices. A more typical embedded system design will contain a Flash device for code storage, SRAM for scratchpad and code execution, and an EEPROM for small parameter storage. Some systems may also use battery-backed SRAM for semipersistent storage.

At the end of this morass of trade-offs is Flash memory. With fast access time, direct execution of code from Flash is possible. Its electrical updateability enables convenient updates for both engineers and end users and solves the inventory and code revision management problems of EPROMs and ROMs. Its byte-level writes enable EEPROM-style data storage when managed by a software data manger.

2.1.4.2 Data Usage Model. From a separate direction, flash has also been heralded, with very limited real-world success, as the successor to disk-based storage devices for all applications that used a floppy or hard disk.

Designer's enthusiasm for disk-based storage in these types of applications has always been mixed. Disk storage is very cheap but is bulky, power consuming, and, worst of all, sensitive to shock and vibration. Flash memory, with its small size, nonvolatile data retention, and solid-state reliability, has been seen as the solution to these storage issues. However, its penetration into this market has been held back by the high cost of Flash storage per megabyte relative to disk. In spite of this, adoption for Flash for data storage has been increasing with the growing popularity of portable digital capture applications such as personal digital assistants, digital cameras, and digital audio recorders. Much of this adoption for data storage has been in removable Flash cards rather than embedded devices.

2.1.4.3 Code+Data Usage Model. Because of these abilities, Flash has largely supplanted other memory devices in embedded and portable applications for code and to some extent data storage. Flash usage models have separated along lines based predominantly on whether the application was either code or data-centric. However, beyond replacing the functions of these traditional memories in traditional usage models, Flash has finally evolved to the point where it is gaining its own usage model, single-chip code+data storage.

2.1.5 Understanding Flash Attributes

While they developed as an extension of EPROMs, over time, Flash memories have gained their own set of attributes that change the way they are used in system applications. (See Table 2.2.) Understanding these attributes is an important first step in grasping the functions and limitations of Flash usage models. This section will look at the most important Flash attributes and how they influence system design. First, device-level features will be discussed, then the growing roles of Flash software and packaging in making a complete system.

TABLE 2.2. Summary of Important Flash Attributes and Benefits

Attribute	Relevant Metrics	What It Enables in Applications
Random-access read time	Nanoseconds (ns); milliseconds (ms)	Code execution (if in nanosecond range, otherwise, data only)
Page-mode interface	Number of words per page, typically 4 or 8 words Page access time (ns)	Faster code execution to support higher performance page-mode CPUs
Burst-mode interface	Number of words per burst, typically 4, 8, 16 words or continuous Maximum zero-wait state clock frequency (MHz)	Faster code execution to support highest performance burst-mode CPUs at zero-wait states
Bus width	Bits, typically 8 or 16	Matching of CPU bus width for higher data throughput per cycle
Density	Megabits (Mb)	Larger code or data sizes leading to more sophisticated applications
Write state machines	Commands supported	Relieves system software from low-level Flash management making writes to Flash routine
Array blocking	Block size in kilobits (kb), typically 8–64 kb	Modular code/data updates without risking entire array contents making writes to Flash routine
Asymmetrical blocking	Block size, quantity, arrangement	Efficient sizing of blocks (boot, parameter, main) according to basic code or code+data application model
Symmetrical blocking	Block size (kb)	Easy integration of multiple devices into a Flash array, enabling total Flash storage capacity larger than the size of a single device
Voltage supplies	Read supply, volts (V)	Matches system supply voltage; lower supply voltages lead to longer battery life
	Program/erase supply, volts (V)	If it matches system supply voltage, then it enables routine writes to Flash enabling in-system code, data updates
	I/O supply, volts (V)	Interface to logic running at voltages below Flash supply voltage, leading to lower power consumption

TABLE 2.2. *Continued*

Attribute	Relevant Metrics	What It Enables in Applications
Cycling	k cycles, typically 100+k	Writes become routine when adequate cycles are available, enabling frequent in-system updates and code+data storage
Suspends	Modes that can be suspended (program, erase, program during erase suspend, etc.)	Suspends must be available when needed for flexibility in servicing interrupts or capturing data when needed; enables software solution for read-while-write
	Operations available when suspended (read, program, etc.)	Performing the needed operations at the required time; enables software solution for read-while-write
	Suspend latency (ms)	Deterministic suspend latency allows software read-while-write to work in real-time systems
Partitioning	Number and size of partitions	Determines memory allocation between code and data
	Functions available during program/erase in other partitions	Adds flexibility in system function if program, erase or other Flash operations can occur simultaneously indifferent partitions; increases code execution and data store throughput
Write protection	Hardware or software control	Hardware control adds additional protection beyond software
Block locking	Individual or group lock/unlock	Individual block locking control removes unnecessary exposure of other blocks, providing more security
	Lock change latency	Low latency allows lock changes to happen in real time
	Hardware interlock	Provides additional layer of protection for code or boot code
Software	Management functions provided	Insulates system software from low-level Flash management and provides higher level interface so development effort can focus on application, not infrastructure

2.1.5.1 Device Attributes. Flash memory possesses its own set of attributes that define the product category and each specific product implements those attributes in a slightly different way. Thus, understanding each attribute is important to be able to measure the usefulness of a particular Flash device for an application.

ACCESS TIME AND READ INTERFACES. Like all memories, flash memory is measured by access time first. Access time is the latency between the time an address location is given to the time the Flash device outputs the stored data from that location. This attribute has obvious implications for system design because it is tied to system performance in both code and data retrieval. For code, it limits the clock frequency at which the CPU can run. For data, it defines data retrieval performance.

Reductions in process geometry have sped up Flash access times, but, at the same time, the trend toward decreasing supply voltage has negated the performance benefits. Consequently, the primary avenues for increasing Flash read performance are now architectural changes to incorporate improved read interfaces such as page-mode or burst-mode read to satisfy the latest low-voltage, high-performance CPUs.

Alternative read interfaces operate similarly to the like-named interfaces commonly seen on DRAMs. Page mode is an asynchronous read mode in which the Flash internally reads a page (usually four or eight words) at a time into a page buffer so that they can be output in sequence very quickly as the lower-order address lines change. Burst mode is a read mode that outputs bursts of sequential data synchronously to the CPU. Burst lengths vary depending on the requirements of the processor. Burst mode is currently the fastest interface being generally used on Flash memory devices.

BUS WIDTH. Bus width is the number of bits that a Flash device provides at one time. Bus widths are commonly 8-bits (×8, "by eight") or 16-bits (×16, "by sixteen"). Some Flash devices have bus widths selectable by an input pin. Bus width is a factor in performance because wider buses can transfer more data in a single cycle. Typically, a system designer will choose a Flash device to match up with the system CPU's bus width. As CPUs have moved from 8- and 16-bit to 32- and 64-bit buses, Flash devices are commonly paired together to supply the needed bus width.

DENSITY. Flash memory density is a synonym for the amount of data stored by the memory and is usually measured in megabits (Mb). The earliest NOR Flash memories were 256 kb (kilobits) and have since grown to 512 Mb in the Intel StrataFlash product line. NAND Flash devices have likewise grown from 16 Mb to 8 Gb (gigabits). Density is directly relevant to an application because the amount of storage it needs for code, data, or code+data must be satisfied by one or more Flash devices.

WRITE STATE MACHINES. The first flash devices followed the precedents set by EPROMs for controlling write operations. The burden was on the system to properly control the voltages and signals needed to correctly perform a program or erase operation. Performing this fundamentally analog process under the control of a digital system was problematic and involved hard to control timeouts and write pulses. The resulting variability in the implementation of these algorithms resulted in reduced reliability of these early Flash devices. In addition, the timing requirements of these algorithms meant the system CPU had to remain fully dedicated to the task, bringing the operation of the system grinding to a halt. All of this meant that in-system write operations were not used routinely by system designers. The solution was to bring control of the program/erase algorithms internal to the Flash device by implementing them in an on-chip write state machine that properly controlled the required algorithms.

The incorporation of write state machines as a standard feature increased device reliability and allowed the CPU to write to the Flash using standard bus cycles. The added ease of use and reliability were the first steps in making in-system write to Flash a regular design practice.

ARRAY BLOCKING. While most NOR-type Flash memory has always been programmable on a single byte or word basis,[1] first-generation Flash devices sported a

---

[1] Some Flash devices require programming on a page basis. While this is usually associated with smaller block sizes on the order of 512 bytes, it increases the overhead and risk required to update single bytes or words.

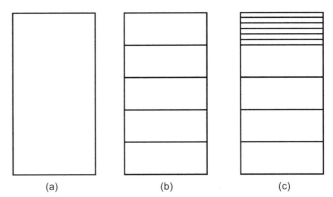

Figure 2.3. Various Flash blocking architectures: (a) bulk-erase, (b) symmetrical, and (c) asymmetrical.

single erasable memory space, much like that of their EPROM predecessors. This is why they are sometimes referred to as *bulk-erase* Flash devices [Fig. 2.3(a)]. To update a bulk-erase Flash, the entire memory array had to be erased and then the whole array reprogrammed. This methodology went against robust system design practices because if the process was interrupted after erasing, but before reprogramming was complete, the system would become nonfunctional. System-level safeguards such as using duplicate Flash devices allowed workarounds but were redundant and inelegant.

While EPROMs had a reason for erasing the whole memory at the same time (their window exposed the entire memory array to the erasing UV light), Flash technology had no window necessitating full array erasure. This led to another major architectural innovation: the division of the memory array into independently erasable blocks [Fig. 2.3(b, c)]. One of the first applications of blocking was in the "boot block" architecture. A small boot block segmented from the rest of the array and separately locked was added so systems could put their fundamental power-up boot code in the boot block and lock it to prevent accidental erasure. Now even if the firmware was somehow erased, the system could bring itself up far enough to go out to the floppy disk and restart the update. The boot block architecture, with its combination of small and large blocks remains to this day one of the most popular Flash devices offered.

Erasing Flash in smaller chunks made management of code and data storage easier and safer. Most wonder why block sizes aren't reduced all the way to the ideal of single byte/word erase. The reason is that the smaller the block, the larger the penalty in transistors and die area, which increases costs. While smaller blocks are easier to use and faster to erase, they are more costly in terms of die size, so every blocking scheme must balance its block sizes with device cost and the needs of its target application.

ASYMMETRICAL VERSUS SYMMETRICAL. The way the Flash memory array is divided into blocks optimizes it for a particular type of storage. Since data tends to change more often, smaller blocks are better for storing data, due to their quicker erase time. On the other hand, code does not change as often, larger blocks can be used for code storage, saving on die size and cost.

Two common approaches to segmenting the array are symmetrical and asymmetrical. Symmetrical blocking [Fig. 2.3(b)] divides the memory array into equally sized blocks. A commonly chosen block size is 64 kB, which is used by the Intel FlashFile architecture. Symmetrical blocking tends to be good for either code or data storage but is not particularly optimized for mixing code+data in the same device since the large block sizes are relatively slow to erase.

Asymmetrical blocking [Fig. 2.3(c)] divides the array into differently sized blocks, small ones for data and larger ones for code, optimizing the block sizes for the code+data requirements. A side effect of asymmetrical blocking is the need for separate top and bottom versions of a Flash device. Since some microprocessors look for boot code at the top of the memory map and others look at the bottom, mirrored versions of the memory map are required. An example of an asymmetrical blocking scheme optimized for code+data is Intel's Advanced+Boot Block architecture, which features 8 kB blocks for small data parameter storage and 64 kB blocks for code and large data storage.

Today, improving Flash erase performance and increasingly sophisticated software data managers are reducing the need for small blocks and asymmetrical blocking schemes. In addition, as embedded devices get connected via networks, Internet, and wireless links, the amount of data they store is increasing, pushing their storage requirements into larger block sizes.

VOLTAGE SUPPLIES. When Flash memories first appeared in system designs, they were direct replacements for ROMs or EPROMs, so they mostly stored stable, unchanging code and only benefited manufacturing processes with improved flexibility. The obvious next step was to provide in-system updates, the ability to reprogram the system Flash device with a completely new code image, similar to how a personal computer's BIOS chip is updated today. To achieve this, however, systems needed to provide a 12-V power supply to the Flash chip's programming voltage pin, V_{pp}, as well as a 5-V supply to the chip's operating voltage pin, V_{cc}. This dual-supply requirement increased the complexity required to implement in-system updates. Since most systems would not leave a 12-V supply on continuously, this requirement meant that writing to the Flash was considered an occasional event, not a routine part of system operation.

While the 12-V differential remains an internal requirement of NOR-type Flash cells (the type most suited for code storage and code+data applications), the requirement is now met by the device itself with internal charge pumps that boost the system-level input voltage for Flash program and erase operations. Today, typical NOR-type Flash chips can read, program, and erase from the single system supply voltage. Some Flash devices continue to allow the temporary application of more efficient, external higher-voltage supplies to speed programming in the manufacturing process.

The current design requirement for Flash operating voltage is to provide read, program, and erase capability at the standard system voltage. This system voltage has been trending down from 5 to 3.3 V and below. Since mobile applications such as wireless phones have been among the largest adopters of Flash memory, Flash operating voltages are being lowered aggressively to 2.5, 1.8 V, and even lower to increase battery life.

CYCLING. Flash devices usually have a maximum erase cycling specification. An erase cycle count is the number of times that a block has been erased. For most

Flash devices, cycles are counted on a block basis and erases on one block does not affect the reliability or performance of the others. The cycle count is relevant because the device's performance specifications, such as block erase time or byte program time, are only guaranteed to hold up to the maximum erase cycling specification. This is because erasing a Flash block puts wear on the oxide layer insulating the floating gate, which stores the information in the chip. A widely held misunderstanding of the cycling spec is that the device will fail or cease operating once the max cycling spec has been reached. This is not true; in most cases, the device will continue operating well after the max cycling spec has been exceeded. The spec signifies the amount of erases for which the manufacturer can guarantee the device's performance specs.

Erase cycling specifications have improved dramatically since the introduction of Flash. The first bulk-erase Flash devices were rated for 1000 cycles, but 10,000-cycle specs were soon commonplace. The addition of embedded write state machines to Flash devices brought tighter control of the erase process and made 100,000 cycles possible. Today, typical devices are rated for 100,000 to 1,000,000 cycles, but improved media management techniques (see Section 2.4.1.7) have decreased the number of erases used in typical data storage processes, so 100,000 cycles is generally enough for most applications.

SUSPENDS. One of the first issues with executing system code directly from Flash and expecting to write to that same Flash device is the *read while write* problem. This problem refers to the fact that a standard single-partition Flash device cannot be read while a write operation (program or erase) is in progress. Since the internal write state machine is manipulating the internal control signals and voltages to perform the write operation on the cells of the Flash array, those same Flash cells cannot be read at the same time. Because the system is usually architected such that CPU instructions are fetched from the Flash, a catch-22 situation is created as follows: Where does the CPU obtain the instructions it needs to control the Flash write operation when those same instructions will become unavailable as soon as the Flash write operation begins? This problem is magnified in a code+data application since writes of data to the Flash occur much more often than code modifications.

The standard method for dealing with the read-while-write issue is to copy the minimal code needed to write to the Flash into RAM and to execute from RAM during the write operation. However, the hole in this method is the inability to execute additional code from the Flash in the case that an interrupt occurs during the Flash write operation. To alleviate this constraint, write suspends were added to the Flash write state machine.

Write suspends allow the CPU to interrupt an in-progress program or erase operation, read from the Flash array, then resume the suspended operation. Recent Flash devices support erase suspend to read, program suspend to read, erase suspend to program, and even the nested erase suspend to program suspend to read. When executing the Flash write code from RAM, an interrupt handler can poll for interrupts and suspend the write operation to allow code execution in order to service an interrupt. This method of interrupt handling is sometimes called *software read-while-write*.

Suspends incur some performance penalties in the area of suspend latency and write performance. Suspend latency is the time the write state machine takes to pause the in-progress write operation and return the Flash array to read mode. The maximum suspend latency specification of a Flash device should be evaluated against the system's maximum interrupt service latency to ensure compatibility. Write performance is impacted by suspends since programming or erasing is paused during a suspend. In a system that interrupts its Flash write operations often, data storage throughput will be degraded.

DUAL PARTITION. A recent architectural development is aimed at relieving the performance compromises associated with program or erase suspends by reducing the need for them, but at the cost of increased silicon area. Dual-partition Flash devices divide the Flash into two partitions that can be operated somewhat independently from each other. For example, the CPU can start an erase or program operation in one partition, while executing code in the other partition. This eliminates the need to execute Flash write code from RAM, since this code can be run from the partition opposite that being written. The general usage model is for code to reside in the larger partition and data in the smaller partition. However, careful planning of code and data location is required to ensure that the required code is always available from one of the partitions when needed. Also, the need for suspends is not completely eliminated by dual-partition devices. If something needs to be read from a partition that is already busy with a write operation, then a suspend will still be necessary.

In the future, Flash memory devices may be further subdivided into more partitions. Multipartition Flash would be able to support independent program, erase, or read operations in each partition. This concept could be taken to its logical limit by allowing autonomous operations in each block.

Designers now have a number of choices for how they handle read-while-write. Systems that can tolerate some interrupt latency and have relatively low data throughput and bus utilization can use standard single-partition Flash to save on component cost. Higher performance systems with less tolerance for interrupt latency or higher throughput and bus utilization can consider dual-partition Flash devices to reduce exposure to performance degrading factors.

WRITE PROTECTION. As Flash devices became write-enabled in the system, the opportunity for data corruption due to accidental writes to the Flash becomes possible. Some of the sources of these accidents are noisy systems or buggy software. To help decrease the vulnerability of the Flash to these incidents, two general approaches to write protection have been developed.

One approach is hardware centric and involves switching off the V_{pp} program/erase voltage source pin when writes to the Flash are not anticipated. Since the contents of the Flash cannot be changed without a programming voltage source, the Flash is shielded from inadvertent writes.

Another approach is software centric and involves extending the length of the command sequence required to initiate an operation that would change the Flash array contents. By making the handshaking sequence longer, any write sequences generated by random system noise would be much less likely to initiate a program or erase operation.

BLOCK LOCKING. As the Flash becomes the central repository for both code and data in embedded system designs, concern increases about the safety of the information stored in the Flash device. Additional protection beyond the broad write protection schemes are needed because the Flash can potentially be programmed or erased by any code running on the CPU. Consequently, designers want the ability to protect that information with various levels of security. Block locking is a feature added to Flash devices to service this need.

One of the first approaches to block locking was the hardware control boot block on Intel's first Boot Block Flash devices. This approach used a hardware pin to control the lock status of the first block in the Flash, which contains the CPU's boot-up code, with the idea that this could protect crucial startup and reprogramming code in case the rest of the Flash was corrupted. However, as today's sophisticated applications deposit more code and more data into the Flash, additional flexibility in the locking scheme is needed.

The newest approach to locking is available on more advanced Flash devices, is sometimes called instant individual block locking, and it protects each block independently with two levels of protection and can be changed with no latency. The first level of protection is useful for protecting blocks that are used for data storage. This first level prevents program or erase to a locked block. Since data is written relatively frequently, this level allows software-only unlocking and relocking for fast locking changes in real time, as access to the block is needed. The second level of protection, called "lockdown," is useful for protecting blocks used for code storage. Since code is not updated as frequently, yet is more crucial to system operation, lockdown incorporates hardware control. Once a block is put into lockdown with a software command, unlocking it for an update requires a hardware input pin to be changed. This additional step provides additional protection for code segments stored in Flash.

2.1.5.2 Software: A Key Player. While not physically part of a Flash device, software is a vital piece of every system using Flash. As such, this section will treat it as an attribute of Flash. In truth, as applications use Flash in more and more sophisticated methods, increasingly sophisticated software is the key to enabling those applications, and in the future, functions once implemented in software may be implemented in the Flash hardware itself.

Software needs vary from the most basic algorithms that put the Flash device in the proper mode to the most sophisticated media managers that assist in managing the Flash device and provide full file systems capabilities. A system will require some or all of this depending on its specific requirements and needs. This section will explore each of these aspects of Flash software separately.

WHY FLASH REQUIRES SOFTWARE. Flash memory has many unique characteristics that mandate the need for software to program or erase information from the Flash device. Software is essential in any system that needs to write information to a Flash device. Inherent characteristics of all Flash memory, regardless of configuration or packaging, make this true. Flash memory is unlike any other memory device in that it provides byte programmability, block erasing, and nonvolatility.

A Flash device powers on in a "read" state and requires a series of commands to put it into another state such as "program" or "erase." A bit in Flash can be changed from the erased state (generally a "1") to a programmed state (generally a "0") but requires a block erasure to change the 0 back to a 1. Since a single byte of information cannot be erased, and block sizes generally range from 8 to 128 kbytes, special software must be used to modify small pieces of information stored on the Flash device. In addition, Flash memory has a limit to the number of times a block can be erased before performance degradation occurs. This is generally in excess of 100,000 erasures, but must be accounted for in a system to ensure optimal operation.

BASIC SOFTWARE FOR FLASH. Basic Flash algorithms allow a user to perform functions such as programming a byte or series of bytes, erasing a block or series of blocks, reading the device ID, or reading the status of the part. Without these algorithms the user can only read information (code or data) from the Flash device since it powers up already in the read state. Each of these algorithms is specific to the device being used although most manufacturers will tend to use many of the same routines across their product line (i.e., the program routine for all Intel devices will generally be the same).

At a minimum, to program information to a Flash device, read, program, erase, and the system software must use status algorithms. For example, to write a person's telephone and address information to a Flash device, a program routine must first be called that sends the program command to the Flash device and then writes the information to the desired location in the Flash. The status routine (which sends the "status" command to the Flash device) must be called between each program command to ensure that the data has been properly written to the device. Once the data has been written, the read command must be sent to the device in order to read information from the device again. If there is initially not enough free space in the Flash device to write this information, the "erase" command must be called first to erase a block of Flash. See Figure 2.4 for an example calling sequence.

These basic Flash algorithms do little more than send a command to the Flash device that puts it into the proper state. These basic algorithms are, however, essential to the operation of a Flash device in a system.

```
Main()
{
if (!(space_available)) {
    Erase_block (block_to_erase);
    }
Program (beginning_address, data, number_bytes_to_program);
Read();
}

Program (beginning_address, data, number_bytes_to_program)
{
Flash_address = beginning address;
while (number_of_bytes_to_program)
    {
    Send_command (Program_command, Flash_address, data_byte);
    while (Flash_status = = not_done) {
      wait;
      }
    number_of_bytes_to_program--;
    Flash_address++;
    }
}
```

<u>Figure 2.4.</u> Example basic algorithm calling sequence.

HIGHER LEVEL SOFTWARE FOR FLASH. Designers accustomed to using other memory devices that allow a memory location to be directly overwritten, such as RAM or EEPROM, may find using Flash strange at first. They are generally used to a piece of data being located at a specific address in the memory map and being able to modify that piece of data at the same location. However, when using Flash, in-place modification is not practical since the entire block contents would have to be backed-up to a temporary location, then the entire block erased to prepare the location for rewriting. This is impractical and a waste of resources. Instead, higher level software (often called a media manager) is necessary to optimize the use of Flash memory.

The most important thing that a media manager does is manage the storage of information and reclaim space used by old, invalid data. Without this function a block of Flash memory (8 to 128 kbytes in size) must be used for every piece of information stored on the device, regardless of its size. This is generally very wasteful since most pieces of information stored in Flash memory are considerably smaller than 8 to 128 kbytes. A robust media manager can make this block size transparent to the system in several different ways, but most use one of two methods—virtual blocking or a linked-list approach.

VIRTUAL BLOCKING. With virtual blocking, the system accesses data by looking a parameter number up in the lookup table that returns the physical location(s) in Flash memory where each piece of the information is stored. When using virtual blocks, all of the information is stored in pieces that can fit into one or more virtual blocks. For example, a piece of information that is larger than a virtual block will span multiple virtual blocks. The information may or may not be contiguous within the Flash memory and all virtual blocks are the same size. See Figure 2.5, which illustrates the use of virtual blocks.

LINKED LIST. When using a linked list, the system uses a lookup table, but this table indicates only where the first piece of information is located. Every piece of information includes a header, which points to where the next piece of information resides. Each piece of memory is not necessarily the same size, and one piece could

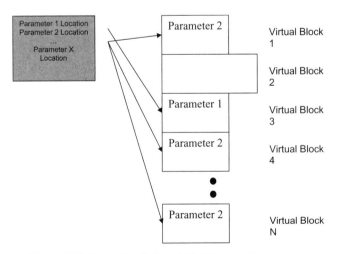

Figure 2.5. Example of virtual blocking media manager.

INTRODUCTION

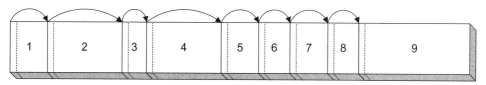

Figure 2.6. Example of linked-list media manager.

span multiple Flash erase blocks. See Figure 2.6, which illustrates the use of a linked list.

ADVANCED MEDIA MANAGEMENT FUNCTIONS. A media manager can also implement advanced features, such as reclamation or wear leveling, which are above the capability of standard Flash drivers.

Reclamation is an essential function of a media manager since it makes portions of the Flash memory usable again by erasing blocks. As information is updated in the Flash memory, the old information is not immediately erased (since you can only erase a block of Flash), but is marked "invalid." When enough of the bits in a block are marked invalid, the media manager will reclaim that block. This method of managing data storage allows many pieces of data to be stored and updated within a block before a reclaim and erase are necessary, which increases the effective block cycling of the Flash device. The steps necessary to reclaim a block of Flash include copying the valid information from the block to be reclaimed to a new block, updating the tables used to translate parameter numbers to a physical location in memory, and finally erasing the old block. While this is occurring, special attention must be paid to power loss recovery. If power were lost to the system in the middle of a reclaim, there is a possibility of losing information if proper power loss mechanisms are not used.

Wear leveling is useful in any system that writes information to a Flash device. If a media manager is not aware of the need for wear leveling, one block can tend to be used exclusively for a particular piece of information. This can wear out a Flash device earlier than it should, especially when a piece of information is updated at a rapid rate. A robust media manager keeps statistics on the number of times each block is erased and will occasionally move data around such that the piece of data being updated at a rapid rate is moved to a block that has not been erased as many times. Generally a wear-leveling algorithm will ensure that the erase counts of the block that has been erased the most number of times and the block that has been erased the least number of times stay close to each other. This extends the amount of time before the Flash reaches its cycling specification.

FLASH FILING SYSTEMS. Finally, the most sophisticated form of Flash software is a filing system. A filing system not only makes Flash's large erase blocks transparent to the system, but allows the system to access data via file names and write, modify, and delete these files. Generally multiple files may be open at one time. A filing system can use a standardized translation protocol such as FTL (Flash translation layer) to transform disk-style file allocation table (FAT) file sectors to locations in the linear Flash memory or a filing system can use a proprietary method to create this translation. With a filing system, requests for files stored on the Flash memory appear to the system as though the files are stored on a standard disk drive. A filing system includes all of the features of a media manager in addition to its file capabilities.

FUTURE OF FLASH SOFTWARE. Flash software continues to expand the usage of Flash and add value to systems. By allowing simultaneous code execution and data storage, Flash software has assisted cellular phone original equipment manufacturers (OEMs) in reducing component count. In doing so, the cellular phone is now lighter weight, smaller, and has a longer battery life. In addition, simultaneous code execution and data storage allows the same component count reduction in handheld personal computers (PCs), micronotebooks, smart phones, web phones, and set-top boxes.

Advances in Flash software for cellular phones will assist in Internet browsing and ultimately over-the-air code updates. Set-top boxes will benefit from Flash software enhancements, which will enable fast, reliable code updates without the need for redundant memory. Finally, all applications using Flash memory for file storage can benefit by using Flash software not only for the file storage but also for their boot code.

As usage of sophisticated software media managers becomes commonplace, certain core functions that are commonly implemented in software may be accelerated by pulling their functionality into the Flash hardware itself. For example, software media managers routinely do "find" operations to search through the Flash contents for a particular piece of data. This involves reading data from the Flash over the memory bus, comparing it to the target data in the CPU, and looping until the data is found. A faster and more efficient method would be to incorporate this into the Flash chip, so that bus and CPU bandwidth are conserved.

2.1.5.3 The Expanding Role of Packaging.

For years, packaging was an afterthought in the Flash market. The Flash device silicon was what was important, and the silicon was put into a package that had enough pins to provide all the necessary signals. Cutting edge technologies such as DIPs (dual in line packages) and PLCCs (plastic leaded chip carrier) were good solutions, both in the omnipresent offline programmer, and in sockets on the prototyping board. Additionally, they offered easy access to the leads for debugging. Fast forward to today where cellular phones in Japan target 60 g for the whole phone and aren't much bigger than the larger pin count DIP devices of yesteryear.

The constant pressure on OEMs to develop smaller products translated into another pressure on silicon vendors to come out with smaller, lighter, and thinner packages.

TYPICAL PACKAGES FOR FLASH MEMORY. In the early years of Flash memory, devices were typically offered in dual-in-line (DIP), and plastic leaded chip carrier (PLCC) packages. While the DIP was the last printed circuit board (PCB) through-hole package, PLCC began the practice of surface mount technology (SMT). Later, these types of packages were followed by a family of surface mount packages called small outline packages (SOP), and most recently ball grid array (BGA) type packages. Some of the most common packages for Flash memory are shown in Figure 2.7.

Over the years, SOP, an established industry standard, has gained momentum primarily due to significantly reducing board space consumed by devices. The SOP family provides flexibility by offering pin-to-pin compatible densities in the same package, making it easier to implement expanded code without the need to redesign the PCB. The SOP family's gull-wing-shaped leads and dual-in-line design provides

INTRODUCTION

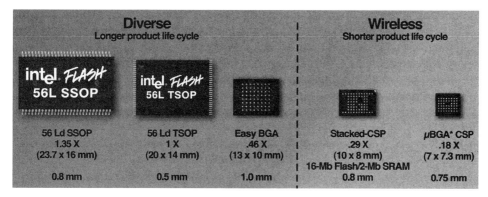

Figure 2.7. Typical Flash memory packages.

improvements over traditional packages. As an example, the SOP family offers improved lead inspection, board routing, and solder joints over other PLCC packages.

The SOP family includes the thin small outline package (TSOP), the shrink small outline package (SSOP), and the plastic small outline package (PSOP). The TSOP has lead pitch of 0.5 mm with a maximum package height of 1.2 mm, while the SSOP and PSOP have 0.8- and 1.27-mm lead pitches and 1.9- and 2.95-mm maximum package heights, respectively.

The TSOP is a multiple supplier, industry standard small form factor package. TSOP provides the easiest Flash memory density upgrade path. The TSOP supports many densities and is ideal for sophisticated functions where size and portability are primary design concerns, such as cellular phones, PC BIOS, networks, Personal Computer Memory Card International Association (PCMCIA) cards, mass storage, fax/modems, arcade games, and printers.

The SSOP is rugged, easy to use, and supports mid to high densities. It is particularly well suited in applications that must withstand extreme temperatures and demand high reliability. It is best suited for the most demanding applications such as automotive, cellular base stations, and industrial embedded applications.

The PSOP offers the same high-reliability characteristics as the SSOP and supports low to mid-density products. It is particularly well suited in applications that must withstand extreme temperatures. It is best used for medical and industrial embedded applications, office settings for printer servers and fax/modems, and Telecom switches and networking.

CHIP SCALE AND BALL GRID ARRAY PACKAGING. Flash memory applications based on PCMCIA cards and other small form factor systems, such as computers and handheld communications (cellular phones, personal communication services (PCS), pagers), are becoming more compact every day. In addition, larger applications such as telephone switching systems and PC network hubs are attempting to add more functionality and memory density into the same size box. Both of these areas are driving the demand for a new type of very small package: the chip scale package (CSP). A package that doesn't exceed the size of the die by greater than 20% falls into the category of CSP.

The market for ever-shrinking consumer products such as cell phones is seeing explosive growth and fierce competition. Just about every handheld consumer product seems to be shrinking at a very rapid rate. The race to sub-handheld-sized products is on, and the race is being judged in millimeters, grams, and additional functionality. At the same time, strong competitive forces are putting pressure on manufacturers to reduce the cost of these products.

Newly designed or redesigned consumer products are being introduced almost daily, and product design cycles are decreasing to meet these accelerated introduction cycles. In the past, aggressive companies may have introduced both Fall and Spring models, but in this new environment, it is not uncommon to see new models being introduced every quarter. The entire market for chip-scale package solutions is seeing the ripple effect of this environment. Suppliers are seeing step-function demand increases for the components. The opportunities for manufacturers and suppliers are great, but so are the challenges to supply components that enable this new wave of consumer products.

In addition, there has been a dramatic shift from traditional leaded packages to BGA-type packages. One of the reasons for this shift is attributed to increased PCB assembly yields due to the lack of coplanarity and bent lead issues and the self-alignment properties of the BGA package, which offset minor component placement inaccuracies.

There are several types of BGA packages that address the needs of a variety of applications. The construction of the packages range from the μBGA package developed by Tessera Corporation that uses a elastomer and polyimide tape interposer, with the back of the silicon die exposed on the top side, to the traditional plastic molded device with either a laminate or polyimide interposer. These packages are also produced with a variety of pitches (distance between balls). The most common pitches are 1.0, 0.8, 0.75, and 0.5 mm. Some companies are also taking integration to a new level by combining more than one silicon die into a single package, typically referred to as a stacked CSP package. For example, Flash memory is being combined in a single package with logic and SRAM.

2.2 CODE STORAGE

Code storage was the first usage and remains a large consumer of Flash memory. The variety of code storage applications is broad since all types of electronic equipment—from personal computers, cellular telephones, laser printers, networking systems, Global Positioning System (GPS) receivers, biomedical equipment, and postal meters to larger machines such as automobiles, aircraft, and farm machinery—use Flash memory for code storage. Virtually all types of microprocessor-based systems use Flash memory. Across this diverse base of applications, two basic usage models have emerged that span most code storage applications: execute in place (XIP) and store and download (SnD). Most applications use either SnD or XIP, although some designs have evolved to use a combination of both.

In this section, we differentiate between these two usage models, discuss the benefits and design considerations pertaining to each, and provide examples of how these usage models are implemented in applications today. In the final portion of this section, we describe some future applications emerging as a result of Flash memory technology.

CODE STORAGE

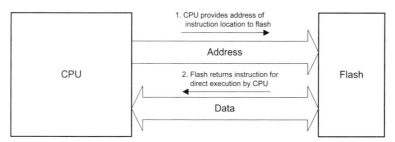

Figure 2.8. XIP model—CPU executes program code directly from Flash memory subsystem.

2.2.1 Execute-in-Place

Execute-in-place, or XIP, refers to the usage model whereby executable code is stored in Flash memory, and the system microprocessor fetches instructions directly from the Flash memory subsystem for execution. Examples of XIP include embedded code storage in industrial control and biomedical equipment, BIOS code storage in personal computers, and engine and drive train control in automobiles (Fig. 2.8). In each of these application examples, several market requirements are key considerations that benefit from the XIP usage model. While the relative ranking of most important to least important attributes may vary between applications, each is essential in meeting cost, performance, and reliability system requirements.

2.2.1.1 Relevant Flash Attributes for XIP Applications. The primary Flash attributes that XIP applications care about are read performance, bus width, and code integrity. While other Flash attributes are used in these types of applications, they are not directly relevant to the core XIP functionality. For example, a Flash device's system voltage write capability, write state machine, and blocking all enable code updates, but these updates happen infrequently relative to the core read functions.

READ PERFORMANCE. The primary attribute required for XIP is read access time or read performance. In order for the system microprocessor to execute code directly from the Flash memory, the memory must output requested code fast enough to keep the instruction queue full, without causing delays that would impact system performance. Typically, Flash memory read performance is directly related to operating voltage; that is, the higher the operating voltage, the faster the read access time. The typical access time of a fully rated 5-V asynchronous Flash device is 60 to 65 ns. This is sufficient to support zero wait state read performance up to a 15-MHz clock speed.

Generally, two factors drive the performance of microprocessor-based systems: clock speed and operating voltage. Interestingly, microprocessor performance is typically inversely proportional to operating voltage since the lower the voltage, the faster the transistors within the microprocessor can switch, because there is less voltage differential between logic levels. As the microprocessor performance increases, the read performance required of the memory subsystem also increases. However, just the opposite effect occurs in Flash memory; the lower the operating voltage, the slower the read performance of the Flash. This contrast in how microprocessor and Flash performance relate to operating voltage presents an inherent performance limitation of the XIP model. When the raw speed of the Flash no longer keeps up with microprocessor performance, the XIP model becomes less viable as a complete memory subsystem, at least when using asynchronous Flash memory devices. The solution is to use page-mode or synchronous-burst read interfaces. This is an important development that warrants its own in-depth discussion, found in Section. 2.2.1.3.

BUS WIDTH. Bus width is another key attribute when selecting a Flash memory for XIP applications, since the memory device's bus width should match that of the CPU for best performance. Flash devices are generally available in 8- or 16-bit widths, and multiple devices can be arranged in parallel to build up the required bus width. Interleaving is another design technique sometimes used to increase memory performance by reading from two devices alternately in a Ping-Pong fashion. However, this use of this technique is decreasing due to page/burst-mode Flash devices, which integrate an optimized version of the same idea directly into the chip.

PROGRAM CODE INTEGRITY. Another very important attribute for code storage is code security. Virtually all systems in today's market require absolute code integrity. Not one bit can be in error or the system may not boot. While error detection and correction can be implemented in SnD usage models (such as hard disk drives), XIP must guarantee zero bit errors to ensure reliable operation. In addition, many Flash memory devices have write protection features to prevent accidental writes to the code storage portion of the memory map. These hardware control inputs must be externally driven to allow code updates. Finally, some Flash memory devices have dedicated program voltage inputs (V_{pp} separate from the V_{cc} input), which can be switched to ground to prevent unintended write operations from altering memory array contents.

NOR-type Flash memory is ideally suited for code storage because it meets both the read performance and security requirements. NOR-type Flash has fast random access, meaning every new address, even nonsequential ones will result in data within the maximum access time. NOR also has extremely high reliability, requiring no error detection and correction, and has data retention of 10 years or more.

By contrast, NAND-type Flash is not well suited for code storage, for two reasons. First, NAND has very slow random access, on the order of 5 to 10 µs (compared to 60 ns for NOR.) Second, NAND may require error detection and correction because some bits may randomly change state. Since code storage requires absolute bit integrity, NAND-type Flash is really most suited for data storage only.

2.2.1.2 Advantages and Limitations of Flash for XIP.
The advantage of using Flash for XIP is the clean memory model. Code is stored in, updated to, and executed from the same place, eliminating the redundant memory that would be needed if code storage and execute took place from separate memories, as it would with store-and-download applications. By reducing memory redundancy, the XIP model reduces the amount of RAM needed in the system. In addition, code does

not need to be copied from nonvolatile memory (Flash or disk) to RAM before the system can begin operating. Thus, systems can boot and get operational faster, which is what users want.

The limitations of Flash for XIP are speed and size. As discussed in Section 2.1.5.1, Flash read performance can directly limit CPU and system performance in an XIP environment. However, burst and page mode reads are now available to address this limitation. With the latest synchronous Flash devices capable of zero-wait state operation at 66 MHz, few systems will be limited by Flash read performance. Code sizes continue to increase as software developers add more and more features. With XIP, all of the code needs to be contained in Flash. Some of the latest Flash devices, such as the Intel StrataFlash cellular memory, pack 64 Mbytes in a single chip, so large amounts of code can be stored for XIP in a very small amount of board space.

2.2.1.3 Advanced Read Interfaces: Page and Burst Modes. Traditional Flash memory employs an asynchronous interface similar to SRAM. In this configuration, the CPU supplies the Flash memory with an address, then some time later (called access time) the Flash memory returns data back to the processor. Typically, 5-V Flash devices have an access time ranging anywhere from 55 to 120 ns; whereas, 3-V components have access times that range from 75 to 150 ns. Lower density components have faster access times. At the 16-Mbit density, the 3-V access times are between 75 and 150 ns. With these access speeds, most high-performance CPUs have to wait for data. For example, operating at 33 MHz, a typical system using an 80-ns Flash memory would have to insert two wait states. During these two cycles, the processor sits idle waiting for information.

The asynchronous read architecture is perfect for nonburst processors because nonburst processors retrieve one piece of data at a time. However, a high-performance burst CPU can access a "group" of data when fetching information from the memory subsystem. This type of memory access, used primarily in code XIP applications, places high read performance demands on Flash memory. System performance degrades further when a burst processor is used with an asynchronous memory. During a burst cycle, the initial memory access takes a minimum of two clock cycles, consisting of an address phase and a data phase. After the processor reads the first piece of data, however, data can be accessed on every clock edge. Using an 80-ns asynchronous Flash memory component again, a typical system would have a wait state profile of 2-3-3-3 (see Fig. 2.9).

To alleviate the performance gap between logic and memory when using asynchronous Flash devices, high-performance systems had to rely on complex and redundant memory configurations such as memory interleaving and code shadowing to overcome Flash memory's read performance shortcomings. Because these techniques increase system cost and decrease system reliability, they are not good solutions.

Clearly the performance gap must be resolved with a clean architectural solution. The brute-force solution, speeding up asynchronous Flash performance, is not possible because the standard techniques used to make CPUs faster, such as lowering operating voltage and increasing transistor density, do not apply to Flash memory. Unlike logic circuits, lower operating voltages and higher densities tend to have a negative impact on Flash memory read performance. A portion of the Flash memory sensing logic consists of analog circuitry, which does not scale efficiently with voltage,

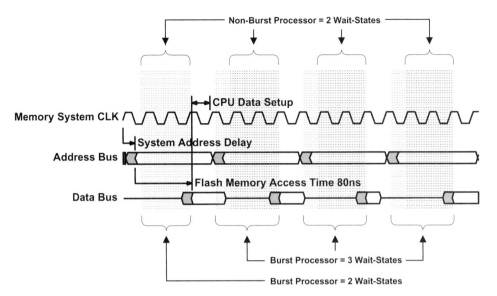

Figure 2.9. Example wait state profile of a burst vs. nonburst processor.

making it difficult to dramatically improve access times in the face of the trend toward lower operating voltages. Density also has a negative impact on the read performance of Flash memory devices. As the Flash memory density increases, the internal capacitance increases slowing down random access speeds.

Instead of tackling the problem with brute force, the issue is best solved by changing the Flash memory read architecture in order to resolve the XIP performance bottleneck. How have other memory technologies dealt with the read performance issue?

In order to keep pace with processor performance, other memory technologies such as ROM, DRAM, and SRAM have implemented interfaces that improve read transfer rates by taking advantage of the processor's burst protocol. To address performance issues faced by the computer industry, for example, DRAM has evolved from fast page mode to extended data out (EDO) to synchronous to direct DRAM interfaces. Likewise, ROM devices have made a similar transformation over time. ROM manufacturers offer asynchronous, page-mode, and synchronous architectures to meet different performance needs. Today, 50% of all ROM components shipped support a page-mode interface while new synchronous ROM devices are being introduced. All of these innovations use the predictable and sequential pattern of the burst protocol to help keep pace with processor performance. Flash memory is following a similar interface and architectural evolution to solve these high-performance application requirements.

Like page-mode ROM, asynchronous page-mode Flash memory provides a high-performance nonclocked memory solution for burst CPUs. Random-access speeds for this architecture are equivalent to traditional, asynchronous Flash memory. However, asynchronous page mode provides access to subsequent information at very high speeds. Typically, the page access time is equal to the speed of the output buffers—20 to 30 ns for Flash memory.

Synchronous Flash memory interfaces provide the highest performance memory solution for burst CPUs. By looking up the next word of data while outputting the current data, the synchronous interface for Flash memory offers enough memory throughput to achieve zero wait-state performance. This allows the burst access time to decrease even further compared to the asynchronous page-mode access time.

The synchronous burst-mode Flash memory interface is very similar to synchronous DRAM (SDRAM) and pseudo static RAM (PSRAM) and, in some cases, can be connected directly to the memory bus through the PSRAM port. To control the Flash memory device in synchronous mode, two new signals are added: address valid and clock. The address valid pin informs the Flash device that a new address is present on the address bus and to start a new burst cycle. The clock signal synchronizes the Flash memory device with the CPU. These signals connect directly to the processor's memory interface unit to minimize system glue logic and system propagation delay.

2.2.1.4 Future Read Interfaces. To continue performance improvements in Flash, new interfaces will be needed, versions of which are showing up in the RAM world such as double data rate (DDR) DRAM and direct Rambus DRAM (RDRAM). The DDR interface is similar to the synchronous burst mode mechanism, but it outputs data on the rising and falling edge of the system clock, which effectively doubles the data transfer rate at a given bus frequency. As system bus frequencies start to exceed the 133-MHz realm, the interface will need to evolve into a source synchronous implementation. The source synchronous approach is required at high frequencies to minimize system flight time (the time it takes a signal to travel from one component to another). With this interface, the signal used to latch data (strobe) is set with the data to the recipient. With a source synchronous interface a frequency of about 400 MHz can be achieved. However, the implementation of these enhanced interfaces is dependent on the adoption of high-performance interfaces on the embedded processors that use Flash. Both the processor and memory need to support the optimized interconnect in order to improve overall system performance.

2.2.1.5 Selecting the Right Interface. Application requirements are changing and so will the Flash memory architecture in the future. Asynchronous Flash memory works well with nonburst processors. Asynchronous page-mode and synchronous burst-mode Flash memory devices help resolve Flash memory performance issues that system designers are faced with today when executing code directly from Flash memory with a burst processor. Asynchronous page-mode Flash memory is ideal for nonclocked memory subsystems and for slow operating frequencies. Synchronous burst-mode Flash provides the high-performance solution for high CPU operating frequencies. All three Flash memory architectures have important characteristics that help address the performance needs for different applications (see Fig. 2.10). To push the Flash memory performance in the future, other interfaces will need to be adopted.

2.2.2 Store and Download

Store and download (SnD) refers to the usage model whereby program code is stored in a nonvolatile media, such as Flash, ROM, or disk, and, upon power-up, downloaded into system RAM for execution (Fig. 2.11). Since nonvolatile memory solutions typically have slower read access time than volatile memories such as

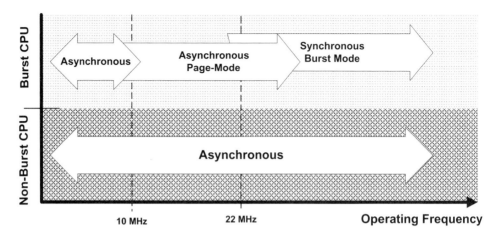

Figure 2.10. Memory solutions for burst and nonburst processors per bus frequency.

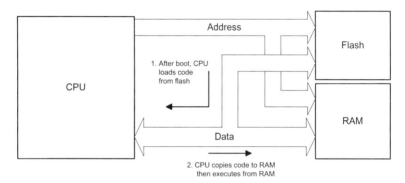

Figure 2.11. Store and download model—program code is copied into system DRAM for faster execution.

DRAM or SRAM, the SnD model is well suited for systems that require faster read performance. Applications that use this approach include boot code and operating system code in data communication networks, switching hubs, network routers, and code stored in telecommunication switches, to name a few. Some applications select SnD for specific needs stemming from their system operation or history. A well-known example is the architecture for personal computers, which stores BIOS code in Flash memory and code on hard disks, but downloads it all to RAM for execution because the PC operation model expects execution to take place from RAM and requires the ability to randomly write to memory. Other examples of system needs driving the memory model toward SnD is wireless local loop base stations, which need to receive over-the-air code updates from their wireless data link. Since such a base station must run code to maintain its wireless transmission protocol to receive a new code update, it can't run that code from Flash at the same time it is writing a new code image to Flash. Thus, SnD is a good fit.

2.2.2.1 Relevant Attributes for SnD. The store and download usage model demands little from the Flash memory. Most of the features and attributes built into Flash memory devices are focused on enabling writing and updateability, and these are not heavily used in SnD applications.

DENSITY. The Flash memory must be large enough to store all of the code needed in the system. Both very small (8 Mb) to very large (1 GB) memories are available to cover all needs, and, since devices do not need to be linearly mapped for SnD, multiple devices are easily configured to provide additional space. However, board space and cost often drive SnD applications to favor the highest density components.

ACCESS TIME. Depending on the system requirements, access time may or may not be important for SnD applications. On one hand, since code will be run from RAM, access time is not a limiter on execution speeds. But on the other hand, the entire code image will be downloaded from Flash at boot time, so faster access time will speed up the boot process for applications that need faster turn-on time. For example, a digital broadcast video camera using SnD will need to get the code downloaded and running as fast as possible, or an important news event may be missed by a cameraman waiting for his camera to boot.

CODE INTEGRITY. While code storage integrity is always important, downloaded code can always be run through error correction algorithms before being placed in RAM for execution; so this is not actually a key requirement. Thus, either NAND (with error correction) or the more reliable NOR can be used to meet the system requirement. However, adding error correction can add cost to the system if done in hardware or slow down the boot time if done in software.

2.2.3 Contrasting Execute-in-Place versus Store and Download

The primary benefits of XIP include cost, power savings, and fewer components. By executing code directly from the Flash device, less DRAM is required. Conversely, the argument for SnD is performance based; by downloading program code from the storage media, Flash, into DRAM for execution, the overall system performance increases (assuming the DRAM has faster access time than Flash.) However, SnD presents a redundant memory model, by having separate memory subsystems, Flash for storage, and DRAM for execution. Even though many XIP systems will have some amount of DRAM, more DRAM is required in an SnD model, which increases cost. The cost/performance benefit is a trade-off that must be optimized based on the performance requirements and cost constraints of each application.

2.2.4 Future Code Storage Applications

The fundamental benefits of Flash technology (including nonvolatile, low power, and high reliability) makes feature-rich small-form-factor battery-powered applications possible. While cellular telephones and handheld GPS receivers existed without Flash memory, both have achieved smaller form factor, increased features, and longer battery life directly as a result of using Flash. A number of new applications are emerging that will also leverage the technological benefits Flash memory provides. For example, Flash memory is being used increasingly in automotive applications. The silicon content in automobiles is growing rapidly. While program code storage for engine and drive train control has relied on Flash memory for years, the fastest growing segment in automotive applications is within the dashboard. Specifically, driver information systems are fast becoming standard in most luxury cars. Automotive electronic suppliers are integrating cellular telephones, automotive stereos, GPS navigation, and wireless digital data transmission into a single compact dashboard system. Flash memory is the ideal storage solution for both the program code and map/route data storage in these driver information systems. Because of their higher price, luxury cars are better able to absorb the price point increase of the new driver information system. Over time, as the volume of these systems increases, the price adder should decrease with the economy of scale, and eventually they will become options to lower end cars, and perhaps standard equipment as car stereos are today.

2.3 DATA STORAGE

All applications need to store some amount of data, but many of today's applications have a need for mass storage of data. These applications can be divided into two broad categories. The first consists of devices that store data specifically for use in performing its function. An example is a network router, which stores large address tables and uses them in order to send network packets to their correct destination. The second category consists of devices whose sole purpose is to capture and store data from the outside world. An example is a digital audio recorder, which stores audio samples for later playback. In both cases, the data storage medium must be nonvolatile.

2.3.1 Why Use Flash to Store Data?

In the rush to digital media, applications that fall into the second category are rapidly increasing. In addition, they tend to be portable, such as digital cameras or audio recorders. However, in portable or handheld devices there are physical size and weight constraints in addition to vibration, shock, reliability, power, and heat dissipation requirements. This type of environment is an ideal place to use Flash memory for storage instead of rotating magnetic media.

Traditional hard disk drives use more power and generate much more heat than Flash memory. Additionally, rotating media is much more susceptible to vibration and shock damage and that makes them less reliable in a portable or handheld system. Applications such as cellular phones and handheld organizers use Flash memory instead of hard disk drives because otherwise they would be big, heavy, fragile, and have short battery life. Another issue is the minimum capacity of disk drives. Hard disk drive capacities are continually increasing but are not readily available in small capacities. Common disk drive sizes are in the gigabyte range while Flash cards are in the one to hundreds of megabyte range. Flash storage may better suit the needs of the application by providing only the necessary capacity. If your application needs only 2 or 100 MB, then a Flash card of that capacity could be more appropriate and available than a hard disk drive.

2.3.2 Architectural Decisions

When engineering a system utilizing Flash for data storage, the first decision must be whether the Flash memory should be removable or nonremovable (resident). In addition, a decision between NOR- and NAND-type Flash must be made since both have pluses and minuses for data storage.

DATA STORAGE

TABLE 2.3. Key Criteria Differentiating between Embedded or Removable Flash Storage

Embedded	Removable Flash Cards
• Combining code and data in the same Flash chip is desired • Data or code is changed frequently • Physical space constraints preclude adding slots for removable cards • A minimum amount of readily available storage is needed • System cannot tolerate an unexpected removal of the storage (as could happen with a removable card) • Weight constraints exist.	• The size of data is greater than what can be reasonably stored locally in embedded Flash memory. • Moving data between systems is necessary. • Flexibility in the amount of storage is needed by the end user.

2.3.2.1 Removable versus Embedded. Nonremovable Flash memory storage refers to a Flash device or group of devices permanently embedded into the system. (See Table 2.3.) A key benefit of embedded Flash is the possibility of code+data. Since most systems that need to store data will also require code to be stored someplace, putting both in the same chip makes sense. For example, a digital camera can have a larger than necessary Flash chip for its code storage and use the extra space to store pictures when no card is available. This can be convenient for the end user for quick, unplanned snapshots. The device may not be able to afford the necessary physical space and weight to add a removable card slot.

Removable Flash memory is placed on a card or other carrying mechanism so that it can be removed from the system in order to transfer data to another system or be swapped with another card for additional storage. Applications that need to store large amounts of data that will not fit in embedded Flash because of cost or size constraints may require removable Flash media. Using multiple removable Flash cards allows the system user to have limitless storage capacity. Flash cards allow for forward compatibility because they provide a standard system interface. The user can increase his storage capacity by buying additional cards or by buying Flash cards with larger capacities. A common Flash card application is a digital camera (Fig. 2.12) or audio recorder. These are good examples of Flash card applications because it is unlikely that the user will have a network connection available when recording the audio or video.

2.3.2.2 NOR or NAND for Data Storage. Unlike code applications, data storage applications are not required to directly interface the storage medium (Flash in this case) with the CPU. Thus, the interface to Flash can be abstracted above the hardware signal timing requirements of the processor. The CPU requests a specified piece of data using a standard protocol and receives back the desired data in the standard format. One of the common abstractions used is the advanced technology attachment (ATA) interface, which will be discussed further in Section 2.3.4.1. By hiding the specifics of the Flash device, this abstraction enables two things: flexibility in the type of Flash device used and manipulations on the retrieved data before being passed to the CPU.

Behind the abstracted interface, the storage medium could be anything. For example, the ATA interface is used as the interface to hard disks, floppy disks, wired networks, Flash memories, CD-ROMs, and even wireless network transceivers. This characteristic can be used to increase flexibility in the type of device providing the actual storage, so either NOR or NAND Flash could be used. A particular application's requirements may demand one of the specific characteristics of NOR (fast random access, bit reliability) or NAND (fast sequential access), but otherwise, designing the interface to support both types of Flash will sensibly increase sources of supply.

The abstracted interface also allows the storage subsystem to massage the retrieved data prior to returning it to the CPU. Two ways that this freedom can be exploited are error correction and compression. However, both compression and error correction will slow down the retrieval of data because the algorithms must decompress and correct the data before returning it.

Error correction uses mathematical algorithms to correct data that has changed through degradation of the Flash cells or other means. Generally, these algorithms require additional bits to be stored with the data in order to provide enough redundancy to fix errors. The more errors are statistically expected from the Flash the more extra bits are needed to correct them. While this decreases the effective storage capacity, error correction can mask flaws in a Flash device, making it possible to tolerate less reliability in the Flash device. In a Flash context, this helps NAND Flash because NAND devices do not guarantee 100% good bits, allowing use of lower cost devices. NOR Flash devices do not benefit from error correction because they do not need it, because all NOR devices currently on the market guarantee 100% good bits. The trade-off is between the lower cost of non-100% NAND devices (partially offset by the required storage of additional bits for error correction) and the full usability of NOR Flash devices.

Compression uses mathematical and symbolic algorithms to represent digital data in fewer bits than the actual data by encoding patterns and repetition. By using an abstracted interface, data can be compressed before being stored into Flash, then, when it is accessed, decompressed before being returned to the CPU. This enables an effectively larger storage device than the actual physical device. Compression should assist both NOR and NAND devices equally, if both devices can deliver 100% good bits. If the error correction is used on NAND devices, the additional correction bits may offset the gain in capacity from compression.

The bottom line for data storage applications is that an abstracted interface blurs the differences between NOR and NAND. It also enables additional design techniques such as correction and compression, whose trade-offs should be carefully considered versus application performance requirements before selecting a Flash technology.

Figure 2.12. Sony Cybershot 8.1 megapixel digital camera with removable flash memory. (*Courtesy of Sony.*)

2.3.3 Embedded Flash Storage

Although an embedded array of Flash could also be used for code or code+data applications, the term *embedded* most commonly refers to an array used for data storage.

Both NOR and NAND Flash can be used for embedded storage. If the Flash is expected to do code storage (e.g., boot code), then the storage integrity and fast random access of NOR makes it the right choice. On the other hand, the fast programming and high density of NAND make it stronger for pure data storage.

Embedded Flash is an excellent solution for applications that need to capture and access data but do not need to transport the data between systems. For example, a fax machine can use embedded Flash to store incoming and output fax transmissions before sending or printing them. Flash provides a good solution in this case because its nonvolatility prevents the loss of faxes if power is lost. In addition, it provides solid-state reliability with no moving parts. Since the data is only transported through the phone line or paper output, removable storage is not necessary.

In other cases, the physical size of the application may preclude usage of a removable Flash card. For example, the smallest and lightest MP3 players are built only with embedded Flash.

2.3.3.1 Relevant Attributes for Embedded Flash Storage.

The relevant attributes for data storage embedded Flash differ from other usage models. Consequently, the needs focus on fitting a large amount of storage into a small physical space and the ability to safely store and update data over the long term.

DENSITY AND PACKAGING. Most applications considering adding embedded Flash need a relatively large amount of storage and want to fit that into as little board space as possible. Density becomes relevant because higher density Flash chips fit more storage into the same space since, in general, new chips pack twice the storage into the same silicon die area. Thus for a given number of chips in an array, upgrading to higher density devices can double the storage capacity. For this reason, designs should always plan for future upgrades by routing the additional address lines needed to support bigger devices in the same layout.

Packaging has also become important as the physical size of applications has shrunk. Two factors should be considered: package size and routing efficiency. Obviously, the size of the package is directly linked to how much capacity can be packed into a given space. However, routing efficiency determines how closely those packages can be placed together on the circuit board and still get all of their signals routed. Some packages offer special pinouts that enable compact routings of multiple devices in an array.

SYMMETRICAL BLOCKING. Because embedded Flash can potentially contain multiple Flash devices wired together, symmetrical blocks that are all the same size make memory management easier. Asymmetrical devices could introduce different block sizes in the middle of the memory map, making management more complex. Thus, symmetrically blocked Flash is preferred for embedded configurations.

CYCLING. Embedded Flash used for data storage may incur more erase cycles compared with code-only Flash because of the potential for constant turnover in the data. For example, in a digital camera, pictures are continually stored and deleted as the user takes pictures, downloads them, and erases them. However, in many cases, embedded Flash will be managed through a software data manager, which can, if properly designed, dramatically decrease erase cycle count since multiple chunks of data can be appended within a block before it needs to be erased.

BLOCK LOCKING. Unlike code, corrupted stored data is not generally fatal to system operations, but data loss can still disable an application's functionality. For example, a network router that stores its network address table in embedded flash could still boot up if that table were lost, but it would not be able to perform its designated function within the network. For this reason, block locking functionality can be used to protect data stored in embedded Flash from corruption or accidental erasure.

SOFTWARE. A software data manager can be very helpful in abstracting the details of the Flash interface so that the system application can store data to Flash using higher level calls that make the Flash store files or emulate a disk. The data manager can also implement sophisticated functions, such as wear leveling, to improve embedded Flash performance.

2.3.4 Removable Media

Applications that capture or store data from the outside world are growing rapidly, as is clear from the numerous digital cameras, digital audio recorders, and MP3 players that line the shelves of consumer electronics retailers. The most salient characteristic of this usage model is that the amount of storage needed by the end users really cannot be quantified since there are an infinite number of pictures out there to be captured. Thus removable media is needed so that the end user can decide how much storage is appropriate. However, this requirement does not preclude the need for built-in Flash chips. Most of these devices have CPUs that need code storage Flash, which can be extended into code+data storage to provide some amount of built-in data space.

The following sections will discuss the issues important to removable Flash and some of the choices available in the market.

2.3.4.1 Relevant Attributes for Removable Flash.
Choosing a type of removable Flash media for a system design introduces a number of new attributes that have not yet been discussed, but keep in mind that inside these cards are Flash memory devices that are subject to the same limitations as discrete components. The following sections address the attributes relevant specifically to removable Flash cards.

NOR AND NAND. Removable Flash memory cards are a good example of a Flash application that can be either NOR- or NAND-type Flash. NOR-based Flash memory is best suited to code or code+data storage, while NAND memory is best suited for data-only storage.

Performance of a Flash card is an important consideration. Performance can be measured by several different metrics. NOR and NAND memory each have their strengths and weaknesses in the areas of random-access read performance, sequen-

DATA STORAGE

tial access read performance, write performance, and power consumption. Depending on what performance metric is most important in an application will help determine which removable Flash media fits best.

INTERFACES: LINEAR AND ATA. Two types of Flash card interfaces exist: linear and ATA. Linear Flash is addressed directly in its memory map, and it is addressed essentially the same way as a memory component by itself. For example, the system will access the 85th word/byte by going to the 85th (55 hex) memory location. No translation circuitry is required in the Flash card, which lowers card manufacturing costs. Also, linear Flash allows for fast data transfers through a 16-bit-wide data bus. Code can be executed in place (XIP) with linear PCMCIA cards.

Software such as Flash Translation Layer (FTL) can be used to manage linear Flash cards. FTL emulates rotating magnetic media (sector emulation). FTL is best suited to emulate removable sector-based drives.

The other interface approach is to use an ATA (advanced technology attachment)/integrated device electronics (IDE) interface to the Flash card. The CompactFlash and SmartMedia technologies discussed shortly take this approach. The ATA (IDE) interface reduces development time by using an industry standard interface that simplifies integration. However, a controller must be used, which adds cost. CompactFlash includes a controller in the card that increases card cost, while SmartMedia requires a controller in the reader. The end result is simpler integration, but this comes at a dollar cost. Additionally, these cards cannot execute code directly out of the card; the code must be downloaded to memory before executing.

REMOVABLE CARD FORM FACTORS. A crucial consideration is the form factor of the Flash card. There are multiple form factors from multiple vendors (Fig. 2.13). Each particular standard tends to be oriented toward either NOR-type Flash or NAND-type Flash. PCMCIA Flash cards, CompactFlash, SmartMedia, SD memory card, and Multimedia Card are different form factors based on accepted standards. Each has its own size, weight, and other strengths or weaknesses. The needs of each specific application will determine which Flash card form factor fits best.

INSERTION METHODS. The insertion method may be a consideration in choosing a Flash card. PCMCIA cards, CompactFlash, SD memory cards, and SmartMedia are all inserted lengthwise into a small slot in the side of a system, much like inserting a floppy disk into a PC. This slot only needs to be the width and thickness of the card. Very little surface area on the outside of the system is required.

Figure 2.13. Examples of various card form factors.

DATA INTERCHANGE. Another consideration in choosing a Flash memory card for a particular application is the way it connects to the rest of the world. Adapters from PCMCIA cards to CompactFlash and SmartMedia are easy to find (Fig. 2.14). These passive adapter cards make it easy to transfer data to and from any Flash card through a PCMCIA slot on a PC. In addition, there are adapters that allow smart media to be read in a standard floppy disk drive or USB drive.

2.3.4.2 *Major Card Standards.* There are several competing standards available in the marketplace. Each standard has strengths and weaknesses, which will determine which is the right choice for a particular system design.

PCMCIA AKA PC CARD. PCMCIA-based Flash cards have 68-pin connectors in two rows of 34 pins. These Flash cards are typically accessed as linear Flash or through an ATA (IDE) interface. Linear Flash is typically based on NOR Flash, and the ATA interface cards are commonly based on NAND memory. PCMCIA cards are a widely accepted open standard that allows for compatibility with almost all notebook computers and many other systems. PCMCIA has cooperated with other standards organizations to make it easy to build adapters to other Flash card standards. Consequently, there are readily available adapters from PCMCIA Flash cards to CompactFlash, and SmartMedia. More information can be found at http://www.pcmcia.org.

One advantage of the PCMCIA cards (Fig. 2.15) is their physical size. The larger card can hold more Flash memory devices, which means they have the capability of greater storage capacities. However, the larger size could be a hindrance in a small handheld device.

SD MEMORY CARD. The SD (Secure Digital) card is a general-purpose expansion card standard that includes both memory and I/O support. It is considerably

Figure 2.14. Flash card adapters to PCMCIA.

DATA STORAGE

Figure 2.15. PCMCIA card with linear Flash.

smaller than the PCMCIA card. Unlike PCMCIA, this standard does not have any pins that could be bent, but rather nine exposed traces on the card. SD cards have become a popular standard in many smaller form factor handheld devices and have become the de facto standard for removable memory in the cellular phone market. In addition to the 3-g SD memory card, there is a shrink of the standard available. This smaller version, the MiniSD, weighs in at only 1 g.

COMPACT FLASH. CompactFlash is based on an open standard that complies electrically with the PCMCIA ATA interface standard. These cards are commonly based on NAND Flash memory. They have an on-board controller to manage error correction and control the interface. The controller adds additional cost but lets the Flash card appear to the system as an ATA (IDE) drive. This allows for easy designs and quick time to market. The fast sequential writes with these cards combined with the small form factor make these devices very well suited for digital cameras. More information can be found at on the Internet at the CompactFlash Association website at http://www.compactflash.org.

SMARTMEDIA (SSFDC). SmartMedia or solid state floppy disk card (SSFDC) is another removable Flash card standard targeting handheld systems. SmartMedia cards are very light-weight devices. They are about 2 g each. SSFDC cards use NAND memory for their storage medium. The design of SmartMedia cards allows for only one die, so the card capacity is limited to the capacity of a single die. SmartMedia host interface chips also use the ATA interface standard allowing for quick time to market with simple designs; however, the trade-off is increased system costs. The fast sequential writes of NAND combined with the small form factor make these devices very well suited for digital cameras. More information can be found on the Internet at the SSFDC Forum homepage at http://www.ssfdc.or.jp/english/.

MULTIMEDIA CARD (MMC). The MultiMediaCard, introduced in November, 1997, is an extremely small Flash card targeted at space-conscious applications such as cellular phones, MP3 music players, and digital camcorders. It weighs less than 2 g and is the size of a postage stamp (32 mm high by 24 mm wide and 1.4 mm thick). The interface to the card is a 7-pin serial interface, which allows small and simple sockets for the card but precludes code execution from an MMC. More information on MMC can be found at http://www.sandisk.com/.

2.4 CODE+DATA STORAGE

Code+data is the only usage model of Flash that has evolved specifically around the strengths of Flash memory. (See Fig. 2.16.) Most of the features that have been added to Flash over time can be seen as steps in the evolution of the code+data usage model. Each advance in Flash architecture has become a building block for the next step in code+data integration. When early Flash memory devices were developed, proponents of the technology advocated it as the eventual final destination of the road to memory consolidation, due to its capability for in-system updates, fast reads, and low cost. However, before this could happen, a number of advances in the Flash architecture and technology would need to take place in order to solve the problems involved in storing both code and data in a single Flash memory component. None of these advances, taken alone, could enable code+data, but each is a vital building block enabling code+data as achieved today. A look at the significant developments in Flash technology and architecture will clarify the issues involved in single-chip code+data and illustrate how today's Flash components are used in this manner.

System functions such as updateable code have become mandatory with today's fast changing environments and rapid development cycles. In addition, due to its strengths across a variety of characteristics, Flash is the logical choice as the recipient of integrated memory functions. Flash is nonvolatile, electrically updateable, fast to read, available in high densities, and relatively low cost. Since other memory types do not provide this combination of functions, these characteristics make including Flash as one of the system memory components nearly mandatory.

If the presence of Flash in a system can be assumed, then what can be eliminated? (See Fig. 2.17.) Direct code execution from Flash, enabled by fast burst or page read modes, eliminates the need for storing code and executing code in different memories, reducing system RAM needs. The EEPROM's parameter storage function and RAM data storage can be incorporated into Flash by using a software data manager. This eliminates the EEPROM from the system and leaves two memory components, the Flash and the RAM. The RAM's size can be greatly reduced because it does not need to store code for execution nor persistent data. The end result is a lower component count that produces lower cost, power consumption, and board space.

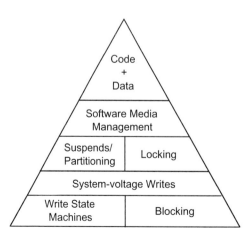

Figure 2.16. Flash developments leading to code+data.

CODE+DATA STORAGE

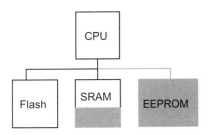

Figure 2.17. Block diagram of memory subsystem showing memory that can be integrated into Flash.

The goal of code+data system architecture is to integrate the various types of memory components. Memory subsystem integration is a clear design trend driven by the immediate benefits of reduction in component count, cost, power consumption, and board space achieved by combining EEPROM functionality into NOR-type Flash memory devices. These devices possess the combination of characteristics that make them the obvious choice for combined code execution and data storage, especially in embedded designs that do not have the burden of legacy architecture. Some of today's important applications that have adopted code+data architectures are wireless phones and handheld computers. Today's Flash memory devices have evolved the required features needed to support integrated code+data architecture.

Unsurprisingly, the applications that are most concerned about size, power, and cost are aggressively pursuing code+data design techniques. The most conspicuous examples are cellular phones and other portable wireless devices. A cellular phone is most useful if carried everywhere and turned on constantly, so cellular phone manufacturers have grabbed onto code+data as a way to maximize functionality while minimizing size and power.

2.4.1 Relevant Attributes for Code+Data

Unsurprisingly code+data applications require a combination of the needs of code and data applications. Specifically, this usage model requires attributes that allow efficient allocation of Flash space between code and data, routine availability of Flash write operations, and methods to protect stored code and data from corruption (see Table 2.4).

2.4.1.1 Array Blocking. The division of the Flash memory array into blocks grants the capability to update code and data without incurring the risk of losing everything by erasing the entire array, enabling more robust system design. This added security gives designers the confidence to write to Flash routinely.

2.4.1.2 Symmetrical Versus Asymmetrical Devices. The type and size of the data being stored in a code+data application determines the Flash block sizes needed to store the data. Small-sized data such as parameters and timers fit better in smaller blocks, while larger data types such as voice recordings or packetized data from the Internet require larger blocks. An asymmetrically blocked device contains both small blocks for parameter storage and big blocks for code storage. A code+data design that only needs to store parameters may find asymmetrically blocked devices an efficient way of meeting both code and data needs. However, if the data types

TABLE 2.4. Selecting a Flash Device for a Code+Data System

Feature	Desirable Characteristics	Keep in Mind
Write state machine	• Compact command set bus cycles • Useful command set extensions, like buffered writes	• Long command sequences increase bus cycles when transferring data sequences
Blocking	• Smaller block sizes erase faster • Asymmetrical blocking can optimize block sizes for your application	• Smaller blocks increase die cost • Software media management reduces the need for small blocks
Voltage	• System-level write voltage • 12-V program operation for faster production lines • Independent read and write voltage pins allow write power supply to be switched for power savings and data protection	• Understanding your application's proportionate use of the Flash power modes (standby, active, powerdown) is key to evaluating Flash power consumption
Write suspends	• Flexible suspend commands: erase/program suspend to read, erase suspend to program, erase suspend to program suspend to read • Deterministic suspend latency • Low suspend latency	• Suspending erase or program too often decreases data throughput
Partitioning	• Code and data partition sizes appropriate for your application	• Fixed partition sizes may not fit your applications data or code sizes • Your application's performance may not need dual partition
Locking	• Ability to change locking in-system • Independent locking control of each block • Fast, real-time locking changes • Different levels of protection for code and data • Hardware control for additional protection	• Locking that can only be defined outside the system • Long locking latencies may mean you are unable to use the locking features
Software	• Ability to target single (software read-while-write) or dual-partition devices from the same code base • Wear leveling • Power loss recovery • Ability to handle your current and future data types • Support and continued development	• Features that limits your design's potential • Media manager architecture nonoptimal for your application (disk-style emulation, etc.) • Licensing or technical restrictions that limit your Flash sources

require more storage space, data may need to be stored in bigger blocks, making symmetrically blocked Flash devices a stronger choice. Cellular telephones provide a good example: basic voice-only phones need to store only small parameters for radio tuning or call timers and are well suited for asymmetrical Flash, whereas high-end phones that record voice or Internet data will need to use the bigger blocks available in both asymmetrical and symmetrical Flash devices. Note that software data managers tend to mix big and small data in the same blocks, reducing the importance of block size in the area of data storage.

2.4.1.3 *Voltage.* Flash devices that require nonstandard voltages to program or erase the Flash make such operations exceptions to be exercised only when the need is crucial, such as a code update to correct a fatal bug. But code+data is about making writes to Flash a routine, integrated part of system operation. Therefore, Flash devices that can program and erase using standard system operating voltages are the best choice for code+data system designs.

2.4.1.4 *Write Suspends and Dual Partition.* Because code+data systems, by their nature, store data and execute code out of the same Flash device, some mechanism is require to either interleave the read and write operations or allow them to take place simultaneously. Program and erase suspend functions in a standard single-partition Flash device enable interleaved read and write operations. If code needs to be executed during a program or erase operation, for example, to service an interrupt, then the write operation can be suspended to read from the Flash. Dual- or multipartition Flash devices take this a step further by allowing simultaneous read and write operations in different partitions. For code+data systems, the decision between using a single or multipartition hinges on the throughput requirements of the system for both data storage and code execution reads. As the data rate of incoming data and code execution performance requirements increase, using suspends on a single-partition Flash device will degrade system performance. Dual- or multipartition devices can alleviate this bottleneck.

2.4.1.5 *Write Protection and Block Locking.* Code+data systems write to their Flash devices often, which engenders the need for write protection or block locking schemes. In addition, systems should protect data segments in Flash because they are not tolerant of corruption. Write protection schemes aim to prevent accidental writes caused by system noise by lengthening the command sequence required to change the Flash contents, but in a code+data system, the longer command sequence may hinder system performance since more bus cycles must be used for the command itself.

Block locking is another approach that is well suited for code+data. A Flash device that allows locking of individual blocks is ideal for code+data since only the affected block need be unlocked when necessary, leaving the rest of the code and data safely locked. Some schemes require groups of blocks to be unlocked together or can only be locked/unlocked outside the system at manufacturing. Another item to look for is the performance of the block locking scheme. If a locking change takes too long, then it will either compromise code storage performance or not be used. Either way, slow locking changes decrease performance and security.

2.4.1.6 *Erase Cycling.* As systems incorporate code and data in the single system Flash device, block erase operations inevitably increase. Whereas a code-only design might accumulate at most a hundred erase cycles over its lifetime of code updates, a code+data design must accommodate those code update cycles in addition to the constant churn of data being stored and updated. Data storage can push the device cycling count into the thousands or more. However, note that using properly designed data manager software dramatically decreases erase cycle counts since multiple pieces of data are stored and updated many times within a block before it needs to be erased.

2.4.1.7 Software. All of the innovations discussed previously provide the hardware building blocks for code+data storage in Flash, but one more piece, software media management, is needed to implement a code+data architecture in a system. A media manager is needed because Flash block sizes generally range from 8 to 128 kB, which is large relative to the size of most data parameters used in a system. Since a single byte of information cannot be erased, erasing a block of Flash every time a new piece of information needs to be written is impractical.

The primary function of a media manager is to make block size transparent to the system by managing the storage of information and the reclaim of space used by old, invalid data.

An example of such a media manager is Intel's Flash Data Integrator (FDI) software. FDI implements two key elements required for code+data designs. The first is a real-time interrupt handler such as the one described in the erase suspend section. This piece of code will poll for interrupts and suspend writes to Flash accordingly and enables code+data applications with single-partition Flash devices. The second is a robust media manager that provides an application programming interface (API) for system application code to store parameters, data streams, and other data types in the Flash. The media manager provides a standard, higher-level way to access data and handles all of the details required to store, retrieve, and protect the data in the Flash. In addition, the media manager provides features such as wear leveling and power-loss recovery to ensure robust code+data system operation in a real-time system.

Selecting the right Flash management software is important because its features and flexibility will define where system development resources are allocated. With a full-featured Flash manager, development can focus on additional system features built on top of the Flash manager, not on the details of setting and looking up bits in the Flash array. While a limited-feature data manager may look simpler to implement, those limitations may hold back the potential of a system.

In the future, if application requirements become more standardized, many of the elements of media management may become incorporated into the Flash hardware. Just as the write state machine abstracted the write pulses needed to program and erase the Flash, a higher-level command interface could incorporate media management functions so that the parameter could be written directly to the Flash. The Flash hardware could take care of space allocation, garbage collection, and more. When looking for the future of Flash hardware, look at the layers of software driving the Flash. Features of that software are likely candidates for incorporation into next-generation Flash chips.

2.4.2 Fitting the Pieces Together for Code+Data

Code+data deserves a more detailed description of how the various pieces of the system interrelate, not just because it is the newest and, most likely, unfamiliar usage model, but because its implementation exercises so many aspects of the system, from the CPU and Flash hardware to the software stack running on the CPU.

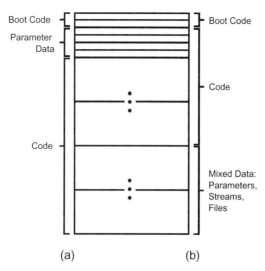

Figure 2.18. Memory map showing allocation of Flash between boot code, data, and code: (a) small data application and (b) large data applications.

2.4.2.1 Allocating Flash Space. A Code+data design needs to allocate space in the Flash for three fundamental categories of information: boot code, code, and data. Boot code needs to be placed either at the top or the bottom of the memory, depending on the processor architecture. Thus, its placement is predetermined. In Figure 2.18, which shows how an asymmetrically blocked Flash device would be allocated, the boot code is shown at the top of the memory map, occupying the first two blocks. Keep in mind that asymmetrically blocked Flash devices are available in top and bottom mirrored configuration, so this diagram can be mirrored vertically to understand bottom boot configurations.

Placement for the rest of the code and for the data is more flexible and adjustable to application needs. Figure 2.18(*a*) shows how memory would be allocated for a code+data design that only needs to store a small amount of data, similar to an EEPROM. In this case, the remaining small blocks of the asymmetrical block architecture can be used for data, and all of the larger blocks can be allocated for code.

On the other hand, Figure 2.18(*b*) shows how memory would be allocated for a code+data design that needs to store a large amount of data. This amount of data will not fit in the small-size blocks, so it needs to be moved to the larger blocks. But should it be placed at the top of the big blocks, adjacent to the small blocks, or at the bottom end of the memory map? Either placement could work, but Figure 2.18(*b*) shows them placed at the bottom. This is done so that the code is placed toward the top of the map, allowing its size to grow toward the bottom without trouble. If the data size needs to expand, then it can be expanded upward. A proper media manager will be able to handle both expansions without problem.

2.4.2.2 System Software Stack. Given a modern Flash component, the key addition to the system architecture for code+data is software media management.

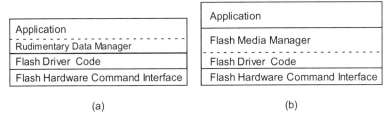

(a)   (b)

Figure 2.19. Comparison of traditional and code+data system software stacks: (a) traditional software stack and (b) code+data software stack.

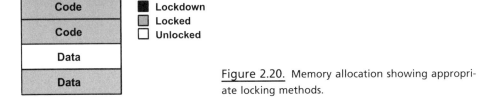

Figure 2.20. Memory allocation showing appropriate locking methods.

A typical software stack in a pre-code+data application might be as shown in Figure 2.19(a). The system application talks to a low-level driver layer, which sends the actual commands to the Flash hardware's command interface. This means that the system application must incorporate media management in order to format and retrieve stored data. Since this Flash media management is usually not a core competency of system-level hardware or software designers, a simplistic media manager may limit data storage into Flash.

Implementing a code+data system requires an additional layer of software between the low-level drivers and the system application, as shown in Figure 2.19(b). This layer of software, the media manager, abstracts the Flash-specific formatting and bit manipulation so that the application software can handle data simply as parameters or data streams to be passed to the media manager for storage.

Figure 2.20 shows how code and data could be mapped in the Flash along with the appropriate locking modes. Starting from the top, the first block contains boot code and has been put into "lockdown," so that both software and hardware changes are required to change the code in that block. The next two blocks contain data and have been locked to prevent changes. However, in the case of a code update, a software command could be used to quickly unlock these blocks. The next two blocks have been designated as data blocks with one unlocked for an update while the other is locked.

2.4.2.3 System Operation. By now, we've discussed all of the elements of the code+data system architecture, but we need to fit it all together and see how it works step by step. The relevant system hardware generally consists of a CPU, Flash, and RAM. The Flash is the primary memory device and contains boot code, data, and application code.

1. The CPU boots from the Flash and begins executing code fetched directly from the Flash device. The boot code takes the system through initialization and then continues execution from the application code in the Flash. The Flash media manager can be considered to be compiled into the application code at this point.
2. At some point, when the application needs to store data to Flash, it will make the appropriate calls to the media manager API. The media manager will format the data and begin writing it to Flash.
3. Assuming for now that the Flash device is single partition, a small portion of code that controls the write process will be copied to RAM.
4. The CPU will begin executing from RAM and start writing data to Flash.
5. If an interrupt is received during the write process, the Flash will be unavailable for code fetches, so the code in RAM should suspend the Flash write process.
6. The CPU can then return to fetching code from the Flash to service the interrupt.
7. When this is completed, then execution returns to RAM.
8. The write process is resumed by the code in RAM until complete.
9. Execution then returns to the Flash.

If the Flash device is a dual-partition device, since code does not have to copied to RAM to control the write process, steps 3 to 9 are simplified as follows:

1. If the Flash device is dual or multipartition, code that controls the write process will begin executing from a partition different from the one containing the data.
2. If an interrupt is received during the write process, the CPU can jump to another code segment in the Flash, as long as that segment is in a different partition from the one containing the data.
3. When the interrupting code is finished, execution returns to the media manager until writing is complete. Code execution never leaves the Flash unless code needs to be executing from a partition that is being written. In that case, then the write process will have to be suspended and handled similar to a single-partition device.

2.4.3 Benefits of Code+Data

Memory subsystem integration is a clear design trend driven by the immediate benefits of reduction in component count, cost, power consumption, and board space achieved by combining EEPROM functionality into NOR-type Flash memory devices. These devices possess the combination of characteristics that make them the obvious choice for combined code execution and data storage, especially in embedded designs that do not have the burden of legacy architecture. Some of today's important applications that have adopted code+data architectures are wireless phones and handheld computers. Today's Flash memory devices have evolved the required features needed to support integrated code+data architecture.

From a broader perspective, code+data is not important just for component and cost reduction, but as the architectural step that provides the foundation for tomorrow's data-rich applications. Traditional system architectures that rely on limited-size EEPROMs will remain locked into a small-data mindset, with their innovation and features limited by the size of their parameter store. Systems that embrace code+data architectures as an enabling technology will be ready to embrace the increasing amount of data flowing in the networked world. Flash device densities continue to increase, and systems designed around them can grow with them, accommodating expanding code and data sizes without limitation.

2.5 CONCLUSION

What future applications will Flash enable? Not that long ago, Flash manufacturers were heralding the impending collapse of the hard disk drive industry as everything but the highest densities were going to convert to Flash by the year 2000. Well the world didn't stop working at the end of the millennium, and the hard disk drive didn't become extinct. It's clear with the density escalation of rotating media that if an application can use a hard disk, it will. The price/megabyte is just too compelling. Flash will continue to expand its use in applications where power consumption, reliability, portability, and space constraints demand solid-state nonvolatile memory.

Phones will integrate the answering machine into the handset. Cell phones, pagers, and digital assistants will merge into a single connectivity device. Chemical film shares the shelves with e-film (Flash). Camcorders will become digital and tape will eventually be displaced with solid-state memory cards. CD-ROMs will eventually be replaced with high-density multilevel NOR or NAND Flash, much in the same way records were replaced by CDs.

The futuristic vision of Star Trek, where people inserted credit card sized devices into systems and listened to messages and music, or had voice-recognition-enabled computers, became reality decades before the show's creators forecasted. No matter what we hypothesize with respect to future applications for Flash, it's clear as applications demand more from memory technology, the direction Flash will take is bigger, faster, and cheaper.

3

MEMORY CIRCUIT TECHNOLOGIES

Giulio G. Marotta, Giovanni Naso, and Giuseppe Savarese

3.1 INTRODUCTION

Flash memories [1] are fairly complex systems from the circuit designer standpoint. A Flash memory chip usually includes the following functional blocks: a large array of nonvolatile memory cells, array decoding circuitry, analog subsystems to generate and regulate the voltages needed for cell programming and erasure, sense amplifiers to retrieve stored data from the memory array, system control logic, and input/output (I/O) interface circuits. Besides, Flash memories incorporate a great amount of circuitry, whose functionality is, so to speak, hidden from the end user, such as, for example, the testability functions and the repair and error correction circuits. In this chapter the design techniques adopted to implement above-mentioned functional blocks will be reviewed along with a summary outline of the most commonly adopted Flash memory array implementations and chip architectural features.

3.2 FLASH CELL BASIC OPERATION

Seen from a circuit designer perspective, the Flash cell is a floating-gate transistor [2, 3], whose current–voltage (I–V) characteristics can be altered by properly controlling the voltages applied to the transistor gate, drain, source, and bulk terminals [see Fig. 3.1(*a*)] for the purpose of storing nonvolatile information.

The peripheral circuitry of a Flash memory chip performs four basic operations on the Flash cell: program, erase, compaction or soft program, and read.

Nonvolatile Memory Technologies with Emphasis on Flash. Edited by J. E. Brewer and M. Gill
Copyright © 2008 the Institute of Electrical and Electronics Engineers, Inc.

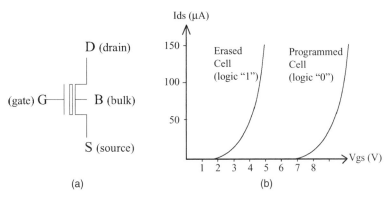

Figure 3.1. (a) Flash cell schematic symbol and (b) programmed and erased cell *IV* characteristics.

3.2.1 Cell Programming

Two basic physical mechanisms are usually exploited to program a Flash cell: channel hot electron (CHE) injection [4] and Fowler–Nordheim (FN) tunneling. Channel hot electron programming consists of applying a relatively high voltage, on the order of 4 to 6 V, to the cell drain D in Figure 3.1, a high voltage (8 to 11 V) to the cell gate G while source S and bulk B are kept at 0 V. With this biasing regime a fairly large current (0.3 to 1 mA) flows in the cell and the hot electrons generated in the channel acquire sufficient energy to jump the gate oxide barrier and get trapped into the floating gate. Most NOR-type Flash architectures adopt CHE programming. The CHE programming process takes a few microseconds to shift the cell threshold voltage from the erased value (about 2 V) to the programmed value (>7 V)—see Figure 3.1(*b*). Conventionally, a programmed cell stores a logic 0 value. FN tunneling programming is used in NAND-type [5], AND-type, and DINOR-type [6, 7] Flash architectures. FN programming is achieved by applying a high voltage of about 20 V to the cell gate while drain, source, and bulk are grounded. FN programming is slower than CHE programming. FN programming time is in the range of a few milliseconds. One important advantage of FN programming is that it requires very small programming current (<1 nA) per cell thus allowing many cells to be programmed at a time.

3.2.2 Cell Erase

Cell erasure is accomplished by FN tunneling. In a NOR-structured Flash memory array, for example, cell gate G is biased at a negative (with respect to bulk B potential) voltage of about −10 V; 4 to 6 V are applied to the drain D, source S is floated, and bulk B is kept at 0 V. With the above biasing regime a high electric field is applied across the gate oxide that pushes electrons out of the floating gate, thus reducing the device threshold voltage. An erased cell conventionally stores a logic 1 value. FN erase usually takes a few milliseconds.

3.2.3 Compaction

The erase process may sometimes overdeplete the cell floating gate, that is, electrons are extracted from the floating gate beyond the neutrality (zero charge) condition, which is equivalent to saying that positive charge (holes) are stored in the floating gate. Upon occurrence of that event, the cell threshold voltage becomes increasingly low (the larger the positive charge the lower the threshold) up to the point the cell conducts current even when the applied gate voltage is 0 V. In NOR-structured arrays the occurrence of overdepletion or overerasure [8] as it is also called, can end up impairing retrieval of the stored information from the memory array. For that reason erase algorithms usually check for the overdepletion state after erase by applying 0 V to the cell gate and reading the cell current. If the cell current is higher than zero by a certain amount, then a compaction [9] procedure is applied to restore floating-gate neutrality to a degree sufficient to ensure memory array functionality. Compaction is basically a soft programming and, similarly to the programming case, either one of the programming methods, CHE or FN tunneling, is used. The name compaction derives from the fact that this method is usually applied either to the entire memory array or to a memory cell column and, after compaction, the distribution of the threshold voltage of the cells belonging to the array (or to the column) is narrower than it was before applying the compaction procedure. In compaction methods based on FN programming a high voltage is applied to cell gate G while drain D and source S are floating and bulk B is at 0 V. The voltage to be applied to the gate G is somewhat lower than the one needed to fully program the cell. Compaction methods based on CHE programming consist of applying high voltage to the gate G, medium voltage to the drain D, with source S and bulk B at 0 V in the same fashion as the CHE programming case. Again, applied voltages have to be lower than the ones needed for programming to prevent a full shift of the cell threshold. The narrowing effect of the compaction over the cell threshold voltage distribution arises from the fact that those cells exhibiting a higher degree of overerasure, that is, a more positive floating-gate potential, tend to have a larger threshold voltage shift because the electric field set up by the positive charge trapped in the floating gate is added to the electric field induced by the applied compaction biasing voltages. In NAND-type Flash, similarly to electrically erasable programmable read-only memory (EEPROM), the cells to be erased are intentionally overdepleted to achieve a negative threshold voltage, and in this case there is, of course, no need for compaction.

3.2.4 Read

In read mode the cell gate G is biased at about 5 V, the source S and the bulk B are at 0 V, and the drain D is at about 1 V. It is very important that reading circuitry be designed in such a way that the cell drain D never experiences a voltage greater than 1 to 1.5 V. If higher voltages were applied to the cell drain during read, then the cell itself would experience read-disturb. Read-disturb can be caused by channel hot electron generation during read. Although read current is small (10 to 50 µA) with respect to the CHE programming current, it has to be considered that a Flash cell is in read mode for most of the time (Flash memories are inherently mostly read devices). Therefore, over the device lifetime even small amounts of read-disturb might impair data integrity.

3.3 FLASH MEMORY ARCHITECTURE

For a better understanding of the topics discussed in the remainder of this chapter, let us refresh some of the basic concepts of Flash memory systems [10].

In contrast to dynamic random-access memory (DRAM) and static RAM (SRAM), Flash memories are "mostly read" devices. That arises from the fact that writing information to a Flash memory array, that is, programming or erasing the Flash cells, requires much longer time than reading the stored data. Programming a byte or a word takes typically a few microseconds, erasing a Flash array takes typically many seconds, while reading a byte or a word takes only 20 to 100 ns. Besides, while programming can be performed at bit, byte, or word level, erasure is a "bulk operation," that is, the entire cell array has to be erased even if only a single bit has to be changed from 0 to 1. After erasure a Flash memory array stores all 1's. Programming operation can selectively change 1's into 0's. Programming 1 to a memory location leaves the location unchanged.

Since, as mentioned above, erasure can be performed at the array level only, large Flash memory arrays are segmented into smaller pieces (subarrays), usually referred to as *blocks* or *sectors*. The size of a block is typically 64 kbytes. In recent years Flash memory chips have been introduced [11] that feature an even smaller block size: one of the many 64-kbyte sectors is further segmented into smaller (typically 8 kbytes) blocks usually referred to as boot blocks or parameter blocks. Besides providing flexibility, the array segmentation makes the sensing and erase circuit design easier because it reduces the array parasitic load.

For the purpose of discussing the architectural features of a Flash memory, let us consider the block schematic of Figure 3.2 in which the basic structure of a Flash memory is shown. The system exchanges signals with the outside world by means of the three sets of signals:

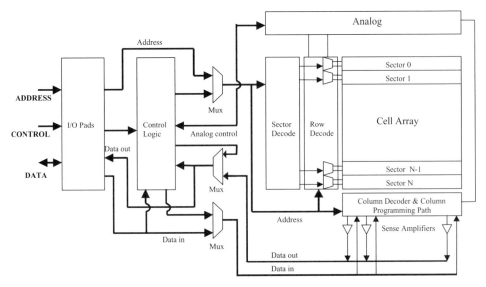

Figure 3.2. Flash memory circuit block diagram.

- **ADDRESS:** A $\log_2$ MS wide input bus, where MS is the *memory size* in bytes or words. The host processor applies to this bus the address of the byte or word to be read or programmed or the address of the sector to be erased.
- **CONTROL:** Chip control input signals. Typical control signals are: reset, chip enable, write enable, and output enable.
- **DATA:** 8- or 16-bit-wide input/output bus. The host processor exchanges data and commands with the memory chip through this bus.

The role of the various circuit blocks can be summarized as follows:

- **I/O Pads:** These are the chip input/output circuits. Other than performing input/output signal buffering the I/O pad circuitry includes special structures aimed at protecting the chip against electrostatic discharge (ESD) events that might damage the chip itself during handling or even when mounted on the final application board.
- **Cell Array:** This is an array of Flash memory cells. As previously mentioned, the cell array is often divided into relatively small blocks or sectors that can be individually erased.
- **Row Decode:** This is the array wordline decoder. It applies to the gate of the cells belonging to the selected wordline, the read and programming voltages depending upon the operation mode. For most Flash architectures in erase mode all the wordlines of the selected sector are driven to a negative voltage.
- **Sector Decode:** Enables the row decoder of the addressed sector in read, programming, and erase mode. The remaining sectors are deselected with all row lines held at 0 V.
- **Column Decoder and Column Programming Path:** In read mode this block connects the addressed columns (or bitlines) to the sense amplifiers. In programming mode it isolates the sense amplifiers from the array and conveys the programming and erase voltages to the memory array.
- **Control Logic:** This is the main chip internal controller. It decodes the commands issued by the host processors and executes the programming and erase algorithms. Besides, it controls the analog circuitry by enabling and disabling the internal programming and erase voltage generators and by determining the duration of the programming and erase voltage pulses.
- **Analog:** This block includes all the analog functions needed to read, write, and erase the memory locations. Namely, it includes voltage multipliers to generate internal high voltages and regulation circuitry.

Architectures have been very recently introduced [12, 13] that feature a double partitioned memory array for the purpose of allowing the host processor to read from one partition while the other one is being programmed or erased. This configuration significantly increases the memory data throughput since it reduces the time the host processor is idle waiting for program/erase operations to be completed. This kind of feature is often referred to as background operation (BGO) or concurrent operation. To further increase data throughput, as required by demanding application platforms such as, for example, wireless communication systems, BGO Flash chips often support page and/or burst, or synchronous [14], read mode. In page architectures, the data read from the memory array are stored in a page buffer. Subsequent accesses within the page address boundary can be performed at a much faster rate. Initial page access takes typically about 100 ns while accesses within the page can be performed in 30 to 40 ns. In burst read a start address is applied to the chip, along with a free running clock signal. After an initial latency time lasting a few clock cycles, the data stored at the start address location are output; subsequent locations are then read at clock signal frequency rate thanks to an internal burst logic that, after the latency period, automatically increments the start burst address at each clock cycle. The general outline of a Flash array structured to support BGO and page/burst access is shown in Figure 3.3.

The physical mechanisms exploited in Flash technology to store logic levels into the memory cell are inherently of analog nature in the sense that the threshold voltage shift the cell experiences during either programming or erase is proportional to the amount of charge injected into the floating gate. This consideration has driven, in the past few years, the development of methods to store more than two logic levels per cell [15, 16]. These developments yielded the introduction on the market of Flash memory chips [17], referred to as multilevel Flash, featuring the storage of two bits per physical cell with consequent doubling of the logic memory size. Multilevel architectures are very well suited for mass storage application characterized by high density, extremely low cost, and a low-medium level of memory speed performance.

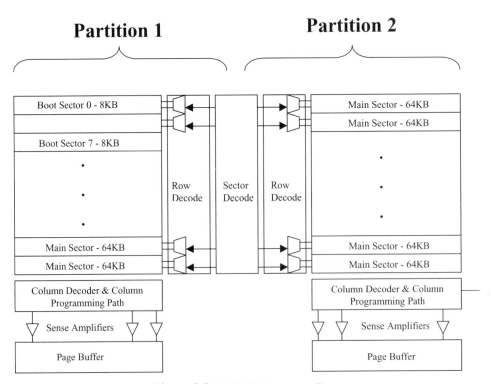

Figure 3.3. BGO Flash array outline.

3.3.1 Memory Cell Array

The most widely adopted Flash memory array architectures can be grouped into four categories: NOR type, divided bitline NOR (DINOR) type, NAND type, and AND type. NOR-type Flash has inherited the array structure used in early ultraviolet (UV) EPROM architectures. Cell programming is usually achieved by hot-electron injection and electrical erase by Fowler–Nordheim tunneling. Figure 3.4(a) shows the basic layout outline of a NOR-type Flash array. Figure 3.4(b) shows the equivalent circuit schematic of the Flash array, and Figure 3.4(c) shows the biasing voltages applied to the memory array in programming, erase, and read mode.

NOR-type Flash [18] is characterized by high-speed random access and is very well suited to program storage applications. Typical sector size is 64 kbytes. At the time this chapter was written, NOR-type Flash chips were available in densities ranging from 4 to 128 Mbits.

DINOR-type (divided bitline NOR) was introduced by Mitsubishi [19, 20]. An outline of DINOR Flash is reported in Figure 3.5.

DINOR-type Flash makes use of triple-well process technology. In the elementary DINOR array structure two polysilicon subbitlines, SBL0 and SBL1, are connected through select gates to an aluminum main bitline (MBL). Typically 64 cells (one block) are connected to each subbitline in parallel similar to a NOR-type array. With this arrangement a roughly 70% cell area reduction can be achieved with respect to the NOR-type configuration. Writing data to the cell is performed by applying about 8 V to the select gateline SGL (deselected SGLs are kept at 0 V). The selected wordline (WL) in the selected block is biased at −11 V (with deselected

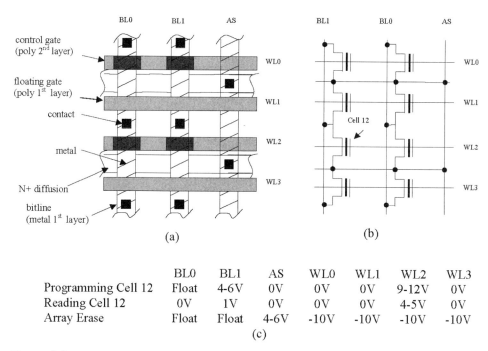

	BL0	BL1	AS	WL0	WL1	WL2	WL3
Programming Cell 12	Float	4-6V	0V	0V	0V	9-12V	0V
Reading Cell 12	0V	1V	0V	0V	0V	4-5V	0V
Array Erase	Float	Float	4-6V	-10V	-10V	-10V	-10V

(c)

Figure 3.4. NOR-type Flash: (a) array layout topology, (b) equivalent schematic, and (c) biasing voltages.

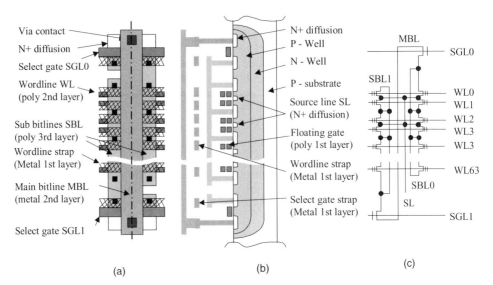

Figure 3.5. DINOR-type Flash: (a) layout outline, (b) cross section, and (c) equivalent schematic.

WLs at 0 V). The main bitline (MBL) potential is raised to about 8 V. With the above biasing voltages, the cell at the crossing of the selected wordline and the selected subbitline sees −11 V on the gate and 8 V on the drain, and, therefore, electrons stored in its floating gate are discharged to the drain by tunneling across the gate oxide. Erase is achieved, again by FN tunneling, by applying about 12 V to the word lines of the selected block, −11 V to the P-well, the source line (SL) and the select gate. Since wordlines are at 12 V and P-well is at −11 V, a high electric field is applied between the floating gate and the P-well, which pushes electrons inside the floating gate. Different than the NOR-type Flash, for the DINOR technology, erase causes an increase in cell threshold voltage and is done at block level, while writing causes a decrease in threshold and is performed at the bit level.

The NAND-type Flash [21–25] array features are summarized in Figure 3.6. Figure 3.6 (a) shows the basic NAND-type cell layout, Figure 3.6 (b) outlines the array cross section, and Figure 3.6 (c) presents the equivalent NAND string circuit schematic. The NAND Flash technology outlined in Figure 3.6 is based on a triple-well process. There are, however, implementations [26] of NAND-type Flash in which the array is built inside P-well regions on an N-type substrate. NAND Flash exploits Fowler–Nordheim tunneling for both programming and erase. With reference to Figure 3.6 (c) let us discuss NAND-type string operation. In erase mode all wordlines (WL) and select lines (SSL) are at 0 V. Select GSL, common-source line (CS) and bitline (BL) are floating; and the P-well is at 20 V. In programming mode the selected WL is at 19 V, deselected WLs are at 12 V, SSL and CS are at 5 V, and P-well and GSL are at 0 V. The BL is at 0 V if the data to be programmed is logic 0 or at V_{cc} if the data is logic 1. Read is performed with BL at 3 V, CS and P-well at 0 V, SSL and GSL at V_{cc}. All WLs are at V_{cc} except the one selected, which is set at 0 V. Erased threshold voltage Vt is about −2 V (overerased state), and programmed Vt is in the 0.6- to 1.7-V range.

FLASH MEMORY ARCHITECTURE

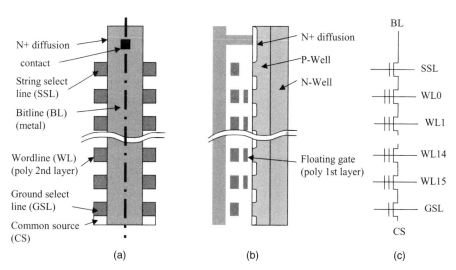

Figure 3.6. NAND-type Flash: (a) array layout, (b) array cross section, and (c) equivalent circuit schematic.

The NAND-type Flash allows high integration densities due to the small number of contacts in the array. The NAND cell size is roughly half that of the NOR-type. NAND Flash is accessed in a serial mode and is widely used in data storage applications, such as Flash memory cards.

The AND-type Flash was introduced by Hitachi [27, 28]. It is a serial access architecture that combines some features of NOR and NAND technologies. It is optimized for data storage applications such as Flash cards and fast silicon disk. AND technology features are summarized in Figure 3.7. Figure 3.7 (a) shows the AND-type Flash array equivalent circuit schematic. The cells are connected in parallel to local diffusion lines. The local diffusion lines can be connected to metal bitlines (BL) through select transistors. Similar to the NAND technology, thanks to the small number of contacts (cells connect to diffusion lines), high integration densities are achieved. Figure 3.7 (b) pictorially summarizes programming and erase mechanisms, which are floating-gate channel and floating-gate drain FN tunneling, respectively.

3.3.2 Analog Blocks

The analog voltages needed for memory array operation are generated on-chip by dedicated circuitry. The typical architecture of the analog block is shown in Figure 3.8.

The charge pumps and voltage regulation circuitry generate all the voltages needed by the memory array: V_{pp} is the wordline voltage in programming mode, V_{hc} is the bitline voltage in programming mode and the source line voltage in erase mode, V_{nn} is the negative voltage applied to the wordlines in erase mode and V_{read} is the wordline voltage in read mode. The reference voltage generator provides the reference voltage V_{ref} for the charge pump regulation circuitry. The oscillators block

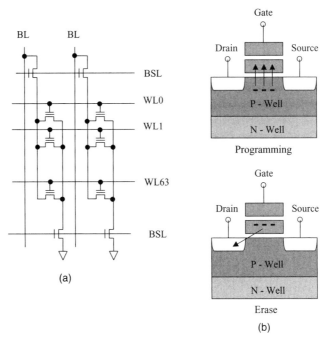

Figure 3.7. AND-type Flash: (a) array equivalent circuit and (b) cell programming and erase mechanism.

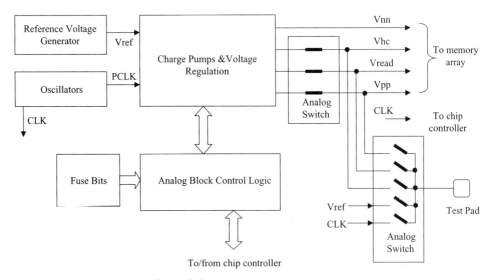

Figure 3.8. Analog block architecture.

generates the charge pump clock signals PCLK and the main chip controller clock CLK. Analog voltages, as well as other relevant internal signals, such as V_{ref} and CLK, for example, can be multiplexed in and out of the chip thanks to the analog switch circuitry. This is particularly useful in the chip development phase either to observe the chip internal functionality or to force internal circuit nodes for debug

FLASH MEMORY ARCHITECTURE 73

purposes. This functionality is also exploited in the manufacturing line to reduce test time by applying analog voltages to the chip via dedicated test pads. The analog block control logic interfaces the chip controller and drives the analog circuitry in the various operation modes. It also reads the information stored in the fuse bits block. This is a bank of nonvolatile memory locations used to configure and trim the voltage generation circuitry. The fuse bits cells are written in the manufacturing line by dedicated test routines.

3.3.3 Control Logic

The numerous functions of a Flash memory chip are controlled by a rather complex controller block. The purpose of the controller is to interface the host processor, to generate the control signals for the memory array and the analog circuits, and to execute the program, erase and verify algorithms.

In a typical application, the Flash memory chip is accessed by the host processor through three main system buses: the address bus, the data bus, and the control bus. The typical configuration of a Flash memory chip controller is shown in Figure 3.9.

The host processor controls the memory chip via the input signals CEB (chip enable, active low), OEB (output enable, active low), WEB (write enable, active low), and RSTB (chip reset, active low). The addresses are applied to the AD lines, and data exchange with the memory chip is performed through the input/output DQ lines. Control signals are interfaced by the I/O logic block. The command user interface (CUI) block interprets the commands the host processor sends to the memory chip. The write state machine (WSM) consists of three functional blocks:

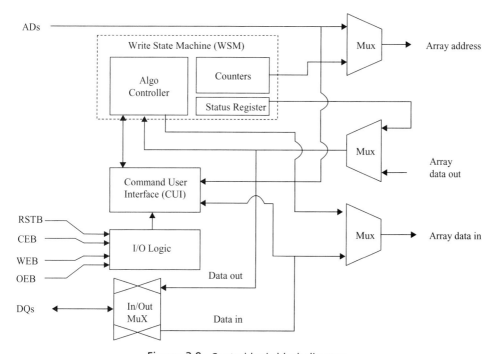

Figure 3.9. Control logic block diagram.

algo controller, which executes the program, erase, and verify algorithms, the counters block, which is a bank of counters used by the algo controller to generate array addresses and to count program and erase retries during algorithm execution, and, finally, the status register.

The status register can be read by the host processor to assess chip status during programming and erase algorithms execution. A typical Flash memory command set includes the following commands:

- **READ ARRAY:** Puts the chip in read array mode. Read array is the default mode at power up and after chip reset.
- **READ STATUS REGISTER:** Upon decoding this command the status register content is put on the DQ data bus.
- **PROGRAM:** This is usually a two-bus cycle command. In the first cycle programming setup is performed, in the following bus cycle the data on the DQ bus are written to the addressed array location.
- **ERASE:** This also is a two-bus cycle command. In the first cycle the erase setup is performed. In the second cycle the erase of the addressed sector is executed.
- **PROGRAM/ERASE SUSPEND:** The host processor can suspend a programming or erase operation by issuing this command to the CUI.
- **PROGRAM/ERASE RESUME:** Previously suspended program/erase operations can be resumed and completed by sending this command to the chip.

Flash programming algorithms are structured according to the following sequence: When a PROGRAM command is written to the chip, the WSM sets the programming pulse counter in the counters block to zero, sets the array in programming mode, and activates the internal charge pumps. Then the data to be programmed are written to the addressed array location (a byte or a word), and the programming pulse counter is incremented. The array is then set in program verify mode (this is usually a read done with a wordline voltage higher than that used in regular read), and the location previously programmed is read back by the WSM. If the read data match the programmed data, then the programming is successful and the ready code is written to the status register. Otherwise the sequence is repeated. If the maximum allowed number of retries is exceeded, then the WSM writes the error code to the status register and exits.

The erase algorithm starts when the ERASE command is decoded by the CUI. The WSM sets the erase pulse counter in the counters block to zero, activates the charge pumps, and, as a first operation, writes 0 (programs) all the locations of the sector to be erased. This operation is called preprogramming and is done to render more uniform the array Vt distribution after erase. An erase pulse is then applied to the addressed sector, and the erase pulse counter is incremented by 1. Erase verify mode is then entered and all locations are read by the WSM (with a read voltage on the wordlines lower than that of the regular read). If all locations are erased, then the WSM performs the so-called depletion check. Depletion check consists in reading the sector locations with all wordlines at about 0 V. Expected data is 0 since no bitline should conduct. If data other than 0 is read, then the WSM applies a soft-programming (compaction) pulse to the failing bitline and performs another verify read. The above sequence is repeated until all the location of the sector being erased pass. If the maximum number of allowed retries is exceeded, then the WSM writes the error code to the status register and exits. Otherwise, it writes ready code and exits successfully.

Usually WSM design follows standard finite state machine design methodologies, such as logic synthesis starting from software modeling and standard cell physical layout, or ROM-based architectures.

3.4 REDUNDANCY

Use of redundancy in Flash memories can dramatically improve the production yield. This is particularly useful during production ramp-up when the yield can be very low due to the learning curve in a new process flow. Yield improvement in production ramp-up allows a reduction in the time to market of new chips and the initial costs. As the memory chips become denser and with a more complex process technology, the yield is expected to decrease. The use of redundancy can help in allowing a production yield much higher than natural yield even when the process has reached its mature stage [29].

Experience over the last 20 years has resulted in the formulation of well-established row and column redundancy design techniques for Flash memories. Criteria for choosing one solution or another are based on the defectivity model associated with the particular array architecture and process.

Final yield after the use of redundancy can be much greater than natural yield, and advanced design techniques to fix some weakness of the traditional solutions for Flash memories have been recently developed that further improve the final yield.

3.4.1 Defectivity and Process Variations

Flash memory array failures due to particles can be of four different kinds: single-bit failures, row failures, column failures, and row/column failures. Single-bit failures are mainly due to small particles and can be repaired either using row redundancy or using column redundancy according to the best strategy to repair the other defects.

A Flash memory array can be more sensitive to row defects or to column defects depending on the particular layout of the cell and on the process technology. For this reason the amount of redundancy to be introduced in a memory and the balance of row redundancy versus column redundancy has to be evaluated according to the expected sensitivity of layout/process to row failures versus column failures.

Redundancy can be used not only to repair defects and increase the yield but also to replace memory cells that are at the edge of process capability for parameters such as length, gate oxide thickness, distribution of implanted ions affecting yield, and performances [30].

3.4.2 Yield Improvement

The use of redundancy can dramatically improve the production yield of a Flash memory, especially during ramp-up when the yield without redundancy could be very low. In Figure 3.10 an example of the possible yield improvement that can be obtained by the use of redundancy is reported for three Flash memories, 8, 4 and

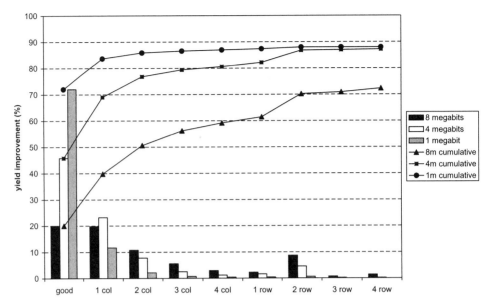

Figure 3.10. Yield improvement using redundancy.

1 Mbits. All the three memory chips are in 0.5-μm technology. Their redundancy scheme is a maximum of two rows per half megabit sector, but no more than four rows per chip; and a maximum of four columns per sector, but no more than four columns per chip.

In case of the 1-Mbit chip, the natural yield (without the use of redundancy) is relatively high, having the value of 72%, and the use of redundancy can only contribute a total increase of 15%. In case of the 8-Mbit chip, the natural yield is very low, having the value of 20%, and the use of redundancy contributes with a total increase of 52%.

The production yield (PY) calculated as

$$PY = \frac{\text{Number of good chips} + \text{number of repaired chips}}{\text{Total number of chips}}$$

is really impressive compared to the natural yield calculated without considering the repaired chips. This is particularly evident for big chips or for nonmature processes having very low natural yield. When developing a redundancy design, it is also necessary to consider the chip area penalty imposed by the added circuitry. This includes both the extra rows/columns as well as the extra logic to manage the redundancy. In the development of a redundancy scheme, the evaluation parameter is the effective yield (EY), defined as [31]

$$EY = PY \times \frac{\text{Chip size with redundancy}}{\text{Chip size without redundancy}} \propto \frac{1}{\text{Manufacturing cost}}$$

Use of a well-designed redundancy scheme makes the difference between a memory chip having acceptable manufacturing costs or having manufacturing costs that are too high to achieve business profitability.

3.4.3 Yield Simulator

From a theoretical point of view it is not easy to evaluate the yield impact due to a particular redundancy scheme. Mathematical techniques have been developed to calculate the yield without redundancy in particular cases where the process defectivity follows specific statistical distributions such as Poisson. For more complex process defectivity there is no closed-form formula to describe the distribution resulting from different types of defectivity such as: single point, single or multiple row, single or multiple column, and row/column. While some specific redundancy architectures have been investigated theoretically [32], the general task of expressing the yield in closed mathematical form becomes far more complicated because redundancy schemes are often based on rules and limitations expressed in words rather than mathematics.

A heuristic method to perform the yield evaluation of a generic redundancy scheme has been developed [31] and is summarized in Figure 3.11. Based on the known statistics of the individual defect processes expressed in closed mathematical form or as experimental tables, defective wafers can be simulated using the Monte Carlo method. Using this technique the natural yield (NY) without redundancy is easily calculated. The particular redundancy scheme to be evaluated is modeled and formalized using a general-purpose software language containing the rules expressing the redundancy coverage per sector and per chip. This software is applied to the simulated defective wafers, and some of the wafers are virtually repaired allowing a yield increase. The area penalty due to the redundancy scheme under investigation can be evaluated from layout architecture planned to be used in implementing the

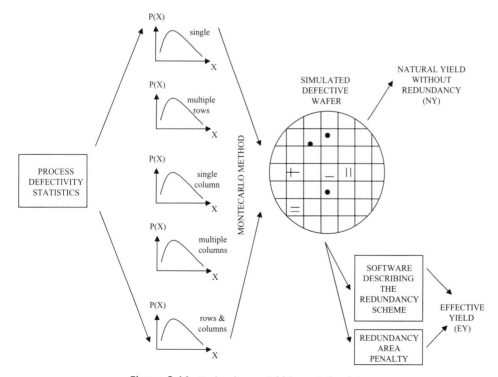

Figure 3.11. Redundancy yield impact simulator.

redundancy scheme. From the yield after repair and the area penalty, the effective yield (EY) can be calculated and the redundancy impact EY/NY evaluated.

3.4.4 Redundancy Fuses Design

Redundancy fuses design is a set of well-established design techniques that have been widely used over the last 20 years. For volatile memories the basic fuse structure is a poly or metal line to be blown by a laser. For Flash memories the basic fuse structure is a Flash cell to be programmed using voltages and timings in accordance with the electrical characteristics of the Flash array used in the chip. Various flavors of redundancy fuses design have been considered over the years, but all the solutions can be functionally represented by the block schematic of Figure 3.12.

The block schematic of Figure 3.12 represents the fuses structure to manage the selection of a redundant row or a redundant column and is composed of three functional units: the enable fuse, the address fuses, and the logic gates that combine the signals associated with all the fuses and generate the signal MATCH. The enable fuse is a Flash cell used to specify that a particular redundant row or a particular output data line has been selected for repair. The address fuses are Flash cells used to select a particular address that corresponds to a defective row or a defective column that must be repaired. Each fuse cell has an associated latch. At power-up, these latches are all preset by the signal PRE_, causing the signals ltout to be low for all the latches. Soon afterward the preset operation signal PRE_ is forced high, and the fuses wordlines (FWL) are forced high to transfer the content of the fuse cells to the latches. If a particular fuse cell is programmed, it will not draw current, its associated latch is not flipped, and the corresponding signal ltout remains low. If

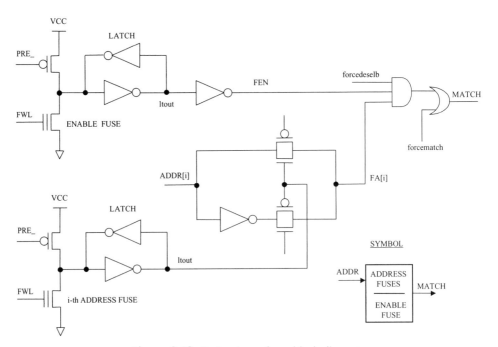

Figure 3.12. Redundancy fuses block diagram.

REDUNDANCY

the fuse cell is erased, it will draw current, causing the latch to flip and force high the corresponding signal ltout. If the enable fuse is programmed, the input FEN to the AND gate is high; otherwise it is low. The effect of each address fuse is to transfer to the input of the AND gate the associated address line ADDR or its inverted value. If an address fuse is programmed, the inverted value of the associated address line is transferred to the input of the AND gate. The signal MATCH at the output of the AND gate will be high if and only if the enable fuse is programmed, and all the addresses associated to the programmed address fuses are low, and all the addresses associated with the erased address fuses are high.

If, for example, four address lines are assumed and the address ADDR[3:0] = 1101 must be detected, it is sufficient to program the enable fuse and the address fuse corresponding to ADDR[1]. In fact, in this case the signals FEN and all the FA[3:0] at the input of the AND gate will be high and the signal MATCH will be forced high.

The fuse control logic reported in Figure 3.12 has also two control signals to override fuse configuration. If forcedeselb is low and forcematch is also low, the MATCH signal is always low even if the fuse enable has been programmed and the particular address matches the address fuses configuration. If forcematch signal is high the MATCH signal is always high no matter how the fuses are programmed.

3.4.5 Row Redundancy Design

Row redundancy is performed in Flash memories by adding one or more wordlines to each sector. In Figure 3.13 a block schematic of such architecture using two redundant wordlines (REDWL1 and REDWL2) is reported.

Redundant wordlines are driven by a level shifter (LS) having the same structure as that used in the regular wordlines WL0, WL1, ..., WLn. Each one of the redundant rows has its own fuse structure programmed to detect a particular row address. If REDWL2 is chosen to repair WL0, the enable fuse associated with REDWL2 must be programmed, and the address fuses associated with REDWL2 must be programmed to detect the WL0 address. When the address of WL0 is applied, the signal MATCH2 is forced high and REDWL2 is selected while all the other regular wordlines are deselected.

REDWL2 will thus replace WL0 for read, program, and verify operations. As far as erase is concerned, all the regular rows WL0, WL1, ..., WLn and REDWL1, REDWL2 are erased together being in the same sector. During the preprogram operation, aimed at raising the threshold voltage before any erase pulse is applied to avoid depletion, the row redundancy replacement must not be effective otherwise the failing row would be erased but not preprogrammed and would go into depletion.

The preprogram operation is performed on all the regular rows, disabling the row redundancy replacement, and it is also performed on all the redundant rows, forcing their MATCH signal high even if the associated fuses are not programmed.

In order to save area, the fuses used to select redundant rows can be shared among sectors. A unique group of fuses can be used to select REDWL1 in a certain number of sectors (eventually all the sectors in the chip), and a unique group of fuses can be used to select REDWL2 in a certain number of sectors (eventually all the sectors in the chip). This implies also an address compression in the fail bit map

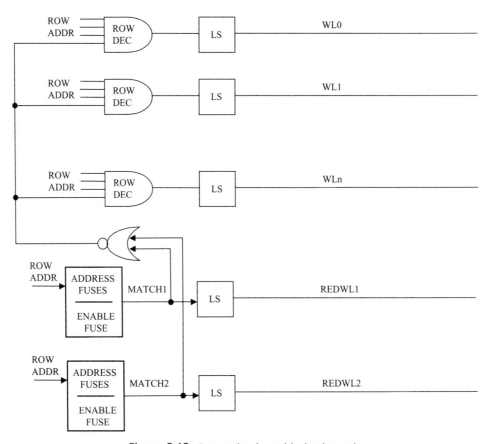

Figure 3.13. Row redundancy block schematic.

allowing the use of a small amount of tester memory to store fail maps. Having common fuse structure for redundant rows in different sectors implies a reduced fail–repair capability. This fuse compression is a result of a compromise between area penalty, test cost reduction, and fail coverage.

3.4.6 Column Redundancy Design

Column redundancy is performed in Flash memories by adding one or more column to each sector. The redundant column will replace the failing column of the failing IN/OUT line DQ. In Figures 3.14 and 3.15 a column redundancy block schematic for the read path and the write path are provided [33].

With reference to the read path of Figure 3.14, each output line $Q0, \ldots, Qn$ is associated to a group of columns selected by a column address (COL ADDR). A similar group of redundant columns can be selected by the same column address, giving the possibility to use different redundant columns for different output lines. In case only one redundant column is used, the redundant column multiplexer (MUX) can be avoided, and only one column of a single output line can be repaired. At the output of the column MUX, the selected regular columns are connected to the sense amplifiers $SA0, \ldots, SAn$ while the selected redundant column is connected to the redundant sense amplifier RSA. Between each sense amplifier,

REDUNDANCY

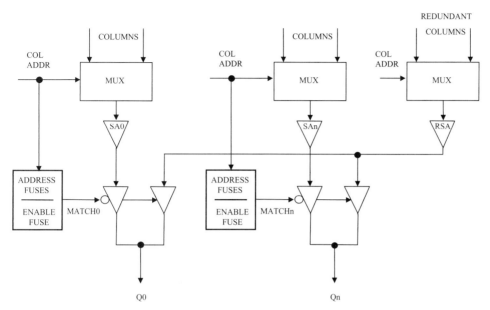

Figure 3.14. Column redundancy block schematic (read path).

SA0, ..., San, and the correspondent output line Q0, ..., Qn there are two tristate buffers having an associated fuses block. If the enable fuse of one block is programmed, then when the column address is consistent with the programmed address fuses configuration, a MATCH signal is generated and the redundant column, instead of the regular column, will be connected to the output line.

The column redundancy program path reported in Figure 3.15 is similar to the read path. Fuse blocks are the same fuse blocks used in the read path. Each input line D0, ..., Dn, after a latch used to internally keep the data value during a program algorithm, can be used to program the selected column or to program the selected redundant column if the correspondent group of fuses generates a match condition.

The discussed column redundancy solution has no speed penalty because the redundant and regular columns are read in parallel and they are MUXed after sensamp. In this situation the MATCH signal is stable when the bit information has completed its propagation from the memory cell to the sensamp [34].

3.4.7 Advanced Redundancy Design

Advanced redundancy design approaches for volatile memories consider flexible redundancy architectures in order to further increase the yield under a wide range of possible fail categories. In the conventional redundancy architecture (intrasector replacement) defective cells can be replaced only by redundant cells that belong to the same sector. This architecture can be efficient only in case of the so-called retention fails, which are numerous but isolated. In presence of hard fails, which are rare but clustered, a distributed globally replaceable (DGR) redundancy scheme is more efficient because redundant cells can repair faults located in any sector of the memory [31]. Other advanced redundancy designs, like the variable domain redun-

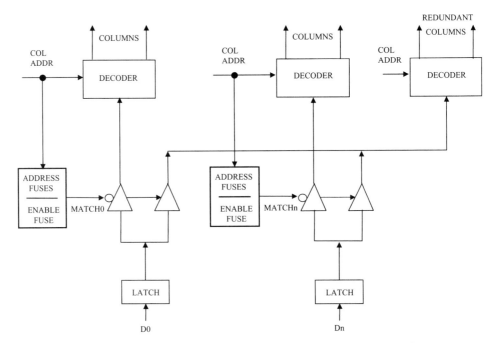

Figure 3.15. Column redundancy block schematics (program path).

dancy replacement (VDRR) technique, can maximize the efficiency of redundancy in the presence of many different kinds of faults [35].

Unfortunately, these flexible redundancy arrangements, which provide good results in volatile memories, are not practical solutions for Flash memories. For Flash memories they impose a big layout and design overhead because of the fact that a sector and its associated redundancy must be erased together as if they were a single sector.

Recent advanced redundancy designs for Flash memories focus attention on improving the efficiency of row redundancy. In fact, in the case of wordline to wordline shorts, or in the case of wordline to bitline shorts, preprogram operation of the failing rows cannot be correctly performed and depletion can easily happen.

Three interesting advanced row redundancy design approaches to avoid depletion or its effects in Flash memories will be considered:

- In case of wordline to bitline shorts, the erase operation of the cells in a failing row can be avoided.
- In case of wordline to wordline shorts, the two adjacent shorted wordlines can be selected at the same time during preprogram operation (using a carry decoder or a Gray code decoder).
- The effects of depleted cells can be avoided by floating the source of the deselected cells during a read operation.

Figure 3.16 shows a row decoder performing a selective wordline erase defeat in a Flash memory array having a short between wordline WL1 and bitline BL1 [36]. In the traditional row decoders and their associated row redundancy, the

REDUNDANCY

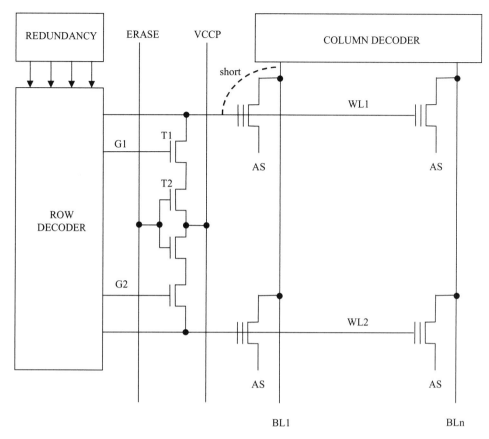

Figure 3.16. Selective wordline erase defeat.

preprogram of the failing row is performed by applying a typical value of 12 V on WL1, a typical value of 6 V on BL1, and connecting AS to ground. The 12-V source is usually a low current pump, while the 6 V is usually a high current pump in order to supply the programming current. Due to the WL/BL short and to the different pump performances, the voltage on WL1 is lowered to about 6 V, and preprogram of the failing row cannot be correctly performed.

The erase operation is performed applying a typical value of −10 V to all the wordlines, a typical value of 6 V to the common array source AS, and all the bitlines are forced to float. For these bias conditions, the failing row is correctly erased. If the preprogram verify operation is skipped, as usually done to save time, all the cells associated to the failing wordline WL1 can very likely go into a depletion state because at every cycle they are poorly preprogrammed but fully erased.

The row decoder shown in Figure 3.16 prevents the cells connected to WL1 to go into depletion because it keeps the wordline WL1 from being erased. During preprogram operation the cells connected to WL1 will be poorly preprogrammed, but during erase the row decoder does not drive the failing wordline WL1. It drives its associated line G1 while all the other good wordlines are correctly driven. On the common auxiliary powerline VCCP a typical voltage of 6 V is applied, and WL1 is biased to 6 V through transistors T1 and T2. All the cells connected to WL1 are not erased, having gate and source connected to the same voltage and drain floating, and cannot go into depletion.

In Figure 3.17 a row decoder performing simultaneous selection of two adjacent shorted wordlines during preprogram is reported [37, 38]. When the row fail is due to wordline to wordline short, the traditional preprogram operation is not effective because the high voltage applied on the failing wordline during preprogram is lowered to ground by the short to the adjacent deselected row having wordline connected to ground. The erase operation will be correctly performed on both the shorted rows because the same negative voltage is applied to all the wordlines and they will go into depletion.

The row decoder reported in Figure 3.17 prevents the depletion on the two shorted wordlines WL_i and $WL_i + 1$ because during preprogram it selects both the wordlines allowing a correct preprogram operation of both. The traditional row decoder is composed by a series of single-row decoders, each one associated to a single wordline, generating the select signals $P_0, P_1, \ldots, P_n$. The $(i + 1)$th single-row decoder can be included in a more completed structure $PG_i + 1$ having one AND gate and one OR gate to generate not only the WL selection signal $P_i + 1$ but also a carry $C_i + 1$ used to force the selection of $WL_i + 2$ in case it is needed. The $PG_i + 1$ block has at its inputs the carry C_i from the PG_i block associated to the WL_i and an enable signal E that is high when the preprogram operation is ongoing and the fuse logic is generating a match condition. If the wordline WL_i has a short with wordline $WL_i + 1$ and is selected during a preprogram operation, signal P_i will be high but also E and C_i will be both high, forcing $P_i + 1$ to be high: both WL_i and $WL_i + 1$ will be correctly preprogrammed.

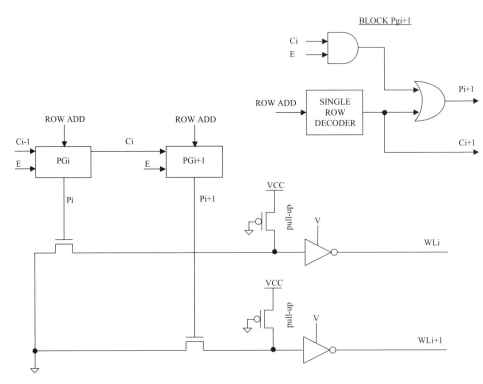

Figure 3.17. Double wordline selection row decoder.

REDUNDANCY

Another possible way to select adjacent shorted rows during preprogram is shown in Figure 3.18 and is based on the Gray row decoder [39, 40]. The Gray row decoder is a decoder where the binary address of any two adjacent wordlines differs by only one bit. In a Gray row decoder it is possible to select two adjacent wordlines by forcing high both the direct and inverted value, A and A_, of a specific address component. For example, in the 8 WL Gray decoder of Figure 3.18, if the address is 000, the inputs to the decoder are A[2] = 0, A_[2] = 1, A[1] = 0, A_[1] = 1, A[0] = 0, A_[0] = 1 and only the WL0 is selected. If both A[0] and A_[0] are forced to be 1, both WL0 and WL1 are selected.

In Figure 3.18 the blocks to control a Gray row decoder are shown. During the preprogram operation a clock ck increments a binary counter to select row addresses to be preprogrammed. The counter contains the uncoded binary version of the address represented by a[n:0]. In the case where there are no failing rows the outputs a[n:0] of the binary counter are converted to a Gray code g[n:0], which will be applied direct (A[n:0]) and inverted (A_[n:0]) to the input of a Gray row decoder to select only one wordline. In the case where there is a failing row, its address is also incremented by 1 to generate the consecutive binary address a'[n:0], which in general will differ from a[n:0] by more than 1 bit. This address is converted to a Gray code g'[n:0], which will differ from g[n:0] by only 1 bit. The two Gray codes g[n:0] and g'[n:0] are compared to detect which bit is different between the two codes. A bus b[n:0] is generated having all the components at 0 except the component corresponding to the different bit in g and g'. This component b[i] = 1 is used to force A[i] = A_[i] = 1 and select both WLi and WLi + 1.

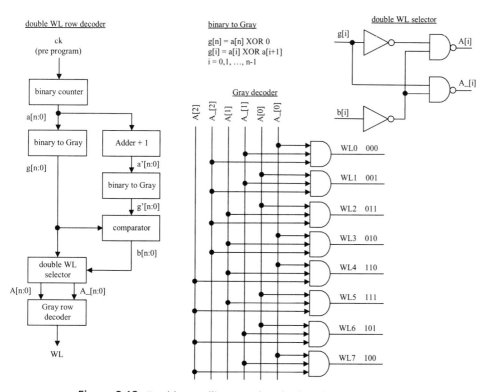

Figure 3.18. Double wordline row decoder based on a Gray code.

In Figure 3.19 a row decoder design that avoids the effects of depleted cells is reported [41]. The concept is to prevent any current flow from bitlines to ground because of deselected wordlines that may be in depletion by floating the memory transistor source terminals during the read operation. The program, erase, and read operating conditions of the row decoder reported in Figure 3.19 are as follows.

Program operation is performed with SCL low and all the transistors S1, S2, ..., SR off. The common source line CSL2 is not connected to any of the source lines SL1, SL2, ..., RSL. The common source line CSL1 is connected to ground. High voltage is applied on the selected wordline WLi and its correspondent pass transistor Ti will connect CSL1 to its source line SLi, thus grounding the source line of the selected row. The deselected wordlines will not have the CSL1 connected to their array sources.

Erase operation is performed with all the wordlines low or negative (T1, T2, ..., TR off), with CSL2 and SCL at a typical value of 6 V (S1, S2, ..., SR on) and the bitlines floating.

Read operation is performed with SCL low (S1, S2, ..., SR off) and with CSL1 connected to ground. The selected ith row will have the read voltage applied to its wordline WLi and the ground connected to its source line SLi through the pass transistor Ti. All the deselected rows will have the wordlines to ground and their source lines floating.

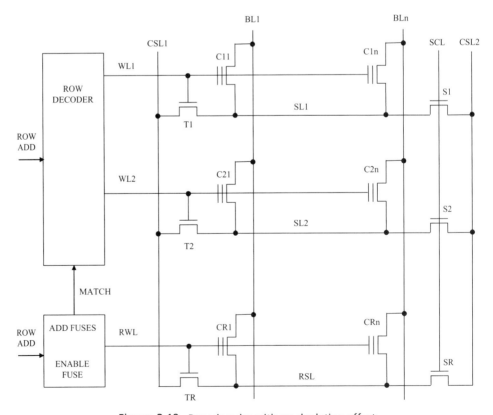

Figure 3.19. Row decoder with no depletion effect.

3.5 ERROR CORRECTION CODING (ECC)

Error correction coding has been an established field of study for over 30 years in communication, computers, and integrated circuits. Error correction coding has been widely adopted in many high-speed memory designs [42] to correct single-bit fails or byte fails and in low-speed mass memories to correct multiple random errors or burst errors.

The main purpose of on-chip ECC in DRAM is to increase reliability of memories affected by alpha particle interactions. The small area of a DRAM cell capacitor is not sufficient to maintain an adequate charge level after the interaction with an alpha particle, and on-chip ECC can recover this kind of soft error. Traditional redundancy repair techniques cannot fix these kinds of errors because they are randomly distributed in position and time.

The main purpose of ECC in mass magnetic memories is to overcome the effects of surface imperfections or random noise error in head reading operations. ECC has been also used in mass integrated memories organized to emulate a computer hard disk. This new type of system, called solid-state disk (SSD), can be implemented using low-cost defective DRAM chips and a unique ECC circuit dedicated to perform the error correction for all the DRAM chips. Such an application can increase wafer production line throughput because even the hard failing chips can be assembled and used [43, 44].

Various SSD designs based on Flash EEPROM and using system ECC have been reported by Intel [45], Micron [46], SanDisk [47, 48], and Toshiba [49, 50].

In addition, various ECC design techniques for DRAM/SRAM have been reported [51]. Special effort has been spent in developing new on-chip ECC concepts such as interlaced vertical parity (IVP) and augmented product code (APC) in order to minimize the area overhead due to the introduction of check memory cells to implement ECC. As an example, an area overhead of 15% and a soft error rate improvement factor of 10^6 has been reported [52] for a 4-Mbit DRAM exposed to an alpha particle flux of $1/cm^2$ per hour.

Error correction coding has been considered for Flash memories in order to increase the endurance and the retention. Recently, the ECC techniques have also been considered for multilevel Flash memories in order to increase the robustness of this new kind of Flash that has reduced noise margins between stored levels versus the traditional Flash memories with two threshold levels per cell [53]. The commonly used architecture of a Flash memory having high-speed parallel on-chip ECC capability is illustrated in Figure 3.20.

Data are written with parallelism of m bits, where m is usually equal to 8 or 16. Associated with the m bits of data are n bits for correction. The n correction bits are calculated by an internal ECC encoder circuit based on the m data bits and on the particular ECC algorithm chosen. The $m + n$ bits will be written into the array through the column path. The read operation will be performed on the $m + n$ bits at the same time. The $m + n$ bits are sent to a syndrome generator circuit that generates m bits, each one corresponding to one data bit. If one syndrome bit is high, the corresponding data bit is to be considered wrong, and it will be inverted by a correction circuit. The correction circuit will then send m corrected data bits to the output buffers.

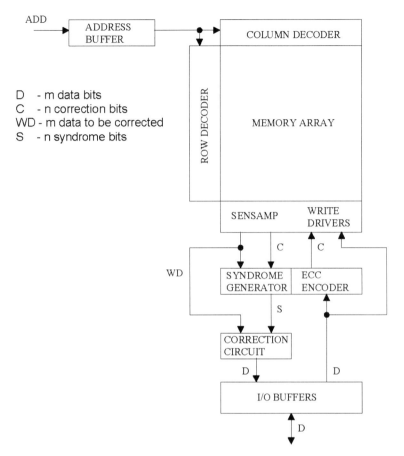

Figure 3.20. On-chip ECC for Flash memory.

3.5.1 On-Chip ECC and Endurance/Retention in Flash Memories

Error correction coding can be used in Flash memories to increase endurance and improve retention. After a certain number of erase/program operations a Flash memory can have defective cells due to oxide degradation or charge trapping. As soon as this happens, and provided that the ECC circuit is able to repair the amount of failing bits, the repair will be automatically performed. The apparent endurance of the chip is thus extended to a greater number of cycling operations.

After a certain lifetime a Flash memory can have a degradation of the cell thresholds due to charge loss or gain and, therefore, can undergo hard or random errors. On-chip ECC can correct these errors as soon as they happen in the application. The apparent data retention of the Flash memory is thus improved.

Redundancy is not useful to fix errors due to endurance or data retention because it can only be set up for fixed locations one time at the factory.

Flash memory ECC can be linked to circuitry to facilitate the performance of tests to screen chips that are weak for endurance [54] or retention. A Flash memory can in fact be provided with an internal address counter and an *internal repairs* counter. In test mode the Flash is fully read, and every time the ECC circuit performs a correction, the repairs counter is incremented. It is possible, therefore, to use the repair count as a criterion for scraping chips that have too many corrections because they are intrinsically weak, or to scrap chips that have too many clustered errors, or to scrap chips that have excessive repairs counter difference after 100 cycles because they are weak for endurance, or to scrap chips that have excessive repairs counter difference after certain time in accelerated conditions because they are weak for retention.

3.5.2 On-Chip ECC and Multilevel Flash Memories

Error correction coding has been considered for multilevel Flash memories in order to reduce the possibility of errors due to data retention.

In traditional two-level Flash memories (1 bit per cell: 0 or 1) the data retention is less critical because the threshold VT0 associated with programmed bits and the threshold VT1 associated with erased bits can be chosen distant enough to guarantee a typical separation of 5 V. If the read operation is performed using a voltage exactly in the middle of the interval VT0–VT1, the read margin is typically 2.5 V. In a multilevel Flash having four levels (2 bits per cells), the read margin is typically 0.8 V. In a multilevel Flash having 16 levels (4 bits per cell), the read margin is typically 0.15 V. Even if a threshold level VT can be programmed using very gradual and precise wordline ramp, thus obtaining very sharp cell distributions around any given VT, a small shift of distribution due to data retention degradation can produce a reading error. As the number of bits per cell in a multilevel Flash memory increases, the probability of errors due to data retention degradation will increase. ECC can help considerably in recovering these errors.

A technique to choose the proper multilevel bit coding has been also devised [55] in order to minimize the number of correction bits in a multilevel Flash memory as shown in Figure 3.21.

If the binary coding $Q0, \ldots, Qz$ associated with two adjacent levels Ei, Ei + 1 is chosen in a way to have only one different bit (Gray coding) instead of the maximum z different bits in the traditional coding (progressive coding), the number of correction bits can be dramatically reduced, and the ECC technique to be used will always be a single error correction (SEC). As an example, in a 4-bit per cell Flash memory having the outputs organized by 16 and a Gray level coding, it will be necessary to use 5 correction bits (31%) and a SEC technique to repair a fail between two adjacent levels. If in the same memory a progressive levels coding is used, it is necessary to use 12 correction bits (75%) and a 4-error correction technique to perform the same repair.

3.6 DESIGN FOR TESTABILITY (DFT)

Continuous advances in memory characteristics, such as reduced minimum features, increased bits per chip, growing numbers of chip I/Os, higher working frequency, and lower power supply voltages, are creating significant challenges to chip testability [56]. In the overall category of semiconductor memories, Flash memories have some peculiar characteristics that contribute further challenges to chip testability [10].

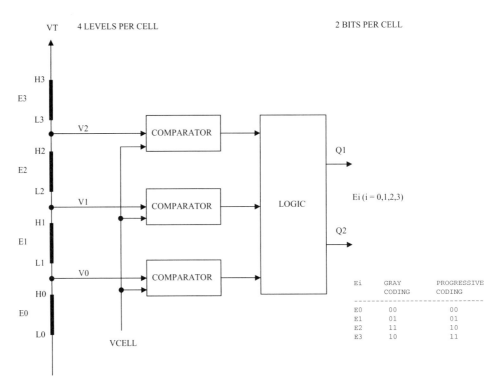

Figure 3.21. Levels coding for ECC in multilevel Flash memories.

Basic challenges in Flash memories DFT can be summarized as follows:

- Chip observability
- Test time reduction
- Chip configurability

Even though the external power supply of present-day Flash memories is low, down to the value of 1.65 V, the operating voltages for program and erase are still very high, in the range of 10 V: This creates a need to monitor and trim these high voltages.

While the working frequency of the peripheral circuits is steadily increasing, some basic operations of a Flash memory such as program and erase continue to require wide pulses that are often repeated many times during control algorithms. This impacts test time and serves as a motivation to find design solutions that can shorten the test time. It has been observed that cost of testing a transistor is becoming higher and higher compared to the cost of manufacturing [56].

Flash memories are used in a fast growing market that has a variety of very different application needs. To accommodate these differences, it is desirable that chip designs allow for easy configurability after manufacturing and avoid having different production splits for different chip versions.

Based on these considerations, fundamental requirements of DFT for Flash memories can be summarized according to the following list:

DESIGN FOR TESTABILITY (DFT)

- Trimmings: Sense amplifier reference trimming and monitor, high voltages trimming, timings trimming and external control
- Internal state machine algorithm skips
- Configuration: Address path configuration, data path configuration
- High voltages external forcing and monitor
- Array direct access and stresses
- Test time reduction: Internal pattern write and verify, data compression

These DFT requirements along with general considerations for DFT will be discussed in detail in the following paragraphs outlining design solutions and specific circuits to implement them.

3.6.1 Test Entry and Organization

A way [57] to meet the basic requirements of Flash EEPROM DFT is to organize the test functions according to Figure 3.22. A command user interface (CUI) recognizes that a special test mode entry request has been sent. A special test mode decoder recognizes which particular test has been requested.

Tests can be of two different kinds:

1. Tests that will act directly on blocks and will complete their action once the chip exits from test mode
2. Tests that will act on particular EEPROM cells used as nonvolatile storage elements to permanently perform trimming operations, skips in the algorithm steps, or chip configuration

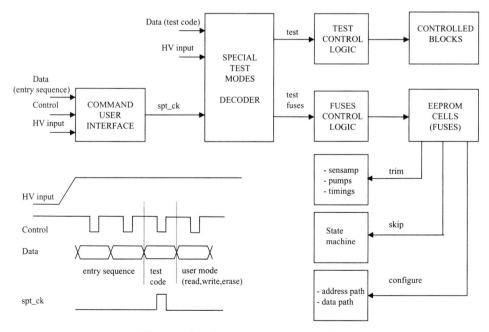

Figure 3.22. Test entry and organization.

The function of these particular EEPROM cells is similar to functions that in volatile memories are implemented using fuses that are blown with a laser or with the injection of very high currents. For this reason also the EEPROM cells used to permanently store information manipulated in test modes are often called fuses.

The test entry must be a safe operation from a user's point of view. In normal customer usage the chips must never enter test mode, and for this reason the ways to enter special test modes make use of voltages that are out of the user specification ranges [high-voltage (HV) inputs] or sequences of signals clearly declared forbidden in the user's specifications [58, 59]. In order to further lower the probability that the user accidentally enters test mode, the use of HV inputs can be associated with entry sequences that are not in the list of commands allowed for the user. The use of HV inputs is impractical in some situations, such as with low-cost testers not having high-voltage capability. In this case a hardware safe way to enter test mode can be implemented by adding in the fuse list an additional fuse (low-voltage command fuse) overriding the high-voltage detector [60]. Once this low-voltage command fuse has been programmed, the test mode is entered even in the absence of high voltage on the pin. The low-voltage command mode fuse must be erased before shipping the memory to the customer.

After the CUI has recognized a test entry request, a clock spt_ck is generated in order to enable the special test mode decoder to proceed in activating the special test mode specified by a particular value (test code) typically present on the data lines [61]. After the particular test mode is entered, the memory can accept all the user commands for read, program, and erase that will be executed having the memory setup in a particular test condition or test fuses configuration.

3.6.2 Fuse Cell

The information to configure and trim a Flash memory is stored in dedicated nonvolatile memory latches referred to as fuse cells. A typical fuse cell implementation [62] is based on a differential configuration. Two Flash cells are programmed with complementary data. The drains of the cells (CELL1 and CELL2) are connected, through isolation transistors, to cross-coupled PMOS transistors as shown in Figure 3.23.

In programming mode the signal ISOLATE is at 0 V to isolate the PMOS transistors from the programming voltage V_{pp}. Complementary data, DATA and DATAB, are then programmed into CELL1 and CELL2. In read mode, ISOLATE is high, DATA and DATAB are both forced at 0 V and cell wordline WL is biased at the read voltage (usually V_{cc}). The fuse cell output FUSEOUT, therefore, flips either to low or to high depending upon the content of CELL1 and CELL2. One big advantage of the above approach is that direct current (dc) in reading mode is zero thanks to the cross-coupled gate connection of the PMOS transistors. Fuse cells are arrayed in a fuse cell bank. They are programmed at the factory to trim analog values and to configure the chip, as well as to store repaired location addresses. Sometimes the fuse cell content is copied in complementary metal–oxide–semiconductor (CMOS) latches, which can be overridden in dedicated special test modes for trimming and repair look-ahead before the final programming of the nonvolatile fuse cells.

Single-ended versions of fuse cells have been also developed [63].

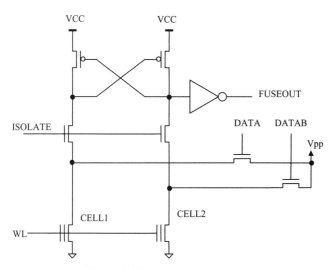

Figure 3.23. Fuse cell schematic.

3.6.3 Sense Amplifier Reference Trimming and Monitor

The EEPROM cell sensing generally makes use of one of two basic techniques. One way is to detect the currents flowing in the EEPROM cell and in an EEPROM ultraviolet erased EEPROM cell, convert them into two voltages, and use a voltage differential amplifier to compare the values [57]. Another way is to detect the current (I_{cell}) flowing in the EEPROM cell to be read and compare it to a fixed reference current (I_{ref}) derived from the V_{cc}. The value of I_{ref} is chosen to be less than the I_{cell} of an erased cell (in this case the sense amplifier output is 1) and greater than the I_{cell} of a programmed cell (in this case the sense amplifier output is 0). The latter technique is particularly suitable when compact array-pitched sensing circuits are required. Sense amplifier reference trimming techniques are different in the two different sensing techniques. The following I_{ref} trimming description is related to the current-sensing scheme illustrated in Figure 3.24.

Sense amplifier reference trimming is necessary for different operating conditions: regular read, read to verify threshold margin for a programmed cell (program verify), read to verify threshold margin for an erased cell (erase verify), and read to perform a depletion check. Each one of these operating conditions is associated with a different value of I_{ref} and to a different range of I_{ref} trimming. A set of sense amplifier reference fuses is associated with each one of the possible four operating conditions: fuserefr for regular read, fuserefpv for program verify, fuserefev for erase verify, and fuserefd for depletion check. Assuming each one of the reference fuse groups has two components, a multiplexer will select only one fuse group providing two fuse lines (fuseref1 and fuseref2) to the sensing circuit according to the active operating condition. Through the connection of resistors RES1, RES2, four currents, greater than the minimum value I_{ref0}, can be forced to flow in the transistor TP1 and mirrored to the transistor TP2 to generate I_{ref}.

In a regular read with WL having a typical value of 5 V, the I_{ref} value can be in the range of 2 to 5 µA, while typically the I_{cell} is around 20 µA. In erase verify with

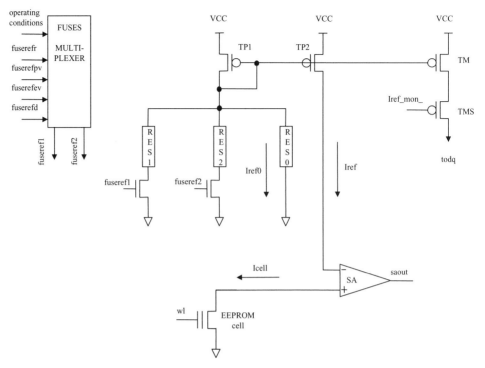

Figure 3.24. Sense amplifier reference trimming and monitor circuit.

WL having a typical value of 5 V, the I_{ref} can be in the range of 5 to 10 μA to guarantee that the cell is erased enough to be read with at least 10 μA of margin. In program verify the cell cannot be read using 5 V on the WL because zero current is driven by the cell and there is no meaning in increasing the I_{ref} to guarantee the margin. A typical value to be used for WL in program verify is 7 V, and the I_{ref} can be in the range of 2 to 5 μA. In a depletion check WL for all of the cells in the selected columns is at 0 V, and all together they should draw no current or a current lower than that used as reference in regular read. The I_{ref} in this case can be in the range of 1 or 2 μA.

The I_{ref} can be further mirrored on transistor TM and connected to an output pad when a particular test sequence is given and signal Iref_mon_ is forced low. Connecting the pad to ground using a resistor of a known value can make it possible to measure I_{ref}.

3.6.4 High Voltages Trimming

Starting from a very high voltage produced by a global charge pump, it is necessary to generate the lower voltages required for the different Flash memories operating conditions. For example, starting from a global high voltage of 12 V, 10 V is generated for the wordlines in program mode or 7 V is generated for the wordlines in program verify mode. Each one of those different voltages must be trimmable using fuse cells in order to optimize the chip performance during two different phases of manufacturing: in production ramp-up to find the right mean values, and in steady

DESIGN FOR TESTABILITY (DFT)

production to reduce the effects of natural processing spreads and avoid array stress, inefficient operations and excessive algorithm looping.

One way [57] to implement trimming of high voltages produced by a charge pump is shown in Figure 3.25. Operating high voltage v_{out} applied to a capacitive LOAD (wordlines) or conductive LOAD (bit lines) is generated from high voltage (hv) produced by a charge pump having different characteristic for different kinds of LOAD. The voltage hv is applied to the drain of a P-channel transistor (TP); switching on and off the transistor TP produces an increment or a decrement in the v_{out} value. The v_{out} value must be trimmable with different values in different operating conditions such as program (pgm) and program verify (vfy).

If the program operation is active, resistive partitioning Rf, Rp0, Rp1, ..., Rpn is selected; if program verify is active, resistive partitioning Rf, Rv0, Rv1, ..., Rvn is selected. When partitioned value v_{part} is greater than the stable reference voltage v_{ref}, the output ctrl of the operational amplifier OPAMP is high and high-voltage hv is applied to the gate of the pass transistor TP, which will be turned off and the value of v_{out} will decrease. When partitioned voltage v_{part} is lower than v_{ref}, the ctrl signal is low and 0 V is applied to the TP gate, which will be turned on and the value of v_{out} will increase.

The regulated value for v_{out} is associated with the condition vpart = vref. In general terms, avoiding using index p (for program) or v (for program verify), the condition vpart = vref under the variation possible using the trimming process produces the following extreme values for v_{out}:

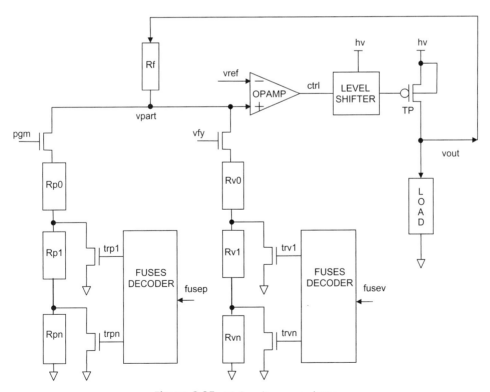

Figure 3.25. High-voltage regulator.

$$\text{Max } v_{out} = (1 + Rf/R0) \times v_{ref}$$
$$\text{Min } v_{out} = (1 + Rf/(R0 + R1 + \cdots + Rn)) \times v_{ref}$$

The first situation corresponds to high values on all trimming signals $tr1, \ldots, trn$, and the second situation corresponds to low values on all trimming signals $tr1, \ldots, trn$. To establish the trimming signals a set of fuse cells are appropriately erased and programmed, and then output from the cells is fed to a decoder. The decoder output is the set of trimming signals $tr1, \ldots, trn$. For example, three fuse cells can be decoded to form eight possible trimming signals $tr1, \ldots, tr8$.

3.6.5 Timings Trimming and External Control

Program and erase operations consist of applying appropriate voltage conditions to the selected EEPROM cells in order to program or erase them. These operations are organized by algorithms to monitor the effects and to proceed with the operation until a satisfactory result is reached. The amount of time that a single operation will last is controlled by a pulse having a width optimized to reach the result with the desired accuracy and with the minimum looping of the algorithm. The typical value for program pulse width is 5 µs while for erase pulse width it is 5 ms. However, processing spread and adjustments require having these typical values trimmable within a certain range around a default value. One way [57] to trim the pulse width is based on a pulse width generator controlled by fuse cells, and this is shown in Figure 3.26.

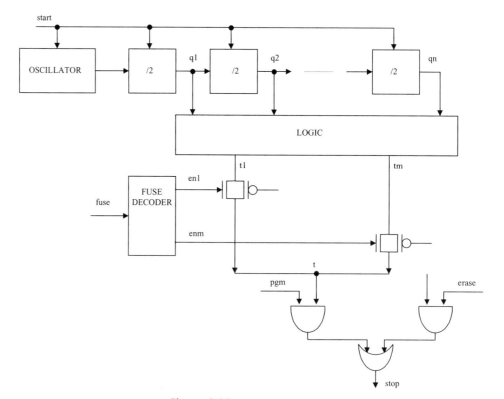

Figure 3.26. Timings trimming.

DESIGN FOR TESTABILITY (DFT)

The pulse width generator is composed of a counter having n stages of by 2 frequency divider flip-flops, and a free-running oscillator producing a clock applied to the first stage of the counter. The number of stages in the counter must be high enough to reach the desired maximum value of pulse width. A logic circuit is designed to generate the desired number of pulse width signals $t1, \ldots, tm$ starting from the $q1, \ldots, qn$ flip-flop outputs. The fuse decoder generates the same number of enable signals $en1, \ldots, enm$ to be applied to the pass gates of a multiplexer, select the proper pulse signal, and thus select the proper value for the pulse width. A logic circuit combines the selected pulse width and the operating signal to apply the pulse width associated with the selected operating condition and to generate the common stop pulse signal. The pulse width generator circuit receives the start signal from the internal state machine when a program or erase operation starts, and will send a stop signal back to the internal state machine to indicate that the operation has been completed.

A test mode can be designed to defeat the internal timer and override the stop signal in order to externally force a program or an erase pulse width [64]. This is a very useful feature, particularly in conjunction with an external high-voltage forcing mode, to program or to erase the array avoiding automatic adjusting mechanisms involved in the algorithms.

3.6.6 Internal State Machine Algorithm Skips and Monitor

The algorithm to erase a Flash memory sector is a complex sequence of operations performed by the internal state machine. The erase algorithm can be divided into three parts: preprogram loop and its verify, erase operation and its verify, and the compaction loop. The sector erase algorithm consists of the following sequence of steps:

- Preprogram and verify:
  1. Preprogram a single word of the selected sector (this is a regular program operation).
  2. Verify the preprogrammed threshold of the word.
  3. Increment word address (row and column) and go to step 1 until a sector has been completed.
- Erase and verify:
  4. Apply an erase pulse to the sector.
  5. Verify the erased threshold of every single word of the sector. If erase verify passes, go to step 6. Otherwise go to step 4 until pass or the maximum number of pulses is reached.
- Compaction:
  6. Perform a depletion check on a single column address (all the wordlines are off).
  7. If depletion check fails, apply a soft program pulse to every single word in the selected column address (soft program is a programming mode having lower wordline and bitline voltages. Its effect is to lightly program the cells having very low threshold and not affect the others).

8. Increment column address and go to step 6 until the full sector has been successfully scanned or a maximum loop count has been reached.

Using fuses cells, it is possible to skip some of the steps in the erase algorithm [65]. Using five fuse cells, the following parts of the erase algorithm can be skipped: preprogram, preprogram verify, erase pulse, erase verify, and compaction. Skipping some of the erase algorithm steps can produce a significant reduction in the erase algorithm execution time without any drawback if the process is mature and well controlled. Considering that the duration of the sector erase algorithm is about 1 s, and that a significant amount of this time is due to the loop of preprogram and its verify, skipping the preprogram verify can significantly reduce the total time necessary to erase a sector.

Erase algorithm skips can be very useful in a special test mode called internal pattern write, designed to automatically write and verify key data patterns into the array as reported in Section 3.6.11.

If the internal state machine is implemented using a microcode programmable machine, the way to implement the fuse cell action is straightforward. Signals from the skip fuses are input to the internal state machine, and they are tested every time the machine is about to begin another step of the erase algorithm. If a skip signal is true, the next algorithm step is not executed and the program jumps to the next step after that one.

Erase and program operations in Flash memories are accomplished using complex algorithms composed of many different steps, requiring control actions to establish timing and involving counters and analog voltages. An algorithm controller manages the execution of the different steps in the erase and program algorithms. Furthermore the erase and program operations on a sector can be suspended to start a read operation in another sector, and the algorithm controller must correctly manage the suspension and its resume. Once an erase/program operation has been requested, and the algorithm is started, it is not possible to say which particular step the algorithm controller is performing at a particular time. In fact the erase and program algorithm is adaptative, and a step is started only when conditions associated with the previous step are satisfied. A way to monitor the very basic activity of an algorithm controller is the status register concept. This has been widely used in Flash design and the register is available to the end user to interface the Flash chip with the board controller.

An alternative to the status register technique is the concept of a programmable ROM-based algorithm controller. Here the activity during the erase and program algorithm can be carefully monitored using test modes. One test mode allows an override of the internal algorithm controller clock using a clock applied to an external pad, thus imposing a specific clock period and synchronization to the algorithm controller. A second test mode allows the controller to follow the evolution of the algorithm, reading ROM address at the output pads [66].

3.6.7 Address Path Configuration

Flash memories have a special sector, called the boot sector, where the kernel instructions the system looks for at power up are stored. The boot sector has special characteristics in terms of size and special protection features. Historically, systems

DESIGN FOR TESTABILITY (DFT)

using a Flash memory look for the boot sector at either the all-1 address or the all-0 address, and to be compatible with this practice two versions of a chip are usually produced. One version has the boot sector addressed with the all-1 address (top boot version), and the second version has the boot sector addressed with the all-0 address (bottom boot version). The two Flash versions can be defined by using a different metal mask, but the availability of fuse cell structures make it possible to design a circuit [57] to electrically configure the Flash memory as top boot or bottom boot.

The boot sector must be designed to be internally decoded by an all-0 address or by an all-1 address. For the sake of simplicity let's suppose that the boot sector is internally decoded by the all-0 address. After each address buffer, an exclusive OR gate is inserted having as the second input a common line coming from a fuse cell associated with the boot function. If the line coming from the fuse is high, address lines are all inverted and a top boot is obtained. If line coming from the fuse is low, address lines are not inverted and a bottom boot is obtained.

3.6.8 Data Path Configuration and Trimming

Chip configurability is important not only for address path but also for data path as reported in Figure 3.27 for the input data path and in Figure 3.28 for the output

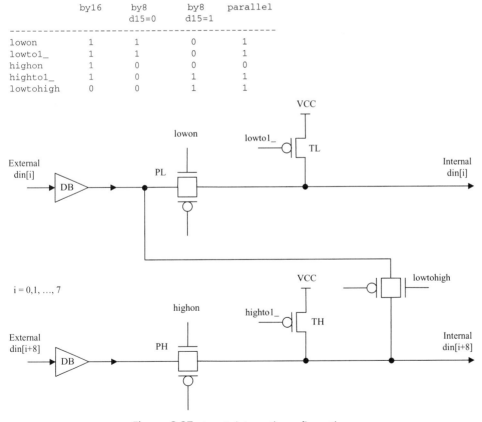

	by16	by8 d15=0	by8 d15=1	parallel
lowon	1	1	0	1
lowto1_	1	1	0	1
highon	1	0	0	0
highto1_	1	0	1	1
lowtohigh	0	0	1	1

Figure 3.27. Input data path configuration.

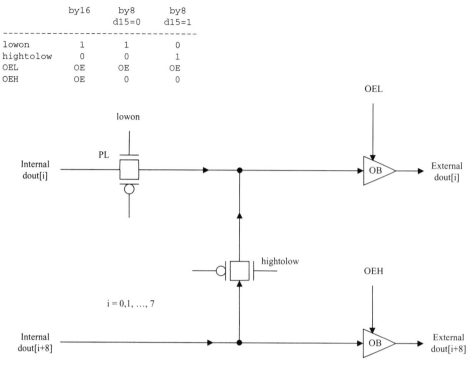

Figure 3.28. Output data path configuration.

data path. Typical data input configurations support by 8 or by 16 input data bussing and use an internal by 16 configuration. The by 8 or by 16 configuration can be important either from the user point of view or from the testing point of view. The by 8 configuration uses fewer data I/O lines but requires two cycles to perform the same operation. Another useful configuration during test mode is the parallel programming configuration where input data is applied only on the external lower byte DQ[7:0], and this byte is stored on the internal lower byte and duplicated on the internal upper byte. Usually when the memory chip is in by 8 mode the data is on the external lower byte DQ[7:0] and the single data line DQ[15] decodes whether the data is to be associated with the internal lower or upper byte.

In Figure 3.27 two generic correspondent ith data lines din[i] and din[i + 8] are drawn for the lower byte and the upper byte. Two fuse cells can be associated with the by 16 mode and with the parallel programming mode, and used to design the logic to generate the control signals in the input data path configuration according to the table given in the figure.

When the memory chip is in by 16 mode, the external lower byte is connected to the internal lower byte and the external upper byte is connected to the internal upper byte. When the memory chip is in by 8 mode and DQ[15] = 0, external lower byte is connected to the internal lower byte and the internal upper byte is forced high to prevent programming. When the memory chip is in by 8 mode and DQ[15] = 1, external lower byte is connected to the internal upper byte and the internal lower byte is forced high to prevent programming. When the memory chip is in parallel programming, the external lower byte is connected to both the internal lower and upper byte.

DESIGN FOR TESTABILITY (DFT)

The output data path configuration, shown in Figure 3.28, is simpler than the input data path configuration. The modes involved are only by 16 or by 8 and are controlled by the same fuses as in the input path configuration.

When the memory chip is in by 16 mode, the internal d_{out} lower byte is connected to the external d_{out} lower byte and the internal d_{out} upper byte is connected to the external d_{out} upper byte; the output enable signals for lower byte (OEL) and for upper byte (OEH) have in this case the same value of OE according to the output enable function.

When the memory chip is in by 8 mode and d15 = 0, the internal d_{out} lower byte is connected to the external d_{out} lower byte having the output enable signal equal to the general OE. The internal d_{out} upper byte is connected to its output buffer (OB), but the output enable signal for the upper byte is forced low and the output buffers are in the 3-state condition.

When the memory chip is in by 8 mode and d15 = 1, internal d_{out} lower byte lines are disconnected from their output buffers, and the internal d_{out} upper byte is connected to the external d_{out} lower byte. Output enable signal for the external lower byte is connected to the general output enable signal, the output enable signal for the external upper byte is forced low, and the external upper byte lines are in the 3-state condition.

Even though simple circuits are needed to implement the output data path configuration, these circuits must be designed and organized with particular care because they influence the read speed path. From the layout point of view there are two basic key points to consider in minimizing the speed penalty due to the output data path configuration: output buffers organization and their data bus routing. If the output buffers are organized having the couples 0 and 8, 1 and 9, ..., 7 and 15 as close as possible, the routing at the output of the pass gates implementing the upper byte to lower byte connections is minimized.

A useful feature that can be introduced in the data path is output buffer strength trimming [67]. In this concept an output driver would be composed of two or more stages arranged to drive the load in parallel. Logic signals derived from fuses or external signals would be used to set the number of parallel stages that would be active, thus determining the available drive current. In this manner the best compromise between speed, overshoot, undershoot, and electromagnetic emissions can be selected at the factory or by the end user.

3.6.9 High Voltages External Forcing and Monitor

Due to process variations, voltages produced by internal charge pumps used in read, program, or erase operations can vary in such a way as to alter the final performance of the chip. The result can be data errors or inefficient operation that increases algorithm looping. For this reason, circuits have been designed to monitor the voltage VREAD applied to the cell gates in read mode, to monitor the voltage HV applied to the cell gates during program or verify operations, and to monitor the voltage HC applied to the cell drains during program or to the cell sources during erase. This monitor function is enabled by entering special test modes to connect each one of the voltages to be monitored to an external pad.

Another useful feature is similar to the voltage monitor in making connections to an external pad, but it acts in the opposite direction. It is possible in fact to disable an internal pump and force the voltage on its output line by applying an external

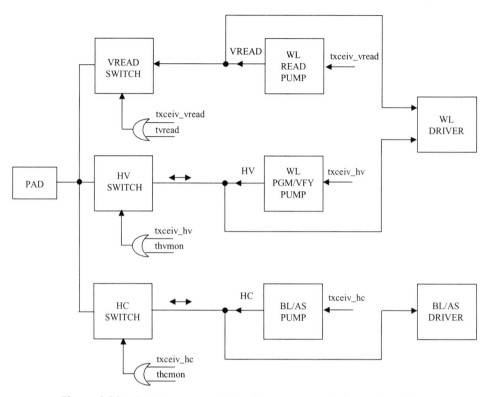

Figure 3.29. Block diagram of high-voltage external forcing and monitor.

voltage to a chip pad. This forcing function can be enabled by entering special test modes to connect each one of the lines to be forced to an external pad. The on-chip charge pumps can only supply a small amount of current, and external voltage forcing can be useful to supply current to the bitlines to speed up a program or erase operation or to have the capability to perform parallel programming or erasing operations [68]. It can also be used to apply voltage ramps on the cell gates to measure the real cell threshold [10].

One possible design scheme to implement the voltage monitor of three voltages VREAD, HV, HC and the forcing function is shown in Figure 3.29. These three monitor operations are activated by three different test modes: VREAD monitor is operated when the test signal tvread is active, HV monitor is associated with the thvmon test signal, and HC monitor is associated with the thcmon test signal. These test signals, which will be activated one at a time, can force the switches to be ON and connect the pumps to the external pad. In Figure 3.29 it is also shown that it is possible to force the lines VREAD, HV, and HC by activating test mode signals txceiv_vread, txceiv_hv, and txceiv_hc. These test signals, applied one at a time, can force the switches to be ON while disabling the correspondent charge pump. VREAD, HV, or HC lines can be forced by connecting a voltage generator to the external pad. Wordline and bitline drivers can thus receive their power supply either by an internal source or by an external source.

A way [69] to implement the switch function is shown in Figure 3.30. When there is no monitor or forcing test mode, the generic test signals tmon and txceiv

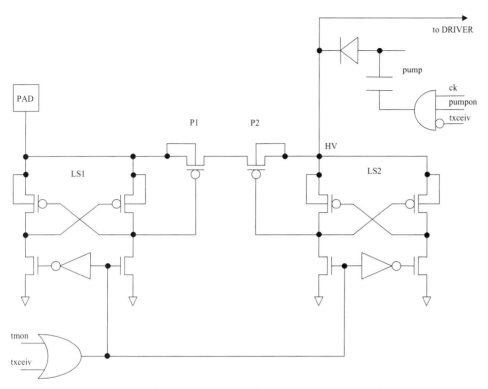

Figure 3.30. Switch circuit for high-voltage external forcing and monitor.

are both low. Level shifter LS1 forces the pass transistor P1 to be off connecting its gate to the pad. Level shifter LS2 forces the pass transistor P2 to be off connecting its gate to HV, the pump is enabled, and the driver can receive the high voltage HV directly from the pump. When tmon is active, level shifter LS1 forces the pass transistor P1 to be on applying 0 V to its gate, and level shifter LS2 forces the pass transistor P2 to be on applying 0 V to its gate, the pump is enabled, and the voltage HV can be applied to the driver and also externally read. When txceiv is active, P1 and P2 are on while the pump is disabled. In this condition the input HV of the driver can be externally forced without affecting the pump having the diode in the off state.

3.6.10 Array Direct Access and Stresses

Array direct access is a key point in the design for testability of Flash memories. In particular the possibility of measuring the real threshold voltage (VT) of the Flash cells gives a powerful means to evaluate the VT array distributions under different circumstances. The usual program and erase verify operations do not give the exact value of VT of a cell. They only give an indication that the VT of the selected cell is below the erase verify value or above the program verify value. A basic way to measure the real VT [69, 70] is to externally apply a voltage to the selected wordline and to connect the drain of the Flash cell directly to an external pad. The first test mode has been already discussed as the external high-voltage forcing test mode, the second mode is a bitline access test mode; combining the two modes it is possible to measure the real VT of a Flash cell. In fact, after the cell has been connected to a pad, an external voltage of typically 0.5 V/1 V is applied to the cell drain through the pad, and the wordline of the cell is ramped up applying an external ramp to pad. When a current greater than a certain conventional amount, typically 1 or 10 µA, will flow through the cell, the corresponding wordline value is considered equal to the cell VT.

A DFT of a Flash memory must include a series of tests to stress the memory cells in order to detect the weak cells where performance is not acceptable or has a significant probability of becoming unacceptable within the lifetime of the device. The Flash cells threshold VT distribution is measured both before and after the stress tests are applied. The VT distribution after the stress is typically shifted to lower values; for every set of stress voltages and duration, criteria about the VT shift can be set in order to define the shift acceptable and consistent with how that particular silicon oxide is expected to perform. If the VT shift is anomalous, a defect in the Flash structure is suspected, and an investigation is required.

There are three kinds of tests to stress an array of Flash cells: gate stress, source stress, and drain stress. The gate stress is a channel oxide stress performed with drain and source connected to ground and control gate at high voltage (positive or negative). The source stress is a stress to check the junction where the erase operation takes place and is performed with the control gate connected to ground, the source at high voltage, and the drain left floating. The drain stress is a stress to check the junction where the program operation takes place and is performed with control gate connected to ground, the drain at high voltage, and the source left floating [10].

Circuits for such stress modes are implemented with traditional techniques to force all the wordlines in all the sectors to ground, to force all the drain (sources) of all the sectors to high voltage, and force all the sources (drain) in all the sectors to be floating. The high-voltage forcing can be performed either using the internal pumps or using the external high-voltage forcing test mode.

When short channel effects are not involved in Flash cells, the band-to-band current during stress is independent from channel length and has the same value if the source (drain) stress is performed with drain (source) floating or grounded. For the short channel length Flash cells significantly worse degradation after stress is observed [71]. VT shift is more evident, and Gm degradation appears with long time stressing. Furthermore, stressing is more effective when lateral junction stresses are performed with the second terminal not floating but connected to ground. Based on this experience, design for testability of short channel Flash memories must include more stress modes than traditionally conceived, including lateral junction stresses without second terminal floating.

For longer channel length, the edge Fowler–Nordheim discharge speed is independent from channel length, but for super short channel Flash cells a super fast discharge speed has been noted. This not only causes device degradation but also causes widened erase distribution due to array channel length variations. Design techniques have been devised to overcome this limitation [71–73] and to help in alleviating what could be considered a fundamental limitation to the scaling of Flash memory cells.

DESIGN FOR TESTABILITY (DFT)

3.6.11 Internal Pattern Write and Verify

Reduction in Flash memories test time is a key issue. Testing memory chips in parallel is a way to reduce significantly the test time. There are some design solutions that make the parallel technique really effective. One such design technique [65] is the internal pattern write: giving just one command from outside the chip, it is possible to write traditional patterns, such as checker boards, vertical or horizontal stripes in a full sector, and to have the possibility to know if the patterns have been written with the appropriate verify margins.

The internal pattern write can also exploit the possibility of skipping steps in the erase algorithm. In fact, if in the erase algorithm the preprogram and preprogram verify operations are kept, while the erase pulses and compaction operations are skipped, the result is that a full sector is written with a pattern having 0 in all the bits; furthermore all the preprogrammed words are verified using the program verify margin and a status bit can be read on the output if the check passes.

This skipping technique is a powerful method, but its limitation is that it is only possible to write the all-0 pattern. By associating the erase algorithm skipping configuration and the design circuits presented in Figure 3.31, it is possible to automatically write a wide range of patterns by giving only the erase command and specifying a starting pattern on the DQ pads.

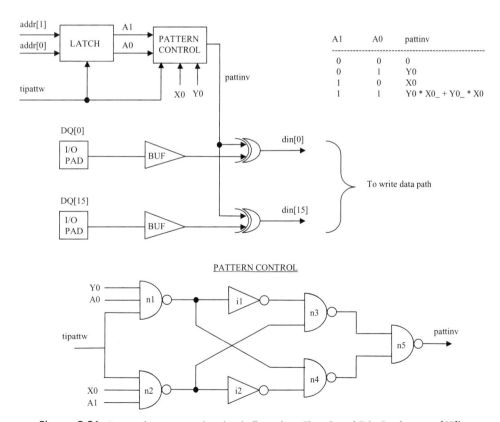

Figure 3.31. Internal pattern write circuit (based on Figs. 6 and 7 in Roohparvar [65]).

The circuit to generate the internal patterns consists of: a LATCH to store two addresses, that is, addr[0] and addr[1], for the selection of one of four possibilities to manipulate the starting pattern; a PATTERN CONTROL; and XOR (exclusive OR) gates to invert or not invert the data coming from the external input pads to the write data path.

If the signal pattinv generated by the PATTERN CONTROL is high, the data coming from the input pads to the write data path are inverted, otherwise they are not inverted. The detailed structure of the PATTERN CONTROL is determined by the gates n1, n2, n3, n4, and n5 and by the inverters i1 and i2. The signal pattinv can be high only if the internal pattern write test mode associated with the signal tipattw is entered using a specific sequence. The signal X0 is the least significant row address, and the signal Y0 is the least significant column address. The signal A0 is the address addr[0] latched at the rising edge of tipattw, exactly when the internal pattern write test mode is entered, and the signal A1 is the address addr[1] latched at the rising edge of tipattw.

As reported in the table included in Figure 3.31, if $A1A0 = 00$, the signal pattinv is low and no inversion will occur; if $A1A0 = 01$, the signal pattinv is equal to Y0 and input data will be inverted only if odd columns are selected; if $A1A0 = 10$, the signal pattinv is equal to X0 and input data will be inverted only if odd rows are selected; if $A1A0 = 11$, the signal pattinv is equal to $Y0 * X0\_ + Y0\_ * X0$ and input data will be inverted only if odd column and even rows or even columns and odd rows are selected.

If consecutive row addresses are physically adjacent and consecutive column addresses are physically adjacent, the internal pattern reported in Figure 3.32 can be automatically written using the internal pattern write and having all 0 as a starting sequence.

3.6.12 Data Compression

Since the cost of testing is becoming a significant component of the total manufacturing cost of memory chips, it is desirable to increase the throughput of the testing

A1	A0	pattern	
		00000000	starting pattern
0	0	00000000 00000000 00000000	all 0
0	1	01010101 01010101 01010101	vertical stripes
1	0	00000000 11111111 00000000	horizontal stripes
1	1	01010101 10101010 01010101	checker board

Figure 3.32. Example of patterns.

DESIGN FOR TESTABILITY (DFT)

operation. Data compression is a special test mode that can represent a way to increase the testing throughput.

Two data compression techniques will be discussed in this section: a technique performing a single internal word compression to a single external nibble [74], which for sake of brevity can be called *I/O compression*, and a technique performing a multiple internal words compression to a single external word, which for sake of brevity can be called *word compression*. Both techniques are summarized in Figure 3.33.

The I/O compression has the purpose of reducing the number of external lines to be read when predefined internal patterns need to be checked. Typical patterns that are used to check the single cell integrity and cell-to-cell isolation are: all zeros, all ones, checker board (1010 1 ... 10) and inverse checker board (0101 ... 01). In normal mode, the 16 lines Q[0:15] at the output of a memory chip carry the information of every single bit of the selected internal word W[0:15]. When a fail happens, the information about the particular failing bit location is complete. In I/O compression mode, the information about the fact that internal word W[0:5] is correctly storing one of the predefined pattern is preprocessed and carried to a reduced set of output lines, namely Q[4:7]. Using this compressed data the tester is still able to extract enough information to detect failing patterns.

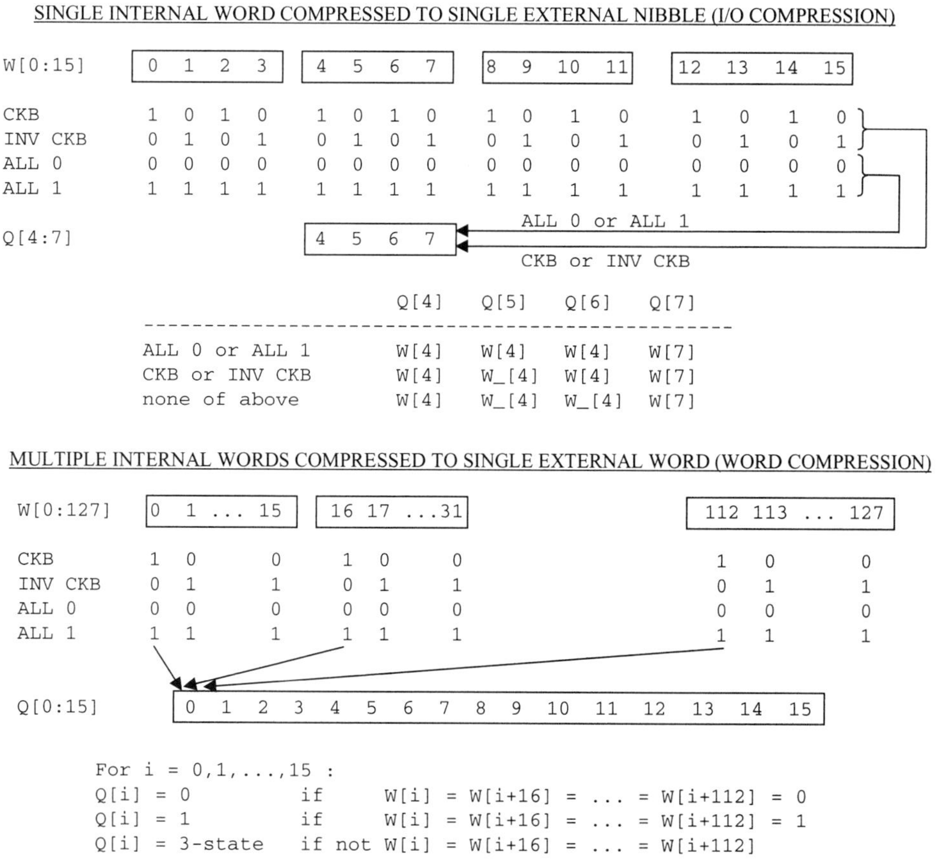

SINGLE INTERNAL WORD COMPRESSED TO SINGLE EXTERNAL NIBBLE (I/O COMPRESSION)

```
W[0:15]      0  1  2  3    4  5  6  7    8  9  10 11   12 13 14 15

CKB          1  0  1  0    1  0  1  0    1  0  1  0    1  0  1  0
INV CKB      0  1  0  1    0  1  0  1    0  1  0  1    0  1  0  1
ALL 0        0  0  0  0    0  0  0  0    0  0  0  0    0  0  0  0
ALL 1        1  1  1  1    1  1  1  1    1  1  1  1    1  1  1  1
                                         ALL 0 or ALL 1
Q[4:7]                     4  5  6  7
                                         CKB or INV CKB

                           Q[4]  Q[5]  Q[6]  Q[7]
             -----------------------------------------
             ALL 0 or ALL 1    W[4]  W[4]  W[4]  W[7]
             CKB or INV CKB    W[4]  W_[4] W[4]  W[7]
             none of above     W[4]  W_[4] W_[4] W[7]
```

MULTIPLE INTERNAL WORDS COMPRESSED TO SINGLE EXTERNAL WORD (WORD COMPRESSION)

```
W[0:127]     0  1 ... 15    16 17 ... 31              112 113 ... 127

CKB          1  0       0    1  0        0             1   0        0
INV CKB      0  1       1    0  1        1             0   1        1
ALL 0        0  0       0    0  0        0             0   0        0
ALL 1        1  1       1    1  1        1             1   1        1

Q[0:15]      0  1  2  3  4  5  6  7  8  9 10 11 12 13 14 15
```

```
For i = 0,1,...,15 :
   Q[i] = 0          if      W[i] = W[i+16] = ... = W[i+112] = 0
   Q[i] = 1          if      W[i] = W[i+16] = ... = W[i+112] = 1
   Q[i] = 3-state    if not  W[i] = W[i+16] = ... = W[i+112]
```

<u>Figure 3.33.</u> Data compression.

With information carried on a lower number of external lines it is possible to test chips in parallel. If during compression mode, the output lines not carrying information are forced into 3-state, it is possible to connect 4 memory chips to a tester having 16 input lines; every group of 4 output lines of the chips is connected to a different set of 4 tester input lines.

A way to preprocess the pattern information is to apply for Q[4:7] the settings reported in Figure 3.33. In this way Q[4:7] = 0000 if W[0:15] is all 0; Q[4:7] = 1111 if W[0:15] is all 1; Q[4:7] = 1010 if W[0:15] is a checker board; Q[4:7] = 0101 if W[0:15] is an inverse checker board, and finally Q[4:7] will be different from any of the reported values if W[0:15] is different from any of the four typical patterns. If a fail is detected, there is no possibility to know which bit is failing, and the column repair must be done either by replacing the entire word with one redundant word or by detecting the position of the failing bit by exiting the compressed test mode.

The word compression technique is different from I/O compression in the following aspects:

1. Word compression does not reduce the I/O lines, and because of this no parallel testing can be performed by wire arrangement to the tester.
2. Word compression performs data compression bit by bit on multiple internal words and is particularly suitable for memory chips that are read with internal parallelism.
3. The testing throughput is increased by collecting, with a single read operation, preprocessed information about all the internal words at the same time.
4. The number of address lines needed to decode the internal words are not significant and the number of tester address drivers can be reduced.

A way to preprocess the pattern information is to apply the settings of Figure 3.33 where a memory chip having eight internal words is reported. The generic output Q[i] will be 1 if all the words have 1 in the same position, or will be 0 if all the words have 0 in the same position, or will be 3-state if neither of the previous conditions is verified. If a fail is detected, the bit position within the word is known, but there is no possibility of knowing in which particular word that bit is failing. The column repair must be done either by replacing the same failing bit location in every single word with a redundant column for every word or by detecting the failing word by exiting the compressed test mode.

3.7 FLASH-SPECIFIC CIRCUIT TECHNIQUES

As outlined in the previous sections, Flash memories need a large variety of voltage values to operate the cell array. These voltages have to be generated on-chip to support single supply operation, which is a strong requirement driven by overall end-product cost reduction. In addition there is an ever increasing demand for lower supply voltages driven by the needs of portable battery-operated equipment, and this has placed enormous pressure on Flash developers to identify technology and circuit design solutions that support low-voltage capability. Scaling of CMOS also

FLASH-SPECIFIC CIRCUIT TECHNIQUES

implies reduction of supply voltages in order to avoid electric field overstress conditions where dimensions have been reduced. This leads to a particularly difficult situation in Flash technology. While chip operating voltage decrease, internal voltages to be generated on-chip do not scale proportionally. Also, the threshold voltage of peripheral transistors cannot be lowered arbitrarily since it would then not be possible to fulfill requirements for low stand-by current due to transistor off-current increases. In this section techniques to generate internal voltages and perform needed regulation and level shifting operations will be outlined and discussed.

3.7.1 Voltage Level Shifting

Voltage level shifting is required wherever I/O or internal digital signals interface to the memory array. Those signal levels are set by the V_{cc} supply level (typically in the 1.8- to 5-V range) while the memory array, as previously seen, inherently requires high voltage for operation. In order to render it possible to interface V_{cc} level signals with the memory array, high-voltage capability components have to be supported by the fabrication process. To fulfill this requirement multiple transistor technologies have been introduced [75, 76]. Usually high-voltage transistors have a thicker gate oxide than low-voltage (LV) V_{cc} transistors. V_{cc} transistor gate oxide thickness is in the 40- to 100 Å range, while high-voltage (HV) transistors have gate oxide thickness in the 150- to 250 Å range. The circuit schematic for a positive voltage level shifter is shown in Figure 3.34.

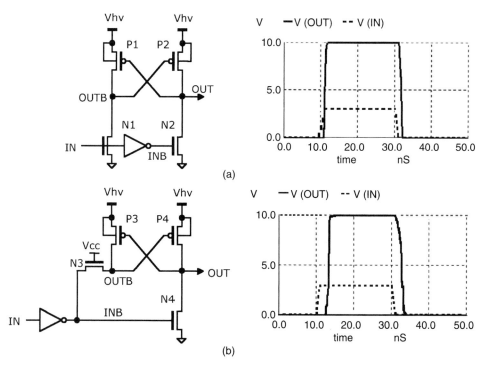

Figure 3.34. Level shifter circuits: (a) cross-coupled PMOS-type and simulated waveforms and (b) feedback PMOS-type and simulated waveforms.

Figure 3.34(a) shows the cross-coupled PMOS-type level shifter. The pair of PMOS transistors P1 and P2 has their sources connected to a high-voltage supply V_{hv}. The gates are connected to nodes OUT and OUTB. The drain of P1 is connected to node OUTB and the drain of P2 to node OUT. Two NMOS transistors, N1 and N2, have the drain connected to OUTB and OUT, respectively. The sources of N1 and N2 are connected to ground. IN is the level shifter input signal, and it is a digital signal with switching voltage levels of 0 V and V_{cc} that drive the gate of NMOS transistor N1. The gate of N2 is driven by signal INB, which is the output of an inverter logic gate. OUT is the output signal of the level shifter. The simulated waveforms in Figure 3.34(a) help to clarify circuit operation. In the simulated case IN switches from 0 to $V_{cc} = 3$ V at time 10 ns. INB is an inverted replica of IN. As IN reaches the 3-V level N1 turns on and drives OUTB to 0 V, which turns P2 on. INB switches to 0 V and turns N2 off. As a consequence OUT switches from 0 to $V_{hv} = 10$ V in the simulated example. Basically, OUT is logically a replica of IN with switching levels 0 and V_{hv} instead of 0 and V_{cc}.

Figure 3.34(b) shows another implementation of the level shifter circuits referred to as PMOS feedback type. Transistors P3, P4, and N4 behave like P1, P2, and N2 of the PMOS cross-coupled type. Differently from N1 of the cross-coupled type implementation, N3 is connected with its gate at V_{cc} and INB drives its source. When INB goes to 0 V N3 turns on and behaves like N1 in the cross-coupled PMOS implementation. By looking at the simulated waveforms of Figure 3.34(b), we can see that the PMOS feedback circuit has somewhat worse switching characteristic. That is due to the current loading at the inverter gate output INB, caused by N3, which results into slowing down INB itself as it is clearly shown by the simulation at time 10 ns. Slower INB switching directly translates into a slower OUT waveform.

The level shifter circuits presented above have been widely used in Flash design to implement the interfacing of digital low-voltage signals with the array high-voltage circuitry. As the V_{cc} supply level scales, the task of performing fast voltage level switching becomes increasingly difficult because of the reduced gate drive of the NMOS transistors. These transistors have thicker gate oxide and longer channel length than peripheral logic transistors in order to tolerate the high voltage and are inherently slow. To manage this situation new schemes have been recently proposed [77] that extend the supply operating range of level shifter circuits to 1.5 V V_{cc} and below. These proposed schemes adopt bootstrapping techniques to increase the driving voltage of NMOS transistors and, therefore, achieve faster switching of the level shifted signal.

Flash memories, as previously discussed, require negative voltage for array operation and, therefore, negative level shifters are needed to interface peripheral logic circuits. Negative voltage switching was one motivation for the introduction of triple-well process technologies. NMOS transistors built on the chip P-type substrate, usually biased at ground potential, cannot have their source or drain N-type terminal connected to negative voltages since the P-N junctions formed by the substrate and the transistor source/drain would be forward biased. A triple-well process allows the fabrication of NMOS transistors inside isolated P-wells. These P-wells can be biased at a negative potential whenever the NMOS transistors source-drain has to be connected to a negative voltage, thus preventing P-N junction forward biasing.

The circuit schematic of a negative level shifter is shown in Figure 3.35(a). Note that to achieve switching of the negative voltage triple-well NMOS transistors (namely N3 and N4) have been used. The structure of a triple-well NMOS transistor is illustrated in Figure 3.35(b). The simulated waveforms of the level shifter are shown in the plot of Figure 3.35(c). In the simulated case, the level shifter of Figure 3.35(a) translates a digital input signal IN, switching between 0 V and V_{cc} = 3 V, into an output signal OUT whose levels are V_{hv} = 5 V and V_{nn} = −10 V, thus achieving both high and negative voltage level shifting. The plot shows how, as IN goes low at about time 10 ns, signal SB goes to 5 V, thus turning PMOS P4 off. Signal S (not plotted) is the logical NOT of SB and, therefore, is driven to 0 V, thus turning PMOS P3 on. When P3 is on and P4 is off, N4 will switch on and N3 will turn off. Having N4 on drives the node OUT to V_{nn} = −10 V. Note that once the switching has completed, at about time 14 ns in the plot, current no longer flows from V_{hv} to V_{nn} since each path from V_{hv} to V_{nn} has at least one transistor, either PMOS or NMOS, in the off state. An important aspect of good level switcher design is to make sure that no direct current path exists that would load internally generated voltages because those voltages are usually the output of charge pump circuits that have limited current capability.

Like positive level shifters, negative level shifters are increasingly difficult to design as V_{cc} voltage scales. Implementation based on bootstrapping techniques has been proposed [77] for negative level shifters, aimed at increasing switching speed and circuit efficiency at low V_{cc}.

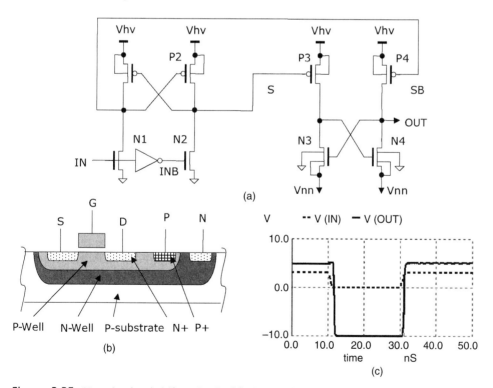

Figure 3.35. Negative level shifter circuit: (a) circuit schematic, (b) structure of a triple-well NMOS transistor, and (c) simulated waveforms of the circuit.

3.7.2 Sensing

Although digital data are stored in a Flash memory array, the operation of reading the information content of a Flash cell is of an analog nature. In fact, as we have seen, what is physically stored in a Flash memory cell is the electric charge trapped in the floating gate. An excess of charge in the floating gate, with respect to the neutrality status, causes the cell threshold voltage Vt to increase and, consequently, the cell drain-source current I_{ds} to decrease at a given gate voltage. The sensing operation consists of comparing the addressed cell current with that of a reference cell. The sensing circuit outputs logic zero or logic one depending on whether the cell current is lower or higher than the reference current.

Figure 3.36 shows the typical arrangement of a Flash sensing scheme, also referred to as a sense amplifier. A Flash sensing circuit consists of four functional blocks: bitline biasing and current-to-voltage converter, reference current generation, differential to single-ended converter, and data latch.

During the sense operation transistors N12 and N13, which are of high-voltage type (thicker gate oxide), are on. Their purpose is to isolate the sensing circuit, implemented with low-voltage-type transistors, from array high voltages during programming and erase operations. The Y selector connects the bitline of the addressed cell to the source of N12 and, therefore, cell current I_{cell} tends to pull to ground potential the source of N2. If I_{cell} is not large enough, then N2 source voltage and, therefore, the voltage of the selected bitline starts increasing until it reaches the threshold voltage of N1. As that happens, N1 starts conducing and, as soon as its current becomes higher than that of P1, lowers the gate voltage of N2. With this

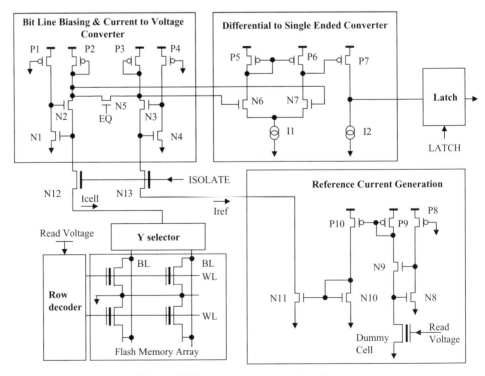

Figure 3.36. Flash memory sensing scheme.

feedback mechanism, the bitline voltage is clamped at about the threshold voltage of N1, which is typically 0.7 to 0.8 V. This circuit ensures that the bitline voltage and, therefore, the cell drain, never reaches values large enough to cause a disturb as discussed in Section 3.2.4. It is important that bitline charging up to the biasing voltage happens in the shortest possible time to speed up data reading. That can be achieved by properly sizing transistors N1, N2, P1, and P2. Cell current I_{cell} flows through diode-connected PMOS P2 and gets converted into a voltage. The same considerations apply to the reference current I_{ref} side of the circuit for symmetry reasons. Transistor N5 is turned on by the EQ signal during the bitline charging phase to equalize the array cell and reference cell sides of the circuit. Also, equalization biases the differential to single-ended stage at the maximum voltage gain point of its transfer function for the purpose of speeding up sensing.

Reference Current Generation. This block is a replica of the bitline voltage biasing generator (compare P8, P9, N9, and N8 to P1, P2, N2, and N1) to ensure that the reference cell and the array cell have exactly the same biasing conditions. The reference cell current is then mirrored by P10, N10, and N11 to generate the reference current. Usually the reference current is a fraction of the reference cell current. The fraction, that is, the mirroring ratio of P9 and P10, is chosen in such a way that the reference current is roughly half the current of an erased array cell.

Differential to Single-Ended Converter. The array cell current I_{cell} and the reference current I_{ref} are converted into voltages by flowing through P2 and P3, respectively. These voltages are applied to the differential stage formed by transistors N6, N7, P5, and P6 and by the current source I1. Transistor P7 and current source I2 form the output stage of the differential to single-ended converter. If the voltage at the gate of N6 is lower than the voltage at the gate of N7, that is, if $I_{ref} > I_{cell}$, then the output switches to 0 V. If $I_{cell} > I_{ref}$, then the output switches to V_{cc}. The signal at the output of the differential to single-ended converter stage is thus a CMOS-level digital signal.

Latch. The output of the differential to single-ended converter is latched by this conventional CMOS latch upon assertion of control signal LATCH and is kept until a new array location is accessed.

The sensing operation is triggered by an address change at the memory array boundary and happens in three phases: a first phase in which EQ signal is asserted and bitline is precharged at the biasing voltage level. The duration of this phase depends basically upon the bitline capacitance and the transistor characteristics. Typical bitline precharge duration is in the 20- to 50-ns range. A second phase is where the sensing happens. In this phase the EQ signal is deasserted and the differential to single-ended converter stage switches either to high or low state, depending upon the data stored in the addressed memory array location. The duration of the sensing phase depends basically upon the differential-to-single-ended converter design for given component features. Typical duration ranges from 10 to 30 ns. In the final phase the LATCH signal is asserted and the read data captured in the latch.

The sensing scheme concepts outlined in the above discussion have been widely adopted to implement Flash memory sense amplifiers. As with everything else in Flash memory technology, sensing circuit design becomes increasingly difficult as

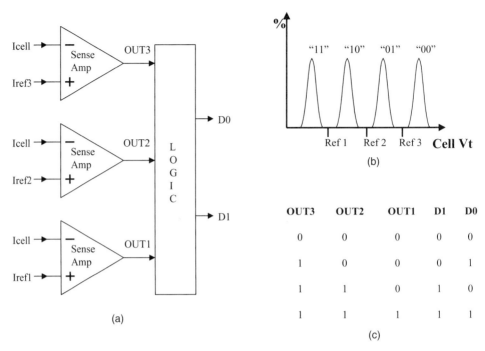

Figure 3.37. Multilevel sensing: (a) multilevel sensing scheme block diagram, (b) cell Vt distribution in multilevel data storage, and (c) logic combination of sense amp output.

supply voltage scales. Novel sensing schemes for very low voltage operation (1.8 to 1.5 V) have been recently proposed [77] that realize bitline precharge by charge sharing between a dummy bitline charged at V_{cc} and the array bitline predischarged at 0 V. Since the dummy bitline has the same capacitance as the array bitline, biasing of the array bitline at roughly half V_{cc} is achieved. Half V_{cc} is a suitable bitline precharge level when V_{cc} is below 2 V.

In multilevel architectures, sensing is implemented by logically combining the output of three sense amplifiers [78]. Each sense amplifier compares the array cell current with a different reference current to determine the cell threshold voltage level. The multilevel sensing concept is summarized in Figure 3.37.

3.7.3 Voltage Multiplication

Flash memories require voltages higher than the supply voltage and negative voltages to program, erase, and read the array. These voltages have to be generated on-chip if single-supply operation is to be supported. The circuits for high and negative voltage generations are referred to as voltage multipliers or charge pumps. The voltage multiplier [79] scheme most frequently adopted in Flash memory is implemented by using MOS transistors as rectifying elements and poly-diffusion or poly-poly capacitors as charge-storage elements. These schemes are often referred to as MOS-C charge pumps. The circuit schematic of a MOS-C charge pump is shown in Figure 3.38(a).

The charge pump circuit is composed of a certain number N of identical stages. Each stage is formed by three NMOS transistors M1, M2, and M3 and two capaci-

FLASH-SPECIFIC CIRCUIT TECHNIQUES

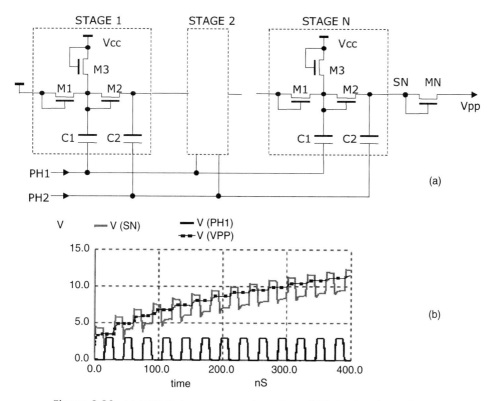

Figure 3.38. (a) MOS-C charge pump schematic and (b) simulated waveforms.

tors C1 and C2. The transistors are connected in a diode configuration, with gate and drain shorted together. In this arrangement the transistor behaves like a rectifying element. The bottom plate of capacitor C1 is connected to clock signal PH1. The bottom plate of capacitor C2 is connected to clock signal PH2. Clock signals PH1 and PH2 have the same period and are nonoverlapping with one another. The number of stages N of a MOS-C charge pump depends upon the operating V_{cc} voltage and the desired level of the output voltage V_{pp}. The lower the value of V_{cc} for given target V_{pp}, the more stages will be required. Usually, for $V_{pp} = 10$–$12\,V$, at $V_{cc} = 3\,V$, 5 to 7 stages are sufficient to provide an output current in the range of 50–100 µA. The theoretical steady-state output voltage is given by

$$V_{out} = V_{in} + N\,\text{Vclk}\,C/(2C+C_S) - (N+1)V_t - N\,I_{out}/f(C+C_S)$$

where V_{in} is the first-stage input voltage (usually V_{cc}), C_s is the parasitic capacitance associated with the top plate of capacitors C1 and C2, C is the value of C1 and C2, Vt is the rectifier threshold voltage, f is the clock frequency, Vclk is the amplitude of the clock signal, and I_{out} the output current.

Figure 3.38(b) shows some simulation waveforms for an MOS-C charge pump circuit made of 6 stages, with C1 = C2 = 10 pF, running at $V_{cc} = 3\,V$ with a clock period of 30 ns. The output voltage V_{pp} reaches about 11 V in 400 ns. Circuit operation is rather straightforward, when PH1 is low and PH2 is high, capacitor C1 of each stage is charged up by C2 of the preceding stage, when PH1 goes high and

PH2 goes low, C1 charges up C2. Each stage contributes $V_{cc} - Vt$ maximum theoretical voltage increment. Transistor M3 initializes capacitor voltage at $V_{cc} - Vt$ for faster pump startup.

The main problem with the MOS-C scheme is related to the increase of transistor Vt due to body-bias effect. Since the bulk of the NMOS transistor is at ground potential (0 V), as voltage increases, the Vt increases, causing the circuit efficiency to decrease. This becomes particularly severe at low V_{cc}. Since, as mentioned previously, the contribution of each stage is $V_{cc} - Vt$, it is apparent that, as V_{cc} decreases, the number of stages has to be increased. It can be calculated that, for a Vt value of about 0.8 V and $V_{cc} = 3$ v, a 5-stage MOS-C pump can output $V_{pp} = 10$ V. At $V_{cc} = 1.8$ V more than 20 stages are needed to achieve the same output voltage. Solutions to this problem based on bipolar junction transistors (BJT) have been proposed by several authors [80–82]. A charge pump scheme based on BJTs is shown in Figure 3.39(*a*). It is similar to the MOS-C scheme, with the sole difference that rectifying elements are realized using *n-p-n* BJT transistors.

In triple-well CMOS technology no additional process steps or masks are required to realize *NPN* BJT devices. The structure of the triple-well *NPN* is shown in Figure 3.39(*b*). The emitter E is formed by an N⁺ diffusion inside an isolated P-well, which is the base terminal B of the device. The base is contacted by a P+ diffusion. The N-well, inside which the P-well is formed, is the collector of the device. The BJT transistor action happens mainly in the region underneath the

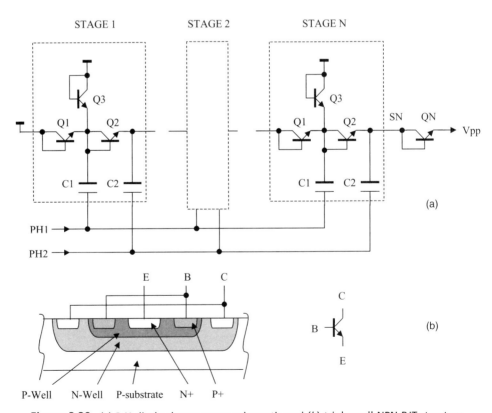

Figure 3.39. (*a*) P-N diode charge pump schematic and (*b*) triple-well NPN BJT structure.

emitter area. For that reason, this type of NPN BJT is sometime referred to as a vertical NPN.

Typical h_{fe} values for triple-well NPN transistors fall in the 20- to 60- range and breakdown voltages are rather high (>12 V), which renders this component well suited for charge pump and, more in general, analog circuitry implementation. The main advantage the BJT transistor offers in charge pump circuits is the absence of body-bias effect and the lower threshold voltage with respect to MOS rectifiers.

Negative voltage charge pump circuits can be realized with the MOS-C configuration shown in Figure 3.38(a) once NMOS are replaced with PMOS transistors. Negative MOS-C voltage multipliers are also affected by the body-bias effect problem. Unfortunately, there is no BJT-based implementation for negative multipliers. The reason why BJTs cannot be used to implement negative charge pumps is that the N-well (collector) cannot be connected to voltages lower than substrate potential without forward biasing the substrate to N-well P-N junction. To overcome this problem in negative voltage multiplier implementation, polysilicon P-N diode [83] structures have been built in 0.25-μm, triple-well technology and adopted for use in low-voltage negative charge pump circuits [84].

Another circuit technique introduced [85–87] to increase the efficiency of voltage multipliers at low supply voltage is the so-called heap charge pump. The heap pump concept is described in Figure 3.40.

The operation principle of the heap charge pump is rather simple. When clock signal CLK is high, all switches SW are in position 1, and assuming no voltage drop across the switch, all capacitors C get charged up to V_{cc}. As CLK goes low in the second half of the clock period, all switches SW are flipped to position 0. In position 0 the capacitors C are connected in series with one another and the bottom plate of the capacitor of the first stage is connected to V_{cc}. Therefore, the voltage at the

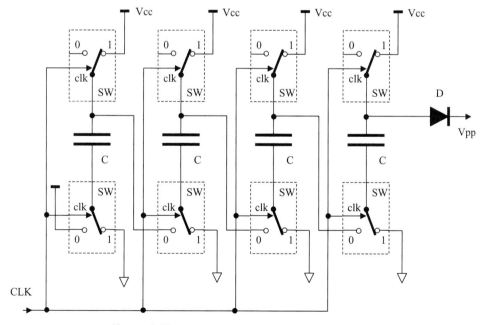

Figure 3.40. Heap charge pump principle schematic.

anode of diode D becomes $5V_{cc}$. The purpose of the diode D is to prevent the discharge of V_{pp} back to V_{cc} when CLK is high. The achievable maximum V_{pp} voltage is $5V_{cc} - V_d$, where V_d is voltage drop across the rectifying diode D. In the general case, if N is the number of stages of the heap charge pump, assuming no voltage drop across the switches SW and neglecting the parasitic capacitance associated with the top plate of capacitors C, the heap charge pump maximum output voltage is given by

$$V_{pp} = (N+1)V_{cc} - V_d$$

In actual circuits, switches SW are realized with MOS transistors, and bootstrapping techniques are adopted to minimize the voltage drop across the switches and the output rectifying element. Diode D is often implemented as an NPN BJT in positive charge pumps. The heap pump concept is very well suited to low-voltage operation.

3.7.4 Reference Voltage Generation

A key element in Flash memory design is the reference voltage generator. The voltages needed for array operations have to be regulated within rather narrow ranges to ensure reliable sensing and narrow programmed and erased cell Vt distributions. Also, well-regulated programming and erase voltages extend cell lifetime, thus resulting in better endurance and data retention performance of the memory chip.

The most widely adopted reference voltage generator is the bandgap reference [88, 89]. An implementation of the bandgap reference circuit is shown in Figure 3.41.

The principle of operation of the bandgap circuit can be outlined as follows: The current mirror formed by PMOS P1 and P2 forces the collector currents of

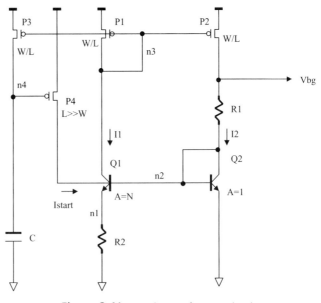

Figure 3.41. Bandgap reference circuit.

NPN BJTs Q1 and Q2, I1 and I2, respectively, to be equal. Since Q1 and Q2 have the same collector current and an emitter area ratio N, the voltage at node n1, which is the voltage across resistor R2 is given by

$$Vn1 = (VbeQ2 - VbeQ1) = \ln N(KT/q)$$

where VbeQ1 is the base-to-emitter voltage of Q1, VbeQ2 is the base-to-emitter voltage of Q2, N is the emitter area ratio of Q1 and Q2, K the Boltzmann constant, T the absolute temperature, and q the charge of the electron.

The value of I1 and I2 is therefore

$$I1 = I2 = Vn1/R2 = (1/R2)\ln N(KT/q)$$

The bandgap circuit output voltage is, hence, given by

$$Vbg = VbeQ2 + I2 \times R1 = VbeQ2 + (R1/R2)\ln N(KT/q)$$

Since the Vbe of a bipolar transistor has a temperature coefficient of about $-2\,mV/°C$ and, since VbeQ2 − VbeQ1 has a positive temperature coefficient, by properly selecting N, R1 and N2 in the design equation for the Vbg voltage, a temperature-independent voltage reference value can be obtained. The typical value of Vbg at which temperature dependence cancellation is obtained is 1.25 V.

Transistors P3, P4 and capacitor C in the schematic of Figure 3.41 have the purpose of providing an initial I_{start} current pulse when the circuit is powered up. In fact, the mesh formed by Q1, Q2, and R2 is in a stable state when I1 = I2 = 0. At power up node n4 voltage is 0 V. Since initially I1 = I2 = 0, the initial current of P3 is equal to 0 also. Therefore, n4 voltage is kept at 0 V by capacitor C until V_{cc} reaches a value high enough for P4 to generate the I_{start} pulse. As I_{start} enters the base of Q1 and Q2, I1 and I2 currents reach their regime value, node n4 is pulled to V_{cc} by P3 drain current, P4 turns off and I_{start} goes to zero, so ending the startup sequence.

Other fundamental requirements for the reference voltage generator are alternating current (ac) stability, good supply noise rejection and, for memory and portable equipment application, low-voltage capability and low power consumption. Power supply rejection can be achieved by increasing the output impedance of PMOS current mirrors (P1, P2, and P3 in the above example). Cascoded current mirror configurations, which exhibit excellent supply rejection, though, are not viable in low-voltage design. So, filtering techniques are sometime adopted. Low-power bandgap circuit implementations have been proposed [90, 91] that adopt dynamic circuit design techniques. Low-voltage architecture based on dynamic threshold MOS transistors has been recently presented [92] based on dynamic threshold MOS transistors, aimed at extending the bandgap circuit operating range below 1 V.

3.7.5 Voltage Regulation

The voltages generated by the internal charge pumps have to be regulated within the ranges suitable for reliable array operation. Voltage regulation techniques can be divided into two main categories: continuous regulation and switch-mode regulation.

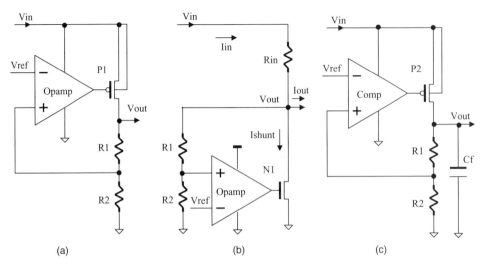

Figure 3.42. Voltage regulation schemes: (a) series regulator, (b) shunt regulator, and (c) switch regulator.

Continuous regulation is, in turn, of two types: series regulation and shunt regulation. These basic concepts are summarized in Figure 3.42. The output voltage V_{out} of the three schemes shown in the figure is given by

$$V_{out} = V_{ref}(1 + R1/R2)$$

but the working principle of the three schemes is different. In the series regulator of Figure 3.42(a), the operational amplifier varies the gate voltage of P1 to modulate its on-resistance in such a way that the output voltage does not change with the regulator output current. In the shunt regulator case [see Fig. 3.42(b)] the op-amp drives the gate of N1 to vary I_{shunt} as a function of I_{out} in a way such that V_{out} stays constant, that is, in a way such that $V_{in} - I_{in}R_{in} = V_{ref}(1 + R1/R2)$. The switch regulator, shown in Figure 3.42(c), uses a voltage comparator instead of an op-amp. When V_{ref} is greater than $V_{out}[R2/(R1 + R2)]$, the comparator turns P2 on and vice versa. So, P2 is switched on and off to regulate the output voltage. The noise on V_{out} caused by the switching action of P2 is smoothed by the filter capacitor Cf.

The switch-mode regulation, also referred to as on–off regulation, is the most frequently adopted technique in charge pump voltage regulation because it is the most efficient method as far as power consumption is concerned. The architecture of a charge pump on–off regulation scheme, similarly to the basic schematic we have just discussed, is based on three main elements: the reference voltage, the voltage comparator, and the feedback loop. The block diagram of the on–off regulated charge pump is shown in Figure 3.43. The charge pump output voltage V_{pp} is divided by the capacitive divider C1, C2 and is sensed by the voltage comparator Comp. The other input of the comparator is the reference voltage V_{ref}, generated by the bandgap circuit. The signal ENABLE enables the charge pump circuit, as well as the other circuit blocks, whenever V_{pp} voltage has to be generated. Note that to ensure precise regulation the capacitor divider is kept in a discharged status when ENABLE is deasserted low. In the disabled state transistors N1 and N2 are on and transistor P1

FLASH-SPECIFIC CIRCUIT TECHNIQUES

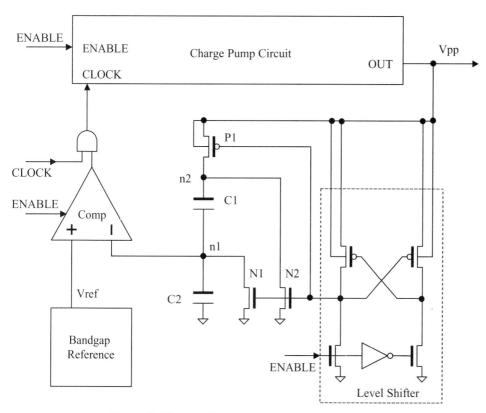

Figure 3.43. On-off charge pump regulation scheme.

is off. Therefore the voltage of circuit nodes n1 and n2 is 0 V, thus ensuring that C1 and C2 are fully discharged. C1 and C2 values must be chosen such that junction leakage is not sufficient to significantly alter node n1 (which is a floating node when ENABLE is asserted) for the entire duration of the programming or erase pulse (typically a few microseconds and a few milliseconds, respectively).

When ENABLE is asserted, the voltage of node n1 is

$$V_{n1} = V_{pp} C1/(C1+C2)$$

If $V_{n1} < V_{ref}$, then the comparator output is high and the clock signal CLOCK propagates through the AND logic gate to activate the charge pump. When $V_{n1} > V_{ref}$, then the comparator output is low and CLOCK is set to zero, causing the charge pump to stop pumping action. With this feedback mechanism, V_{pp} voltage is kept constant at the nominal value of

$$V_{pp} = V_{ref}(1+C2/C1)$$

The above-shown regulation scheme does not draw any direct current from the charge pump output since there is no dc path from V_{pp} to ground. For that reason it is very well suited to regulate weak charge pumps (50 to 100 μA maximum output current) like those employed to generate the programming voltage to be applied to the cell gate and the negative erase voltage. High current charge pumps (5 to 10 mA maximum output current), like those generating the cell drain programming voltage in Flash technologies exploiting the CHE programming mechanism, and the source erase voltage are often regulated with a scheme similar to the one shown in Figure 3.43 with the capacitor divider replaced by a resistor divider, which does not need to be initialized when the pump is disabled. Shunt regulation is usually avoided in Flash memory design for its poor power efficiency. Series regulation is used especially when regulated voltages lower than V_{cc} are needed or when voltages lower than V_{pp} have to be generated when V_{pp} is active.

3.7.6 I/O Signal Buffering

A Flash memory chip interfaces the external system bus by means of three categories of signals: address, data, and control signals. Address and control lines are input signals for the chip, data lines are input–output, or I/O, signals. The design of input and I/O buffering circuits deserves special attention since the I/O circuit performance directly influences several data sheet parameters, as well as some crucial aspects of the chip reliability. Data sheet parameters directly influenced by I/O circuit design are memory access time, I/O switching levels, I/O leakage, and output current drive capability. Reliability aspects encompass electrostatic discharge (ESD) and latch-up immunity. Although treatment of these implications in detail is beyond the scope of this chapter, the general aspects of I/O buffering circuitry will be discussed in this section.

The schematics for an input buffer and an output buffer circuit are shown in Figure 3.44(a) and 3.44(b) respectively.

The input buffer circuit of Figure 3.44(a) consists of two inverter stages: the first inverter stage is formed by transistors P2 and N1. A third PMOS, P1, is used to disable the input buffer when the chip is deselected, that is, when enable signal CEB,

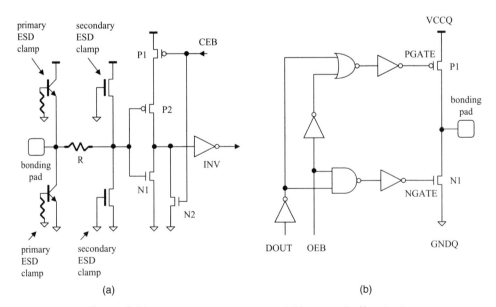

Figure 3.44. (a) Input buffer circuit and (b) output buffer circuit.

active low, is deasserted high. The disable function prevents the internal circuitry from burning power when the chip is deselected. Transistor N2 is on when the chip is deselected (CEB = high) to default to high the output of logic gate INV, which is the second inverting stage of the buffer circuit. The input bonding pad is connected to the gates of N1 and P2 via resistor R. The role of resistor R is to protect the gate oxide from ESD. The primary ESD clamps are large bipolar structures whose purpose is absorbing the energy of the ESD pulse at the beginning of the discharge event. The voltage at the pad rises, due to the ESD current flowing in the primary ESD clamp, up to the trigger point of the secondary ESD clamp. Once the secondary ESD clamp is conducting, the IR drop across resistor R protects the transistor gates. The value of resistor R is in the 0.5- to 1.5-kΩ range depending upon the characteristic of the ESD clamp. Its size is the result of a design trade-off since big resistors values, although improving ESD protection effectiveness, might impair circuit speed.

Figure 3.44(*b*) illustrates the structure of an output buffer. The output buffer has three possible states: logic high, logic low, and high impedance. When output enable signal OEB is high, the output pad is driven either high or low, depending upon the logic value of signal DOUT (data to be output). When OEB is asserted low, both P1 and N1 are off and the output pad exhibits high impedance. Note that P1 and N1 are connected to VCCQ and GNDQ, respectively. VCCQ and GNDQ are dedicated supply lines for the output buffers. Separating the power supplies prevents switching noise from being injected into the "clean" VCC and GND supply rails. Transistors N1 and P1 have to be sufficiently large to drive 30 to 50 pF of bus capacitance with less than 5 ns rise and fall time. Bi-directional I/O buffer circuit is implemented by merging an input buffer and an output buffer. I/O buffer circuitry is usually surrounded by multiple guard rings, made of N-type and P-type diffusion connected to VCC and GND, to achieve sufficient latch-up immunity.

REFERENCES

1. R. Bez, E. Camerlenghi, A. Modelli, and A. Visconti, "Introduction to Flash Memory," *Proc. IEEE*, Vol. 91, No. 4, pp. 489–502, Apr. 2003.
2. D. Kahng and S. M. Sze, "A Floating-Gate and Its Application to Memory Devices," *Bell Syst. Tech. J.*, Vol. 46, No. 4, pp. 1288–1295, 1967.
3. S. T. Wang, "On the I-V Characteristics of Floating-Gate MOS Transistors," *IEEE Trans. Electron Devices*, Vol. ED-26, pp. 1292–1294, Sept. 1979.
4. K. Robinson, "Endurance Brightens the Future of Flash," *Electronic Component News*, pp. 167–169, Nov. 1989.
5. S. Aritome, S. Satoh, T. Maruyama, H. Watanabe, S. Shuto, G. J. Hemink, R. Shirota, S. Watanabe, and F. Masuoka, "A 0.67 μM^2 Self-Aligned Shallow Trench Isolation Cell (SA-STI Cell) for 3V-Only 256Mbit NAND Eeproms," *Electron Devices Meeting, 1994 Tech. Dig.*, pp. 61–64, 11–14 Dec. 1994.
6. S. Kobayashi, H. Nakai, Y. Kunori, T. Nakayama, Y. Miyawaki, Y. Terada, H. Onoda, N. Ajika, M. Hatanaka, H. Miyoshi, and T. Yoshihara, "Memory Array Architecture and Decoding Scheme for 3V Only Sector Erasable DINOR Flash Memory," *IEEE J. Solid-State Circuits*, Vol. 29, No. 4, pp. 454–460, Apr. 1994.
7. H. Onoda, S. Kobayashi, H. Nakai, Y. Kunori, T. Nakayama, Y. Miyawaki, and Y. Terada, "Improved Array Architecture of DINOR for 0.5 μm 32M and 64Mbit Flash Memories," *IEICE Trans. Electron.*, Vol. E77-C, No. 8, pp. 1279–1286, Aug. 1994.

8. A. Chimenton, P. Pellati, and P. Olivo, "Overerase Phenomena: An Insight Into Flash Memory Reliability," *Proc. IEEE*, Vol. 91, No. 4, pp. 1454–1458, Apr. 2003.
9. G. Santin, "Circuit and Method for Erasing EEPROM Memory Array to Prevent Over-Erased Cells," U.S. Patent No. 5,122,985, 1992.
10. G. Campardo, M. Scotti, S. Scommegna, S. Pollara, and A. Silvagni, "An Overview of Flash Architectural Developments," *Proc. IEEE*, Vol. 91, No. 4, pp. 523–536, April 2003.
11. "28F160B3-3 Volt Advanced Boot Block Flash Memory Data Sheet," Intel, July 1998.
12. "MT28F321P2FG Page Flash Memory Data Sheet," Micron Technology, Apr. 2002.
13. "28F320D18—Dual Plane Flash Memory data Sheet," Intel, Oct. 1999.
14. D. Mills, M. Bauer, A. Bashir, R. Fackenthal, K. Frary, T. Gullard, C. Haid, J. Javanifard, P. Kwong, D. Leak, S. Pudar, M. Rashid, R. Rozman, S. Sambandan, S. Sweha, and J. Tsang, "A 3.3 V 50 MHz Synchronous 16 Mb Flash Memory," *IEEE ISSCC Tech. Dig.*, pp. 120–121, 1995.
15. G. J. Banks, "Electrically Alterable Non-Volatile Memory with N-Bits per Memory Cell," U.S. Patent No. 5,394,362, 1995.
16. A. Fazio, G. Atwood, and J. Q. Mi, "Method and Circuitry for Storing Discrete Amounts of Charge in a Single Memory Element," U.S. Patent No. 5,440,505, 1995.
17. "StrataFlash Wireless Memory Data Sheet," Intel, Aug. 2005.
18. T. Tanzawa, A. Umezawa, T. Taura, H. Shiga, T. Hara, Y. Takano, T. Miyaba, N. Tokiwa, K. Watanabe, H. Watanabe, K. Masuda, K. Naruke, H. Kato, and S. Atsumi, "A 44-mm^2 Four-Bank Eight-Word Page-Read 64-Mb Flash Memory with Flexible Block Redundancy and Fast Accurate Word-Line Voltage Controller," *IEEE J. Solid-State Circuits*, Vol. 37, No. 11, pp. 1485–1492, Nov. 2002.
19. H. Onoda, Y. Kunori, S. Kobayashi, M. Ohi, A. Fukumoto, N. Ajika, and H. Miyoshi, "A novel cell structure suitable for a 3 V operation, sector erasable Flash memory," *IEDM Tech. Dig.*, pp. 599–602, Dec. 1992.
20. S. Kobayashi, H. Nakai, Y. Kunori, T. Nakayama, Y. Miyawaki, Y. Terada, H. Onoda, N. Ajika, M. Hatanaka, H. Miyoshi, and T. Yoshihara, "Memory Array Architecture and Decoding Scheme for 3 V Only Sector Erasable DINOR Flash Memory," *IEEE J. Solid-State Circuits*, Vol. 29, No. 4, pp. 454–460, Apr. 1994.
21. Y. Itoh, R. Shirota, Y. Iwata, R. Nakayama, R. Kirisawa, T. Tanaka, S. Aritome, T. Endoh, K. Ohuchi, and F. Masuoka, "An Experimental 4 Mb CMOS EEPROM with a NAND Structured Cell," *ISSCC Dig. Tech. Papers*, pp. 134–135, Feb. 1989.
22. R. Shirota, R. Nakayama, R. Kirisawa, M. Momodomi, K. Sakui, Y. Itoh, S. Aritome, T. Endoh, F. Hatori, and F. Masuoka, "A 2.3 mm^2 Memory Cell Structure for 16 Mb NAND EEPROMS," *IEDM Tech. Dig.*, pp. 103–106, Dec. 1990.
23. K. Suh, B. Suh, Y. Um; J. Kim, Y. Choi, Y. Koh, S. Lee, S. Kwon, B. Choi, J. Yum, J. Choi, J. Kim, and H. Lim, "A 3.3 V 32 Mb NAND Flash Memory with Incremental Step Pulse Programming Scheme," *IEEE ISSCC Tech. Dig.*, pp. 128–129, 1995.
24. J. Lee, H. Im, D. Byeon, K. Lee, D. Chae, K. H. Lee, S. W. Hwang, S. Lee, Y. Lim, J. Choi, Y. Seo, J. S. Lee, and K. Suh, "High-Performance 1 Gb NAND Flash Memory with 0.12 µm Technology," *IEEE J. Solid-State Circuits*, Vol. 37, No. 11, pp. 1502–1509, Nov. 2002.
25. H. Takahiko, T. Hara, K. Fukuda, K. Kanazawa, N. Shibata, K. Hosono, H. Maejima, M. Nakagawa, T. Abe, M. Kojima, M. Fujiu, Y. Takeuchi, K. Amemiya, M. Morooka, T. Kamei, H. Nasu, K. Chi Ming Wang; Sakurai,1 N. Tokiwa, H. Waki, T. Maruyama, S. Yoshikawa, M. Higashitani, T. D. Pham, T. Yupin Fong, and T. Watanabe, "A 146-mm^2 8-Gb Multi-Level NAND Flash Memory with 70-nm CMOS Technology," *IEEE J. Solid-State Circuits*, Vol. 41, No. 1, pp. 161–169, Jan. 2006.
26. M. Momodomi, T. Tanaka, Y. Iwata, Y. Tanaka, H. Oodaira, Y. Itoh, R. Shirota, K. Ohuchi, and F. Masuoka, "A 4 Mb NAND EEPROM with Tight Programmed Vt Distribution," *IEEE J. Solid-State Circuits*, Vol. 26, No. 4, pp. 492–496, Apr. 1991.

27. H. Kume, M. Kato, T. Adachi, T. Tanaka, T. Sasaki, T. Okazaki, N. Miyamoto, S. Saeki, Y. Ohji, M. Ushiyama, J. Yugami, T. Morimoto, and T. Nishida, "A 128 mm^2 Contactless Memory Cell Technology for 3 V-only 64Mbit EEPROM," *IEDM Tech. Dig.*, pp. 991–993, Dec. 1992.
28. A. Nozoe, T. Yamazaki, H. Sato, H. Kotani, S. Kubono, K. Manita, T. Tanaka, T. Kawahara, M. Kato, K. Kimura, H. Kume, R. Hori, T. Nishimoto, S. Shukuri, A. Ohba, Y. Kouro, O. Sakamoto, A. Fukumoto, and M. Nakajima, "A 3.3 V High-Density AND Flash Memory with 1 ms/512 B Erase & Program Time," *IEEE ISSCC Tech. Dig.*, pp. 124–125, 1995.
29. I. Koren and Z. Koren, "Defect Tolerance in VLSI Circuits: Techniques and Yield Analysis," *Proc. IEEE*, Vol. 86, No. 9, pp. 1819–1835, Sept. 1998.
30. R. Aitken, "Redundancy—It's Not Just for Defects Anymore," paper presented at the IEEE International Workshop on Memory Technology, Design and Testing, pp. 117–120, San Jose, California, 2004.
31. T. Yamagata, H. Sato, K. Fujita, Y. Nishimura, and K. Anami, "A Distributed Globally Replaceable Redundancy Scheme for Sub-Half-Micron ULSI Memories and Beyond," *IEEE J. Solid-State Circuits*, Vol. 31, No. 2, pp. 195–201, Feb. 1996.
32. S. Matarrese, and L. Fasoli, "A Method to Calculate Redundancy Coverage for FLASH Memories," *Proc. IEEE Workshop on Memory Technology, Design and Testing*, San Jose, California, pp. 41–44, 2001.
33. A. Silvagni, G. Fusillo, R. Ravasio, M. Picca, and S. Zanardi, "An Overview of Logic Architecture Inside Flash Memory Devices," *Proc. IEEE*, Vol. 91, No. 4, pp. 569–580, Apr. 2003.
34. F. Roopharvar, Micron Technology, "Parallel Processing Redundancy Scheme for Faster Access Time and Lower Die Area," U.S. Patent No. 5,808,946, 1988.
35. T. Kiriata, G. Daniel, J. M. Dorta, and K. P. Pfefferl, IBM, Siemens, "Method of Making a Memory Device Fault Tolerant Using a Variable Domain Redundancy Replacement Configuration," U.S. Patent No. 5,881,003, 1999.
36. R. R. Lee and F. Gonzales, Micron Technology, "Flash Memory Having Transistor Redundancy," U.S. Patent No. 5,559,742, 1996.
37. C. M. Golla and M. Olivo, SGS-Thomson Microelectronics, "Double-Row Address Decoding and Selection Circuitry for an Electrically Erasable and Programmable Non-Volatile Memory Device with Redundancy, Particularly for Flash EEPROM Devices," U.S. Patent No. 5,581,509, 1996.
38. J. Brennan, Intel Corporation, "Row Redundancy for Flash Memories," U.S. Patent No. 5,233,559, 1993.
39. M. Mihara, T. Nakayama, M. Ohkawa, S. Kawai, Y. Miyawaki, Y. Terada, M. Ohi, H. Onoda, N. Ajika, M. Hatanaka, H. Miyoshi, and T. Yoshihara, "Row-Redundancy Scheme for High-Density Flash Memory," *ISSCC Dig. Tech. Papers*, pp. 150–151, 1994.
40. T. Futatsuya, M. Mihara, Y. Terada, T. Nakayama, Y. Miyawaki, S. Kobayashi, and M. Ohkawa, Mitsubishi Electric, "Nonvolatile Semiconductor Memory Device with a Row Redundancy Circuit," U.S. Patent No. 5,602,778, 1997.
41. M. Higashitani, Fujitsu, "Word Line Redundancy Nonvolatile Semiconductor Memory," U.S. Patent No. 5,426,608, 1995.
42. T. R. N. Rao and E. Fujiwara, *Error-Control Coding for Computer Systems*, Prentice-Hall, Upper Saddle River, NJ, 1989.
43. G. Cardarilli, M. Di Zenzo, P. Pistilli, and A. Salsano, "A High Speed Reed-Solomon Encoder-Decoder for Fault Tolerant Solid State Disk," *Proc. IEEE International Workshop on Defect and Fault Tolerance in VLSI Systems*, 1993, Venice, Italy, pp. 33–34, 1993.
44. M. Di Zenzo, Texas Instruments Italy, and P. Pistilli and A. Salsano, Rome University Italy, "Method and Apparatus for Mapping Memory as to Operable and Faulty Locations," U.S. Patent No. 5,541,938, 1996.

45. M. Christopherson, S. Wells, G. Atwood, M. Bauer, A. Fazio, and R. Hasbun, Intel, "Error Management Processes for Flash EEPROM Memory Arrays," U.S. Patent No. 5,475,693, 1995.
46. R. D. Norman, Micron, "Write Reduction in Flash Memory System Through ECC Usage," U.S. Patent No. 5,754,567, 1998.
47. J. S. Mangan, R. D. Norman, J. Craig, R. Albert, A. Gupta, J. D. Stai, and K. M. J. Lofgren, SunDisk, "Flash EEPROM Array Data and Header File Structure," U.S. Patent No. 5,471,478, 1995.
48. E. Harari, R. D. Norman, and S. Mehrotra, SanDisk, "Multi-State Flash EEPROM System on a Card that Includes Defective Cell Substitution," U.S. Patent No. 5,877,986, 1999.
49. H. Sukegawa, Toshiba, "Storage System with a Flash Memory Module," U.S. Patent No. 5,559,956, 1996.
50. H. Sukegawa, Y. Maki, and T. Inagaki, Toshiba, "Semiconductor Disk System Having a Plurality of Flash Memories," U.S. Patent No. 5,603,001, 1997.
51. A. Sharma, "Semiconductor Memories-Technology, Testing, and Reliability," in *Memory Error-Detection and Correction Techniques*, Chapter 5.6, IEEE Press Wiley, Hoboken, NJ, 2002.
52. P. Mazumder, "An On-Chip ECC Circuit for Correcting Soft Errors in DRAM's with Trench Capacitors," *IEEE J. Solid-State Circuits*, Vol. 27, No. 11, pp. 1623–1633, Nov. 1992.
53. S. Gregori, A. Cabrini, O. Khouri, and G. Torelli, "On-Chip Error Correcting Techniques for New-Generation Flash Memories," *Proc. IEEE*, Vol. 91, No. 4, pp. 602–616, Apr. 2003.
54. T. Nakayama, Y. Terada, M. Hayashikoshi, K. Kobayashi, and Y. Miyawaki, Mitsubishi, "Semiconductor Memory Device Having Error Correction Function," U.S. Patent No. 5,233,610, 1993.
55. F. Roohparvar, Micron, "Method and Apparatus for Performing Error Correction on Data Read From a Multistate Memory," U.S. Patent No. 5,864,569, 1999.
56. G. Singer, Intel, "The Future of Test and DFT," *IEEE Design and Test of Computers*, pp. 11–14, July–Sept. 1997.
57. F. Roohparvar, Micron Technology, "Memory System Having Non-Volatile Data Storage Structure for Memory Control Parameters and Method," U.S. Patent No. 5,880,996, 1999.
58. "TMS4C2972 Field Memory Data Sheet," Texas Instruments, Oct. 1997.
59. G. Naso, E. D'Ambrosio, P. Pistilli, and C. Palazzo, "Design for Testability in TMS4C2972/3," Texas Instruments Asia Pacific Technical Conference, pp. 131–134, Singapore, 1995.
60. F. Roohparvar, Micron Technology, "Low Voltage Test Mode Operation Enable Scheme with Hardware Safeguard," U.S. Patent No. 5,825,700, 1998.
61. G. Naso, and E. D'Ambrosio, "Test Mode Decoder in a Flash Memory," U.S. Patent No. 6,785,162, Aug. 2004; U.S. Patent No. 6,977,410, Dec. 2005.
62. F. Roohparvar, Micron Technology, "Non-Volatile Data Storage Unit and Method of Controlling Same," U.S. Patent No. 5,864,499, 1999.
63. G. Santin and G. Naso, "Flah Cell Fuse Circuit," U.S. Patent No. 6,654,272, Nov. 2003; U.S. Patent No. 6,845,029, Jan. 2005; U.S. Patent No. 7,002,828, Feb. 2006.
64. F. Roohparvar, Micron Technology, "Apparatus for Externally Timing High Voltage Cycles of Non Volatile Memory System," U.S. Patent No. 5,901,108, 1999.
65. F. Roohparvar, Micron Technology, "Non Volatile Memory System Including Apparatus for Testing Memory Elements by Writing and Verifying Data Patterns," U.S. Patent No. 5,825,782, 1998.
66. G. Naso, P. Pistilli, L. De Santis, and P. Conenna, "ROM-Based Controller Monitor in a Memory Device," U.S. Patent No. 6,977,852, Dec. 2005.

67. G. Gallo, G. Marotta, and G. Naso, "Output Buffer Strength Trimming," U.S. Patent No. 7,064,582, June 2006.
68. G. Naso, G. Santin, and P. Pistilli, "Flash Memory Sector Tagging for Consecutive Sector Erase or Bank Erase," U.S. Patent No. 6,717,862, Apr 2004; U.S. Patent No. 6,909,641, June 2005.
69. F. Roohparvar, Micron Technology, "Memory Circuit for Performing Threshold Voltage Tests on Cells of a Memory Array," U.S. Patent No. 5,790,459, 1998.
70. F. Roohparvar, Micron Technology, "Memory Circuit with Switch for Selectively Connecting an I/O Pad Directly to a Non Volatile Memory Cell and Method for Operating Same," U.S. Patent No. 5,706,235, 1998.
71. J. Chen, J. Hsu, S. Luan, Y. Tang, D. Liu, and S. Haddad, "Short Channel Enhanced Degradation During Discharge of Flash EEPROM Memory Cell," *IEDM Tech. Dig.*, pp. 331–334, 1995.
72. T. Endoh, H. Iizuka, S. Aritome, R. Shirota, and F. Masuoka, "New Write/Erase Operation Technology for Flash EEPROM Cells to Improve the Read Disturb Characteristics," *IEDM Tech. Dig.*, pp. 603–606, Dec. 1992.
73. T. Endoh, K. Shimizu, H. Iizuka, R. Shirota, and F. Masuoka, "A New Write/Erase Method to Improve the Read Disturb Characteristics Based on the Decay Phenomena of Stress Leakage Current for Flash Memories," *IEEE Trans. Electron Devices*, Vol. 45, No. 1, pp. 98–104, Jan. 1998.
74. F. Roohparvar, Micron Technology, "Output Data Compression Scheme for Use in Testing IC Memories," U.S. Patent No. 5,787,097, 1998.
75. S. Atsumi, A. Umezawa, M. Kuriyama, H. Banba, N. Ohtsuka, N. Tomita, Y. Iyama, T. Miyaba, R. Sudoh, E. Kamiya, M. Tanimoto, Y. Hiura, Y. Araki, E. Sakagami, N. Arai, and S. Mori, "A 3.3 V only 16 Mb Flash Memory with Row-Decoding Scheme," *ISSCC Dig. Tech. Papers*, pp. 42–43, Feb 1996.
76. Y. Iwata, K. Imamiya, Y. Sugiura, H. Nakamura, H. Oodaira, M. Momodomi, Y. Itoh, T. Watanabe, H. Araki, K. Narita, K. Masuda, and J.-I Miyamoto, "A 35 ns Cycle Time 3.3 V Only 32 Mb NAND Flash EEPROM," *IEEE J. Solid State Circuits*, Vol. 30, pp. 1157–1164, Nov. 1995.
77. N. Otsuka and M. Horowitz, "Circuit Techniques for 1.5 V Power Supply Flash Memory," *IEEE J. Solid State Circuits*, Vol. 32, pp. 1217–1230, Aug. 1997.
78. "Intel Strata Flash Memory Technology," AP-677 Application Note, Intel, Dec. 1998.
79. J. F. Dickson, "On-Chip High-Voltage Generation in MNOS Ics Using an Improved Voltage Multiplier Technique," *IEEE J. Solid-State Circuits, Vol. SC-11*, pp. 374–378, June 1976.
80. S. Kobayashi, M. Mihara, Y. Miyawaki, M. Ishii, T. Futatsuya, A. Hosogane, A. Ohba, Y. Terada, N. Ajika, Y. Kunori, K. Yuzuriha, M. Hatanaka, H. Miyoshi, T. Yoshihara, Y. Uji, A. Matsuo, Y. Taniguchi, and Y. Kiguchi, "A 3.3 V-Only 16 Mb DINOR Flash Memory," *ISSCC Dig. Tech. Papers*, pp. 122–123, Feb. 1995.
81. Santin, G. Marotta, and M. C. Smayling, "High Efficiency, High Voltage, Low Current Charge Pump" U.S. Patent No. 5,815,026, 1998.
82. Marotta, G. Santin, P. Piersimoni, and M. C. Smayling, "Low Voltage, High Current Charge Pump for Flash Memory" U.S. Patent No. 5,874,849, 1999.
83. M. Dutoit and F. Sollberger, "Lateral Poly Silicon p-n Diodes," *J. Electrchem. Soc. Solid-State Science and Technology*, Vol. 125, No. 10, pp. 1648–1651, 1978.
84. Y. Miyawaki, O. Ishizaki, Y. Okihara, T. Inaba, F. Niita, M. Mihara, T. Hayasaka, K. Kobayashi, T. Omae, H. Kimura, S. Shimizu, H. Makimoto, Y. Kawajiri, M. Wada, H. Sonoyama, and J. Etoh, "A 29 mm^2, 1.8 V-Only, 16 Mb DINOR Flash Memory with Gate-Protected-Poly-Diode (GPPD) Charge Pump," *IEEE J. Solid-State Circuits*, Vol. 34, No. 11, pp. 1551–1556, Nov. 1999.

85. S. Menichelli, "Multiplier Improved Voltage," U.S. Patent No. 5,831,469, 1998.
86. M. Mihara, Y. Terada, M. Yamada, "Negative Heap Pump for Low Voltage Operation Flash memory," *Symp. VLSI Circuits, Dig. Tech. Papers*, pp. 76–77, 1996.
87. S. Menichelli, "Negative Voltage Charge Pump Particularly for Flash EEPROM Memories," U.S. Patent No. 5,994,949, 1999.
88. R. J. Widlar, "New Developments in IC Voltage Regulators," *IEEE J. Solid-State Circuits*, Vol. 6, pp. 2–7, Feb. 1971.
89. K. E. Kuijk, "A precision Reference Voltage Source," *IEEE J. Solid-State Circuits*, Vol. 8, pp. 222–226, June 1973.
90. H. Tanaka, Y. Nakagome, J. Etoh, E. Yamasaki, M. Aoki, and K. Miyazawa, "Sub-1 µA Dynamic Reference Voltage for Battery-Operated Drams," *Symp. VLSI Circuits Dig. Tech. Papers*, pp. 87–88, June 1993.
91. T. Kawahara, T. Kobayashi, Y. Jyouno, S. Saeki, N. Miyamoto, T. Adachi, M. Kato, A. Sato, J. Yugami, H. Kume, and K. Kimura, "Bit-line Clamped Sensing Multiplex and Accurate High Voltage Generator for Quarter-Micron Flash Memories," *IEEE J. Solid-State Circuits*, Vol. 31, No. 11, pp. 1590–1599, Nov. 1996.
92. A. J. Annema, "Low-Power Bandgap Reference Featuring Dtmost's," *IEEE J. Solid-State Circuits*, Vol. 34, No. 7, pp. 949–955, July 1999.

4

PHYSICS OF FLASH MEMORIES

J. Van Houdt, R. Degraeve, G. Groeseneken, and H. E. Maes

4.1 INTRODUCTION

Flash memory emerged in the late 1980s as a combination of erasable programmable read-only memory (EPROM) and electrically erasable programmable ROM (EEPROM) features to form a "best of two worlds" solution for high-density nonvolatile storage. A lot of the know-how established for these "older" technologies was readily exploited in Flash memory research and development (R&D) programs. This had three major implications for researchers and engineers worldwide:

1. A firm base of experimental tools and know-how was available to support Flash memory development (Section 4.2). The basic floating-gate principle and technology introduced in 1967 for EPROMs was firmly established. The basic characteristics of EPROM/EEPROM such as transient programming and erasing characteristics, soft-write and disturb effects, retention, and endurance, as well as the basic models for these devices could be easily translated to the field of Flash memory.
2. The physics of electron transport through a dielectric layer (Section 4.3) was fairly well known at that time, which meant that research could be mainly concentrated on new concepts (such as source-side hot-electron injection) and improvement with respect to existing concepts.
3. Flash reliability aspects (Section 4.4) also did benefit from the experience built up in EPROM/EEPROM resulting in a sharper learning curve and a considerable time-to-market reduction.

Nonvolatile Memory Technologies with Emphasis on Flash. Edited by J. E. Brewer and M. Gill
Copyright © 2008 the Institute of Electrical and Electronics Engineers, Inc.

This has led to the fast introduction of Flash memory products for code storage, that is, low-cycle EPROM replacement applications. However, when Flash memory became more and more mature, specifications were pushed toward a higher number of cycles, larger densities, smaller sectors, and the like. These evolutions envisaged a more aggressive market expansion that also affected other types of memories, such as byte-erasable EEPROM, static random-access memory (RAM), and even hard drive technology. Doing so, the reliability aspects became increasingly critical. Typical examples are the detrimental effects of band-to-band tunneling-induced hot holes and stress-induced leakage currents in thin oxides.

The purpose of this chapter is to present the basic concepts and physics of operation as well as some basic reliability physics that are of great importance in the context of Flash memory development.

4.2 BASIC OPERATING PRINCIPLES AND MEMORY CHARACTERISTICS

4.2.1 Floating-Gate Principle

The basic operating principle of floating-gate nonvolatile semiconductor memory devices is the storage of electrons on a floating polysilicon gate that is isolated from the rest of the device by high-quality dielectrics (Fig. 4.1). Since the floating gate controls at least part of the underlying transistor channel, the charge on this gate will directly influence the current in the channel. In other words, by storing electrons on the floating gate, the current can be switched off, leading to a first logical state (e.g., a logical 0). On the other hand, by removing the electrons from the floating gate, the channel will be able to transport current from drain to source and a second logical state is obtained (e.g., a logical 1). For external current control, a second gate, referred to as the control gate, is used. This implies that the external threshold voltage of the memory cell has to be defined from this control gate. In summary, the external threshold voltage of the memory transistor (V_T) can be modified to switch between two distinct values, while the readout or sensing operation uses a readout voltage that lies somewhere between these two threshold voltages (Fig. 4.2).

Two major remarks have to be added here for completeness:

1. In practice, a wide variety of memory devices are being used, and depending on the particular device structure, the definition of the *external threshold voltage* has to be refined.

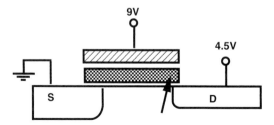

Figure 4.1. Cross section of an exemplary Flash memory device with typical (hot-electron) programming voltages.

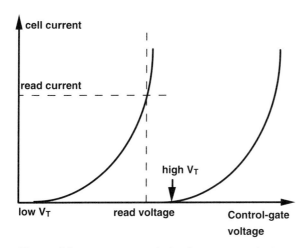

Figure 4.2. Sensing principle for floating-gate devices.

2. It is clear from this basic concept that the information stored inside the device is essentially analog information since the threshold voltage can be continuously varied between two extreme values. This explains the recent interest in multilevel memories where more than two states are used in order to store more than one bit per cell (2^n states is equivalent to n bits). Such a concept reduces the cost per bit by a factor of n without requiring a smaller cell. On the other hand, readout becomes more complicated since one has to distinguish between different cell current values, and also reliability issues become much more critical because of the reduction of the available window between two (neighbor) states that has to account for process variations, charge loss, and degradation effects (see also Section 4.4).

4.2.2 Basic Definitions with Examples

4.2.2.1 Programming and Erase Operation. The operation that corresponds to the transfer of information from the outside world into the Flash memory is referred to as the programming or write operation. The operation that establishes the initial condition of the cell array is called the erase operation. Since, generally speaking, Flash memory is not byte alterable, the difference between programming and erase is more essential than in, for example, EEPROMs. Indeed, since Flash memory is erased in blocks or in "sectors," the erase operation need not be as fast as in EEPROMs. Typical erase times are, therefore, in the order of 100 ms to 1 s. On the other hand, the selective programming operation is usually performed in a byte-by-byte or word-by-word manner, which requires short programming times. Therefore, for a given device using a set of two opposite conduction mechanisms for transporting electrons to and from the floating gate, the faster mechanism will be used as the programming operation while the slower one as the erase mechanism. From the discussion of the different cell concepts throughout this book, it will become clear that this can lead to confusion when comparing different cell concepts: When hot electrons are being used, the programming operation corresponds to the increase of the threshold voltage in Figure 4.3, while some tunneling-based Flash concepts will have the opposite convention.

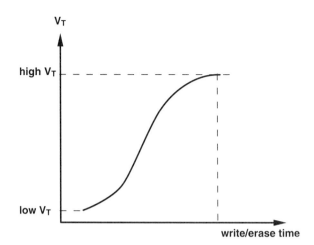

Figure 4.3. Transient characteristic of a floating-gate device.

4.2.2.2 Threshold Voltage Window. The threshold voltage window of a memory cell is the range of threshold voltage values swept through during programming and erasing. The threshold voltage is usually defined from the external control gate as the voltage that has to be applied to this gate to obtain a certain (arbitrary but low) channel current through the cell when keeping the other terminals at the voltages that correspond to the readout operation. As an example, the threshold voltage of the simple device in Figure 4.1 is the control gate voltage, which corresponds to, for example, a channel current of $1\,\mu A$ when applying $1\,V$ to the drain while grounding source and substrate (assuming the drain voltage during readout is $1\,V$). In some cases, especially when additional device terminals are being used (e.g., triple-gate structures and cells using a substrate or source biasing scheme), this definition can become quite complicated.

In any case, one should keep in mind that any (positive) applied voltage during readout would have an impact on the threshold voltage window for a given charge on the floating gate. The most basic example is the drain coupling effect known from EPROMs that can initiate "drain turn-on." This is a lowering of the cell's threshold voltage caused by the partial coupling of the drain readout voltage to the floating gate during readout.

As a conclusion, it is sufficient to mention here that the set of readout voltages used has an impact on the threshold voltage of the memory cell. The difference between the threshold voltage in the programmed or erased state and the gate readout is often referred to as the programming or erase margin, respectively. These margins should stay sufficiently large taking statistics, endurance, and charge loss mechanisms into account.

It should be noted that the floating-gate device, as operated from the control gate and the drain, acts as a metal–oxide–semiconductor (MOS) transistor, however, with slightly different characteristics. The main reason is that the external control gate of the cell is not directly controlling the transistor channel. The floating gate, however, has a more direct impact on the transistor channel while it is biased only through capacitive coupling from other terminals (mainly the control gate).

A main parameter for characterizing memory cells is, therefore, the gate coupling ratio, which is defined as the ratio of the control-gate-to-floating-gate capacitance and the overall floating-gate capacitance. Basically, it quantifies the percentage of control gate voltage that is coupled from the control gate to the floating gate in any operation mode. The subthreshold slope of the memory cell as measured from this control gate is smaller than that of the floating-gate transistor by a factor equal to this gate-coupling ratio. A mathematical treatment will be given in Section 4.2.3.

4.2.2.3 Readout Operation. In contrast to, for example, dynamic random-access memories (DRAMs), recovering the information from the memory is done by drawing a certain current from the memory cell that is subsequently detected by a sense amplifier. This circuit converts the (analog) current value into a digital bit.

4.2.2.4 Transient Characteristics. Programming and erase characteristics are usually presented as the evolution of the threshold voltage as a function of time. These characteristics (see Fig. 4.3) are sometimes referred to as the transient characteristics since they essentially describe the transition from one state to another. In some cases, the readout current as a function of time is also used to monitor the change in memory cell content. This is particularly interesting if

1. The relationship between external threshold voltage and readout current is highly nonlinear.
2. The cell structure and operation is such that it does not allow a "hard-off" state, which implies a small leakage current through the cell even in case of a high threshold voltage (e.g., split-gate cells that use polyoxide tunneling for erasure; see Section 4.3.2).
3. A multilevel approach is envisaged that requires the separation of different current levels in the sense amplifier.

4.2.2.5 Endurance. The *write/erase cycling endurance* (or *endurance* for short) of a floating-gate device is the major reliability characteristic because it describes the gradual degradation of the cell and, therefore, its lifetime in terms of the number of write/erase cycles that can be applied before failure. A general definition of endurance could therefore read as:

Endurance is "the ability to perform according to the specifications when subjected to repeated write/erase cycling" [1].

When looking at a single cell, the *intrinsic* endurance is usually monitored by measuring the threshold voltage window as a function of the number of applied write/erase cycles (see Fig. 4.4). As long as the high threshold voltage is higher than the readout voltage, the leakage current in this state will be smaller than the channel current at which the cell's threshold is defined (see Section 4.2.2.2). On the other hand, the low threshold voltage is usually a measure for the readout current of the cell and, depending on the type of cell considered, this threshold voltage value should not increase above a certain value in order to guarantee a sufficient readout current.

In some cases, the readout current is monitored versus the number of cycles. For the high threshold voltage, the leakage current through the cell is often a more suitable monitor for degradation since the subthreshold slope is decreased by the gate-coupling ratio (see Section 4.2.2.2). In the case of devices that use a low gate-coupling ratio on the order of 10 to 20% (as, e.g., the split-gate cells that use polyoxide tunneling for erasure; see Section 4.3.2), the subthreshold slope becomes very

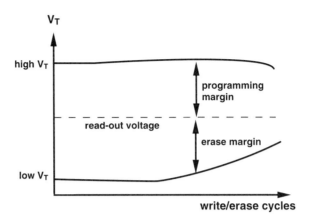

Figure 4.4. Endurance characteristics of a floating-gate device.

small and a considerable leakage will flow through the cell even when the programmed threshold voltage is higher than the readout voltage. This problem becomes even worse after cycling the cell because the local charges trapped in the oxide at the injection point tend to decrease the subthreshold slope even further [2]. On the other hand, the readout current degradation is often more relevant than the actual window closing, especially in those cases where the relationship between both is nonlinear.

An important remark is that the cell-level endurance test always underestimates the intrinsic lifetime of the cell because a large number of cycles are applied to the cell in a very short period in order to reduce test time. In practice, there will be more time in between cycles, and the degradation will be relaxed because of recovery effects such as, for example, charge detrapping from the oxide.

The *extrinsic* endurance is monitored at the circuit level and can be very different from the cell-level endurance depending on the envisaged technology and/or application. Typical examples are failures because of breakdown events in the cell and failures of the peripheral circuitry. Both examples are discussed below to illustrate the difference between circuit-level and cell-level endurance.

First, cells can fail after cycle accumulation considerably less than that predicted by the intrinsic endurance. Especially when high oxide fields are being used (devices using tunneling for programming), it is known that the failure rate is almost constant with the number of applied cycles [3]. This means that the larger the memory size, the lower the number of cycles until the first cell failure. Along with other factors, this was one of the main reasons why full-feature EEPROM was not easily "upgraded" toward the megabit level. An alternative approach [3], which showed a more intrinsic failure rate, was the textured poly floating-gate (TPFG) technology that used enhanced tunneling through a polyoxide.

Important for Flash memory is the fact that technologies using tunneling for programming will have to take into account a nonzero failure rate even before the product lifetime is reached. This can partially be solved by redundancy schemes and error correction codes (ECC), but these measures can never completely rule out eventual failures. In some applications, for example, image storage in digital camera's and voice recording, a nonzero failure rate can be accepted while in code storage it cannot.

In general, one could say that low-oxide-field injection mechanisms such as hot-electron injection and (slow) tunneling-induced erase suffer much less from the problem and the circuit-level endurance should be fairly close to the cell-level characteristic (currently typically 100,000). In recent products, the ramp of the erase pulse is controlled in such a way that the very high peak field typical of tunnel erase is avoided, and as a result, the endurance is considerably enhanced [4].

A second example where circuit-level endurance can differ significantly from cell-level tests illustrates the importance of the readout circuitry. It is the trip point of the sense amplifier that determines the current level at which an erased cell will be detected as having been "written." This current may or may not be the same current used for cell-level tests. Thus, endurance-induced changes in the cell window will have an observable effect only if the combination of the characteristic changes and the readout circuitry is unfavorable. Moreover, one could correct this by lowering the trip point, but this can in some cases lead to increased access time for detecting truly written cells.

In general, the access time increases as the readout current decreases, and the rate at which this will happen is also important for defining the circuit-level endurance. In some Flash technologies, the window is forced to stay open as long as possible by increasing the erase time and/or erase voltage. In such a scheme, the circuit-level endurance is maximized to approach the wear-out of the oxide.

4.2.2.6 Retention.
The most basic requirement for a nonvolatile memory of the floating-gate type is the ability to contain its information in the absence of an external power supply. Therefore, charge retention can be defined as "the ability to retain valid data over a prolonged period of time under storage conditions."

On the cell level, the so-called intrinsic retention is the ability to retain charges in the absence of applied external bias voltages. This is usually characterized by monitoring the threshold voltage window versus storage time at high temperature. In practice, *temperature-accelerated tests* are used, which means that the charge loss is enhanced by using elevated testing temperatures (typically 250 to 300°C). The bake time corresponding to a given threshold voltage shift is displayed as a function of the inverse of the absolute temperature. Figure 4.5 shows such a log (time) versus $1/T$ result, which is known as an *Arrhenius plot*. From practical experience, it is known that this relationship describes the temperature dependence of the charge loss mechanism quite well. This allows data taken at high temperatures to be extrapolated to predict what will happen at room or operating temperatures. If the abscissa is $1/kT$ (with k Boltzmann's constant), the slope of this curve is equal to the activation energy E_A of the dominant charge loss mechanism in the considered testing temperature range. Mathematically:

$$t_{\Delta V_t} \sim \exp(E_A / kT) \tag{4.1}$$

Table 4.1 summarizes some typical charge loss mechanisms and their associated activation energies. The mechanisms listed in Table 4.1 are now briefly discussed. For more details, we refer to the associated references. From the most commonly observed charge loss mechanisms, *intrinsic charge loss* has the highest activation energy. In this context, *intrinsic charge loss* refers to the fact that the stored electrons actually jump over the physical barrier between polysilicon gate and (tunnel) oxide. This typically shows an activation energy of about 1.4 eV, which is sometimes

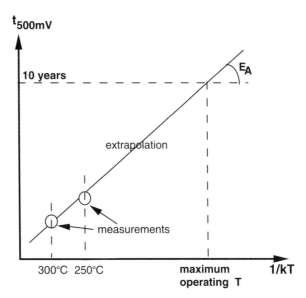

Figure 4.5. Arrhenius diagram of a floating-gate device.

TABLE 4.1. Selected Charge Loss Mechanisms and Associated Activation Energies as Reported in Literature

Failure Mode	Activation Energy (eV)	Reference
Intrinsic charge loss	1.4	Shiner-IRPS80 [5]
Oxide defects	0.6	Shiner-IRPS80 [5]
Tunnel oxide breakdown	0.3	Baglee-IRPS84 [6]
ONO	0.35	Wu-IRPS90 [7]
Ionic contamination	1.2	Shiner-IRPS80 [5]
Cycling induced charge loss	1.1	Verma-IRPS88 [8]

observed in, for example, single-poly floating-gate cells that have only a limited number of possible leakage paths. This high activation energy implies that the extrapolation down to room temperature usually yields a very long retention time (order of centuries and longer).

More often, the charge loss occurs through imperfections in an oxide layer that yields an activation energy of only 0.6 eV. For example, devices using a polyoxide for floating-gate-to-control-gate isolation often exhibit this activation energy in retention tests. If tunnel oxide breakdown is the main charge loss mechanism, the activation energy is even smaller (0.3 eV), due to the low-temperature dependence of tunneling. Many floating-gate devices use a composite oxide–nitride–oxide (ONO) layer for floating-gate-to-control-gate isolation. In this case, a number of complex charge transport mechanisms come into play as described in Mori et al. [9]. Indeed, apart from tunneling through the bottom oxide and Poole–Frenkel current in the nitride, also hole injection from the top electrode into the top oxide can lead to charge loss. In most cases, the mechanisms observed in ONO layers show a low activation energy of 0.35 eV, which means that these mechanisms have to be very well controlled in order not to affect the device at operating temperature.

In some cases, ionic impurities (mobile ions) are released from the interlevel dielectric layers or from the passivation layer, which leads to severe charge losses at high temperature. The mechanism is partially field driven and partially diffusion driven, which leads to dedicated techniques in order to identify this mechanism [10]. Usually small positive ions (like alkali metals) migrate from the upper layers of the chip down to the silicon surface where they are attracted by the cells that have a high threshold voltage (excess negative charge on the floating gate). Attracted by the associated electric field, they will move toward the floating gates of these cells and partially compensate the (negative) floating-gate charge, which leads to an apparent electron loss as measured from the control gate. When the device is erased, a second bake will result in an apparent charge gain in these cells that were previously showing a charge loss. This is due to the diffusion-driven withdrawal of the positive ions away from these previously "attacked" floating gates.

It is clear that single-poly cells and cells having a large portion of their floating gate uncovered by the control gate are more susceptible to this kind of charge loss. The associated activation energy is about 1.2 eV, and the mechanism is, therefore, very much temperature dependent. Moreover, it is clear that ion contamination is becoming more critical with device scaling since the number of electrons on the floating gate is scaling with the cell dimensions. It has been found that—apart from the reduction of the mobile ion contamination level—clustering of these ions (e.g., in a phosphorous-rich borophosphosilicate glass (BPSG) layer [11]) is an efficient way to cope with this problem. Finally, it is interesting to mention that cycling can sometimes activate new charge loss mechanisms. A typical example is given in Mielke et al. [3], where hole detrapping from the oxide after cycling is reducing the threshold voltage of a typical cell during a bake test. The activation energy is essentially linked to the bandgap of silicon (1.12 eV).

The main drawbacks of temperature-accelerated testing are listed below [1]:

1. The higher temperature only enhances *certain* leakage mechanisms. Therefore, it is never clear whether the dominating mechanism at the high test temperature is also dominant at room or operating temperatures.
2. Extrapolation of data down to operating temperature is of limited validity since it is based on few data. Indeed, in order to keep the test within practical limits, the temperature can in most cases not be decreased below 200°C, which means one has to extrapolate over a range that is considerably larger than the considered testing temperature range.
3. Retention tests after cycling are not always relevant because of partial recovery of damage at high temperature.
4. High-temperature tests without bias voltages are not realistic since the presence of a low electric field inside the device due to readout voltages can largely influence the actual retention time (see also Section 4.2.2.7).

Therefore, some authors have claimed that *field-accelerated testing* is to be used, which means that the charge loss is enhanced by increasing the field somewhere inside the device [12]. However, by increasing the electric field across a specific dielectric layer, one is "amplifying" one particular charge loss mechanism, which will inevitably also mask part of the potentially hazardous retention problems. For example, when increasing the tunnel oxide field, the mechanism will change from direct tunneling to Fowler–Nordheim tunneling, which means that extrapolating from such tests to low fields is not relevant for the device reliability (see Lanzoni et al. [12]). Therefore, both tests are necessary: unbiased temperature-accelerated tests (referred to as retention tests) and field-accelerated tests (more device dependent and further referred to as disturb tests; see Section 4.2.2.7).

Again, the circuit-level (or extrinsic) retention is the retention behavior of the worst cell in an array, and, therefore, essentially defect related. Enhanced charge loss in so-called erratic cells, especially after cycling, is a limiting factor in most practical situations.

4.2.2.7 Soft-Write and Disturb Effects. Disturb effects occur when the threshold voltage of a memory cell is altered unintentionally under the influence of a combination of operating voltages. A general definition of *immunity* with respect to disturb effects, therefore, reads: "disturb immunity is the ability to retain valid data over a prolonged period of time under *operating* conditions."

Note that we have kept the same wording as for the definition of retention in the previous section, which highlights the main difference with the retention issue: Disturb effects occur only when operating the device, and, therefore, are inherently different from retention problems. In other words, while retention refers essentially to the nonvolatility of the memory cell in the absence of a supply voltage, disturb effects refer to unwanted changes in the memory cell's content when operating (reading, writing, or erasing) a part of the memory.

SOFT-WRITE EFFECTS. On the cell level, only one disturb effect is of particular importance in most practical cases, that is, the slow writing (increase of the threshold voltage) while reading a memory cell. This is also referred to in literature as the *soft-write effect* [13]. Since the floating-gate cell is designed to withstand high programming and erase voltages, the much lower readout voltages do not usually cause significant charge loss.

However, for cells that use a single transistor channel, such as the one depicted in Figure 4.1, the channel-hot-electron programming scheme and the readout scheme differ only in the *level* of the applied voltages. Typical conditions for the cell in Figure 4.1 are: programming with 8 to 9 V on the control gate and 4.5 V on the drain (typical hot-electron conditions; see also Section 4.3.3), and readout with 3.3 V on the control gate and 1 V on the drain. The programming conditions should establish a threshold voltage shift from 2 to 7 V in about 10 µs (typical Flash programming time), while the readout condition should not cause a significant shift (e.g., <0.5 V) in 10 years (product lifetime). This implies a difference in the injection levels for both conditions of more than 14 orders of magnitude! In order to realize this margin the drain readout voltage will have to be limited.

The soft-write effect is even worse if the cell has been optimized for fast programming, and it is clear that the readout voltages have to be scaled together with the programming voltages in order to maintain the necessary margin. In some cells, the problem is circumvented by using a split-gate device [13]. In this type of design, the voltage drop is concentrated in two different channels depending on the operation mode, virtually removing the soft-write phenomenon and decoupling the readout voltages from this reliability constraint.

DISTURB EFFECTS. Most disturb effects are circuit related in the sense that they depend on the way the memory cells are connected in the memory circuit. One can classify these effects into read-disturb, program-disturb, and erase-disturb. Essentially all of these effects stem from the fact that voltages applied to accomplish readout, programming, or erasure are also partially applied to nonselected cells.

Depending on the node of the cell that is suffering from the disturb effect, one can distinguish between gate disturb and drain disturb. Because the drains of the cells are connected to a common bitline and the gates to a common wordline, terms such as *wordline disturb* and *bitline disturb* are sometimes used. This will be explained in more detail further on in this section.

In the case of the readout voltages, associated disturb effects such as the soft-write effect mentioned above, have to be tolerated for the full 10-year typical lifetime of the device. In contrast, the program- and erase-disturb effects have to be tolerated for only a short amount of time, which is a function of the programming/erase time, the number of applied programming/erase cycles, and the array size and organization.

Disturb effects depend strongly on the detailed cell and circuit arrangement, and the nature of disturb effects peculiar to specific Flash technologies are presented in other chapters. In this chapter the simple structure of Figure 4.1 will be used to point out some general aspects of the disturb phenomena.

In full-feature (byte-erasable) EEPROMs, program- and erase-disturb effects are avoided by providing select transistors inside the memory array to limit the influence of the associated voltages as much as possible. In the ideal case, voltages would be applied only to the addressed word or byte. In the case of Flash memory, however, the high density requires that the select device is left out and the disturb effects become more critical. For example, when programming the cell in Figure 4.1 with 12 V at the control gate and 5 V at the drain, all cells on the same wordline will be subjected to this 12-V bias, which can cause slow programming of erased cells or slow erasing of programmed cells due to the nonzero tunneling currents inside the device (so-called gate disturb). In addition, the devices on the same bitline as the cell being programmed will experience the 5-V bias, and these cells could be erased partially due to tunneling of electrons from the floating gate to the drain junction (so-called drain disturb).

Assuming that the sectors of the memory are organized as a set of wordlines, it can already be concluded that gate and drain disturb effects have to be tolerated for a different time. In the case of the gate voltage of 12 V, the *gate* disturb time (i.e., the maximum time this effect can occur) is limited by the programming time and by the number of words in a sector. On the other hand, the drain voltage will affect all cells on the same bitline due to the absence of select transistors. Consequently, this *drain* disturb effect will appear for a time that is the product of the programming time, the number of wordlines, and the number of applied write/erase cycles. This last factor is typically on the order of 100,000, which implies that the drain disturb effect during programming has to be cleared for a much longer time than the gate disturb effect.

In most cases, the erase-disturb effects are avoided by a proper array organization and/or by using inhibit voltages, that is, additional voltages are applied to the nonselected sectors during erase in order to reduce the corresponding disturb effects. The main reason for this is the considerably long erase time that is used in Flash memories. For example, the device in Figure 4.1 is typically programmed in 10 μs and erased in 0.1 to 1 s, resulting in 100,000 longer disturb times during erase as compared to programming. The use of erase inhibit voltages also strongly depends on the particular properties of the considered technology, and, therefore, is beyond the scope of this chapter.

4.2.3 Basic Equations and Models

4.2.3.1 Capacitor Model. The main difference between floating-gate (FG) cells and classical MOS devices is the presence of a floating gate containing a certain amount of charge that can be changed through programming and erase operations. In order to describe such a device, an adequate capacitor model is required. The essential difference between the floating-gate storage node and a set of capacitors as known from electrostatics is the fact that the floating node is now allowed to contain a nonzero charge that is achieved by moving charges through the dielectric of one of the surrounding capacitors (Fig. 4.6). Indeed, in a classical electrostatic situation, the net charge on the floating node is zero ($\Sigma Q_{fgi} = 0$) because it can only be induced by the displacement of charges in accordance with external influences. However, if one of the surrounding capacitors is allowed to "leak" net charge onto this node, an additional term appears in the charge balance equation:

$$\sum Q_{fgi} = Q_{fg} = \sum C_i(\Psi_{fg} - \Psi_i) \tag{4.2}$$

in which C_i is the capacitance from node i to the FG and Ψi is the corresponding *electrostatic* potential. The value of the electrostatic potential of the FG is then found as

$$\Psi_{fg} = \sum c_i \Psi_i + Q_{fg}/C_{tot} \tag{4.3}$$

where $C_{tot} = \Sigma C_i$ is the total FG capacitance and $c_i = C_i/C_{tot}$ are the coupling ratios of the different device nodes with respect to the FG. Physically, the coupling ratio of node i is the relative impact of that external node on the FG potential (thus $\Sigma c_i = 1$ by definition).

In order to transform Eq. (4.3) to externally applied voltages, the following relationship has to be introduced:

$$\Psi = V_{app} \pm \Phi_f \tag{4.4}$$

where V_{app} is the externally applied voltage and

$$\Phi_f = \frac{kT}{q} \ln\left(\frac{N}{n_i}\right) \tag{4.5}$$

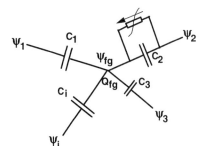

Figure 4.6. Principle of the charge balance equation. The "leaky capacitor" is represented by a nonlinear resistor in parallel with C_2.

BASIC OPERATING PRINCIPLES AND MEMORY CHARACTERISTICS

in which k is Boltzmann's constant, T is the absolute temperature, q is the electron charge ($kT/q = 25.8\,\text{mV}$ at room temperature), N is the doping level, and n_i is the intrinsic electron concentration ($=1.07 \times 10^{10}/\text{cm}^3$ at room temperature). The positive sign in Eq. (4.4) has to be used in case of an N-type node (N^+ polysilicon gate/ junction or inversion of a P-type substrate) while the minus sign accounts for a P-type node (accumulation of a P-type substrate). The transformation (4.4) is necessary because of the fact that the charge balance equation (4.2), by definition, makes use of electrostatic potentials while externally applied voltages are to be treated as quasi-Fermi potentials (in inversion the electron quasi-Fermi level is used, in accumulation the hole quasi-Fermi level is used). The difference between both types of potentials is due to band bending in the silicon (if applicable) and due to the work function of the considered node with respect to midgap. The term $\Phi_f \approx E_g/2q$ (half the silicon bandgap) for highly doped nodes such as N^+ gates. In cases where the substrate is in the depletion regime, the formula is no longer valid since the difference between quasi-Fermi level and electrostatic potential then varies between $-\Phi_f$ and $+\Phi_f$.

This calculation is illustrated hereafter for the case of the device in Figure 4.1. Writing down Eq. (4.3) for the programming regime and introducing Eq. (4.4) yields

$$V_{fg} + E_g/2q = cg(V_{cg} + E_g/2q) + ch(V_{ch} + \Phi_f) + d(V_d + E_g/2q) + Q_{fg}/C_{tot} \quad (4.6)$$

in which cg, ch, and d are the control gate, channel, and drain coupling ratios, respectively; V_{cg} and V_d are the control gate and drain voltages applied during programming; while V_{ch} is the average channel potential assuming the entire channel is inverted from source to drain. A last, yet significant, simplification in Eq. (4.6) is the assumption that the field regions are accumulated and that the bulk (or P-well) is grounded. If these assumptions would not be fulfilled (e.g., when using a bulk bias as in the case of secondary impact ionization initiated hot-electron programming; see Section 4.3.6), an additional term would appear taking into account the (small) contribution of the field-to-FG coupling.

In practice, it is more convenient to perform calculations in terms of the external (memory cell's) threshold voltage. Therefore, the same equation is written down a second time but now for the regime that corresponds to the threshold voltage measurement condition. In most cases, a small current (e.g., $1\,\mu\text{A}$) is forced into the source junction while an analog feedback circuit adjusts the control gate voltage accordingly until equilibrium is reached. The necessary control gate voltage is—by definition—the threshold voltage of the memory cell:

$$V_{th} + E_g/2q = cg(V_T + E_g/2q) + ch(V_{chmeas} + \Phi_f) + d(V_{dread} + E_g/2q) + Q_{fg}/C_{tot} \quad (4.7)$$

in which V_{th} is the *internal* threshold voltage as measured on the floating gate, V_T is the external threshold voltage as measured on the control gate, V_{dread} is the drain voltage used during readout, and V_{chmeas} is the average channel potential in the V_T measurement mode. The impact of the average channel potential can usually be neglected since at these low current levels, the main voltage drop across the channel appears close to the drain junction and, therefore, its contribution to the FG potential is very small.

From Eqs. (4.6) and (4.7) one can derive the following important relationship between the FG voltage shift, the V_T shift, and the charge on the FG:

$$\Delta Q_{fg} = C_{tot} \Delta V_{fg} = C_g \Delta V_T \tag{4.8}$$

where C_g is the control gate capacitance. Equation (4.8) is valid only when assuming that the channel potential does not significantly change during programming. In that case, the change in FG voltage is directly proportional to the associated threshold voltage shift:

$$\Delta V_{fg} = \text{cg} \Delta V_T = \Delta Q_{fg} / C_{tot} \tag{4.9}$$

Or alternatively:

$$\Delta V_T = \Delta V_{fg}/\text{cg} = \Delta Q_{fg}/C_g \tag{4.10}$$

Equation (4.9) shows that a small gate-coupling ratio corresponds to a small charge transfer to or from the FG, since, in practice, the external threshold voltage window is fixed based on reliability considerations (see Section 4.2.2.5). Equation (4.10) illustrates the fundamental physical reason for the high scalability of Flash memory: The external "signal" (ΔV_T) is only a function of the ratio between the stored charge and the coupling capacitor. When scaling down the cell, this capacitor is scaled approximately with F^2 (where F is the feature size of the process). At the same time, the charge on the FG can be scaled with F^2 as well *without* affecting the margins on the memory cell operation. Thus, apart from some specific effects such as ion contamination, one could say that the charge on the FG is scaled proportionally to the storage capacitor without truly affecting the operability and reliability of the cell. This is in strong contrast to, for example, DRAMs, where the capacitor cannot be scaled to the same extent, since the stored charge determines the refresh time of the memory. This explains why DRAM technology has evolved toward vertical capacitor integration and new high-dielectric-constant materials, while Flash technologies are still using virtually the same device concepts.

Equations (4.6) and (4.7) allow elimination of the (unknown) FG charge in order to obtain a relationship between the externally applied voltages, the coupling ratios, and the threshold voltage of the memory cell:

$$V_{fg} = V_{th} + \text{cg}(V_{cg} - V_T) + \text{ch}(V_{ch} - V_{chmeas}) + d(V_d - V_{dread}) \tag{4.11}$$

Usually, the contribution of the channel coupling term can be neglected and the third term on the right-hand side of Eq. (4.11) can be omitted. This equation is often used to calculate the FG voltage evolution starting from the threshold voltage values as recorded during a transient measurement. To illustrate the difference between electrostatic and quasi-Fermi potentials, consider the case of the erase operation for the device depicted in Figure 4.1. Erasure is typically achieved by applying a negative voltage to the control gate and a positive voltage to the source junction. Repeating the same exercise that led to Eq. (4.11) for this case yields

$$V_{fg(erase)} = V_{th} + \text{cg}(V_{cg(erase)} - V_T) + \text{ch}(-2\Phi_f - V_{chmeas}) + sV_{s(erase)} - dV_{dread} \tag{4.12}$$

PHYSICS OF PROGRAMMING AND ERASE MECHANISMS

where s is the source-coupling ratio. In this case, the difference between both kinds of potentials leads to an additional channel-coupling term. The reason is that the additional Φ_f terms in Eq. (4.4) are no longer automatically canceled out when eliminating the FG charge between Eqs. (4.6) and (4.7), due to the fact that the channel is in the accumulation regime during erase.

Considering again Eq. (4.8) and taking the derivative with respect to time results in:

$$I_g = dQ_{fg}/dt = C_{tot} dV_{fg}/dt = C_g dV_T/dt \quad (4.13)$$

Since the transient measurement of programming/erase behavior yields the threshold voltage versus time, the gate current can be calculated from Eq. (4.13). Together with Eq. (4.11) or (4.12) this allows calculation of the gate-current characteristic $I_g(V_{fg})$ for a given technology.

4.2.3.2 Transient Equation. When electrons are injected onto or from the floating gate, the floating-gate potential will continuously change, therefore altering the magnitude of the gate current. Thus, in order to calculate the threshold voltage shift as a function of time, the gate current has to be integrated as follows:

$$I_g(t) = I_g[V_{fg}(t)] = F\left[V_{fg0} - \frac{1}{C_{tot}} \int_0^t I_g(t)\,dt\right] \quad (4.14)$$

In this equation, the function F describes the gate-current characteristic $I_g(V_{fg})$. The argument of the function F rewrites the FG voltage as an initial value V_{fg0} at the beginning of the programming/erase operation and an integral that represents the change of this initial value due to the collected charge on the FG. In Eq. (4.14), the implicit convention has been used that injecting current on the FG corresponds to a positive gate current. This equation is the most general description of the transient behavior of a floating-gate memory device. The only assumption made is that changes in FG potential are *entirely* attributed to changes in the FG charge. This is not always the case since in some erase schemes the cell's channel can evolve from accumulation to inversion during the operation that adds a correction term to the FG potential due to channel coupling. Equation (4.14) is essentially a differential equation in Q_{fg} [or alternatively in ΔV_{fg} or ΔV_T; see Eq. (4.8)]:

$$\frac{dQ_{fg}(t)}{dt} = F\left[V_{fg0} - \frac{1}{C_{tot}} \Delta Q_{fg}(t)\right] \quad (4.15)$$

The complexity of the function F will determine whether an analytical solution is possible.

4.3 PHYSICS OF PROGRAMMING AND ERASE MECHANISMS

Electrical conduction through thin dielectric layers has been studied extensively in the past. It is generally understood that the behavior of the charge transport mechanisms in dielectrics can be divided into two main classes: *bulk-limited conduction* and *electrode-limited conduction*. In the former class, the current is determined mainly by the characteristics of the dielectric itself and is independent of the electrodes from where the current originates. On the other hand, in the class of electrode-limited current, the conduction is determined by the characteristics of the electrodes, that is, the interface from where the current originates.

Many dielectrics, such as silicon nitride (Si_3N_4) or tantalum oxide (Ta_2O_5) belong to the bulk-limited conduction class. The current through silicon nitride is determined by Schottky emission from trapping centers in the nitride bulk and is commonly referred to as Poole–Frenkel conduction [14]. Thin (<30 nm) nitride layers, however, also show a strong electrode-limited contribution.

In silicon dioxide, the current is mainly determined by the electrode characteristics, more specifically by the characteristics of the injecting interface. This is because SiO_2 has a large energy bandgap of about 9 eV, as compared to 5 eV for Si_3N_4, and a high-energy barrier at its interfaces with aluminum or silicon. For example, the barrier of SiO_2 is about 3.2 eV for electrons in the conduction band of silicon, and 4.8 eV for holes in the valence band, as compared to 2 eV for holes and electrons in Si_3N_4. In currently used Flash technologies, silicon dioxide is still the basic dielectric through which electrons are transported for programming and erasing, and, therefore, the remainder of this section will concentrate on this case only.

Currently, several injection mechanisms are being used in commercially available Flash memories. When the Flash concept first emerged, channel hot-electron injection (CHEI) was the most popular programming mechanism as being well-known from EPROMs, while the Fowler–Nordheim (FN) tunneling was chosen for the slow erase operation. However, when Flash matured and scaling proceeded, other combinations were implemented and a lot of them found their particular place on the Flash market.

First, the EEPROM concept was "translated" to a Flash concept by a number of companies, which led to low-power Flash programming since FN tunneling was used for both programming and erasing. On the other hand, the knowledge built up on polyoxide conduction for EEPROMs has led to Flash memories that use CHEI for programming and (low-field) polyoxide conduction for erasing. In the meantime, relatively "new" mechanisms have been explored such as substrate hot-electron injection (SHEI) and source-side hot-electron injection (SSI), both aiming at an improvement of the low injection efficiency of conventional CHEI, sometimes referred to as *drain injection*. In recent publications, a lot of attention is focused on secondary impact ionization (SII) effects that could increase the CHEI efficiency considerably when the supply voltage drops below the magic barrier of 3.2 V. This corresponds to the potential that an electron must possess in order to have a reasonable probability of crossing the silicon dioxide energy barrier.

Thus, during the last two decades, various mechanisms for charge injection into the oxide have been considered as listed below including some relevant references:

- Fowler–Nordheim tunneling through thin oxides (<10 nm) [15, 16]
- Enhanced tunneling through polyoxides [17, 18]
- Channel hot-electron injection [19, 20]
- Substrate hot-electron injection [21, 22]
- Source-side injection [23–25]
- Secondary impact ionization initiated hot-electron injection [26]

PHYSICS OF PROGRAMMING AND ERASE MECHANISMS

The first two are based on a quantum mechanical tunneling mechanism through an oxide layer, while the last four are based on injection of carriers that are heated in a large electric field in the silicon, followed by injection over the energy barrier of SiO_2. These mechanisms will be treated in more detail in the following sections.

4.3.1 Fowler–Nordheim Tunneling

One of the most important injection mechanisms used in floating-gate devices is the so-called Fowler–Nordheim (or FN) tunneling, which is in fact a field-assisted electron-tunneling mechanism [27]. When a large voltage is applied across a polysilicon–SiO_2–silicon structure, its band structure has a shape as indicated in Figure 4.7. Due to the high electric field, the electrons in the silicon conduction band see a triangular energy barrier of which the width is dependent on the applied field. The height of the barrier is determined by the electrode material and the band structure of SiO_2.

At sufficiently high fields, the width of the barrier becomes sufficiently small for electrons to tunnel through the barrier from the silicon conduction band into the oxide conduction band. This mechanism was already demonstrated by Fowler and Nordheim for the case of tunneling of electrons through a vacuum barrier and was later also described by Lenzlinger and Snow for oxide tunneling. The Fowler–Nordheim current density is given by [27]

$$J = \alpha E_{inj}^2 \exp\left(\frac{-E_c}{E_{inj}}\right) \quad (4.16)$$

with

$$\alpha = \frac{q^3}{8\pi h \phi_b} \frac{m}{m^*} \quad (4.17)$$

$$E_c = \frac{4\sqrt{2m^*}\phi_b^{3/2}}{3\eta q} \quad (4.18)$$

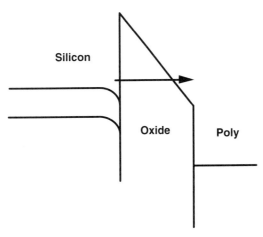

Figure 4.7. Energy band representation of Fowler–Nordheim tunneling through thin oxides: The injection field equals the average thin oxide field. Electrons in the silicon conduction band tunnel through the triangular energy barrier.

where h = Planck's constant
- ϕ_b = energy barrier at the injecting interface (3.2 eV for Si–SiO$_2$)
- E_{inj} = electric field at the injecting interface
- q = charge of a single electron (= 1.6 × 10^{-19} C)
- m = free electron mass (= 9.1 × 10^{-31} kg)
- m* = effective mass of an electron in the bandgap of SiO$_2$ (0.42 m according to [27])
- $\eta = h/2\pi$

Equation (4.16) is the simplest form for the Fowler–Nordheim tunnel current density and is quite adequate for use with nonvolatile memory devices. A complete expression for the tunnel current density takes into account two second-order effects: the image force barrier lowering and the influence of temperature.

The image force lowers the effective barrier height due to the electrostatic influence of an electron approaching the interface. Two correction factors, $t(\Delta\phi_b)$ and $v(\Delta\phi_b)$, have to be introduced in Eq. (4.16), both of which are tabulated elliptic integrals and slowly varying functions. The reduction in energy barrier height ($\Delta\phi_b$) given by [27]

$$\Delta\phi_b = \frac{1}{\phi_b}\sqrt{\frac{q^3 E_{inj}}{4\pi\varepsilon_{ox}}} \tag{4.19}$$

Although tunneling is essentially independent of temperature, the number of electrons in the conduction band, available for tunneling, is depending on the temperature. This dependence can be taken into account by a correction factor $f(T)$, given by [27]

$$f(T) = \frac{\pi ckT}{\sin(\pi ckT)} \tag{4.20}$$

with

$$c = \frac{2\sqrt{2m^*}t(\Delta\phi_b)}{hqE_{inj}} \tag{4.21}$$

Taking these two corrections into account, the expression for the Fowler–Nordheim tunnel current density becomes

$$J = \alpha E_{inj}^2 \frac{1}{t^2(\Delta\phi_b)} f(T) \exp\left[\frac{-E_c}{E_{inj}} v(\Delta\phi_b)\right] \tag{4.22}$$

The influence of the correction factors is small, however, and, for most practical calculations, the basic Eq. (4.16) is sufficiently accurate.

The Fowler–Nordheim tunnel current density shows an almost exponential dependence on the applied field. This dependence is shown in Figure 4.8(a) for the monocrystalline silicon–SiO$_2$ interface. The Fowler–Nordheim current is usually plotted as J/E^2 vs. $1/E$, which should yield a straight line with a slope proportional to the oxide barrier, as shown in Figure 4.8(b).

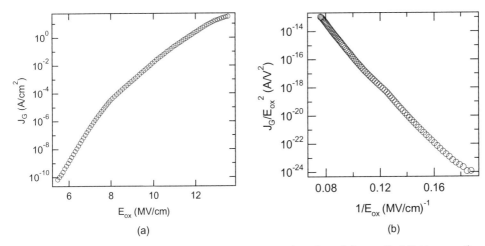

Figure 4.8. (a) Fowler–Nordheim tunneling current as a function of the applied field across the oxide. The current is exponentially dependent on the field. (b) Fowler–Nordheim plot: J/E^2 as a function of $1/E$, extracted from (a). A straight line is obtained.

In practice, an injection field of 10 MV/cm leads to a current density of approximately 10^{-2} A/cm². This high value of injection field is of the order of that needed across the oxide during the programming of a nonvolatile memory device. The breakdown field of these oxides should be, of course, significantly larger than this value. In order to reach these high field values, and limit the voltages needed during programming, very thin tunnel oxides are used; an injection field of 10 MV/cm is attained by applying a voltage of 10 V across an oxide of 10 nm thickness. If one were to reduce the programming voltages, the tunnel oxide would have to be made even thinner. A thickness of 6 nm, however, is the lower limit for good retention behavior. However, these oxides are difficult to grow with low defect densities, as is required for floating-gate devices. Moreover, below this thickness value other injection mechanisms, such as direct tunneling, can become important.

It should be noted that the tunnel current density is totally controlled by the field at the injecting interface, and not by the characteristics of the bulk oxide. Once the electrons have tunneled through the barrier, they are traveling in the conduction band of the oxide with a rather high saturated drift velocity of about 10^7 cm/s [28].

For the calculation of the injection field at the silicon–SiO$_2$ interface, however, the flatband voltage has to be taken into account as follows:

$$E_{inj} = \frac{V_{app} - V_{fb}}{t_{ox}} \quad (4.23)$$

where V_{app} = voltage applied across the oxide
V_{fb} = flatband voltage
t_{ox} = thickness of the oxide

When voltages are applied so that the silicon is driven into depletion, a voltage drop in the induced depletion layer must be accounted for in the calculation of the oxide field.

The basic advantages of Fowler–Nordheim tunneling are summarized below:

- Low current programming/erasing and, hence, low power operation in non-volatile memories
- Good endurance due to the uniform current conduction in the oxide
- Good controllability of the tunneling current due to the one-dimensional character of the mechanism

Major disadvantages are:

- Large voltages required
- Importance of tunnel oxide quality (yield issue)
- Constant failure rate during cycle test, which limits endurance on circuit level (reliability issue; see, e.g., Mielke et al. [3])
- Generation of stress-induced leakage currents (see Section 4.4.2.5)

From the point of view of memory cell operation, it can be concluded that the steep FN tunneling characteristic leads to slowly programming/erasing cells (strong decrease of the current as the oxide field decreases). Of course, the peak current must be controlled because of reliability and yield concerns.

Today virtually all byte-erasable EEPROMs and some Flash memories use Fowler–Nordheim tunneling for both programming and erase. FN tunneling is also the major erase mechanism for Flash memories programmed by another mechanism.

4.3.2 Polyoxide Conduction

Fowler–Nordheim tunneling requires injection fields on the order of 10 MV/cm to narrow the Si–SiO$_2$ energy barrier so that electrons can tunnel from the silicon into the SiO$_2$ conduction band, as discussed in the previous section.

In oxides thermally grown on monocrystalline silicon, the injection field is equal to the average field in the SiO$_2$. Therefore, thin oxides have to be used to achieve large injection fields. Oxides thermally grown on polysilicon, so-called polyoxides, however, show an interface covered with asperities due to the rough texture of the polysilicon surface [29, 30]. This has led to the name *textured polyoxide*. These asperities give rise to a local field enhancement at the interface and an enhanced tunneling of electrons [31, 32]. In polyoxides the field at the injecting interface is, therefore, much larger than the average oxide field. Consequently, the band diagram of a polysilicon–polyoxide interface is as shown schematically in Figure 4.9. Average oxide fields of the order of 2 MV/cm are sufficient to yield injection fields of the order of 10 MV/cm. This has the major advantage that large injection fields at the interface can be obtained at moderate voltages using relatively thick oxides, which can be grown much more reliably than the thin oxides necessary for Fowler–Nordheim injection from monocrystalline silicon.

A quantitative analysis of the tunnel current–voltage relations for polyoxides is rather complicated. Though the tunnel mechanism itself is described by the same formula (4.16), discussed in the previous section, the difficulty lies in the accurate determination of the injection fields to be used. One can no longer use a single value

PHYSICS OF PROGRAMMING AND ERASE MECHANISMS

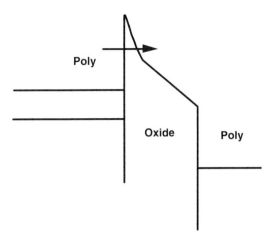

Figure 4.9. Energy band representation of Fowler–Nordheim tunneling through oxides thermally grown on polysilicon: The injection field is much higher than the average oxide field. The high injection field is due to local field enhancement at polysilicon–oxide interface asperities.

for this injection field because of the nonuniformity of the field enhancement over the injecting interface. Indeed, the field enhancement factor is not uniform over the surface of one asperity bump [33]: The factor is maximum at the top of the asperity and decreases strongly down the slope of the bump surface. Additionally, variations of the bump shape may be another cause for the nonuniformity.

In the past, attempts have been made to model the current through the polyoxide by use of some mean field enhancement factor [34], but as was proven in Ref. [33], this method always leads to incorrect results. Therefore, a complete model for the current conduction must be based on:

- The Fowler–Nordheim expression for tunnel current density
- A model for the distribution of the field enhancement factors over the total injecting area
- A model for the charge trapping behavior of the oxide under current injection conditions

The latter has to be taken into account because charge trapping is of much more importance in polyoxides than in oxides grown on monocrystalline material. This is again due to the strong nonuniform field enhancement. Initially, the injection current originates almost completely from the regions of maximum field enhancement. Extremely large current densities occur at these injection points, leading to strong local trapping of electrons near these sites. This trapping reduces the injection locally, by which the current is taken over by regions with a slightly lower field enhancement. This process proceeds gradually, such that the injection current, which is initially extremely localized, becomes increasingly uniform and decreases continuously. Whereas in conventional Fowler–Nordheim injection the trapping occurs only after some critical current level has been reached, charge trapping and current injection are occurring simultaneously over the whole current range during polyoxide injection. The nonuniform field enhancement at the polysilicon–polyoxide interface makes it impossible to use a closed analytical expression for the polyoxide conduction. A complete model for the polyoxide conduction, based on the principles indicated above, can be found in Ref. [33].

Polyoxide conduction has the advantage that considerable current levels can be obtained at moderate average oxide fields and thus moderate applied voltages. The need for thin polyoxides is therefore not so stringent. In addition, from a reliability point of view, this is an advantage since the oxides are not stressed at large fields during programming, such that dielectric breakdown failures and stress-induced leakage current are avoided [3, 35]. On the other hand, the growth of textured polyoxides has to be carefully controlled in order to obtain the desired interface features (shape, size of the asperities) that determine the injection current and reliability characteristics. For this reason, reproducibility may be a problem for this kind of injection mechanism. Another disadvantage is that the injection is asymmetrical with respect to polarity. For injection from a top polysilicon layer, the currents are much smaller. Finally, the strong change in injection currents due to the decrease in the mean enhancement factor during current injection can pose severe constraints on the number of programming cycles that can be allowed in a memory cell.

From the point of view of the memory cell operation, it can be concluded that the polyoxide tunneling mechanism offers a viable alternative for conventional FN tunneling, primarily because of the lower susceptibility to oxide breakdown and stress-induced leakage currents. The major drawback is the controllability of the tunneling current over a large amount of cells, and the larger decrease of the current with stress time, both leading to larger programming windows and large voltages.

Nevertheless, optimized process technologies supported by adequate design solutions have enabled the use of polyoxide conduction mostly as an erase mechanism in some Flash memories [36, 37].

4.3.3 Channel Hot-Electron Injection (CHEI)

At sufficient drain bias, the minority carriers that flow in the channel of a MOS transistor are heated by the large electric fields seen at the drain side of the channel and their energy distribution is shifted higher. This phenomenon gives rise to impact ionization at the drain, by which both minority and majority carriers are generated. The highly energetic majority carriers are normally collected at the substrate contact and form the so-called substrate current. The minority carriers, on the other hand, are collected at the drain. A second consequence of carrier heating occurs when some of the minority carriers gain enough energy to allow them to surmount the SiO_2 energy barrier. If the oxide field favors injection, these carriers will be injected over the barrier into the gate insulator and give rise to the so-called channel hot-carrier injection gate current [38, 39]. The corresponding energy band diagram is schematically represented for the case of an N-channel transistor in Figure 4.10.

The electrons gain enough energy from the lateral field at the drain side of the channel to surmount the energy barrier between the silicon and the oxide.

An important difference with respect to the two previously discussed mechanisms is that CHEI can only bring electrons onto the (floating) gate. It cannot be used to remove them again. Channel hot-hole injection is not feasible due to the very low gate current levels (higher energy barrier of 4.8 eV) and the large degradation associated with hole trapping (see also Sections 4.4.2.2 and 4.4.2.4).

Several models have been proposed to describe the gate current due to channel hot-electron injection. In contrast to the Fowler–Nordheim tunneling case, no

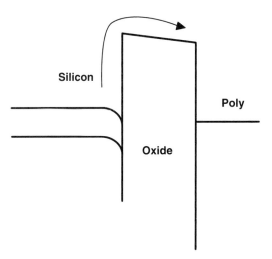

Figure 4.10. Energy band representation of channel hot-electron injection (CHEI) in an N-channel MOS transistor.

closed-form analytical expression exists for the channel hot-electron injection current due to the complex two-dimensional nature of the phenomenon and many unknown physical parameters. Therefore, the models are merely qualitative. They can be divided into three main categories: the lucky electron models, the effective electron temperature models, and the physical models.

The lucky electron model, or LEM [40, 41], assumes that an electron is injected into the gate insulator if it can gain enough energy in the lateral field without undergoing a collision, by which energy could be lost. By phonon scattering, the electrons are then redirected toward the Si–SiO$_2$ interface. If these electrons can reach the interface and still have enough energy to surmount the Si–SiO$_2$ energy barrier (and eventually also a repulsive field), they will be injected into the gate insulator. The LEM formula is given hereafter:

$$I_g = C \left(\frac{\lambda E_m}{\phi_b} \right)^2 \exp\left(\frac{-\phi_b}{\lambda E_m} \right) \qquad (4.24)$$

where ϕ_b = (zero oxide field) energy barrier at the interface (3.2 eV)
E_m = peak lateral electric field at the drain junction
C = preexponential coefficient
λ = electron mean free path

The effective electron temperature model [42] assumes that the electrons are heated and become an electron gas with a Maxwellian distribution with an effective temperature, T_e, that is dependent on the electric field. The gate current can then be calculated as the thermionic emission of heated electrons over the interface energy barrier.

The physical models [43] attempt to calculate the gate currents based on a more physical treatment and an accurate solution of the two-dimensional electric field distribution at the drain side of the channel. Then, the gate current is calculated based on an injection efficiency that is dependent on the interface barrier energy and the lateral electric fields.

For all above-mentioned models, one always has to keep in mind that there exists a difference between the number of injected electrons and the number of electrons actually reaching the gate. Indeed, due to a repulsive oxide field, all or part of the injected electrons can be repelled into the silicon [44].

Qualitatively, it can be stated that the gate current is determined on the one hand by the number of hot electrons and their energy distribution (which is largely dependent on the electric fields occurring in the channel of the transistor) and on the other hand by the oxide field (which determines the fraction of hot electrons that can actually reach the gate). In practice, this means that the gate current versus gate voltage characteristic will rise to a significant level only as the oxide field becomes favorable for electron injection, and then will decrease according to the corresponding decrease in the lateral field with increasing gate voltage. This gives the typical curve of Figure 4.11, which shows a maximum around $V_g = V_d$.

The gate current, thus, essentially peaks very close to the zero-oxide-field condition.

Due to this typical gate voltage dependence of the CHEI current, the gate voltage during programming of a nonvolatile memory cell, using CHEI, has to be chosen carefully in relation to the applied drain voltage. For completeness, it should be mentioned that the maximum in the CHEI characteristic disappears for submicron Flash devices that have been optimized for high (drain) injection efficiency [45]. Instead, the gate current becomes a constant as a function of the gate voltage, due to the increasing overlap between carrier concentration and electric field peak at higher gate voltages, which compensates for the decrease of this field peak [46].

In order to evaluate the dependence of the CHEI current on processing and geometrical parameters, a simplified expression for the lateral electric field in the channel can be used [47]:

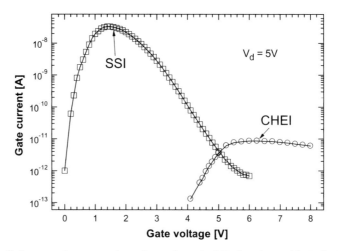

Figure 4.11. Gate current versus gate voltage characteristic for channel hot-electron injection (CHEI) showing a maximum gate current around $V_g = V_d$. For comparison, also the gate current for source-side injection is depicted (see Section 4.3.5).

PHYSICS OF PROGRAMMING AND ERASE MECHANISMS

$$E_m \approx \frac{V_d - V_{dsat}}{l} \quad (4.25)$$

with

$$l \approx 0.22 t_{ox}^{1/3} x_j^{1/2} \quad (4.26)$$

and with V_{dsat} expressed as [48]

$$V_{dsat} = \frac{(V_g - V_t)L_{eff} E_{sat}}{V_g - V_t + L_{eff} E_{sat}} \quad (4.27)$$

where E_{sat} is the electric field at which the electron velocity saturates.

From these formulas, it can be concluded that the gate current increases with thinner gate oxides, shallower junctions, smaller effective channel lengths, and higher substrate doping levels (through the influence on the threshold voltage at the drain through the body effect).

From the point of view of the memory cell operation, it can be concluded that the CHEI mechanism provides a fast programming method for Flash memory, however, requiring a considerable channel current resulting in a high power consumption. On the other hand, because of the small oxide field, cells relying on CHEI for programming are less susceptible to oxide breakdown and stress-induced leakage currents.

4.3.4 Substrate Hot-Electron Injection (SHEI)

With the CHEI mechanism, discussed in the previous section, hot electrons are injected near the drain region only, and the injected current is determined by a combination of many factors. The substrate hot electron injection (SHEI) technique, originally proposed by Verwey [49] and later improved by others [21, 50], allows injecting hot electrons into the oxide in a uniform manner, at a fixed oxide field, with independently controlled current density.

A typical setup and a schematic band diagram, illustrating the SHEI mechanism, are shown in Figures 4.12 and 4.13, respectively, for the case of an N-MOS transistor. The transistor is biased in inversion with source and drain grounded, and electrons are injected in the substrate from an additional PN junction (the injector junction), which can be formed either around the MOS field-effect transistor (FET) (as in the figure) or as an underlying N-well to P-well junction. A fraction of the injected electrons reach the depletion region of the transistor by diffusion and are accelerated toward the gate, the source or drain junction. If the electrons gain sufficient energy in the depletion region, they become "hot," and they are injected into the gate oxide giving rise to a gate current. The remaining electrons flow in the channel to the source or to the drain.

With this electron injection technique, the oxide field is determined by the gate to source/drain voltage, while the silicon field and, thus, the energy of the injected carriers are determined by the substrate-to-source/drain voltage and the substrate doping. The injected current can be controlled by the forward current from the injector diode. For the injection of holes, a similar setup is used, but in this case,

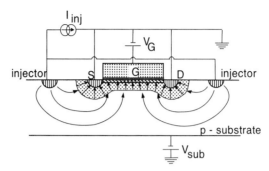

Figure 4.12. Schematic representation of the SHEI mechanism. The oxide field is only determined by the gate voltage, the gate current can be controlled by tuning the amount of electrons at the injector junction.

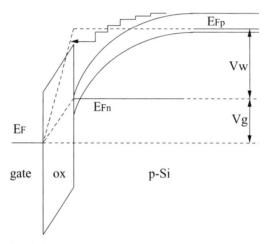

Figure 4.13. Energy band diagram corresponding to the bias conditions of Figure 4.12. Also shown is the trajectory of a "hot" electron able to surmount the Si–SiO$_2$ interface barrier. E_{Fn}/E_{Fp} are the quasi-Fermi levels for electrons and holes, respectively, while V_w and V_g are the voltages applied to the well (or substrate) and to the gate, respectively.

P-channel devices instead of N-channel devices have to be used [51, 52]. Substrate hot-hole injection (SHHI) is especially interesting for experimental study of degradation effects (see Section 4.4.2) but is of no practical use for memory cells (see also Section 4.3.3).

The physical description of SHEI is analogous to that of CHEI, and the current can be described by Eq. (4.24). The main difference is that the energy of the electrons is now determined by the electrostatic field in the silicon and not by the peak electric field near the drain.

From the point of view of the memory cell operation, the SHEI mechanism provides an efficient programming method for Flash memory, however, requiring a nearby electron source, which, in practice, urges for the use of triple-well technology. The main advantage is the unique combination of uniform injection (in contrast to CHEI) and low oxide field (in contrast to FN tunneling).

4.3.5 Source-Side Injection (SSI)

The main disadvantage of the CHEI mechanism for programming a nonvolatile element stems from its low injection efficiency and, consequently, its high power consumption. This is due to the incompatibility of having a high lateral field *and* a high vertical field, favorable for electron injection, at fixed bias conditions, as explained in the Section 4.3.3. Indeed, the lateral field in a conventional MOS device is a decreasing function of the gate voltage, while the vertical field increases with the gate voltage. Therefore, in order to generate a large amount of hot electrons, a low gate voltage is required, combined with a high drain voltage. However, for electron injection and collection on the floating gate of the memory device, a high gate voltage and a low drain voltage are required (see Fig. 4.14). In practice, both gate and drain voltages are kept high as a compromise. The main drawback is clearly the high drain current (on the order of $100\,\mu A$ to $1\,mA$) and the correspondingly high power consumption.

Therefore, a novel injection scheme, now commonly referred to as source-side injection (SSI), was proposed to overcome this problem [25]. In most cases, the MOS channel between the source and drain regions is split into two "subchannels" controlled by two different gates. The gate on the source side of the channel is biased at the condition for maximum hot-electron generation, that is, somewhat above the threshold voltage of this channel. The gate at the drain side, which is the floating gate of the cell, is capacitively coupled to a potential that is comparable to or higher than the drain voltage in order to establish a vertical field component that is favorable for hot-electron injection in the direction of the floating gate. The latter condition can be accomplished either by implementing an additional gate with a high coupling ratio toward the floating gate [4, 53] or by using a high drain-coupling ratio [23]. As a result, the drain potential is entirely or partially extended toward the region between the gates that control the MOS channel. This effect is referred to as the *virtual drain effect* since the inversion layer under the floating gate merely

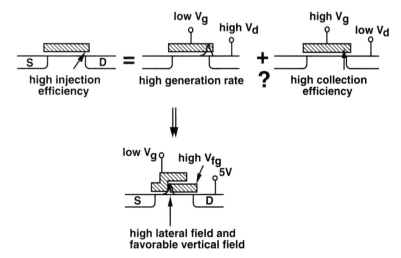

Figure 4.14. Schematic representation of the problem of low injection efficiency for CHEI and the principle of source-side injection to overcome this problem.

acts as a drain extension, while the effective transistor channel is formed by the subchannel at the source side of the device [25]. Consequently, a high lateral field peak is obtained in the gap between both subchannels (Fig. 4.14). The hot electrons are thus generated inside the MOS channel and not at the drain junction of the cell. Because of the high floating-gate potential, the vertical field at the injection point is favorable for electrons, and most of the generated hot electrons that overcome the potential barrier between the channel and the oxide layer are effectively collected on the floating gate.

The main advantage of this mechanism is the much higher injection efficiency (on the order of 10^{-3} and higher), which allows for fast 5-V-only and 3.3-V-only operation combined with a low power consumption [24, 53]. This is illustrated in Figure 4.11 where the gate currents of the conventional channel hot-electron injection and the source-side injection mechanisms are shown for comparable devices (same channel width to length ratio W/L, same drain voltage, same technology). It is clear that the SSI mechanism provides a gate current that is more than 3 orders of magnitude higher than conventional hot-electron injection [25]. At the same time, the drain current in the SSI case is also reduced by a factor of 40 with respect to the conventional case [54].

The gate current tends to saturate with increasing floating-gate voltage because the virtual drain potential (i.e., the channel potential at the injection point) approaches the externally applied drain voltage [25]. Since the floating-gate voltage changes during programming, the maximum observed in the conventional gate current characteristic (Fig. 4.11) is no longer relevant in the SSI case: The gate current decreases monotonically, and only slightly, while programming an SSI cell. In addition, the SSI mechanism is not a function of the drain profile and only a smooth linear function of the channel length of the device [25]. This is in strong contrast to the conventional hot-electron injection where the injection is strongly dependent on both drain profile and channel length.

4.3.6 Secondary Impact Ionization Initiated Channel Hot-Electron Injection

As explained in the previous sections, the conventional drain side hot-electron injection suffers from two fundamental scaling problems: the drain voltage can not be decreased below ~4 V due to the low efficiency of the mechanism and the barrier of 3.2 eV that has to be surmounted at zero-oxide-field conditions. Consequently, the drain voltage has to be supplied from a bitline charge pump that can only provide a limited current. Moreover, the channel length will not be easily scaled without scaling this drain-to-source voltage.

The power consumption is very high, which makes the concept even less attractive for the deep-submicron era that requires low-voltage low-power operation.

A new mechanism has been reported that could provide an interesting alternative for deep-submicron low-voltage Flash memories. When applying a small negative voltage to the substrate of a conventional stacked-gate memory cell, a large increase in gate current is observed that cannot be explained from the conventional channel hot-electron theory [26]. This phenomenon has been attributed to an impact ionization feedback mechanism that allows secondary particles (i.e., holes generated from the primary impact ionization caused by channel electrons) to gain sufficient energy from the vertical field in the silicon in order to cause secondary impact ionization deeper in the substrate of the device (Fig. 4.15).

The resulting electrons are partially injected into the gate, creating a much higher contribution to the gate current than the conventional gate current. The latter is true under the assumption that the vertical silicon field is high enough to generate a large amount of hot electrons by means of the secondary impact ionization effect. Indeed, these electrons are generated much deeper in the substrate, which decreases the probability for reaching the interface considerably. Therefore, for this condition to occur, a negative voltage must be applied to the substrate or P-well in order to enhance the vertical electric field. On the other hand, the mechanism only becomes appreciable for deep-submicron technologies where shallow junctions are used in combination with very thin gate oxides.

The most interesting advantages of the mechanism are listed below:

- The injection current increases by several orders of magnitude as compared to the conventional (primary) hot-electron current, which allows further voltage scaling, and thus also further channel length reduction.
- A significant injection is still observed when the gate voltage drops below the drain voltage, which allows for low power programming similar to source-side injection.

Disadvantages are:

- The necessity of a triple-well technology in order to allow a negative P-well biasing scheme
- The fact that drain engineering is still required in order to obtain true 2.5-V-only operation

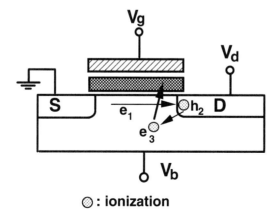

Figure 4.15. Schematic representation of the secondary impact ionization effect.

4.4 PHYSICS OF DEGRADATION AND DISTURB MECHANISMS

4.4.1 Band-to-Band Tunneling

When using a nonuniform erase scheme [FN tunneling of electrons from the floating gate to the (N^+) source or drain junction of the memory cell], band-to-band tunneling will inevitably occur at the junction–oxide interface (Fig. 4.16).

Since the gate is biased at high negative voltage while a positive voltage is applied to the junction, a very high vertical silicon field appears in the gate-to-junction overlap region where the doping is sufficiently low. Because holes are drained away by the junction-to-substrate field, the N-type junction cannot be inverted and the device enters the deep-depletion mode. This implies that the energy band bending becomes larger than the energy gap, which allows valence electrons to directly tunnel to the conduction band of the N-type region. These electrons are collected at the junction due to the positive junction bias. On the other hand, the corresponding generated holes are drawn to the substrate to form a large bulk current (typically a few hundred nanoamperes for a state-of-the-art memory cell; see Fig. 4.17).

This implies that the current to be supplied from the junction supply voltage can grow significantly large when erasing a large number of cells simultaneously. When using a large junction voltage (in order to lower or even avoid the negative gate voltage), this current has to be supplied from a charge pump, which creates limitations for the memory sector size. In addition, the large substrate current can cause significant potential drops in the substrate of the memory array and increase the risk for forward biasing of other junctions, or even cause parasitic bipolar triggering and latch-up.

Moreover, some of the holes are heated by the junction field and are injected into the gate oxide, which is known to lead to severe reliability problems:

- The impact of hole trapping on the interface quality is orders of magnitude larger than that of electrons (see Section 4.4.2.2).

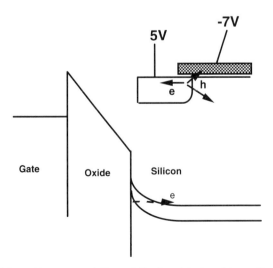

Figure 4.16. Schematic representation of the band-to-band tunneling phenomenon.

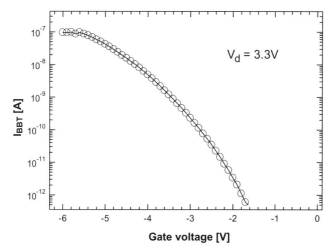

Figure 4.17. Typical band-to-band tunneling current as a function of the gate voltage for a 0.35-μm HIMOS technology [35].

- Endurance is lower since holes generate traps that can subsequently be filled by electrons. Also, erratic erase (anomalous behavior of a few tail bits in large memories during cycling) is assumed to be correlated to the presence of trapped holes over the tunneling region [55].
- Some disturb effects (e.g., stress-induced leakage current) are enhanced by hole trapping [8].

The band-to-band tunneling effect in Flash devices is not easily controlled since the junction profile basically affects only the exact position of the deep depletion condition, while the magnitude of the current is more or less independent of this profile. Switching from drain or source-erase to channel-erase or to polyoxide-erase removes the problem at the expense of more complicated processing.

4.4.2 Oxide Degradation

From the discussion of the different programming and erase mechanisms in Section 4.3, it is clear that the reliability of the gate oxide layer is crucial for a reliable operation of the memory cell. In this section the physical phenomena related to oxide degradation caused by injection of electrons are described.

When electrons tunnel through the silicon/oxide potential barrier or gain sufficient energy to overcome that barrier (hot electrons), the properties of the oxide film gradually deteriorate because of structural damage generated in the oxide. When the accumulated damage becomes too large, the oxide abruptly looses its insulator properties, which is defined as oxide breakdown.

The oxide degradation and breakdown are mainly studied on capacitor and MOSFET structures. During oxide stressing several phenomena can be observed that are indicators of the degradation process or are related to its mechanism. The most important phenomena are the generation of a hole fluence (Section 4.4.2.1), interface trap creation (Section 4.4.2.2), neutral electron trap creation (Section

4.4.2.3), negative and/or positive charge trapping (Section 4.4.2.4), and the generation of a stress-induced leakage current (SILC) (Section 4.4.2.5).

In Section 4.4.2.6, the different physical mechanisms for the generation of oxide damage (i.e., trap generation) proposed in the literature are reviewed.

4.4.2.1 Hole Fluence. When the gate oxide of an N-MOSFET is stressed with a positive gate voltage, while source and drain are grounded, electrons tunneling through the oxide are injected from the transistor channel and provided from the source and drain. In this configuration, a positive current can be measured at the substrate (charge separation technique) [56, 57]. The substrate current density has similar oxide field dependence as the FN current density, as is shown in Figure 4.18. However, it should be remarked that the curves in Figure 4.18 are not parallel: The ratio between gate and substrate current depends on the oxide field.

A well-known and widely accepted explanation for the physical origin of this substrate current is given in the literature [56, 59, 60], and is schematically illustrated in Figure 4.19. When the injected electrons enter the anode (the poly-Si gate), they lose their energy by creating highly energetic holes—possibly through the excitation of some intermediate state [61]—and the holes can then be injected back in the oxide. The hole flow reaches the cathode and is measured as a positive substrate current, J_p. This *anode hole injection mechanism* is often used to explain the generation of both interface traps and bulk oxide traps (see Section 4.4.2.6). However,

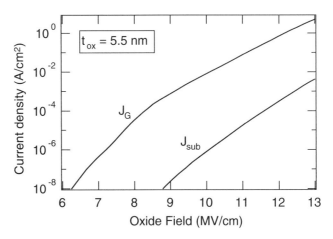

Figure 4.18. Gate and substrate current as a function of the oxide field. The ratio α between the substrate and the gate current is field dependent. α increases with increasing field.

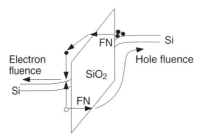

Figure 4.19. Schematic illustration of the anode hole injection model. Injected electrons reach the anode with high energy and can generate hot holes that can tunnel back to the cathode.

another possible explanation for the substrate current is the creation of holes in the cathode by photons generated in the anode [62]. The photons result from energy relaxation of the injected electrons, and they can be easily observed with emission microscopy [63, 64].

The hole current density can be related to the electron current density, J_n, as follows:

$$J_p = \alpha(E_{ox})J_n \quad (4.28)$$

with $\alpha(E_{ox})$ the field-dependent hole generation efficiency. In the anode hole injection model, α is interpreted as the probability for a tunneling electron to generate an anode hole, which is injected back into the oxide toward the cathode [60]. When a constant voltage is applied, α is found to be almost constant as a function of time [65], while for a constant current stress, α slightly increases. However, for thin oxides the difference between the initial and final values of α is very small. Therefore, with good approximation, an equivalent relation as Eq. (4.28) also holds between the integrated values of J_p and J_n, the hole fluence, Q_p, and the electron fluence, Q_n, respectively:

$$Q_p = \alpha(E_{ox})Q_n \quad (4.29)$$

In Chen et al. [56] it was observed for the first time that the hole fluence reaches a critical value at breakdown: $Q_{p,crit}$. This result has been confirmed in Satake and Toriumi [66] and Degraeve et al. [58]. In Figure 4.20, $Q_{p,crit}$ and the charge-to-breakdown Q_{BD}, measured in a wide field range from 8 to 14 MV/cm, are plotted. Clearly, $Q_{p,crit}$ remains constant in the entire field interval, while Q_{BD} decreases with increasing field. A satisfactory physical explanation for the experimentally observed invariance of $Q_{p,crit}$ cannot be found in the literature, but the various suggested

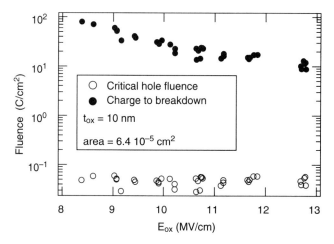

Figure 4.20. The Q_{BD} and the hole fluence at breakdown as a function of the electric field. On the one hand, the hole fluence reaches a critical value at breakdown, $Q_{p,crit}$, independent of the stress field. On the other hand, the charge-to-breakdown decreases continuously with the field [58].

possibilities are discussed in detail in Section 4.4.2.6. It should be remarked that in Satake and Toriumi [66], $Q_{p,\text{crit}}$ was observed to be constant only at 300 K. For lower temperatures, it decreases as a function of the field. This temperature effect indicates that the hole fluence at breakdown is possibly not the factor determining the triggering of breakdown.

4.4.2.2 Interface Trap Creation. When a thin oxide film is stressed either in FN (or direct tunneling) regime, or by hot carrier injection, interface traps are created at both the cathode and the anode interface. The density of interface traps D_{it} at the substrate–oxide interface can be extracted by several techniques. In a capacitor structure, the capacitance–voltage (CV) method [67] is best known. In transistors, D_{it} can also be measured with the charge-pumping (CP) technique [68] or by the subthreshold slope method [67]. The part of the bandgap that is assessed depends on the specific technique, so in general the results will not be identical. The CP technique is today the most extensively used method to determine the interface trap density in MOSFETs. It has a much higher resolution than CV analysis, and it can even be used to characterize single traps in small transistors [69]. CP has the additional advantage that information on the trap localization under nonuniform stress conditions can be obtained [70].

To study the impact of electron and hole injection separately, substrate hot carrier techniques (Section 4.3.4) have been used. The most important features of electron-induced interface trap generation are summarized in Figure 4.21 [50]. This plot compares the interface trap generation ΔD_{it} (at a fixed fluence) at two injection temperatures, 295 and 77 K, as a function of oxide field. At 295 K, the interface trap generation rate is seen to exhibit approximately exponential field dependence over a wide oxide field range. This strong field dependence suggests a correlation with electron heating in the oxide field [71]. At 77 K, curve A is the direct generation during injection, while curve B is the total generation after injection and subsequent warmup to 295 K (at zero bias). Hot-electron-induced interface trap generation is clearly a strongly temperature-activated process: Going from 295 to 77 K, the gen-

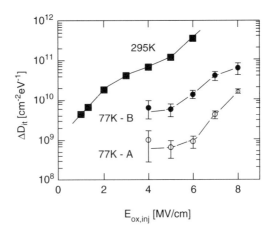

Figure 4.21. Total interface trap generation after injection of $6.25 \times 10^{18}\,\text{cm}^{-2}$ electrons (with SHE injection) as a function of oxide field, at 295 and 77 K. Curves A and B are after injection at 77 K and after warmup to 295 K, respectively [50].

eration efficiency drops by more than 2 orders of magnitude. Curve B shows that during injection at 77 K also some latent damage is formed.

Similar experiments have been conducted with substrate hot-hole injection to study the interface trap generation by hole injection. In contrast to the bulk oxide hole trap generation *interface trap generation is much more efficient when injecting holes than when injecting electrons.* Typical generation efficiencies are in the range of 10^{-3} to 10^{-2} [51, 72, 73]. As is also the case for hole trapping, the generation process was observed to be independent of silicon field [72]) and only weakly dependent on the oxide field [72, 73], in contrast to the exponential oxide field dependence of interface trap generation by electron injection.

At 77 K the interface trap generation efficiency is also higher than at 295 K, clearly in strong contrast to electron-induced D_{it} generation, as discussed previously.

4.4.2.3 Electron Trap Generation.
During oxide stress, apart from interface traps, electron traps are also generated in the bulk of the oxide. Although many researchers have related this trap creation process to the breakdown process, direct measurements of the oxide trap density D_{ot} as a function of the applied stress conditions are rare. Indeed, most of these traps remain neutral during a high field stress because of the high detrapping probability for electrons at these conditions (see next section). Moreover, the electron trapping probability depends on the oxide field and therefore the negative charge density measured directly after stress is not a good monitor to measure the density of neutral traps [74].

In order to measure this degradation phenomenon, the neutral traps have to be made electrically visible. This can be accomplished by interrupting the stress periodically and by filling the generated traps with electrons under controlled conditions without creating any additional traps. A technique ideally suited for this purpose is uniform substrate hot-electron injection (SHEI) (see Section 4.3.4).

In Figure 4.22 [75], the increase of the neutral trap density as a function of the injected electron fluence is shown during electron injection. Degradation and

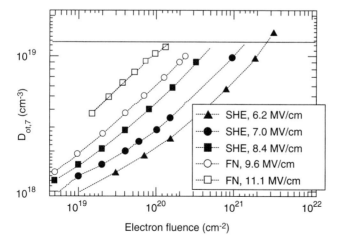

Figure 4.22. Occupied electron traps at a filling field of 7 MV/cm, as a function of the injected electron fluence measured during oxide stressing at fields between 6.2 and 11.1 MV/cm [75]. Oxide stress was performed either by FN (open symbols) or SHE injection (closed symbols). For each stress condition, the mean breakdown value is indicated by an asterisk.

breakdown in the field range 6 to 8.5 MV/cm with SHEI and 8.5 to 11 MV/cm with constant voltage FN injection are presented. The traps were filled with SHE at a fixed oxide field of 7 MV/cm, resulting in a fixed filling fraction for all stress [74]. In the considered field range and apart from small statistical fluctuations, breakdown occurs when D_{ot} reaches a critical value. This result supports the idea that a critical electron trap density is necessary to trigger breakdown and is the base of the breakdown model discussed in Section 4.4.3.

4.4.2.4 Oxide Charge Trapping. During FN and hot carrier injection, both holes and electrons can become trapped in the oxide. During FN injection at constant voltage, this trapping can be easily observed in capacitor structures. In oxides with thickness larger than 4 nm, typically an initial increase of the current density is observed, followed by a current decrease. The current changes are caused by trapping of holes (generated as described in Section 4.4.2.1), followed by electron trapping. The trapped charge leads to an oxide field distortion and subsequent change of the tunnel current density. Other possible ways to measure the trapped charge are by monitoring the shift of the flatband voltage or the threshold voltage.

The holes are only trapped in traps existing before the current injection started (as no additional hole traps are formed during stress), but electron trapping occurs in both prestress existing traps as well as in new traps created during stress. Therefore, the net charge density becomes negative as the oxide degradation proceeds.

In sub-4-nm oxides, the negative charge built up almost completely disappears. Typically, a very small increase of the stress current during constant voltage stress is measured, which is attributed to positive charge trapping and the gradual generation of a stress-induced leakage current [76].

Hot carrier trapping has been studied under controlled conditions using the substrate hot carrier injection technique (Section 4.3.4). From these experiments, it is concluded that hot electrons injected into the gate oxide have a certain, though small, probability (about 10^{-7}) of being trapped in electron traps present in the oxide bulk [71]. Low-field electron trapping has been found to be independent of the silicon field (i.e., the energy at injection). At 77 K electron trapping is one to two orders of magnitude more effective due to additional trapping in energetically shallow traps [77]. Field detrapping from shallow traps is strongly promoted at higher oxide field [77]. For this reason, most electron traps remain neutral (i.e., unoccupied) after a high field stress (see also in previous section).

The trapping of hot holes injected into the gate oxide occurs with several orders of magnitude higher efficiency than that of electrons: The trapping efficiency reaches values as high as 10 to 20% [51, 72]. Hole trapping is quite insensitive to the oxide field conditions. Also, no dependence on silicon field nor injected current density has been observed, an observation that is consistent with the assumed fast thermalization of injected holes in the oxide [51, 72]. At 77 K hole trapping is enhanced, the efficiency increasing to values up to 70% [72]. Analysis of the trapping kinetics at 295 and 77 K has demonstrated that the enhanced trapping at 77 K is not due to additional trapping in shallow hole traps, but rather to a larger effective capture cross section at low temperatures [72].

In some publications [78], it is claimed that the positive charge trapping in the oxide is responsible for triggering the oxide breakdown event. During, for example, a constant current stress (CCS), a locally enhanced charge trapping will not influence the total current density in the capacitor, but leads to a local current density

PHYSICS OF DEGRADATION AND DISTURB MECHANISMS

increase, resulting in an increased stress, which in turn leads to an increased positive charge trapping. In this way a positive feedback mechanism is initiated that finally results in breakdown.

Two arguments oppose this idea. First, in ultrathin sub-4-nm oxides, the measurable positive charge trapping is extremely small and yet the oxides break down. Second, the possible role of the negative charge in the oxide is completely ignored, while the net charge trapped at breakdown is negative.

Other publications [79, 80] claim that the net negative trapped charge in the oxide exceeds a critical threshold value at breakdown. In Vincent et al. [80], the authors related their observation to charge trapping kinetics, without assuming any additional trap creation. This is in contradiction with most publications on oxide degradation, which clearly demonstrate an increase of the electron trap density in the oxide.

4.4.2.5 Stress-Induced Leakage Current.
Another important phenomenon that occurs during oxide degradation is the generation of a leakage current through the gate oxide. This is the stress-induced leakage current (SILC). SILC is illustrated in Figure 4.23, where the I_g-V_g curve is shown on a fresh sample and on stressed samples. The SILC is easily observed in the field range below the onset of the FN tunneling current. SILC rises continuously with injected fluence and applied oxide field.

When the SILC after a given stress is continuously monitored as a function of time, two components can be distinguished [81]. Initially, a decaying transient component is observed, leading to a steady-state SILC after some time. Both components depend on the oxide thickness. Thick oxides have a large transient component and low steady-state component, while very thin oxides have a very small transient component and a large steady-state component.

In nonvolatile memories, SILC is the main limitation for tunnel oxide scaling. When the electrons stored on the floating gate are leaking off, the threshold voltage of the cell shifts and the stored information will be lost after some time. In some

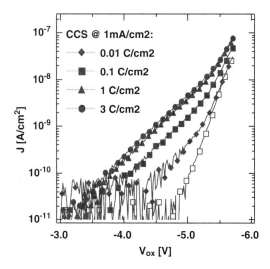

Figure 4.23. Gate current density vs. the applied oxide voltage for a fresh device (open symbols) and after a high field stress has been applied. A gradually increasing leakage current appears.

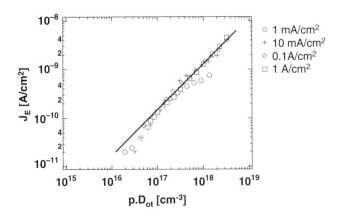

Figure 4.24. Steady-state SILC measured at a fixed field E (here 5 MV/cm) plotted as a function of the generated density of neutral traps. Independent of the stress condition, a one-to-one relation is observed [83].

cases, the transient component is the limiting factor for reliability [82]. This is the case when thick oxides are used and when the tolerated threshold voltage shift is small. For thinner oxides and with a more relaxed reliability criterion, the steady-state component will determine the reliability of the cell [83].

Stress-induced leakage current is directly related to the density of electron traps in the oxide as is illustrated in Figure 4.24. To construct this figure, the FN stress was periodically interrupted to measure the leakage current density J_E at a fixed low oxide field E and, on the same device, the density of neutral oxide traps was determined with the technique explained in Degraeve et al. [75] (see also Section 4.4.2.3). Plotted versus one another, a stress-independent one-to-one relation is revealed between the steady-state SILC increase and the generation of neutral traps [83].

Steady-state SILC can be explained as a trap-assisted tunneling of electrons from cathode to anode. Directly after stress, the traps near the interfaces are charged causing the transient component. In thick oxides, the probability that an electron has sufficient traps that can be used as "stepping stones" for tunneling is very small, resulting in a small steady-state component. In thinner oxides, fewer traps are necessary for the electrons to move from cathode to anode and, therefore, the steady-state component will dominate. The physical modeling of the trap-assisted tunneling process causing SILC has been the topic of several publications [84, 85].

Since the density of neutral traps reaches a critical value at breakdown, it is obvious from the relation in Figure 4.24 that also the SILC reaches a critical value at breakdown [86] and can therefore be used as a degradation monitor and predictor [76, 87].

4.4.2.6 Trap Generation Mechanism. In Sections 4.4.2.1 to 4.4.2.5, the degradation of thin oxide layers during tunneling stress or hot carrier stress was described. The main phenomenon was trap creation and all other parameters were related to this. In this section, several literature models describing the electron trap generation are compared. In summary, three trap generation models are discussed: the anode hole trap generation model, the hydrogen release model, and the electric field energy model.

The anode hole injection mechanism has already been explained in Section 4.4.2.1 (see Fig. 4.19). It has been proposed that the generation of traps in the oxide is related to the holes tunneling back to the cathode. The physics of the trap creation process are still speculative. Indeed, there have been several studies demonstrating that the interaction of electrons and holes in an oxide results in trap creation [88, 89]. However, the precise role of electrons and holes in the trap creation process and the details of the microscopic mechanism of the trap creation are still uncertain. The most important difficulty in studying this effect is the inability of many techniques to separately control the hole and the electron injection.

In DiMaria et al. [62], it is demonstrated that below 7.8 V the dominant source of the hole current is photoexcitation of valence band electrons in the cathode by light generated in the anode. If there exists a voltage limit to anode hole injection, it could be questioned whether this model will correctly predict the oxide degradation at low voltage.

When the density of oxide traps is plotted versus the hole fluence, a unique correlation is found, as illustrated in Figure 4.25. According to the anode hole trap generation model, this relation is interpreted as a *causal* relation, i.e., the holes are necessary to create the traps. This line of thought is outlined in Figure 4.26(*a*). However, other explanations are possible as well. As is outlined in Figure 4.26(*b*), the energy release of the incoming electrons at the anode can, besides hole creation, also activate some other mechanism that is responsible for neutral electron trap generation. This line of thought suggests that the hole fluence and the trap creation have a "common origin", that is, the energy released by the electrons at the anode.

A serious candidate for the other mechanism of Figure 4.26(*b*) is *hydrogen release at the anode*. Most older publications [71, 90] on this phenomenon deal with interface trap creation at the cathode, but the model's basic ideas can easily be extended to bulk oxide traps. In the hydrogen release model, the electrons tunnel through the oxide potential barrier and reach the anode with sufficient energy to release hydrogen from the anode–oxide interface. This hydrogen is always present

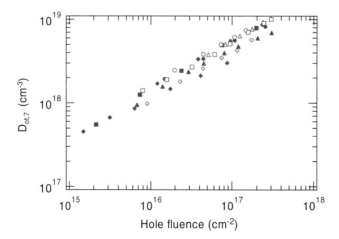

Figure 4.25. Generated neutral electron traps as a function of the hole fluence for different oxide thicknesses (range 7 to 14 nm), different stress types [constant voltage stress (CVS) and constant current stress (CCS)] and different oxide fields (9.5 to 12 MV/cm). A unique relation is observed [75].

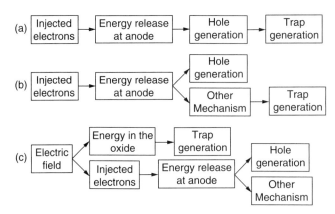

Figure 4.26. Outline of the three lines of thought on neutral electron trap generation that can be found in the literature.

in sufficient amounts because of interface annealing applied to reduce the initial interface trap density. The released hydrogen diffuses through the oxide and can generate electron traps.

Again, as for the anode hole injection model, the precise physical details of the microscopic trap generation mechanism remain speculative. An interesting argument in favor of the hydrogen release model is provided by the experiments performed in Cartier and Stathis [91]: An oxide (without polysilicon gate) on a silicon substrate is exposed to a remote hydrogen plasma, and it is observed that the trap creation at the oxide–silicon interface evolves identically to the interface trap creation induced by a hot-electron stress.

Some authors have even suggested a third possible route to explain the generation of electron traps, as is illustrated in Figure 4.26(c). According to this approach, the electric field itself induces sufficient energy directly into the oxide to cause electron trap creation (= *electric field energy model*). With this interpretation, all processes that are related to the energy release of the injected electrons at the anode are independent of the trap generation mechanism [92–97]. It has been proposed that the oxide can be treated as a collection of dipoles, and the free energy of activation for the breakdown reaction can be expressed as a series expansion of E_{ox}, keeping only the first term [95, 96].

4.4.3 Oxide Breakdown

When the oxide degradation continues, the oxide abruptly looses its insulating properties, which is commonly defined as "breakdown." There is an almost general consensus that breakdown is triggered when the accumulated damage (= generated traps) exceeds a critical value [66, 75, 76, 86, 92, 98–104].

Based on this idea, in the beginning of the 1990s, Suñé et al. [100] presented a two-dimensional "weakest link" breakdown model. In work that is more recent a new three-dimensional weakest link model has been proposed that can accurately describe the intrinsic breakdown distribution. This model is based on the principles of percolation theory. The use of the percolation concept for modeling of oxide breakdown has been suggested [105] and has been thoroughly elaborated in the literature [77, 103, 104]. The model is schematically illustrated in Figure 4.27.

PHYSICS OF DEGRADATION AND DISTURB MECHANISMS

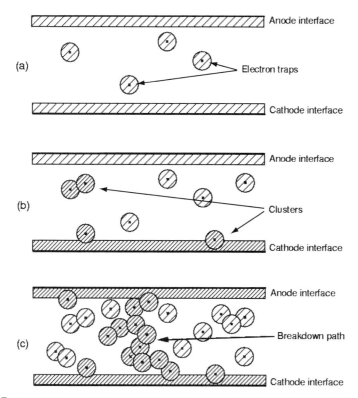

Figure 4.27. Percolation model for oxide breakdown explained step by step. As the density of neutral electron traps increases, conductive clusters of traps are formed ultimately leading to the creation of a conductive breakdown path from anode to cathode [103].

It is assumed that electron traps are generated inside the oxide at random positions in space. Around these traps, a sphere is defined with a fixed radius r, which is the only parameter of the model [Fig. 4.27(a)]. If the spheres of two neighboring traps overlap, conduction between these traps becomes possible by definition. Also, the two interfaces are modeled as an infinite set of traps [Fig. 4.27(b)].

The mechanism of trap generation continues until a conducting path is created from one interface to the other, which is the defining breakdown condition [Fig. 4.27(c)]. In a computer simulation, the total critical electron trap density, $D_{ot,crit}$, can now be calculated. It is found that the simulated $D_{ot,crit}$ distribution can be fitted with a Weibull expression. This distribution can be easily associated with the experimentally obtained time-to-breakdown distribution [103].

The percolation model for breakdown is able to explain quantitatively two important experimental observations: (1) as the oxide thickness decreases, the density of oxide traps needed to trigger breakdown decreases [77, 103, 104], and (2) as the oxide thickness decreases, the Weibull slope of the breakdown distribution decreases, that is, a larger spread on the t_{BD} values is observed [77, 103, 104, 106]. Both effects are illustrated in Figures 4.28 and 4.29, respectively.

An important consequence of the decreasing Weibull slope for thinner oxides is the increased area dependence of the time to breakdown or charge to breakdown [77, 106]. Indeed, based on the random nature of the breakdown position, it has

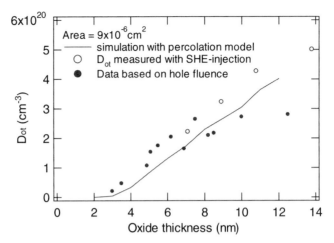

Figure 4.28. Critical density of electron traps in the oxide needed to trigger breakdown as a function of the oxide thickness. Both direct measurements using the trap filling technique (Section 4.4.2.3) as well as indirect measurements based on critical hole fluence and applying the relation shown in Figure 4.25 are presented [103]. The percolation model predicts the observed oxide thickness dependence very well.

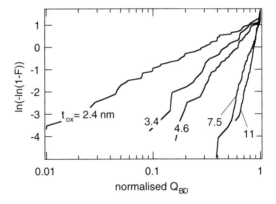

Figure 4.29. Normalized Q_{BD} distributions for different oxide thicknesses. The decrease of the Weibull slope with oxide thickness is experimentally observed [103].

been shown that for the scale factors η_1 and η_2 of two Weibull distributions (either t_{BD} or Q_{BD} distributions) of capacitors with identical oxide thickness, but area A_1 and A_2, respectively, the following relationship holds [107, 108]:

$$\frac{\eta_1}{\eta_2} = \left(\frac{A_2}{A_1}\right)^{(1/\beta)} \quad (4.30)$$

As β decreases, the area dependence becomes stronger. It can be concluded that for thick oxides ($t_{ox} > 10$ nm), the intrinsic Q_{BD} value is almost in the range of capacitor areas commonly available for experimental purposes, but for thin oxides, the Q_{BD} can no longer be considered area independent. This has important implications when the Q_{BD} value is used as the figure of merit for oxide quality. It is meaningless to specify a Q_{BD} value without specifying the area, and it is mandatory to specify the area when Q_{BD} values of different oxide thicknesses are compared.

4.5 CONCLUSION

In this chapter, an overview has been given of the relevant physical mechanisms for Flash memory operation and reliability. After discussing the basics of floating-gate devices and the corresponding characteristics, a general mathematical treatment of this class of devices has been presented (Section 4.2). The physics of programming and erase mechanisms have been reviewed by covering the most commonly used transport mechanisms in SiO_2. These have been treated from a general physics point of view, while the particular implementation and the corresponding trade-offs are discussed in the more specific technology chapters.

As most important mechanisms, Fowler–Nordheim tunneling, polyoxide tunneling, channel hot-electron injection, and substrate hot-electron injection have been treated in Sections 4.3.1 to 4.3.4. It is concluded that Fowler–Nordheim tunneling requires high voltages, leading to extensive oxide degradation, which is particularly detrimental with respect to stress-induced leakage current (SILC). Especially when used as a programming mechanism, very high oxide fields are required. Moreover, due to the very steep tunneling I-V characteristic, the corresponding programming times are rather long. Polyoxide conduction offers a viable and SILC-free alternative, especially as an erase mechanism, however, at the expense of a larger trapping rate.

Channel hot-electron injection is by far the most popular programming mechanism due to the lower voltages as compared to tunneling and the faster programming times. However, the most crucial drawback is the high power consumption that is a direct consequence of the low injection efficiency. Substrate hot-electron injection has been proposed as a low-power alternative, however, requiring a more complicated cell technology, which has limited its application in Flash memories.

More recently developed mechanisms have been discussed in Sections 4.3.5 and 4.3.6. Source-side injection offers a very attractive alternative for channel hot-electron injection as a high-efficiency programming mechanism offering low-voltage, low-power capabilities. Therefore, cells using the source-side injection mechanism for programming can more easily be scaled down to deep-submicron dimensions. An alternative programming mechanism also providing higher injection efficiency is the channel hot-electron injection mechanism that makes use of secondary impact ionization. This also results in lower programming voltages and a lower programming power consumption, however, requiring a triple-well technology.

In Section 4.4, the basic reliability physics have been presented, focusing on the oxide-related physics that are of particular importance for Flash technology. After a short description of the band-to-band tunneling mechanism, which is of particular importance for a large class of Flash devices, the physics of oxide degradation have been reviewed. Hole and electron trapping, interface state, and oxide trap generation, as well as stress-induced leakage current and oxide breakdown have been discussed. Especially SILC has been recognized as the main limitation for tunnel oxide scaling.

REFERENCES

1. J. Van Houdt, "Reliability Issues of Flash Memory," short course of the 14th IEEE Nonvolatile Semiconductor Memory Workshop, Monterey, CA, 1995.
2. D. Wellekens, J. Van Houdt, L. Faraone, G. Groeseneken, and H. E. Maes, "Write/Erase Degradation in Source Side Injection Flash EEPROMs: Characterization Techniques and Wearout Mechanisms," *IEEE Trans. Electron Devices*, Vol. 42, No. 11, p. 1992, 1995.
3. N. Mielke, A. Fazio, and H.-C. Liou, "Reliability Comparison of FLOTOX and Textured Polysilicon EEPROMs," *Proc. IRPS*, p. 85, 1987.
4. R. Bez, D. Cantarelli, L. Moioli, G. Ortolani, G. Servalli, C. Villa, and M. Dallabora, "A New Erasing Method for A Single-Voltage Long-Endurance Flash Memory," *IEEE Electron Device Lett.*, Vol. 19, No. 2, p. 37, 1998.
5. R. Shiner, J. Caywood, and B. Euzent, "Data Retention in EPROMs," *Proc. IRPS*, p. 238, 1980.
6. D. A. Baglee, "Characteristics and Reliability of 100 Å Oxide," *Proc. IRPS*, p. 152, 1984.
7. K. Wu, C.-S. Pan, J. J. Shaw, P. Freiberger, and G. Sery, "A Model for EPROM Intrinsic Charge Loss through Oxide-Nitride-Oxide (ONO) Interpoly Dielectric," *Proc. IRPS*, p. 145, 1990.
8. G. Verma, N. Mielke, "Reliability Performance of ETOX Based Flash Memories," *Proc. IRPS*, p. 158, 1988.
9. S. Mori, Y. Kaneko, N. Arai, Y. Ohshima, H. Araki, K. Narita, E. Sakagami, and K. Yoshikawa, "Reliability Study of Thin Interpoly Dielectrics for Nonvolatile Memory Application," *Proc. IRPS*, p. 132, 1990.
10. N. Mielke, "New EPROM Data Loss Mechanism," *Proc. IRPS*, p. 106, 1983.
11. G. Crisenza, G. Ghidini, and M. Tosi, "Floating Gate Memories Reliability," *Proc. ESREF*, p. 505, 1991.
12. M. Lanzoni, R. Menozzi, C. Riva, P. Olivo, and B. Riccò, "Evaluation of E^2PROM Data Retention by Field Acceleration," *Proc. ESREF*, p. 329, 1990.
13. J. F. Van Houdt, D. Wellekens, G. Groeseneken, and H. E. Maes, "Investigation of the Soft-Write Mechanism in Source-Side Injection Flash EEPROM Devices," *IEEE Electron Device Lett.*, Vol. 16, No. 5, p. 181, 1995.
14. S. M. Sze, "Current Transport and Maximum Dielectric Strength of Silicon Nitride Films," *J. Appl. Phys.*, Vol. 38, p. 2951, 1967.
15. W. Johnson, G. Perlegos, A. Renninger, G. Kuhn, and T. Ranganath, "A 16 Kb Electrically Erasable Nonvolatile Memory," *ISSCC Tech. Dig.*, p. 152, 1980.
16. J. Yeargain and K. Kuo, "A High Density Floating Gate EEPROM Cell," *IEDM Tech. Dig.*, p. 24, 1981.
17. R. Klein, W. Owen, R. Simko, and W. Tchon, "5-V-Only, Nonvolatile RAM Owes It All to Polysilicon," *Electronics*, p. 111, Oct. 11, 1979.
18. G. Landers, "5-V-Only EEPROM Mimics Static RAM Timing," *Electronics*, p. 127, June 30, 1980.
19. B. Rössler and R. Müller, "Electrically Erasable and Reprogrammable Read-Only Memory Using the n-Channel SIMOS One-Transistor Cell," *IEEE Trans. Electron Devices.*, Vol. ED-24, p. 806, 1977.
20. D. Guterman, I. Rimawi, T. Chiu, R. Halvorson, and D. McElroy, "An Electrically Alterable Nonvolatile Memory Cell Using a Floating Gate Structure," *IEEE Trans. Electron Devices*, Vol. ED-24, p. 806, 1977.
21. T. H. Ning, "Hot-Electron Emission from Silicon into Silicon Dioxide," *Solid-State Electronics*, Vol. 21, p. 273, 1978.

22. N. Tsuji, N. Ajika, K. Yuzuriha, Y. Kunori, M. Hatanaka, and H. Miyoshi, "New Erase Scheme for DINOR Flash Memory Enhancing Erase/Write Cycling Endurance Characteristics," *IEDM Tech. Dig.*, p. 53, 1994.
23. M. Kamiya, Y. Kojima, Y. Kato, K. Tanaka, and Y. Hayashi, "EPROM Cell with High Gate Injection Efficiency," *IEDM Tech. Dig.*, p. 741, 1981.
24. A. Wu, T. Chan, P. Ko, and C. Hu, "A Novel High-Speed, 5-V Programming EPROM Structure with Source-Side Injection," *IEDM Tech. Dig.*, p. 584, 1986.
25. J. Van Houdt, P. Heremans, L. Deferm, G. Groeseneken, and H. E. Maes, "Analysis of the Enhanced Hot-Electron Injection in Split-Gate Transistors Useful for EEPROM Applications," *IEEE Trans. Electron Devices*, Vol. ED-39, p. 1150, 1992.
26. J. D. Bude, "Gate Current by Impact Ionization Feedback in Sub-Micron MOSFET Technologies," *Symp. VLSI Technol. Dig. Tech. Papers.*, p. 101, 1995.
27. M. Lenzlinger and E.H. Snow, "Fowler–Nordheim Tunneling into Thermally Grown SiO_2," *J. Appl. Phys.*, Vol. 40, p. 278, 1969.
28. R. C. Hughes, "High Field Electronic Properties of SiO_2," *Solid State Electronics.*, Vol. 21, p. 251, 1978.
29. D. J. DiMaria and D.R. Kerr, "Interface Effects and High Conductivity in Oxides Grown from Polycrystalline Silicon," *Appl. Phys. Lett.*, Vol. 27, p. 505, 1975.
30. R. M. anderson and D. R. Kerr, "Evidence for Surface Asperity Mechanism of Conductivity in Oxide Grown on Polycrystalline Silicon," *J. Appl. Phys.*, Vol. 48, p. 4834, 1977.
31. H. R. Huff, R. D. Halvorson, T. L. Chiu, and D. Guterman, "Experimental Observations on Conduction through Polysilicon Oxide," *J. Electrochem. Soc.*, Vol. 127, p. 2482, 1980.
32. P. A. Heimann, S. P. Murarka, and T. T. Sheng, "Electrical Conduction and Breakdown in Oxides of Polycrystalline Silicon and Their Correlation with Interface Texture," *J. Appl. Phys.*, Vol. 53, p. 6240, 1982.
33. G. Groeseneken and H. E. Maes, "A Quantitative Model for the Conduction in Oxides Thermally Grown from Polycrystalline Silicon," *IEEE Trans. Electron Devices*, Vol. ED-33, p. 1028, 1986.
34. R. D. Jolly, H. R. Grinolds, and R. Groth, "A Model for Conduction in Floating Gate EEPROMs," *IEEE Trans. Electron Devices.*, Vol. ED-31, p. 767, 1984.
35. D. Wellekens, J. Van Houdt, J. De Blauwe, L. Haspeslagh, L. Deferm, and H. E. Maes, "A Low Voltage, High Performance 0.35 µm Embedded Flash EEPROM Cell Technology," paper presented at the 16th IEEE Nonvolatile Semiconductor Memory Workshop, Monterey, CA, pp. 106–108, 1998.
36. D. J. Lee, R. A. Cernea, M. Mofidi, S. Mehrotra, E. Y. Chang, W. Y. Chien, L. Goh, J. H. Yuan, A. Mihnea, G. Samachisa, Y. Fong, D. C. Guterman, and R. D. Norman, "An 18Mb serial Flash EEPROM for Solid-State Disk Applications," *Symp. VLSI Circ. Dig. Tech. Papers*, p. 59, 1994.
37. S. Kianian, A. Levi, D. Lee, and Y.-W. Hu, "A Novel 3 Volts-Only, Small Sector Erase, High Density Flash E^2PROM," *Symp. VLSI Technol. Dig. Tech. Papers*, p. 71, 1994.
38. P. E. Cottrell, R. R. Troutman, and T. H. Ning, "Hot Electron Emission in n-Channel IGFET's," *IEEE J. Solid-State Circuits*, Vol. SC-14, p. 442, 1979.
39. B. Eitan and D. Frohman-Bentchkowsky, "Hot Electron Injection into the Oxide in n-Channel MOS-Devices," *IEEE Trans. Electron Devices.*, Vol. ED-28, p. 328, 1981.
40. C. Hu, "Lucky Electron Model of Hot Electron Emission," *IEDM Tech. Dig.*, p. 22, 1979.
41. P. Ko, R. Müller, and C. Hu, "A Unified Model for Hot Electron Currents in MOSFET's," *IEDM Tech. Dig.*, p. 600, 1981.

42. E. Takeda, H. Kume, T. Toyabe, and S. Asai, "Submicrometer MOSFET Structure for Minimizing Hot Carrier Generation," *IEEE Trans. Electron Devices*, Vol. ED-29, p. 611, 1982.
43. S. Tanaka and M. Ishikawa, "One-Dimensional Writing Model of n-Channel Floating Gate Ionization-Injection MOS (FIMOS)," *IEEE Trans. Electron Devices*, Vol. ED-28, p. 1190, 1981.
44. K. R. Hoffmann, C. Werner, W. Weber, and G. Dorda, "Hot-Electron and Hole Emission Effects in Short n-Channel MOSFET's," *IEEE Trans. Electron Devices*, Vol. ED-32, p. 691, 1985.
45. M. Melanotte, R. Bez, and G. Crisenza, "Nonvolatile Memories—Status and Emerging Trends," *Proc. ESSDERC, reprinted from Microelectron. Eng.* Vol. 15, p. 603, 1991.
46. J. Van Houdt, "Physics and Characteristics of the High Injection MOS (HIMOS) Transistor: Novel Fast-Programmable Flash Memory Device," Ph.D. Dissertation, Catholic University of Leuven, Belgium, 1994.
47. T. Y. Chan, P. K. Ko, and C. Hu, "Dependence of Channel Electric Field on Device Scaling," *IEEE Electron Device Lett.*, Vol. EDL-6, p. 551, 1985.
48. C. Hu, "Hot Electron Effects in MOSFET's," *IEDM Tech. Dig.*, p. 176, 1983.
49. J. F. Verwey, "Nonavalanche Injection of Hot Carriers into SiO_2," *J. Appl. Phys.*, Vol. 44, p. 2681, 1973.
50. G. Van den Bosch, G. Groeseneken, and H. E. Maes, "Direct and Post-Injection Oxide And Interface Trap Generation Resulting from Low-Temperature Hot-Electron Injection," *J. Appl. Phys.*, Vol. 74, No. 9, pp. 5582–5586, 1993.
51. A. v. Schwerin, M. M. Heyns, and W. Weber, "Investigation on the Oxide Field Dependence of Hole Trapping and Interface State Generation in SiO_2 Layers Using Homogeneous Nonavalanche Injection of Holes," *J. Appl. Phys.*, Vol. 67, pp. 7595–7601, 1990.
52. G. Van den Bosch, G. Groeseneken, and H.E. Maes, "Critical Analysis of the Substrate Hot-Hole Injection Technique," *Solid-State Electron.*, Vol. 37, pp. 393–399, 1994.
53. J. Van Houdt, L. Haspeslagh, D. Wellekens, L. Deferm, G. Groeseneken, and H. E. Maes, "HIMOS—A High Efficiency Flash EEPROM Cell for Embedded Memory Applications," *IEEE Trans. Electron Devices*, Vol. ED-40, p. 2255, 1993.
54. J. Van Houdt, D. Wellekens, L. Faraone, L. Haspeslagh, L. Deferm, G. Groeseneken, and H. E. Maes, "A 5 V-Compatible Flash EEPROM Cell with Microsecond Programming Time for Embedded Memory Applications," *IEEE Trans. CPMT*, Vol. 17, No. 3, Part A, pp. 380–389, 1994.
55. T. C. Ong, A. Fazio, N. Mielke, S. Pan, N. Righos, G. Atwood, and S. Lai, "Erratic erase in ETOX™ Flash Memory Array," *Symp. VLSI Technol. Dig. Tech. Papers*, p. 83, 1993.
56. I. C. Chen, S. Holland, K. K. Young, C. Chang, and C. Hu, "Substrate Hole Current and Oxide Breakdown," *Appl. Phys. Lett.*, Vol. 49, No. 11, pp. 669–671, 1986.
57. Z. A. Weinberg and M. V. Fischetti, "SiO_2-Induced Substrate Current and Its Relation to Positive Charge in Field-Effect Transitors," *J. Appl. Phys.*, Vol. 59, No. 3, pp. 824–832, 1986.
58. R. Degraeve, J. L. Ogier, R. Bellens, Ph. Roussel, G. Groeseneken, and H. E. Maes, "A New Model for the Field Dependence of Intrinsic and Extrinsic Time-Dependent Dielectric Breakdown," *IEEE Trans. Electron Devices.*, Vol. 45, No. 2, pp. 472–481, 1998.
59. D. J. DiMaria, "Hole Trapping, Substrate Currents, and Breakdown in Thin Silicon Dioxide Films," *IEEE Electron Device Lett.*, Vol. 16, No. 5, pp. 184–186, 1995.
60. K. F. Schuegraf and C. Hu, "Metal-Oxide-Semiconductor Field-Effect-Transistor Substrate Current During Fowler–Nordheim Tunneling Stress and Silicon Dioxide Reliability," *J. Appl. Phys.*, Vol. 76, No. 6, pp. 3695–3700, 1994.

REFERENCES

61. M. V. Fischetti, "Model for the Generation of Positive Charge at the Si–SiO$_2$ Interface Based on Hot-Hole Injection from the Anode," *Phys. Rev. B*, Vol. 31, No. 4, pp. 2099–2113, 1985.
62. D. J. DiMaria, E. Cartier, and D. A. Buchanan, "Anode Hole Injection and Trapping in Silicon Dioxide," *J. Appl. Phys.*, Vol. 80, No. 1, pp. 304–317, 1996.
63. C.-L. Chiang and N. Khurana, "Imaging and Detection of Current Conduction in Dielectric Films by Emission Microscopy," *IEDM Tech. Dig.*, pp. 672–675, 1986.
64. E. Cartier, J. S. Tsang, M. V. Fischetti, and D. A. Buchanan, "Light Emission During Direct and Fowler–Nordheim Tunneling in Ultra Thin MOS Tunnel Junctions," *Microelectron. Eng.*, Vol. 36, Nos. 1–4, pp. 103–106, 1997.
65. K. F. Schuegraf and C. Hu, "Hole Injection SiO$_2$ Breakdown Model for Very Low Voltage Lifetime Extrapolation," *IEEE Trans. Electron Devices*, Vol. 41, No. 5, pp. 761–767, 1994.
66. H. Satake and A. Toriumi, "Substrate Hole Current Generation and Oxide Breakdown in Si MOSFETs under Fowler–Nordheim Electron Tunneling Injection," *IEDM Tech. Dig.*, pp. 337–340, 1993.
67. E. H. Nicollian and J. R. Brews, *MOS Physics and Technology*, Wiley, New York, 1982.
68. G. Groeseneken, H. E. Maes, N. Beltràn, and R. F. De Keersmaecker, "A Reliable Approach to Charge-Pumping Measurements in MOS Transistors," *IEEE Trans. Electron Devices*, Vol. 31, No. 1, pp. 42–53, 1984.
69. G. Groeseneken, I. De Wolf, R. Bellens, and H. E. Maes, "Observation of Single Interface Traps in Submicron MOSFETs by Charge Pumping," *IEEE Trans. Electron Devices*, Vol. 43, p. 940, 1996.
70. P. Heremans, J. Witters, G. Groeseneken, and H. E. Maes, "Analysis of the Charge Pumping Technique and Its Application for the Evaluation of MOSFET Degradation," *IEEE Trans. Electron Devices*, Vol. 36, p. 1318, 1989.
71. D. J. DiMaria and J. W. Stasiak, "Trap Creation in Silicon Dioxide Produced by Hot Electrons," *J. Appl. Phys.*, Vol. 65, No. 6, pp. 2342–2356, 1989.
72. G. Van den bosch, G. Groeseneken, H. E. Maes, R. B. Klein, and N. S. Saks, "Oxide and Interface Degradation Resulting from Substrate Hot-Hole Injection In Metal-Oxide-Semiconductor Field-Effect Transistors at 295 and 77 K," *J. Appl. Phys.*, Vol. 75, No. 4, pp. 2073–2080, 1994.
73. Q. D. M. Khosru, N. Yasuda, K. Taniguchi, and C. Hamaguchi, "Oxide Thickness Dependence of Interface Trap Generation in a Metal-Oxide-Semiconductor Structure during Substrate Hot-Hole Injection," *Appl. Phys. Lett.*, Vol. 63, No. 18, pp. 2537–2539, 1993.
74. R. Degraeve, G. Groeseneken, I. De Wolf, and H. E. Maes, "Oxide and Interface Degradation and Breakdown under Medium and High Field Injection Conditions: A correlation Study," *Microelectron. Eng. (Proc. INFOS)*, Vol. 28, Nos. 1–4, pp. 313–316, 1995.
75. R. Degraeve, G. Groeseneken, R. Bellens, J. L. Ogier, M. Depas, Ph. Roussel, and H. E. Maes, "New Insights in the Relation between Electron Trap Generation and the Statistical Properties of Oxide Breakdown," *IEEE Trans. Electron Devices*, Vol. 45, No. 4, pp. 904–911, 1998.
76. T. Nigam, R. Degraeve, G. Groeseneken, M. M. Heyns, and H. E. Maes, "A Fast and Simple Methodology for Lifetime Prediction of Ultra-Thin Oxides," *Proc. IRPS*, pp. 381–388, 1999.
77. T. Nishida and S. E. Thompson, "Oxide Field and Temperature Dependences of Gate Oxide Degradation by Substrate Hot Electron Injection," *Proc. IRPS*, pp. 310–315, 1991.
78. I. C. Chen, S. Holland, and C. Hu, "A Quantitative Physical Model for Time-Dependent Breakdown in SiO$_2$," *Proc. IRPS*, pp. 24–31, 1985.

79. C. Monsérié, C. Papadas, G. Ghibaudo, C. Gounelle, P. Mortini, and G. Pananakakis, "Correlation between Negative Bulk Oxide Charge and Breakdown, Modeling and New Criteria for Dielectric Quality Evaluation," *Proc. IRPS*, pp. 280–284, 1993.
80. E. Vincent, C. Papadas, and G. Ghibaudo, "Electric Field Dependence of Charge Build-Up Mechanisms and Breakdown Phenomena in Thin Oxides during Fowler–Nordheim Injection," *Proc. ESSDERC*, pp. 767–770, 1996.
81. R. Moazzami and C. Hu, "Stress-Induced Current in Thin Silicon Dioxide Films," *IEDM Tech. Dig.*, pp. 139–142, 1992.
82. M. Kato, N. Myamoto, H. Hume, A. Satoh, T. Adachi, M. Ushiyama, and K. Kimura, "Read-Disturb Degradation Mechanism Due to Electron Trapping in Tunnel Oxide for Low-Voltage Flash Memories," *IEDM Tech. Dig.*, pp. 45–48, 1994.
83. J. De Blauwe, J. Van Houdt, D. Wellekens, R. Degraeve, Ph. Roussel, L. Haspeslagh, L. Deferm, G. Groeseneken, and H. E. Maes, "A New Quantitative Model to Predict SILC-Related Disturb Characteristics in Flash E^{2}PROM Devices," *IEDM Tech. Dig.*, pp. 343–346, 1996.
84. B. Ricco, G. Gozzi, and M. Lanzoni, "Modeling and Simulation of Stress-Induced Leakage Current in Ultrathin SiO$_2$ Films," *IEEE Trans. Electron Devices*, Vol. 45, p. 1554, 1998.
85. J. Wu, L. F. Register, and E. Rosenbaum, "Trap-Assisted Tunneling Current Through Ultra-Thin Oxide," *Proc. IRPS*, pp. 389–395, 1999.
86. M. Depas and M. M. Heyns, "Relation between Trap Creation and Breakdown during Tunneling Current Stressing of Sub-3 nm Gate Oxide," *Microelectron. Eng.*, Vol. 36, Nos. 1–4, pp. 21–24, 1997.
87. K. Okada, H. Kubo, A. Ishinaga, and K. Yoneda, "A New Prediction Method for Oxide Lifetime and Its Application to Study Dielectric Breakdown Mechanism," *Symp. VLSI Technol. Dig. Tech. Papers*, pp. 158–159, 1998.
88. H. Uchida and T. Ajika, "Electron Trap Center Generation Due to Hole Trapping in SiO$_2$ under Fowler–Nordheim Tunneling Conditions," *Appl. Phys. Lett.*, Vol. 51, No. 87, pp. 433–435, 1987.
89. H. Satake, S. Takagi, and A. Toriumi, "Evidence of Electron-Hole Cooperation in SiO$_2$ Dielectric Breakdown," *Proc. IRPS*, pp. 156–163, 1997.
90. D. J. DiMaria, E. Cartier, and D. Arnold, "Impact Ionization, Trap Creation, Degradation, and Breakdown in Silicon Dioxide Films on Silicon," *J. Appl. Phys.*, Vol. 73, No. 7, pp. 3367–3384, 1993.
91. E. Cartier and J. H. Stathis, "Atomic Hydrogen-Induced Degradation of the Si/SiO$_2$ Structure," *Microelectron. Eng.*, Vol. 28, Nos. 1–4, pp. 3–10, 1995.
92. D. J. Dumin, J. R. Maddux, R. Scott, and R. S. Subramoniam, "A Model Relating Wearout to Breakdown in Thin Oxides," *IEEE Trans. Electron Devices*, Vol. 41, No. 9, pp. 1570–1580, 1994.
93. R. S. Scott, N. A. Dumin, T. W. Hughes, D. J. Dumin, and B. T. Moore, "Properties of High Voltage Stress Generated Traps in Thin Silicon Oxides," *Proc. IRPS*, pp. 131–141, 1995.
94. J. W. McPherson and D. A. Baglee, "Acceleration Factors for Thin Gate Oxide Stressing," *Proc. IRPS*, pp. 1–5, 1985.
95. B. Schlund, C. Messick, J. S. Suehle, and P. Chaparala, "A New Physics-Based Model for Time-Dependent-Dielectric-Breakdown," *Proc. IRPS*, pp. 84–92, 1996.
96. J. W. McPherson and H. C. Mogul, "Disturbed Bonding States in SiO$_2$ Thin Films and Their Impact on Time-Dependent Dielectric Breakdown," *Proc. IRPS*, pp. 47–56, 1998.
97. M. Kimura, "Field and Temperature Acceleration Model for Time-Dependent Dielectric Breakdown," *IEEE Trans. Electron Devices*, Vol. 46, No. 1, pp. 220–229, 1999.
98. E. Avni and J. Shappir, "A Model for Silicon-Oxide Breakdown under High Field and Current Stress," *J. Appl. Phys.*, Vol. 64, No. 2, pp. 743–748, 1988.

REFERENCES

99. P. P. Apte and K. C. Saraswat, "Modeling Ultrathin Dielectric Breakdown on Correlation of Charge Trap-Generation to Charge-to-Breakdown," *Proc. IRPS*, pp. 136–142, 1994.
100. J. Suñé, I. Placencia, N. Barniol, E. Farrés, F. Martín, and X. Aymerich, "On the Breakdown Statistics of Very Thin SiO_2 Films," *Thin Solid Films*, Vol. 185, pp. 347–362, 1990.
101. N. Shiono and M. Itsumi, "A Lifetime Projection Method Using Series Model and Acceleration Factors for TDDB Failures of Thin Gate Oxides," *Proc. IRPS*, pp. 1–6, 1993.
102. D. J. DiMaria and J. H. Stathis, "Explanation for the Oxide Thickness Dependence of Breakdown Characteristics of Metal-Oxide-Semiconductor Structures," *Appl. Phys. Lett.*, Vol. 70, No. 20, pp. 2708–2710, 1997.
103. R. Degraeve, G. Groeseneken, R. Bellens, M. Depas, and H. E. Maes, "A Consistent Model for the Thickness Dependence of Intrinsic Breakdown in Ultra-Thin Oxides," *IEDM Tech. Dig.*, pp. 863–866, 1995.
104. J. H. Stathis, "Quantitative Model of the Thickness Dependence of Breakdown in Ultra-Thin Oxides," *Microelectron. Eng.*, Vol. 36, Nos. 1–4, pp. 325–328, 1997.
105. H. Z. Massoud and R. Deaton, "Percolation Model for the Extreme-Value Statistics of Dielectric Breakdown in Rapid-Thermal Oxides," *extended abstracts of the ECS Spring Meeting*, pp. 287–288, 1994.
106. G. M. Paulzen, "Qbd Dependencies of Ultrathin Gate Oxides on Large Area Capacitors," *Microelectron. Eng.*, Vol. 36, Nos. 1–4, pp. 321–324, 1997.
107. T. Nigam, R. Degraeve, G. Groeseneken, M. M. Heyns, and H. E. Maes, "Constant Current Charge-to-Breakdown: Still a Valid Tool to Study the Reliability of MOS Structures?" *Proc. IRPS*, pp. 62–69, 1998.
108. D. R. Wolters and J. F. Verwey, Chap. 6 in G. Barbottin and A. Vapaille, (Eds.), *Instabilities In Silicon Devices*, Elsevier Science, Amsterdam, 1986, pp. 332–335.

5

NOR FLASH STACKED AND SPLIT-GATE MEMORY TECHNOLOGY

Stephen N. Keeney, Manzur Gill, and David Sweetman

5.1 INTRODUCTION

Flash memories were first conceived in the early 1980s as a trade-off between the low cost of the ultraviolet electrically programmable read-only memory (UV-EPROM) and the higher functionality of the in-system byte-alterable electrically erasable PROM (EEPROM). The high-volume production for both device types used a stacked floating-gate N-metal–oxide–semiconductor (NMOS) technology, with various enhancements appropriate to the wide variety of manufacturers. The initial Flash EEPROM was a device that could be both erased and programmed electrically in-system; however, the erase size was a sector or entire device, not a byte. Programming and reading still occurred at the byte level. Thus, a reasonable compromise between cost and functionality was achieved.

Several means were available to either eliminate or mitigate the effects of overerase, an intrinsic function of a nonvolatile single floating-gate memory transistor. Whether hot-electron injection or Fowler–Nordheim tunneling was used for erase and/or programming, a single memory transistor could be overerased, which would effectively short-circuit some portion of the array. This problem could not occur with UV-EPROMs, due to the nature of how UV light removed electrons from the floating gate. The EEPROM technology eliminated the problem by using a select transistor that isolated the memory transistor from the array, which also made byte alterability an easily implemented function. However, the multiple transistors for a single memory cell for an EEPROM significantly added to the cost compared to the single transistor in the memory cell of a UV-EPROM.

Two approaches rapidly dominated the initial Flash market. The first used a split-gate approach; however, this was not successful until the advent of SuperFlash.

Nonvolatile Memory Technologies with Emphasis on Flash. Edited by J. E. Brewer and M. Gill
Copyright © 2008 the Institute of Electrical and Electronics Engineers, Inc.

The ETOX-type approach uses a software algorithm to prevent overerase. Both approaches allow the successful production of a Flash memory that is reliable and cost effective.

The initial stacked or split floating-gate Flash devices were direct derivatives of design and process experience and knowledge gained from floating-gate UV-EPROMs and EEPROMs. The ETOX memory transistor most closely evolves from the UV-EPROM memory transistor, while the SuperFlash memory transistor represents a hybrid of both design and layout of the UV-EPROM memory transistor and the EEPROM cell (memory and select transistors).

5.2 ETOX FLASH CELL TECHNOLOGY

5.2.1 Introduction

ETOX (Erase Through Oxide) is a trade mark of the Intel Corporation, and the acronym emphasizes the transition from UV erase to electrical erase.

5.2.2 Cell Structure

The ETOX cell's ancestor was the EPROM cell that used UV light to erase the device. The EPROM cell used a floating polysilicon gate sandwiched between the transistor channel and a top gate to store charge. The Flash EEPROM device has basically the same structure but the oxide between the floating gate and the channel was thinned down to become the tunnel oxide during the electrical erase operation.

Figure 5.1 shows the key elements of a Flash cell. It looks similar to a MOS transistor with the addition of a floating gate between the channel and the control gate. The floating gate is the key to nonvolatile memory storage. Since this floating polysilicon region is surrounded by an insulating dielectric, any charge that is placed on the floating gate will remain there even in the absence of electrical power for a very long period of time (typically 10 to 100 years). The presence or absence of charge will cause a threshold voltage shift in the transistor that can be detected as a 1 or 0. The Flash cells are normally built into an array of devices with the drain contacts connected by a metal bitline and the control gates forming a wordline. The source contact is normally common to all the devices in a given block. Figure 5.2 shows how cells are formed into an array.

Figure 5.2 shows how drain contacts are shared by two cells, and all the drain contacts for a given column are connected to a common metal line called the bitline.

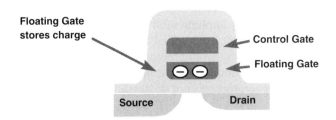

Figure 5.1. Sketch of a basic Flash EEPROM device.

ETOX FLASH CELL TECHNOLOGY

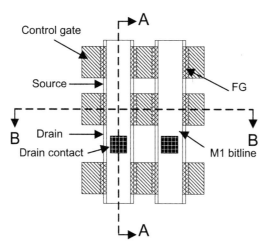

Figure 5.2. Top view of a flash EEPROM array showing cut lines for cross sections depicted later.

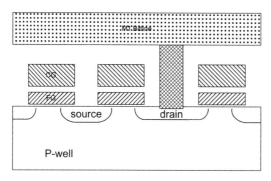

Figure 5.3. Cross section along the A–A cut line.

Typically, anywhere from 128 to 2048 cells share the same bitline. Bitlines are repeated in parallel to each other to build out the array to the required size. The diagram also shows the polysilicon wordlines that form rows of devices. Typically, 128 to 2048 cells share the same wordline, which is again repeated to build out the array. Notice that the space between the wordlines is irregular since a drain contact is required for each pair of cells, whereas the source is not contacted as frequently (since it is common to all cells in the array). It is not shown in this diagram, but the source regions are joined together for all the cells by diffusion, which is periodically strapped with metal. Viewing the cross sections illustrated by the cut lines in Figure 5.2 will help clarify the cell structure in the array. The view along the A–A cut line taken along the bitline is shown in Figure 5.3.

In this cross section we can see the source, drain, and gates of the Flash cell. The drain is shown making contact to the metal 1 bitline through a plug. The source and drain junctions may not be the same since they are often optimized for different operating conditions that will be described later. A tunnel oxide exists between the floating gate (FG) and the channel, which is typically in the 80- to 100-Å range. A

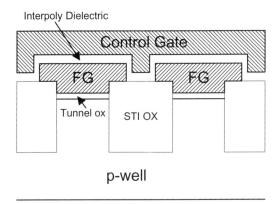

Figure 5.4. Cross section along the B–B cut line.

different dielectric is typically used between the floating gate and control gate and is often a composite of three layers consisting of silicon dioxide–silicon nitride–silicon dioxide. The electrical equivalent thickness of this dielectric is typically about 150 Å.

Figure 5.4 shows how the cells are isolated from one another along the wordline. In this case a shallow trench isolation (STI) scheme is shown. The wordline is continuous for all cells along the wordline, but each cell has its own unique floating gate to store the information for that cell location. The control gate is shown to wrap around the floating gate, and this helps improve the coupling to the floating gate from the control gate. We will see later how this is important for cell operation.

5.2.3 Read (Sensing)

The sensing operation occurs to "read" the contents of a cell. Information is stored in the cell by adjusting the threshold voltage. Changes in the cell threshold voltage will cause the current flowing through the cell to change, which can be compared with a reference current. The cell current will lie above or below the reference depending on the data stored. The cell threshold voltage is adjusted by changing the amount of charge stored on the floating gate.

In Figure 5.5 the cell with the low threshold voltage is the erased cell and the cell with the high threshold voltage is the programmed cell. In the next sections we will describe the program and erase operations that allow the user to adjust the threshold voltage of the Flash cell.

During the sensing operation, it is desirable to have a large current difference between the programmed and erased cell to make the sensing operation faster and more robust to noise. In order to achieve this, the transconductance (or G_m) of the cell should be as large as practical. For this reason it is important that the coupling between the control gate and floating gate be high. It is important to keep the interpoly dielectric thin (i.e., the dielectric between the floating and control gate). However, there is a constraint on how thin this can be made since we must avoid any significant charge loss from the floating gate.

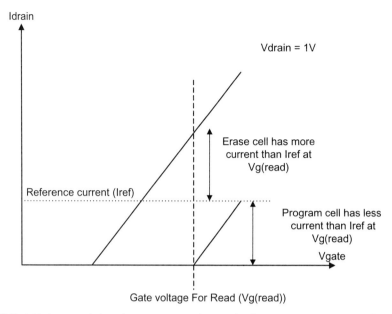

Figure 5.5. I–V characteristics of a program and erased cell. Currents are compared to a reference at a given gate voltage to determine a 1 or 0.

5.2.4 Programming

The programming operation is the process of placing a surplus of electrons on the floating gate. For most ETOX cells this is achieved by channel hot-electron (CHE) injection. A large V_{gs} (typically ~10 V) and a moderate V_{ds} (~4 V) are applied to the cell, which will bias the structure in a way that creates significant hot electrons in the channel while creating a favorable bias condition for gate current (see Figure 5.6). This operation, while not very efficient, is very fast with cells programming in about 1 μs. The physics of channel hot-electron injection (CHEI) has been covered in Section 4.3.3.

5.2.5 Erasing

The erasing operation is the process of removing electrons from the floating gate that were placed there during the programming operation. For EPROM devices UV light was used to remove any residual charge from the floating polysilicon gate since the photons of the UV light had sufficient energy to allow the electrons to surmount the energy barrier separating the electrons from the floating gate to either the top gate or bulk silicon below. For ETOX-type memories it is not unusual to program an array first before erasing to help maintain a consistent erase distribution with cycling. Erase is achieved by Fowler–Nordheim (FN) tunneling of electrons from the floating gate to the source junction. This can be achieved by grounding the gate and applying a large voltage (~11 V) to the source junction with the drain floating or a negative voltage can be applied to the wordline (~−10 V), allowing the source voltage to drop to about 4 V during the erase operation. This technique can also be

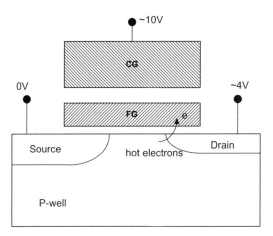

Figure 5.6. High electric fields at the drain junction edge create hot electrons that can contribute to gate current. These electrons can be injected into the floating gate.

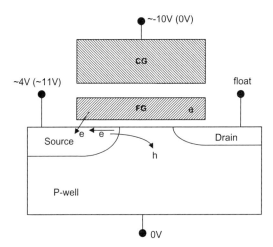

Figure 5.7. During erase Fowler–Nordheim tunneling of electrons occurs from the floating gate to the source. There is also band-to-band tunneling occurring near the silicon surface at the source to P-well junction.

modified to apply the positive voltage to the substrate if the Flash cell lies in a well that can be biased to a positive voltage. The first bias scheme is commonly referred to as *positive source erase*, the second is known as *negative gate erase*, and the third is known as *channel erase*. Most modern stacked-gate Flash memories use a *channel erase* scheme since it allows more aggressive scaling of the memory device through source junction scaling.

Figure 5.7 shows the voltages that are applied to the cell during erase, showing the negative gate erase conditions and in brackets the positive source erase conditions. The advantages of the negative gate erase condition is the ability to lower the source voltage, which in turn makes the source junction engineering easier since it now only needs to support 4 V instead of 11 V. This can also have implications for cell reliability and cell scaling.

ETOX FLASH CELL TECHNOLOGY

From Figure 5.8 one can see the erase operation gets exponentially slower as the Flash cell threshold voltage drops. This is due to the linear drop in the tunnel oxide electric field as the floating gate discharges and the floating gate voltage becomes more positive. The physics of FN tunneling and band-to-band tunneling (BBT) have been covered in Sections 4.3.1 and 4.4.1, respectively.

The channel erase technique puts the positive voltage on the body of the Flash cell with the source/drain junctions either biased to the same potential or floating. One key advantage of this technique is that it eliminates the BBT current and so reduces substantially the power required to do the erase. It should also help in device scaling but does require the addition of another well for the Flash cell, which would increase processing cost.

Figure 5.9 shows a Flash cell in a channel erase bias condition. With a negatively biased gate and positive P-well, a high electric field is established between the floating gate and P-well, which causes electrons to move from the floating gate to the well via Fowler–Nordheim tunneling.

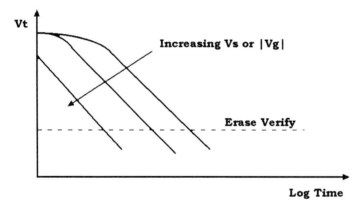

Figure 5.8. Flash cell erasing characteristic showing how the Flash cell threshold voltage changes with time under fixed erase bias conditions.

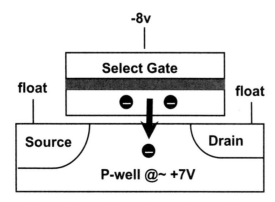

Figure 5.9. Cell channel erase bias condition.

5.2.6 Array Operation

5.2.6.1 Disturbs. Let us first introduce the concept of a *disturb* and then see how it relates to cell operation in an array. A disturb can be defined as a voltage condition where the contents of a cell (the charge stored on the floating gate) are altered unintentionally. In general, it is difficult to avoid disturbing cells in an array since they share wordlines, bitlines, and source terminals. There are two key categories of disturbs: (1) interblock and (2) intrablock. Interblock disturb means that an intentional operation on one block disturbs the charge stored in some other block. Intrablock disturb means that an intentional operation on a given block disturbs charge stored in that block. We will focus mainly on the latter since the former can normally be dealt with by appropriate choice of array blocking architecture. Blocking is the term used to describe how the memory is divided into individually erasable portions, which are also known as sectors. If voltage signals are not shared between blocks, then there are no interblock disturbs to be concerned with. In general, one might try and share signals between blocks to save die area and reduce cost. However, the concerns here are very similar to the intrablock disturbs so we will only cover those with any detail.

5.2.6.2 Intrablock Disturbs. The Flash cells are configured into an array for standard memory operation. It is important to consider what happens to the neighboring cells in an array during any operation that may be in progress. Table 5.1 and the array schematic in Figure 5.10 summarize some typical operating voltages for both the cell being accessed and for the neighboring cells. We will examine some of the concerns with cell operation in an array.

All cells in a given block share the source terminal. This is often referred to as an erase block or sector. All the cells in this block will be erased together in an erase

TABLE 5.1. Flash Cell Bias Conditions in an Array Configuration

Operation	Cell X (Cell being accessed)			Cells A and B (Common wordline)			Cells C and D (Common bitline)		
Terminal	V_{d1}	V_{g1}	V_s	$V_{d0/2}$	V_{g1}	V_s	V_{d1}	$V_{g0/2}$	V_s
Read	1 V	4 V	0 V	0 V	4 V	0 V	1 V	0 V	0 V
Program	5 V	10 V	0 V	0 V	10 V	0 V	5 V	0 V	0 V
Erase	Float	−10 V	3 V	Float	−10 V	3 V	Float	−10 V	3 V

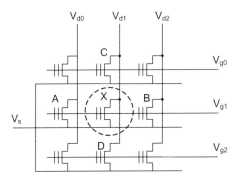

Figure 5.10. Schematic of an ETOX Flash cell array where cell X is being addressed.

operation and all see the same voltages applied so there are no disturbs to be concerned with.

During the program operation cells A and B see the high gate voltage (~10 V). If cell A is an erased cell, it will see what is called a direct current (dc) program stress. This stress is present whenever a cell on the same wordline is being programmed so the stress duration can be thousands of times longer than the program operation of a single cell. If the tunnel oxide quality is poor and the stress is for an extended duration, this erased cell can "soft" program and look like a programmed cell. Once the contents of the cell are changed the memory becomes corrupt and no longer contains valid data without the use of error-correcting techniques. Cell B sees the same stress voltage, but if it was already a programmed cell, it too may lose charge if the interpoly dielectric has poor integrity. This cell is said to see a dc erase disturb. If the cell V_t drops low enough it will look like an erased cell and the memory will again become corrupt. These disturbs need to be well understood to ensure that the disturb stress voltages do not change the contents of the memory inadvertently. Appropriate choice of the array architecture and high-quality tunnel oxide and interpoly dielectrics can ensure that this is not a problem in ETOX memories. This is one of the reasons that the interpoly dielectric is often made up of a three-layer sandwich of silicon dioxide–silicon nitride–silicon dioxide. A multilayer dielectric will prevent problems with pin holes, and the nitride with its higher dielectric constant will allow a thicker dielectric film to be used without effecting the coupling of the floating gate to the control gate. This, as pointed out earlier, is important to maintain the gain of the cell. There has also been considerable work done in improving the tunnel oxide quality with many works published.

Also during the program operation cells C and D see the programming drain voltage (~5 V), although they don't see the gate voltage. This stress voltage is present when any cell on the bitline is being programmed, which may be thousands of cells. If the tunnel oxide quality is poor, a programmed cell can lose charge due to the field present between the drain and floating gate. It is important therefore to ensure a good-quality tunnel oxide even after many program and erase cycles. An erase cell on the same bitline is not usually a disturb concern unless its V_t becomes very low and it starts to conduct current in the subthreshold region. In this case, holes can be accelerated in the channel and cause considerable damage to the tunnel oxide, which may later result in a drain disturb problem when this cell has been programmed.

5.2.7 Erase Threshold Control

Control of the erase distribution is important for ETOX memories since unlike other approaches there is no select transistor for each Flash cell, and because many cells share a common bitline it is important that the cells do not leak when deselected. Furthermore, since the erase operation occurs on many cells in parallel (typically an entire block or sector), it is important that uniform behavior is observed in all cells. Figure 5.11 shows a typical erase distribution often characterized by two populations: a normal population and a tail population of fast to erase bits. There are two key elements that need to be controlled here: (1) the sigma of the intrinsic normal population should be minimized and (2) the tail population should be carefully controlled and monitored. Since we would always like to maximize the current from an erased cell to improve read performance and provide immunity to noise,

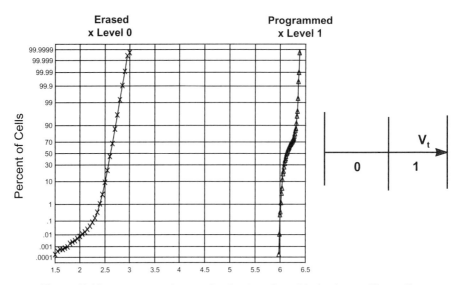

Figure 5.11. Program and erase distributions for a block of 0.5 million cells.

process variations and circuit offsets, it is desirable to erase the cells to as low a threshold as practical. There are other constraints related to erase performance that may determine this as well, but it is desirable to keep the intrinsic sigma small. The fast tail should also be controlled, but a few fast bits can be tolerated since this can be repaired by self-convergence or soft programming techniques at the end of the erase operation.

The erase of the Flash cell is an exponential function of the electric field in the tunnel oxide between the floating gate and the erase terminal. This can be seen from the FN equation that describes this tunneling process:

$$J = AE^2 \exp(-B/E)$$

where J is the tunneling current density, A and B are constants, and E is the electric field in the tunnel oxide. To better understand the source of the erase distribution, it is worth considering the effect the key cell parameters have on the electric field in the tunnel oxide. This is also dependent on the erase scheme chosen, which can place different weighting factors on these variables.

Let us consider in some detail the three key cell parameters that have the most influence on erase distribution control. From Yoshikawa et al. [1] they are: (1) tunnel oxide thickness, (2) electrical width W, and (3) source junction under lap (with the exception of channel erase).

Tunnel oxide thickness plays an obvious role in the erase distribution since it directly influences the electric field E in the tunnel oxide. A linear change in T_{ox} will cause a linear change in E that will result in an exponential change in J, the tunnel oxide current density. Therefore, it is important to have precise control of the oxide thickness in the manufacturing of Flash cells.

The electrical width of the device and the junction under lap (except in the case of channel erase) determine the tunneling area during the erase operation. Any change in the tunneling area will cause a cell to erase faster or slower. The isolation process determines the width of the device that in turn is controlled by the isolation lithography and etches. Good critical dimension (CD) control is very important during the lithography and etching operations, which can be quite a challenge at small dimensions. The junction under lap is determined by the junction engineering; too shallow a junction will cause considerable tunneling area variations, and too deep a junction will limit the ability to scale the cell in the channel length direction. These trade-offs are part of the device optimization made by the device engineer.

The fast tails at the low end of the erase distribution are often repaired or recovered by some soft programming techniques. Several self-convergent schemes have been proposed that normally rely on hot-electron or hole injection that occurs when a device is biased with a high V_{ds} and a low V_{gs}. Essentially, cells with a low threshold voltage will program faster than cells with a higher threshold voltage since they have higher subthreshold current flowing in the channel and thus have more electrons available for carrier heating in the high electric field near the junction with the high voltage.

A fundamental problem of a Flash erase operation is how to prevent cells from becoming overerased. In a NOR architecture, if a cell V_t is too low, then the cell on an unselected wordline (biased at 0 V) will leak, causing undesired current on the column, which can be viewed as a noise on the read signal. If many cells on a column leak, their currents will accumulate and can easily swamp the read current.

A number of measures are taken to prevent this. First, if a block containing random data is subjected to a series of erase pulses, by the time the high-V_t cells have erased to their target, the lower V_t cells may have overerased. To prevent this, before the bulk erase operation, the entire block is preprogrammed. This operation, known as *preprogramming* or *preconditioning* will help prevent cells from becoming overerased. Second, after the bulk erase operation is completed (defined when the highest V_t cell in the block has reached a predetermined erase target), any cells with a very low V_t are searched and soft-programmed up to within the main erase distribution. This operation is known as *posterase repair* (PER). (see Fig. 5.12)

Soft programming means programming with lower than normal gate voltages. Finding very low V_t cells is problematic since one or a few very low V_t cells could cause an entire column to leak. Since the standard Flash sensing is capable only of sensing column current, it is difficult to determine exactly which cells are leaking. The solution is to search for leaking columns (sense current with all wordlines grounded) and soft program every cell on leaky columns with gate voltages that are

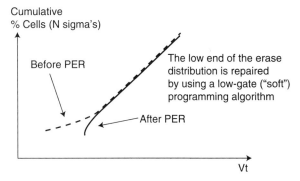

Figure 5.12. Post erase repair of the erase distribution.

high enough to recover the most leaky cells but low enough to not overprogram the highest V_t cells. Such voltages can usually be found through empirical characterization of the technology. More complex algorithms have been considered that include multiple passes between erase pulse and soft programming phases with the goal of further tightening the final erase distribution.

5.2.8 Process and Scaling Issues

Scaling the dimension of the Flash cells is a very important component in the cost reduction roadmap of semiconductor memories. As a general rule each Flash memory technology generation shrinks the size of the Flash cell by 2×. However, Flash cells offer a few unique restraints when it comes to scaling since they operate at low voltages during read operations but at high voltages during program and erase operations. Let us consider three key aspects of Flash cell scaling: (1) feature size, (2) film thickness, and (3) architectural techniques.

5.2.8.1 Feature Size. Conceptually, feature size is one of the more straightforward ways to scale a Flash cell. With each new generation of equipment the engineer is able to print and etch smaller dimensions with greater control. New materials and equipment that improve gap fill at high aspect ratios also contribute to this scaling process. Referring to Figures 5.1, 5.3, and 5.13, this allows the contact to be drawn smaller, the contact to gate dimension to shrink, smaller isolation lines between cells, a smaller source region, and a smaller active cell area (both channel length and width) to be drawn. However, the channel length scaling is not necessarily dictated by the ability to print and control the gate length since the cell must be able to support the large (~4 V) V_{ds} used during programming without punch-through occurring. If the length is not scaled, then it is difficult to scale the electrical width as well since this would result in a drop in the cell gain, which is undesirable from a design perspective as pointed out before. Channel length scaling needs to be coupled with appropriate junction scaling for both the source and drain as well as engineering of the channel doping concentration for adequate punch-through performance.

In early Flash scaling (early 1990s) cell scaling was largely achieved through reduction of inactive areas of the cell, such as the contact, through pure lithography

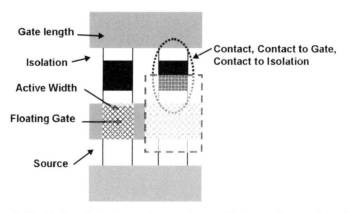

Figure 5.13. Flash cell structure where regions subject to scaling are identified.

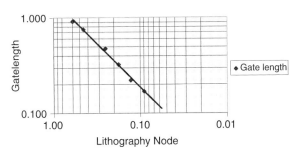

Figure 5.14. Gate length scaling as a function of the lithography technology node dimension.

advances. After that, more advanced scaling tricks were played on inactive areas, such as self-aligned poly and unlanded contacts, and active area scaling became necessary. This led to electrical device scaling challenges, such as wider program and erase distributions, which in turn drove advances in algorithms, such as more sophisticated erase PER and multilevel cell (MLC) state placement techniques.

Gate length scaling (Fig. 5.14) has been achieved through a variety of means. The requirements for the source junction (depth, doping profile) have been eased by going from positive-source erase to negative-gate erase to channel erase. Further reduction has been made possible through tunnel oxide scaling, improvement in litho control, and other junction engineering. Self-aligned techniques have reduced to a minimum the number of registration requirements in the cell.

The future of lithographic scaling will likely involve a variety of resolution enhancement tricks such as phase shift and other mask technologies.

5.2.8.2 Film Thickness. The tunnel oxide and interpoly dielectric have scaled slowly over the last five Flash technologies. The main reason for this is the requirement to maintain a 10-year data retention specification, which is much more difficult with thinner dielectrics. This has resulted in a slow drop in the voltage scaling for both program and erase operations. Therefore, although the external voltage supply has been scaling with each technology, pumps have been employed on-chip to provide the high voltages still required by the Flash cells.

Dielectric scaling in nonvolatile memories (NVM) is reaching the point where new approaches will be required to meet the reliability and performance requirements of future products. For both the tunnel oxide and the interpoly dielectric (IPD), high-K materials are being explored as possible candidates to replace the traditional SiO_2 and ONO (oxide/nitride/oxide) films used today. New storage node concepts are also becoming attractive as an alternative approach to address some of the dielectric scaling limitations.

DIELECTRIC REQUIREMENTS. The tunnel oxide has the most stringent requirements in a Flash cell with properties summarized by the following: (1) It must provide a good interface to the silicon channel for reliable transistor operation. (2) It should provide efficient charge transport either through tunneling or hot carrier injection during the programming and erase operations, which allow the data to be changed in the memory cell. (3) It should enable years of retention of the charge on the floating gate. Figure 5.15 summarizes the requirements with an electric field

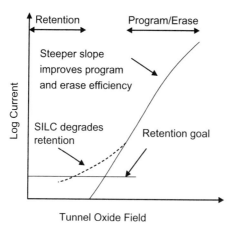

Figure 5.15. Electric field versus current characteristic of a tunnel dielectric showing the dielectric requirements.

versus current plot for a tunnel dielectric. It is desirable to have a large slope to this characteristic to give fast programming and erase at high fields but very low leakage at low fields to provide good retention.

It is also important to study the characteristic after stress since SILC (stress-induced leakage current) can be the limiter for retention since it increases leakage current at low fields. The IPD is not expected to transport charge, but it must block any leakage during the programming and erase operations. It should also deliver years of retention of charge on the floating gate and provide good capacitive coupling of the control gate to the floating gate to lower the voltages needed for read, program, and erase operations.

TUNNEL OXIDE SCALING. SiO_2 thickness limitations have been reported, with 6 nm being the lower limit due to direct tunneling, but a more practical limit of >8 nm has been proposed [2] due to postcycling SILC limitations. In actual fact the International Technology Roadmap for Semiconductors (ITRS) tunnel oxide scaling roadmap has been made more conservative in the 2001 edition due to the challenges in going below 8 nm. It is becoming clear that to enable a significant reduction below 8 nm equivalent oxide thickness (EOT) there needs to be a significant change in the approach, which until now has focused on SiO_2 and nitridation techniques using NH_3 and N_2O [3]. Chapter 10 discusses tunnel dielectrics in detail. Several alternatives have been presented in the literature with two of them selected here for further discussion, namely JVD (jet vapor deposition) nitride [4–6] and tunneling barrier engineering [7, 8].

JVD Nitride. Traditional chemical vapor deposition (CVD) nitride is a poor choice for a tunnel oxide due to the high trap density leading to Poole–Frenkel (PF) conduction (trap-assisted transport) resulting in poor retention under low field bias along with poor interface quality. A JVD nitride has been shown to have low trap density and interface states, making it a possible candidate for the Flash cell tunnel oxide. This is attributed to the low hydrogen incorporation in the film as a result of the deposition conditions [4]. Data has been reported [5] showing good endurance and reduced SILC, which will improve retention and enable a thinner EOT. There is the additional benefit of a lower barrier height for electrons shown in Figure 5.16. (2.1 eV versus 3.1 eV for SiO_2) leading to more efficient electron injection resulting in improved performance or scaled voltages. (see Fig. 5.17)

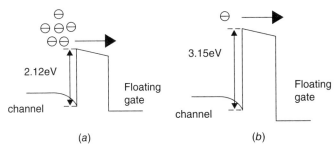

Figure 5.16. Lower electron barrier for a JVD nitride tunnel dielectric (a) compared to SiO$_2$ shown in (b) [5].

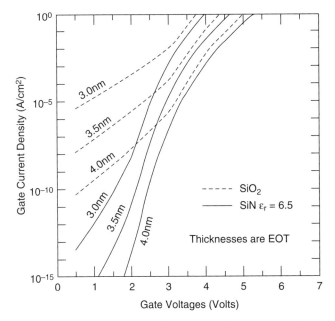

Figure 5.17. Dielectric tunneling characteristic showing the *theoretical* improvement in low field leakage with the higher dielectric constant nitride film [6].

Similarly, for holes the barrier is only 1.9 eV versus 4.9 eV for SiO$_2$, making hot-hole injection a candidate for discharging the floating gate. Although the single-cell fundamentals look encouraging, there has been little data reported yet on the dielectric performance in a memory array. For this to be a viable replacement for SiO$_2$ the statistical performance of the film needs to be studied in terms of erase uniformity and data retention on a large sample of cells.

Tunnel Barrier Engineering. Barrier engineering approaches [7, 8] are applicable to FN tunneling program and erase operations. With SiO$_2$, or any material with a uniform barrier, you cannot achieve high transparency at electric fields necessary for fast program and erase operations (~10 MV/cm) along with low transparency at fields (1 to 3 MV/cm) where good retention is required. The reason for this is that the barrier height remains fixed regardless of the applied electric field, and

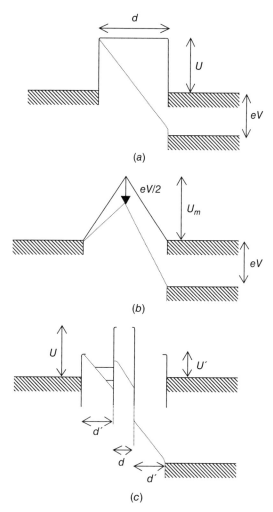

Figure 5.18. Classical single barrier (a) shows tunneling modulated by the barrier width. With barrier engineering it is possible to have a lower barrier under high field and a higher barrier under low field (b) and (c) [7].

field emission occurs due to thinning of the barrier width [Fig. 5.18 (a)]. If a triangular or crested barrier approach is used [Fig. 5.18 (b)], then the barrier height reduces under the high field condition when charge transport is desired but increases again when a low field is present, reducing the low field leakage current and improving retention. The challenge here is to find the material combination that meets these barrier needs. One possible solution is a trilayer stack of $Si_3N_4/AlN/Si_3N_4$, which is discussed in Likharev [7]. It is necessary to provide a very low trap nitride, otherwise low-level leakage will be high due to PF conduction. There is a possible opportunity here for a JVD nitride approach, but there may be alternative material options that are more attractive. Other such approaches reported by Govoreanu et al. [8] explore high-K dielectrics in combination with SiO_2. In this study Al_2O_3, Y_2O_3, ZrO_2, and HfO_2 are studied with promising results. Integration of these dielectrics into nonvolatile memory devices is underway, which will help address the next level of concerns with these approaches.

INTERPOLY DIELECTRIC (IPD). The key requirements from this dielectric are simple: Provide good capacitive coupling to the floating gate from the control gate and minimize any leakage through the dielectric. Unlike the tunnel oxide, this dielectric stack is not expected to support charge transfer during programming and erase operations. For most of the last decade ONO scaling has been achieved by thinning the dielectrics while maintaining control of the thickness. This scaling has also been enabled by improving the quality of the oxide on the floating gate, which can be aided by reducing the poly surface roughness. Nitridation techniques can also be deployed for the oxide portions of this stack, similar to those used to improve the tunnel oxide, and this will likely enable scaling into the 10-nm EOT thickness [9]. Although it is difficult to predict when this approach will no longer meet the needs due to direct tunneling scaling limitations, there would be a clear advantage for voltage scaling if higher capacitive coupling could be achieved. One such approach would be to replace the IPD with a high-K material. To maintain the high capacitive coupling improvement, a metal gate would be required to reduce the series capacitance from gate depletion. Since this film is not expected to transport charge (as is the case with the tunnel oxide), it will be easier to find a candidate to meet the cell requirements. Replacing the entire ONO stack with a single high-K material may be impractical due to the difficulties in depositing a high-K film directly on silicon, so a SiO_2 interfacial layer may still be required. This bottom oxide will block electron injection, but a high-K material along with the top gate electrode must be chosen in order to block hole injection. Possible options include Al_2O_3 with a TaN gate, which has been proposed for a silicon oxide nitride oxide silicon (SONOS)-type memory device [10], but could in principle be deployed in a floating-gate device as well. However, the gate electrode should be chosen consistent with the high-K film to make sure the energy barrier is engineered to reduce carrier injection. Going forward much of the development in this area will be engineering the right combination of dielectric and gate materials.

ALTERNATIVE FLOATING NODES. Rather than solving the scaling challenges of the tunnel oxide, alternative floating-node concepts are being explored. The basic idea here is to limit the source of electrons rather than trying to maintain the insulation with thinner dielectrics. There has been substantial activity in this area, which includes two basic approaches. The first is to replace the polysilicon gate with silicon nitride [11]; the second looks to use nanocrystals as storage nodes [12, 13]. In principle, these approaches can also adopt the IPD scaling options discussed in the previous section.

Nitride Floating Gates. Section 13.3 treats this topic in detail. The key concept is that a nitride gate has low conductivity so any tunnel oxide pin hole will not allow discharge of the entire floating gate, as would happen with doped semiconductors such as polysilicon. This local storage effect can also be taken advantage of to provide multibits per cell storage. There are also barrier height changes that effect conduction and retention. Several variations on this theme have been reported in the literature, but key differences center around the programming and erasing techniques and the film thicknesses. Figure 5.19 shows the basic structure of the cell

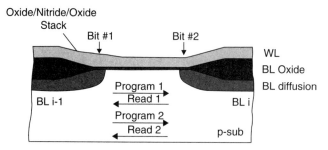

Figure 5.19. Nitride storage layer replaces the floating gate allowing tunnel oxide scaling [11].

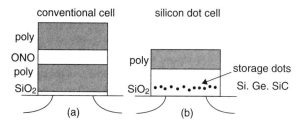

Figure 5.20. Conventional floating gate cell (a) and silicon nanocrystal memory cell (b) [13].

with an ONO film between the gate and the channel. In this case storage occurs in the nitride film above either bitline junctions, so two bits of information can be stored in the device, which alternates the source and drain to read out the information. Some approaches use hot carrier injection and tunneling while others are focused on tunneling only. In general, this approach integrates easily into standard complementary metal oxide semiconductor (CMOS) technology. For those approaches that employ hot carrier injection, charge redistribution in the nitride is the primary mode for retention failure (threshold voltage shifts in the Flash cell) rather than conduction through the surrounding oxide layers. It is not clear at this time if these limitations can be addressed sufficiently to make this a viable nonvolatile memory.

Nanocrystal Storage Nodes. This approach replaces the floating gate with a sea of nanocrystals embedded in an insulator, as shown in Figure 5.20. Several types of crystals have been explored with most of the literature focused on silicon or metal structures [12, 14] with a good overview published in De Blauwe [13]. The key challenge with this approach is to create small, uniform, high-density nanocrystal films. These requirements are important for the following reasons: (1) Need to have a large number of nanocrystals per device so statistical fluctuations from device to device are small, and (2) need small uniform size crystals so that we get equivalent electron storage per node and can rely on Coulomb repulsion to self-saturate the charge stored on each node. However, it has been observed that as we take this approach to the limit (very small, very dense nanocrystals) they begin to behave like a single floating gate since we can get charge transfer between the crystals; thus, this would represent the scaling limit for this approach [15]. Most development focus is centered on the nanocrystal growth techniques and retention.

ETOX FLASH CELL TECHNOLOGY

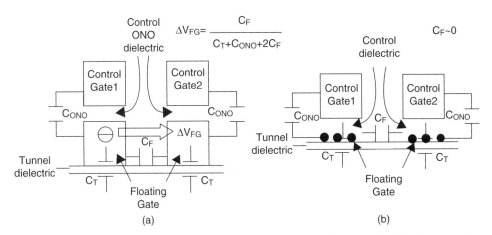

Figure 5.21. Shows the capacitive coupling (C_F) that occurs between the floating gates in diagram (a). Using nanocrystal memory, this coupling (C_F) can be substantially reduced (b) [16].

FLOATING-GATE TO FLOATING-GATE ISOLATION. So far we have discussed dielectric scaling with regards to isolation of the floating gate from the channel or control gate. One other key concern for NVM scaling is the isolation of floating gates from each other. The main concern is not charge transport between floating gates but capacitive coupling (Fig. 5.21). As device dimensions shrink, this can become a limitation since coupling between the floating-gate structures can cause a V_t shift on one device when the charge on the neighboring device is changed (as would occur when this cell would be programmed). Both storage node approaches previously discussed to address tunnel oxide scaling will potentially address this concern since coupling is reduced when thin layers of nitride or nanocrystals are deployed [16]. However, with the high-K approaches where the polysilicon floating gate remains, this coupling can limit some of the scaling opportunities and needs to be carefully accounted for during the cell design.

SUMMARY. The industry is facing some significant challenges in the area of dielectric scaling for nonvolatile memories. Several approaches for both tunnel oxide scaling and interpoly dielectric scaling have been reported along with alternative Flash cell structures that reduce some of the traditional limitations. Many challenges remain before any of these approaches are ready for production, and these solutions will compete with alternative memory storage approaches being investigated.

5.2.8.3 Architectural Techniques.
These methods attempt to provide new scaling opportunities above those provided by the feature size reduction or enable further feature size reduction. Several key architectural techniques have been reported over the last few years including self-aligned source (SAS), shallow trench isolation (STI), self-aligned contact (SAC), and self-aligned poly (SAP).

SELF-ALIGNED SOURCE (SAS). This technique is outlined in Figure 5.22, and it is shown to reduce the cell height. Previously, the source region was defined by a mask so that when the isolation was formed there was still a continuous diffusion

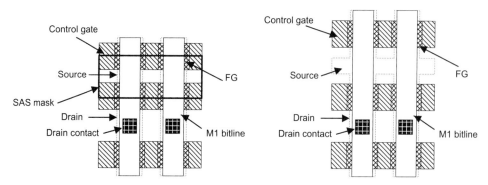

Figure 5.22. Demonstration of the SAS mask to shrink cell size.

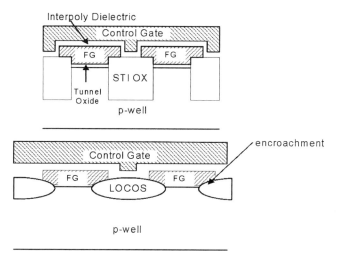

Figure 5.23. Comparison of STI and LOCOS isolation.

link between the sources of all the cells. It is necessary to keep some distance from the source diffusion edge and the gate edge between the cells to ensure adequate isolation. This spacing is what is eliminated when we adopt an SAS technique and allow the cell height to shrink.

With the SAS approach there is no diffusion link formed between the cells when the isolation is formed because after the gate formation an oxide etch occurs to remove the isolation along the source area between the cells. This etch is self-aligned to the gate edge so there are no layer-to-layer registration concerns here.

SHALLOW TRENCH ISOLATION (STI). Local oxidation of silicon (LOCOS) has been the mainstay isolation scheme for many technology generations. With this approach a nitride pattern is formed and an oxidation step is performed. Areas that are not covered by nitride can oxidize easily, whereas those covered by nitride cannot. The problem with this approach occurs when both narrow active and isolation regions are required. The LOCOS approach has a lateral encroachment, commonly called a bird's beak, which is difficult to control or minimize. As illustrated in Figure 5.23, the shallow trench scheme does not have this restriction since the isolation is deposited and not grown, allowing a sharp transition from the active to isolated region. This comes at a cost of increased process complexity, which has been well documented for the STI process.

SELF-ALIGNED CONTACT (SAC). Looking at the layout for the Flash cell in Figure 5.2, it can be seen that a considerable portion of the cell area is consumed by the contact and the spaces required around the contact to ensure that the contact does not short to the gate or land on the isolation when all the misregistrations are accounted for.

The SAC approach tries to eliminate the diffusion overlaps of the contact as well as the contact to gate spacing. This is achieved by self-aligning the contact etch to the gate and isolation edge. Essentially, this is achieved by coating the gate and isolation with a nitride layer to act as an etch stop when the contact etch (through oxide) is being performed. This is shown schematically in Figure 5.24 where the contact is shown to be drawn bigger than the actual space to land the contact. Similarly, the nitride acts as an etch stop on the isolation in a similar manner to the gate so that the contact can be drawn bigger than the active area width without concerns of etching the oxide isolation during the contact etch.

SELF-ALIGNED POLY (SAP). Self-aligned poly is an architecture feature that allows the width of the flash cell to be scaled more aggressively. The floating polysilicon gates are self-aligned to the isolation, which removes the need to build in a poly overlap of the active area to account for any registration error between the

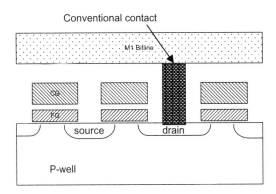

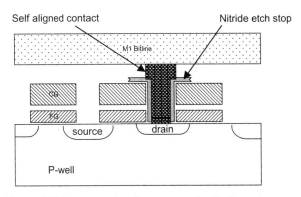

Figure 5.24. Conventional contact and self-aligned contact.

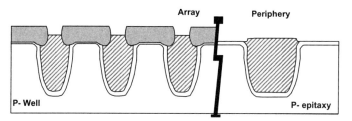

Figure 5.25. Wafer cross section with poly gates formed by SAP (self-aligned poly).

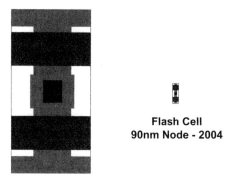

Flash Cell - 0.8μm Node - 1991

Flash Cell 90nm Node - 2004

Figure 5.26. Comparison of Flash cell size in 1991 to size in 2004.

poly layer and the isolation layer. The self-alignment is achieved by using the topography on the wafer from the isolation processing along with CMP (chemical mechanical polishing) to leave polysilicon above the active areas, which were recessed from the isolation areas. In Figure 5.25 it is worth noting that the isolation is now left recessed from the top of the polysilicon to give better coupling to this gate once the top gate is deposited.

SUMMARY. The overall effectiveness of all the scaling approaches is shown in Figure 5.26 where the size difference achieved over a 13-year period is documented.

5.2.9 Key Circuits and Circuit/Technology Interactions

This section will survey some of the key Flash memory circuit building blocks. Chapter 3 treats the topic of Flash circuits in greater detail.

5.2.9.1 High-Voltage Handling. Flash memories require high voltage for all the internal operations, read, program, and erase so there is a need to shift from V_{dd} logic levels to higher voltages. Figure 5.27 shows a typical level shifter implementation. This is a cross-coupled level shifter. The logic stage drives MN1 and MN2, which are designed to be much stronger than MP1 and MP2. One N-channel device will be off while the other is driven by V_{dd}, so one side of the cross-coupled latch pulls down to GND while the other goes up to the V_{pp} (high voltage) rail. The circuit burns no static power but uses six transistors.

ETOX FLASH CELL TECHNOLOGY

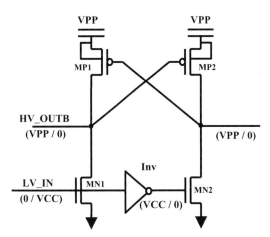

Figure 5.27. High-voltage level shifter.

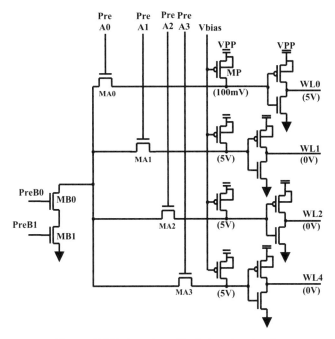

Figure 5.28. Row (wordline) decoder circuit.

5.2.9.2 Wordline Driver. A row (wordline) decoder shown in Figure 5.28 depicts the decoding of four rows. This is a special case of a ratioed level shifter. Pre-A0 through Pre-A3 along with the shared Pre-B0 and Pre-B1 form a pulldown path that is ratioed against the pullups MPU biased by V_{bias}. The final stage is simply a high-voltage inverter. This design burns static power but only on the one selected wordline and uses only an average of 4.25 transistors per wordline. Since there are hundreds of thousands of these per chip that must fit in the tight cell pitch, this design is favored over the static level shifter approach shown in Figure 5.27 for a wordline decoder, but the static approach is more commonly used in random periphery circuits where many may be active at once and the power is more of a concern.

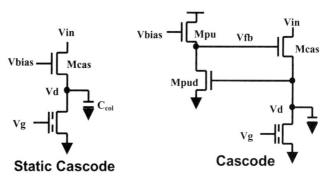

Figure 5.29. Cascode column bias circuits.

5.2.9.3 Read Path: Column Bias. During read a Flash array column must maintain a bias of around 1 V. A higher voltage will result in a disturb condition—cell V_t will move up. The column is typically biased with a cascode as shown in Figure 5.29. There are two basic ways of doing this; either with a static (fixed) cascode bias (V_{bias}) or with a feedback scheme. The feedback cascode maintains a tight cascode bias and will amplify V_{gs} through the feedback gain, which helps charge the column faster when V_d is below its final target. The cascode gate bias is also less dependent on cell current since any drop at the drain will be compensated for by a rising gate bias. The feedback scheme thus has the advantages of faster charging time and less V_d variation dependence, but has the drawback of higher power dissipation due to the feedback leg.

5.2.9.4 Read Path: Column Load. The column bias circuits shown in Figure 5.29 activate a Flash cell in the read mode and cause a current to flow through the Flash cell. The cell current is a function of the charge on the floating gate and indicates the data stored in the cell. The next step in the sensing scheme is to convert this current into a voltage via a "column load." Conceptually, the column load is a resistor to the supply rail and can be implemented in a variety of ways, such as a diode-connected P- or N-channel device, or a current mirror driven by a bias current. The goal is to provide maximum voltage swing at the V_{in} node, as this is what is ultimately sensed.

The resistor implementation is not commonly used in practice since it requires a precision resistor in the technology and also tends to be large in order to get an acceptable voltage swing with small cell currents. The diode-connected MOS device implementations become difficult as V_{cc} power supply scales due to the column load V_t drop. The cascode forming the drain bias must remain in saturation to keep the drain at a constant bias voltage, which places a floor on the V_{in} window. The remaining voltage window is available for sensing. The current-mirror load provides the most voltage gain and allows more voltage window since V_{in} can go higher and keep the P-channel load in saturation.

5.2.9.5 Read Path: Differential Sensing. A critical piece of any sensing scheme is the referencing technique, that is, how a decision is made between a 1 and a 0. Figure 5.30 shows a simple inverter connected to V_{in}. This is typically not used in practice. The decision between a 1 and a 0 in that circuit is dictated by the trip

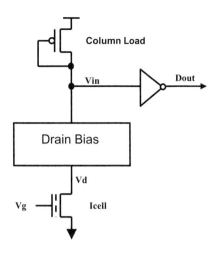

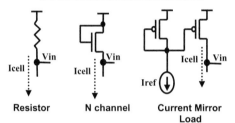

Figure 5.30. Column load implementation alternatives.

point of the inverter (the implied "reference voltage" is just the inverter trip point). However the flash cell IV characteristic variations and other variations of electrical properties of the cell versus temperature, noise response, and the like does not necessarily track well with the inverter trip point.

The best way to produce a reference (threshold decision) is with another Flash cell and differential sensing. Using a Flash cell as a reference cell achieves common mode tracking of Flash cell variations with temperature, voltage, and process (Fig. 5.31). The read path to the reference cell is matched (same column parasitics, matched drain bias circuit, matched column load), and a differential sense amplifier is used to compare the V_{in} of the main cell's path to that of the reference. The reference cell is carefully trimmed (during wafer sort) to a V_t between that of a programmed cell and an erased cell.

5.2.9.6 Read Path: Timing. A waveform diagram with the key signals and phases of sensing is shown in Figure 5.32. During the first phase, the columns of both the main array and matched reference array are charged. Optionally, the two differential amplifier inputs (R_{in} and S_{in}) shorted together thus ensuring identical common-mode voltage before the differential development stage. In the second phase, the equalization and column charging is released and the main and reference cells' currents move the column voltage one way or the other. Ideally, the common mode (precharge) voltage remains, and during signal development the cell only needs to discharge the column enough to overcome noise and offsets in the differential amplifier (100 mV is generally sufficient). In the case of a low V_t cell,

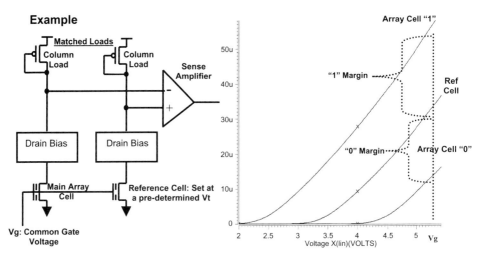

Figure 5.31. Sensing reference circuit using a Flash cell.

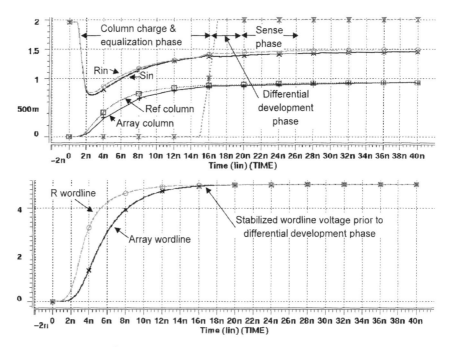

Figure 5.32. Read waveforms showing timing.

only the cell needs to charge. In the case of a high V_t cell, the column load needs to discharge. After sufficient differential is developed, the output of the differential amplifier is latched and translated to logic levels.

5.2.9.7 Program Path. The key components of the program path are shown in Figure 5.33. Level shifters are used to drive high voltages to the N-channel column select and data select devices, in order to reduce their IR drop from the large programming current. The wordline is biased to a voltage much greater than

ETOX FLASH CELL TECHNOLOGY

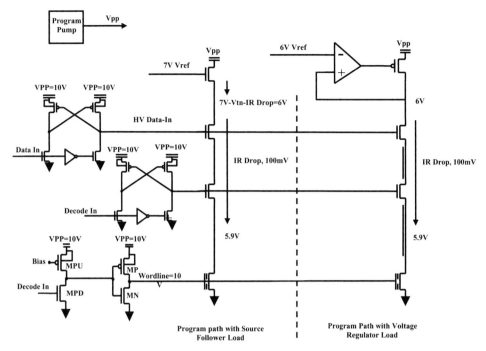

Figure 5.33. Key components of the programming path.

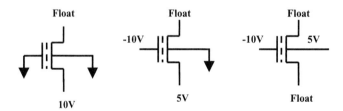

Figure 5.34. Alternative terminal voltage arrangements for erase.

in read mode, as much as 10 V. The diagram shows as examples two ways to bias the column (drain) either through a source follower (e.g., 7 V on the gate with 7 V $- V_{tn} - $ IR $= \sim 6$ V on the column), or through a regulation scheme, where a feedback op-amp with a 6-V reference forces the column to 6 V. This latter method provides a more predictable voltage at the expense of higher power due to the op-amp.

5.2.9.8 Erase. Various erase techniques have been employed in NOR Flash memories over the last decade (Fig. 5.34). All are variations on the basic mechanism of Fowler–Nordheim tunneling.

The earliest method placed a high voltage on the source and ground on the gate. This created a high field between gate and source and caused electrons to tunnel off into the source. This method required greater source overlap and actually caused some amount of source junction breakdown thus requiring a graded source junction. This was common in the early 1990s and was feasible with the cell sizes of those days. The advent of negative-gate-erase (about −10 V on the gate with a smaller ~5 V on the source) maintained the same vertical field while relieving some of the source junction engineering, which helped with cell size scaling, however, required negative voltage handling with triple-well NMOS transistors in the periphery circuits. The latest advent is channel erase, where the entire array is placed in a biased P-well. This further alleviates source junction engineering issues and reduces damage in the tunnel oxide by allowing erase to occur across the entire channel instead of only at the source junction.

5.2.10 Multilevel Cell Technology Circuits

Multilevel cell (MLC) technology is an important aspect of modern nonvolatile memory technology in that it is an architectural scheme that greatly enhances the cost effectiveness of the product. Chapter 12 provides an introduction and overview of MLC fundamentals together with references to the literature.

5.3 SST SUPERFLASH EEPROM CELL TECHNOLOGY

5.3.1 Introduction

The field-enhancing tunneling injector SuperFlash EEPROM cell from Silicon Storage Technology, Inc. (SST) [17] is a single-transistor, double-poly, split-gate memory cell using poly-to-poly Fowler–Nordheim tunneling for erasing and source-side channel hot-electron (SS-CHE) injection for programming. Poly-to-poly tunneling is from the field-enhancing tunneling injector formed on the floating gate using industry standard CMOS process oxidation and dry etching techniques. SS-CHE injection has an injection efficiency of $\approx 10^{-3}$ (compared to $\approx 10^{-6}$ for stacked gate); thus, allowing the use of small on-chip charge pumps from a single voltage power supply, for example, 2.7 or 1.8 V. Cells are normally erased prior to programming. The split-gate memory cell has the channel between the source and drain controlled by the series combination of the select gate and the floating gate. The select gate isolates the floating gate from the drain, thus avoiding the overerase phenomena of stacked-gate cells.

The SST SuperFlash technology typically uses a simpler process with fewer masking layers, compared to other flash EEPROM approaches. The fewer masking steps significantly reduce the cost of manufacturing a wafer. Reliability [18] is improved by reducing the latent defect densities, that is, fewer layers are exposed to possible defect causing mechanisms.

The SST split-gate memory cell is comparable in size to the single-transistor stacked-gate cell (for a given level of technology), yet provides the performance and reliability benefits of the traditional two-transistor byte-alterable E^{2}PROM cell. By design, the SST split-gate memory cell eliminates the stacked-gate issue of "overerase" [19] by isolating each memory cell from the bitline. "Erase-disturb" [19] cannot occur because all bytes are simultaneously erased in the same erase element (e.g., page, sector, block, and bank), and each erase element is completely isolated from every other erase element during any high-voltage operation.

The split-gate memory cell size is comparable to traditional stacked-gate memory cells using the same process technology. This is possible because:

1. The tunneling injector cell does not need the extra spacing to isolate the higher voltages and currents required for programming the stacked-gate array.
2. Floating-gate extensions are not needed to achieve the required stacked-gate coupling ratios.

Additionally, the simplicity of the structure eliminates many of the peripheral logic functions needed to control erasing (e.g., preprogramming, erase threshold) of the stacked-gate device. Memory arrays may use either random-access or sequential-access peripheral architectures.

5.3.2 Cell Cross Sections and Layout

A top view and a cross-sectional view along the wordline are presented in Figures 5.35 and 5.36 (note drawings are not to scale).

A cross-sectional view along the bitline and a scanning electron microscope (SEM) cross section are presented in Figures 5.37 and 5.38. Polysilicon or polysilicon with silicide or salicide is used to connect control gates along the wordline (row). Metal is used to connect the drain of each memory cell along the bitline (column). A common source is used for each page, that is, each pair of bits sharing a common source along a row pair (even plus odd row). A single wordline is referred to as a row; the combination of the even and odd rows forms a page or sector or block, which is erased as an entity. Programming may be either byte by byte individually or for a number of bytes in parallel (including all bytes) within the same page simultaneously. Note that although byte will be used for convenience, the same principles apply for a word.

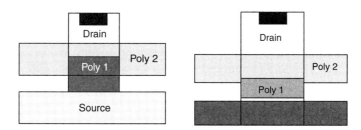

Figure 5.35. Top view of the non-self-aligned cell (left) and self-aligned cell (right).

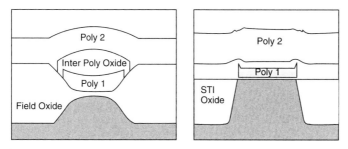

Figure 5.36. Cross sections along the wordline (NSA cell and SA cell).

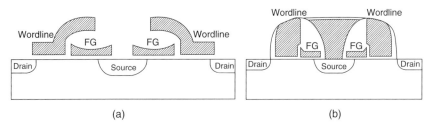

Figure 5.37. Cross section along the bitline: (a) First-generation SuperFlash non-self-aligned cell for low density and (b) second-generation SuperFlash self-aligned cell for high density.

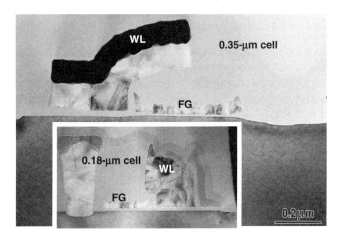

Figure 5.38. Cross section TEM pictures for 0.35- and 0.18-μm cells (same scale).

The drain region consists of an N + source/drain (S/D) diffusion, which is aligned with the edge of the poly-2 control gate and uses the same process as the base logic S/D formation. The source region consists of an N + S/D diffusion, which overlaps the floating poly. A cell implant beneath the floating gate is used to control the intrinsic cell threshold (V_T) and the punch-through voltage. The same oxide is used to separate the select gate from the channel and from the sidewall of the floating gate. The oxide thickness is 25 and 15 nm for 0.35 and 0.18 μm, respectively.

The floating gate is separated from the channel and source diffusion by a thermally grown 8- to 9-nm gate oxide. The tunneling injector on the floating gate is formed by oxidation of the polysilicon, similar to the formation of the field oxide "bird's beak" on single-crystal silicon, followed by a reactive ion etching of polysilicon. A silicide or polycide can be formed on the control gate to reduce the poly wordline resistance.

5.3.3 Charge Transfer Mechanisms

Table 5.2 illustrates typical voltages applied to various nodes of the memory cell to execute the read, erase, and program operations. Note that, the wordline is connected to the "control gate," the bitline is connected to the drain, and the source line is connected to the source, as illustrated in Figure 5.37. A schematic of the cell array is illustrated in Figure 5.41.

TABLE 5.2. Operating Conditions for the SST SuperFlash Cell

Node	Operation	Erase 1	Program 0	Read
Word line	Selected	≈ 11–12 V	V_T	V_{REF1}
	Unselected	0 V	0 V	0 V
Bitline	Selected	0 V	≈ V_{DD} → 1 ≈ V_{SS} → 0	V_{REF2}
	Unselected	0 V	≈ V_{DD}	0 V
Source Line	Selected	0 V	≈ 8–10 V	0 V
	Unselected	0 V	0 V	0 V

Table 5.2 gives the conditions for the memory cell terminals during the erase, program, and read operations. V_{DD} is the power supply, V_{SS} is ground, and V_T is the memory cell threshold for a given technology. V_{REF} are the reference voltages (a fraction of V_{DD} for a given technology) used to access the memory cell during the read cycle. The high voltages on the wordline during erase and on the source line during programming are generated by an on-chip charge pump.

5.3.4 Erase

The cell erases using floating gate to control gate Fowler–Nordheim tunneling. The floating-gate poly oxidation process provides a uniform field-enhanced tunneling injector along the edges of the floating gate. This repeatable manufacturing process provides consistent oxide integrity that minimizes endurance-induced degradation, that is, charge trapping or oxide rupture.

During erasing, the source and drain are grounded and the wordline is raised to ≈11 to 12 V. The conditions for erasing are in Table 5.2; reference Figure 5.37 for identification of terminals. The low coupling ratio between the control gate and the floating gate provides a significant ΔV across the interpoly oxide, which is the same everywhere between poly 1 and poly 2. A local high electric field is generated primarily along the edge of the tunneling injector. Charge transfer is very rapid and is eventually limited by the accumulation of positive charge on the floating-gate. This positive charge raises the floating-gate voltage until there is insufficient ΔV to sustain Fowler–Nordheim tunneling. An extensive investigation of the Fowler–Nordheim electron injection from field-enhancing floating-gate injector is presented in Yoshikawa et al. [1].

See Figure 5.39 for an example of how the 0.18-μm cell current changes as a function of erase time for a range of erase voltages.

The removal of negative charge leaves a net positive charge on the floating gate. The positive charge on the floating gate decreases the memory cell's threshold voltage, such that the memory cell will conduct ≈35 to 80 μA (depending on process and design) when the reference voltage is applied to the memory cell during a read cycle. The reference voltage is sufficient to turn on both the select transistor and the erased memory transistor in the addressed memory cell. Erasing is by fixed erase pulses generated by an internal timer, whose value does not change during the useful life of the device, for example, with accumulated erase/program cycles. Since the split-gate cell does not overerase, additional erase pulses do not cause any problems with the device.

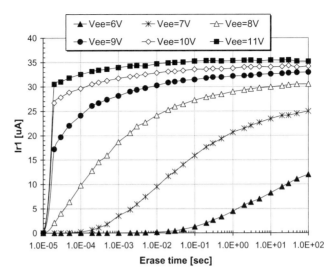

Figure 5.39. Erase characteristics, 0.18-μm cell current as a function of erase time.

5.3.4.1 Erase-Disturb. The enhanced field-tunneling injector devices are internally organized by pages (pairs). Each page consists of an even and odd row of memory cells. Each row pair (page) shares a common source line and each row pair has the wordline at the same voltage potential during erasing. Thus, all bytes are simultaneously erased along the common wordlines. All other wordlines (pages) do not receive the erasing high voltage. Since erase is between the floating gate and selected wordline, not to the substrate or other node common to the array, erase-disturb is not possible. Note, an erase element, for example, page, sector, block, bank, will contain one or more pages. The column leakage phenomenon caused by "overerase" in stacked-gate cells is not possible because the split gate provides an integral select gate to isolate each memory cell from the bitline.

5.3.5 Programming

The cell programs use high-efficiency SS-CHE injection. The conditions for programming are in Table 5.2; reference Figure 5.37 for identification of terminals. The intrinsic (i.e., UV erased) floating-gate threshold is positive; thus, the memory cell is essentially nonconducting, with the wordline at the reference voltage during a read cycle.

During programming, a voltage approximately 0.6 to 0.8 V above the threshold V_T of the select transistor is placed on the control gate, via the word line. This is sufficient to turn on the channel under the select portion of the control gate. The drain is at $V_{dp} \approx V_{SS}$ (actually small voltage above V_{SS}), if the cell is to be programmed. If the drain is at $\approx V_{DD}$, programming is inhibited. The drain voltage is transferred across the select channel because of the voltage on the control gate. The source is at $V_{sp} \approx 8$ to $10\,V$. The source-to-drain voltage differential (i.e., $V_{SD} = V_{sp} - V_{SS}$) generates CHEs. The source voltage is capacitively coupled to the floating gate. The field between the floating gate and the channel very efficiently ($\cong 100\%$) sweeps to the floating gate those CHEs that cross the Si-SiO$_2$ barrier height of $\approx 3.2\,eV$.

SST SUPERFLASH EEPROM CELL TECHNOLOGY

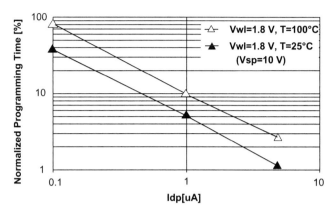

Figure 5.40. Time to program cell as a function of programming current (0.35-μm cell data).

The programming effect is eventually self-limiting as negative charge accumulates on the floating gate. The programming source–drain current (I_{dp}) is very low. (See Figure 5.40 for a quantification of current required.) Thus, the voltage on the source and the associated I_{dp} can be generated by a small charge pump on the die. The program time is fast because of the high efficiency of SS-CHE. The addition of negative charge to the floating gate neutralizes the positive charge generated during erasing; thus, the cell is nonconducting when the reference voltage is applied during a read cycle. Programming is by fixed program pulses generated by an internal timer whose value does not change during the useful life of the device, for example, with accumulated erase/program cycles.

5.3.5.1 Program-Disturb. The memory cells are arranged in a true crosspoint array, using a wordline and bitline for address location selection; thus, unselected cells within a page will see the programming voltages. There are two types of possible program-disturbs with the field-enhanced tunneling injection cell, both of which are described in the following paragraphs. Both mechanisms are preventable by proper design and processing. Defects are screenable with testing [20]. Devices with this memory architecture do not have program-disturb caused by accumulated erase/programming cycles because each page is individually isolated. Each cell is only exposed to high voltage within the selected page (along the row or source line); there is no high voltage on the bitline during an operation.

5.3.5.2 Reverse Tunnel Disturb. Reverse tunnel disturb [21] can occur for unselected erased cells within the page sharing a common source line, but on the mirrored row of the selected page to be programmed; thus, the wordline is grounded. The source voltage is capacitively coupled to the floating gate of the unselected erased cell. If there is a defect in the oxide between the control gate and the floating gate, Fowler–Nordheim tunneling may occur. This could program the unselected erased cell. Proper design and processing assures the reverse tunnel voltage is significantly higher than any applied voltage. Normally, reverse tunneling cannot occur since the electric field required to tunnel electrons from the wordline to the floating gate is significantly higher than any naturally applied electric field. The reverse tunneling requires a higher electric field because of the retarding effects of the tunneling injector tip, when the field is applied in the reverse direction. Defects are eliminated by reverse tunnel voltage screening in the 100% testing operations. Forward tunneling is defined as occurring when electrons are transferred from poly 1 (the floating gate) to poly 2 (the control gate), thereby erasing the cell. Reverse tunneling is defined as occurring when electrons are transferred from poly 2 to poly 1, thereby programming the cell.

5.3.5.3 Punch-through Disturb. Within a page, punch-through disturb can occur for erased cells in the adjacent inhibited wordline, that share a common source line and bitline with the cell being programmed. An inhibited wordline is grounded to prevent normal CHE injection. If there is a defect that reduces channel length and creates punch-through along the select gate channel, there could be hot electrons available to program the inhibited erased cell. Proper design and processing assures the punch-through voltage is significantly higher than any applied voltage. Defects are eliminated by punch-through voltage screening in the 100% testing operations.

5.3.6 Cell Array Architecture and Operation

Figure 5.41 represents a section of a typical cross-point memory array, arranged as eight memory cells in two columns (bitlines), two source lines, and four wordlines (rows). Note that the wordline is split into an even and odd row, which isolates the source line from all other source lines. Figure 5.42 is an equivalent memory cell, showing how the split-gate cell provides the logical equivalent of a select transistor

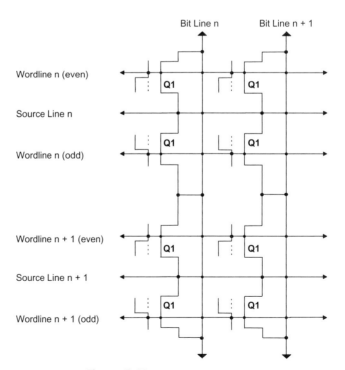

Figure 5.41. Cell array schematic.

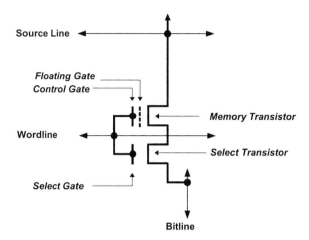

Figure 5.42. Equivalent memory cell structure for Q1.

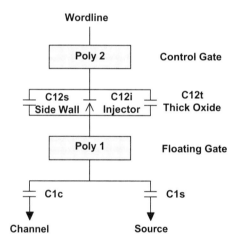

Figure 5.43. Equivalent capacitive coupling circuit.

and a memory transistor. The voltage applied to each terminal during normal operations is listed in Table 5.2. An equivalent circuit representation used to illustrate capacitive coupling is presented in Figure 5.43.

For the split-gate memory cell, the channel between the source and drain is split and controlled by the series combination of the select gate transistor and the memory gate transistor. The memory transistor is either in negative threshold or charge neutral (enhancement mode) state depending on the amount of stored electric charge on the floating gate.

During the read operation, this reference voltage is applied to the control gate and the select gate, via the wordline. The reference voltage will "turn on" the select gate portion of the channel. If the floating gate is programmed (high threshold state), the memory transistor portion of the channel will not conduct. If the floating gate is erased (low or negative threshold state), this memory cell will conduct. The conducting state is output as logic 1, the nonconducting state is logic 0.

Referring to Figure 5.43:

$$C10 = C1c + C1s \qquad C12 = C12s + C12i + C12t$$

Coupling ratios (CRs) are defined as:

1. CR10 = CR (poly 1 to substrate)
   = C10/(C10 + C12)
2. CR12 = CR (poly 1 to poly 2)
   = C12/(C10 + C12)
3. CR10 + CR12 = 1

During erasing, the channel is in inversion due to the wordline voltage. This increases the value of C1c. During programming the channel is in depletion; thus, C1c is negligible. Therefore, the coupling ratios are different during erasing and programming.

During programming, the coupling capacitance ratio between the source and the floating gate is approximately 50 to 60%. This means approximately 50 to 60% of the voltage at the source will be coupled to the floating gate, for example, if the source is at 10 V, the floating gate will be at 5 to 6 V, given no charge on the floating gate.

5.3.7 Erase Threshold Control and Distribution

The tunneling injector memory cell is a split-gate cell; therefore, the concerns with "overerase" are eliminated by design. The control gate portion of the cell isolates the memory cell from the bitline. Thus, there can only be conduction through the memory cell when the control gate is selected.

Unlike stacked-gate cells, the tunneling injector split-gate cell does not require control of the erase threshold within a narrow range with limits on both ends. The SST SuperFlash cell only has to be erased, so that the minimum erase threshold is greater than the level necessary to have the proper cell current; the maximum level is not applicable. In general, the cell is erased to saturation, such that for a given erase time, there is no change in the characteristics of the cell in the specified useful life of the device.

Erase V_t spread or margin, that is, the difference in V_t (or erased cell current) between the weakest and strongest bit is optimized in order to reduce erase time. The tunneling injector split-gate cell erases significantly faster than stacked-gate cells, for example, tens of milliseconds versus seconds; therefore, rigid control is not required in order to have an array that is much faster to erase than the equivalent array using stacked-gate technology.

5.3.8 Process Scaling Issues

The use of thicker charge transfer oxides greatly extends the vertical scalability of the SuperFlash oxides compared with the thin oxides of the stacked-gate cell. There is no limitation based on the cell of the horizontal scalability of the cell, that is, within the current and near future wafer process limitations.

The SST SuperFlash cell has high voltage on only two nodes, unlike the stacked-gate cell with high voltage on three nodes (during program or erase). No additional isolation is then required on the drain node, which never sees a high voltage.

SST SUPERFLASH EEPROM CELL TECHNOLOGY

TABLE 5.3. Comparison of Process Technology, Cell Size, and Die Area Efficiencies

	SST	AMD	SST
Density/part number	16 Mbit s 39VF160	6 Mbits 29F016	16 Mbits 39VF160
Sector size	2048 words	65,536 bytes	2048 words
Technology (μm)	0.35	0.35	0.18
Cell size (μm^2)	1.71	1.21	0.38
Die size (mm^2)	43.62	48.28	10.3
Efficiency (%)	65.8%	42%	62%

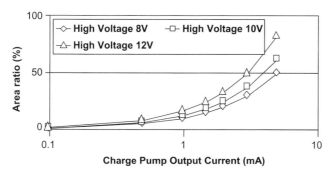

Figure 5.44. Charge pump area ratio as a function of charge pump output.

Since eventually even the SST relatively thicker oxides will approach the physical limits of SiO$_2$, alternative insulation and charge transfer materials will be required.

Table 5.3 compares SST first and second cell generations, and SST with Advanced Micro Devices (AMD) stacked-gate devices on similar process geometry technologies.

5.3.9 Key Circuit Interactions

As with any floating-gate reprogrammable nonvolatile technology, the key circuits are for the generation and transfer of the high voltage used for erasing and programming. The tunneling injector split-gate cell uses Fowler–Nordheim tunneling for erase from the injector, which significantly reduces the average voltage across the tunneling oxide. The high-voltage circuit for erase requires little current from the internal charge pump. Lower current requirements allow use of a smaller circuit. Figure 5.44 illustrates the increase in charge pump size, and thus die area required to provide more current for programming or erasing.

Since source-side injection is extremely efficient, little current is required from the internal charge pump, thus, allowing the use of a smaller charge pump circuit, relative to drain CHE technology.

Since no high voltage is ever on the drain of the memory cell, special doping or isolation circuits are not required. The high voltage applied to the gate during erase or the source during program is transferred using standard high-voltage transistors.

5.3.10 Multilevel Cell Implementation

The SST tunneling injector split-gate cell may be used for MLC (multilevel cell) arrays for the following reasons:

1. The split-gate cell eliminates the issue of overerase and vastly minimizes the need for tight erase Vt control.
2. Source-side injection is a low current (relative to drain-side CHE injection) and fast (relative to Fowler–Nordheim tunneling cells) charge transfer mechanism; therefore, control of programmed Vt is simplified.
3. SST cell has superior data retention reliability showing no noticeable cell current drift during extended bake as can be seen in the Figure 5.46.

Thus, the circuitry to produce different programmed V_t's is reduced, and there is still adequate separation between V_t's for signal detection. Note, the amount of charge on the floating gate, for a given technology, is about the same for the SST cell as stacked-gate cells; therefore, a given MLC implementation will require less circuitry for the SST cell than alternatives.

5.4 RELIABILITY ISSUES AND SOLUTIONS

5.4.1 Oxide Integrity

All oxides are subject to time-dependent dielectric breakdown (TDDB), that is, for a given oxide and electric field, eventually the oxide will break down. The lower the electric field and the less time the field is applied, the longer the time to breakdown. For oxides used in normal logic voltage level circuits, this time is essentially infinite; however, in Flash memories that use high voltages, the time of oxide exposure to high electric fields can contribute to the intrinsic device reliability.

The SST memory cell uses an approximate 6- to 7-MV/cm electric field during erasing. This value is significantly lower than the $\approx$10 MV/cm used by stacked-gate Flash approaches or the $\approx$11 MV/cm used by the thin oxide E^{2}PROM and NAND Flash approaches. Since the oxide time-dependent breakdown rate is an exponential function of the field strength, the SST memory cell intrinsically has a much lower failure rate than stacked-gate cell for oxide breakdown. Note the SST cell is exposed to the lower electric field for significantly less time during erase, compared with stacked-gate approaches. The charge trapping occurs in accordance with a very repeatable and predictable distribution, for a given technology. Therefore, intrinsic charge trapping may be measured on each cell and an endurance reliability prediction for the expected number of erase/program cycles per cell may be established in the 100% testing operation.

Figure 5.45 shows how the erase voltage increases for a fixed erase time as a function of accumulated erase/program cycles for different cell generations. It can be seen that the use of thinner interpoly oxides result in less cycling-induced change of the cell's erase voltage. This is a combined effect of less charge trapping (due to thinner oxide) and higher oxide capacitance.

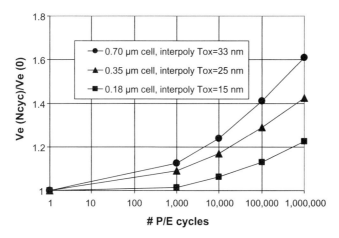

Figure 5.45. Normalized erase voltage as a function of accumulated erase/program cycles.

5.4.2 Contact Integrity

All memory arrays contain metal-to-silicon contacts, typically from the metal bitline to the diffused drain of the memory cell. Stacked-gate and the SST memory cells use a standard cross-point array, whereby a contact is shared by every two memory cells; thus, there are many contacts in a large memory array, for example, a 16-Megabit chip contains over 8 million contacts. Contacts must have a very low failure rate because there are so many of them. Contacts and associated metal lines are subject to failure based on the current density passing through the contact and metal line. The lower the current density, the lower the potential failure rate due to contact damage or electromigration mechanisms.

The SS-CHE injection current used in programming SST cells is significantly lower than the drain-side CHE injection current used in programming stacked-gate cells. During programming, SST cells use approximately 1 µA of source–drain current; this is much less than the read cell current. In contrast, a stacked-gate cell requires 500 to 1000 µA of source–drain current during programming, which is much higher than the read cell current. The high programming-current density in stacked-gate cells results in a higher probability of failure due to contact damage or electromigration. Since the programming current for the SST cell is much lower than the read current, there is no increase in the reliability failure rate due to programming-induced current density failure mechanisms.

Fowler–Nordheim (FN) tunneling used for erase is intrinsically a low current operation. Therefore, both the SST and stacked-gate cells are not measurably affected by current density during the erase operation.

5.4.3 Data Retention

The field enhancing tunneling injector cell uses relatively thick charge transfer oxides, compared with other E^{2}PROM or Flash EEPROM cells; therefore, intrinsic data retention is robust. The thicker charge transfer oxides minimize initial and latent oxide defects, thus, improving yield and oxide integrity. The lower voltages

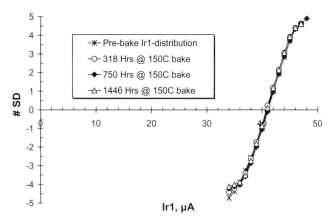

Figure 5.46. Normal probability plot for 4-Mb Ir1 current distribution. No noticeable Ir1 shift was found during 60 days 150°C bake for device subjected to 1 M P/E cycles before bake.

used for erase and programming combined with the relatively thicker charge transfer oxides reduce the endurance-related extrinsic data retention failure rate. In fact, oxide leakage mechanism, SILC (stress-induced leakage current), which is known to be a reliability concern in stacked-gate cell, so far has not been observed in the relatively thick interpoly oxide of the SuperFlash cell. Next advantage is that a thin 8- to 10-nm oxide underneath the FG of the SuperFlash cell is never exposed to high FN fields, thus showing substantially higher resistance to the SILC as compared to the FG oxide in stacked-gate cell. Figure 5.46 shows no noticeable shift in bit current distribution over 4 Mb 0.35-μm memory array during 2 months at 150°C bake after 10^6 program/erase cycles.

5.4.4 Endurance

Since the field enhancing tunneling injector cell uses a relatively thick oxide for the Fowler–Nordheim tunneling transfer oxide, the primary endurance limitation [22] is due to charge trapping in the interpoly oxide. Since both erasing by tunneling and the SS-CHE programming utilize relatively weaker electric fields across the poly-1 insulating oxides, the oxide rupture failure rate is low.

Trapping [23] occurs mainly in an approximate 20-Å-thick shallow region adjacent to the tunneling injector. Within this distance, direct tunneling detrapping occurs in the quiescent times between erase/program cycles. In practice, this means the endurance of the device in real-world applications will be greater than the endurance demonstrated in a stress environment, where the device is being erase/program cycled at the maximum possible frequency.

Charge trapping results in single bits that are unable to erase. These bits are detectable immediately after the erase operation and, therefore, are repairable with either extended erase time or with error code correction (ECC) circuits or software. There is no catastrophic failure such as the loss of a column as with overerase or a field failure due to charge loss from a leaky or slightly ruptured thin oxide.

Weibull distribution adequately depicts endurance data. The good thing about Weibull distribution is that endurance data can be easy scaled to any memory array size. Let's say there is a chip endurance data (4-Mb devices). Sector endurance data

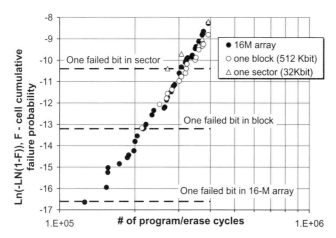

Figure 5.47. Weibull plot for 16 Mb array endurance data. The memory array consists of 32 blocks; each block includes 16 × 32 Kb sectors.

will be parallel shifted to chip data by $\ln(32\,\text{Kb\_sector}/4\,\text{Mb\_chip}) = \ln(1/128) = -4.85$ in y direction. This is because Weibull function $\ln[-\ln(1-F)]$ can be approximated by $\ln(F)$ if $F \ll 1$. Scaling can be done in both directions: from chip to cell and backward.

To verify the parallel-scaling law, the endurance failures have been monitored in different portions of the same 16-Mb memory array. As shown in the Figure 5.47, endurance data from any portion (in terms of cell failure probability) appears to land on the same unique line. This remarkable result allows:

1. In one stroke to evaluate endurance for arbitrary-chosen array and any desired failure rate if Weibull function has been found.
2. Develop express technique for endurance evaluation based on miniarray cycling.

5.4.5 Disturbs

A major concern of reprogrammable nonvolatile memories is that of disturb phenomena, that is, where a different location than the one being erased or programmed is altered. Disturbs can occur whenever a high voltage is applied to the gate, source, or drain of a memory cell that is not being intentionally erased or programmed. The SST cell has several design advantages to reduce the possibilities for a disturb:

1. There is no high voltage placed on the bitline, as is common for stacked-gate approaches. In addition, the split-gate cell isolates each memory storage node from all other nodes along the bitline. Thus, a disturb via the bitline (connected to the drain) is not possible.
2. The device uses a page, sector, or block erase, whereby all bytes in the page/sector/block are erased simultaneously, that is, see the same high voltage at the same time. Since each page/sector/block is isolated from every other page/sector/block by the wordline selection circuitry, disturbs along the wordline (connected to the gate) during erasing are not possible.

3. The device uses a unique source line for each page, unlike most stacked-gate devices that have the source line common to large sectors or the entire array. This limits exposure to disturb conditions to only the cells within a page during the time that page is being programmed. This greatly reduces the probability of a disturb and eases the detection, that is, only the page being programmed need be verified after any programming operation.

5.4.6 Life Test (Dynamic Burn-in)

The field enhancing tunneling injector cell uses standard CMOS technology in both the periphery and memory array; therefore, the life test results will be comparable to other devices built with the same process technology. As with all floating-gate reprogrammable nonvolatile memories, life test results for a given technology will generally be better than other memories, for example, static random-access memory (SRAM), built with the same technology because of the standard endurance and data retention infant mortality screening. [24, 25]

5.5 APPLICATIONS

The SST SuperFlash memory cell can be used for any reprogrammable nonvolatile circuit application. The limitations are strictly with the peripheral circuits. The memory cell is small and flexible enough to compete with volatile random-access memories [e.g., dynamic RAM (DRAM)] or sequential access nonvolatile memories (e.g., NAND) or random-access nonvolatile memories (e.g., NOR) for all code and most data applications. Some data applications cannot use floating-gate reprogrammable nonvolatile memory because of either endurance or write time constraints. The SST tunneling injector split-gate cell is well suited for stand-alone memories, especially low- or medium-density devices (relative to the maximum practical density for a given technology) where the memory array circuit efficiency must be high. See Table 5.3 for a comparison of die area efficiencies by technology. For embedded memories, the SST tunneling injector split-gate cell is well suited to applications where the memory density demands low peripheral circuitry overhead or the process cannot tolerate the addition of many new modules.

REFERENCES

1. K. Yoshikawa, S. Yamada, J. Miyamoto, T. Suzuki, M. Oshikiri, E. Obi, Y. Hiura, K. Yamada, Y. Ohshima, and S. Atsumi, "Comparison of Current Flash EPROM Erasing Methods: Stability and How to Control," *IEDM Tech. Dig.*, pp. 595–598, 1992.
2. S. Lai, "Tunnel Oxide and ETOX™ Flash Scaling Limitation." paper presented at the Nonvolatile Memory Technology Conference, 1998 Proceedings, Seventh Biennial IEEE, pp. 6–7, June 1998.
3. J. Kim, J. D. Choi, W. C. Shin, D. J. Kim, H. S. Kim, K. M. Mang, S. T. Ahn, and O. H. Kwon, "Scaling Down of Tunnel Oxynitride in NAND Flash Memory: Oxynitride Selection and Reliabilities," *IEEE Reliability Physics Symposium*, pp. 12–16, Apr. 8–10 1997.
4. D. Wang, T.-P. Ma, J. W. Golz, B. L. Halpern, and J. J. Schmitt, "High-Quality MNS Capacitors Prepared by Jet Vapor Deposition at Room Temperature," *IEEE Electron Device Lett.*, Vol. 13, No. 9, pp. 482–484, Sept. 1992.

5. M. She, T.-J. King, C. Hu, W. Zhu, Z. Luo, J.-P. Han, and T.-P. Ma, "Low-Voltage, Fast-Programming P-Channel Flash Memory with JVD Tunneling Nitride," *2001 Semiconductor Device Research Symposium*, pp. 641–644, Dec. 2001.

6. X. W. Wang, Y. Shi, T. P. Ma, G. J. Cui, T. Tamagawa, J. W. Golz, B. L. Halpen, and J. J. Schmitt, "Extending Gate Dielectric Scaling Limit by Use of Nitride Or Oxynitride," *VLSI Technol. Dig Tech. Papers*, pp. 109–110, June 1995.

7. K. K. Likharev, "Riding the Crest of a New Wave in Memory NOVORAM," *IEEE Circuits Devices Mag.*, Vol. 16, No. 4, pp. 16–21, July 2000.

8. B. Govoreanu, P. Blomme, M. Rosmeulen, J. Van Houdt, and K. De Meyer, "VARIOT: A Novel Multilayer Tunnel Barrier Concept for Low-Voltage Nonvolatile Memory Devices," *IEEE Electron Device Lett.*, Vol. 24, No. 2, pp. 99–101, Feb. 2003.

9. S. Mori, Y. Y. Araki, M. Sato, H. Meguro, H. Tsunoda, E. Kamiya, K. Yoshikawa, N. Arai, and E. Sakagami, "Thickness Scaling Limitation Factors of ONO Interpoly Dielectric for Nonvolatile Memory Devices," *IEEE Trans. Electron Devices*, Vol. 43, No. 1, pp. 47–53, Jan. 1996.

10. C. H. Lee, K. I. Choi, M. K. Cho, Y. H. Song, K. C. Park, and K. Kim, "A Novel SONOS Structure of SiO/sub 2//SiN/Al/sub 2/O/sub 3/ with TaN Metal Gate for Multi-Giga Bit Flash Memories," *IEEE IEDM Tech. Dig.*, pp. 26.5.1–26.5.4, Dec. 2003.

11. B. Eitan, P. Pavan, I. Bloom, E. Aloni, A. Frommer, and D. Finzi, "NROM: A Novel Localized Trapping, 2-bit Nonvolatile Memory Cell," *IEEE Electron Device Lett.*, Vol. 21, No. 11, pp. 543–545, Nov. 2000.

12. H. I. Hanafi, S. Tiwari, and I. Khan, "Fast and Long Retention-Time Nano-Crystal Memory," *IEEE Trans. Electron Devices*, Vol. 43, No. 9, pp. 1553–1558, Sept. 1996.

13. J. De Blauwe, "Nanocrystal Nonvolatile Memory Devices," *IEEE Trans. Nanotechnol.*, Vol. 1, No. 1, pp. 72–77, Mar. 2002.

14. M. Takata, S. Kondoh, T. Sakaguchi, H. Choi, J.-C. Shim, H. Kurino, and M. Koyanagi, "New Non-Volatile Memory with Extremely High Density Metal Nano-Dots," *IEEE IEDM Tech. Dig.*, pp. 22.5.1–22.5.4, Dec. 2003.

15. B. De Salvo, C. Gerardi, S. Lombardo, T. Baron, L. Perniola, D. Mariolle, P. Mur, A. Toffoli, M. Gely, M. N. Semeria, S. Deleonibus, G. Ammendola, V. Ancarani, M. Melanotte, R. Bez, L. Baldi, D. Corso, I. Crupi, R. A. Puglisi, G. Nicotra, E. Rimini, F. Mazen, G. Ghibaudo, G. Pananakakis, C. M. Compagnoni, D. Ielmini, A. Lacaita, A. Spinelli, Y. M. Wan, and K. van der Jeugd, "How Far Will Silicon Nanocrystals Push the Scaling Limits of NVMs Technologies?" *IEEE IEDM Tech. Dig.*, pp. 26.1.1–26.1.4, Dec. 2003.

16. S. J. Baik, S. Choi, U.-I. Chung, and J. T. Moon, "High Speed and Nonvolatile Si Nanocrystal Memory for Scaled Flash Technology Using Highly Field-Sensitive Tunnel Barrier," *IEEE IEDM Tech. Dig.*, pp. 22.3.1–22.3.4, Dec. 2003.

17. S. Kianian, A. Levi, D. Lee, and Y.-W Hu, "A Novel 3 Volts-Only, Small Sector Erase, High Density Flash E^{2}PROM", *1994 IEEE Symposium of VLSI Technology*, pp. 71–72, June 1994.

18. D. Sweetman, "Reliability of Reprogrammable Nonvolatile Memories", *1998 IEEE Nonvolatile Memory Technology Conference, Albuquerque*, pp. 101–108, June 1998.

19. IEEE Std 1005–1998, *IEEE Standard Definitions and Characterization of Floating Gate Semiconductor Arrays*, Feb. 1999.

20. D. Sweetman, "Guardbanding VLSI EEPROM Test Programs," *1991 IEEE VLSI Test Symposium*, pp. 155–160, April 1991.

21. K. Kotov, A. Levi, Y. Tkachev, and V. Markov, "Tunneling Phenomenon in SuperFlash Cell," *2002 IEEE Nonvolatile Memory Technology Symposium*, pp. 110–115, Nov. 2002.

22. V. Markov, A. Kotov, A. Levi, T. Dang, and Y. Tkachev, "SuperFlash Memory Program / Erase Endurance," *2003 IEEE Nonvolatile Memory Technology Symposium*, pp. 231–234, Nov. 2003.

23. T. Tkachev, X. Liu, A. Kotov, V. Markov, and A. Levi, "Observations of Single Electron Trapping / Detrapping Events in Tunnel Oxide of SuperFlash Memory Cell," *2004 IEEE Nonvolatile Memory Technology Symposium*, pp. 45–50, Nov. 2004.
24. D. Sweetman, "EEPROM Endurance—Technology, Design, Screens," *presentation at 1991 IEEE Nonvolatile Semiconductor Memory Workshop, Monterey, CA*, Feb 1991.
25. D. Sweetman, "Endurance and Retention Specification," *presentation at Nonvolatile Memory Technology Conference, Linthicum, MD*, June 1987.

6

NAND FLASH MEMORY TECHNOLOGY

Koji Sakui and Kang-Deog Suh

6.1 OVERVIEW OF NAND EEPROM

Memory that can retain data when the power is shut down is an indispensable part of a modern electronic system. System designers benefited greatly when the electrically erased and programmed nonvolatile EEPROM (electrically erasable programmable read-only memory) became available. However, the large cell area and high cost per bit of the device relative to dynamic random-access memory (DRAM) prevented its widespread usage. A lower cost alternative appeared on the scene when in 1987 Fujio Masuoka of Tohoku University invented the NAND-structured EEPROM cell, which arranged eight memory transistors in series to form a small EEPROM memory chip [1]. Later work, as summarized in Table 6.1, produced NAND EEPROM devices of increasing density using progressively finer process dimensions.

Because the NAND-structured cell shares resources among several cells, this technology achieves a small cell area using conventional self-aligned stacked polysilicon gate processes without scaling down the device dimensions. Figure 6.1 compares a NAND-structured cell and a NOR EEPROM cell. The cell area of the NAND EEPROM includes the contact, select gate, and source-wiring areas. Therefore, the areas allocated to a single bit consists of one memory transistor area, one-eighth of a select transistor area, and one-thirty-second of a contact hole area.

The NAND EEPROM uses a uniform Fowler–Nordheim (FN) tunneling between the substrate and floating gate for both erasing and programming. No huge hot-electron injection current is required, and no band-to-band current occurs. The power consumption for programming does not significantly increase even when the number of memory cells to be programmed is increased. As a result, the NAND

Nonvolatile Memory Technologies with Emphasis on Flash. Edited by J. E. Brewer and M. Gill
Copyright © 2008 the Institute of Electrical and Electronics Engineers, Inc.

TABLE 6.1. NAND EEPROM Development

Device	Source	Technology	Reference
Original NAND	Toshiba	—	[1]
4-Mbit NAND	Toshiba	1.0 μm	[2]
16-Mbit NAND	Toshiba	0.7 μm	[3]
32-Mbit NAND	Toshiba	0.5 μm	[4]
32-Mbit NAND	Samsung	0.5 μm	[5]
128-Mbit Multi-NAND	Samsung	—	
64-Mbit NAND	Samsung/Toshiba	0.4 μm	[6]
256-Mbit NAND	Toshiba	0.25 μm	[7]

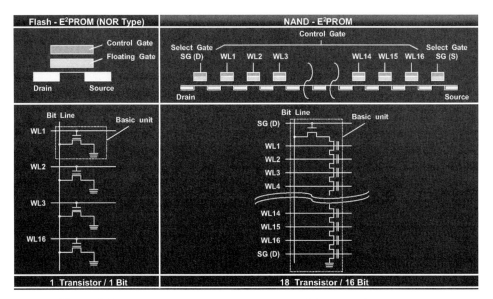

Figure 6.1. NAND EEPROM and Flash EEPROM cell (NOR-type) comparison.

EEPROM can be easily programmed in a long page (on the order of 16 kbits) so that the programming time per byte can be quite short (typically 200 μs). A 56-nm 8-Gb NAND EEPROM has introduced 8-kbyte page programming to achieve 10-MB/s program throughput with multilevel cell (MLC), in which 2 bits are srored per cell [8].

On the other hand, the random-access speed is slow. It can take several tens of microseconds (typically 25 μs) to access a particular address because the resistance of memory cells and select transistors connected in series typically limit the read cell current to 1 or 2 μA. NAND EEPROM designs incorporate fast page buffers to work around the read delay. For example, one current design has decreased the serial cycle time of the page buffers from 50 to 30 ns [9].

Figure 6.2 summarizes NAND- and NOR-type EEPROMs. The features of NAND EEPROM and NOR EEPROM are quite complementary. NAND is a block-oriented memory, and NOR is a bit-oriented memory. NAND EEPROM is used for applications that require reading and programming blocks of data. One high production volume example is a digital camera, which is required to quickly

OVERVIEW OF NAND EEPROM

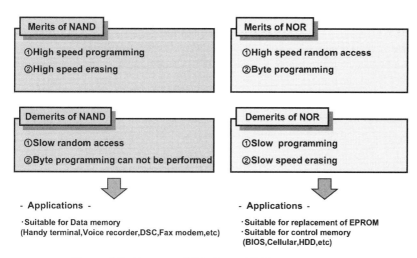

Figure 6.2. Features of NAND- and NOR-type EEPROMs.

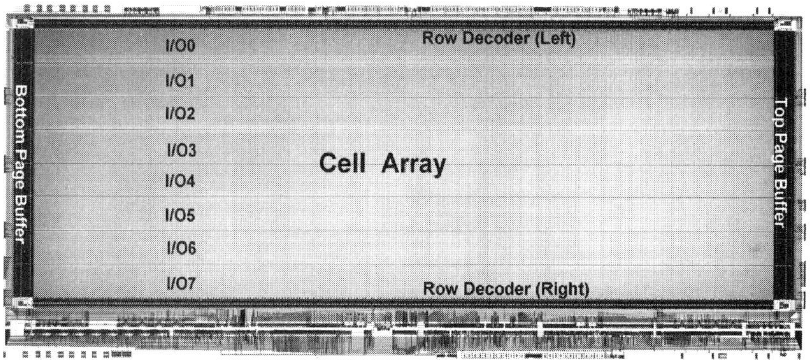

Figure 6.3. Microphotograph of a 3.3-V-only 64-Mbit NAND EEPROM.

program and read one-half megabit blocks. In contrast, NOR EEPROM might be used in the control system for a mobile telephone where fast random access is of value.

Figure 6.3 presents a microphotograph of a 3.3-V-only 64-Mbit NAND EEPROM designed by Samsung and Toshiba [6], which is fabricated using a 0.4-µm, single-metal complementary metal–oxide–semicondutor (CMOS) technology resulting in a 122.9-mm^2 die size and a 1.1-µm^2 effective cell size. The key device features are summarized in Table 6.2.

Figure 6.4 is a block diagram for the core circuits. To minimize the die size, 69,206,016 cells are organized in a single 16-kbit row by (4 kbit + 128 bit) column array. One block corresponds to a row of NAND strings, and a NAND string includes 16 cells serially connected between two string select transistors. An extra 128 bits are added to each row for spare data storage, which are typically used to store system data and/or error correction code (ECC) data.

Row decoders are symmetrically split into left and right sections to accommodate the wordline pitch of 0.76 µm. Due to a very tight bitline pitch of 1.2 µm, page

TABLE 6.2. Key Device Features of 64-Mbit NAND EEPROM

Chip size	122.0 mm^2 (16.74 × 7.34 mm)
Cell size	1.10 μm^2
Organization	8 Mbits × 8
Page size	528 bytes (512 + 16)
Block size	(8192 + 256) Bytes = 16 pages
Supply voltage	3.3 V ± 0.3 V
Page program	200 μs
Block erase	2 ms
Random access	5 μs
Serial access	35 ns
Process	400 nm CMOS, double poly, wordline WSi, 1 metal

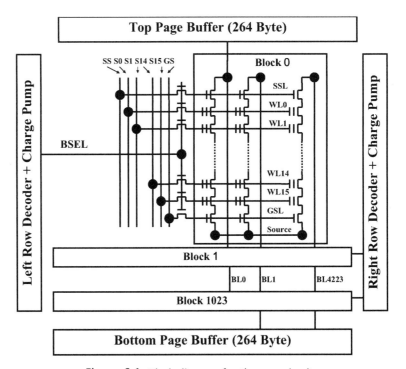

Figure 6.4. Block diagram for the core circuits.

buffers are also split into top and bottom banks. Each sense and latch unit in the page buffer is connected to a bitline, and the page buffers in the top bank are connected to even-numbered bitlines, while those in the bottom banks are connected to odd-numbered bitlines. The split page buffer scheme supports an interleaved data path for fast sequential read operations. Program and read operations are performed in a page unit of 512 + 16 bytes, and a page corresponds to a row of cells. The erase unit is a block, and each block corresponds to 16 pages or 8192 + 256 bytes.

6.2 NAND CELL OPERATION

This section introduces the basic NAND cell operations of erase, program, and read [10–13].

6.2.1 Cell Structure

Figure 6.5 illustrates the equivalent circuit diagram and top view of a NAND-structured cell. Sixteen stacked cells are serially connected between two select transistors. By changing the memory cell configuration from NOR type to NAND type, the memory cell pitch can be defined by the polysilicon wiring lithography, which is the easiest for scaling in the large-scale integration (LSI) process so that its cell size can be the smallest. This NAND structure is suitable not only for miniaturizing the cell size but also for operating with a single supply voltage because it can be programmed and erased by FN tunneling current. Therefore, NAND EEPROM can offer many advantages, which are beyond the simple logical connection of the memory cells. As a result, NAND EEPROM has a different market than NOR EEPROM with respect to device characteristics, power consumption, and cost.

6.2.2 Erase Operation

Erase operations can be performed in a block unit of 8 kbytes [6]. The control gate voltage is set to 0 V and the P-well of the cell array is biased at about 20 V, as shown in Figure 6.6. This causes FN tunneling of electrons off the floating gates in the selected block and into the P-well so that the threshold voltage of the stacked-gate memory cell transistor is shifted from positive to negative. The memory cell can be deliberately overerased to as low as −3 V with a single erase pulse because overerasure is of no concern for NAND EEPROM.

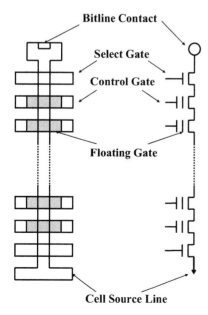

Figure 6.5. Equivalent circuit diagram and top view of NAND structured cell.

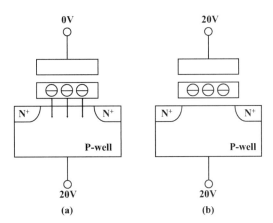

Figure 6.6. Schematic description of NAND EEPROM electrical erase by Fowler–Nordheim tunneling of electrons from the floating gate to the substrate: (a) erase and (b) erase-inhibit.

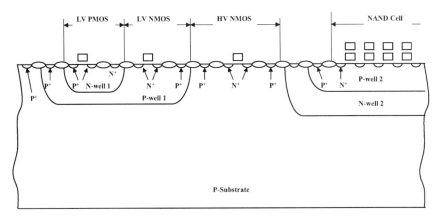

Figure 6.7. Cross-sectional view of NAND EEPROM.

In order to realize this sophisticated erase operation on-chip, the P-well for the memory cell area isolates the peripheral circuit P-well, which can be grounded for the erase operation, as illustrated in Figure 6.7. For an erase inhibit block, the control gates are biased, or boosted up to about 20 V by the capacitive coupling of the control gates and P-well so as to prevent FN tunneling from occurring. In the erase verify operation, all the wordlines of the selected block are set to 0 V, allowing all 16×4096 cells to be verified simultaneously.

6.2.3 Program Operation

The program operation is divided into three procedures. The first is data load in which the data is serially transferred from input/output (I/O) buffers to the data registers.

The second part is program in which all 528×8 cells connected to the same wordline are programmed simultaneously. The channels of the NAND-structured cells are charged or boosted up to 8 V for 1-programming (inhibiting) and are held

NAND CELL OPERATION

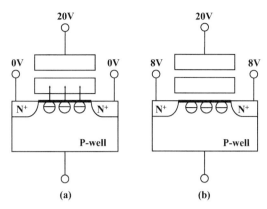

Figure 6.8. Schematic description of NAND EEPROM electrical program by Fowler–Nordheim tunneling of electrons from the channel to the floating gate: (a) 0-program and (b) 1-program (Inhibit).

to ground for 0-programming. The program is accomplished by applying 20 V to the selected control gate while 10 V is applied to the unselected control gates. In 0-programming, electrons are injected to the floating gate from the inversion layer by FN tunneling, and the threshold voltage of the memory cell changes from the negative value of the erased state to positive value, as shown in Figure 6.8. In 1-programming, the memory cell is kept at the same state as before the program operation because the voltage over the tunnel oxide under the floating gate is too low to cause a tunnel current.

After the program operation, a verify-read operation is performed. In the NAND string, the memory cells are connected in series so that the threshold voltage of each programmed memory cell must be kept between the selected control gate voltage and the pass gate voltage for reading. If the threshold voltage of any pass cell in the NAND string exceeds the pass control gate voltage, the other cells in the same NAND string cannot be read out. In order to suppress overprogramming and to obtain a tight threshold distribution for high-speed page-read, a program verification is required.

6.2.4 Program Disturb

In the NAND cell array when a particular memory cell is programmed the other unselected cells may experience program disturb modes in which several volts are applied between the NAND cell channel and floating gate. There are two types of the program disturb modes, A and B, as illustrated in Figure 6.9. The intermediate level, V_M, which is applied to the unselected control gate during programming, should be set at a value to minimize these two disturbance modes. If V_M were increased to 20 V, the program-inhibit cell would be programmed. At the other extreme, suppose the V_M should be at a lower level. In case that the programmed cell threshold voltage is defined as $V_{th}\,0$, and the bitline voltage is 7 V, the channel voltage for cell B will be the smaller value between 7 V and $(V_M - V_{th}\,0)$. If the latter is smaller, the voltage between the control gate and channel becomes $20\,V - (V_M - V_{th}\,0)$. And if this condition continues for a long time, then programming will occur in cell B. Figure 6.9 shows an experimental result when the program-disturb stress voltage, which is 100 times as long as the normal operation, is applied to a worst-

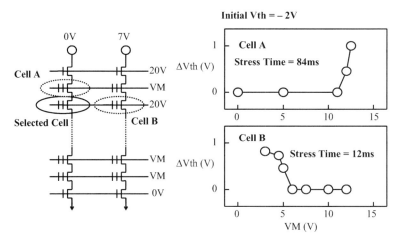

Figure 6.9. Program-disturb modes.

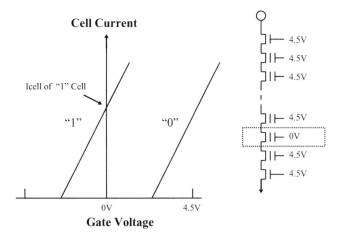

Figure 6.10. Reading of the NAND EEPROM.

case process variation cell. Looking at the data plots, it is evident that a substantial margin can be attained when V_M is set between 6 and 11 V.

6.2.5 Read Operation

The reading method is essentially the same as that of the NAND-type mask ROM, as shown in Figure 6.10. In this case, 0 V is applied to the gate of the selected memory cell, while V_{read} of 4.5 V, which is higher than $V_{\text{cc}} = 3.3$ V, is applied to the gate of the other cells. Therefore, all of the other memory transistors, except for the selected transistor, serve as transfer gates. As a result, in the case where 1 is being written, the memory transistor is in the depletion mode and current flows. Conversely, current does not flow in the case where 0 is written because the memory transistor is in the enhancement mode. The state of the cell is detected by a sense amplifier, which is connected to the bitline. The difference between 0 and 1 is determined by whether negative charge is stored in the floating gate. If negative charge is stored in the floating gate, the threshold voltage becomes higher so that the memory transistor is in the enhancement mode.

6.3 NAND ARRAY ARCHITECTURE AND OPERATION

6.3.1 Staggered Row Decoder

Since the wordline pitch of the NAND cell is determined by the photolithography limit of repeating polysilicon control gates, the narrow control gate pitch makes row decoder layout very difficult. In the previous design [2–5], row decoders are asymmetrically spilt into main and subdecoders, as shown in Figure 6.11(a). The main-row decoder and the subrow decoder have the same pitch as a NAND string. In case of 64-Mbit NAND EEPROM, row decoders are staggered, as shown in Figure 6.11(b), and each row decoder occupies the pitch of two NAND strings [6, 14]. Details of the conventional row decoder and those of the staggered row decoder are illustrated in Figures 6.12 and 6.13, respectively.

Even-numbered wordlines of a unit NAND string in the conventional row decoder are accessed through the combination of BSEL_M and even-numbered common gate lines (S0, S2, S4, ..., S14), while odd-numbered wordlines are accessed through the combination of BSEL_S and odd-numbered common gate lines (S1, S3, ..., S15). The main-row decoder includes a decoding circuit and a local charge pump circuit to supply high enough gate voltage to pass transistors, while the subrow decoder has a local charge pump only. The decoding information of the main-row decoder is transferred to the subrow decoder through a string select line (SSL), and the SSL signal activates the BSEL_S signal. This causes a signal delay from the main-row decoder to the subrow decoder, and the transition time difference between even-numbered wordlines and odd-numbered wordlines is 600 ns.

In the staggered row decoder scheme, the transition time difference between even and odd wordlines is removed through the combination of BSEL and the common gate lines (S0–S15) as shown in Figure 6.13. The staggered row decoder

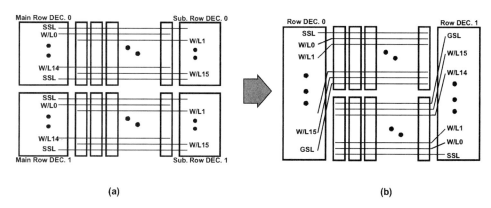

Figure 6.11. Block diagram of (a) conventional row decoder and (b) staggered row decoder.

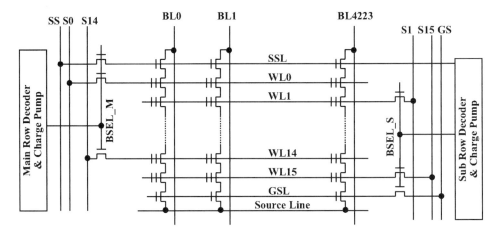

Figure 6.12. Conventional row decoder.

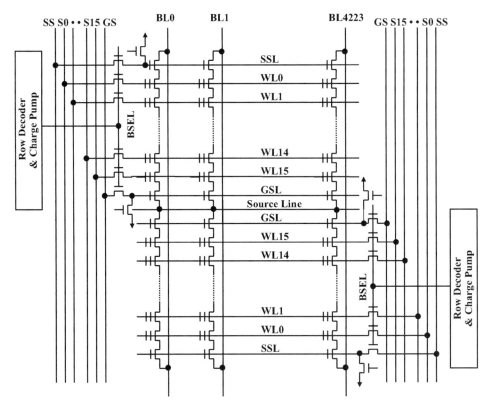

Figure 6.13. Staggered row decoder.

scheme improves the wordline delay by 30% in comparison with the conventional design. Furthermore, the driver circuit, composed of high-voltage transistors, which are the transfer gates for the control gates, is provided in a pitch of two NAND strings so that the layout limitations can be relaxed.

NAND ARRAY ARCHITECTURE AND OPERATION

TABLE 6.3. Bias Conditions for Erase, Program, and Read Operation

Array Nodes	Erase	Program	Read
Selected block			
SSL	Vera × β	Vcc	Vread
WL0, WL2-15	0 V	Vpass	Vread
WL1	0 V	Vpgm	0 V
GSL	Vera × β	0 V	Vread
Unselected block			
SSL	Vera × β	0 V	0 V
WL0, WL2-15	Vera × β	L-floating	L-floating
WL1	Vera × β	L-floating	L-floating
GSL	Vera × β	0 V	0 V
BL0: 0	Vera – Vbi	0 V	Vpre
BL1: 1	Vera – Vbi	Vcc	0 V
SL	Vera – Vbi	Vcc	0 V
Vwell	Vera	0 V	0 V

Table 6.3 summarizes the bias conditions for erase, program, and read operations, which will be described below.

6.3.2 Self-Boosted Erase Inhibit Scheme

The erase operation is performed in a block unit. The control gate (CG) voltage is set to 0 V, and the P-well is biased at about 20 V, causing FN tunneling of electrons from floating gates in the selected block to the P-well. Since a NAND Flash array has many blocks on a common P-well, CGs of unselected blocks should be biased to the same voltage as that of the P-well to inhibit erasing during the block erase operation. Since a CG can have two different voltage levels depending on the selected or the unselected during erase operation, the block decoders are split into two parts for supplying the high erase inhibit voltage. Splitting the block decoders, however, costs large layout area and it makes the layout too difficult. Another issue of splitting the block decoders is that supplying the erase inhibit bias to unselected CGs needs an extra large charge pump as well as the charge pump driving the P-well substrate.

The above problems are readily solved by adopting a self-boosted erase inhibit scheme [5] instead of directly applying the erase-inhibited voltage to the control gate in the unselected NAND blocks. The self-boosted erase inhibit scheme is to utilize the capacitive coupling of cells between CGs and the P-well. As shown in Figure 6.14, before the P-well is biased to the erase voltage, all CGs of unselected blocks are floated by shutting off the pass gate for the selecting signal (0 V) in the decoder while those of selected blocks are grounded. Then, all CGs of unselected blocks are coupled up as the P-well is rising to the erase voltage so that the unselected blocks are automatically erase inhibited. Adopting the self-boosted erase inhibit scheme can make layout of the block decoder easy and the layout area becomes compact. And also it reduces power consumption by eliminating the extra charge pump.

The block erase operation is explained by the miniarray of two blocks, as shown in Figure 6.15. Then 0 V is applied to the common gate lines (S0, S1, S2, S3, ..., S15)

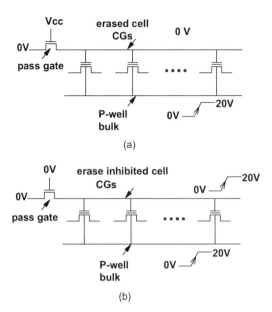

Figure 6.14. Self-boosted erase-inhibit scheme. (a) The bias condition of erase block and (b) the bias condition of erase-inhibited blocks.

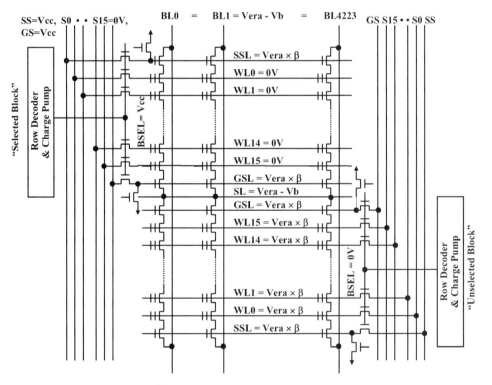

Figure 6.15. Block-erase operation.

NAND ARRAY ARCHITECTURE AND OPERATION

and V_{cc} is applied to the block select node, BSEL, which is the gate node of the transfer gates so that the selected wordlines (control gates) are grounded. Conversely, BSEL in the unselected block is 0 V and their transfer gates are in the off state. When P-well is biased to the erase voltage of V_{era}, those unselected wordlines, which are floating, are self-boosted to $V_{era} \times \beta$. Here, β is the capacitive coupling ratio of cells between CGs and the P-well, and $\beta \cong 0.8$. All the string select lines of SSL and GSL are also self-boosted to $V_{era} \times \beta$ so as not to put a high electric field to the thin gate oxide of the select gates.

6.3.3 Self-Boosted Program Inhibit Scheme

All 528-byte cells connected to the same wordline are programmed simultaneously by changing some cells from erased states to programmed states or by maintaining the other cells' erased states. While the channels of the NAND-structured cells for the cells to be programmed should be biased at 0 V, those of program-inhibited cells should be biased by a program-inhibiting voltage to prevent FN tunneling.

In the previous NAND Flash memory [4], a high program inhibit voltage (e.g., 8 V) was supplied to the NAND string channel directly through the bitlines. However, there are several disadvantages with this method:

1. A large bitline charge pump is required to supply the high voltage on the highly capacitive bitlines. This bitline charge pump will occupy a large chip area.
2. Time and extra current are required to set up the bitline to program inhibit voltage.
3. The page buffer size is increased due to the high-voltage input path and increased transistor size to handle high voltages.

On the other hand, the self-boosted inhibit scheme [5] provides the necessary program inhibit voltage of approximately 8 V even though bitlines are only biased to V_{cc}. The bias conditions for supplying program inhibit voltages to the channel of cells are shown in Figure 6.16.

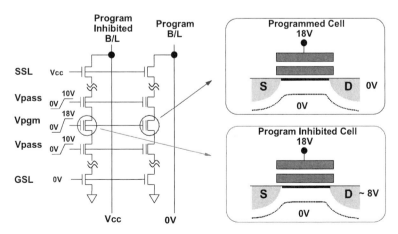

Figure 6.16. Self-boosted program-inhibit scheme.

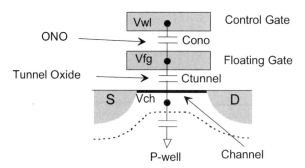

Figure 6.17. Capacitance model for self-boosted program-inhibit scheme.

With the SSL transistors turned on and the GSL transistors turned off, the bitline voltages for cells to be programmed are set to 0 V, while the bitline voltages for cells to be program inhibited are set to V_{cc}. When the program voltage is applied to the control gate of the selected cell, the large potential difference between the floating gate and channel causes FN tunneling of electrons for the programming cells. In program-inhibited cells, the V_{cc} bitline initially precharges the associated channel. When the wordlines of the unit NAND string rise (selected wordline to the program voltage and unselected wordlines to the pass voltage), the series capacitance through the control gate, floating gate, channel, and bulk are coupled and the channel potential is boosted automatically. Assuming a single boosted pass cell, and the model of Figure 6.17, the boosted channel voltage, V_{ch}, can be estimated as follows:

$$V_{ch} = \frac{C_{ins}}{C_{ins} + C_{channel}} V_{cg}$$

where C_{ins} is $C_{ono} \| C_{tunnel}$.

It is calculated that the floating channel voltage rises to approximately 80% of the control gate voltage. Thus, channel voltages of program-inhibited cells are boosted to approximately 8 V when program and pass voltages are raised to 10 V. This self-boosted channel voltage is enough to prevent the program-inhibited cell from FN tunneling if the junction leakage current is lower than 1 nA.

The page program operation is explained by the miniarray of two blocks, as shown in Figure 6.18. At first, the 0-programmed bitlines are precharged to V_{cc}, and the 1-inhibited bitlines become 0 V as forced by the load data in the page buffers. In this case, BL0 and BL1 stand for 0-programmed bitline and 1-inhibited bitline, respectively. In case that the wordline, WL1, in the selected block is being programmed, the common gate line, S1, becomes the program voltage, V_{pgm}, and the other common gate lines (S0, S2, S3, ..., S15) become the pass voltage, V_{pass}. The block select node, BSEL, is boosted to over $V_{pgm} + V_{th}$ by the local charge pump circuit, which is allocated in each block decoder. Here, V_{th} is the threshold voltage for the transfer gates of the wordlines so that the transfer gates are turned on in the linear region and V_{pgm} is smoothly applied to WL1. The other wordlines in the unselected blocks are "low" floating because the block select node, BSEL, is 0 V, however, the string select lines of SSL and GSL in the unselected blocks become 0 V by the grounded transistors of Tr1 and Tr2.

PROGRAM THRESHOLD CONTROL AND PROGRAM V_t SPREAD REDUCTION

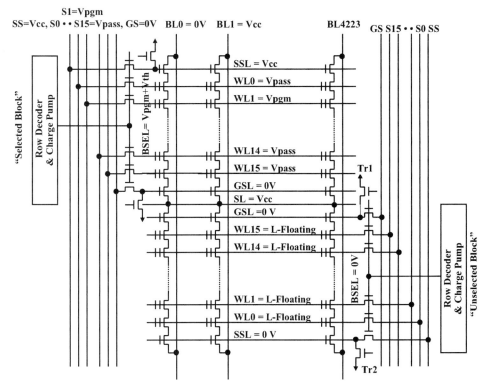

Figure 6.18. Page program operation.

6.3.4 Read Operation

In case of the read operation, the selected wordline and the pass wordlines in the selected block become 0 V and V_{read}, respectively, via the common gate lines, as shown in Figure 6.19. V_{read} is slightly higher than V_{cc} of 3.3 V. In the 64-Mbit NAND EEPROM, V_{read} is set to 4.5 V. The other wordlines in the unselected blocks are "low" floating, however, the string select lines of SSL and GSL in the unselected blocks are set to 0 V by the grounded transistors of Tr1 and Tr2. Therefore, the unselected NAND string doesn't flow any current.

6.4 PROGRAM THRESHOLD CONTROL AND PROGRAM V_t SPREAD REDUCTION

6.4.1 Bit-by-Bit Verify Circuit

In the NAND EEPROM, 528-byte memory cells are simultaneously programmed. If the memory cells are programmed with the same program time, the threshold voltages have a wide distribution resulting from the differences in the program characteristics of the memory cells. This wide program threshold voltage distribution causes difficulties in operating the NAND EEPROM with 3 V. The new verify circuit, which is composed of only two transistors, results in a simple intelligent program algorithm for 3-V-only operation [15, 16].

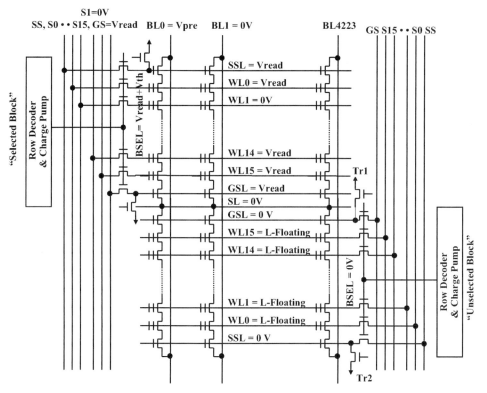

Figure 6.19. Read operation.

Figure 6.20 illustrates a circuit diagram for a bit-by-bit verify circuit and a read/write circuit. A bit-by-bit circuit, composed of only two transistors, is connected to each bitline. Two bitlines and bit-by-bit verify circuits share a common read/write circuit like in an open-bitline architecture of a DRAM. The read/write circuit acts as a flip-flop-type differential sense amplifier in the read operation and as a data latch circuit in the program operation. Figures 6.21, 6.22, and 6.23 show the clock timing diagrams of each operation of program, read, and verify-read in the case that a control gate of CG4 in the array in Figure 6.20(a) is selected.

In the program operation, the initial program data are loaded by a page sequence and latched into the read/write circuits. The memory cells, which share the same control gate, are programmed simultaneously. For 1-programming, the bitlines are charged up to 8 V for the power supply voltage of V_{rw} for the read/write circuits, which is pumped up from the external power supply of V_{cc}. For 0-programming, the bitlines are grounded. The program is accomplished by applying 20 V to the selected control gate of CG4 while 10 V is applied to the unselected control gates of CG1-3, 5-8, as shown in Figure 6.21.

In the read operation, the selected bitlines of BLai are precharged to $\frac{3}{5}V_{cc}$ and the dummy bitlines of BLbi are precharged to $\frac{1}{2}V_{cc}$ at t1 in Figure 6.22. After precharging, the unselected control gates of CG1-3, 5-8 are raised to V_{cc} while the selected control gate of CG4 is grounded <t2>. If the V_{th} of the selected memory cell is smaller than 0 V, a read current flows and the selected bitline voltage decreases below $\frac{1}{2}V_{cc}$ (1-read). Conversely, if the V_{th} is larger than 0 V, no cell read current

PROGRAM THRESHOLD CONTROL AND PROGRAM V_t SPREAD REDUCTION

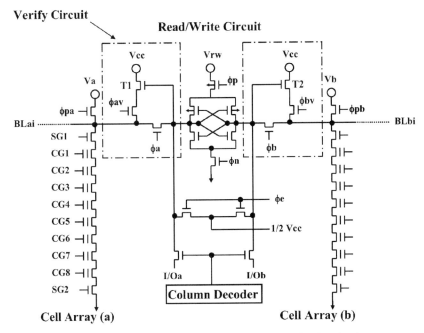

Figure 6.20. Bit-by-bit verify circuit and a read/write circuit.

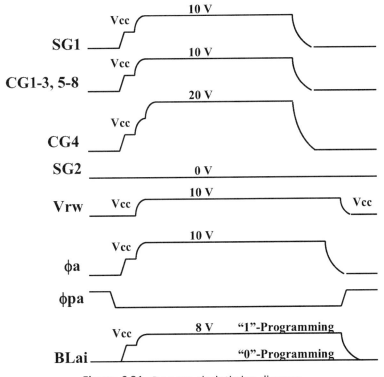

Figure 6.21. Program clock timing diagram.

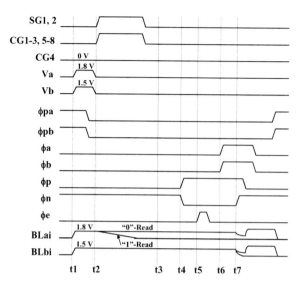

Figure 6.22. Read clock timing diagram.

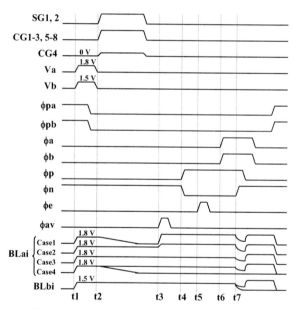

Figure 6.23. Verify-read clock timing diagram.

flows and the bitline voltage is kept at $\frac{3}{5}V_{cc}$ (0-read) <t3>. After all control gates turn low, the read/write circuits are reset <t4> and equalized <t5>, and the bitlines are connected to the read/write circuits by turning the clock of ϕa and ϕb high <t6>. The voltages of the bitlines are sensed by the open-bitline differential sensing manner <t7>.

After the program operation, a verify-read operation is performed to detect the memory cells that require more time to reach the 0-programmed state. In the verify-

TABLE 6.4. Data Modification Rule

	Case 1	Case 2	Case 3	Case 4
Program data (latch data)	1	1	0	0
Cell data	1	0	1	1
Reprogram data (relatched data)	1	1	0	1

read operation, the program data latched in the read/write circuit is modified to the reprogram data, according to the data modification rule, as shown in Table 6.4. As a result, a reprogram operation is performed only on the memory cells that have not yet reached the 0-programmed state. Thereby, the program time is optimized bit by bit. The verify-read operation is similar to the read operation except that the selected control gate of CG4 is raised to the verify voltage of 0.4 V at t2 in Figure 6.23 and the bit-by-bit verify circuits are enabled by turning the clock of ϕav high <t3>. The read/write circuits act as program data latch circuits until $t4$ and as sense amplifiers after $t4$. The sensed data are latched as the reprogram data. In the case that the read/write circuit latches the program data 1, the state of the transistor T1 in the bit-by-bit verify circuit is ON. The bitline after 1-programming is recharged over $\frac{1}{2}V_{cc}$ by the bit-by-bit verify circuit <t3>. Therefore, the latched reprogram data is 1 independent of the memory cell data (Case 1,2). In the case that the read/write circuit latches the program data 0, the state of the transistor T1 is OFF. So, the bitlines are not recharged by the bit-by-bit verify circuits even if the clock of ϕav turns high. If the memory cell has been successfully programmed 0, the bitline voltage after 0-programming is over $\frac{1}{2}V_{cc}$. Conversely, if the memory cell does not reach the 1-programmed state, the bitline voltage decreases below $\frac{1}{2}V_{cc}$. The latched reprogram data is 1 for the memory cell, which is in the 0-programmed state (Case 3). The reprogram data is 0 for the memory cell, which has not yet reached the 0-programmed state (Case 4). By using the bit-by-bit verify circuit, the program data are automatically and simultaneously modified to the reprogram data according to Table 6.4.

Figure 6.24 compares a new quick page-programming algorithm with the conventional one. At first, the initial program, the first verify-read is performed. If the verify-read data is all 0, the program operation is completed. Otherwise, the reprogram and the verify-read are repeated until the verify-read data becomes all 0. The verify-read data of all 0 indicates that there is no memory cell that requires more time to reach the 0-programmed state because it indicates that the reprogram data is all 1. The new quick algorithm has no complicated modify operation and no data reload in comparison with a conventional one [17], as shown in Figure 6.24(b). Therefore, utilizing the quick algorithm, the total page-programming time is significantly reduced.

Under the conditions that (1) the page size is 528 bytes, (2) the internal serial access time is 30 ns, (3) the external serial cycle time is 50 ns, (4) the random access time is 3 μs, (5) the programming pulse width is 25 μs, (6) the programming cycle time is 30 μs, (7) the number of verify-read and reprogram cycles is 5, and (8) the operating frequency of the external processor unit is 50 MHz, by the conventional algorithm, the total program time is

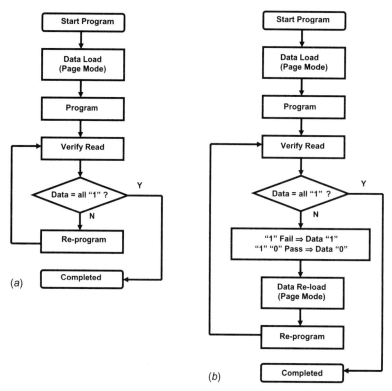

Figure 6.24. Page-programming algorithm. (a) quick page programming; (b) conventional page programming.

{Verify-read + data modification + data reload + program} × 5 cycles = page program time

{(3 μs + 527 bytes × 50 ns) + (528 bytes × 20 ns) + (528 bytes × 50 ns) + 25 μs} × 5 = 456.6 μs

By the quick algorithm,

Verify-read + {program × 5 cycles} = page program time
30 μs + 30 μs × 5 = 180 μs

Thus, the total page-programming time by the quick algorithm can be reduced to 39% of that by the conventional one. Moreover, no external processor is needed for the data modification and the data reload.

6.4.2 Sophisticated Bit-by-Bit Verify Circuit

A sophisticated bit-by-bit verifying scheme, which is able to realize a tight programmed threshold voltage distribution of 0.8 V, has been proposed for NAND EEPROM's [18]. A new bit-by-bit verifying circuit is composed of a conventional sense amplifier and a dynamic latch circuit with only three transistors, increasing the chip size of the 64-Mbit NAND EEPROM [6] less than 1%.

PROGRAM THRESHOLD CONTROL AND PROGRAM V_t SPREAD REDUCTION

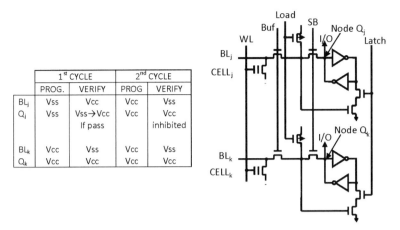

Figure 6.25. Conventional bit-by-bit verifying scheme according to a page of load data.

Figure 6.25 shows the conventional bit-by-bit verifying scheme according to a page of load data. If the memory cell, Cell j, is being programmed, the node of a page buffer, Q_j, is set to the source voltage, V_{ss}. After each programming attempt, a verify-read is performed by applying a reference voltage, V_{ref}, to the wordline to verify whether the threshold voltage of the programming cell surpasses V_{ref} or not. If the threshold voltage of Cell j exceeds V_{ref} in the first programming cycle, the voltage of Q_j is charged up to V_{cc} from V_{ss} by the load transistor so that the page buffer is flipped, and then Cell j is not programmed after that. Once the voltage of Q_j becomes V_{cc}, the page buffer cannot be flipped back even though the bitline voltage of Cell j is discharged down to V_{ss} from V_{cc} in the next verify-read operation. Conversely, Q_k for the nonprogramming cell is set to the power supply voltage, V_{cc}, and keeps it. When every voltage Q in the 528-byte page buffer eventually becomes V_{cc}, the program operation is stopped.

As shown in Figure 6.26, the real NAND cell architecture has a parasitic small resistance [19, 20]. In the worst case that each cell current of the NAND string is 10 µA, and the total resistance of the cell source wiring and the driver transistor is 10 Ω, ΔV_s is calculated to be 0.4 V. In this verify-read operation, a voltage rise, ΔV_s, of the NAND string source from the ground level, significantly influences the comparison of the threshold voltages with V_{ref}. For a rather fast programming cell, the equivalent wordline voltage is regarded as $V_{ref} - \Delta V_s$. Thereby, when the threshold voltage becomes higher than $V_{ref} - \Delta V_s$, the verify-read result for a rather fast programming cell passes. ΔV_s approaches 0 V from 0.4 V as the number of programmed cells increases because the total cell current, Icell_sum, decreases. As a result, some programmed threshold voltages go below V_{ref}, as shown in Figure 6.27. This distribution tail prevents V_{ref} from being lowered below 1 V, which is a significant disadvantage toward a future low-voltage operation.

Figure 6.28 compares the page-programming algorithms of the conventional and proposed bit-by-bit verifying schemes. In the proposed algorithm, when the page data are loaded into the page buffers, the initial load data are simultaneously stored in the dynamic latch circuits in order to detect whether the page buffers are programming or not. The initial load data are then recalled to the page buffers

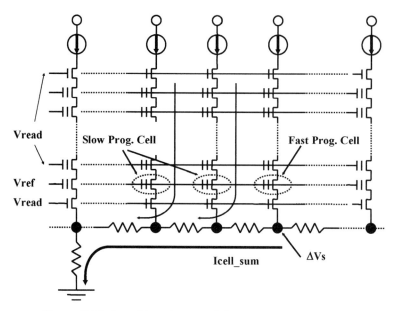

Figure 6.26. Parasitic resistance of the NAND cell architecture.

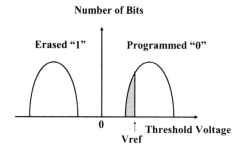

Figure 6.27. Threshold voltage distribution for the conventional bit-by-bit verifying scheme.

before the verify-read operation so that all programming page buffers are verified, independent of the previous verification results.

Figure 6.29 presents the proposed bit-by-bit verifying circuit, which adds only three transistors, M1, M2, and M3, for a dynamic latch circuit to the conventional sense amplifier. A timing diagram of the proposed circuit is illustrated in Figure 6.30. In the data load operation, the node of N_j for the programming page buffer stores V_{cc}. It can distinguish the programming page buffer from the nonprogramming page buffer, even if Q becomes V_{cc} after the verify-read operation. Before each verify-read operation, every programming page buffer can be reset according to the initial load data. Therefore, even if Q_j turns to V_{cc} from V_{ss} in the first programming cycle, Cell j can be programmed again in the $(n + 1)$th programming cycle when the nth verify operation fails due to a decrease in ΔV_s.

The programming page buffer reset takes less than 100 ns, which is a negligibly short increase in a programming cycle time of 35 µs. With respect to the area penalty, the dynamic latch circuits occupy only 1 mm², which is less than a 1% increase of the chip size for the 122.9-mm² 64-Mbit NAND EEPROM [6].

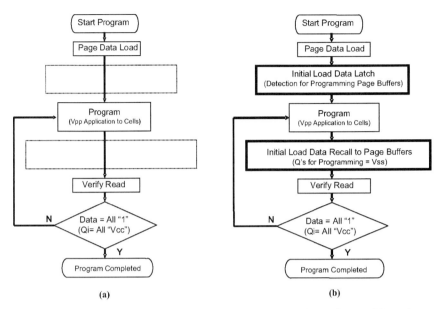

Figure 6.28. Page programming algorithm for (a) conventional bit-by-bit verifying scheme and (b) proposed bit-by-bit verifying scheme.

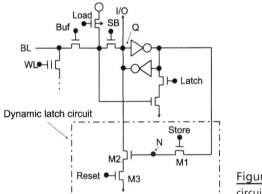

Figure 6.29. Proposed bit-by-bit verifying circuit.

The conventional and proposed verifying schemes are experimentally compared by using the 64-Mbit NAND EEPROM, as shown in Figure 6.31. In this case, V_{ref} is set to 1.0 V. However, the threshold voltage distribution of the programmed cells in the conventional verifying scheme results in voltages below 1.0 V. Conversely, in the proposed verifying scheme, the distribution has been successfully maintained between 1.1 and 1.9 V. Therefore, V_{ref} can be lowered from 1.0 to 0.5 V by utilizing the proposed verifying scheme, so that a wider margin in the read operation can be attained for the wordline voltage of the pass memory cell, V_{read}, of 4.5 V.

Moreover, it is important to note that the programming voltage can also be lowered by the proposed scheme. When V_{ref} is lowered by 0.5 V, the programming voltage, V_{pgm}, can be practically lowered by 0.5 V from 18.5 to 18 V. The pass wordline voltage for the unselected cells, V_{pass}, which boosts the channel of the

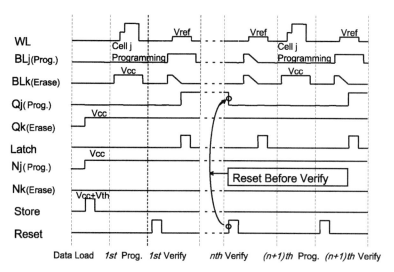

Figure 6.30. Timing diagram for the proposed bit-by-bit verify.

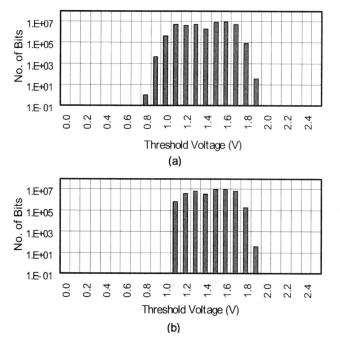

Figure 6.31. Measured programmed threshold voltage distribution of 64-Mbit NAND EEPROM for (a) conventional bit-by-bit verifying scheme and (b) proposed bit-by-bit verifying scheme.

nonprogramming NAND string, can then be lowered by 1.0 V from 12 to 11 V due to reductions in both the programmed threshold voltage and programming voltage. As a result, the program disturbance has been suppressed, and the area for the high-voltage circuits, such as the charge pumps, row decoders, and memory arrays, has been reduced. It is therefore essential that a high reliability NAND EEPROM be attained with the sophisticated bit-by-bit verifying scheme.

PROGRAM THRESHOLD CONTROL AND PROGRAM V_t SPREAD REDUCTION

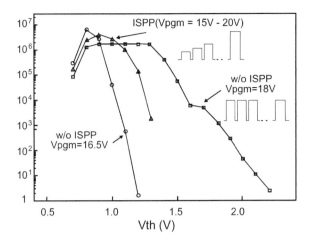

Figure 6.32. Programmed cell V_{th} distribution of 32-Mbit NAND Flash memory with ISPP and without ISPP.

As described earlier, 528 byte cells (one page) are programmed simultaneously by FN tunneling in the NAND Flash memory. FN tunneling current is strongly dependent on the electric field between the floating gate and the channel. The electric field is determined by the control gate voltage, the tunneling oxide thickness, and the cell coupling ratio. Normally, cell program times can vary widely due to nonuniformity in the process. When a page is simultaneously programmed by the constant programming voltage pulse using the bit-by-bit verification scheme, the page program speed is determined by the slowest programming cell within the page. The program speed could be improved by increasing program voltage, but this can result in an overprogramming problem because of fast programming cells.

Incremental step pulse programming (ISPP) [5] shown in Figure 6.32 achieves fast program performance under process and environment variations while still keeping a tight programmed cell V_{th} distribution. ISPP effectively covers process variation by allowing "easy-to-program" cells to be programmed with a lower program voltage and "difficult-to-program" cells to be programmed with a higher program voltage. After an initial 15-V program pulse, each subsequent pulse (if required) is incremented in 0.5-V steps up to 20 V. Since sufficiently programmed cells are automatically switched to the program inhibit state in the verification step, easily programmed cells are not affected by the higher program voltage.

Incremental step pulse programming has the effect of increasing pulse width without actually increasing the program time by dynamically optimizing program voltage to cell characteristics. Figures 6.32 and 6.33 show the distribution of V_{th} and the distribution of the number of program cycles for all 8 K pages in a 32-Mbit NAND Flash memory with and without ISPP, respectively. With ISPP, all of pages are programmed within 10 program pulses, keeping a narrow-cell V_{th} distribution.

6.4.3 Overprogram Elimination Scheme

In the NAND string, 16 memory cells are connected in series so that 15 cells serve as pass transistors when one selected cell is read. Therefore, a NAND string failure occurs if 1 of 16 memory cells should be programmed to more than 4.5 V, which is

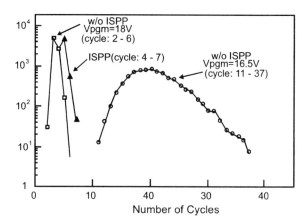

Figure 6.33. Distribution of program cycles in 32-Mbit NAND Flash memory.

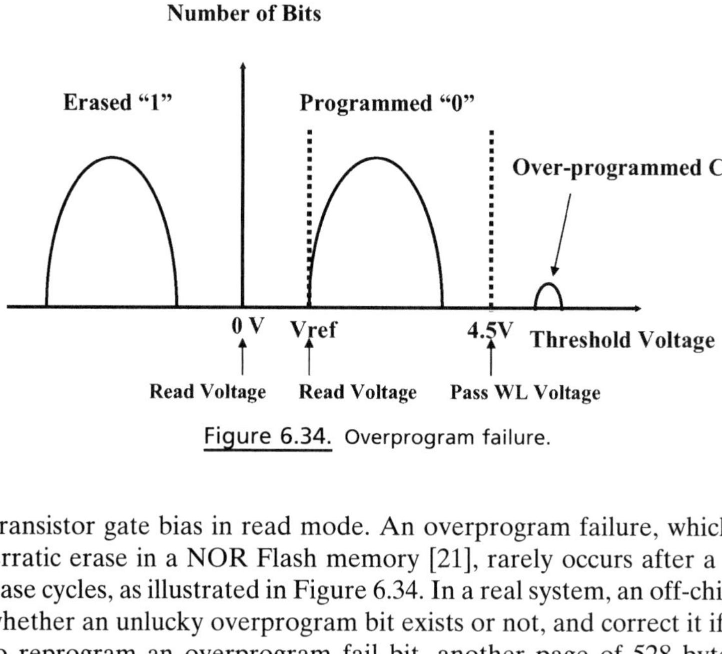

Figure 6.34. Overprogram failure.

a pass transistor gate bias in read mode. An overprogram failure, which is similar to the erratic erase in a NOR Flash memory [21], rarely occurs after a number of write/erase cycles, as illustrated in Figure 6.34. In a real system, an off-chip ECC can check whether an unlucky overprogram bit exists or not, and correct it if it does. In order to reprogram an overprogram fail bit, another page of 528 byte registers, which can store the initial program data, is needed either on- or off-chip because the initial program data latched in one page of registers eventually turns to all 0 after the bit-by-bit program verify completely finishes.

Figure 6.35 shows the flowchart of the conventional program and program-verify sequence, and Figure 6.36 illustrates the data pattern of the page buffers and selected page of memory cells step by step. Previous to the program, the data are loaded to the page buffers (i). These data are programmed to a selected page simultaneously (ii). Fast programmed cells due to the process variations are programmed during one program pulse, and then the program-verify follows automatically (iii). The register data for the sufficiently programmed cells, the V_{th}'s of which become higher than the reference voltage of 1 V, are changed to the program inhibit status of 1. In the next programming cycle, these modified data are used. Thereby, the suf-

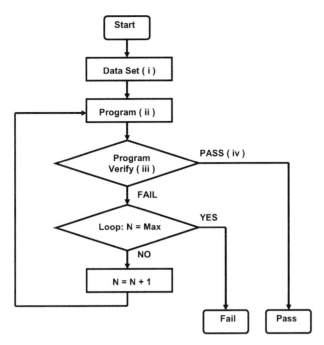

Figure 6.35. Flowchart of the conventional program and program-verify sequence.

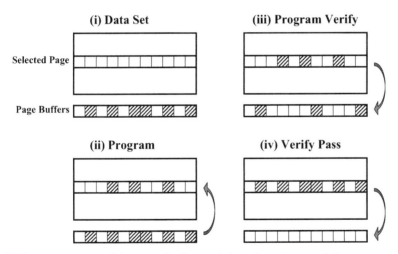

Figure 6.36. Data pattern of the page buffers and the selected page of the memory cells for the conventional sequence.

ficiently programmed cells are not programmed any more. When the register data turn to all 1, the program and program-verify sequence finishes, even if an overprogram fail bit exists (iv). In the conventional scheme, the fail page address, which includes an overprogrammed cell, cannot be found out from 16 pages in a block. When an overprogram failure occurs, the entire block of 16 pages must be erased and reprogrammed. Therefore, the user must preserve the 16-page data.

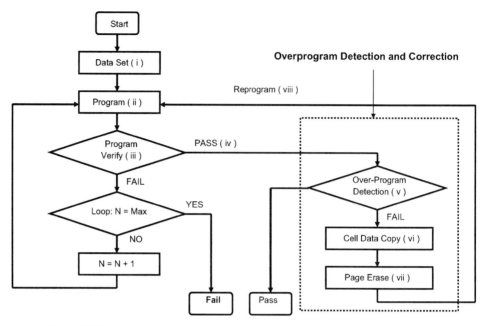

Figure 6.37. Flowchart of the overprogram detection and correction sequence.

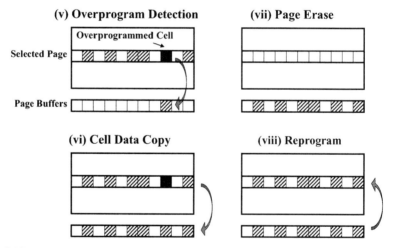

Figure 6.38. Data pattern of the page buffers and the selected page of the memory cells for the overprogram detection and correction sequence.

Figure 6.37 presents the proposed verify-sequence with overprogram detection and correction [22]. Figure 6.38 (v) to (viii) schematically show the overprogram detection and correction sequence followed by the conventional program-verify. In the overprogram detection mode, all 16 wordlines in the NAND string are biased to 4.5 V, which is the same voltage as the pass transistor gate bias in read mode. In case that an overprogrammed cell, the V_{th} of which is higher than 4.5 V, is detected, the bitline voltage becomes high because no cell current flows. In the detection sequence, the page buffer corresponding to the fail bit is reversed (v). This reversed

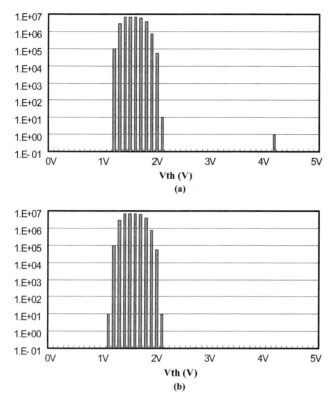

Figure 6.39. V_{th} distribution for 64-Mbit NAND EEPROM. The programmed V_{th} distribution of 32-Mbit cells for the checker-pattern is plotted: (a) intentional overprogram failure and (b) after overprogram correction.

datum represents the programming status of 0. In case that an overprogrammed cell is detected, the overprogram correction sequence starts. The rest of data are copied back from the memory cells and the correction data are prepared (vi). Next, the page data are erased (vii). After that, the program through program-verify sequence, (ii) and (iii), is repeated until the copy-back data are completely programmed.

The proposed overprogram detection and correction scheme has been experimentally confirmed by using the 64-Mb NAND Flash memory. Figure 6.39 shows the V_{th} distribution of memory cells in a chip, where a checker pattern is programmed. Here, the positive V_{th} distribution for 32-Mbit cells is plotted. To confirm the proposed scheme, one bit cell is intentionally overprogrammed [Fig. 6.39(a)]. Consequently, the fail page, which contains an overprogrammed cell, is erased and reprogrammed after the overprogram detection sequence. As a result, the V_{th} of the overprogrammed cell is absorbed into the main distribution without changing the shape of the distribution [Fig. 6.39(b)]. This kind of failure is erratic and seldom occurs in the next program cycle. Figure 6.40 shows the time-consumption penalty of overprogram detection and correction. The overprogram detection penalty is only 3.6%. When an overprogram failure occurs in one page, the total penalty for overprogram detection and correction is 81.1%. For the actual usage, the frequency of failure is extremely low.

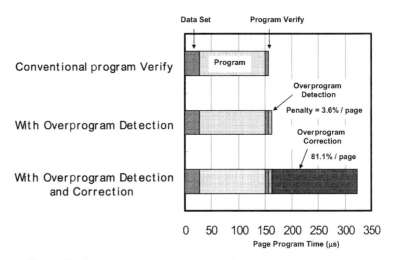

Figure 6.40. Time-consumption penalty for one page programming.

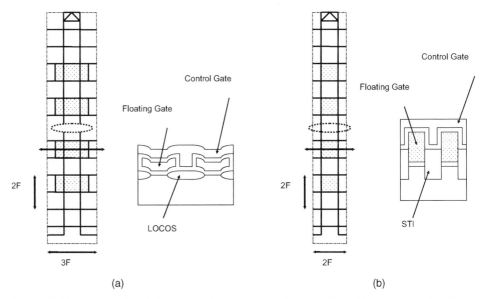

Figure 6.41. SA-STI cell and the conventional LOCOS cell comparison: (a) LOCOS NAND cell and (b) STI NAND cell.

6.5 PROCESS AND SCALING ISSUES

6.5.1 Shallow Trench Isolation NAND Technology (256-Mbit NAND)

For an ultra-high-density NAND structure cell, a self-aligned shallow trench isolation (SA-STI) technology is introduced [23]. Figure 6.41 compares the cross-sectional and top views of the SA-STI cell with those of the conventional LOCOS cell. A small cell size is achieved by the shallow trench isolation to separate neighboring bits and the self-aligned floating gate, which eliminates the floating-gate wings. Even though the floating-gate wings are eliminated, a high coupling ratio of 0.65 can be obtained by using the side walls of the floating gate to increase the coupling ratio.

However, the high aspect ratio of the gate space has made it difficult to control the planarization process of the trench isolation by chemical mechanical polishing (CMP). To overcome this problem, a stacked floating-gate structure is applied [24, 25]. The first thin poly-Si gate is self-aligned with the active area of the cells to control the channel width precisely. A global planarization by CMP process is very controllable due to the reduction of the gate space aspect. The second poly-Si gate is formed on the first poly-Si gate to achieve a high coupling ratio (0.6) for the cell. The second poly-Si gate is patterned with spacing of 0.15 µm by a novel SiN spacer process. This process has made it possible to realize 0.55-µm pitch isolation.

The process flow of the proposed NAND STI cell is illustrated in Figure 6.42. The active area is isolated by the STI formation using a self-aligned mask of a thin poly-Si gate [Fig. 6.42(a)]. After chemical vapor deposition (CVD) SiO$_2$ deposition and planarization by CMP, the second poly-Si gate is deposited on the exposed first poly-Si layer, resulting in the stacked floating-gate structure [Fig. 6.42(b)]. The second poly-Si layer is striped with spacing of 0.15 µm, which is smaller than a design rule by a novel SiN spacer process as follows. A SiN mask is patterned at spacing of 0.25 µm, and a 50-nm-thick spacer SiN is then deposited. By etching the SiN mask back, a stripe mask pattern with 0.15-µm spacer is obtained [Fig. 6.42(c)]. After removal of the SiN mask, an interpoly dielectric (ONO) and the control gate are successively deposited [Fig. 6.42(d)]. The control gate and the floating gate are continuously patterned, followed by deposition of a barrier SiN layer and an interlayer.

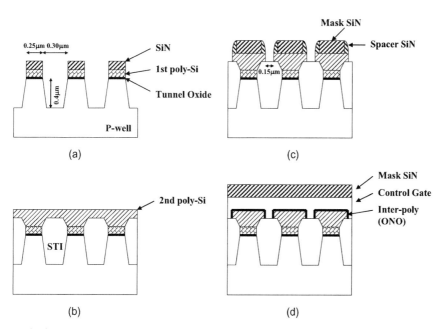

Figure 6.42. Process flow of the NAND STI cells. (a) Trench etching, (b) LPCVD SiO$_2$ fill-in and planarization by CMP, 2nd poly-Si gate deposited, (c) floating-gate formation by SiN spacer process, and (d) ONO and the control gate formation.

The SiN layer covering the control gate prevents a short circuit between the gate and the borderless contacts. Finally, a doped poly-Si is filled in with the bitline contact and source line contact and etched back, followed by the metallization. The memory cells and the peripheral transistors can be formed simultaneously without additional process steps. Before the first poly-Si layer for the floating gate is deposited, 40-nm-thick gate oxides for the high-voltage transistors, and 9-nm-thick gate oxides for the cells and the low-voltage transistors are formed. Therefore, a gate electrode does not overlap the trench corner so that a high reliability of the gate oxide is realized.

Figure 6.43 shows subthreshold characteristics of low-voltage peripheral transistors as a function of the well voltage. No hump is observed because gate overlap with the STI corners has been avoided. Figure 6.44 shows a junction breakdown voltage and a threshold voltage of a parasitic field transistor. The isolation ability

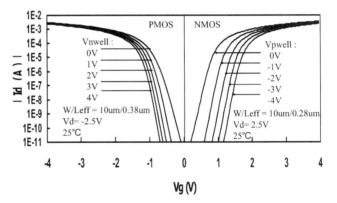

Figure 6.43. Subthreshold characteristics of the peripheral transistor (NMOS and PMOS) as a function of well voltage.

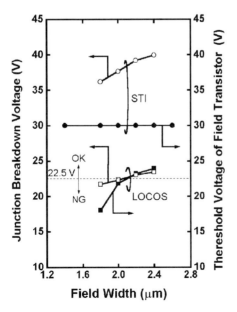

Figure 6.44. Punch-through voltage and a threshold voltage of parasitic field transistor for high-voltage isolation.

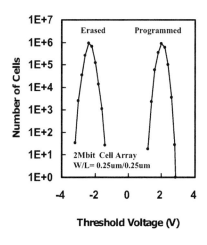

Figure 6.45. Cell threshold voltage distribution in one program and one erase pulse (without verify). Programming and erasing are carried out by 17 V, 10 μs and 18 V, 1 ms, respectively.

of the STI exceeds the required value (>22.5 V) by an adequate margin. Therefore, the self-aligned STI process is compatible with both the peripheral transistor and the memory cell.

A threshold voltage distribution of programmed and erased cells is evaluated by measuring 2-Mbit cell array, as shown in Figure 6.45. Both the programming and the erasing are performed by Fowler–Nordheim tunneling of electrons. A tight distribution of about 2.0 V is realized, although the programming and the erasing are carried out by one pulse without verification, because of a good uniformity of the channel width in the memory cell by using the self-aligned STI structure.

This STI cell has been applied to a 256-Mbit NAND EEPROM [7]. STI has shrunk the bitline pitch by 73% in comparison with the conventional LOCOS isolation, so that an ultrasmall cell of 0.29 μm^2 can be attained by using a 0.25-μm design rule. Therefore a small chip area of 129.76 mm^2 has been accomplished by using NAND-type memory cell and STI.

Figure 6.46 shows a top view of the memory cells. A cross point of an active area line (horizontal line) and a polycide wordline (vertical line) forms the memory cell. Sixteen memory transistors are arranged between two select transistors so that both the bitline and source line contacts are arranged every 18 vertical lines. The memory cells are fabricated with the self-aligned STI and have the stacked floating-gate structure. The stacked floating gate gives a high coupling ratio of 0.6 for the cells. The STI can minimize the isolation width so that the bitline pitch becomes 2.2F, while it was 3F with the conventional LOCOS, where F is the feature size. As a result, 0.55 × 8.54 μm/16 Flash cells can be realized. In order to reduce the fabrication process steps, the peripheral transistors are also made by the same process as the memory cells. They have a stacked gate structure; however, the control gate layer is floating and the floating-gate layer is controlled as a gate electrode. The control gate layer is removed only at the gate contact area.

In the 256-Mbit NAND EEPROM, the cell array is divided into two planes. Each cell array consists of 8k bitlines and 16k wordlines. In order to decrease the chip size, the bitline should not be divided further. The long bitline, which has 1k NAND strings, causes large bitline capacitance, and suppression of the bitline swing is necessary to allow high-speed operation. In order to reduce bitline swing, the floating bitline sensing scheme with a shield is adopted. In comparison with the case of using a bitline load transistor, the required swing is reduced from 1.1 to 0.4 V, and

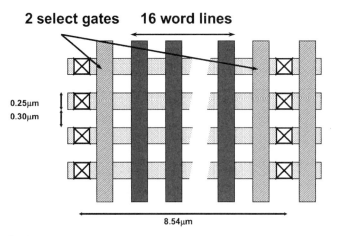

Figure 6.46. Top view schematic of 256 Mbit NAND EEPROM memory cell array.

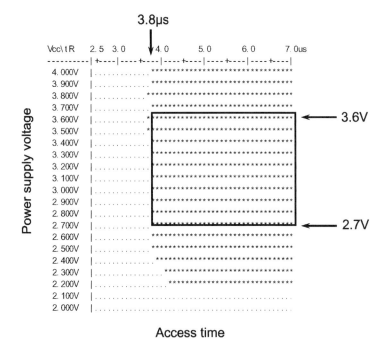

Figure 6.47. Shmoo of first access time.

thus the bitline noise is drastically reduced. A fast readout time of 3.8 μs was achieved in spite of the large bitline capacitance. A shmoo for the first access time from memory cells to data registers is shown in Figure 6.47.

6.5.2 Booster Plate Technology

NAND Flash memory using FN tunneling for cell programming needs a relatively higher voltage (~20 V) than NOR Flash memory using the hot carrier injection. The

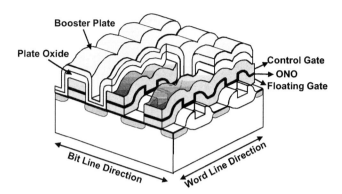

Figure 6.48. Bird's eye view of NAND Flash memory cells with a booster plate.

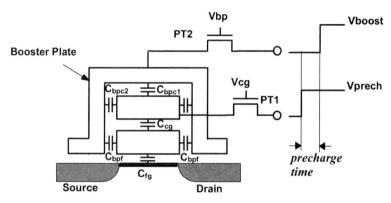

Figure 6.49. Capacitive model of the booster plate cell.

use of higher voltages will pose a serious limitation to the reduction of the device dimensions because the layout of high-voltage transistors in a tight bitline or wordline pitch becomes more difficult as a device is scaled down. Therefore, the reduction of the need for high-voltage levels is very important in high-density NAND Flash memories.

A booster plate technology for NAND Flash cells was introduced in Choi et al. [26]. A bird's eye view of NAND Flash memory cells with a booster plate is shown in Figure 6.48. The booster plate cell has an additional polycide layer that completely covers the top and side of the control gate and the side of the floating gate in the cell transistors of a whole block.

The capacitive model of the booster plate cell with the double boosting program is shown in Figure 6.49. The successive boosting using the booster plate bias after the precharging control gate is a key technique to increase the floating-gate potential. Figure 6.50 shows the simulated result. It can be seen that the floating-gate voltage, V_{fg}, could be raised up to 11 V, which is enough to enable FN tunneling, by the control voltages of 9 V. Table 6.5 summarizes the bias conditions of the booster cell for program, erase, and read operations. By using the booster plate, the high-voltage peripheral circuitry could be operated below 15 V maintaining the tunnel oxide thickness of 100 Å.

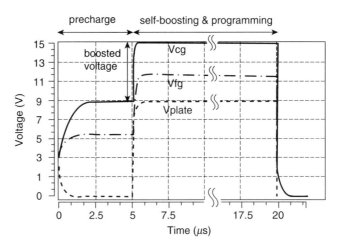

Figure 6.50. Simulation results of the double-boosting program in the booster plate cell.

TABLE 6.5. Comparison of Bias Conditions with/without the Booster Plate

Mode Node	With Booster Plate			Without Booster Plate		
	Program	Erase	Read	Program	Erase	Read
B/L	0	F	1.0	0	F	1.0
SSL	Vcc	F	Vcc	Vcc	F	Vcc
Booster plate	9–12 V	0	Vcc	—	—	—
W/L (select)	9–12 V	0	0	15–20 V	0	0
W/L (unselect)	7 V	F	Vcc	10 V	F	Vcc
GSL	0	F	Vcc	0	F	Vcc
CSL	0	F	0	0	F	0
BULK	0	14 V	0	0	20 V	0

6.5.3 Channel Boost Capacitance Cell

Because of a low channel boost ratio [27] program disturb becomes a significant issue in highly scaled and fast programming memory cells. In the CBC (channel boost capacitance) cell, a high channel boost ratio is obtained by forming a channel boost capacitance between control gates and a cell source/drain diffusion area. In the CBC cell the channel boost ratio increases with decreasing design rules [28] and acts to suppress program disturb. The number of allowable program cycles is 15 times longer than that of the conventional cell.

The cross-sectional view and layout of the CBC cell are shown in Figures 6.51 and 6.52, respectively. There are three key points in the CBC cell. First, a channel boost capacitance (C_{cb}) is formed around the control gates, and the channel boost capacitance gate (CBC gate) contacts with one of the source/drain diffusion areas in the NAND strings. Second, a large C_{cb} is obtained by capacitance oxide thickness (d) and side-wall height (H). H can be controlled by etch-back of the interlayer oxide after planarization process. Third, the CBC cell realizes a drastic increase of the channel boost ratio (C_r) by C_{cb}, without an additional gate control.

PROCESS AND SCALING ISSUES

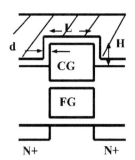

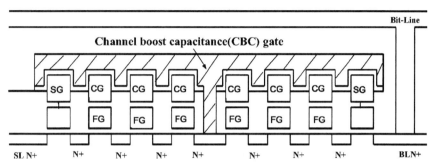

Figure 6.51. Proposed NAND Flash memory cell. Channel boost capacitance (C_{cb}) gate contacts with the source/drain diffusion area. A high channel boost ratio is obtained by C_{cb}. C_{cb} strongly depends on L, H, and d.

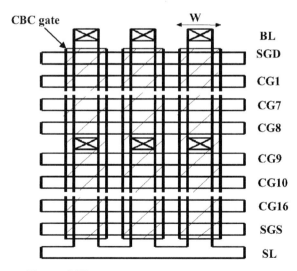

Figure 6.52. Layout of CBC NAND EEPROM cells.

The equivalent circuit of the CBC memory cell is shown in Figure 6.53. Channel voltage (V_{ch}), which is the inhibit voltage for 1 data ($V_{th} < 0$) programming, is raised by capacitive coupling with V_{pass} and V_{pgm}. V_{ch} is mainly affected by C_r. C_r is shown as a function of d and H for a 0.2-μm rule in Figure 6.54. C_r strongly depends on d

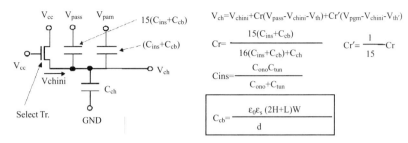

Figure 6.53. Equivalent circuit of CBC memory cell. Channel boost ratio, Cr, is significantly increased by C_{cb}. V_{th} and $V_{th'}$ (are unselected and selected cell's threshold voltage, respectively. C_{ins} is a cell capacitance, C_{ch} the total channel, and the source/drain diffusion area capacitance [28].

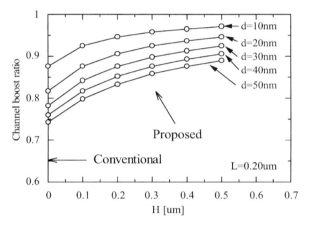

Figure 6.54. Channel boost ratio of the CBC memory cell. The channel boost ratio strongly depends on H and d and is increased to more than 0.85 by C_{cb}.

and H. Thereby, very high V_{ch} can be obtained by C_{cb}. Therefore, the CBC cell drastically improves the program disturbance.

The program disturbance of V_{pgm}-stressed cells and V_{pass}-stressed cells are shown in Figure 6.55 [5]. V_{pgm} is applied to the selected wordline, and V_{pass} is applied to the unselected wordlines. V_{cc} is applied to the program inhibit bitlines to cut off the select gate transistors. In the fast programming, the maximum V_{pgm} and V_{th} become higher than those in the slow programming (V_{th} is strictly controlled).

The program disturbance of V_{pgm}-stressed cells is increased by the V_{th}'s of the unselected cells, as shown in Figure 6.56(a), and a higher V_{pgm} causes larger program disturbance, as shown in Figure 6.56(b). Moreover, the C_r decreases with decreasing design rule [27]. In order to compensate for this effect, V_{pass} must be increased, resulting in a large V_{th} shift of V_{pass}-stressed cells. Therefore, a very high C_r is essential for realizing highly scaled and fast programming NAND Flash memories.

The minimum channel voltage to reduce the V_{th} shifts to less than 0.1 V is shown in Figure 6.57(a). V_{ch} is required to exceed 8.5 V (6.4 V in strict V_{th}-controlled cells). Figure 6.57(b) shows V_{ch} as a function of C_r. V_{ch} exceeds 8.5 V when C_r exceeds 0.85. This is impossible for gate-length-scaled cells because a highly doped channel profile is necessary to improve the cutoff ability of cell and select gate transistors, and this increases channel capacitance (C_{ch}) and decreases C_r. The CBC cell is the key to

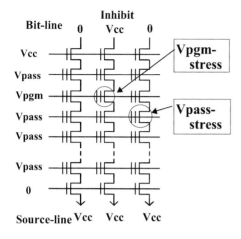

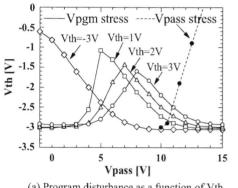
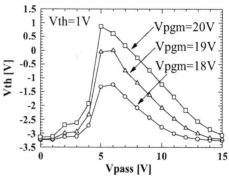

(a) Program disturbance as a function of Vth. (b) Program disturbance as a function of Vpgm.

Figure 6.55. Schematic diagram of the self-boosted programming method [5]. Inhibit V_{ch} is raised by capacitive coupling with V_{pass} and V_{pgm}. Program disturbance of V_{pass}-stressed cells and V_{pgm}-stressed cells has occurred.

Figure 6.56. Measured program disturbance of conventional memory cell. Program disturbance is fatally increased by unselected cells V_{th} and program voltage, V_{pgm}: (a) program disturbance as a function of V_{th} and (b) program disturbance as a function of V_{pgm}.

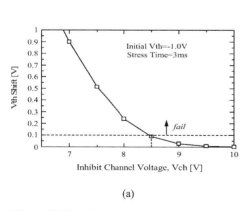
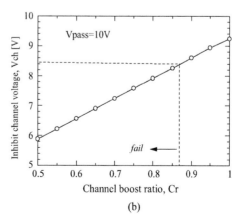

(a)   (b)

Figure 6.57. Inhibit channel voltage to prevent the program disturbance. V_{ch} exceeding 8.5 V and Cr exceeding 0.85 are necessary for small V_{th} shifts (less than 0.1 V): (a) V_{th} shifts versus V_{ch} and (b) V_{ch} versus Cr.

overcome this problem because high C_r can be obtained by C_{cb} when the feature size is scaled.

The most remarkable point of the CBC cell is that C_r is strictly controlled by d and H. A high C_r exceeding 0.85 is obtained for $d = 30$ nm and $H = 0.15$ μm, as was shown in Figure 6.54. Due to this high channel boost ratio, the program disturbance is drastically improved in an accelerated test, as shown in Figure 6.58. The number of allowable additional program cycles is 15 times larger than that of a conventional memory cell. Therefore, the program disturbance doesn't cause any problem when the CBC cell is used.

As mentioned above, the program disturbance is the largest problem in the scaled memory cell. That is to say, as the cell feature size decreases, the number of allowable program cycles is fatally decreased [27]. Figure 6.59 shows the channel boost ratio as a function of design rule. Even when the gate length and gate width are scaled, there is no need to scale d and H, so that the channel boost ratio increases with decreasing design rule. A very high channel boost ratio (exceeding 0.9) is real-

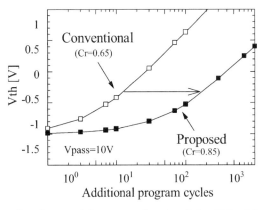

Figure 6.58. Program disturbance of the conventional and the CBC cell in accelerated test. The number of allowable additional program cycles is 15 times larger than that of a conventional memory cell.

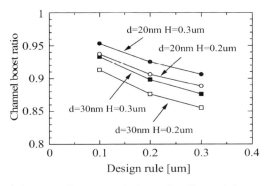

Figure 6.59. Channel boost ratio versus design rule. Channel boost ratio increases with decreasing design rule due to the channel boost capacitance, when the CBC cell is used. Program disturbance must be improved significantly in scaled (especially less than 0.20 μm) NAND Flash memory.

ized when the design rule is less than 0.2 μm. Therefore, the program disturbance must be improved significantly in highly scaled NAND Flash memory.

6.5.4 Negative V_{th} Cell

The concept of the negative V_{th} cell is shown in Figure 6.60 [29, 30]. Both the erased 1 and the programmed 0 state have negative V_{th}. The thermal equilibrium threshold voltage, V_{thi} is also decreased. As a result, the program or erase time of the proposed cell is the same as the conventional one, as shown in Figure 6.61.

In the floating programming scheme, the key issue to reduce the program disturb is to achieve a higher channel voltage for program inhibit cells, V_{ch}. In a conventional cell, 0 cells of unselected wordlines are not turned on until the wordlines are raised to $V_{th0} + V_{chi}$, where V_{th0} is the V_{th} of the 0 state and V_{chi} is the transferred bitline voltage via a select transistor connecting to a bitline. After 0 cells are turned on, the channel is raised by the capacitive coupling with wordlines. Therefore, V_{ch} of a conventional cell is as follows [27]:

```
Vch_conv.=Vchi + 15γ(Vpass - Vth0 - Vchi) + γVpgm -
Tp*Ileak/Ctotal
γ = Cono*Cox/{16Ctotal*(Cono + Cox)}
```

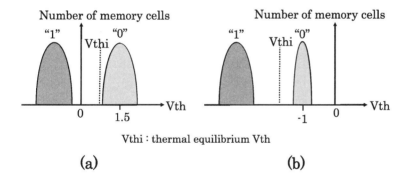

Figure 6.60. Concept of the negative V_{th} cell: (a) conventional cell and (b) proposed cell.

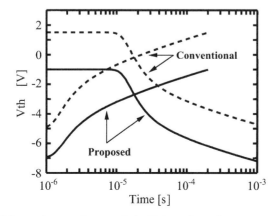

Figure 6.61. Write and erase characteristic. Write voltage is 17 V. Erase voltage is 21 V.

C_{ono}, C_{ox}, and C_{total} are the ONO capacitance, the tunnel oxide capacitance, and the total channel capacitance, respectively. T_p is the program pulse width and I_{leak} is the field leakage current. Conversely, in the negative V_{th} cell, V_{th} of the 0 state is negative and 0 cells of unselected wordlines are turned on even when the wordlines are 0 V. Therefore, V_{ch} of the negative V_{th} cell is as follows:

```
Vch proposed = Vch conv. + 15γ(Vth0 + Vchi) = Vch conv. + 2.4 V
```

Much higher V_{ch} of the proposed cell reduces the electric field across the tunnel oxide and provides a much wider V_{pass} window, as shown in Figure 6.62. Figure 6.63 shows the allowable program-disturb time of the LOCOS cell so as not to cause an incorrect operation. As the LOCOS width, that is, the channel length of the field transistor, is decreased, the threshold voltage of the field transistor decreases due to the short channel effect. As a result, the field leakage current increases, which

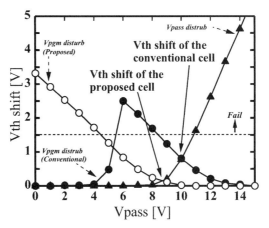

Figure 6.62. V_{th} shift after V_{pgm} and V_{pass} disturb measured at 85°C. The LOCOS width is 0.8 μm.

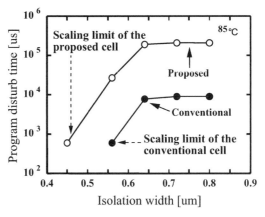

Figure 6.63. Scaling limit of the LOCOS width. The program-disturb time is defined as when V_{th} shifts by 1.5 V. As the LOCOS width decreases, the field leakage current increases and degrades the program-disturb.

decreases V_{ch} and eventually affects the program disturb [27]. The program-disturb time of the negative V_{th} cell is 20 times longer, in comparison with the same LOCOS width. The scaling limit of the LOCOS width decreases from 0.56 μm [27] to 0.45 μm, which leads to 20% isolation width reduction.

To realize the negative V_{th} cell operation, there are two circuit issues. One is to read a negative V_{th}. In NAND Flash memories, the negative voltage bias for a wordline is not preferred because high-voltage NMOS transistors should be fabricated on a P-substrate to lower the manufacturing costs. Thus, a new sensing method is required. The other issue is to realize a bit-by-bit program verify to tighten the V_{th} distribution.

To overcome these problems, a PMOS drive column latch and a V_{cc} bitline shield sensing method are introduced, as shown in Figures 6.64 and 6.65. In a conventional read, a bitline is precharged to V_{cc} and cell current flows from the bitline to a grounded source line. In the proposed scheme, a source line bias method is used. After the bitline is precharged to 0 V, V_{cc} is applied to the source line. The selected and unselected wordlines are 0 and 2 V. As a result, the cell current flows from the source line to the bitline, and then the bitline voltage becomes the absolute value of V_{th}. After N_{sense} is reset to 0 V, the bitline voltage is sensed by the PMOS.

As the bitline pitch is scaled down, the inter-bitline capacitive coupling noise becomes more significant. To suppress the noise, a shielded bitline sensing method has been proposed [16, 31], where neighboring bitlines are alternately selected, and the unselected bitlines are kept grounded. However, if used together with the source line bias method, a cell current continuously flows throughout the read operation

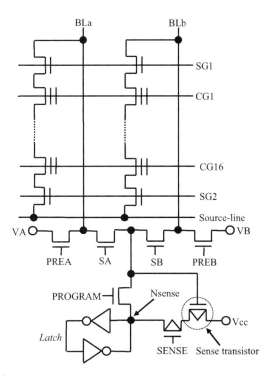

Figure 6.64. PMOS drive column latch and a V_{cc} bitline shield-sensing circuit.

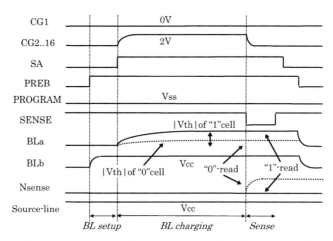

Figure 6.65. Read operation timing diagram. Selected BLa is charged from a source line. Unselected BLb is kept V_{cc} from VB.

from the source line (V_{cc}) to the unselected bitlines (0 V). As the number of unselected bitlines is 4k, the total current increase is as much as 100 mA. To overcome this problem, a V_{cc} bitline shield sensing method is developed, where the unselected bitlines are connected to V_{cc}. In this case, after the unselected bitlines are charged to V_{cc}, no current flows through the unselected bitlines. Therefore, both the inter-bitline capacitive coupling noise and a large current consumption can be successfully eliminated.

By using the new column latch, a bit-by-bit program verify successfully operates and a narrow V_{th} distribution is realized. The program data is latched in the column latch. During program, the bitline voltage is changed according to the program data by activating the PGM signal. The verify read is the same as the read except that the selected CG1 is 0.3 V and the reset of N_{sense} is not performed. The reprogram data is modified such that programming is executed only on memory cells that have not been successfully programmed.

The V_{th} of a memory cell fluctuates due to array noise during a program verify [19, 20]. The array noise is caused by the high resistance of a diffused source line. The source line bounces in a conventional cell and drops in a proposed cell, as illustrated in Figure 6.66. The gate–source voltage, V_{gs}, of a conventional cell decreases and the cell read current decreases, as shown in Figure 6.67. As a result, the V_{th} fluctuates in both positive and negative direction.

The source line resistance and the V_{th} shift decrease by increasing metal bypass lines that connect to a diffused source line. However, this increases the cell area because the bypass lines are not counted as part of the capacity. The total V_{th} fluctuation which is a sum of the positive and negative V_{th} shift, is 0.7 V with 2.5% area penalty [19, 20], as shown in Figure 6.68. If the step voltage of program pulses is 0.5 V, the V_{th} distribution is 1.2 V. Conversely, in the negative V_{th} cell, even if the source line drop decreases the gate–drain voltage, V_{gd}, the cell read current does not change. Therefore, the V_{th} shift is strongly reduced. The V_{th} fluctuation decreases to 0.1 V and a very narrow V_{th} distribution of 0.6 V can be realized.

PROCESS AND SCALING ISSUES 267

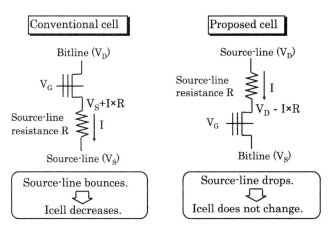

Figure 6.66. Source line drop (proposed) and source line bounce (conventional) during a verify-read.

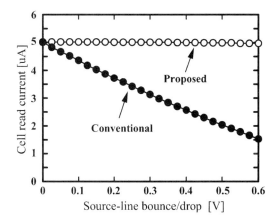

Figure 6.67. Measured cell read current change due to the source line drop (proposed) and source line bounce (conventional).

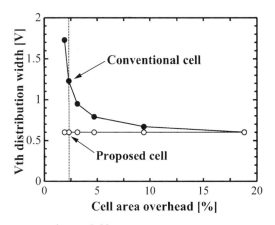

Figure 6.68. V_{th} distribution width.

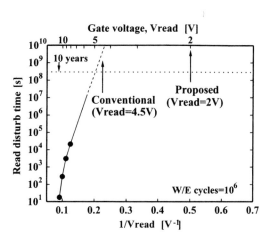

Figure 6.69. Improvement of read disturb. The read-disturb time is defined as when V_{th} reaches $-3.5\,V$ (proposed) and $-1\,V$ (conventional).

In a read operation, the read voltage, V_{read}, is applied to the control gates of the unselected cells in series with a selected cell and V_{read} must be higher than the maximum V_{th} to make the unselected cells act as transfer gates. A read disturb occurs to the unselected cells. The electric field across the tunnel oxide is [32]:

```
Eox = {Cono(Cox + Cono)}(Vread + Vthi - Vth)Tox
```

where T_{ox} is a tunnel oxide thickness. Thus, the tunneling current that causes the read disturb decreases by lowering $V_{read} + V_{thi} - V_{th}$. Here, $V_{thi} - V_{th}$ of the proposed and the conventional cell are the same. In the proposed cell, as V_{th} is negative and the V_{th} distribution is 0.6 V narrower, the maximum V_{th} is 3.1 V lower than the conventional one. As a result, V_{read} decreases from 4.5 to 2 V, and the read-disturb time increases by more than 3 orders of magnitude, as shown in Figure 6.69.

6.5.5 Free Wordline Spacing Cell

In the NAND structure cell, the space between neighboring stacked control gates can be reduced to zero theoretically because neither contact hole nor source line is required between wordlines, and also the voltage between the drain and source of the cell is 0 V for programming. A novel wordline pattering process has reduced this idle space from 0.6 to 0.3 µm with a 0.6-µm ground rule, as illustrated in Figure 6.70 [33].

Figure 6.71 shows the major processing steps to fabricate a 0.3-µm space self-aligned double polysilicon stacked gate. After depositing SiN on the second polysilicon layer, SiN stripe patterning is formed [Fig. 6.71(a)]. Next, the resist is patterned between the SiN lines, where the SiN and resist lines have the same line width of 0.6 µm, and a spacing between the SiN and resist lines is 0.3 µm [Fig. 6.71(b)]. Then, the stacked double polysilicon layers with the interpoly dielectric layer are etched simultaneously by using the SiN and resist mask [Fig. 6.71(c)]. Finally, the resist is removed, and the second and first polysilicon layers serve as the control gate and floating gate, respectively [Fig. 6.71(d)].

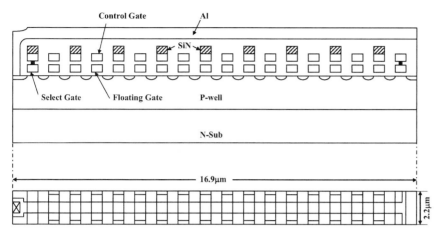

Figure 6.70. Top and cross-sectional views of the NAND-free wordline spacing cells.

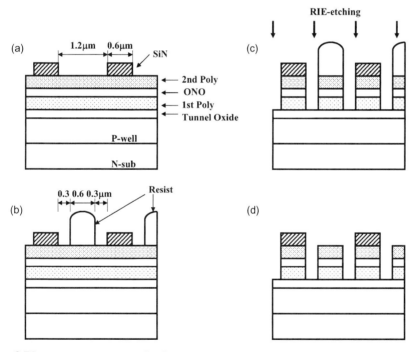

Figure 6.71. Major process steps for fabricating a 0.3-μm space self-aligned double polysilicon stacked gate.

A row decoder design for this cell becomes more difficult due to the tightened wordline pitch. However, it is important to note that the NAND structured cell requires neither control gate spacing nor source and drain implantation of the memory transistor.

6.6 KEY CIRCUITS AND CIRCUIT/TECHNOLOGY INTERACTIONS

6.6.1 Shielded Bitline Sensing Method

As the integration density of the NAND Flash memory has increased, the bitline–bitline (BL–BL) coupling noise has become a serious problem during the read operation. The coupling noise is increased as the bitline pitch becomes smaller. In this section, the BL–BL coupling noise will be investigated, and a bitline architecture to reduce this noise will be proposed [16, 31].

A simplified model for the coupling noise is shown in Figure 6.72(*a*). Where the capacitance of C1 is defined as the coupling capacitance to the adjacent bitline, and C2 is defined as the bitline capacitance excluding $2 \times$ C1. The capacitance of C2 consists of the capacitances between the bitline and the P-well, select gates, control gates, and the third polysilicon wiring. In the NAND-type cell array, the capacitance between the bitline and the P-well is small because only $\frac{1}{16}$ contact hole per memory is used for the case of the eight-NAND-cell string. The measured capacitances of C1 and C2 for the cell array are, respectively, 300 and 950 fF. The bitline length is 1.6 μm × 130 NAND strings.

In the read operation, the coupling noise of ΔVBL is generated in the bitline of BL1 when the select gates and the control gates are activated and a signal voltage appears in the adjacent bitlines of BL0 and BL2. Figure 6.72(*b*) represents the worst case in which 1 data appears on the concerned bitline of BL1 and 0 data on the neighboring bitlines of BL0 and BL2.

The bitline of BL1 voltage should stay at the precharged voltage of 1.8 V. However, the bitline of BL1 voltage is decreased through the BL–BL capacitive coupling as the voltages of both neighboring bitlines of BL0 and BL2 are decreased from 1.8 to 0 V. The maximum coupling noise of ΔVBL is

```
ΔVBL = 1.8 V × [2 × C1 / (2 × C1 + C2)]
```

Under the condition that C1 = 300 fF and C2 = 950 fF, the noise of ΔVBL is 697 mV.

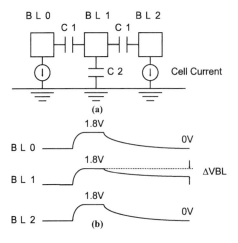

Figure 6.72. Maximum bitline–bitline coupling noise. (*a*) Simplified model of unshielded bitline. (*b*) Bitline waveforms in read operation.

In the read operation, the precharged voltage difference between the selected bitline and the dummy bitline must be larger than this coupling noise. Therefore, the large noise causes a malfunction of the device or results in a long read access time.

In order to reduce the BL–BL capacitive coupling noise, the shielded bitline architecture is proposed. Figure 6.73 illustrates the circuit configuration of the shielded bitline architecture. Four bitlines (BLai, BLai+1, BLbi, BLbi+1) share the common Read/Write sense amplifier. In a read operation, the bitlines are alternately selected, and the unselected bitlines are grounded. If BLai is selected, BLbi becomes the dummy bitline while BLai+1 and BLbi+1 are grounded. In this case, the clocks of $\phi1$, $\phi pa2$, and $\phi pb2$ are activated and the voltages of Va2 and Vb2 are 0 V.

Figure 6.74 shows a simplified model to estimate the BL–BL coupling noise in the case of the shielded bitline. A capacitance of C3 is defined as a BL–BL coupling capacitance between a bitline and its second nearest bitline. The maximum coupling noise becomes

```
ΔVBL = 1.8 V × [2 × C3 / (2 × C3 + 2 × C1 + C2)]
```

The measured capacitance of C3 is 15 fF, so that the noise of ΔVBL decreases drastically down to 5%, from 700 to 35 mV.

In a program operation, unselected bitlines are raised to 8 V to prevent 1 programming. If BLai is selected, BLai+1 is raised to 8 V while BLbi and BLbi+1 are grounded.

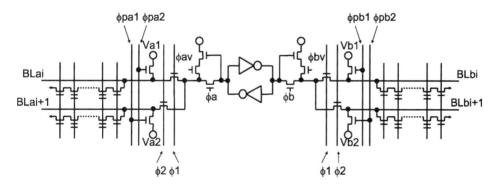

Figure 6.73. Shielded bitline architecture.

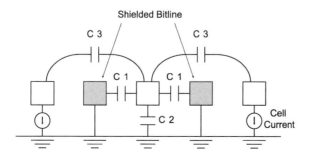

Figure 6.74. Simplified model of shielded bitline.

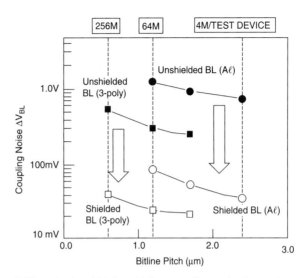

Figure 6.75. Calculated bitline–bitline coupling noise in read operation.

The BL–BL coupling noise in a read operation was calculated as a function of the bitline pitch with two types of bitline materials in Figure 6.75. In a 4-Mb NAND Flash memory, having an aluminum bitline pitch of 2.4 μm, the coupling noise of ΔVBL becomes 700 mV. This noise causes the random access time of over 10 μs in the 4-Mb device. With the shielded bitline architecture, the coupling noise is suppressed by a factor of 20, and the noise is reduced to only 35 mV.

In the case of the third polysilicon bitline, the thickness of the bitlines can be decreased to lower the BL–BL coupling noise. Even in a 256-Mb device and beyond, where the bitline pitch is less than 0.6 μm, the noise is suppressed to 40 mV by adopting the third polysilicon shielded bitline architecture.

The shielded bitline architecture also suppresses a control gate coupling noise, as illustrated in Figure 6.76. When the cell data are read out to the bitlines, the selected NAND strings are charged. If the memory cell is in the programmed state, in the enhancement mode, the selected control gate of 0 V is raised due to the drain–gate capacitive coupling. If the memory cell is in the erased state, in the depletion mode, the selected control gate is also raised due to channel–gate capacitive coupling [Fig. 6.76(a)]. When the selected control gate is raised from 0 V, a programmed cell, which has a rather lower threshold voltage, is turned on. However, every other bitline is grounded for the shielded bitline architecture. The coupling noise is reduced to half that of the unshielded bitline architecture, as shown in Figure 6.76(c), because the coupling ratio of the drain–gate and channel–gate capacitors to the total wordline capacitors becomes a half [Fig. 6.76(b)].

Therefore, the bitline shield technology prevents the programmed cells from being read out due to the control gate coupling noise.

6.6.2 Full Chip Burst Read Operation

Since the NAND Flash memory is used for applications requiring fast bulk data transfer with the simple device interface, the NAND Flash memory uses an internal address counter to support sequential read over multiple pages or over a whole chip.

KEY CIRCUITS AND CIRCUIT/TECHNOLOGY INTERACTIONS

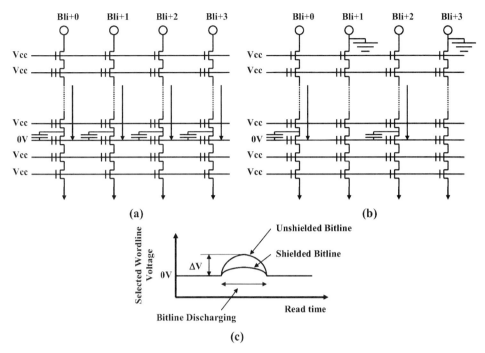

Figure 6.76. Wordline coupling noise suppression. (a) Channel-gate capacitive coupling for the conventional sensing method. (b) Channel-gate capacitive coupling for the shielded bitline sensing method. (c) Coupling noise comparison between the conventional and shielded bitline sensing methods.

During this sequential read operation, the page sensing latency of tR, ~10μs, exists in conventional NAND Flash memories for every new page access. Thus, tR latency is one of the limiting factors in achieving high read throughput. There is an FCBR (full chip burst read) operation that eliminates tR latency except the first page sensing, as shown in Figure 6.77 [6].

In order to implement FCBR capability, each I/O cell array is divided into the left-half page (LHP) and the right-half page (RHP). Read control circuits are also split into two groups according to the cell array, and the read control circuits corresponding to RHP are controlled independently from those of LHP. As soon as a sequential read operation is invoked, the LHP cells of the starting page (nth page) are sensed and latched in the LHP during T1 period. While the stored data in the LHP are read sequentially during T2 period, the RHP cells of the nth page are sensed and latched to RHP concurrently so as to be accessed in the next period T3. Since the sequential read operation of the half page and the sense-and-latch operation of the next half page are performed simultaneously, tR latency between successive page is eliminated completely.

6.6.3 Symmetric Sense Amplifier with Page Copy Function

Figure 6.78 shows the symmetric sense amplifier with page copy function [34, 35]. A flip-flop is connected to one end of each BL. The voltage source of the flip-flop is connected to V_{cc} during the read operation, or to V_m, which is a program inhibit

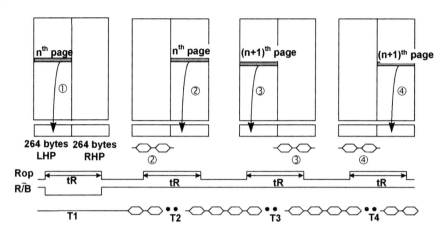

Figure 6.77. Full chip burst read (FCBR) operation.

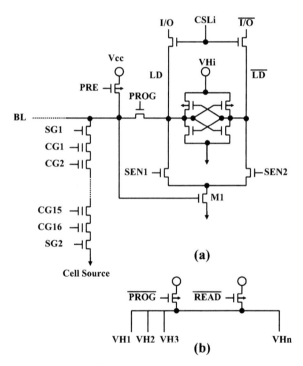

Figure 6.78. Symmetric sense amplifier with page copy function. (a) A single sense amplifier. (b) Connection of the flip-flop voltage source.

voltage, during the program operation, as shown in Figure 6.78. Figure 6.79 shows the timing diagram of program operation, and Figure 6.80 shows the timing diagram of the read and program verify–read operations. In both cases, the control gate, CG3, is selected.

In the program operation, the initial data are loaded by a page sequence and latched into the data registers in the sense amplifiers. The memory cells, which share

KEY CIRCUITS AND CIRCUIT/TECHNOLOGY INTERACTIONS 275

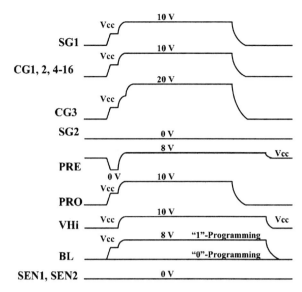

Figure 6.79. Program operation timing diagram.

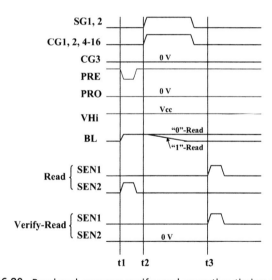

Figure 6.80. Read and program verify-read operation timing diagram.

the same control gate, are simultaneously programmed. The 0 and 1 programming BL's are set to 0 V and V_{cc}, respectively. The selected control gate, CG3, and the unselected control gate, CG1, CG2, CG4–CG16, are boosted to 20 and 10 V, respectively. The threshold voltages of 0 programming memory cells are positively shifted, while those of 1 programming cells are kept negative.

In the read operation, the BL's are precharged to V_{cc} and SEN2 goes high at <t1>. LD is initialized to 0 V. Next, BL is left in a floating state and the selected control gate, CG3, is set to 0 V, while the unselected control gates, CG1, CG2, CG4–CG16, and the select gates, SG1 and SG2, are raised to V_{cc} at <t2>. Finally, BL is discharged to the L level if the selected cell is conducted for 1 data, or is kept high if it is programmed for 0 data. By raising SEN1, LD is inverted at <t3> if BL is high. The BL voltages are detected by the transistor M1. As a result, the cell data are latched in the data register. The readout of the cell data to the register is performed simultaneously for the cells connected to the same control gate.

After the program operation, a verify-read operation is performed in order to modify the program data latched in the register. The read and verify-read operations are performed with the same signal timing except for SEN2. In the verify-read operation, the register data are not initialized at the beginning of the operation. Using the verify-read operation changes the flip-flop state when a cell is programmed. Once LD is set to high, it is kept high until the program operation ends, so that the reprogram data are automatically and simultaneously set.

Figure 6.81 depicts a typical "garbage collection" operation, which is used in most memory systems. A simple and fast page copy operation is required in order to achieve a high garbage collection performance. The page copy function is generally utilized when the data from one page is moved to another page and is one of the most attractive features of the NAND Flash architecture.

Figure 6.82 illustrates the inverted operation timing diagram. Contrary to the normal read operation, the inverted data are stored in the flip-flops in the inverted-

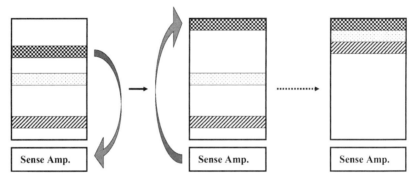

Figure 6.81. Garbage collection operation.

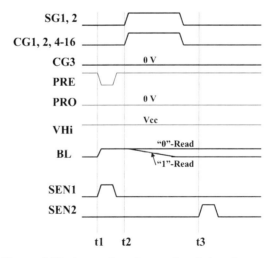

Figure 6.82. Inverted-read operation timing diagram.

KEY CIRCUITS AND CIRCUIT/TECHNOLOGY INTERACTIONS   277

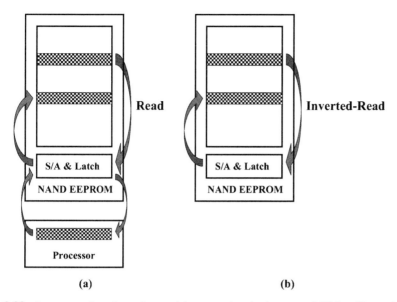

Figure 6.83. Page copy function scheme: (a) conventional scheme and (b) intelligent function scheme.

read operation, in which the timings of SEN1 and SEN2 are exchanged (relative to the read operation). The sense amplifier in Figure 6.78 consists of a symmetric flip-flop in order to detect BL voltages with the same sensitivity during both the read and the inverted-read operations.

Figure 6.83 shows two page copy function schemes: the conventional scheme and the proposed intelligent function scheme. The external processor and extra buffers outside the chip, which are required in the conventional page copy function, are not necessary in the intelligent page copy function. It is not necessary for the new function scheme to modify or invert the read data.

The readout data in the inverted-read operation can be utilized for programming because the LD states for 1 and 0 data programming are the same as these for 1 and 0 data readout in the inverted-read operation as summarized in Table 6.6. In the page copy function from page 1 to page 2, the data that are copied to page 2 are transferred to the sense amplifier in only one inverted-read operation in this scheme. Two additional operations of data reloading must be performed after the read operation in the conventional scheme.

Figure 6.84 outlines the algorithm of the intelligent page copy function. The intelligent copy function is performed on the chip in the background after command and address input sequences. During the background function, the external processor is free and available for other functions. If we assume a page size of 528 bytes, a random-access time of 10 μs, a cycle time of 50 ns for data input and output, and a total required program time of 200 μs, then the intelligent page copy function time is 210 μs. However, when an external processor is used, it requires 263 μs. The page copy time can thus be reduced by 20% by the symmetric sense amplifier. Moreover, it should be noted that neither central processing unit (CPU) resource nor extra buffers are required for this on-chip page copy function, and the external processor is free and available for other functions during execution of the page copy function.

TABLE 6.6. LD State in Program, Read, and Inverted-Read Operations

	0 Dara	1 Data
Program	H (8 V)	L (0 V)
Read	L (0 V)	H (V_{cc})
Inverted-Read	H (V_{cc})	L (0 V)

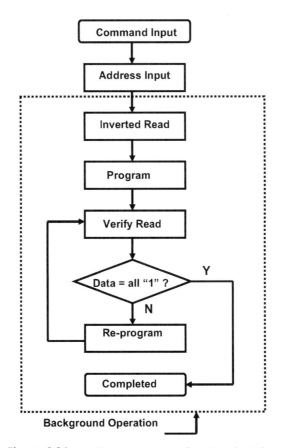

Figure 6.84. Intelligent page copy function algorithm.

6.6.4 Source Line Programming Scheme

For low-voltage operation, this section proposes a new programming scheme, which drastically reduces the program disturbance and realizes a high reliability [36].

Figure 6.85 shows the conventional V_{cc}-bitline [5] (Conv.1), the conventional boosted bitline [37] (Conv.2), and the proposed programming scheme. Table 6.7 summarizes the performance comparison. The erased (1) and the programmed (0) states have a negative and a positive V_{th}, respectively. In Conv.1, a program bitline is grounded and a program-inhibit bitline is V_{cc} [Fig. 6.85(a)]. In Conv.1, the program disturb becomes worse at lower V_{cc}. The V_{cc} minimum of Conv.1 is 2.0 V. In Conv.2,

KEY CIRCUITS AND CIRCUIT/TECHNOLOGY INTERACTIONS

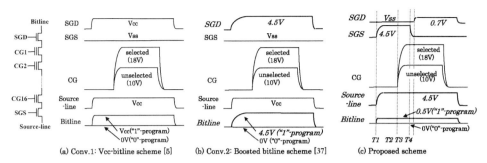

Figure 6.85. Program operation of (a), (b) the conventional scheme, and (c) the proposed scheme.

TABLE 6.7. Performance Comparison ($V_{cc} = 1.4\,\text{V}$)

	Conv.1 [5]	Conv.2 [37]	Proposed
V_{cc} (minimum)	2.0 V	None	None
Programming time	178 µs/page	500 µs/page	192 µs/page
Chip size overhead	None	5%	None
Additional process steps	None	High-voltage PMOS req.	None
Energy dissipation	0.49 µJ	1.13 µJ	0.53 µJ

the program-inhibit bitline and SGD are raised to 4.5 V to reduce the program disturb [Fig. 6.85(b)].

In NAND Flash memories, 4000 memory cells are programmed at the same time to increase the program throughput. In Conv.2, the huge capacitance, as much as 17 nF, of the 4000 bitlines is boosted to 4.5 V. The operating current and the rise time of the bitline boosting increase at lower V_{cc}. In short, although Conv.2 shows excellent program-disturb characteristics, the program time and the power dissipation increase. Moreover Conv.2 is subject to circuit area and manufacturing cost overheads.

In the proposed scheme a higher program-inhibit voltage, for example, 4.5 V is applied from the source line to the channel so as to reduce the program disturb. At T1 in Figure 6.85(c), the source line and SGS are biased to 4.5 V. SGD is 0 V. At T2, V_{pgm} and V_{pass} are applied to the selected and the unselected control gates, respectively. By the capacitive coupling with the control gates, the channels of both the 0 and the 1 program cells are raised to V_{ch}, 8 V. At T3, SGS is discharged to 0 V and then SGD is applied to 0.7 V at T4. As a result, the channel voltage is changed or kept according to the program data. Here, SGD must be higher than 0.7 V to turn on the select transistor connected to the grounded bitline in the case of the 0 program. The channel of the 0 program cell is discharged to 0 V and electrons are injected from the channel to the floating gate. On the other hand, in case of the 1 program, the bitline should be higher than 0.5 V to turn off the select transistor. The channel of the 1 program cell is kept at V_{ch} and the electron injection is prohibited.

In a 16-NAND cell array, one source line is arranged every 32 wordlines, whereas the bitlines cover all memory cells. Therefore, the source line capacitance is much smaller than the bitline capacitance. The source line capacitance is 2 nF and about one-tenth of the bitline capacitance. Due to the small capacitance of the source line, the source line boosting in the proposed scheme is performed during the setup of the bitline and does not degrade the program speed. The time added from T2 to T4 increases the program time by only 2%. As a result, highly reliable and high-speed programming can be achieved. In the proposed scheme, the bitline voltage can be reduced from V_{cc} to 0.5 V because a high program-inhibit voltage is applied from the source line. This drastically decreases the bitline charging current. Therefore, despite the source line boosting, the power consumption of the proposed scheme is almost the same as Conv.1 and a very low power operation can be realized.

The program disturb of Conv.1 is illustrated in Figure 6.86. To reduce the V_{pgm} disturb, the key issue is to achieve a higher channel voltage of program-inhibit cells, V_{ch} [27]. In Conv.1, 0 cells of unselected control gates are not turned on until the control gates are raised to Vth0 + Vchi. Vth0 is the V_{th} of the 0 state. V_{chi} is the voltage transferred from a bitline via a select transistor and is expressed as V_{cc} − Vthsg, where Vthsg is the V_{th} of the select transistor.

After 0 cells are turned on, the channel is raised by the capacitive coupling with control gates. V_{ch} of Conv.1 is [27]

$$V_{ch\_conv.1} = V_{chi} + 15\gamma(V_{pass} - V_{th0} - V_{chi}) + \gamma V_{pgm}$$
$$= (1 - 15\gamma)(V_{cc} - V_{thsg}) + 15\gamma(V_{pass} - V_{th0}) + \gamma V_{pgm}$$
$$\gamma = C_{ono} \times C_{ox}/\{16 C_{total} \times (C_{ono} + C_{ox})\} = 0.04$$

where C_{ono}, C_{ox}, and C_{total} are the ONO capacitance, the tunnel oxide capacitance, and the total channel capacitance. In Conv.1, as V_{cc} decreases, V_{ch} also decreases, and the V_{pgm} disturb becomes worse.

Conversely, in both Conv.2 and the proposed scheme, the program-inhibit voltage is raised to 4.5 V. In this case, V_{ch} is as follows:

$$V_{ch\_proposed} = V_{ch\_conv.1} + (1 - 15\gamma)(4.5 - V_{cc})$$

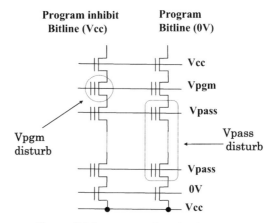

Figure 6.86. Program disturb of Conv. 1.

KEY CIRCUITS AND CIRCUIT/TECHNOLOGY INTERACTIONS

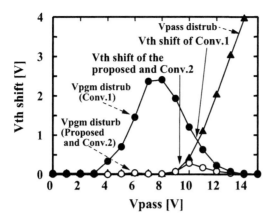

Figure 6.87. V_{th} shift after V_{pgm} and V_{pass} disturb measured at 85°C. V_{cc} = 1.8 V.

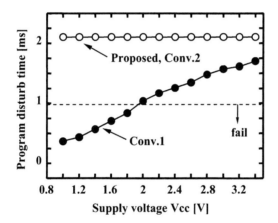

Figure 6.88. Allowable program-disturb time measured at 85°C. The program disturb time is defined as when V_{th} shifts by 1.5 V.

Much higher V_{ch} reduces the electric field across the tunnel oxide and reduces the V_{pgm} disturb.

Figure 6.87 shows the measured V_{th} shift after the program disturb. The V_{pass} disturb of the proposed scheme is the same as Conv.1 and Conv.2. To decrease the V_{pass} disturb, V_{pass} must be decreased. To decrease the V_{pgm} disturb, V_{pass} has to be large enough. The intersection of the V_{pass} and the V_{pgm} disturb is where the program disturb is minimized. The V_{th} shift of the proposed scheme and Conv.2 is smaller than that of Conv.1.

Figure 6.88 shows the allowable program-disturb time of the STI cell so as not to cause an incorrect operation. The V_{cc} minimum is 2.0 V in Conv.1. Conversely, in Conv.2 and for the proposed scheme a highly reliable operation is realized irrespective of V_{cc}.

In the proposed scheme, the source line is boosted by the read charge pump, which generates an unselected wordline voltage during a read operation. No additional charge pump is required. Figure 6.89 shows the program time per page. At each V_{cc}, the number of stages and the capacitance of the read charge pump are

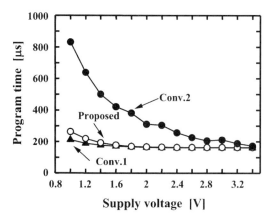

Figure 6.89. Programming time per page.

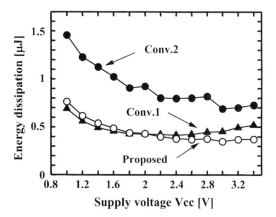

Figure 6.90. Energy dissipation per programming pulse.

optimized so that the wordlines rise within a predetermined period during a read operation. In Conv.2, the program time is prolonged to 500 μs/page at 1.4 V to charge the huge bitline capacitance. In the proposed scheme, a fast programming of 192 μs/page can be realized.

In Conv.1, a voltage higher than V_{cc} is applied only to a row decoder and peripheral circuits. The high-voltage (HV) circuit is composed of only HV-NMOS. HV-PMOS is not used to lower the manufacturing cost. In Conv.2, the bitline is biased to 4.5 or 0 V during a program. Therefore, the column latch must be made of HV-NMOS and HV-PMOS, which occupy much larger circuit areas than low-voltage transistors. As the number of column latches is 4000, the chip area overhead is as much as 5%. In short, Conv.2 is subject to manufacturing cost and circuit area overheads. Contrarily, in the proposed scheme, the source line voltage is applied via a peripheral HV-NMOS switch whose circuit area is just 0.01% of the chip area and negligibly small. Then, an HV-PMOS is unnecessary. Therefore, the proposed scheme can be realized without any circuit area or manufacturing cost overhead.

Figure 6.90 shows the energy dissipation per program pulse. In Conv.2, as V_{cc} decreases, the energy dissipation greatly increases to raise the bitlines to 4.5 V.

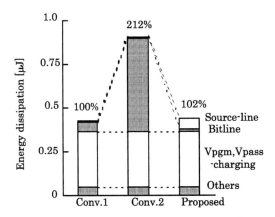

Figure 6.91. Comparison of energy dissipation per programming pulse at $V_{cc} = 1.8\,\text{V}$.

Conversely, in the proposed scheme, the small bitline swing decreases the power consumption (Fig. 6.91). The energy dissipation is almost the same as Conv.1 and a very low power operation of 22 mW at 1.4 V can be realized. In both Conv.1 and the proposed scheme, an increase in the energy dissipation at lower V_{cc} is caused by the V_{pgm} and V_{pass} boosting.

The proposed architecture shows excellent program-disturb characteristics without circuit area, manufacturing cost, program-speed, or power consumption overhead. This architecture is a promising candidate for the prospective low-voltage operation NAND Flash memories.

6.7 MULTILEVEL NAND

6.7.1 Multilevel Circuit Technology

The multilevel cell concept [38], which doubles or triples the memory density without increasing the number of physical cells, combined with advantages of the NAND Flash memory can be an ideal solution for many mass storage applications. In this section, key technologies of the multilevel NAND Flash memory are introduced, based on four cell states (2 bits/cell) [37].

Controlling the threshold voltage distribution of programmed cells is a key issue for multilevel flash memories. Target threshold voltage distributions for the four cell states are shown in Figure 6.92. To achieve the V_{th} distributions for multilevel Flash memory, the program scheme must be able to precisely control cell V_{th} levels. As described in Section 6.4.2, the ISPP scheme has advantages for achieving the tight threshold voltage distributions. In the ISPP scheme, a smaller increment (ΔV_{pgm}) of the program voltage can result in a tighter V_{th} distribution. ΔV_{pgm} of 0.2 V is needed to achieve the target V_{th} distribution of 0.4 V.

Also, to maintain the narrow cell V_{th} distribution, the effect of background pattern dependency (BPD) should be considered because the programmed V_{th} can be varied by the program sequence in the NAND cell string. To avoid V_{th} shift due to BPD, a sequential programming restriction is imposed. Pages of a block are programmed in a fixed sequence, beginning with cells at the bottom of the string (closest to array ground line) and then moving to the top cells (closest to the bitline contact).

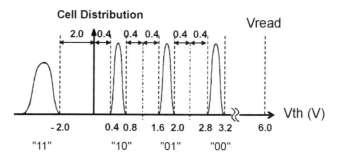

Figure 6.92. Target threshold voltage distribution for 2 bits/cell.

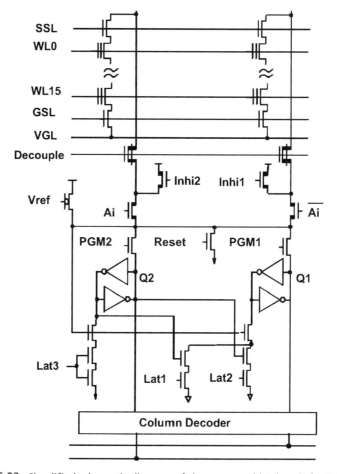

Figure 6.93. Simplified schematic diagram of the sense-and-latch unit for 2 bits/cell.

This minimizes BPD since the selected cell V_{th} is self-adjusted during the program-verify operation to the previously programmed cells below the selected one. The V_{th} shift by BPD can be controlled within 0.05 V by following the program sequence.

The simplified circuit schematic diagram of a sense-and-latch (SL) unit in the page buffer for the 2 bits/cell NAND Flash memory is shown in Figure 6.93. The latches in the page buffer are tied to adjacent I/O lines such that the 2 bit data associated with each cell maps to two data-out pins. For simple circuit design, the page buffer in Figure 6.93 employs a wordline sweeping read (WSR) scheme during page-read and program-verify operations. The WSR scheme identifies one of four different cell states by sequentially changing the wordline voltage.

Figure 6.94 shows the timing diagram for page-read operation using WSR. During the first phase after page buffer reset, the wordline voltage is set to 2.4 V to sense 00 state cells and only Q1 is changed after sensing. During the second phase, the wordline voltage is set to 1.2 V to sense 01 state. Since Q2 value for 00 and 01 is V_{cc}, Q2 state is changed during the second phase by pulsing Lat3 signal. Once Q2 changes its state from 0 V to V_{cc} during the Phase 2, Q1 state is not affected from the final phase. The 10 state is sensed during the third phase. Once the page-read operation is finished, data stored in the latches can be accessed with a burst read cycle. This WSR scheme eliminates the need for complicated multilevel sensing reference circuits within each SL circuit. In case of using four cell states, however, WSR scheme results in a random page access time that is three times longer than that of a single-bit-per-cell device.

This page buffer scheme allows the programmed cell V_{th} to be optimized on a cell-by-cell and state-by-state basis even though the cells in a page are programmed simultaneously. Since the SL unit shares two bitlines, only one of two bitlines operates during page program operation and the unselected bitlines are program inhibited. Program inhibit operation on unselected bitlines is performed by supplying V_{cc} using the Inhi1 or Inhi2 signal in Figure 6.93. Each cell state in a page is programmed sequentially from the 11 state (erased state) to the 00 state, and the program inhibit is initiated as soon as the cell is programmed to a desired state. Figure 6.95 shows the timing diagram for page program operation.

After setting all the page buffer latches according to the input data, program operation is carried out in three phases, one for each of the programmed states. The 10 state cells in a page are programmed during the Phase 1 in Figure 6.95, and 01, 00 cells are programmed during the Phase 2 and the Phase 3, respectively. Within each phase, cells are programmed through repeated program cycles consisting of a program pulse and verification. The program voltage is uniformly incremented in

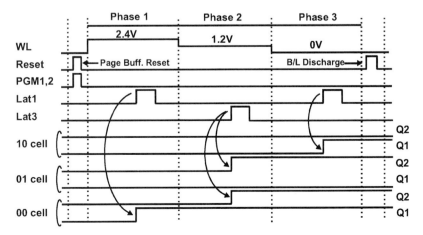

Figure 6.94. Signal timing diagrams in page read for 2 bit/cell operation.

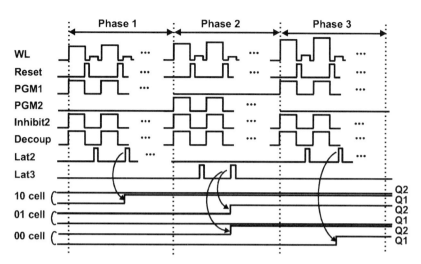

Figure 6.95. Signal timing diagrams in page program for 2 bit/cell operation.

each phase from the starting voltage of each phase. The starting program voltage of the second and the third phases are set to be lower than the final program voltage of the previous phase to improve program speed while preventing overprogramming.

When a cell is programmed to the target V_{th} of its respective state, the associated page buffer latches are automatically reset in the verification cycle to inhibit further programming. The wordline voltage for each phase is controlled differently during the verify operation to distinguish three programmed states sequentially. During the first phase, for example, the wordline voltage is set to 0.4 V which is the lowest V_{th} voltage of 01 state as shown in Figure 6.92. The PGM1 and PGM2 signals in Figure 6.93 are used to multiplex Q1, Q2 to the bitline for proper program verify and inhibit operations.

6.7.2 Array Noise Suppression Technology

In a conventional NAND-type cell array, each NAND cell unit shares a common N^+ diffused source line, which is shunted with metal bypass lines parallel to the metal bitlines, as shown in Figure 6.96. The resistance of the diffused source line becomes higher as the metal bypass line pitch increases and causes source line noise during a bit-by-bit program-verify operation. In this section, it is explained how the source line noise adversely affects the threshold voltage control and causes both negative and positive threshold voltage shifts [19, 20].

A simplified model of the noise is shown in Figure 6.97. In this model, a shielded bitline sensing method is used [16, 31]. Therefore, half of the memory cells sharing the same control gates are programmed and verified simultaneously. In the case that half of the memory cells, except Cell B, stay in the erased state after the first program pulse, a large current I flows through the erased cells during the first program verify [Fig. 6.97(a)]. As a result, the N^+ diffused source line bounces up to $\Delta V_{SL} = I \times R$. The source voltage of Cell B is even higher than ΔV_{SL}, due to the resistance of the series-connected memory cells. In this case, the cell read current through Cell B is

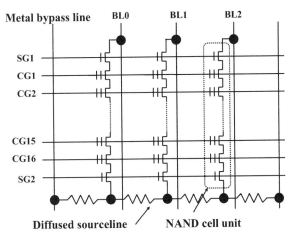

Figure 6.96. Conventional NAND-type cell array. Each NAND cell unit shares 16 control gates, two select gates, and a diffused source line that is shunted with metal bypass lines.

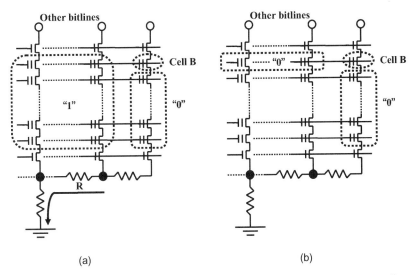

Figure 6.97. Model for the threshold shift in the negative direction due to source line noise.

reduced and Cell B is read out to be successfully programmed. However, after that, if the other memory cells that share the same control gates with Cell B are also programmed [Fig. 6.97(b)], the ground bounce of the source line is reduced. In that case, a larger cell read current flows through Cell B than in the case of the first program-verify operation [Fig. 6.97(a)].

In this worst case, the threshold voltage of Cell B is $\Delta V_{th}(\text{NEG})$ below the reference voltage, V_{vrfy}, as shown in Figure 6.98. The calculated threshold voltage shift in the negative direction, $\Delta V_{th}(\text{NEG})$, is shown in Figure 6.99. As the metal bypass line pitch decreases, source line bounce is reduced and $\Delta V_{th}(\text{NEG})$ decreases, however, the cell area overhead increases.

Figure 6.100 illustrates the mechanism, of the threshold voltage shift in the positive direction. The program cycle starts from the source line side cell and ends

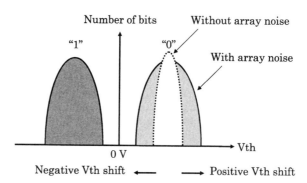

Figure 6.98. Schematic threshold voltage distribution. Due to array noise, the threshold voltages of memory cells shift in both positive and negative direction.

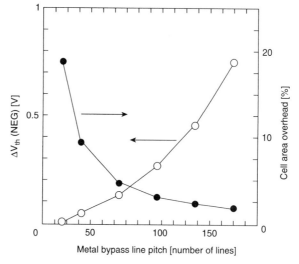

Figure 6.99. Calculated threshold voltage shift in the negative direction due to source line noise.

at the bitline side cell to prevent interference between selected and unselected cells. Therefore, in the case that Cell C is programmed, the series-connected cells, except Cell C, are still in the erased state [Fig. 6.100(a)], and the series resistance is low. After programming Cell C, and after the series-connected cells are also programmed [Fig. 6.100(b)], the series resistance of the NAND cell unit is higher. Therefore, the cell read current through Cell C is reduced compared with the case of Figure 6.100(a). In this worst case, the threshold voltage of Cell C is $\Delta V_{th}(POS)$ higher than is expected without noise. Figure 6.101 shows the calculated threshold voltage shift in the positive direction, $\Delta V_{th}(POS)$. $\Delta V_{th}(POS)$ is also enhanced as the source line bounce is increased.

Figure 6.102 shows the threshold voltage distribution for a 4-level NAND Flash memory cell. In this calculation, the inter-bitline capacitive coupling noise is eliminated by adopting a shielded bitline sensing scheme, as discussed above. As the metal bypass line pitch increases, the threshold voltage distribution of each level,

MULTILEVEL NAND

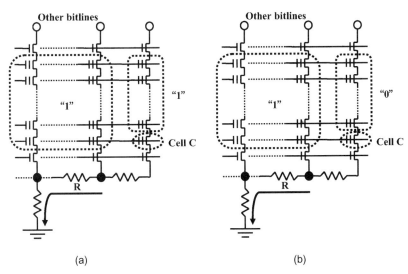

Figure 6.100. Model for the threshold shift in the positive direction due to source line noise.

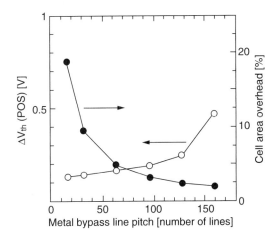

Figure 6.101. Calculated threshold voltage shift in the positive direction due to source line noise.

ΔV_{th}, is increased. The threshold voltage distribution without source line noise, ΔV_{th0}, is assumed to be 0.5 V, which can be realized by using staircase programming pulses combined with a bit-by-bit program verify [5, 39].

Figure 6.103 illustrates the read voltage, V_{cgr}. Without array noise, the read voltage can be 4 V. Enlarged V_{cgr} degrades the read-disturb characteristics. Furthermore, the large difference in the threshold voltage between the 0 and the 3 states will also reduce the retention time when no read voltage is applied to the memory cell since the electric field strengths in the cell will be higher because of the larger amounts of stored charge. Source line noise is a dynamic noise, when dynamic differential sensing is used during program verify. Therefore, if the access time is increased, the threshold voltage shift can be reduced. On the other hand, in the case

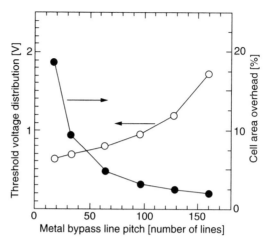

Figure 6.102. Calculated threshold voltage distribution. The threshold voltage distribution without source line noise is assumed to be 0.5 V.

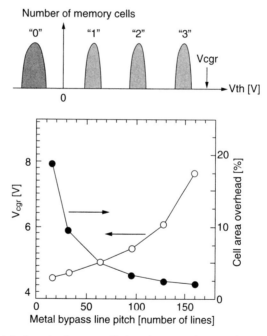

Figure 6.103. Calculated read voltage applied to the control gates of the unselected cells during a read operation in a conventional 4-level cell array.

of the current sensing method [40], the source line noise is continuous and causes a large threshold voltage shift, irrespective of the access time.

A *double-level V_{th} select gate array architecture* is illustrated in Figure 6.104, and the operating conditions are summarized in Table 6.8. Each NAND cell unit is composed of an E-type (V_{th} = 2.0 V) select gate, an I-type (V_{th} = 0.5 V) select gate and 16 memory cells. I-type and E-type select gates are fabricated by respectively

MULTILEVEL NAND

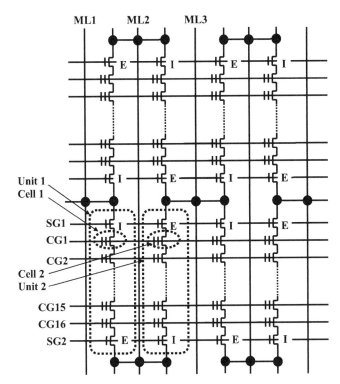

Figure 6.104. Double-level V_{th} select gate array architecture. Each NAND cell unit is composed of an E-type select gate (V_{th} = 2.0 V), an I-type select gate (V_{th} = 0.5 V), and 16 memory cells.

TABLE 6.8. Operation Condition

	ML1	ML2	CG1	CG2–16	SG1	SG2	P-well
Prog. Cell 1	Vcc/0	Vcc	19	10	1.5	1.5	0
Prog. Cell 2	Vcc	Vcc/0	19	10	1.5	1.5	0
Read. Cell 1	1.8	0	Vsense	Vcgr	1.5	Vcc	0
Read. Cell 2	0	1.8	Vsense	Vcgr	Vcc	1.5	0
Verify Cell 1	1.8	0	Vvrfy	Vcgr	1.5	Vcc	0
Verify Cell 2	0	1.8	Vvrfy	Vcgr	Vcc	1.5	0
Erase	Open	Open	0	0	20	20	20

masked and unmasked channel implantation. The metal lines are used both as bitlines and source lines. For example, in the case where ML1 acts as a bitline, the neighboring ML2 acts as a source line. In this structure, the high-resistance diffused source line is removed. Consequently, the source line noise is eliminated. In comparison with a conventional array, the cell area is reduced because neither additional select transistors nor source line shunts are needed. The bitline contact pitch is relaxed, which increases the process margin when the isolation width between memory cells is reduced with trench isolation.

In the case that Cell 1 is selected [Fig. 6.105(a)] during read, the select gate line SG1 is raised to 1.5 V, to turn on the I-type select gates, while the E-type select gates

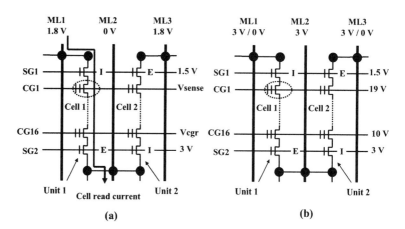

Figure 6.105. Operation principles: (a) read and (b) program.

are turned off. The select gate line SG2 is connected to V_{cc} (3 V), so that the E-type select gates are biased to the ON states. Therefore, a cell read current may flow through Unit 1, according to the stored data in Cell 1. On the other hand, Unit 2 is unselected because the E-type select gates are turned off. The selected bitline ML1 is connected to a sense amplifier, while the unselected bitline ML2 is grounded. In this operation, ML2 acts as a low-resistance source line, forming a shield, between the selected ML1 and ML3. As a result, both source line noise and inter-bitline capacitive coupling noise are eliminated. Due to the excellent noise immunity, a fast and stable read operation can be achieved. On the other hand, if Cell 2 in Unit 2 is selected, the select gate lines SG1 and SG2 are, respectively, raised to 3 and 1.5 V. Therefore, a cell read current may flow from the precharged ML2 through Unit 2 to the grounded ML3. In this case, Unit 1 is unselected because the E-type select gate of Unit 1 is turned-off.

Figure 6.105(b) explains the program operation. Cell 1 is selected. The selected control gate CG1 is raised to about 19 V, and the unselected control gates CG2–CG16 are connected to 10 V; and 1.5 V is applied to both SG1 and SG2, in order to turn on the I-type select gates of both Unit 1 and Unit 2 and to turn off the E-type select gates of both Unit 1 and Unit 2. The program data for the selected Unit 1 is applied via the selected bitline ML1. If the program data is 1, 2 or 3, ML1 is grounded. In this condition, electrons are injected from the grounded channel into the floating-gate by the FN tunneling mechanism. On the other hand, in case of 0 programming, ML1 is connected to 3 V. The channel voltage during 0 programming is raised to about 8 V by the capacitive coupling with the control gates [5] and electron injection is inhibited. The unselected bitline ML2 is also connected to V_{cc} to prevent the unselected Cell 2 from 1, 2 or 3 programming.

The characteristics of the proposed cell array are summarized in Table 6.9. The cell read current paths of each NAND cell unit, from a bitline via a selected cell to a grounded bitline, are completely independent. So the cell read current is not affected by the cell read current that flows through neighboring cells. Thus the threshold voltage shift in the negative direction, $\Delta V_{th}(\text{NEG})$, is eliminated, independent of the value of the bitline resistance. The threshold voltage shift in the positive direction, $\Delta V_{th}(\text{POS})$, is significantly reduced because the source line bounce is

MULTILEVEL NAND

TABLE 6.9. Characteristics of the Proposed 4-Level Cell Array

	Ideal	Conventional	Proposed
V_{th} negative shift	None	0.46 V	None
V_{th} positive shift	None	0.24 V	0.03 V
V_{th} distribution	0.5 V	1.2 V	0.53 V
V_{cgr} (V_{cgr} during read)	4 V	6.1 V	4.1 V
cell area overhead	None	2.5%	None

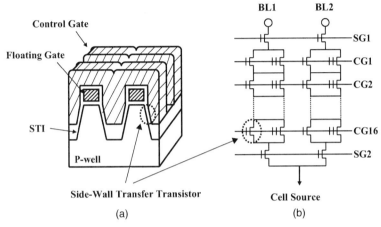

Figure 6.106. Schematic view and equivalent circuit of the side-wall transfer transistor cell (SWATT cell). A transfer transistor is located at the side wall of the shallow trench Isolation (STI) region and is connected in parallel with the floating-gate transistor.

removed. Due to the high noise immunity, a precise threshold voltage control can be achieved, as shown in Table 6.9.

The threshold voltage fluctuation, which is the sum of the threshold voltage shift in the negative and positive direction, is as low as 0.03 V. As a result, the threshold voltage distribution can be 0.53 V if the threshold voltage distribution without noise, ΔV_{th0}, is assumed to be 0.5 V, and a nearly ideal performance can be obtained. In a conventional array, the threshold voltage fluctuation is 0.7 V. As a result, the threshold voltage distribution would be increased to 1.2 V. In the proposed array, the read voltage, V_{cgr}, for a 4-level cell is as low as 4.1 V and a reliable operation of a multilevel Flash memory can be realized. In a conventional array V_{cgr} has to be 6.1 V, as a result, the read-disturb time would be decreased by about 2 orders of magnitude [10] to an unacceptably low level. Moreover, the cell area is reduced in the proposed array, in comparison with the conventional array.

6.7.3 Side-Wall Transfer Transistor Cell

The schematic view and equivalent circuit of the side-wall transfer transistor (SWATT) cell are shown in Figure 6.106 [41, 42]. One cell consists of both a floating-gate transistor and a transfer transistor, which is located at the side-wall of the shallow trench isolation (STI) region. These two transistors are connected in

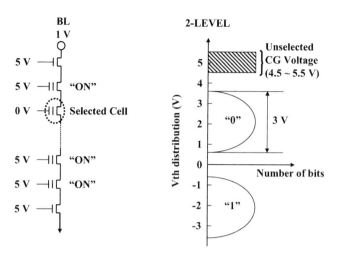

Figure 6.107. Read condition of a conventional NAND cell. The V_{th} distribution of the cells in the programmed state must be narrow with a width of 3.0 V or less, because the unselected cells must function as pass transistors for a control gate voltage of 5 V.

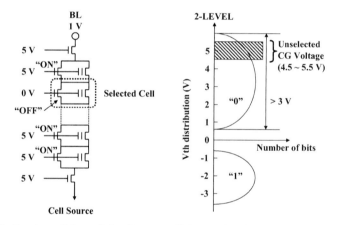

Figure 6.108. Read condition of the SWATT cell for a 2-level scheme. The side-wall transfer transistor functions as a pass transistor. Therefore, the V_{th} distribution of the floating-gate transistor in the programmed state is allowed to be very wide with a width > 3.0 V for 2-level operation.

parallel. Furthermore, 16 cells are connected in series between two select transistors to form a NAND structure cell.

The read conditions of a conventional NAND cell and a SAWTT cell for the conventional 2-level scheme are compared in Figures 6.107 and 6.108. In a conventional cell, 0 V is applied to the gate of the selected memory cell, while 5.0 V is applied to the gates of the other cells. All the memory cells, except for the selected cell, serve as transfer gates. Thereby, for the conventional NAND cell, the threshold voltage (V_{th}) of the programmed cell must be lower than the 4.5 to 5.5 V, which is the unselected control gate (CG) voltage. Thus, the V_{th} distribution of the cells in the programmed state must be narrower than 3.0 V for 2-level operation, as shown

MULTILEVEL NAND

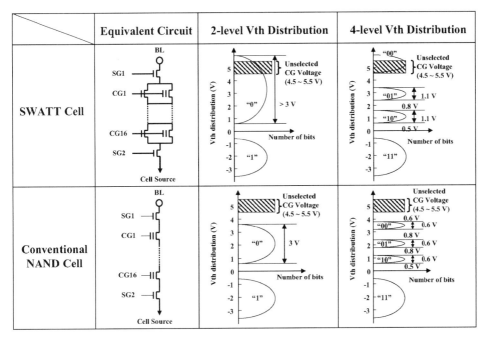

Figure 6.109. Cell threshold voltage distribution of the SWATT cell and the conventional NAND cell for 2-level and 4-level operation. In the SWATT cell, a wide threshold distribution of 1.1 V is allowed for 4-level operation, in comparison with a 0.6-V distribution that is required for the conventional NAND cell. The range of the unselected CG (control gate) voltage is limited because of read-disturb.

in Figure 6.107. Using the SWATT cell, the side-wall transfer transistor functions as a pass transistor instead of as a floating-gate transistor, as illustrated in Figure 6.108. Therefore, the threshold voltage of the floating-gate transistor does not have to be lower than the unselected CG voltage of 4.5 to 5.5 V. As a result, the programmed V_{th} distribution can be much wider than 3.0 V in contrast to the conventional NAND cell.

The threshold voltage distributions of the 2-level and 4-level scheme are compared in Figure 6.109. In a conventional NAND cell, the V_{th} distribution of the cells in the programmed states (10, 01, and 00) must be as narrow as 0.6 V because the cells connected in series must function as pass transistors. Conversely, in the SWATT cell, the V_{th} distribution of the floating-gate transistor in the programmed 10 and 01 states is allowed to be as wide as 1.1 V. The V_{th} distribution in the programmed 00 state is allowed to be even wider than 1.1 V. This wide threshold voltage distribution results in a high programming speed [39] (reducing the number of program/verify cycles [15, 16]) and good data retention characteristics.

The fabrication of the SWATT cell is simple and uses only the conventional techniques. The process sequence of the SWATT cell is shown in Figure 6.110. First, a stacked layer of the gate oxide, the floating-gate polysilicon and the cap oxide are formed. Next, the trench isolation region is defined by patterning these three layers, followed by the trench etching, trench bottom boron implantation and filling with LP-CVD SiO_2 [Fig. 6.110(a)]. Subsequently, the LP-CVD SiO_2 is etched back until

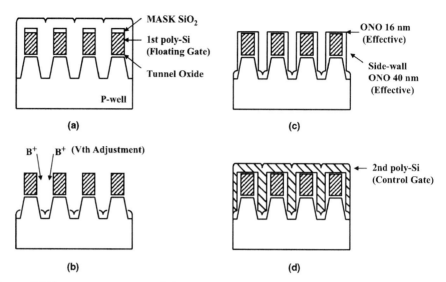

Figure 6.110. Process sequence of the SWATT process. (a) Trench etching, LP-CVD SiO$_2$ fill-in, (b) oxide etchback and B$^+$ implantation of the V_{th} adjustment of the side-wall transfer transistor, (c) ONO formation, and (d) control gate formation. The thermal-oxide of the STI side wall is about 2 times thicker as that on the polysilicon due to oxidation enhancement at the STI side walls.

the side wall of the STI is exposed [Fig. 6.110(b)]. Boron (B$^+$) ion implantation (60 keV, 2E12/cm^2) is carried out for V_{th} adjustment of the side-wall transfer transistor. After that, the interpoly dielectric (ONO) and transfer transistor gate oxide are formed at the same time [Fig. 6.110(c)]. Then, the control gate polysilicon is deposited, followed by the stacked-gate patterning [Fig. 6.110(d)]. In this process, the thermal oxide of the STI side wall is about two times thicker than that on the polysilicon due to oxidation enhancement at the STI side walls. As a result, the breakdown of the control gate does not occur even if a high voltage of about 20 V is applied to the control gate during the programming operation.

An accurate control of the threshold voltage of the side-wall transfer transistor is important for the SWATT cell. The range of the threshold voltage is determined as follows. The side-wall transfer transistor must be in the ON state when the unselected CG voltage (4.5 to 5.5 V) is applied to the control gate. Thereby, the upper limit of the V_{th} of the side-wall transistor is 4.5 V. Conversely, the side-wall transistor must be in the OFF state when the read voltage (about 3.9 V) between the 01 and 00 state for 4-level operation is applied to the control gate. Therefore, the threshold voltage of the side-wall transfer transistor must be in the range from 3.9 to 4.5 V for 4-level operation (from 0 to 4.5 V for 2-level). The important statistical parameters of V_{th} of a side-wall transfer transistor are the boron concentration in the channel region and the side-wall gate oxide thickness. The boron concentration is well controlled by B$^+$ implantation, as shown in Figure 6.110(b). And also, the oxide thickness of the STI side wall is controlled within 10% variation. Therefore, the narrow range of the threshold voltage of the side-wall transfer transistor can be adjusted.

6.7.4 Three-Level NAND

A 4-level 128-Mbit NAND Flash memory has reduced the die size to about 50% of a 2-level chip, however, it has increased the 512-byte programming time to 900 μs. That is a 0.57-Mbyte/s programming throughput. A 3-level NAND Flash memory has been introduced that achieves a 3.4-Mbyte/s programming throughput while reducing a die size per bit to 61% [43].

The 3-level memory cell has data 0, 1 and 2, as shown in Figure 6.111. A 0 state corresponds to a threshold voltage of lower than −1 V. A pair of memory cells stores 3-bit data. The 528-byte data, which includes parity-check bits and several flag data, are simultaneously transmitted from or to 2816 memory cells through 2816 compact intelligent 3-level column latches.

The compact intelligent 3-level column latch is illustrated in Figure 6.112. The 3-level data to be read out from or programmed into the memory cell is temporarily stored in a pair of flip-flop circuits, FF1 and FF2, as an intermediate code.

Figure 6.113 shows a timing chart of a read operation in the case that bitlines BLa's are selected. FF1 and FF2 detect whether the memory cell stores data 0 and data 2, respectively. For a precise sensing of the threshold voltage, the bitlines are precharged to 1.3 V and a power supply voltage applied to the flip-flop is clamped at 2 V during a sensing period.

The timing chart of a program operation, in the case that the bitlines BLa's are selected, is shown in Figure 6.114. The program pulses of 16.5 to 19.3 V are applied to the selected control gate, CG. In order to get a programming speed of 1 programming to be close to that of 2 programming, the bitline voltage of 1 programming is raised to 1.6 V.

After each program operation, a program-verify operation is carried out. Figure 6.115 shows the timing chart of the program-verify operation. The intermediate codes stored in the column latches are modified, such that 1 or 2 programming are, respectively, executed on only memory cells in which data 1 or 2 have not been successfully programmed.

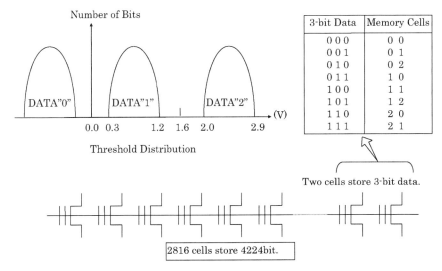

Figure 6.111. Three-level NAND Flash data structure.

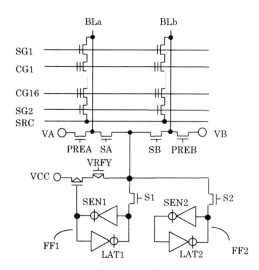

Figure 6.112. Three-level NAND Flash architecture.

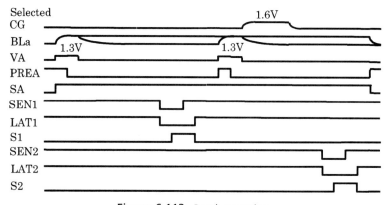

Figure 6.113. Read operation.

The program pulses are gradually raised up by 0.7 V/10 μs, as illustrated in Figure 6.116. The first pulse causes the memory cells to be programmed in a saturation region [39]. Therefore, the threshold voltages of the memory cells are shifted by 0.7 V/10 μs, as shown in Figure 6.117. The threshold voltage of the fastest memory cell after the first pulse is raised to 0.7 V in the 1 programming and 2.3 V in the 2 programming. After the fourth pulse, the threshold voltage of the slowest memory cell reaches 0.8 V in the 1 programming and 2.4 V in the 2 programming. The program pulse margin has a value of 0.9 V. Taking a 0.2-V source line noise into account [39], the programmed threshold voltage distribution is controlled as shown in Figure 6.111.

Four intermediate codes, that is, 6-bit data, are loaded within a 25-ns cycle time. During the data load, a charge pump generates a high voltage for the first program pulse. The setup time of the high voltage is 5 μs. The first pulse duration is 20 μs, and each of the subsequent pulses has a 10-μs duration. A program recovery time and

MULTILEVEL NAND

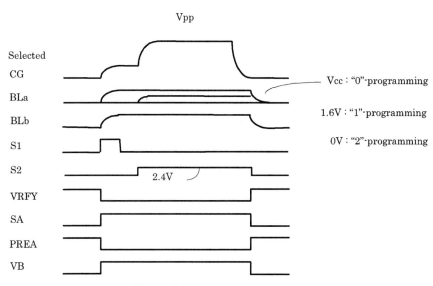

Figure 6.114. Program operation.

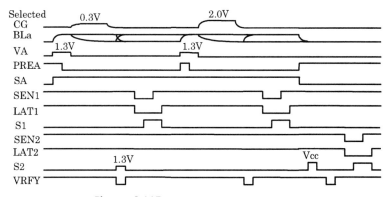

Figure 6.115. Program-verify operation.

the program-verify times are 1 and 16 μs, respectively. The total program time for 528-byte data is as follows:

$$704 \times 25\,\text{ns} + 20\,\mu\text{s} + 1\,\mu\text{s} + 16\,\mu\text{s} + (5\,\mu\text{s} + 10\,\mu\text{s} + 1\,\mu\text{s} + 16\,\mu\text{s}) \times 3 = 150.6\,\mu\text{s}$$

The typical program throughput is 3.4 Mbyte/s and 68% of the 2-level NAND Flash, as shown in Figure 6.118(a).

The die size is estimated on the assumption that, in case of the 2-level flash memory, the memory cells and the column latches occupy 66% of the die size. The number of the memory cells and the area of the column latches are increased by 133.3% when the memory capacity is doubled. The die size of the 3-level Flash memory chip is increased by 122%. Therefore, the die size per bit is reduced to 61%, as shown in Figure 6.118(b).

Figure 6.119 illustrates a 3-level NAND Flash system configuration. Between the 3-level NAND Flash memories and a Flash interface, the intermediate codes

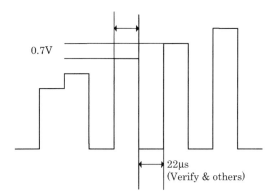

Figure 6.116. Program pulse.

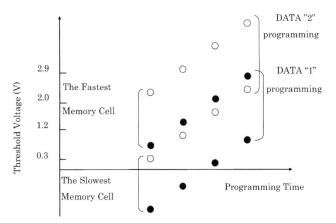

Figure 6.117. Program characteristics.

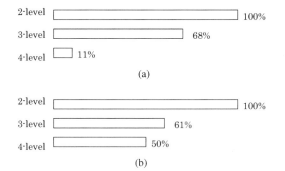

Figure 6.118. (a) Program throughput and (b) die size per bit.

are transmitted. In the Flash interface, the intermediate codes are translated from or to 528-byte data including the parity-check bits and flag data. In the 3-level NAND Flash system, there is a possibility that a 3-bit burst error is induced even if an error occurs in only one memory cell. An ECC engine generates 36-bit parity-check bits so as to be capable of correcting two 9-bit symbol errors of 528-byte data

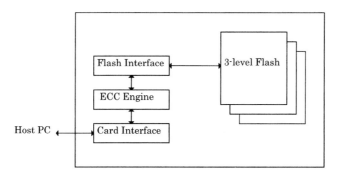

Figure 6.119. Three-level Flash system.

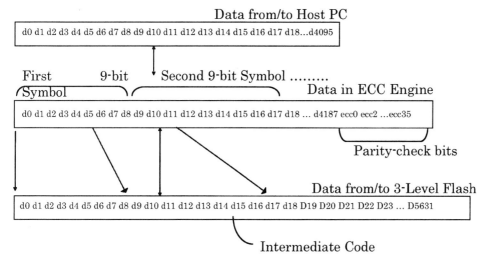

Figure 6.120. Data format for 9-bit symbol architecture.

on the basis of 9-bit symbol Reed–Solomon code, as shown in Figure 6.120. The 9-bit symbol architecture minimizes the length of the parity-check bits for 528-byte data.

6.7.5 High-Speed Programming

Figure 6.121 shows the conventional and the proposed 4-level cell [44, 45]. Two-bit data stored in a conventional cell correspond to two Y addresses (Y1, Y2). These 2-bit data belong to the same page. Also, three levels (the 1, 2, and 3 level) are programmed during the same operation [37, 38, 46]. On the other hand, the proposed cell contains two "pages." Here, a page means the group of memory cells that are programmed simultaneously. In other words, the 2-bit data of each cell correspond to two X addresses, X1, X2, and the programming of X1 and that of X2 are performed during different operations.

As shown in Figure 6.122, at the first page program, the memory cells are programmed to the 1 state. After that, the 0 or 1 cells are, respectively, programmed to the 2 or 3 state during the second page program.

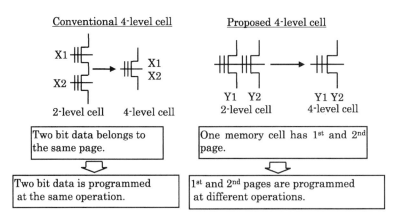

Figure 6.121. (a) Conventional 4-level cell. (b) Proposed 4-level cell.

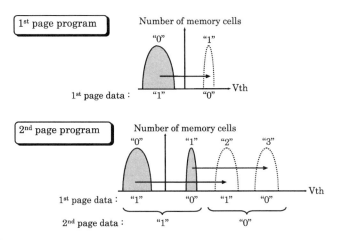

Figure 6.122. Program operation of the proposed architecture: (a) First page program and (b) second page program.

Figure 6.123 shows the V_{th} distribution. In the conventional method, the ΔV_{th} of each programmed state is the same [37, 38, 46], as shown in Figure 6.123(a). Conversely, 1 program is controlled more precisely than 2 or 3 program in the proposed method, as described below [Fig. 6.123(b)].

Figure 6.124(a) shows the program operation of the conventional method with simultaneous multilevel program. During the data load, 2-bit program data are input to the column latch circuit, as seen in Figure 6.125(a). During the program, the channel voltage of the memory cell, V_{ch}, is changed according to the stored data in the column latch. Then, three verify read sequences (1, 2, 3 verify) are subsequently carried out, where the control gate is applied to verify voltages, 0.5, 1.9, and 3.3 V, respectively. Ideally, V_{ch} is 0, 1.4, and 2.8 V for the 3, 2 and 1 program where 1.4 V (2.8 V) corresponds to the V_{th} difference between the 3 and 2 (1).

The simulated write characteristics of the slowest cell are shown in Figure 6.126(a). As can be seen in Figure 6.123(a), ΔV_{pp} is 0.3 V to make ΔV_{th} 0.6 V.

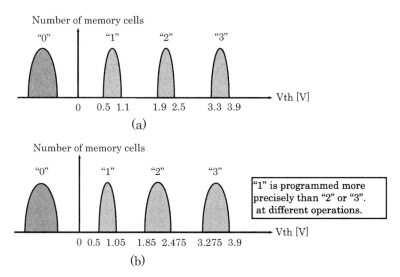

Figure 6.123. Threshold voltage distribution of a 4-level NAND cell: (a) conventional and (b) proposed.

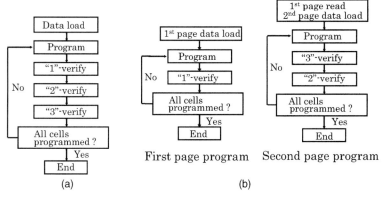

Figure 6.124. Program algorithm: (a) Conventional and (b) proposed.

The three levels are programmed at the same speed. Including the V_{ch} variation, δV_{ch}, the number of program pulses, N_p, and the program time per page, T_p, are as follows:

$$N_p = 1 + (\Delta V_{th0} + \delta V_{pp} + \delta V_{ch} / \Delta V_{pp} = 10$$
$$T_p = T_{load} + (T_{pulse} + 3 \times T_{vfy}) \times N_p = 395\,\mu s$$

where δV_{ch} is 0.1 V. The verify-read time, T_{vfy}, is 7.5 μs. In comparison with the 2-level cell, the increased verify sequences, which occupy as much as 57% of the program time, seriously slow down the program speed, as shown in Figure 6.127.

Actually in the case of the 0 program, the bitline and the select gate connected to the bitline are applied to V_{cc} and the select gate transistor is turned off [5]. If the minimum V_{cc} is 2.8 V, V_{ch} must be lower than 1.5 V so as to turn on the select gate

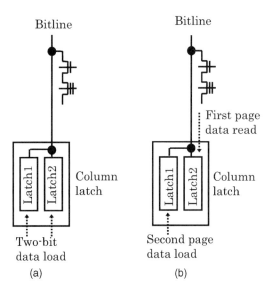

Figure 6.125. Data load operation of (a) the conventional method and (b) the second page program of the proposed method.

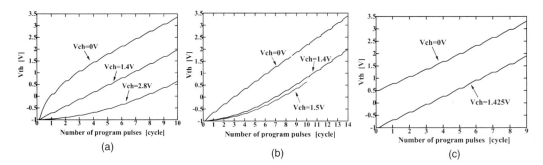

Figure 6.126. Simulated program characteristics of the slowest cell. (a) Ideal simultaneous multilevel program: V_{pp0} = 18.3 V; δV_{pp} = 0.3 V. (b) Actual simultaneous multilevel program: V_{pp0} = 17 V; ΔV_{pp} = 0.3 V. (c) Second page program of the proposed method; V_{pp0} = 18.3 V; ΔV_{pp} = 0.325 V. The couple factor of the cell was 0.57, and the tunnel oxide thickness was 10 nm.

transistor during the 1, 2 or 3 program. Thus, V_{ch} is 0 V for the 3 program, 1.4 V for the 2 program, and 1.5 V for the 1 program. The initial value for V_{pp}, V_{pp0}, must be ΔV_{bl} lower than the ideal case, where ΔV_{bl} is the V_{ch} difference between the actual and ideal case during the 1 program. If this is not done, then the fastest cell would be programmed higher than the 1 state. As a result, the 2 and 3 program become slower than the ideal case, as shown in Figure 6.126(b) and the number of program pulses, N_p, definitely increases:

$$N_p = 1 + (\Delta V_{th0} + \delta V_{pp} + \delta V_{ch} + \delta V_{bl})/\Delta V_{pp} = 14$$

$$T_p = T_{load} + (T_{pulse} + 3 \times T_{vfy}) \times N_p = 545\,\mu s$$

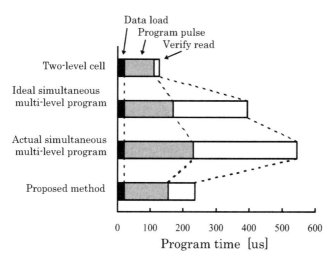

Figure 6.127. Program time comparison.

where δV_{bl} is 1.3 V. The programming is entirely too slow because the number of program pulses, N_p, increases due to the limitation of V_{ch}, as seen in Figure 6.127.

The first page program is performed just like a 2-level cell. As described above, ΔV_{th} of the 1 state is reduced down to 0.55 V by decreasing ΔV_{pp} to 0.25 V. As a result, the number of program pulses, N_{p1}, and the program time per page, T_{p1}, are expressed as follows:

$$N_{p1} = 1 + (\Delta V_{th0} + \delta V_{pp})/\Delta V_{pp} = 11$$

$$T_{p1} = T_{load} + (T_{pulse} + T_{vfy}) \times N_{p1} = 234.5\,\mu s$$

In case of the 1 program the bitline is grounded, so there is no bitline voltage fluctuation, that is, δV_{ch} is 0 V. The verify-read time, T_{vfy}, decreases to 4.5 µs, due to the reduced bitline capacitance. In comparison with a conventional 4-level cell, the program time decreases because there is no V_{ch} variation and only one verify sequence is required.

At the second page program, the channel voltage of the selected memory cell is 0 V for the 3 program and 1.425 V for the 2 program, while the control gate is applied to V_{pp}. Figure 6.126(c) shows the simulated write characteristics of the slowest cell. The 3 and 2 program are performed at the same speed. As a result, the number of program pulses, N_{p2}, decreases. To make the maximum V_{th} the same as that of the conventional method, ΔV_{th} of the 2 and the 3 state is increased to 0.625 V, as shown in Figure 6.126(b). Thus, ΔV_{pp} can be increased to 0.325 V, which further decreases N_{p2} and drastically improves program performance:

$$N_{p2} = 1 + (\Delta V_{th0} + \delta V_{pp} + \delta V_{ch})/\Delta V_{pp} = 9$$

The verify sequences decrease to two sequences (2 and 3 verify), in comparison with the three sequences in the conventional method. This also decreases the program time per page, T_{p2}.

$$T_{p2} = T_{\text{load}} + (T_{\text{pulse}} + 2 \times T_{\text{vfy}}) \times N_{p2} = 236\,\mu s$$

The first page has already been programmed before the second page program. Therefore, the program time of the second page is T_{p2}, not $T_{p1} + T_{p2}$. Consequently, whichever page is selected, any page can be programmed within 236 µs. Due to the reduced program pulses and verify sequences, a high programming speed of 236 µs/512 bytes or 2.2 Mbyte/s can be obtained, which is as much as 2.3 times faster than the conventional method, as shown in Figure 6.127.

In NAND Flash memories, the program operation is performed in a page unit to increase program throughput. The page size is kept at 512 bytes in both 2-level and 4-level cells. A schematic chip overview of the conventional [37] and the proposed 128-Mb Flash memory is shown in Figure 6.128. Table 6.10 explains the core circuit configuration.

In both the conventional and the proposed 4-level cell array, each column latch includes two latch circuits to store 2-bit data, which are read out from or programmed into a 4-level cell. Two adjacent bitlines share one column latch so as

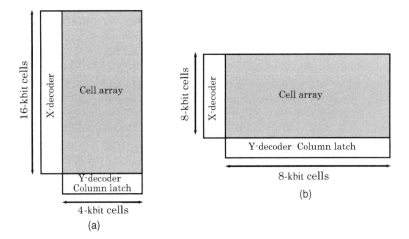

Figure 6.128. Cell array organization of a 128-Mbit NAND Flash memory: (a) Conventional [37] and (b) proposed

TABLE 6.10. Core circuit configuration of the 64-Mbit 2-level NAND Cell Array [6], the Conventional 128-Mbit 4-level NAND Cell Array [37], and the Proposed 4-level NAND Cell Array

	64-Mb 2-level Cell [6]	Conventional 128-Mb 4-level cell [37]	Proposed 128-Mb 4-level Cell
Page size	4 kbit	4 kbit	4 kbit
Column latch circuit: Na	4000	2000	4000
Latch circuits per one column latch: Nb	1	2	2
Total latch circuits: Na x Nb	4000	4000	8000
X-address decoder (Block selector)	1000	1000	512

to reduce the column latch area. In both conventional and proposed 4-level cell array, two bitlines sharing a column latch belong to different pages. And program or read is performed in quite different operations, because two latch circuits are needed to program or read one memory cell connecting to one of the neighboring bitlines. In the conventional 128-Mb cell array, 4000 cells share the same control gate and 2000 column latches are activated during a read or program operation. Although each column latch needs two latch circuits, the number of column latch is halved. Under the same page size condition, the total number of latch circuits in the conventional 128-Mb 4-level cell array is the same as the conventional 64-Mb 2-level cell array and the circuit area does not increase. On the other hand, in the proposed 4-level cell array, 8000 cells share the same control gate so as to make the page size 512 bytes and 4000 column latches are included. In comparison with the conventional 128-Mb 4-level cell array and the 64-Mb 2-level cell array under the same page size condition, there is no circuit area penalty because the number of X-decoders is halved, whereas the circuit area of column latch is doubled, as seen in Table 6.10.

In NAND Flash memories, the cell read current is as small as $1\,\mu A$. Therefore, the read access time is determined by the bitline capacitance, and the delay time of the wordline boosting has little effect on the access time. In the proposed array, as the bitline capacitance is halved, the (verify) read is accelerated.

REFERENCES

1. F. Masuoka, M. Momodomi, Y. Iwata, and R. Shirora, "New Ultra High Density EPROM and Flash EEPROM with NAND Structured Cell," *IEDM Tech. Dig.*, pp. 552–555, Dec. 1987.
2. Y. Itoh, M. Momodomi, R. Shirota, Y. Iwata, R. Nakayama, R. Kirisawa, T. Tanaka, K. Toita, S. Inoue, and F. Masuoka, "An Experimental 4 Mb CMOS EEPROM with a NAND Structured Cell," *ISSCC Dig. Tech. Papers*, pp. 134–135, Feb. 1989.
3. M. Yatabe, K. Kanazawa, and R. Kirisawa, "16 M-Bit NAND EEPROM," *Toshiba Rev.*, Vol. 48, No. 7, pp. 555–558, 1993.
4. K. Imamiya, Y. Iwata, Y. Sugiura, H. Nakamura, H. Oodaira, M. Momodomi, Y. Itoh, T. Watanabe, H. Araki, K. Masuda, and J. Miyamoto, "A 35 ns-Cycle-Time 3.3 V-Only 32 Mb NAND Flash EEPROM," *ISSCC Dig. Tech. Papers*, pp. 130–131, Feb. 1995.
5. K.-D. Suh, B.-H. Suh, Y.-H. Lim, J.-K. Kim, Y.-J. Choi, Y.-N. Koh, S.-S. Lee, S.-C. Kwon, B.-S. Choi, J.-S. Yum, J.-H. Choi, J.-R. Kim, and H.-K. Lim, "A 3.3 V 32 Mb NAND Flash Memory with Incremental Step Pulse Programming Scheme," *ISSCC Dig. Tech. Papers*, pp. 128–129, Feb. 1995.
6. J.-K. Kim, K. Sakui, S.-S. Lee, Y. Itoh, S.-C. Kwon, K. Kanazawa, K.-J. Lee, H. Nakamura, K.-Y. Kim, T. Himeno, J.-R. Kim, K. Kanda, T.-S. Jung, Y. Oshima, K.-D. Suh, K. Hashimoto, S.-T. Ahn, and J. Miyamoto, "A 120 mm² 64 Mb NAND Flash Memory Achieving 180 ns/Byte Effective Program Speed," *Symp. VLSI Circuits Dig. Tech. Papers*, pp. 168–169, June 1996.
7. K. Imamiya, Y. Sugiura, H. Nakamura, T. Himeno, K. Takeuchi, T. Ikehashi, K. Kanda, K. Hosono, R. Shirota, S. Aritome, K. Shimizu, K. Hatakeyama, and K. Sakui, "A 130 mm² 256 Mb NAND Flash with Shallow Trench Isolation Technology," *ISSCC Dig. Tech. Papers*, pp. 112–113, Feb. 1999.
8. K. Takeuchi, Y. Kameda, S. Fujimura, H. Otake, K. Hosono, H. Shiga, Y. Watanabe, T. Futatsuyama, Y. Shindo, M. Kojima, M. Iwai, M. Shirakawa, M. Ichige, K. Hatakeyama, S.

Tanaka, T. Kamei, J.-Y. Fu, A. Cernea, Y. Li, M. Higashitani, G. Hemink, S. Sato, K. Oowada, S.-C. Lee, N. Hayashida, J. Wan, J. Lutze, S. Tsao, M. Mofidi, K. Sakurai, N. Tokiwa, H. Waki, Y. Nozawa, K. Kanazawa, and S. Ohshima, "A 56 nm CMOS 99 mm^2 8 Gb Multi-Level NAND Flash Memory with 10 MB/s Program Throughput," *ISSCC Dig. Tech. Papers*, pp. 144–145, Feb. 2006.

9. D.-S. Byeon, S.-S. Lee, Y.-H. Lim, J.-S. Park, W.-K. Han, P.-S. Kwak, D.-H. Kim, D.-H. Chae, S.-H. Moon, S.-J. Lee, H.-C. Cho, J.-W. Lee, J.-S. Yang, Y.-W. Park, D.-W. Park, D.-W. Bae, J.-D. Choi, S.-H. Hur, and K. Suh, "An 8 Gb Multi-Level NAND Flash Memory in a 63 nm CMOS Process," *ISSCC Dig. Tech.* pp. 46–47, Feb. 2005.

10. S. Aritome, R. Shirota, G. J. Hemink, T. Endoh, and F. Masuoka, "Reliability Issues of Flash Memory Cells," *Proc. IEEE*, Vol. 81, No. 5, pp. 776–788, May 1993.

11. R. Kirisawa, S. Aritome, R. Nakayama, T. Endoh, R. Shirota, and F. Masuoka, "A NAND Structured Cell with a New Programming Technology for High Reliable 5 V-Only Flash EEPROM," *Symp. VLSI Technol. Dig. Tech. Papers*, pp. 129–130, June 1990.

12. S. Aritome, R. Kirisawa, T. Endoh, N. Nakayama, R. Shirota, K. Sakui, K. Ohuchi, and F. Masuoka, "Extended Data Retention Characteristics after More Than 10^4 Write and Erase Cycles in EEPROM's," *IEEE Proc. IRPS*, pp. 259–264, 1990.

13. S. Aritome, R. Shirota, R. Kirisawa, T. Endoh, R. Nakayama, K. Sakui, and F. Masuoka, "A Reliable Bi-Polarity Write/Erase Technology in Flash EEPROM," *IEDM Tech. Dig.*, pp. 111–114, Dec. 1990.

14. K. Sakui, H. Nakamura, T. Tanaka, M. Momodomi, F. Masuoka, and T. Hasegawa, "Semiconductor Memory Device," U.S., Patent No. 5,517,457, May 14, 1996.

15. T. Tanaka, Y. Tanaka, H. Nakamura, H. Oodaira, S. Aritome, R. Shirota, and F. Masuoka, "A Quick Intelligent Page-Programming Architecture 3 V-Only NAND- EEPROMs," *Symp. VLSI Circuits Dig. Tech. Papers*, pp. 20–21, June 1992.

16. T. Tanaka, Y. Tanaka, H. Nakamura, K. Sakui, R. Shirota, K. Ohuchi, and F. Masuoka, "A Quick Intelligent Page-Programming Architecture and a Shielded Bitline Sensing Method for 3 V-Only NAND Flash Memory," *IEEE J. Solid-State Circuits*, Vol. 29, No. 11, pp. 1366–1373, Nov. 1994.

17. M. Momodomi, T. Tanaka, Y. Iwata, Y. Tanaka, H. Oodaira, Y. Itoh, R. Shirota, K. Ohuchi, and F. Masuoka, "A 4-Mbit NAND EEPROM with Tight Programmed Vt Distribution," *IEEE J. Solid-State Circuits*, Vol. 26, No. 4, pp. 492–496, Apr. 1991.

18. K. Sakui, K. Kanda, H. Nakamura, K. Imamiya, and J. Miyamoto, "A Sophisticated Bit-by-Bit Verifying Scheme for NAND EEPROM's," *Symp. VLSI Circuits Dig. Tech. Papers*, pp. 236–237, June 1998.

19. K. Takeuchi, T. Tanaka, and H. Nakamura, "A Double-Level-V_{th} Select Gate Array Architecture for Multilevel NAND Flash Memories," *Symp. VLSI Circuits Dig. Tech. Papers*, pp. 69–70, June 1995.

20. K. Takeuchi, T. Tanaka, and H. Nakamura, "A Double-Level-V_{th} Select Gate Array Architecture for Multilevel NAND Flash Memories," *IEEE J. Solid-State Circuits*, Vol. 31, pp. 602–609, Apr. 1996.

21. T. C. Ong, A. Fazio, N. Mielke, S. Pan, N. Righos, G. Atwood, and S. Lai, "Erratic Erase in ETOXTM Flash Memory Array," *Symp. VLSI Technol. Dig. Tech. Papers*, pp. 83–84, June 1993.

22. Y. Sugiura, Y. Iwata, K. Imamiya, H. Nakamura, H. Oodaira, and K. Sakui, "An Over-Program Detection and Correction Scheme for NAND Flash Memory," *IEEE Nonvolatile Semiconductor Memory Workshop*, Monterey, CA, pp. 120–123, Aug. 1998.

23. S. Aritome, S. Satoh, T. Maruyama, H. Watanabe, S. Shuto, G. J. Hemink, R. Shirota, S. Watanabe, and F. Masuoka, "A 0.67 μm^2 Self-Aligned Shallow Trench Isolation Cell (SA-STI Cell) for 3 V-Only 256 Mbit NAND EEPROM's," *IEDM Tech. Dig.*, pp. 61–64, Dec. 1994.

24. K. Shimizu, K. Narira, H. Watanabe, E. Kamiya, Y. Takeuchi, T. Yaegashi, S. Aritome, and T. Watanabe, "A Novel High-Density 5F2 NAND STI Cell Technology Suitable for 256 Mbit and 1 Gbit Flash Memories," *IEDM Tech. Dig.*, pp. 271–274, Dec. 1997.

25. Y. Takeuchi, K. Shimizu, K. Narita, E. Kamiya, T. Yaegashi, K. Amemiya, and S. Aritome, "A Self-Aligned STI Process Integration for Low Cost and Highly Reliable 1 Gbit Flash Memories," *Symp. VLSI Technol. Dig. Tech. Papers*, pp. 102–103, June 1998.

26. J.-D. Choi, D.-J. Kim, D.-S. Jang, J. Kim, H.-S. Kim, W.-C. Shin, S. T. Ahn, and O.-H. Kwon, "A Novel booster Plate Technology in High Density NAND Flash Memories for Voltage Scaling-Down and Zero Program Disturbance," *Symp. VLSI Circuits Dig. Tech. Papers*, pp. 238–239, June 1996.

27. S. Satoh, H. Hagiwara, T. Tanzawa, K. Takeuchi, and R. Shirota, "A Novel Isolation-Scaling Technology for NAND EEPROMs with the Minimized Program Disturbance," *IEDM Tech. Dig.*, pp. 291–294, Dec. 1997.

28. S. Satoh, K. Shimizu, T. Tanaka, F. Arai, S. Aritome, and R. Shirota, "A Novel Channel Boost Capacitance (CBC) Cell Technology with Low Program Disturbance Suitable for Fast Programming 4 Gbit NAND Flash Memories," *Symp. VLSI Technol. Dig. Tech. Papers*, pp. 108–109, June 1998.

29. K. Takeuchi, S. Sato, T. Tanaka, K. Imamiya, and K. Sakui, "A Negative Vth Cell Architecture for Highly Scalable, Excellently Noise Immune, and Highly Reliable NAND Flash Memories," *Symp. VLSI Circuits Dig. Tech. Papers*, pp. 234–235, June 1998.

30. K. Takeuchi, S. Sato, T. Tanaka, K. Imamiya, and K. Sakui, "A Negative V_{th} Cell Architecture for Highly Scalable, Excellently Noise Immune, and Highly Reliable NAND Flash Memories," *IEEE J. Solid-State Circuits*, Vol. 34, pp. 675–684, May 1999.

31. K. Sakui, H. Nakamura, M. Momodomi, R. Shirota, and F. Masuoka, "Non-Volatile Semiconductor Memory Device," U.S. Patent No. 5,453,955, Sept. 26, 1995.

32. A. Kolodny, S. Nieh, B. Eitan, and J. Shappir, "Analysis and Modeling of Floating-Gate EEPROM Cells," *IEEE Trans. Electron Devices*, Vol. ED-33, pp. 835–844, 1986.

33. R. Shirota, R. Nakayama, R. Kirisawa, M. Momodomi, K. Sakui, Y. Itoh, S. Aritome, T. Endoh, F. Hatori, and F. Masuoka, "A 2.3 μm2 Memory Cell Structure for 16 Mb NAND EEPROM's," *IEDM Tech. Dig.*, pp. 103–106, Dec. 1990.

34. H. Nakamura, J. Miyamoto, K. Imamiya, and Y. Iwata, "A Novel Sense Amplifier for Flexible Voltage Operation NAND Flash Memories," *Symp. VLSI Circuits Dig. Tech. Papers*, pp. 71–72, June 1995.

35. H. Nakamura, J. Miyamoto, K. Imamiya, Y. Iwata, Y. Sugiura, and H. Oodaira, "A Novel Sensing Amplifier for Flexible Voltage Operation NAND Flash Memories," *IEICE Trans. Electron.*, Vol. E79-C, No. 6, pp. 836–844, June 1996.

36. K. Takeuchi, S. Satoh, K. Imamiya, Y. Sugiura, H. Nakamura, T. Himeno, T. Ikehashi, K. Kanda, K. Hosono, and K. Sakui, "A Source-Line Programming Scheme for Low Voltage Operation NAND Flash Memories," *Symp. VLSI Circuits Dig. Tech. Papers*, pp. 37–38, June 1999.

37. T.-S. Jung, Y.-J. Choi, K.-D. Suh, B.-H. Suh, J.-K. Kim, Y.-H. Lim, Y.-N. Koh, J.-W. Park, K.-J. Lee, J.-H. Park, K.-T. Park, J.-R. Kim, J.-H. Lee, and H.-K. Lim, "A 3.3 V 128 Mb Multi-Level NAND Flash Memory for Mass Storage Applications," *ISSCC Dig. Tech. Papers*, pp. 32–33, Feb. 1996.

38. M. Bauer, R. Alexis, G. Atwood, B. Baltar, A. Fazio, K. Frary, M. Hensel, M. Ishac, J. Javanifard, M. Landgraf, D. Leak, K. Loe, D. Mills, P. Ruby, R. Rozman, S. Sweha, S. Talreja, and K. Wojciechowski, "A Multilevel-Cell 32 Mb Flash Memory," *ISSCC Dig. Tech. Papers*, pp. 132–133, Feb. 1995.

39. H. G. Hemink, T. Tanaka, T. Endoh, S. Aritome, and R. Shirota, "Fast and Accurate Programming Method for Multilevel NAND Flash EEPROM's," *Symp. VLSI Technol. Dig. Tech. Papers*, pp. 129–130, June 1995.

40. A. Baker, R. Alexis, S. Bell, V. Dalvi, R. Durante, E. Baer, M. Fandrich, O. Jungroth, J. Kreifels, M. LAndgraf, K. Lee, H. Pon, M. Rashid, R. Rotzman, J. Tsang, K. Underwood, and C. Yarlagadda, "A 3.3 V 16 Mb Flash Memory with Advanced Write Automation," *ISSCC Dig. Tech. Papers*, pp. 146–147, Feb. 1994.
41. S. Aritome, Y. Takeuchi, S. Sato, H. Watanabe, K. Shimizu, G. J. Hemink, and R. Shirota, "A Novel Side-Wall Transfer-Transistor Cell (SWATT Cell) for Multi-Level NAND EEPROM's," *IEDM Tech. Dig.*, pp. 275–278, Dec. 1995.
42. S. Aritome, Y. Takeuchi, S. Sato, H. Watanabe, K. Shimizu, G. J. Hemink, and R. Shirota, "A Side-Wall Transfer-Transistor Cell (SWATT Cell) for Highly Reliable Multi-Level NAND EEPROM's," *IEEE Trans. Electron Devices*, Vol. 44, pp. 145–152, Jan. 1997.
43. T. Tanaka, T. Tanzawa, and K. Takeuchi, "A 3.4-Mbyte/sec Programming 3-level NAND Flash Memory Saving 40% Die Size per Bit," *Symp. VLSI Circuits Dig. Tech. Papers*, pp. 65–66, June 1997.
44. K. Takeuchi, T. Tanaka, and T. Tanzawa, "A Multi-Page Cell Architecture for High-Speed Programming Multi-Level NAND Flash Memories," *Symp. VLSI Circuits Dig. Tech. Papers*, pp. 67–68, June 1997.
45. K. Takeuchi, T. Tanaka, and T. Tanzawa, "A Multi-Page Cell Architecture for High-Speed Programming Multilevel NAND Flash Memories," *IEEE J. Solid-State Circuits*, Vol. 33, No. 8, pp. 1228–1238, Aug. 1998.
46. M. Ohkawa, H. Sugawara, N. Sudo, M. Tsukiji, K. Nakagawa, M. Kawata, K. Oyama, T. Takeshima, and S. Ohya, "A 98 mm² 3.3 V 64 Mb Flash Memory with FN-NOR Type 4-Level Cell," *ISSCC Dig. Tech. Papers*, pp. 36–37, Feb. 1996.

BIBLIOGRAPHY

T. Cho, Y. Lee, E. Kim, J. Lee, S. Choi, S. Lee, D. Kim, W. Han, Y. Lim, J. Lee, J. Choi, and K. Suh, "A 3.3 V 1 Gb Multi-Level NAND Flash Memory with Non-Uniform Threshold Voltage Distribution," *ISSCC Dig. Tech. Papers*, pp. 28–29, Feb. 2001.

T. Hara, K. Fukuda, K. Kanazawa, N. Shibata, K. Hosono, H. Maejima, M. Nakagawa, T. Abe, M. Kojima, M. Fujiu, Y. Takeuchi, K. Amemiya, M. Morooka, T. Kamei, H. Nasu, K. Kawano, C.-M. Wang, K. Sakurai, N. Tokiwa, H. Waki, T. Maruyama, S. Yoshikawa, M. Higashitani, and T. Pham, "A 146 mm2 8 Gb NAND Flash Memory in 70 nm CMOS," *ISSCC Dig. Tech. Papers*, pp. 44–45, Feb. 2005.

J. Lee, H. Im, D. Byeon, K. Lee, D. Chae, K. Lee, Y. Lim, J. Lee, J. Choi, Y. Seo, J. Lee, and K. Suh, "A 1.8 V 1 Gb NAND Flash Memory with 0.12 µm Process Technology," *ISSCC Dig. Tech. Papers*, pp. 104–105, Feb. 2002.

J. Lee, S. S. Lee, O.-S. Kwon, K.-H. Lee, K.-H. Lee, D.-S. Byeon, I-Y. Kim, Y.-H. Lim, B.-S. Choi, J.-S. Lee, W.-C. Shin, J.-H. Choi, and K.-D. Suh, "A 1.8 V 2 Gb NAND Flash Memory for Mass Storage Applications," *ISSCC Dig. Tech. Papers*, pp. 290–291, Feb. 2003.

S. Lee, Y.-T. Lee, W.-G. Hahn, D.-W. Kim, M.-S. Kim, H. Cho, S.-H. Moon, J.-W. Lee, D.-S. Byeon, Y.-H. Lim, H.-S. Kim, S.-H. Hur, and K. Suh, "A 3.3 V 4 Gb Four-Level NAND Flash Memory with 90 nm CMOS Technology," *ISSCC Dig. Tech. Papers*, pp. 52–53, Feb. 2004.

R. Micheloni, R. Ravasio, A. Marelli, E. Alice, V. Altieri, A. Bovino, L. Crippa, E. Di. Martino, L. D'Onofrio, A. Gambardella, E. Grillea, G. Guerra, D. Kim, C. Missiroli, I. Motta, A. Prisco, G. Ragone, M. Romano, M. Sangalli, P. Sauro, M. Scotti, and S. Won, "A 4 Gb 2 b/cell NAND Flash Memory with Embedded 5 b BSH ECC for 36 MB/s System Read Throughput," *ISSCC Dig. Tech. Papers*, pp. 142–143, Feb. 2006.

H. Nakamura, K. Imamiya, T. Himeno, T. Yamamura, T. Ikehashi, K. Takeuchi, K. Kanda, K. Hosono, T. Futatsuyama, K. Kawai, R. Shirota, K. Shimizu, N. Arai, F. Arai, K. Hatakeyama, H. Hazama, M. Saito, H. Meguro, K. Conley, K. Quader, and J. Chen, "A 125 mm^2 1 Gb NAND Flash Memory with 10 MB/s Program Throughput," *ISSCC Dig. Tech. Papers*, pp. 106–107, Feb. 2002.

7

DINOR FLASH MEMORY TECHNOLOGY

Moriyoshi Nakashima and Natsuo Ajika

7.1 INTRODUCTION

DINOR (divided bitline NOR) is a Flash cell that is programmed and erased by Fowler–Nordheim tunneling [1]. Using this approach a low-power dissipation Flash cell can be realized that has a fast random-access time and can be erased and programmed using low voltages. DINOR Flash memory is mainly applied and well accepted in code storage applications for mobile products, currently dominated by cellular phone systems.

Mobile products require higher system performance and higher density, and DINOR Flash memory, as will be described below, has been supporting these needs by incorporating background operation (BGO) functionality, and by offering a smaller package area.

7.2 DINOR OPERATION AND ARRAY ARCHITECTURE

7.2.1 DINOR Operation

DINOR is basically a NOR-type cell that has two main features. The first feature is the use of Fowler–Nordheim (FN) tunneling phenomena for both programming and erasing. The operating current for FN tunneling is small, and low-power dissipation and low-voltage operation are easily realized. Also page-mode programming and fast erasure are enabled. Figure 7.1 shows the circuit arrangement for DINOR operation.

Programming is the same as the gate negative erasing operation for a conventional NOR. It causes FN electron ejection from a floating gate to the drain region

---

Nonvolatile Memory Technologies with Emphasis on Flash. Edited by J. E. Brewer and M. Gill
Copyright © 2008 the Institute of Electrical and Electronics Engineers, Inc.

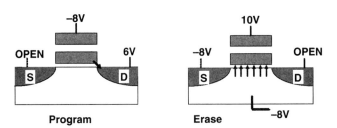

Figure 7.1. DINOR operation schematic: (a) program and (b) erase.

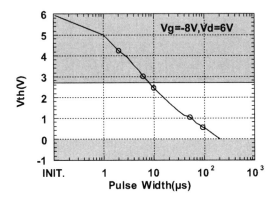

Figure 7.2. Program characteristics of the DINOR operation. Programming condition of $V_g = -8\,\text{V}$ and $V_d = 6\,\text{V}$ realizes program time of 100 μs.

through the gate/drain overlapped area. Erasing causes FN electron injection from a whole channel region to a floating gate. In DINOR operation, to reduce the absolute value of the operating voltage applied to a transistor to less than 10 V, both positive and negative biases are used.

The second DINOR feature is the definition of the threshold voltage (V_{th}) of the programmed and the erased state. In the case of DINOR operation, programming means to set the cell threshold voltage to a low state, and erasing means to set the cell threshold voltage to a high state. This is opposite to the conventional NOR. By defining the low V_{th} as a programmed state, the DINOR operating scheme has the advantage of allowing the V_{th} distribution to be tight because of the bit-by-bit verify programming technique. This makes it possible to employ a low-voltage read without applying a wordline boosting scheme.

7.2.2 DINOR Cell Characteristics

Figures 7.2 and 7.3 show typical program characteristics and erase characteristics for DINOR operation, respectively. The white area of both figures indicates the acceptable range of low state or high state in the situation of $V_{cc} = 3.3\,\text{V}$.

From the figures, the programming time of a single cell is faster than 100 μs at the drain voltage of 6 V and the gate voltage of −8 V, and erase time is less than 1 ms. Erase time of about 1 ms is more than 100 times faster than conventional NOR. This is one of the advantages of the DINOR, especially for the BGO function, which will

DINOR OPERATION AND ARRAY ARCHITECTURE

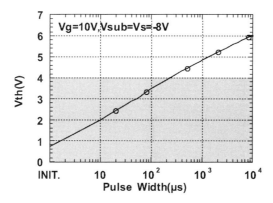

Figure 7.3. Erase characteristics of the DINOR operation. Erase condition of $V_{sub} = V_s = -8\,V$ and $V_g = 10\,V$ realizes erase time of 1 ms.

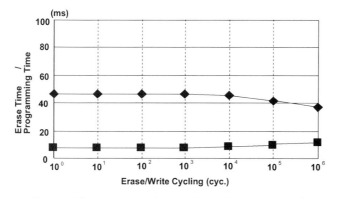

Figure 7.4. Endurance characteristics of the DINOR cell.

TABLE 7.1. DINOR Operating Conditions

	Bitline	Control Gate	Source Line	Substrate
Write	6 V	−8 V	Floating	0 V
Erase	Floating	10 V	−8 V	−8 V
Read	1 V	3 V	0 V	0 V

be mentioned in Section 7.5. The programming time of 100 μs seems slow compared with the conventional NOR. But it is practical to program a large number of cells simultaneously (e.g., 256 bytes) in the DINOR operating mode because of the high efficiency of FN programming. This makes the effective programming time per byte almost the same as that of conventional NOR. The operating conditions for the DINOR cell are summarized in Table 7.1.

Measured endurance characteristics of a device are shown in Figure 7.4. It can be seen that the endurance of the DINOR cell is greater than 10^6 erase/write cycles.

Figure 7.5 shows the measured threshold voltage distribution of a device. Simultaneous programming of plural bits is used. A very narrow distribution width of the programmed state, around 0.6 V, is realized by using a bit-by-bit verify technique.

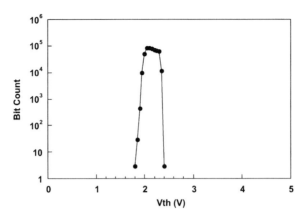

Figure 7.5. V_{th} distribution.

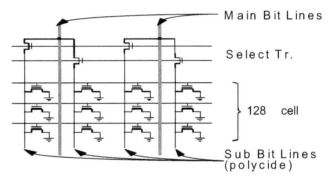

Figure 7.6. Schematic circuit diagram of the cell array architecture. An aluminum main bitline is attached to every two subbitlines, through double-stage select transistors.

7.2.3 DINOR Array Architecture

DINOR uses a hierarchical bitline structure where one main bitline is applied to every two subbitlines through double-stage select gates to every 128 cells. By using these double-stage select gates as a subdecoder, the number of main bitlines can be reduced to one half of that of the subbitlines. A schematic circuit diagram of the array structure is shown in Figure 7.6.

A schematic cross-sectional drawing of the cell is shown in Figure 7.7. The basic structure is similar to that of a conventional NOR, with the above exceptions.

7.2.4 DINOR Advanced Array Architecture

In this section application of the DINOR operation to a virtual ground array that can realize the ideal smallest cell area of $4F^2$ (F = minimum feature size) is discussed.

Among the many innovative technologies and cell structures for high-density Flash memories, contactless array technology is notable because of the possibility of cell area reduction without further scaling a design rule [2]. Contactless cell array with FN tunneling erase/write operation has also been reported [3–5]. However,

DINOR OPERATION AND ARRAY ARCHITECTURE

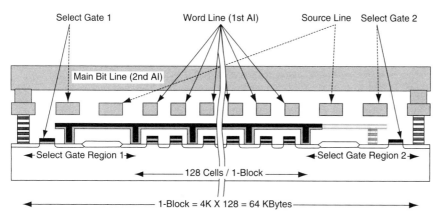

Figure 7.7. Schematic cross-sectional diagram of the DINOR. Double-stage select gates are applied to every 128 cells.

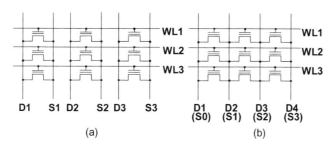

Figure 7.8. Schematic drawings of the virtual ground array: (a) conventional contactless array with FN operation and (b) simple virtual ground array.

these contactless arrays require two diffusion layers per cell to make use of FN tunneling erase/write operation as shown in Figure 7.8(a), and/or the fabrication steps and cell structure are rather complicated and specialized.

DINOR can achieve a simple virtual ground array [Fig. 7.8(b)] with an asymmetrical offset source/drain structure that can program and erase by FN tunneling. The concept of this cell has been tested and verified using a conventional stacked gate test cell with an asymmetrical offset source/drain structure [6].

7.2.5 VGA-DINOR Device Structure and Fabrication

Schematic cross sections of the proposed virtual ground array cell (VGA-DINOR cell) structure are shown in Figure 7.9. One side of the drain region has an offset against the floating gate, and the other side overlaps it. During programming, FN tunneling current only flows to the overlapped drain region from the floating gate that is the selected cell. On the other hand, because of the existence of offset between the drain region and the adjoining unselected cell, programming does not occur for the unselected cell [Fig. 7.9(b)].

The process flow to make the asymmetrical offset source/drain structure is very simple as shown in Figure 7.10. After the tunnel oxide formation, first polysilicon film is deposited and etched with a photoresist mask to form long continuous slots. Next, conventional 7° oblique implantation of arsenic ions is performed perpendicular to the slots without removing the photoresist mask. Because of the oblique ion

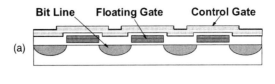

Figure 7.9. Schematic cross sections of the virtual ground array: (a) conventional virtual ground array that cannot use FN mechanism and (b) novel virtual ground array using FN mechanism with offset source/drain structure.

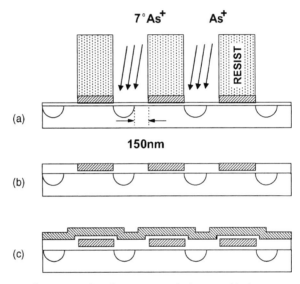

Figure 7.10. Process flow to make the asymmetrical source/drain structure: (a) 7° oblique implantation of arsenic ion with the photoresist mask, (b) CVD SiO2 deposited and etched back and/or CMP, and (c) interpoly dielectric and control gate formation.

implantation, the implanted N^+ regions have an offset of about 150 nm against the polysilicon layers [Fig. 7.10(a)]. Afterward, chemical vapor deposition (CVD) oxide is deposited to fill in the slots and etched back or chemical mechanical polishing (CMP) planarized until the surface of the polysilicon is exposed [Fig. 7.10(b)]. This is followed by the standard contactless array process. In this manner a novel virtual ground array is accomplished [Fig. 7.10(c)].

7.2.6 Characteristics of the Cell with Asymmetrical Offset Source/Drain Structure

Program and erase operations are the same as those of the DINOR approach (Table 7.2). FN programming current ratio between adjoining selected and unselected cells is more than 1000, as shown in Figure 7.11.

TABLE 7.2. Operating Conditions for the Virtual Ground Array DINOR (VGA-DINOR)

	Bitline	Control Gate	Source Line	Substrate
Write	5 V	−9 V	Open	0 V
Erase	−9 V	10 V	−9 V	−9 V
Read	1.5 V	3 V	0 V	0 V

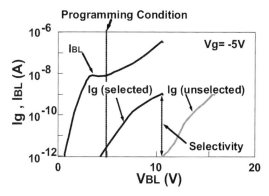

Figure 7.11. FN tunneling current ratio between the adjoining selected and unselected cells.

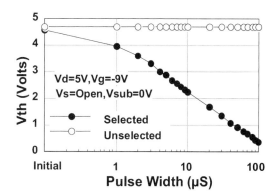

Figure 7.12. Program characteristics of selected and unselected cells that are adjoining each other.

Figure 7.12 shows the programming characteristics of selected and unselected cells that are adjacent to each other. The threshold voltage (V_{th}) of the selected cell is reduced to the acceptable program range at the pulse width of about 100 μs. On the other hand, the V_{th} of the unselected cell is maintained at the high value (erase state) because of the existence of the offset region.

Figure 7.13 shows the erase characteristics at the parameter of applied voltage V_{sub}. An erase time of less than 1 ms is obtained. These characteristics are acceptable for 3-V operations.

Another advantage of this asymmetrical offset source/drain structure is fast data access speed. The virtual ground array uses a buried N^+ line as a bitline and also as a source line for each adjoining cell, so that the electrical resistance and parasitic capacitance of those lines are essentially high. Moreover, the adjoining

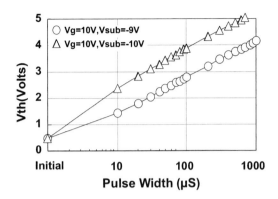

Figure 7.13. Erase characteristics for specific applied voltage V_g and V_{sub}.

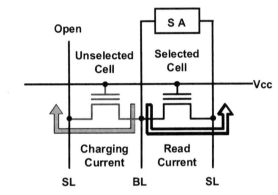

Figure 7.14. Besides the read current of the selected cell, current for charging an unselected source line flows through the adjoining unselected cell. Some delay time is needed to wait until the unselected source line is completely charged.

unselected cell is turned on unwillingly during a read operation, so that a delay time for charging an unselected source line is needed to sense the data (Fig. 7.14). In the case of the proposed cell, the unselected cell has an offset at the source side, so that the unselected cell is kept almost turned off.

Figure 7.15 shows that the cell current (I_{ds}) versus drain voltage (V_d) characteristics of nonoffset and offset source structures. These correspond to selected and unselected cells, respectively. It is shown that the cell current ratio of the selected and the unselected cell is more than 10 at $V_d = 1.5\,\text{V}$ and $V_{fg} = 1.5\,\text{V}$. This means that there is no need to wait for unselected source line charging. That is, high-speed sensing is possible compared with the conventional virtual ground array cell.

7.3 DINOR TECHNOLOGY FEATURES

7.3.1 Low-Voltage Read

In Flash memory, the V_{th} level of a memory cell is treated as representing an information bit, and V_{th} is controlled by electron emission from/injection to the floating gate of the memory cell. In a NOR-type memory array individual memory cells

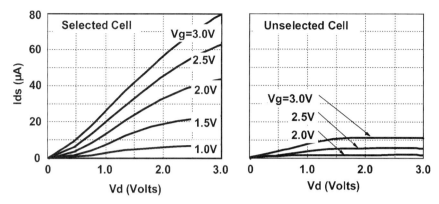

Figure 7.15. I_{ds} versus V_d characteristics of selected and unselected cells of VGA-DINOR. This data is measured on a test structure where gate electrodes are coupled to floating gates.

share one bitline, and because of this, when electrons are emitted from the memory cell, over emission has to be prevented. Since V_{th} of the memory cell can be lowered to 0 V or less, the memory cell is conductive, even if it is not selected. Therefore, precise control of V_{th} for electron emission operation is needed.

As shown in the previous section, in DINOR Flash memory, programming is accomplished using electron emission from the floating gate of the memory cell and erasing is accomplished using injection of electrons. This relation is opposite that of the NOR mechanism. Because of this, V_{th} control of memory cells can be accomplished in a bit-by-bit manner, so that tight distribution of V_{th} of memory cells can be achieved (Fig. 7.5). This means that it is easy to lower the center of the V_{th} distribution, and a relatively lower operating voltage can be used during the read operation. Therefore, DINOR Flash memory has an advantage in that the wordline voltage for reading can be lowered.

7.3.2 Fast Read Access

In DINOR Flash memory, hierarchical bitline architecture is used as shown in Figures 7.6 and 7.7. The memory array is divided so that bitlines are constructed from main bitlines and subbitlines. Using select gate transistors the parasitic capacitance of bitlines are reduced for the sense amplifiers.

The current required for program and erase operations using FN tunneling is small. Because of this, interconnect material with higher resistance can be used for subbitlines. Also lower resistance material can be used for wordline shunts, which means that it is possible to reduce the wordline delay time. Therefore, DINOR architecture is suitable for reducing sensing delay and achieving faster access time.

7.4 DINOR CIRCUIT FOR LOW-VOLTAGE OPERATION

7.4.1 High-Voltage Generation [7]

7.4.1.1 Charge Pump Circuit Without Body Effect. High voltages are required for all Flash memory, and charge pumps are used to raise the external logic supply voltage to the levels required for program and erase. Innovative charge pump

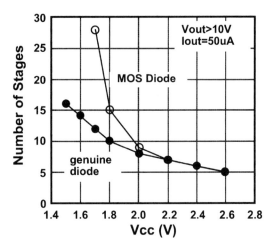

Figure 7.16. Conventional charge pump and diode pump.

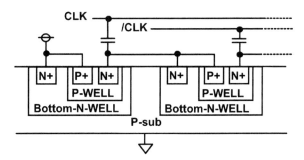

Figure 7.17. Charge pump with P-N junction diode.

design is indispensable when the logic voltages are low as in the case of the DINOR cell. Efficiency of the high-voltage generation is the key issue.

Conventional charge pumps use transistors intended to operate as diodes. Due to the body effect, the output voltage of a metal–oxide–semiconductor (MOS) diode pump is lower than that of a genuine diode pump. Some schemes have been proposed [8, 9] to reduce the threshold voltage of the transistors in the chain, but these approaches are not effective at low voltages. Simulated results of a conventional MOS charge pump [9] and a diode pump are shown in Figure 7.16. The conventional charge pump demands more stages than genuine diode charge pumps. This characteristic becomes more pronounced as the V_{cc} becomes lower than 1.8 V.

To compensate the body effect, a charge pump with a P-N junction diode [10] has been proposed as shown in Figure 7.17. The junction between the P-well and the N^+ diffusion area acts as the P-N diode. The P-well is connected to the N-well in order to avoid bipolar action. However, the N-well cannot be negative because the P-substrate is grounded. This leads to the conclusion that neither a P-N junction diode nor an MOS transistor diode is available for low-voltage operation charge pump circuits. To overcome this problem, a diode element using the N-type polysilicon of the floating gate [11] was fabricated.

DINOR CIRCUIT FOR LOW-VOLTAGE OPERATION

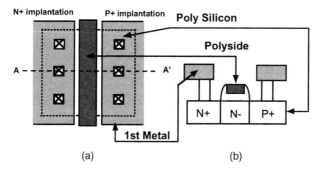

Figure 7.18. Gate-protected polysilicon diode (GPPD) element: (a) top view and (b) cross-sectional view.

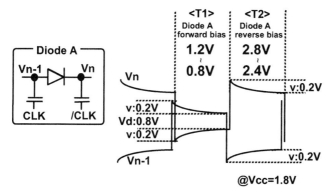

Figure 7.19. Waveforms of the GPPD charge pump.

The structure of a gate-protected polysilicon diode (GPPD) element is shown in Figure 7.18(a) in a top view. Figure 7.18(b) is a cross-sectional view taken along the line A–A' in Figure 7.18(a). The floating-gate material is employed as the material for the diode. P^+ implantation for the peripheral PMOS transistor is employed for forming P^+ electrode of one. N^+ implantation for the peripheral NMOS transistor is employed for reducing the resistance of an N^+ electrode. The gate material of the MOS transistor protects the P-N junction interface from process damage. Because the GPPD is a discrete element, we can utilize the GPPD pump for both positive and negative voltage generation. Moreover, as understood from the above description, we can fabricate the GPPD element without extra process steps.

Figure 7.19 shows the waveforms for the GPPD charge pump. In the period T1, the diode A is forward biased. Initially the diode is biased at $V_d + 2\,\text{V}$. After the charge is transferred, the biased voltage becomes V_d. Assuming that V is 0.2 V, and the V_d is 0.8 V, the forward voltage swings from 1.2 to 0.8 V. In the period T2, the diode A is reverse biased. At the beginning, the diode is biased at $2\,V_{cc} - V_d$. The former capacitance receives the charge, and the voltage increases by V. The latter capacitance loses the charge, and the voltage decreases by V. From this, the reverse voltage is reduced to $2\,V_{cc} - V_d - 2\,\text{V}$. Therefore, the reverse voltage swings from 2.8 to 2.4 V. As long as the reverse current at maximum reverse voltage (2.8 V) is smaller than the forward current at minimum forward voltage (0.8 V), the pump up operation is performed successfully.

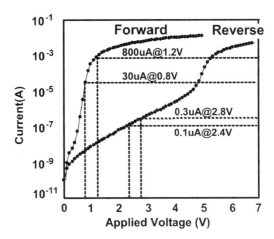

Figure 7.20. Electrical characteristics of the GPPD element.

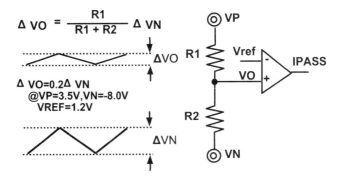

Figure 7.21. Conventional negative voltage detect circuit.

Figure 7.20 illustrates the electrical characteristics of the GPPD element. The reverse current is about 0.13 µA when the reverse voltage is 2.8 V. The forward current is about 820 µA at 1.2 V forward bias, and 30 µA at 0.8 V forward bias. The performance of the GPPD element is poor compared to the genuine diode approach, but the reverse current is small enough compared to the forward current. Therefore, the performance of the GPPD element is adequate to perform a pump up operation. Utilizing this GPPD pump, low-voltage operation is achieved free from the body effect restriction. This results in a reduction of the number of required stages and a reduction in the layout area.

7.4.1.2 Accurate Voltage Detector. The high voltages generated by charge pump circuits are subject to change with the operating conditions, and voltage regulation is required. The conventional negative voltage detect circuit [10] is shown in Figure 7.21. Referring the Figure 7.21, the potential of VO is given by

$$VO = VN + R2(VP - VN)/(R_1 + R_2)$$

The comparator receives the VO and the reference voltage V_{ref}. When VO goes lower than V_{ref}, the detect signal/PASS is at level L to indicate that VN reaches the predetermined level. The voltage conversion factor $\Delta VO/\Delta VN$ is $R_1/(R_1 + R_2)$.

If you set the VN to be −8 V, the conversion factor is 0.2 where VP = 3.5 V and V_{ref} = 1.2 V, and then if you regulate the ΔVN is smaller than 0.3 V, and ΔVO is set less than 0.06 V. This requirement is difficult to realize, and the accuracy of the detection is low.

To overcome this problem, a new negative voltage detect circuit shown in Figure 7.22 can be used. The operation of the proposed detector is as follows. The feedback circuit that controls the voltage applied to the resistance R_1 is identical to the input voltage V_{ref}. Therefore, the current flowing through the resistance R_1 is $I = V_{ref}/R_1$. This is a current mirror circuit, and the same current flows through the resistance R_1 and R_2. The potential VO is given by

$$VO = VN + (R_2)I = VN + V_{ref}(R_2/R_1)$$

The VO is also compared with the reference voltage V_{ref}. The voltage conversion factor ΔVO/ΔVN is 1, and the accuracy of detection is increased. This method can also be applied to the positive voltage detection function.

Figure 7.23 shows the measured waveforms of the output voltages generated by positive and negative charge pumps. Each charge pump generates high voltage using a V_{cc} of 1.8 V.

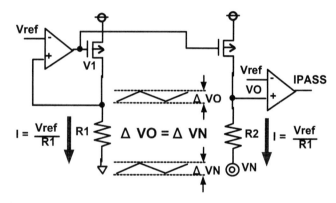

Figure 7.22. Negative voltage detect circuit with feedback control.

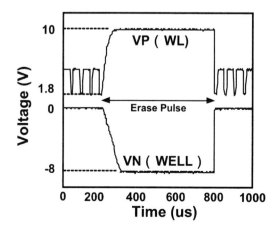

Figure 7.23. Output voltage of GPPD charge pump.

The detect circuit regulates the output voltage up to 10 V. On the other hand, the negative pump generates −8 V, which is also regulated by the detector. The voltage of 10 V is applied to the wordline (WL) through the switching circuit. The voltage of −8 V is applied to the well of the memory array through the switching circuit. Under this condition, the array will be erased.

7.4.2 Wordline Boost Scheme

For low-voltage operation, WL boost is needed. Furthermore, for background operation (BGO), high voltages have to be switched as well as the boosted voltages. BGO allows program or erase in one bank while the reading in the other bank. This BGO feature is suitable for mobile, personal computing, and communication products. Figure 7.24 shows the block diagram of the switching circuit or distributor.

In Figure 7.24 bank 1 is in the read mode and bank 2 is in the erase mode. In bank 1, the boosted voltage is connected to the WL through the WL distributor and the WL decoder. In bank 2, the output of the negative charge pump is connected to the well through the well distributor and the well decoder. Bank 1 also has a well distributor that is not disclosed in Figure 7.24. Similarly, bank 2 also has a WL distributor that is not disclosed in Figure 7.24. Using these distributors, the BGO is developed.

PMOS transistors cannot be used for the switching function because the allowed N-well voltage varies from operating mode to operating mode. Only NMOS transistors can be used for this function, but the body effect limits the voltage that can be transferred. To solve this problem a "heap-pump" circuit [12] is used.

Figure 7.25 shows the circuit diagram of the heap-pump for WL boost. The operation of the switching circuit is as follows. In the first step, three pumping capacitors C_1, C_2, and C_3 are charged up to the V_{cc} level. In response to the address transition detection (ATD) pulse, C_1 and C_2 are connected in series. By heaping C_2 on C_1, a high voltage is achieved. This high voltage is supplied to the gate of the transfer transistor (Tr.1) through the level shift circuit. At the same time C_3 is also boosted. In consequence, the boosted voltage on the C_3 is successfully transferred to the load. This is because the high gate voltage on the transistor Tr.1 compensates for the body effect. In the erase mode, the output of the positive charge pump is connected to the VP1. At this time the gate voltage of the Tr.1 is grounded so as to disconnect the VP1 from the boost circuit.

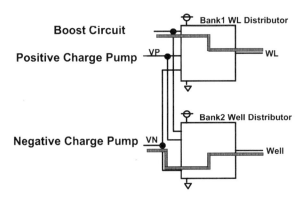

Figure 7.24. Block diagram of distributor.

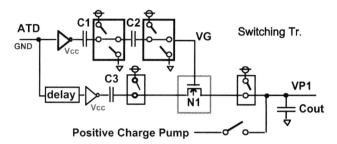

Figure 7.25. Circuit diagram of heap pump for WL boost.

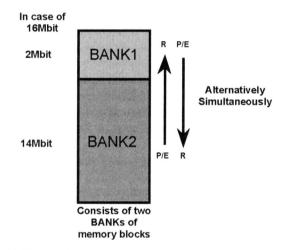

Figure 7.26. BGO (background operation) and two-bank architecture.

7.5 BACKGROUND OPERATION FUNCTION

7.5.1 Background Operation and DINOR

Modern Flash memory incorporates a capability called back ground operation (BGO) that allows a portion of a chip to be reading while erase/program operations are performed elsewhere in the chip. BGO Flash memory can erase/program itself without a need for additional software or external storage to control the process. The control routines are stored within the Flash device. In a two-bank architecture, the data in one bank can be altered (erased and programmed) while the data in the other bank is accessed (read), as illustrated in Figure 7.26.

When DINOR Flash is employed, this is an efficient operation because of its fast erasing time. BGO system overhead time is greatly reduced using the DINOR architecture.

7.5.2 Emulating Electrically Erasable Programmable Read-Only Memory (EEPROM) and Static Random-Access Memory (SRAM)

Background operation Flash memory can avoid the need for embedded EEPROM and SRAM and the associated silicon area penalty. For example, bank 1 of a BGO

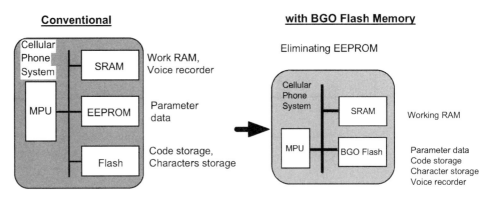

Figure 7.27. EEPROM emulation using BGO flash.

device could contain parameter data that would otherwise be stored in EEPROM while bank 2 might contain voice-recording data that would normally be stored in SRAM. Figure 7.27 illustrates this example by showing how a conventional cellular phone system composed of a microprocessor unit (MPU), Flash, EEPROM, and SRAM might be simplified to just BGO Flash and SRAM. This, of course, has advantages in reduced cost, mounting area, bus area, and weight.

7.5.3 Background Operation Fast Erase

Background operation enables reading a bank while erasing/programming a block in another bank. However, in the case of a continuous erase and program within the same bank a fast erase is an important factor in assuring enhanced system performance. Flash memory erase time is a major portion of the time required for data alteration, and it adds significantly to the system overhead time.

Since high-speed erasing is possible with DINOR Flash memory, as shown in Section 7.2.2, this degradation of system performance can be minimized by using the DINOR approach. This feature becomes more important in case of the "data reclaim" operation, in which erase and program operations are repeated within a short period.

7.6 P-CHANNEL DINOR ARCHITECTURE

7.6.1 Introduction

An electron injection scheme for DINOR Flash memory is described in this section, where band-to-band tunneling-induced hot electrons (BBHE) are employed in a P-channel cell. This method ensures the realization of high program efficiency, high scalability, and hot-hole injection-free operation. It has been demonstrated that a high-speed programming rate of 60 ns/byte can be achieved with a leakage current less than 1 mA by utilizing 512-byte parallel programming. This P-channel DINOR Flash memory can be used to realize low-voltage, high-performance, and high-reliability Flash memory [13–17].

7.6.2 Band-to-Band Hot-Electron Injection Cell Operation

A schematic illustration of the proposed BBHE injection method in a P-channel cell is shown in Figure 7.28. When a negative drain voltage and a positive control gate voltage are applied to the cell, electron-hole pairs are generated by band to band tunneling (BTBT) in the drain region [18, 19]. The electrons are accelerated by a lateral electric field toward the channel region and some of them obtain high energy. The injection of such hot electrons [20] into the floating gate through the tunnel oxide is used as an electron injection method.

The major advantage of this injection method is its high efficiency. Figures 7.29(a) and 7.29(b) show the V_d dependence of I_d, I_g and I_g/I_d (injection efficiency).

To clarify the advantage of the proposed P-channel cell, the characteristics of a typical N-channel cell using the edge FN are shown in Figures 7.30(a) and 7.30(b).

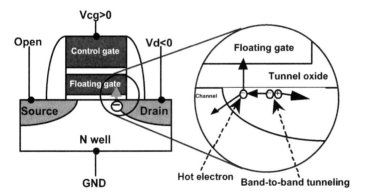

Figure 7.28. Schematic illustration of band-to-band tunneling-induced hot-electron (BBHE) injection method in a P-channel cell. Due to the positive bias to the control gate, holes are not injected into the tunnel oxide.

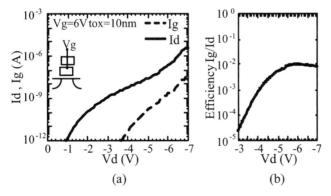

Figure 7.29. (a) V_d dependence of I_d, I_g. (b) V_d dependence of injection efficiency (I_g/I_d) of the P-channel cell. I_d is the band-to-band tunneling leakage current and I_g is the BBHE injection current. A high injection efficiency, which is defined as the ratio of the injection current (I_g), of about 10–2 is obtained.

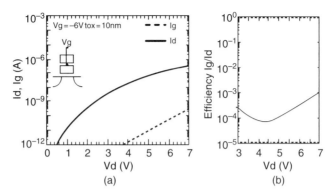

Figure 7.30. (a) V_d dependence of I_d and I_g. (b) V_d dependence of efficiency (I_g/I_d) of the conventional N-channel cell. I_d is the band-to-band tunneling leakage current and I_g is the FN tunneling current from the floating-gate edge to the drain region (edge FN). The efficiency (I_g/I_d) is seen to be less than 10–3. It can be seen that the efficiency of the BBHE injection (Fig. 7.29) is one to two orders of magnitude higher than that of the edge FN.

The maximum injection efficiency of the BBHE injection is 10^{-2}, which is one to two orders of magnitude higher than that of the edge FN. Such high efficiency is achieved because electrons, generated by BTBT, are efficiently accelerated by the lateral electric field in the drain depletion region. In the case of channel hot electron (CHE) injection, lower vertical electric field strength at the electron injection point compared with the BBHE injection is responsible for quite low injection efficiency (10^{-4} to 10^{-5}). Channel-hot-hole-induced hot-electron-injection in a P-channel cell [21] enables low-voltage operation, but the efficiency is only 10^{-4}.

The second advantage of BBHE injection is the elimination of the hot-hole injection into the tunnel oxide. During the BBHE injection operation holes generated by BTBT drift toward the P$^+$ region of the drain, where many holes exist. Because of carrier–carrier scattering, the holes cannot gain high energy. Even if the holes do obtain high energy due to the positively biased control gate, the electric field in the tunnel oxide keeps them from being injected. Therefore, the tunnel oxide is free from degradation due to the hot-hole injection, which occurs in the case of the edge FN operation (Fig. 7.31 [22]).

The third advantage of BBHE injection is the reduction of the vertical electric field strength in the tunnel oxide at the injection point. Device simulation of the BBHE injection operation was performed as shown in Figure 7.32. Calculations included electron temperature ($T_{electron}$), BTBT generation rate (GBTBT), electric fields in the Si substrate (E_{Si}) and in the tunnel oxide (E_{OX}) during BBHE injection operation by device simulation (Avant Corporation MEDICI). Simulated results of $T_{electron}$, GBTBT, and E_{Si} profile at the Si surface in the drain region and E_{OX} profile in the drain region are shown in Figure 7.32. It should be noted that although E_{OX} at the maximum BTBT generation point is 10 MV/cm, E_{OX} of the electron injection point is a low 7.5 MV/cm at the bias condition of $V_d = -6$ V, $V_g = 6$ V ($I_g = 2$ nA/cell). On the other hand, E_{OX} exceeding 13 MV/cm is necessary to obtain the same I_g in the case of the edge FN in the N-channel cell. Therefore, the BBHE injection achieves a 40% reduction of E_{OX} at the point of electron flow as compared with the edge FN. The second and third advantages mentioned above contribute to the improvement of oxide reliability.

P-CHANNEL DINOR ARCHITECTURE

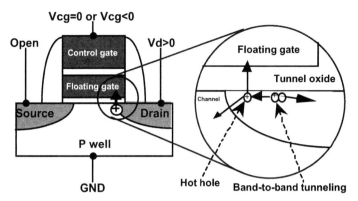

Figure 7.31. Schematic illustration of hot-hole injection into tunnel oxide in the conventional N-channel NOR and DINOR cell. During edge FN operation, hot holes are injected into the tunnel oxide.

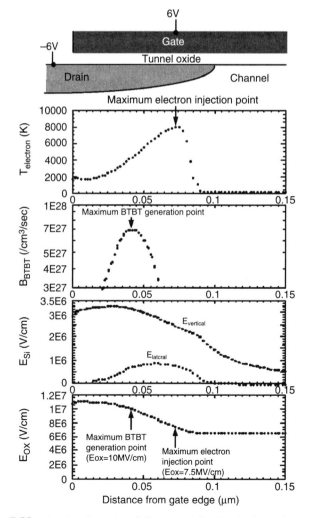

Figure 7.32. Simulated results of the BBHE injection in the P-channel cell.

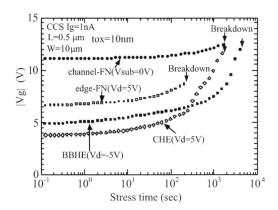

Figure 7.33. Constant-gate-current-stress (CCS) characteristics for BBHE, edge FN, channel FN and CHE. CCS characteristics of BBHE, edge FN, channel FN and CHE were investigated. Compared with the other methods, the lifetime of the tunnel oxide of the BBHE is the longest.

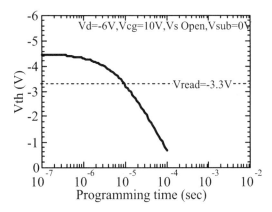

Figure 7.34. Programming characteristics of the P-channel DINOR cell. A programming time of less than 30 μs is obtained.

As shown in Figure 7.33, constant-gate–current-stress characteristics were investigated for the BBHE injection, the CHE injection, the edge FN, and the FN injection from the channel region to the gate (channel FN). The BBHE injection shows high reliability, having the longest lifetime compared with the other methods. This superiority is assumed to be due to the low E_{OX} value at the point of electron flow and to the hot-hole injection-free operation, as discussed above. From the viewpoint of reducing oxide damages, the combination of the BBHE injection and the channel FN could be the best choice for the program and erase method.

7.6.3 DINOR BBHE Programmed Cell

The BBHE injection method can be applied to DINOR cells. The proposed cell is based on a P-channel MOSFET with an N+ polysilicon floating gate. In the erase operation, channel FN ejection from the floating gate is used. The programming characteristics of the cell are shown in Figure 7.34. A programming time of less than 30 μs is obtained, as seen in this figure. Because of the high efficiency of the BBHE injection, a high effective programming speed can be obtained. For example, a high-speed programming rate of 60 ns/byte can be achieved with a leakage current of less than 1 mA by utilizing 512-byte parallel programming. As the $|V_{th}|$ for the programmed state is set at the lower side for P-channel DINOR, same as the conventional N-channel DINOR, a deplete cell problem does not occur with a bit-by-bit verify programming technique.

Figure 7.35 shows the erase characteristics of the cell. An erase time of less than 5 ms is obtained, almost the same as the conventional DINOR. All kinds of disturb characteristics are confirmed to be acceptable for the memory array operation.

Endurance characteristics of the P-channel DINOR cell and its G_m variation during the cycling are shown in Figure 7.36. As is seen in the figure, window narrowing does not occur and no emergence of G_m degradation is observed. This is because the hot-electron injection point is not-located at the channel region but at the drain in the case of the BBHE, as shown in Figures 7.31 and 7.32. Furthermore, the buried channel structure of the cell contributes to its high immunity to G_m degradation, in spite of the utilization of the channel FN ejection for the erase operation, which might damage the whole channel region.

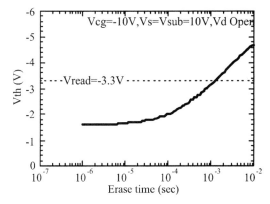

Figure 7.35. Erase characteristics of the P-channel DINOR cell. The erase time is less than 5 ms, which is about the same as the conventional DINOR cell.

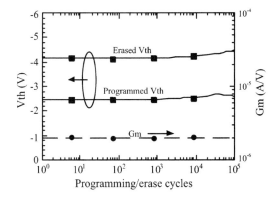

Figure 7.36. Endurance characteristics of the P-channel DINOR cell and its G_m variation during the programming/erase cycling. Window narrowing does not occur after 10^5 programming/erase cycles. Furthermore, the degradation of G_m is not seen after 10^5 cycling.

7.6.4 P-Channel DINOR Summary

An electron injection scheme for Flash memory using BBHE in a P-channel cell has been described. A high injection efficiency of 10^{-2} and hot-hole-injection-free and oxide-damage-reduced operation are realized by this method. Furthermore, a simple symmetrical lightly doped source/drain structure can be adopted, so that high scalability of the cell is achieved. The application of this method to DINOR programming was also described. By utilizing 512-byte parallel programming a high-speed programming rate of 60 ns/byte can be obtained with a leakage current of less than 1 mA. All other characteristics were also confirmed to be acceptable for device operation. The DINOR Flash memory applying the described methods offers high-performance, high-programming speed, high reliability and low-voltage operation.

REFERENCES

1. H. Onoda, Y. Kunori, K. Yuzuriha, S. Kobayashi, K. Sakakibara, M. Ohi, A. Fukumoto, N. Ajika, M. Hatanaka, and H. Miyoshi, "Improved Array Architectures of DINOR for 0.5 um 32M and 64Mbit Flash Memories," *IEICE Trans. Electron.*, Vol. E77-C, No. 8, p. 1279, 1994.
2. B. J. Woo, T. C. Ong, A. Fazio, C. Park, G. Atwood, M. Holler, S. Tam, and S. Lai, "A Novel Memory Cell Using Flash Array Contactless EPROM (FACE) Technology," *IEDM Tech. Dig.*, p. 91, 1990.
3. S. D'Arrigo, G. Imondi, G. Santin, M. Gill, R. Cleavelin, S. Spagliccia, E. Tomassetti, S. Lin, A. Nguyen, P. Shah, G. Savarese, and D. McElroy, "A 5V-Only 256k Bit CMOS Flash EEPROM," *IEEE ISSCC Dig. Tech. Papers*, p. 132, 1989.
4. H. Kume, M. Kato, T. Adachi, T. Tanaka, T. Sasaki, T. Okazaki, N. Miyamoto, S. Saeki, Y. Ohji, M. Ushiyama, J. Yugami, T. Morimoto, and T. Nishida, "A 1.28 um^2 Contactless Memory Cell Technology for a 3V-Only 64 Mbit EEPROM," *IEDM Tech. Dig.*, p. 991, 1992.
5. Y. Hisamune, K. Kanamori, T. Kubota, Y. Suzuki, M. Tsukiji, E. Hasegawa, A. Ishitani, and T. Okazawa, "A High Capacitive-Coupling Ratio (HiCR) Cell for 3V-Only 64Mbit and Future Flash Memories," *IEDM Tech. Dig.*, pp. 19, 1993.
6. M. Ohi, A. Fukumoto, Y. Kunori, H. Onoda, N. Ajika, M. Hatanaka, and H. Miyoshi, "An Asymmetrical Offset Source/Drain Structure for Virtual Ground Array Flash Memory with DINOR Operation," *Symp. VLSI Tech. Dig.*, p. 57, 1993.
7. M. Mihara, Y. Miyawaki, O. Ishizaki, T. Hayasaka, K. Kobayashi, T. Omae, H. Kimura, S. Shimizu, H. Makimoto, Y. Kawajiri, M. Wada, H. Sonoyama, and J. Etoh, "A 29mm2 1.8V-Only 16Mb DINOR Flash Memory with Gate-Protected Poly-Diode Charge Pump," *ISSCC Dig. Tech. Papers*, pp. 114–115, Feb. 1999.
8. T. Jinbo, H. Nakata, K. Hashimoto, T. Watanabe, K. Ninomiya, T. Urai, M. Koike, T. Sato, N. Kodama, K. Oyama, and T. Okazawa, "A 5-V-Only 16-Mb Flash Memory with Sector Erase Mode," *IEEE J. Solid-State Circuits*, Vol. 27, No. 11, pp. 1547–1553, Nov. 1992.
9. A. Umezawa, S. Atsumi, M. Kuriyama, H. Banaba, K. Imamiya, K. Naruke, S. Yamada, E. Obi, M. Oshikiri, T. Suzuki, and S. Tanaka, "A 5-V-Only Operation 0.6-um Flash EEPROM with Row Decoder Scheme in Triple-Well Structure," *IEEE J. Solid-State Circuits*, Vol. 27, No. 11, pp. 1540–1546, Nov. 1992.
10. S. Kobayashi, M. Mihara, Y. Miyawaki, M. Ishii, T. Futatsuya, A. Hosogane, A. Ohba, Y. Terada, N. Ajika, Y. Kunori, K. Yuzuriha, M. Hatanaka, H. Miyashi, T. Yoshihara, U. Uji, A. Matsuo, Y. Taniguchi, and Y. Kigushi, "A 3.3V-Only 16Mb DINOR Flash Memory," *ISSCC Dig. Tech. Papers*, pp. 122–123, Feb. 1995.

11. M. Dutoit and F. Sollberger, "Lateral Poly Silicon p-n Diodes," *J. Electrochem. Soc.*, Vol. 125, pp. 1648–1651, 1978.
12. M. Mihara, Y. Terada, and M. Yamada, "Negative Heap Pump for Low Voltage Operation Flash Memory," IEEE Symposium on VLSI Circuits, pp. 76–77, June 1996.
13. T. Ohnakado, K. Mitsunaga, M. Nunoshita, H. Onoda, K. Sakakibara, N. Tsuji, N. Ajika, M. Hatanaka, and H. Miyoshi, "Novel Electron Injection Method Using Band-to-Band Tunneling Induced Hot Electron (BBHE) for Flash Memory with a P-channel Cell," *IEDM Tech. Dig.*, p. 279, 1995.
14. O. Sakamoto, H. Onoda, T. Katayama, K. Hayashi, N. Yamasaki, K. Sakakibara, T. Ohnakado, H. Takada, N. Tsuji, N. Ajika, M. Hatanaka, and H. Miyoshi, "A High Programming Throughput 0.35 um P-channel DINOR Flash Memory," paper presented at the Symposium on VLSI Technology, p. 222, 1996.
15. T. Ohnakado, H. Takada, K. Hayashi, K. Sugahara, S. Satoh, and H. Abe, "Novel Self-Limiting Program Scheme Utilizing N-channel Select Transistors in P-channel DINOR Flash Memory," *IEDM Tech. Dig.*, p. 181, 1996.
16. T. Ohnakado, H. Onoda, K. Hayashi, H. Takada, K. Kobayashi, K. Sugahara, S. Satoh, and H. Miyoshi, "1.5 V Operation Sector-Erasable Flash Memory with Bipolar Transistor Selected (BITS) P-channel Cells," paper presented at the Symposium on VLSI Technology, p. 14, 1998.
17. T. Ohnakado, H. Onoda, O. Sakamoto, K. Hayashi, N. Nishioka, H. Takada, K. Sugahara, N. Ajika, and S. Satoh, "Device Characteristics of 0.35 um P-channel DINOR Flash Memory Using Band-to-Band Tunneling Induced Hot Electron (BBHE) Programming," *IEEE Trans. Electron Devices*, Vol. 46, No. 9, pp. 1866–1871, Sept. 1999.
18. T. Y. Chan, J. Chen, P. K. Ko, and C. Hu, "The Impact of Gate-Induced Drain Leakage Current on MOSFET Scaling," *IEDM Tech. Dig.*, p. 718, 1987.
19. C. Chang and J. Lien, "Corner-Field Induced Drain Leakage in Thin Oxide MOSFETs," *IEDM Tech. Dig.*, p. 714, 1987.
20. M.-J. Chen, K.-C. Chao, and C.-H. Chen, "New Observation and the Modeling of Gate and Drain Currents in Off-State P-MOSFET's," *Electron Devices*, Vol. 41, No. 5, p. 734, 1994.
21. C. C.-H. Hsu, A. Acovic, L. Dori, B. Wu, T. Lii, D. Quinlan, D. DiMaria, Y. Taur, M. Wordeman, and T. Ning, "A High Speed, Low Power P-Channel Flash EEPROM Using Silicon Rich Oxide as Tunneling Dielectric," *Ex. Abstr. SSDM*, p. 140, 1992.
22. K. T. San, C. Kaya, and T. P. Ma, "Effects of Erase Source Bias on Flash EPROM Device Reliability," *Electron Devices*, Vol. 42, No. 1, p. 150, 1995.

BIBLIOGRAPHY

J. De Blauwe, J. Van Houdt, D. Wellekens, R. Degraeve, P. Roussel, L. Haspeslagh, L. Deform, G. Groeseneken, and H. E. Maes, "A New Quantitative Model to Predict SILC-Related Disturb Characteristics in Flash E2PROM Devices," *IEDM Tech. Dig.*, pp. 343–346, 1996.

M. Kato, N. Migamota, H. Kume, A. Sato, T. Adachi, M. Ushigama, and K. Kimura, "Read-Disturb Degradation Mechanism due to Electron Trapping in the Tunnel Oxide for Low-Voltage Flash Memories," *IEDM Tech. Dig.*, pp. 45–48, 1994.

K. Sakakibara, N. Ajika, and H. Miyoshi, "Influence of Holes on Neutral Trap Generation," *IEEE Trans. Electron Devices*, Vol. 44, pp. 2274–2280, 1997.

K. Sakakibara, N. Ajika, and H. Miyoshi, "On a Universal Parameter of Intrinsic Oxide Breakdown Based on Analysis of Trap-Generation Characteristics," *IEEE Trans. Electron Devices*, Vol. 45, pp. 1336–1341, 1998.

K. Sakakibara, N. Ajika, K. Eikyu, K. Ishikawa, and H. Miyoshi, "A Quantitative Analysis of Time-Decay Reproducible Stress-Induced Leakage Current in SiO_2 Films," *IEEE Trans. Electron Devices*, Vol. 44, pp. 1002–1008, 1997.

K. Sakakibara, N. Ajika, M. Hatanaka, and H. Miyoshi, "A Quantitative Analysis of Stress Induced Excess Current in SiO2 Films," paper presented at the International Reliability Physics Symposium, pp. 100–107, May 3–6, 1996.

K. Sakakibara, N. Ajika, M. Hatanaka, H. Miyoshi, and A. Yasuoka, "Identification of Stress-Induced Leakage Current Components and the Corresponding Trap Models in SiO_2 Films," *IEEE Trans. Electron Devices*, Vol. 44, pp. 2267–2273, 1997.

N. Tsuji, K. Sakakibara, N. Ajika, and H. Miyoshi, "Microscopic and Statistical Approach to SILC Characteristics—Exponential Relation between Distributed Fowler Nordheim Coefficients and Its Physical Interpretation," *Symp. VLSI Tech. Dig.*, p. 196, 1998.

S. Yamada, K. Amemiya, T. Yamane, H. Hazama, and K. Hashimoto, "Non-Uniform Current Flow through Thin Oxide after Fowler-Nordheim Current Stress," *IEEE Proc. IRPS*, pp. 108–112, 1996.

8

P-CHANNEL FLASH MEMORY TECHNOLOGY

Frank Ruei-Ling Lin and Charles Ching-Hsiang Hsu

8.1 INTRODUCTION

Flash memory, due to its nonvolatility, is an indispensable product in portable electronics. Flash memory evolved from EPROM (electrically programmable read-only memory) and possessed the feature of being electrically field programmable similar to EEPROM (electrically erasable programmable read-only memory). For the past decades, researchers and engineers have explored different materials, reduced tunnel oxide thickness, created different structures, innovated more efficient mechanisms to inject or eject carriers, and so forth to produce high-density and high-speed Flash memory. Inheriting from EPROM not only in technology but also in market share, N-channel Flash memory, especially the stacked-gate Flash memory, has played an important role in the development of Flash memory. P-channel Flash memory, introduced by Hsu et al. in 1992 [1], was proposed as an alternative for producing high-density and high-speed Flash memory. The significant advantages of high-speed programming and low-power operation were the major driving force for development of the P-channel Flash memory.

P-channel Flash memory, as illustrated in Figure 8.1(a), resembles N-channel Flash memory, as shown in Figure 8.1(b), in terms of structures and materials but differs in source/drain dopant type. Compared to N-channel Flash memory, P-channel Flash memory can achieve large programming gate current at low power consumption and provides high-speed programming. As very large scale integration (VLSI) technologies progress and battery-powered electronics products prevail, the advantages of P-channel Flash memory become more and more important because of the requirement for low power.

Nonvolatile Memory Technologies with Emphasis on Flash. Edited by J. E. Brewer and M. Gill
Copyright © 2008 the Institute of Electrical and Electronics Engineers, Inc.

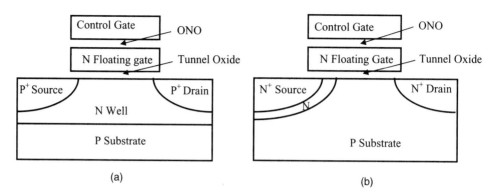

Figure 8.1. Cross sections of (a) P-channel and (b) N-channel Flash memory.

8.2 DEVICE STRUCTURE

The device structure of a stacked-gate P-channel Flash memory is illustrated in Figure 8.1(a). It differs from the N-channel Flash memory in the dopant types of the substrate, source, and drain, while the control gate and floating gate remain unchanged. In addition, the operational ranges of threshold voltages are different, which will be discussed in the next section. The source engineering of the P-channel device is the pure P$^+$ diffusion, while that of N-channel device requires double diffusions, that is, lightly doped N surrounding heavily doped N$^+$, due to reliability considerations. Consequently, the P-channel device possesses advantages in device sizing from the viewpoint of source engineering.

8.3 OPERATIONS OF P-CHANNEL FLASH

Table 8.1 lists the published operations of P-channel Flash memory, where mechanisms and recommended voltages are as shown. The electron injection mechanisms include channel hot-hole-induced hot-electron (CHHIHE) injection, band-to-band tunneling-induced hot-electron (BTBTIHE) injection, and Fowler–Nordheim (FN) tunneling injection, while the electron ejection relies on the FN tunneling.

According to the tunneling paths, erase operations can be categorized into two types: (a) channel FN erase between floating gate and channel/source/drain and (b) interpoly erase between floating gate and control gate. Moreover, the read operation utilizes the channel hole current and operates in the linear region. Hot carriers have to be inhibited against read disturbance (soft programming), and 10-year retention should be guaranteed. For those unselected cells, drain disturbance and gate disturbance are of concern. The tolerable disturbance time (the time to achieve a designated threshold voltage shift) determines whether byte programming or page programming can be applied. For example, if byte programming is used and one page (one wordline) contains 64 bytes, the tolerable disturbance time has to sustain 63 disturbances. If not, the wordline length has to be cut into smaller pieces. Similarly, tolerable drain disturbance time determines the maximum bitline lengths. The

TABLE 8.1. P-Channel Flash Memory Operations

Mode	Mechanism	Operating Voltage Levels (V)					Time or Current
		NOR-Type Control Gate	NAND-Type Control Gate	Drain	Source	N-Well	
Program	CHHIHE	−5 to 0	−3 to 3	−6 to −8	0	0	20–50 μs
	BTBTIHE	8 to 12	10 to 14	−6 to −9	Floating	0	5–20 μs
	FN tunneling	18 to 20	20 to 22	0	0	0	1–10 ms
Erase	FN tunneling (from floating gate to control gate)	7 to 9	9 to 11	0	0	0	10–100 ms
	FN tunneling (from floating gate to channel)	−18 to −20	−16 to −18	0	0	0	10–100 ms
Read	Channel holes	−1 to −3	0	−1 to −2	0	0	20–40 mA
Program Disturb	CHHIHE	0	0	−6 to −8	0	0	0.1–10 s
	BTBTIHE	0	0	−6 to −8	Floating	0	0.1–10 s
	FN tunneling	8 to 12	0	0	0	0	0.1–10 s
Read Disturb	CHHIHE	−1 to −3	0	−1 to −2	0	0	10 years
	BTBTIHE	0	0	−1 to −2	0	0	10 years

resulting array consists of global wordlines and subwordlines, as well as global bitlines and subbitlines.

The programming characteristics of experimental P-channel Flash memory are shown in Figure 8.2. First, the CHHIHE injection gate current shape of the P-channel transistor is a bell centered at the threshold voltage [Fig. 8.2(a)]. The programming speed is 40 to 60 μs [Fig. 8.2(b)]. Next, for BTBTIHE injection the gate current monotonically increases as the gate voltage increases [Fig. 8.2(c)]. The gate current can be increased to 1 nA with 7-V floating-gate voltage. The programming time can be improved to around 40 to 60 μs [Fig. 8.2(d)]. The FN tunneling also monotonically increases as the gate voltage increases [Fig. 8.2(e)]. The tunneling electrons are cool electrons, not hot carriers. The current is proportional to the tunneling probability due to quantum effect. The programming time using FN tunneling is as long as 1 to 10 ms [Fig. 8.2(f)].

The erase operation employs FN tunneling, and the erase time is 100 ms, as shown in Figure 8.3. Oxide electric fields are held to 10 to 12 MV/cm for the sake of reliability, thus complying with erase specifications and still achieving moderate tunneling speeds.

The cross sections and band diagrams shown in Figures 8.4 and 8.5, respectively, are helpful to understand the programming and related gate current. The CHHIHE programming is performed by applying low gate voltage (e.g., −2 V), grounded source, and high drain voltage (e.g., −6 V), as in Figure 8.4(a). The AA' cross section illustrates the lateral band diagram during programming. The channel holes are accelerated by the lateral high field induced by drain voltage and causes electron–hole pair generation through impact ionization. Owing to the large drain electric

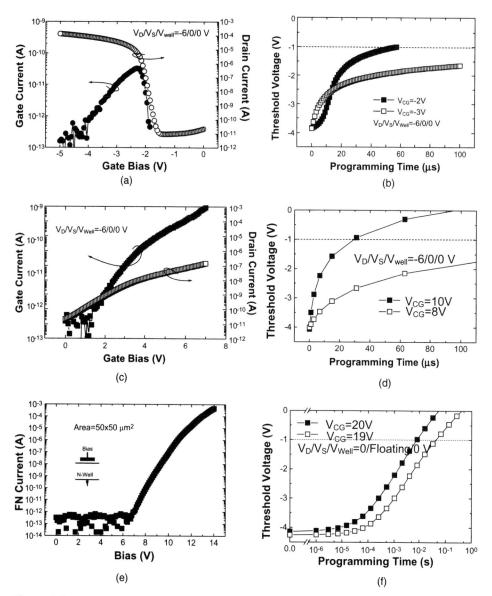

Figure 8.2. (a) CHHIHE gate current, (b) CHHIHE programming trend, (c) BTBTIHE gate current, (d) BTBTIHE programming trend, (e) FN gate current, and (f) FN programming trend. (From 1998 NVSM Workshop [2].)

filed, the electrons are accelerated and gain sufficient energy (i.e., 3.2 eV) to inject into the floating gate, as shown in Figure 8.4(c).

As for BTBTIHE programming, the programming is performed by applying positive control gate voltage (e.g., 10 V), floating source voltage, and high drain voltage (e.g., −6 V). The cross section and band diagrams of BTBTIHE programming are shown in Figures 8.5(a) and 8.5(c), respectively. Due to the high field at such bias conditions, the band-to-band tunneling at the drain/gate overlap regions

OPERATIONS OF P-CHANNEL FLASH

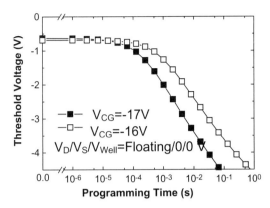

Figure 8.3. Erase characteristics. (From 1998 NVSM Workshop [2].)

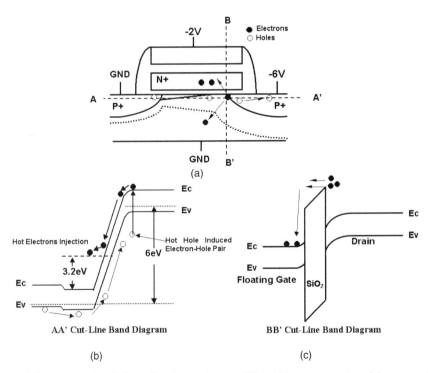

Figure 8.4. Channel hot-hole induced-hot-electron (CHHIHE) programming: (a) cross section, (b) AA' cut-line band diagram, and (c) BB' cut-line band diagram.

generate electron–hole pairs, so-called gate-induced drain leakage (GIDL), as shown in Figure 8.5(b). The generated electrons flow through the large electric field at the drain–gate overlap region and some of the electrons gain sufficient energy (i.e., 3.2 eV) to inject into the floating gate [Fig. 8.5(c)].

The third type is the FN tunneling mechanism for programming. The cross section and related band diagrams are shown in Figures 8.6(a) and 8.6(b), respectively. The N-type substrate is piled up with electrons and portions of those electrons

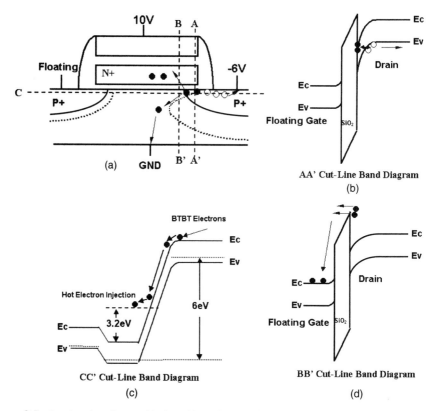

Figure 8.5. Band-to-band tunnel-induced hot-electron (BTBTIHE) programming: (a) cross section, (b) AA' cut-line band diagram, (c) BB' cut-line band diagram, and (d) CC' cut-line band diagram.

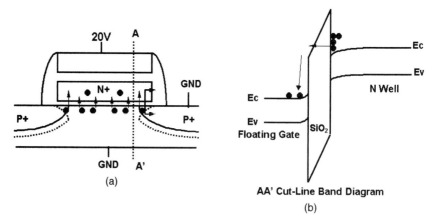

Figure 8.6. FN tunneling programming: (a) cross section and (b) AA' cut-line band diagram.

ARRAY ARCHITECTURE OF P-CHANNEL FLASH

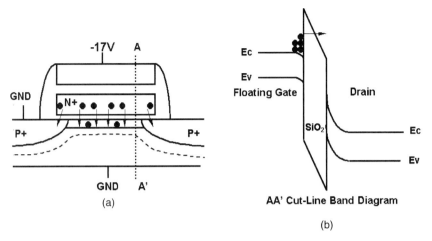

Figure 8.7. FN tunneling erase: (a) cross section and (b) AA' cut-line band diagram.

are tunneled into the floating gate with the well-known FN tunneling mechanism [Fig. 8.6(b)].

As for erasing the P-channel cell, FN tunneling is utilized to eject electrons out of the floating gate, as shown in Figures 8.7(a) and 8.7(b). Threshold voltage is reduced toward a more negative value. On tunneling through the thin oxide, the created damages on the tunneling paths, such as trap generation and trapping, will degrade the subsequent tunneling speed. Therefore, after program/erase cycles, the program/erase speed may be changed.

8.4 ARRAY ARCHITECTURE OF P-CHANNEL FLASH

The array architecture of P-channel Flash memory can be categorized into two types: NOR array and NAND array. The NOR-type array is preferable when random access at high speed is required, while the NAND-type array features high density and a reduced usage of contacts with the penalty of reduced access speed.

8.4.1 NOR-Type Array Architecture

Owing to the significant drain disturbance (explained later), P-channel Flash memory has to adopt a modified NOR-type array architecture, where select transistors have to be inserted to isolate the bitline voltage from the cell's drain. Figure 8.8 shows one feasible layout and schematic diagram for a NOR-type array. Each cell consists of a select transistor and a stacked-gate transistor. Wordlines are formed by connecting the select transistor gates, while the bitlines are made by connecting the select transistor drains. Programming is achieved by applying high voltages to the bitline and control gate, while the select transistor, operated in linear region, passes the bitline voltage to the drain of the stacked-gate transistor. The control gates are all connected to simplify decoding. Obviously, because the unit cell is composed of two transistors, the cell size is too large for high-density applications and is most suitable for memory capacities smaller than 16 Mb, such as embedded applications, PC basic input/output system (BIOS), network switching hubs, and the like.

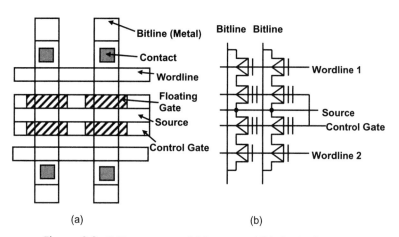

Figure 8.8. NOR-type array: (a) layout and (b) circuit diagram.

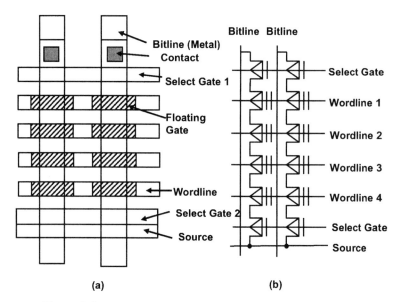

Figure 8.9. NAND-type array: (a) layout and (b) circuit diagram.

8.4.2 NAND-Type Array Architecture

A NAND-type array is shown in Figure 8.9. The memory cells are connected in series, sharing source and drain with each other and forming a string of cells. Each string is gated by top and bottom select transistors for proper operation. The roles of the two select transistors are different. Top select transistor is in charge of passing bitline voltage to the cell in the string, while the bottom select transistor isolates the cells from common source to achieve floating source programming.

To program one cell in the NAND string, the unselected cells have to operate in the linear region to pass the bitline voltage to the drain of the selected cell. The selection process requires the cooperation of the unselected cells involved in passing

EVOLUTION OF P-CHANNEL FLASH

bitline voltage. Complicated timing and decoding are inherently a drawback of the NAND-type array. Furthermore, the series resistance of the long string reduces the read current available for sensing, which makes the access speed slower than that of a NOR-type array. It is even worse that the mobility of holes in a P-channel cell is about one-half of that of an N-channel cell. The read current of a P-channel cell is much reduced. Therefore, the number of cells in series is a trade-off between memory density and memory performance. There is no denying that the array area is smaller in NAND type than in NOR type. For those applications that demand large memory (larger than 16 Mb) and moderate access speed, the NAND-type array is very suitable.

8.5 EVOLUTION OF P-CHANNEL FLASH

The historical evolution of P-channel Flash presented in Table 8.2 is documented by a group of research studies related to the technology. (Also see Lin et al. [3] for an overview of the technology.) These eight studies [1, 2, 4–9] discuss approaches for improving the devices, arrays, and operations of P-channel Flash memories. In order to bring out the innovative features of the technology, the essential message of each study is summarized below.

8.5.1 Hsu et al. [1]

In 1992 C. C. H. Hsu et al. published "A high speed, low power P-channel Flash EEPROM using silicon rich oxide as tunneling dielectric" [1]. CHHIHE was, for the first time, proposed to program P-channel Flash memory. The erase was performed by tunneling through the interpolysilicon-rich oxide. Figure 8.10 shows the cross sections and band diagrams of both N- and P-channel Flash memory operating in the channel hot-electron mode for programming. From the viewpoint of vertical field of the tunneling oxide, the large field in P-channel cell favors electron injection, and hence the probability of electron injection is considerably large in P-channel cell.

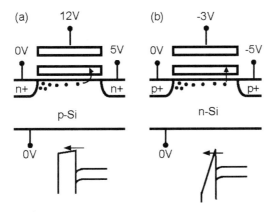

Figure 8.10. Cross section and band diagram of (a) N- and (b) P channel Flash memory [1].

TABLE 8.2. History of P-Channel Flash Memory

Year/Source and Company	Research Topic
1992/SSDM [1] IBM	The first P-channel Flash memory using channel hot-hole-induced hot-electron (CHHIHE) injection and FN tunneling through silicon-rich oxide (SRO)
1995/IEDM [4] Mitsubishi	The first P-channel Flash memory using band-to-band tunneling-induced hot-electron (BTBTIHE) injection
1996/IEDM [5] Mitsubishi	A new self-convergent and self-limiting method to precisely control threshold voltage using N-channel select transistor and alternating control gate pulse oscillating between programming and reading
1997/IEDM [6] National Tsing Hua University	A new self-balanced method that creates dynamic balance between injection and ejection for precisely controlling threshold voltage in multilevel storage
1997/IEDM [7] National Chiao Tung University	Compare degradation and performance of channel hot-hole-induced hot-electron (CHHIHE) and band-to-band tunneling-induced hot-electron (BTBTIHE)
1998/NVSM [8] PMT, PMC	A new single-poly P-channel Flash structure for embedded applications
1998/NVSM [9] National Tsing Hua University	A new programming scheme using ramping control gate voltage to achieve flexible analog/multilevel programming
1998/VLSI Technology [2] Mitsubishi	A new P-channel Flash memory structure with parasitic NPN bipolar transistor at the source side for amplifying read current and suppression of drain disturbance
2000/TED [10] National Tsing Hua University	Multilevel P-channel Flash by using the self-aligned dynamic source
2001/TED [11] Hyundai	A new single-poly P-channel memory with Si nanocrystal charge media
2004/IPFA [12] eMemory	A new single-poly P-channel developed for embedded memory
2005/IEDM [13] MXIC	A new P-channel Flash with 2-bits/cell capability under NAND Flash architecture, suitable for high-density application
2005/VLSI [14] Seoul University	A novel P-channel hot-hole injection programming SONOS memory; high scalability to 50 nm
2006/VLSI [15] GENUSION	A P-channel SONOS memory using back-bias-assisted band-to-band tunneling-induced hot-electron injection
2006/NVSM [16] eMemory	A novel P-channel SONOS memory, very suitable for low cost, embedded Flash memory

Figure 8.11 compares the gate current and drain currents of the two types of Flash memory. The maximum gate current of P-channel cell occurs when gate voltage is biased near its threshold voltage, while that of N-channel cell occurs when gate voltage is equal to drain voltage.

Rearranging the gate (I_G) and drain (I_D) current to be a ratio representing the injection probability (IG/ID), the P-channel cell possesses a dominant 10^{-4} probability, rather than the 10^{-9} of the N-channel cell (Fig. 8.12).

Furthermore, the power consumption of a P-channel cell is much less than that of an N-channel cell because the P-channel cell is biased at the threshold voltage region and the N-channel cell is operated at the saturation region, as shown in Figure 8.13. Obviously, a P-channel cell is more energy saving than an N-channel cell.

Another excellent idea from the study is the silicon-rich oxide (SRO) for the tunneling erase. The tunneling current is greatly improved due to tunneling enhancement by the embedded silicon grain, as shown in Figure 8.14. For the same bias, such as 4 V, 1 µA is generated in SRO while 0.4 nA is flowing through thermal oxide.

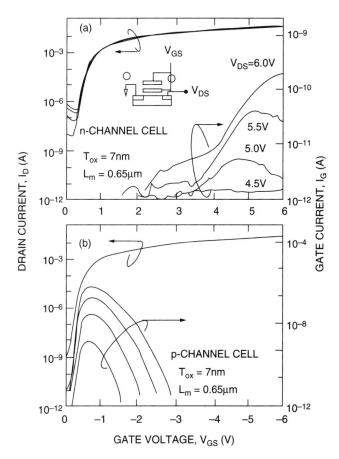

Figure 8.11. Gate current and drain current characteristics of (a) N- (b) P-channel MOSFETs [1].

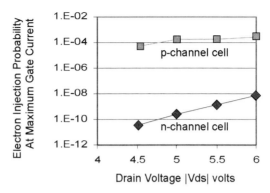

Figure 8.12. Comparisons of injection probability [1].

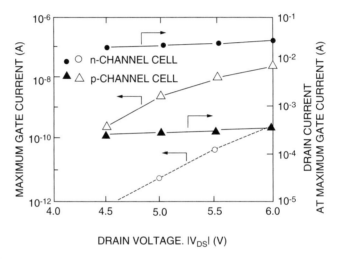

Figure 8.13. Comparison of gate and drain currents between N- and P-channel MOSFETs [1].

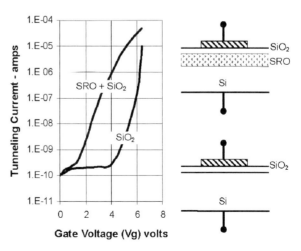

Figure 8.14. Comparison of tunneling current between SRO and thermal oxide [1].

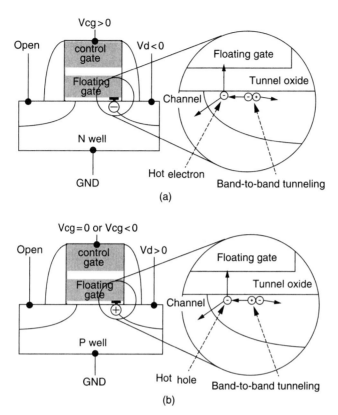

Figure 8.15. Cross sections of (a) P-channel and (b) N-channel Flash memory [4].

8.5.2 Ohnakado et al. [4]

In 1995 T. Ohnakado et al. published a "Novel electron injection method using band-to-band tunneling induced hot electron (BBHE) for Flash memory with a P-channel cell" [4], as shown in Figure 8.15. Band-to-band tunneling-induced hot-electron (BTBTIHE) was, for the first time, applied to the programming of P-channel Flash memory. The positive floating-gate voltage forces the electrons in the valence band to tunnel into the conduction band.

The induced electron–hole pair generates the substrate current, as shown in Figure 8.16. Due to the lateral electric field between drain and substrate (Fig. 8.17), the electrons gain sufficient energy and become "hot." Some of those hot electrons are injected into the floating gate. If the same voltages with reverse polarity are applied to N-channel Flash memory, the band-to-band-tunneling (BTBT) current still occurs, and the generated hot holes are expected to be heated and then injected into the floating gate in a manner similar to the P-channel cell (Fig. 8.15). However, because the barrier height for holes (i.e., 4.8 eV) is larger than that of electrons (i.e., 3.2 eV), the injected hole current is much reduced. The injection efficiencies (I_G/I_D) are also compared in Figure 8.16. The drain current of the N-channel cell is 100 times larger than that of the P-channel cell, and the injection efficiency of the P-channel cell is 100 times larger than that of the N-channel cell. Consequently, the

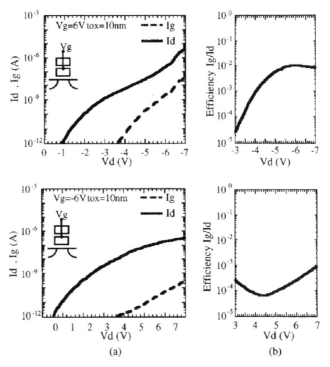

Figure 8.16. Comparisons of drain current, gate current, and injection efficiency between (a) P-channel and (b) N-channel MOSFETs [4].

programming time is greatly reduced from 1 ms to 10 μs if a P-channel cell is adopted rather than a N-channel cell.

This BTBTIHE injection possesses not only high injection efficiency but also abundant advantages, such as ease of downsizing, hot-hole-free, improved reliability, and the like. In addition, the maximum BTBT generation point is separated from the maximum injection point (Fig. 8.17). This result suggests that the injected electrons enter the floating gate through a lower vertical oxide field, and less damage is caused by the accelerated electrons in the oxide.

8.5.3 Ohnakado et al. [5]

In 1996 T. Ohnakado et al. published a "Novel self-limiting program scheme utilizing N-channel select transistor in P-channel DINOR Flash memory" [5]. This study discussed the precise control of threshold voltage applied in divided-bitline NOR arrays (DINOR), as shown in Figure 8.18. The select transistor is a N-channel transistor, instead of a P-channel transistor. The BTBTIHE injection is used for low-power programming. The drain current consumption for one cell is around 100 nA.

The operation comparisons of different select transistors are shown in Figure 8.19. To supply both negative voltage and small current, a N-channel transistor is a better choice than a P-channel transistor. In other words, the N-channel transistor

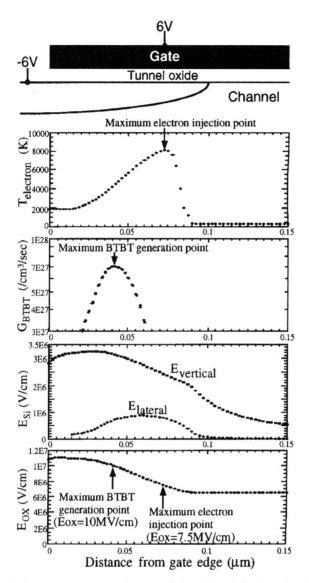

Figure 8.17. Carrier temperature, generation rate, lateral electric field, and vertical electric field at drain edge [4].

is suitable for forming the current source for negative voltages that ideally has high internal impedance.

If the external circuit demands larger current than what the N-channel transistor can supply, the drain-to-source voltage of the N-channel transistor will be increased. As shown in Figure 8.20, a control gate pulse sequence oscillating between programming voltage and reading voltage is used to automatically discharge the bitline voltage as the threshold voltage reaches the designated target.

As the programming proceeds, the threshold voltage changes toward more positive values. The reading voltage of control gate voltage tries to access the read

8. P-CHANNEL FLASH MEMORY TECHNOLOGY

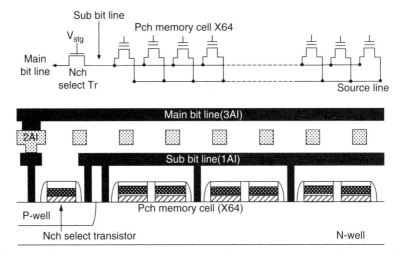

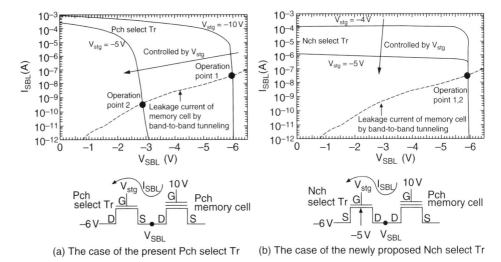

Figure 8.19. Comparisons of operations using (a) P-channel and (b) N-channel select transistors [5].

current for discharging the high-impedance subbitline, which is controlled by the N-channel select transistor. If the threshold voltage reaches the level at which the subbitline is discharged, no further programming proceeds even if the consecutive programming pulses continue, as shown in Figure 8.21.

The programming speed is very sensitive to the subbitline voltage, theoretically an exponential dependence. As a result, with the subbitline voltage discharge phenomenon, the programmed threshold voltage can be self-limited to the designated value controlled by the reading voltage. Since the discharging process stops the threshold voltage shift, the programming is insensitive to the subbitline voltage and to the programming control gate voltage.

EVOLUTION OF P-CHANNEL FLASH

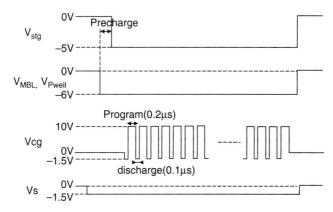

Figure 8.20. Programming waveform for self-limiting method [5].

8.5.4 Shen et al. [6]

In 1997 S. J. Shen et al. published "Novel self-convergent programming scheme for multilevel P-channel Flash memory" [6]. Multilevel storage using P-channel Flash memory was, for the first time, introduced. The programming and erase methods were revolutionarily modified. The BTBTIHE injection method is used for erase (Fig. 8.22), while the programming method uses alternating pulses oscillating between the programming voltage and soft-erase voltage (Fig. 8.23). The erase speed is very fast due to large BTBTIHE gate current. The alternating programming pulse creates a dynamic balance between electron injection and electron ejection.

Threshold voltage shift ceases changing if the balance is reached, as shown in Figure 8.24. Adjusting either programming voltage (VgFN) or soft-erase voltage (VgCHEI) can change the balance point, resulting in changing the convergent values of threshold voltage, as shown in Figure 8.25. The results are very suitable for multilevel applications, as illustrated in Figure 8.26.

8.5.5 Chung et al. [7]

In 1997 Steven S. Chung et al. published "Performance and Reliability Evaluations of P-channel Flash Memories" [7]. The damage to tunnel oxide by using BTBTIHE and CHHIHE was compared, as shown in Figure 8.27 and Table 8.3. Evaluating the gate current degradation stressed by BTBTIHE and CHHIHE, the BTBTIHE damages the tunnel oxide more severely than CHHIHE does, as shown in Figure 8.28. The programming time needs to be increased to obtain the same threshold voltage shift if the degradation is faster, as shown in Figure 8.29. The results suggest that the BTBTIHE injection features high-speed programming for a fresh cell, but the programming speed degrades rapidly after cycling. Therefore, the BTBTIHE is very suitable for high-speed programming and low-power programming at the expense of reduced endurance (i.e., the lifetime of Flash memory).

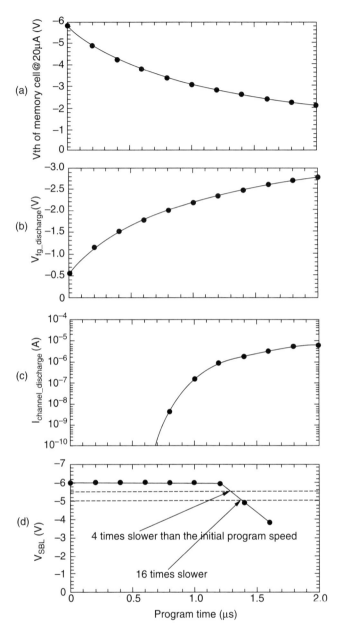

Figure 8.21. Self-limiting programming characteristics [5].

8.5.6 Sarin et al. [8]

In 1998 Vishal Sarin et al. published "A Low Voltage, Low Power P-channel Memory for Embedded and System-on-a-Chip Application" [8]. P-channel Flash memory can be produced using the commercial single-poly complementary metal–oxide–

EVOLUTION OF P-CHANNEL FLASH

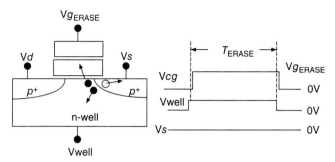

Figure 8.22. Cross section and pulses for band-to-band tunneling erase [6].

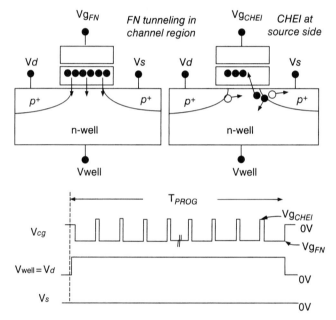

Figure 8.23. Self-convergent programming between injection and ejection [6].

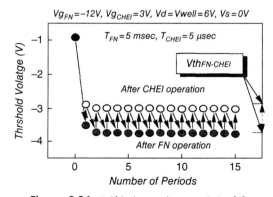

Figure 8.24. Self-balance characteristics [6].

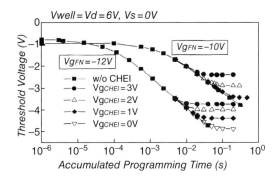

Figure 8.25. Self-convergent programming characteristics [6].

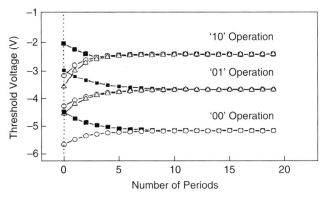

Figure 8.26. Multilevel programming characteristics [6].

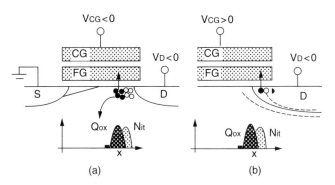

Figure 8.27. Schematic diagram of trap distribution for different gate current stresses: (a) CHE and (b) BTB [7].

semiconducter (CHOS) process by using diffusions as the control gate, as shown in Figure 8.30. The cell cross section is shown in Figure 8.31, where the control gate and the erase gate consist of N diffusion inside the shallow P-well. The fattened cell structure is very suitable for low-cost system-on-a-chip technology.

EVOLUTION OF P-CHANNEL FLASH

TABLE 8.3. Comparison of BTB and CHE

Characteristics	Programming Scheme	
	BTB	CHE
Injection efficiency	Large	Small
Operation window	Large	Small
$\Delta I_G / I_G$	Serious	Moderate
Programming speed	Fast	Slow
Speed retardation	Serious	Moderate

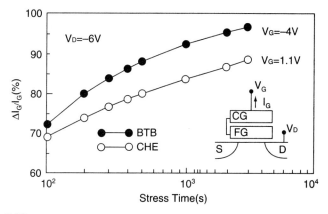

Figure 8.28. Degradation of gate current for different gate current stresses [7].

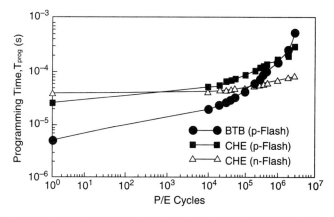

Figure 8.29. Degradation of programming time for different programming mechanisms [7].

8.5.7 Wang et al. [9]

In 1998 Y. S. Wang et al. published "New Programming Scheme for P-channel Flash Memory" [9]. The control gate voltage is modified from one pulse [Fig. 8.32(a)] into an incremental step pulse, as shown in Figures 8.32(b) and 8.32(c). The gate current

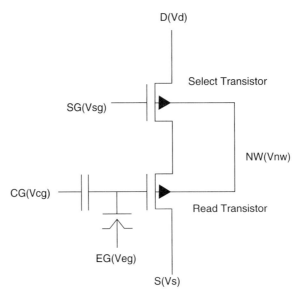

Figure 8.30. Circuit diagram of single-poly P-channel Flash memory [8].

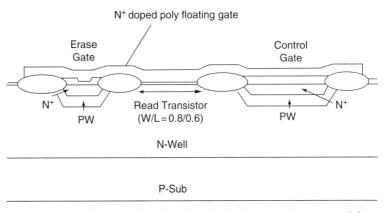

Figure 8.31. Cross section of single-poly P-channel Flash memory [8].

during programming will be controlled to be dynamically constant by the special ramping-gate waveform, as shown in Figure 8.33. The gate current of programming cells with different initial threshold voltages will be converged to the same gate current after a period of transition. After transient state, the programming enters a dynamic balanced situation where nearly constant gate current is charging the floating gate such that the threshold voltage is proportional to the programming time as well as the control gate voltage, as shown in Figure 8.34. By adjusting the programming time, threshold voltages can be tuned in a flexible manner. Therefore, the incremental stair-shape pulse is very suitable for analog storage, such as voice recording.

EVOLUTION OF P-CHANNEL FLASH

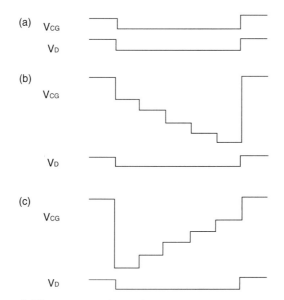

Figure 8.32. Pulse waveform of ramping gate programming [9].

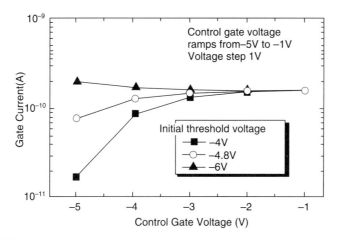

Figure 8.33. Constant gate current programming using ramping control gate pulse [9].

8.5.8 Ohnakado et al. [2]

In 1998 Ohnakado et al. published "1.5V Operation Sector-Erasable Flash Memory with Bipolar Transistor Selected (BIST) P-channel Cells" [2]. The source of the P-channel transistor is modified into a parasitic NPN bipolar select transistor, as shown in Figure 8.35(*a*). The schematic diagram of the modified NOR-type array is illustrated in Figure 8.35(*b*).

It is noted that each cell has its own parasitic bipolar at the source side, with the emitters of those bipolar transistors being connected to form the source line. The P-channel transistor feeds base current to the bipolar transistor such that the

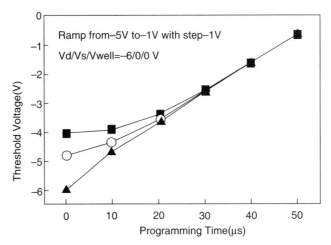

Figure 8.34. Linear increasing threshold voltage using ramping control gate programming [9].

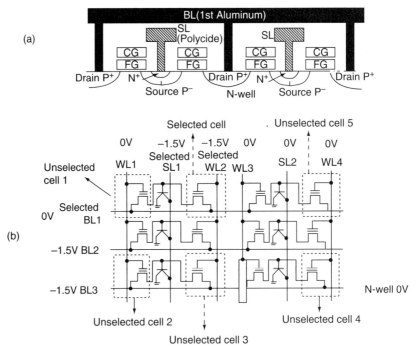

Figure 8.35. (a) Cross section and (b) circuit diagram of BIST P-channel Flash memory structure [3].

EVOLUTION OF P-CHANNEL FLASH

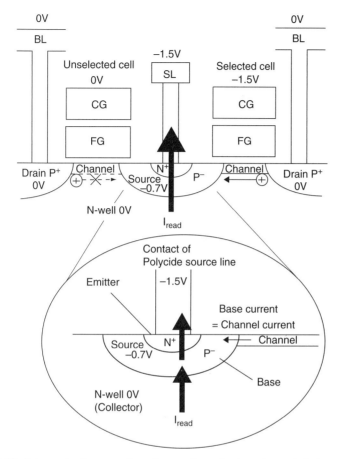

Figure 8.36. Schematic diagram of read operation of BIST P-channel Flash memory [3].

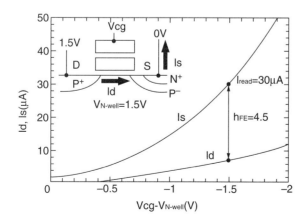

Figure 8.37. Read current characteristics [3].

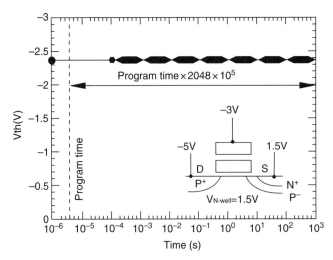

Figure 8.38. Drain disturbance characteristics [3].

channel current is amplified by the bipolar with its current gain, as shown in Figure 8.36. The new device structure improves the small read current of conventional P-channel Flash memory by a factor of hFE (e.g., 4.5). Therefore, the drawback of small hole mobility is overcome by the bipolar transistor's gain, as shown in Figure 8.37. In addition, the new structure allows the negative control gate voltage to suppress the BTBTIHE injection, as shown in Figure 8.38. Negative gate voltage relaxes the vertical oxide field at the drain junction, resulting in the suppression of BTBTIHE gate current. Although channel is conductive, the reverse bias of base and emitter blocks the current flowing into the common source line. As a result, the negative control gate does not introduce undesired leakage in unselected cells. The new structure is a breakthrough innovation in P-channel Flash memory and gives P-channel Flash memory a promising future.

8.5.9 For Further Study

Three major topics relative to P-channel Flash technology are discussed in this section: technology shrink feasibility, multibit storage, and embedded memory.

8.5.9.1 Technology Scalability. Floating-gate-type nonvolatile memory has some specific scaling problems. The thickness of the dielectrics in the gate stack cannot shrink further due to the charge retention and SILC (stress-induced leakage current) issues. This means simple geometric scaling cannot be applied to the nonvolatile memory cell. In addition, with the thick dielectrics, the programming and erasing voltages are nonscaling. And another problem is the HV high-voltage module in the NVM (nonvolatile memory) process. When the feature sizes are reduced in NVM cells, the high-voltage operations are still necessary. Of course, the HV module is itself a nonshrinkable part, and it could impact customer convenience.

The literature contains many ideas for solving these problems by using new structures or new device operation methodologies. Han et al. in 2001 [11] introduced the single-poly P-channel cell using a silicon nanocrystal solution. In this type of structure the HV operation problem can be reduced in comparison with the floating-gate technology.

Sim et al. in 2005 [14] reported the band-to-band tunneling initiated avalanch injection (BAVI) cell on 50-nm single-poly p-channel SONOS technology. SONOS technology stores charge in nitride traps and is compatible with the CMOS process. Band-to-band tunneling initiated avalanche injection programming, and FN electron injection erasing keeps the BAVI cell free from the lateral S/D punch-through, which is a problem in nanoscale devices when using the channel hot carrier injection mechanism. Obviously, this novel operating scheme makes it feasible to scale down to at least 50 nm.

Another new concept for scalability on HV devices can be found by the invention from H. M. Lee et al. in 2006 [16]. Single-poly NeoFlash uses a two-transistor (2T) P-channel SONOS NVM cell. It employs CHE programming and FN erasing. The 2T memory cell has good resistance to programming punch-through. In the NeoFlash cell erasing operation, the well can be held at +6 V while the gate is at −6 V to achieve 12-V FN tunneling. In this scheme the HV operations do not exceed 6 V relative to ground. Therefore, standard input/output (I/O) devices in the CMOS logic process can serve the program/erase (P/E) functionality, and no special HV module has to be installed in the technology.

To solve the lateral programming punch-through problem, Shukuri et al. in 2006 [15] invented a new programming operation using back-bias-assisted band-to-band tunneling-induced hot-electron injection (B4-Flash). In conjunction with a single-poly P-channel SONOS device, ultralow drain-to-source bias was demonstrated for NOR Flash devices scaling down to 60 nm.

8.5.9.2 Multibit Storage. Chapter 12 contains an introduction to multilevel cell (MLC) technology. These concepts have been applied to P-channel nonvolatile memory and are an effective way to multiply the memory density.

A multilevel P-channel Flash memory using a new self-convergent programming scheme was proposed by S.-J. Shen et al. in 1997 [6]. In 2000, R.-L. Lin et al. [10] also introduced a new "self-adjustable dynamic source" programming operation to achieve reliable multilevel P-channel stacked-gate-type Flash.

Another approach to achieve multiple-bits per cell memory density is to have multiple physical storage sites in a cell. This concept is described in Section 13.3 for the NROM technology. A novel P-channel Flash device using nitride trapping was proposed by H.-T. Lue in 2005 [13]. Combining NAND-type array architecture with 2-bit/cell operation, it is suitable for data Flash applications with high programming throughput.

8.5.9.3 Embedded Memory. Embedded memory has become a crucial subject in the semiconductor industry due to the requirement of highly integrated SoC (system-on-chip) application. Chapter 9 treats embedded memory in detail. An ideal SoC should have the capability of integrating a mixed signal/radio frequency module, a dynamic/static random-access memory (DRAM/SRAM) module, and a nonvolatile memory module, into the logic process. However, cost criteria, process complexity, extra thermal budget introduced, and technology porting flexibility will limit this integration process.

Some innovative P-channel embedded memory has been invented. The P-channel NVM cell features low-voltage and low-power operation when compared to N-channel NVM one. That is, the P-channel NVM cell is more feasible for embedded memory because it requires less effort to deal with the HV system circuit design. S.-C. Wang et al. in 2004 [12] introduced single-poly 2T P-channel Neobit for logic

TABLE 8.4. NOR-Type Process Flowchart

Process Steps	Mask No.	Description
1 Starting wafer		P substrate
2 Deep N-well	1	Deep N-well masking/implant/drive-in
3 Shallow P-well	2	Shallow P-well inside deep N-well
4 Shallow N-well	3	Shallow N-well in P-substrate
5 Active definition	4	Active region of devices
6 P-field implant	5	Field threshold voltage adjustment
7 Field oxidation		Isolation oxide growth
8 N-channel V_{TH} adjustment	6	Adjust N-channel threshold voltage
9 P-Channel V_{TH} adjustment	7	Adjust P-channel threshold voltage
10 Cell V_{TH} adjustment	8	Adjust cell threshold voltage
11 Tunnel oxidation		Tunnel oxide growth
12 Poly-I definition	9	Deposit poly-I and pattern cell floating gate
13 Poly oxidation		Bottom oxide growth
14 Nitride deposition		Deposit nitride
15 HTO deposition		High-temperature top oxide deposition
16 ONO removal	10	Remove peripheral ONO
17 Thick gate oxidation		Gate oxide growth of high-voltage transistor
18 Thick oxide etching	11	Pattern thin oxide region
19 Thin oxidation		Gate oxide growth of low-voltage transistor
20 Poly-II deposition		Deposit poly-II
21 Stacked-gate etching	12	Pattern floating-gate cells
22 Self-aligned source (SAS)	13	Remove field oxide on cell sources
23 Source P$^+$ implant	14	Cell source implant
24 Poly-II etching	15	Pattern peripheral gate
25 N-DD implant	16	N-LDD implant
26 P-LDD implant	17	P-LDD implant
27 Oxide spacer		Deposit oxide and etch back
28 N$^+$ implant	18	N$^+$ implant
29 P$^+$ implant	19	P$^+$ implant
30 Salicide		Self-aligned silicide
31 Oxide deposition		Oxide deposition
32 Contact opening	20	Contact opening
33 Contact plug metallization		Metal contact plug
34 Metal-I definition	21	Metal-I sputtering and patterning
35 Oxide deposition		Oxide deposition
36 1st via opening	22	Via opening
37 1st via plug metallization		Metal via plug
38 Metal-II definition	23	Metal-II sputtering and patterning
39 Passivation		Passivation oxide
40 PAD	24	PAD opening

EVOLUTION OF P-CHANNEL FLASH

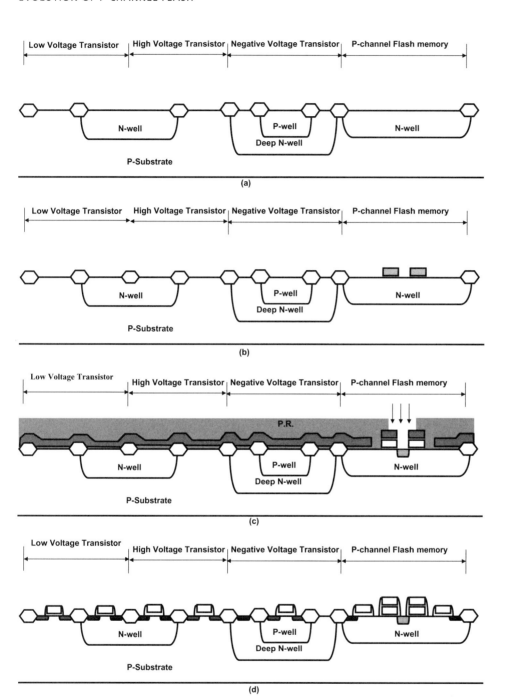

Figure 8.39. Process flow of NOR-type array.

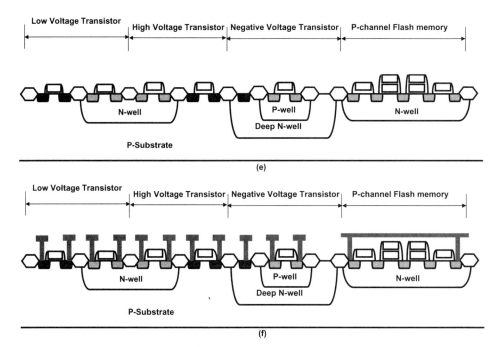

Figure 8.39. *Continued*

nonvolatile memory. Neobit enables the incorporation of general-purpose embedded memory, such as fuse/ROM code replacement, SRAM/DRAM redundancy, security code, RC trimming in analog circuits, and configuration setting for low endurance applications. It meets the low cost, no additional masking layers, easy porting requirements of logic NVM. And it also helps customers by adding design flexibility, more security, and more product features than conventional fuse or ROM methods.

Another logic NVM for higher endurance was also presented by H.-M. Lee, et al. 2006 [16] as mentioned above. This single-poly 2T P-channel SONOS cell (named NeoFlash.) with channel hot-electron programming and FN tunneling erasing requires 2 to 3 additional masking layers. It thus has little impact on logic devices, and this makes the technology easy to transfer to logic process derivatives (such as mixed mode, RF or HV process etc). Single-poly-embedded memory with no HV process module and the same verified ONO module also avoids very difficult logic process turning in the fabrication plant.

8.6 PROCESSING TECHNOLOGY FOR P-CHANNEL FLASH

Flash memory is CMOS compatible, but it requires consideration of some specific issues. Because of high voltages during write operations, high-voltage devices with thicker gate oxide are necessary. Another special concern is the use of negative voltages, which requires well isolation using triple well technology. In this section, the typical process flows for NOR-type arrays and NAND-type arrays will be introduced.

8.6.1 NOR Type Array Architecture

A typical triple-well process flow with 24 masks is presented in Table 8.4. As shown in Figure 8.39(a), a deep N-well is formed at the beginning, followed by the shallow N-well and the P-well. Next, the active regions are defined with field oxide growth. The threshold voltages of N-channel transistors, P-channel transistors, and Flash cells are adjusted with different implantations, respectively. After surface clean, the tunnel oxide is grown, followed by poly-I deposition and poly implantation, as illustrated in Figure 8.39(b). Upon poly-I, the interpoly dielectric ONO is formed. In

TABLE 8.5. NAND-Type Process Flowchart

Process Steps	Mask No.	Descriptions
1 Starting wafer		P substrate
2 Deep N-well	1	Deep N-well masking/implant/drive-in
3 Shallow P-well	2	Shallow P-well inside deep N-well
4 Shallow N-well	3	Shallow N-well in P-substrate
5 Active definition	4	Active region of devices
6 P-field implant	5	Field threshold voltage adjustment
7 Field oxidation		Isolation oxide growth
8 N-channel V_{TH} adjustment	6	Adjust N-channel threshold voltage
9 P-channel V_{TH} adjustment	7	Adjust P-channel threshold voltage
10 Cell V_{TH} adjustment	8	Adjust cell threshold voltage
11 Tunnel oxidation		Tunnel oxide growth
12 Poly-I definition	9	Deposit poly-I and pattern cell floating gate
13 Poly oxidation		Bottom oxide growth
14 Nitride deposition		Deposit nitride
15 HTO deposition		High-temperature top oxide deposition
16 ONO removal	10	Remove peripheral ONO
17 Thick gate oxidation		Gate oxide growth of high-voltage transistor
18 Thick oxide etching	11	Pattern thin oxide region
19 Thin oxidation		Gate oxide growth of low-voltage transistor
20 Poly-II deposition	12	Deposit poly-II
21 Stacked-gate etching	13	Pattern floating-gate cells
22 Poly-II etching	14	Pattern peripheral gate
23 N-LDD implant	15	N-LDD implant
24 P-LDD implant	16	P-LDD implant
25 Oxide spacer		Deposit oxide and etch back
26 N$^+$ implant	17	N$^+$ implant
27 P$^+$ implant	18	P$^+$ implant
28 Salicide		Self-aligned silicide
29 Oxide deposition		Oxide deposition
30 Contact opening	19	Contact opening
31 Contact plug metallization		Metal contact plug
32 Metal-I definition	20	Metal-I sputtering and patterning
33 Oxide deposition		Oxide deposition
34 1st via opening	21	Via opening
35 1st via plug metallization		Metal via plug
36 Metal-II definition	22	Metal-II sputtering and patterning
37 Passivation		Passivation oxide
38 PAD	23	PAD opening

order to be compatible with the CMOS, ONO layer is removed in the peripheral regions. Next, thick gate oxide is grown for high-voltage transistors, such as charge pumping circuits and wordline drivers, followed by the regional removal of thick oxide in the thin oxide regions. Then thin oxide is grown for high-performance transistors. The second polysilicon layer is deposited, followed by floating-gate transistor definition. The memory sources are reconnected by self-aligned source (SAS) etching and source implantation, as illustrated in Figure 8.39(c). Following SAS is the peripheral gate definition. The lightly doped drain and source are formed at N-channel and P-channel transistors, as well as at the select transistors in the memory array. Spacers are then generated at the transistors' sidewalls and memory cells' gate edges and serve as the implant barriers, as illustrated in Figure 8.39(d). Source and drain are implanted with heavy dosage in N-channel and P-channel devices for reduced transistor resistance and contact resistance as shown in Figure 8.39(e). The gate, source, and drain react with metal to form silicide. The isolation oxide is deposited and the back-end processes of interconnection are performed in sequence, as shown in Figure 8.39(f).

8.6.2 NAND-Type Array Architecture

As listed in Table 8.5, NAND-type array fabrication requires 23 masks, rather than 24 masks for a NOR-type array. The reduction is due to the removal of the self-aligned source etching procedure. The process flow is very similar to that of the NOR-type array. At the beginning, the triple wells are formed as shown in Figure 8.40(a), followed by tunnel oxide growth and floating-gate patterning, as illustrated in Figure 8.40(b). After sequentially stacking ONO layers, poly-II is deposited and

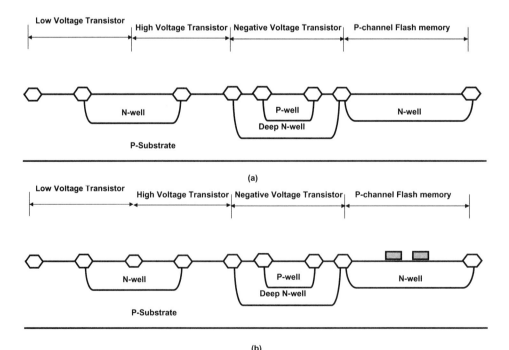

Figure 8.40. Process flow of NAND-type array.

PROCESSING TECHNOLOGY FOR P-CHANNEL FLASH

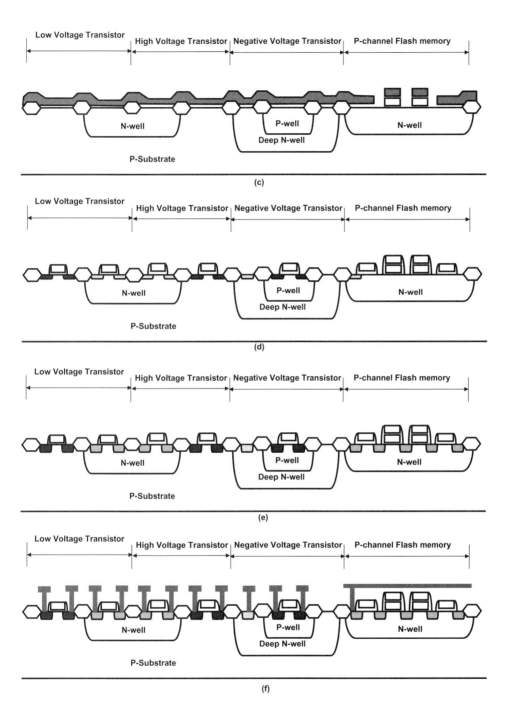

Figure 8.40. *Continued*

patterned, as shown in Figure 8.40(c). Next, the peripheral transistors are defined, followed by lightly doped drain (LDD) implantation and subsequent spacer formation, as shown in Figure 8.40(d). Then source and drain regions are doped with heavy dosage to reduce resistance of diffusion lines and contacts, as shown in Figure 8.40(e). Finally, the oxide deposition, subsequent contacts, and first metals are carried out, as shown in Figure 8.40(f).

REFERENCES

1. C. C.-H. Hsu, A. Acovic, L. Dori, B. Wu, T. Lii, D. Quinlan, D. DiMaria, Y. Taur, M. Wordeman, and T. Ning, "A High Speed, Low Power p-Channel Flash EEPROM Using Silicon Rich Oxide as Tunneling Dielectric," paper presented at the Conference on Solid State Devices and Materials, pp. 140–141, 1992.
2. T. Ohnakado, N. Ajika, H. Hayashi, H. Takada, K. Kobayashi, K. Sugahara, S. Satoh, and H. Miyoshi, "1.5 V Operation Sector-Erasable Flash Memory with Bipolar Transistor Selected P-Channel Cells," *Symp. VLSI Technol. Dig. Tech. Papers*, pp. 14–15, 1998.
3. R. L. Lin, Y. S. Wang, and C. C.-H. Hsu, "P-Channel Flash Memory," paper presented at the Nonvolatile Semiconductor Memory Workshop, Monterey, CA, pp. 27–34, 1998.
4. T. Ohnakado, K. Mitsunaga, M. Nunoshita, H. Onoda, K. Sakakibara, N. Tsuji, N. Ajika, M. Hatanaka, and H. Miyoshi, "Novel Electron Injection Method Using Band-to-Band Tunneling Induced Hot Electron (BBHE) for Flash Memory with a p-Channel Cell," *IEDM Tech. Dig.*, pp. 279–282, 1995.
5. T. Ohnakado, H. Takada, K. Hayashi, K. Sugahara, S. Satoh, and H. Abe, "Novel Self-Limiting Program Scheme Utilizing N-Channel Select Transistors in P-Channel DINOR Flash Memory," *IEDM Tech. Dig.*, pp. 181–184, 1996.
6. S. J. Shen, C. S. Yang, Y. S. Wang, and C. C.-H. Hsu, "Novel Self-Convergent Programming Scheme for Multilevel P-Channel Flash Memory," *IEDM Tech. Dig.*, pp. 287–290, 1997.
7. S. S. Chung, S. N. Kuo, C. M. Yih, and T. S. Chao, "Performance and Reliability Evaluations of P-Channel Flash Memories with Different Programming Schemes," *IEDM Tech. Dig.*, pp. 295–298, 1997.
8. V. Sarin, A. Yu, V. kowshik, R. Yang, T. Chang, N. Radjy, K. Huang, B. Ly, M. Chen, A. Kwok, and A. Wang, "A Low-Voltage, Low Power P-Channel Memory for Embedded and System-on-a-Chip Applications," paper presented at the Nonvolatile Semiconductor Memory Workshop, pp. 116–119, 1998.
9. Y.-S. Wang, S.-J. Shen, C.-S. Yang, and C. C.-H. Hsu, "New Programming Scheme for P-Channel Flash Memory," paper presented at the Nonvolatile Semiconductor Memory Workshop, Monterey, CA, pp. 88–91, 1998.
10. R.-L. Lin, T. Chang, A.-C. Wang, and C. C.-H. Hsu, "New Self-Adjusted Dynamic Source Multilevel p-Channel Flash Memory," *IEEE Trans. Electron Devices*, Vol. 47, No. 4, pp. 841–847, Apr. 2000.
11. K. Han, I. Kim, and H. Shin, "Characteristics of p-Channel Si Nanocrystal Memory," *IEEE Trans. Electron Devices*, Vol. 48, No. 5, pp. 874–879, May 2001.
12. R. S.-C. Wang, R. S.-J. Shen, and C. C.-H. Hsu, "Neobit[r]—High Reliable Logic Non-Volatile Memory (NVM)," paper presented at the Physical and Failure Analysis of Integrated Circuits, IPFA 2004, Proc. of the 11th Inter. Sym., pp. 111–114, July 5–8, 2004.
13. H.-T. Lue, S.-Y. Wang, E.-K. Lai, M.-T. Wu, L.-W. Yang, K.-C. Chen, J. Ku, K.-Y. Hsieh, R. Liu, and C.-Y. Lu, "A Novel p-Channel NAND-Type Flash Memory with 2-Bit/Cell Operation and High Programming Throughput (>20 MB/sec)," *IEEE IEDM Tech. Dig.*, pp. 331–334, Dec. 2005.

14. J. S. Sim, I. H. Park, H. Park, S. Cho, T. H. Kim, K. W. Song, J. Kong, H. Shin, J. D. Lee, and B.-G. Park, "BAVI-Cell: Novel High-Speed 50 nm SONOS Memory with Band-to-Band Tunneling Initiated Avalanche Injection Mechanism," *Symp. VLSI Technol. Dig. Tech. Papers*, pp. 122–123, June 2005.
15. S. Shukuri, N. Ajika, M. Mihara, K. Kobayashi, T. Endih, and M. Nakashima, "A 60 nm NOR Flash Memory Cell Technology Utilizing Back Bias Assisted Band-to-Band Tunneling Induced Hot Electron Injection (B4-Flash)," *Symp. VLSI Technol. Dig. Tech. Papers*, pp. 15–16, June 2006.
16. H.-M. Lee, S.-T. Woo, H.-M. Chen, R. Shen, C.-D. Wang, L.-C. Hsia, and C. C.-H. Hsu, "NeoFlash—True Logic Single Poly Flash Memory Technology," paper presented at the Non-Volatile Semiconductor Memory Workshop, IEEE NVSMW 2006, Monterey, CA, pp. 15–16, Feb. 2006.

BIBLIOGRAPHY

T. S.-D. Chang, "Non-Volatile Electrically Erasable Memory with PMOS Transistor NAND Gate Structure," U.S. Patent No. 5,581,504, Dec. 3, 1996.

T. S.-D. Chang, "PMOS Flash Memory Cell Capable of Multi-Level Threshold Voltage Storage," U.S. Patent No. 5,666,307, Sept. 9, 1997.

T. S.-D. Chang, "PMOS Memory Cell with Hot Electron Injection Programming and Tunneling Erasing," U.S. Patent No. 5,687,118, Nov. 11, 1997.

9

EMBEDDED FLASH MEMORY

Chang-Kiang (Clinton) Kuo and Ko-Min Chang

9.1 INTRODUCTION

Combining the high-density, low-cost and electrical programmability of erasable programmable read-only memory (EPROM) with the electrical erasability of electrically erasable programmable read-only memory (EEPROM), Flash EEPROM, or Flash memory has widely been accepted as the nonvolatile memory of choice for many applications. Flash memory has been used extensively to replace EPROM for program storage to realize the added advantage of being electrically erasable. Flash memory, being block or array erasable, has also been used to replace the byte-erasable EEPROM for data storage to achieve a significant cost reduction. While initial development and applications of Flash memory were devoted to stand-alone Flash memory products, it was recognized early on that the embedded Flash memory could offer many added advantages over the stand-alone Flash memory. Therefore, the development and introduction of embedded Flash memory products followed closely the stand-alone Flash memory [1]. In this chapter, embedded Flash memory is defined as the Flash memory that is physically and electrically integrated into the host logic device on a monolithic silicon substrate. The host logic device can itself be or contain cores of a microcontroller (MCU), digital signal processor (DSP), application-specific integrated circuit (ASIC), application-specific standard product (ASSP), and/or programmable logic device (PLD).

There is a very wide range of applications for the embedded Flash memory. In general, all applications suitable for the stand-alone Flash memory are also suitable for the embedded Flash memory, except for those applications requiring a very large or very small array size to be cost effective. By integrating a Flash memory into a host logic device to form an embedded Flash memory, many added advantages can

---

Nonvolatile Memory Technologies with Emphasis on Flash. Edited by J. E. Brewer and M. Gill
Copyright © 2008 the Institute of Electrical and Electronics Engineers, Inc.

be realized and new applications created. In this chapter, embedded Flash memory applications have been categorized in several different ways including the type of host logic devices in which the Flash memory is embedded and the end products apply the host logic device with the embedded Flash memory.

While a stand-alone Flash memory product is typically developed to achieve maximum density at the lowest cost possible, an embedded Flash memory product must meet the special performance and feature requirements for the host logic device. The special feature and performance requirements can include special array size and memory organization, very high data throughput, very fast access time, very low voltage, and/or low power operation. For this reason, the embedded Flash memory can be considered custom or semicustom from the memory design or cell selection aspect.

The criteria for selecting a Flash memory cell for embedded applications can be different from those for a stand-alone Flash memory. Beside the need of meeting specific performance and competitive cost requirements, the cell along with the process selected for the embedded Flash memory will also require ease of design and ease of manufacturing in order to serve a broader base of embedded Flash memory products providers.

Cell size, while important, is less critical to an embedded application than to a stand-alone Flash memory due to the fact that an embedded Flash memory occupies only a fraction of the total chip area, and therefore cell size has less direct impact on chip size. For these reasons, an application-optimized multiple-cell approach can be suitable for embedded Flash memory cell selection, despite the desirability of a "one-size-fits-all" single-cell approach. The 1.5T (1.5-transistor or split-gate cell) and 2T (2-transistor cell) Flash memory cells, in spite of their larger cell areas when compared to a 1T cell (single transistor cell), can be suitable for embedded applications to take advantage of their specific attributes. Figure 9.1 shows basic structures for 1T, 1.5T, and 2T cells. A 1T cell offers the highest density possible, a 1.5T cell eliminates the over program or over erase problem and offers good manufacturability, and a 2T cell offers the capability of very low voltage operation. In this chapter, examples of embedded Flash memories are given, showing 1T, 1.5T, and 2T Flash memory cells optimized for specific application spaces.

The speed/power performance of the host logic device, in which the Flash memory is embedded, can be very high and is expected to be maintained after integration of the Flash memory. For this reason, an embedded Flash memory process often is the combination of a high-performance logic process and a Flash memory process that contains high-voltage devices to support program and erase operations. The embedded Flash memory process can be either logic process based

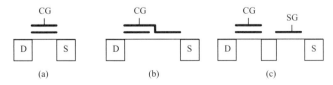

Figure 9.1. Three basic Flash cell structures for optimized embedded applications: (a) 1T-single transistor for high density, (b) 1.5T (split-gate) structure for manufacturability, and (c) 2T (two transistor) for very low voltage.

or Flash memory process based, depending on a company's primary product family being the host logic device or stand-alone Flash memory. Either way, an embedded Flash memory process must contain high-performance logic transistors to support high-speed host logic cores, in addition to Flash memory.

Design of an embedded Flash memory, especially a Flash memory core (FMC), is quite similar to design of a stand-alone Flash memory. However, there are significant differences in features, performance, design environment, and applications between the embedded and the stand-alone Flash memory, which require special consideration from cell selection, process integration, and design aspects. For an embedded Flash memory, design emphasis is on performance, variable array size, flexible array architecture, very low voltage operation, short derivative design time, and design reuse. In this chapter, special areas of design emphasis required for embedded Flash memory will be discussed. These special areas of interest to the embedded Flash memory design include system architecture, array architecture, cell selection, process consideration, circuit techniques, and low-voltage design.

9.2 EMBEDDED FLASH VERSUS STAND-ALONE FLASH MEMORY

A stand-alone Flash memory is essentially self-contained and uses the total chip area for Flash-memory-related operations. An embedded Flash memory, on the other hand, performs the same basic stand-alone Flash memory functions, but it is integrated into a host logic device to help accomplish intended system functions. Stand-alone Flash memory products offered by semiconductor manufacturers can have the same array size, array configuration, pinout, input/output (I/O) timings, and power supply voltage, and are considered standard, discrete, or commodity products. The Flash memory module embedded in the host logic device can have varied array size, array configuration, access time, and power supply voltage and can be considered a custom or semicustom design.

High density, low cost, and high speed are typically the primary goals for the stand-alone Flash memory. Small chip area or small cell size with simple wafer process is key to achieving low cost. For this reason, small cell area is critical to the stand-alone Flash memory for both density and cost. High performance and low cost are the primary goals for the embedded Flash memory. High-performance requirements for the embedded Flash memory are met by taking advantage of the custom or semicustom nature of the design. Cell size, while important, is not as critical to the embedded Flash memory as to the stand-alone Flash memory.

9.2.1 Advantages of Embedded over Stand-Alone Flash Memory

The embedded Flash memory possesses essentially all the advantages applicable to the stand-alone Flash memory. These advantages include fast system code development, reduced time to market, on-board or in-system reprogramming for code correction, product optimization, or product reconfiguration for different applications, and inventory flexibility. Additionally, the embedded Flash memory offers the following extra advantages over the stand-alone Flash memory.

Higher System Speed Two main reasons account for the faster system speed possible for utilizing the embedded Flash memory. The first is the elimination of I/O buffers in the stand-alone Flash memory and direct interface with internal host logic device on the same chip. The second is the customized array size and array configuration possible to achieve a higher speed Flash module design.

1. *Lower Power* By eliminating the I/O buffers and capacitive load for the bus connecting the stand-alone Flash memory and the host logic device, a significant reduction in dynamic power and overall system power dissipation can be achieved.
2. *Higher Density (Physical)* By eliminating the stand-alone Flash memory package and the associated buses connecting the package and the host logic device, the multilevel personal computer (PC) board area is reduced to give a higher density. In addition, reduced power dissipation achievable for the embedded Flash memory offers a tighter packing possibility.
3. *Higher Reliability* Reduced package count, reduced I/O pads, reduced bus connections, and reduced power dissipation can all contribute to higher reliability for the system with the embedded Flash.
4. *Reduced RFI/EMI* By eliminating the I/O buffers and associated large transient currents for charging and discharging the external capacitive loads for the stand-alone flash memory, RFI (radio-frequency interference) and EMI (electromagnetic interference) are reduced for the embedded Flash memory.
5. *Improved Security* Program code stored in the embedded Flash memory is inherently more secure because of less accessible internal buses. In addition, the embedded Flash memory can be made more secure to prevent any unauthorized access by implementing built-in hardware and/or software security features that may not be practical for implementation in a stand-alone Flash memory.
6. *Design Flexibility* The embedded Flash memory is considered a custom or semicustom design in many aspects and offers the flexibility of optimized array size, array configuration, cell type, access time, and/or supply voltage to meet special performance needs for the host logic device. This flexibility is unavailable to stand-alone or commodity Flash memory products.
7. *System-on-Chip (SOC) Capability* By integrating the Flash memory into the host logic device, a system on a chip can be realized, which is not possible with the stand-alone Flash memory. System-on-chip can make good use of the increased integration capability provided by a new generation of scaled technology and reap the benefits of increased functions and performance at reduced cost.
8. *Lower System Cost* Reduced package count, reduced PC board area, reduced power dissipation, reduced product inventory, and reduced total silicon area can lead to reduced overall system cost for the embedded Flash memory over the stand-alone Flash memory.

9.2.2 Disadvantages of Embedded over Stand-Alone Flash Memory

While the embedded Flash memory offers many advantages over the stand-alone Flash memory, some potential disadvantages exist that may or may not be significant depending on particular applications involved. These potential disadvantages include:

1. *Nonstandard Products* Unlike a stand-alone Flash memory that is readily available from the marketplace, an embedded Flash memory is typically custom or semicustom and requires a special design that takes time and resources to complete. The added nonrecurring engineering cost can limit the applications to those with relatively high volume products.
2. *Single Source* Because of possible unique host logic cores and the nonrecurring engineering cost involved for an embedded Flash memory product, multiple source for the product, while possible, is more difficult to accomplish. However, with the multiple production plant capability of most semiconductor companies, and with the availability of wafer foundries, the potential second source problem can be minimized.
3. *Increased Process Complexity* To accommodate both high-performance host logic cores and high-performance Flash memory on the same silicon chip, process complexity for the embedded Flash memory is increased. As a result, the number of wafer fabrication plants that are capable of manufacturing the device may be limited.
4. *Density Limitations* With added process complexity due to required dual-transistor types to support high-performance host logic cores, very low and very high memory density may not be cost effective for the embedded Flash memory. The maximum density for the embedded Flash memory may trail the stand-alone Flash memory by one generation or a factor or 2 to 4.
5. *Increased Test Cost* Test cost for the embedded Flash memory may be higher because of two possible reasons. The first is increased test time due to possible added test features required for the host logic device. The second is the higher tester cost due to high speed and high pin-count requirements for the host logic device.

9.3 EMBEDDED FLASH MEMORY APPLICATIONS

In general, all applications suitable for stand-alone Flash memory are also suitable for embedded Flash memory. Possible exceptions are situations that require a very large or very small array size because that may not be cost effective for the embedded Flash memory. Many applications are suitable only for the embedded Flash memory. These cases may require a special array size, array configuration, speed, and/or special supply voltage for a host logic device. There are many different ways to categorize embedded Flash memory applications. The applications can be defined by the host logic devices in which the Flash memory is embedded, by the electrical functions performed by the embedded Flash memory in the host logic device, by the end products in which the host logic device with the embedded Flash memory is applied, or by the ways in which the host logic device with the embedded Flash memory is utilized.

9.3.1 Applications by Device Type

The type of host logic device in which the flash memory is embedded can be used to categorize embedded Flash memory applications. The host logic device types include microcontroller (MCU), digital signal processor (DSP), application-specific integrated circuit (ASIC), application-specific standard product (ASSP), and programmable logic device (PLD).

Application of microcontrollers with the embedded Flash memory, ranging from 4 bits to 8 bits, 16 bits, and 32 bits, is very pervasive in many of today's electronic products ranging from handheld wireless products to automotive engine and transmission controls. MCU was the first host logic device to employ the embedded Flash memory for program storage. From the beginning of the 1990s, the embedded Flash memory has been widely used to replace ROM and EPROM for code storage. A typical microcontroller (MCU) may contain a central processing unit (CPU) core, static random-access memory (SRAM) modules, embedded Flash memory modules, system integration module, and peripheral modules including timer, analog-to-digital converter (ADC), serial communication, and networking. Figure 9.2 shows a block diagram of the Motorola SPS (now Freescale) MPC555 RISC MCU with 448 kbytes of embedded Flash memory. Figure 9.3 shows a microphotograph of the corresponding MPC555 chip.

The DSP is another popular host logic device that contains peripherals, as well as embedded Flash memory, for signal processing and controls applications. The products range from audio and video equipment to wireless communication products. The key applications of the embedded Flash memory in a DSP include program storage and coefficient storage as shown in a DSP block diagram in Figure 9.4. While the array size of the embedded Flash memory for DSP applications is typically smaller than that for MCU, very high speed operation of the Flash memory is generally required.

A Flash memory module can be designed as a self-contained memory core and used as a macrocell block for ASIC or ASSP design applications. While a Flash module contains all necessary circuits to operate as a Flash memory, it requires the design of a bus interface unit to provide the necessary interface between the Flash memory core and the bus on the ASIC chip.

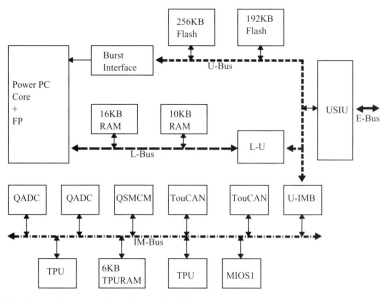

Figure 9.2. Block diagram of a Motorola SPS (Freescale) RISC MCU MPC555 with embedded Flash memory modules and peripheral modules connected to the internal bus systems.

EMBEDDED FLASH MEMORY APPLICATIONS

Figure 9.3. Microphotograph of the RISC Motorola SPS (Freescale) MPC555 MCU with 448 kbytes of embedded Flash memory, corresponding to the same MCU shown in Figure 9.2.

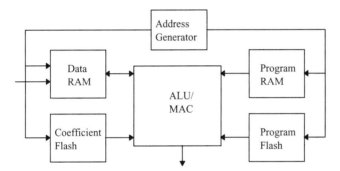

Figure 9.4. Applications of the embedded Flash memory in a DSP core.

A PLD that includes field-programmable gate array (FPGA) and standard cells uses the embedded Flash memory to configure a logic device or gate array. The embedded Flash memory can be programmed by the user to implement a specific logic function. The nonvolatile nature of the Flash memory allows the PLD or FPGA to remain programmed after power-off. For some high-speed applications, a PLD may use the embedded Flash memory as a medium for storing a program and use a higher speed register or SRAM array for executing the program. In this case, the embedded Flash memory serves as a shadow register or shadow SRAM. The code or program stored in the Flash memory is transferred during the power-up reset into registers or SRAM array.

9.3.2 Applications by Function

The embedded Flash memory can perform many different functions inside the host logic device. The two main categories are data storage and program or code storage.

For data storage, the embedded Flash memory is used to store parameters, coefficients, and lookup tables. It can also be used to emulate an EEPROM to store data that requires a frequent update such as the odometer in an automobile. For program storage applications, the embedded Flash memory can be used to store BIOS (basic input–output system) for computers ranging from palmtop to desktop. Other functions for the code storage include boot vector and boot code storage for the host logic device of a MCU or DSP. Additional applications for code storage include the configuration register to configure a host logic device to implement specific types of operations and firmware for carrying out a predefined set of functions.

Another embedded Flash memory function is to store reset machine states or reset configuration information for the host logic device. An independent section of the Flash memory array can be reserved to store this information, which is retrieved by the host logic device at power-up or reset.

Another possible function that can be served by the embedded array is the shadow array. The shadow array stores the nonvolatile logic equations that are transferred into a high-speed SRAM array during the power-up so that the PLD can operate at a higher speed than what an embedded Flash memory can provide.

9.3.3 Applications by End Product

The applications of the embedded Flash memory can be defined by the end products that utilize the embedded Flash memory. This is the conventional and most common way of defining embedded Flash memory applications. These applications can be categorized by automotive, consumers, communications, office automation, and industrial. Among these end products, automotive applications play a pivotal role in promoting and expanding the speed, density, and performance envelopes for the embedded Flash memory. This is because the need for high-density and high-throughput nonvolatile memories to meet the higher fuel efficiency and tighter emission regulations imposed by the California Air Resources Board (CARB) and the Environmental Protection Agency (EPA) can best be met by a high-performance MCU with embedded Flash memory.

9.3.3.1 Automotive. The MCU is the most prevalent host logic device for automotive applications. Starting from the engine spark timing control for powertrain to antilock brake system (ABS) and airbag for vehicle safety, the microcontroller has been utilized extensively for automotive applications. The number of MCUs used in a vehicle has increased steadily to as high as dozens of units per vehicle. Many of these MCUs contain an embedded Flash memory to provide needed reprogrammability for program and data storage. The array size of the embedded Flash memory for automotive applications has been increasing steadily, due to a larger program required to support higher engine diagnostic capability, stricter emission control, added safety features, and driver information system. Automotive applications of the embedded Flash memory are summarized in Table 9.1.

9.3.3.2 Communication. The embedded Flash memory is being applied extensively to many communication products, both wired and wireless. DSP along with MCU are the prominent host logic devices for most of these applications. A reprogrammable embedded Flash memory offers the flexibility of reconfiguring communication products such as cellular phones for different customers with cus-

TABLE 9.1. Typical 2006 Embedded Automotive Flash Memory Applications

Specific Circuit Function	Host Logic Device	Typical Array Size Range	Operating Frequency Range
Powertrain (engine/transmission)	MCU	1 MB–4 MB	80–200 MHz
Safety system (ABS/air bag)	MCU	128 kB–1 MB	16–80 MHz
Body electronics	MCU	32 kB–512 kB	8–40 MHz
Instrumentation	MCU	32 kB–512 kB	4–40 MHz
Driver information system	DSP/MCU	512 kB–4 MB	80–200 MHz

TABLE 9.2. Typical 2006 Embedded Communication Flash Memory Applications

Specific Circuit Function	Host Logic Device	Typical Array Size Range	Operating Frequency Range
Cellular phones	DSP/MCU	1 MB–4 MB	8–120 MHz
Pager	MCU/DSP	64 kB–512 kB	1–16 MHz
Two-way radio	MCU/DSP	128 kB–1 MB	8–60 MHz
Answering machine	MCU/DSP	16 kB–1 MB	8–25 MHz
Phones	MCU/DSP	32 kB–1 MB	8–40 MHz
Cable TV set-top box	MCU/DSP	256 kB–8 MB	16–250 MHz

TABLE 9.3. Typical 2006 Embedded Consumer Flash Memory Applications

Specific Circuit Function	Host Logic Device	Typical Array Size Range	Operating Frequency Range
Digital still camera	MCU/DS	64 kB–2 MB	4–60 MHz
Camcorder	MCU/DSP	64 kB–4 MB	8–60 MHz
Voice recorder	MCU/DSP	16 kB–2 MB	1–16 MHz
Video games	MCU	32 kB–256 kB	8–200 MHz
Smart cards	MCU	16 kB–1 MB	1–16 MHz
Musical systems	MCU/DSP	16 kB–2 MB	16–100 MHz
Spell checkers	MCU	16 kB–512 kB	1–16 MHz
Remote-controlled toys	MCU	8 kB–256 kB	8–40 MHz

tomized features. Table 9.2 provides a list of communication products with the key host logic device, typical ranges of Flash memory size, and operating frequency.

9.3.3.3 Consumer. Low voltage and low power are the key features for many of these handheld consumer products. The embedded Flash memory is used for audio or video storage. High speed and high throughput are typically not a requirement except for video games. MCU and DSP are the main host logic devices for most consumer products some examples of which are listed in Table 9.3.

9.3.3.4 Office Automation. Relatively high density and high speed characterize the embedded Flash memory for most office automation applications. MCU is the primary host logic device for the office automation products listed in Table 9.4.

TABLE 9.4. Typical 2006 Embedded Office Automation Flash Memory Applications

Specific Circuit Function	Host Logic Device	Typical Array Size Range	Operating Frequency Range
Printers	RISC/MCU	128 kB–4 MB	16–200 MHz
Copiers	MCU	32 kB–4 MB	16–80 MHz
Hard disk drives	MCU/DSP	32 kB–8 MB	16–60 MHz
Fax machines	MCU	32 kB–2 MB	16–60 MHz
PC peripherals	MCU	16 kB–4 MB	8–100 MHz
Routers-Enet	MCU	8 kB–4 MB	8–60 MHz

TABLE 9.5. Typical 2006 Embedded Industrial Flash Memory Applications

Specific Circuit Function	Host Logic Device	Typical Array Size Range	Operating Frequency Range
Robotics/servo controls	MCU/DSP	128 kB–2 MB	10–100 MHz
Bar code reader	MCU	32 kB–2 MB	10–60 MHz
Motor control	MCU	128 kB–2 MB	10–100 MHz
Point-of-sale terminals	MCU	64 kB–4 MB	10–150 MHz
Instrumentation	MCU/DSP	16 kB–4 MB	10–150 MHz
Zigbee	MCU/DSP	16 kB–256 kB	8–25 MHz

9.3.3.5 Industrial. For industrial applications, the embedded Flash memory requires a relatively large Flash memory size operating at a high system clock frequency. MCU is the primary host logic device followed by DSP. Table 9.5 shows some key industrial applications for the embedded Flash memory.

9.3.4 Applications by Usage

Operations and performance of a host logic device with an embedded Flash memory typically can be met by the same host logic device with an embedded read-only memory (ROM). The host logic device with an embedded ROM lacks the versatility of nearly instant program alteration capability of the embedded Flash memory, but offers the lowest cost possible for the device because of reduced silicon area and process complexity. Any ROM code changes will involve a mask layer change resulting in additional weeks to complete. For this reason, usage or application of a product with an embedded Flash memory is dependent on the cost premium for the embedded Flash memory over ROM, the value of reprogrammability for the system, and the volume usage of the host logic device. The application of the embedded Flash memory product can range from relatively low volume for system code development, to increased volume for prototyping, to high volume for pilot production, and to even higher volume for full production. The value of reprogrammability or the cost premium acceptable for the embedded Flash memory ranges from nearly "cost-no-object" for initial system development to a very low value for full production, unless reprogrammability is an inherent requirement for the product.

9.3.4.1 System Development. During the initial stage of system development, changes of program code are unavoidable and can be frequent. For that reason, an embedded Flash memory is essential for achieving fast turn-around time for code and system development. The value that the embedded Flash memory brings to system development is so high that the cost premium for the embedded Flash memory is practically a nonissue. This is especially true when considering that the volume of the device involved is relatively small and the improvement gain on time to market is very high.

9.3.4.2 Prototyping. Prototypes in quantity of 10s or 100s are built to demonstrate design integrity and product reliability. At this stage of product development, some modifications to the program developed are often necessary, and, therefore, the embedded Flash memory is still the most desirable media for program or code storage. Since volume for prototyping is still relatively low, the acceptable cost premium for the embedded Flash memory can still be relatively high.

9.3.4.3 Pilot Production. Once performance and reliability for a system are demonstrated from the prototypes, the product can be introduced into production. At the early stage of production or pilot production, cost of the host logic device or the price premium for the embedded Flash memory has become important. The embedded Flash memory is still desirable as occasional changes to the program may still take place.

9.3.4.4 Full Production. Full production commences once a product has been fully proven from the pilot production; no further changes in system design or program are expected. Cost of the host logic device or the cost premium for the embedded Flash memory is a sensitive and critical factor for high-volume production. Usage of the embedded Flash memory may only be justifiable if the cost premium for the embedded Flash memory is low or modification of code is an inherent requirement through the operation life of the device. An example of this is warranty recall, where returned devices may be updated with new software. A read only memory (ROM) or one-time programmable memory (OTP) would require expensive rework that could be done in seconds with a Flash.

Figure 9.5 illustrates the cost or the acceptable cost premium of the embedded Flash memory products as a function of the usage applications or volume. At very high volume for full production, the acceptable cost premium for the embedded Flash memory over the embedded ROM is diminished, unless reprogrammability is an inherent requirement for the product.

9.4 EMBEDDED FLASH MEMORY CELLS

9.4.1 Special Requirements and Considerations

Although the general requirements for features and reliability are no different between embedded Flash and stand-alone flash, there are some subtleties, which will influence the design of the Flash memory cells. It is of utmost importance to recognize that there are other circuits sharing the same piece of silicon, usually in rather large portion, with the Flash module. The end goal is to maximize economic

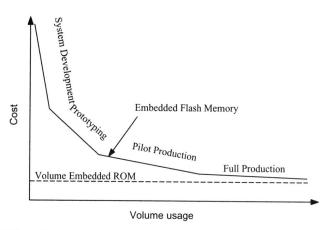

Figure 9.5. Embedded Flash memory applications categorized by volume usage.

return while minimizing the cost of the product. The die cost, not counting assembly and test costs, is a function of wafer cost, die size, and yield. It is advisable to balance the engineering tendency of pushing the smallest possible memory cell with considerations of process complexity, yield, and other cost-sensitive parameters.

9.4.1.1 Cell Size Versus Module Size Versus Chip Size.
In a stand-alone Flash memory product, the total bit cell area usually occupies more than half of the chip with the rest of the area devoted to simple state machine, decoders, level shifters, sense amplifiers, and charge pumps—the essential circuitry for the proper operation of the Flash memory array. In contrast, even in the most advanced microcontroller or system chip with embedded Flash memory, the Flash module size is seldom more than half of the chip size. This is due to the fact that it is the complexity of the chip that requires a rich set of functions, and those functions determine the amount of Flash memory and not vice versa.

Based on this understanding, it is not difficult to see that the Flash cell size in an embedded Flash is not as critical in determining the die size as that in a stand-alone Flash. This is illustrated in an example by comparing chip A with 32% chip area in Flash module and chip B with 80% chip area in Flash module. It is assumed the Flash module is of 16-Mbit density with a cell size of $0.40\,\mu m^2$ and an array efficiency of 40%. The Flash module size therefore is $16\,mm^2$. The die size for chip A is then $50\,mm^2$ and the die size for chip B is $20\,mm^2$. Now assume the cell size is reduced by 25% to $0.30\,\mu m^2$, and let's generously assume the array efficiency stays at 40%, the Flash module size is also reduced by 40% to $12\,mm^2$. The die size for the new chip A, say chip A1, is now 46 (50 − 16 + 12) mm^2 and the new chip B1 is now 16 (20 − 16 + 12) mm^2. For a 40% cell size reduction, the Flash memory intensive chip B saw 20% die size reduction while the typical embedded chip A saw a mere 8% die size reduction.

The obvious question for chip A is: Is it worth the effort to push the reduction in cell size by 25% in order to get 8% die size reduction in return?

9.4.1.2 Cell Size and Process Complexity Trade-off.
With the technology moving rapidly into sub-90nm regime, the task of reducing cell size by squeezing

the last nanometer of design rules is daunting. For self-aligned cells, the challenge is in the floating-gate length reduction. The scaling of the floating-gate length is mainly limited by the overlap of the source junction, drain junction, and the punch-through-free channel length. The difficulty in scaling the tunnel oxide thickness further complicates the task. Self-aligned source is a commonly used process technique to reduce the source bus width to poly-poly spacing to compensate for the ineffectiveness of scaling in the cell height direction. For non-self-aligned cells, including many of the split-gate cells, the challenge is in the budgeting of registration tolerance while managing the punch-through characteristics of the split gate.

There are very few simple options available to the Flash technologists for scaling the cell without invoking additional process complexity. In fact, attention should be diverted to the scaling of the high-voltage transistors used in the decoding and level shifting circuits. This is especially true for systems needing Flash modules with less density. The array efficiency is usually less for smaller Flash modules.

It is again very important to realize that the cost of any additional process devoted to scaling the Flash cell is levied on the rest of the chip that does not benefit from it. Therefore, any tendency to trade process complexity for smaller cell size has to be weighed against the potential gain of the action.

9.4.1.3 Logic-Based Versus Memory-Based Processes. With computing and networking applications pushing the envelope for performance, and the wireless personal productivity applications pushing the envelope for low-power bandwidth [million instructions per second per milliwatt, (MIPS/mW)], the complementary metal–oxide–semiconductor (CMOS) logic technology has advanced to a point where the device transistors have much higher gain than those employed by the stand-alone Flash memory makers. In fact, some process modules associated with Flash memory technology can no longer deliver the features and performance required. New logic process modules such as retrograde wells, shallow trench isolation, ultra-thin gate dielectrics, salicide, local interconnect, full chemical-mechanical polishing (CMP) metal interconnects, and even Cu/low-K backend have now become part of the process portfolio.

Choices must be made in the Flash cell design whether to bring in specialty process modules to achieve design goals. For example, the popular self-aligned source (SAS) module for NOR 1T Flash cells cannot be directly plugged into the logic process without modification because the field oxide profile is steeper with shallow trench isolation. A possible replacement for SAS is to use regular contacts for source pick up and devote a metal layer to connect them [2].

9.4.2 Cell Selection for Embedded Applications

One of the most important steps in developing an embedded Flash memory is the selection of a Flash memory cell. Depending on the scope of the development, the selection can come in two levels. The top-level selection is more global and holistic. The goal is to determine whether to use one cell or more to meet the technology requirements of a wide variety of product specifications. The major advantage of having just one cell is in the consolidation of resources and the ability to focus. The major advantage of having more than one cell is in the ability to optimize for different applications. The next level selection is much more specific with a goal to match the cell attributes with the application space of choice.

TABLE 9.6. Some Published Flash Memory Cells for Embedded Applications

Cell Name	Key Features	Source	Reference
SCSG	Split gate, source erase	Motorola SPS (Freescale)	1,3
2TS	Two transistor, source select	Freescale, Philips	4,5
2TS	0.9 V Flash	Matsushita	6
MoneT	Drain FN program, channel erase	Freescale	7
1T	Drain HCI program, channel erase	Freescale	2
HIMOS	Split-gate, source-side injection, additional program gate	IMEC	8–10
SuperFlash	Split-gate, source-side injection, poly-poly erase	Silicon Storage Technology (SST)	11
SPIN	Sidewall gate, source-side injection, nitride storage	Freescale	12
CHISEL	Secondary electron programming	Lucent	13
PFlash	P-channel, single poly	Programmable Microelectronics Corp. (PMC)	14
Pflash	P-channel, double-poly	John Caywood	15
UCPE	Uniform channel program and erase	Freescale, Infineon	16–18
NeoFlash	P-channel, single-poly, nitride storage	eMemory	19
TFS	Silicon nanocrystal memory	Freescale	20
SG-MNOS	Split-gate, top erase MNOS	Renesas	21

The top-level selection is highly product portfolio and system dependent. Therefore a thorough treatment is outside the scope of this section. Selective examples are given below to illustrate this point.

9.4.2.1 Examples of Embedded Flash Cells. During the past 20 years many Flash cell concepts have been published in journals and conference digests, and quite a few of those were intended for embedded applications. Certainly even more cell proposals have been discussed in engineering war rooms and corporate boardrooms. Table 9.6 captures some of the major published Flash cell options [1–21].

In the following subsections, three distinctively different cells were chosen to illustrate the specific matching of cell attributes to application space. They are the source-coupled splite-gate (SCSG) cell for manufacturability, the 2TS cell for low-voltage operation, and the MoneT cell for high-density and high-speed operation.

9.4.2.2 The SCSG Cell for Manufacturability. The source-coupled split-gate (SCSG) [3] cell is a Flash EEPROM cell developed exclusively for embedded applications. It employs a double-poly floating-gate transistor on the drain side and a single-poly select transistor on the source side in a split-gate configuration. The control gate of the floating-gate transistor and the gate of the select transistor are merged by sharing the same second-level polysilicon. The programming of the cell is performed by the conventional hot-electron injection into the floating gate near the drain junction. The erasure of the cell is achieved by electron extraction using Fowler–Nordheim tunneling through a tunnel region overlapping the source line, therefore the term source coupled.

SPLIT-GATE CONFIGURATION. The main purpose of incorporating a select transistor in the SCSG cell is to overcome the well-documented "overerase" problem seen in most single-transistor (1T) Flash EEPROM cells. In an ETOX array, for example, the need to ensure the bit with the least read current, the slow bit, to deliver the specified read performance dictates a minimum erase time. However, the erase speed distribution in a rather large sector can push the threshold voltage of a handful of "fast bits" to near or below 0 V, resulting in excessive leakage in the bitline to either overwhelm the sense amplifier or to cause significant push out of the access time. The probability of occurrence of these potentially catastrophic fast bits is on the order of less than 1 ppm. In addition, these fast bits exhibit an erratic, that is, nonpredictable, behavior that cannot be screened out easily. Therefore, a tremendous amount of engineering overhead, including system, device, process, design, test, and reliability, has to be invested in order to contain these fast bits.

Conversely, in a properly designed split-gate Flash, also referred to as 1.5T Flash, the erased cell threshold voltage (V_{te}) is largely controlled by the threshold voltage of the select transistor (V_{tsg}) instead of by the threshold voltage of the floating-gate transistor (V_{tfg}). This is achieved by erasing the floating-gate transistor sufficiently hard so that V_{tfg} is much lower than V_{tsg}. The read current distribution in an array will therefore be relatively insensitive to the distribution of the erased V_{tfg}. Because the threshold voltage of the select transistor is set and well controlled by the process, the dependency of the access time on slow bits is much reduced. Furthermore, the off-leakage of the cells is also determined by the threshold voltage of the select transistor, rendering the "overerase" problem a nonissue for the SCSG cell. Therefore, less circuit overhead can be had for the SCSG cell.

SOURCE COUPLING. In order to simplify the manufacturing process and to avoid unnecessary circuit complexity, erasure by Fowler–Nordheim tunneling with high positive source voltage is used. However, the incorporation of a split gate on the source side of the cell poses a challenge for the erase operation because the source end of the floating gate is now separated from the source line. To overcome this, as shown in Figure 9.6(a), a piece of floating gate is extended over the field region into the source line to allow the extraction of electrons from the floating gate to take place. The tunnel region is defined in a self-aligned manner as the source region where the floating gate overlaps. It is doped heavily to eliminate any hot-hole injection due to band-to-band tunneling. Charge loss and charge gain after program/erase cycling have been attributed to stress-induced leakage current (SILC), which has in turn been linked to hot-hole injection and trapping over the tunneling area. The elimination of band-to-band current during erase also allows an efficient charge pump to be used to further boost the erase voltage to enhance the erase speed to around 1 ms.

MANUFACTURABILITY AND AREA TRADE-OFF. The process steps used in fabricating the SCSG cell were designed to be compatible with the base logic process technology. All the Flash EEPROM-specific steps were carried out before logic devices were built in order to preserve their electrical characteristics. This allows functional modules to be incorporated without having to recharacterize existing circuit modules. Conventional process modules for high-density stand-alone Flash memories, such as self-aligned stack etch and self-aligned source, were not employed to allow multiple existing logic fabrication plants to manufacture the same product. This is an important consideration for embedded memories.

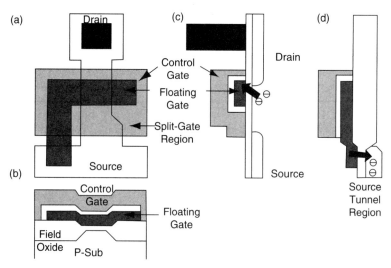

Figure 9.6. SCSG cell structure: (a) top view, (b) side view along the wordline through the floating gate, (c) side view along the bitline, and (d) side view along the poly1 erase strip.

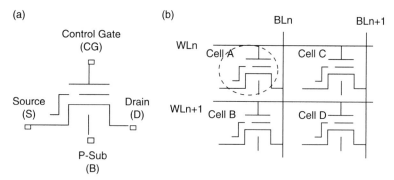

Figure 9.7. SCSG cell and array schematics: (a) five-terminal SCSG cell schematic and (b) a 2 × 2 array electrical schematic.

The penalty for choosing the modular approach and being logic process compatible is the apparent increase in cell size and the resultant die size. However, for most embedded systems with average density, the trade-off is usually in favor of process simplicity over cell size reduction.

PHYSICAL STRUCTURE OF CELL. The top view of an SCSG cell can be seen in Figure 9.6(a). The cell size is $18\,\mu m^2$ at $0.8\,\mu m$ technology. Figure 9.6(b) shows the side view cut along the floating-gate wordline. Figure 9.6(c) shows the side view cut along the bitline. Figure 9.6(d) shows the side view cut along the poly1 erase strip.

CELL OPERATION. The electrical schematic of the SCSG cell is shown in Figure 9.7(a). Table 9.7 contains the typical bias conditions for read, program, and erase.

ARRAY ARCHITECTURE AND OPERATION. The array schematic is shown in Figure 9.7(b) in a 2 × 2 mini array fashion. Cell A is the selected bit to be programmed.

EMBEDDED FLASH MEMORY CELLS

TABLE 9.7. Operating Conditions for the SCSG Cell

Operation	P-Sub	Source	Bitline	Wordline	Bitline + 1	Wordline + 1
Read	0 V	0 V	1 V	V_{dd}	0 V	0 V
Program	0 V	0 V	7 V	9 V	0 V	0 V
Erase	0 V	12 V	Float	0 V	Float	0 V

Cell B shares the same bitline with cell A and is subject to drain disturb. Cell C shares the same pair of wordlines and is subject to control gate disturb. Cell D is undisturbed during programming.

Each array segment (erase block) can have up to 256 rows and 512 columns. Every bit in one array segment has to be erased simultaneously as a block. For a typical programming time of 20 μs, the drain disturb time for cell B is up to 5.1 ms and the control gate disturb time for cell C is up to 1.26 ms.

9.4.2.3 2TS Cells for Low-Voltage Operation.

The two-transistor source-select cell, 2TS cell [4], was developed exclusively as an embedded Flash for low-voltage operations where power consumption is critical. The 2TS cell allows a Flash module be designed into a battery-powered system with one to two single-cell batteries. With a battery end-of-life output of 0.9 V, the 2TS cell was designed to operate as a single-supply Flash where V_{DD} ranges from 0.9 to 3.6 V with an optimal point of 1.8 V.

Two key attributes were specified at the outset of the development. One was to achieve low operating power, both active and standby. The other was to have enhanced read access at low V_{DD} while maintaining low power consumption.

LOW OPERATING POWER. One common practice in NOR Flash design is to maintain a high wordline voltage during read access, either by using a separate voltage source or by boosting the wordline, to ensure a high enough cell current. In most cost-sensitive applications, single supply with wordline boost is the preferred approach. Boosting, however, consumes more power and introduces access latency. At the cell design level, the most logical move is to shift the program V_t (low V_t state) to below 0 V. Thus, a wordline voltage of V_{DD} without boosting can be employed while producing high enough cell current to tip the balance of sense amplifier in the specified access time budget. To allow a reliable and expedient sensing of a selected bit, all other unselected bits sharing the same bitline must be shut off. This is achieved by inserting a select transistor between the floating-gate transistor and the source connection. This also ensures zero standby current from the Flash memory core.

The power consumption during rewrite is deemed secondary because most applications don't demand a high number of program/erase cycles. Even so, Fowler–Nordheim tunneling was used for both programming and erasure to minimize peak rewrite current demand. Besides, charge pumps that are much more area efficient can be realized especially at the low V_{DD} range for which this cell is designed.

ENHANCED READ SPEED. The placement of the select transistor on the source side of the floating-gate transistor has a major advantage over other cells: higher speed at low V_{DD}. The read access selection is controlled by the select transistor, which, being on the source side, is shielded by the floating-gate transistor from all the high voltages during program and erase operations. The fact that the select transistor sees no voltages higher than V_{DD} opens the door for the use of higher transconductance devices for speed optimizations at low V_{DD}. In a typical embodiment, the select transistor is constructed the same way as the floating-gate transistor only to have its poly1 tied to poly2, a higher gain device than the high-voltage devices used in the periphery. It is also important to point out that during read access the control gate voltage of the floating-gate transistor does not toggle and is kept at a low voltage, typically V_{DD}. The wordline select time during read access is determined solely by the select gate of the select transistor.

PHYSICAL STRUCTURE OF CELL. The top view of a self-aligned 2TS cell can be seen in Figure 9.8(a). The cell size is 4.1 μm² at 0.45 μm technology, roughly 20 square features. Figure 9.8(b) shows the side-view cut along the floating-gate wordline. Figure 9.8(c) shows the side-view cut along the bitline.

CELL OPERATION. The electrical schematic of the 2TS cell is shown in Figure 9.9(a). Table 9.8 contains the typical bias conditions for read, program, and erase.

ARRAY ARCHITECTURE AND OPERATION. The array schematic is shown in Figure 9.9(b) in a 2 × 2 mini array fashion. Cell A is the selected bit to be programmed. Cell B shares the same bitline with cell A and is subject to drain disturb. Cell C shares the same pair of wordlines and is subject to control gate disturb. Cell D is undisturbed during programming.

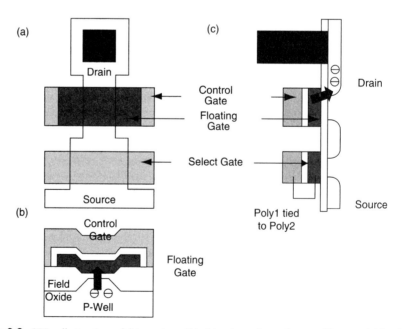

Figure 9.8. 2TS cell structure: (a) top view, (b) side view along the wordline, and (c) side view along the bitline.

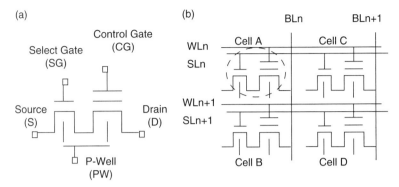

Figure 9.9. 2TS single cell and 2 × 2 array schematics. (a) Five-terminal single-cell schematic and (b) a 2 × 2 array schematic. During the program operation, cell A is the selected cell. Cell B is subject to drain stress and cell C is subject to gate stress. Cell D is not subject to program disturbs.

TABLE 9.8. Operating Conditions for the 2TS Cell

Operation	P-well	Source	Bitline	Wordline	Source Line	Bitline + 1	Wordline + 1	Source Line + 1
Read	0 V	0 V	1 V	1 V	V_{dd}	0 V	1 V	0 V
Program	0 V	0 V	5 V	−9 V	0 V	0 V	3.5 V	0 V
Erase	−5 V	−5 V	−5 V	12 V	0 V	−5 V	12 V	0 V

Each array segment (erase block) can have up to 128 rows and 2048 columns. Every bit in one array segment has to be erased simultaneously as a block. Each row is divided into 8 pages with a page size of 256 bits. For a typical page programming time of 1 ms, the drain disturb time for cell B is up to 127 ms and the control gate disturb time for cell C is up to 7 ms. Erase block to erase block isolation is achieved using a block select transistor. If the block select transistor is not employed, a wordline voltage of 3 V can be applied to all unselected rows to counteract the effect of drain disturb.

9.4.2.4 MoneT Cell for High Density and High Speed. As embedded systems get more sophisticated, the demand for on-chip code storage increases sharply. The Motorola one–transistor cell, or MoneT (pronounced as in Claude Monet, the French Impressionist), was created to meet such a demand. The lineage of MoneT can be traced back to FETMOS (floating-gate electron tunneling MOS) [22], an EEPROM cell that's in the heart of nearly every 68HC11, Motorola's highly successful 8-bit microcontroller.

MoneT is a NOR-type Flash cell. It was developed with two objectives in mind. One is to have a power-efficient Flash module that can operate at 3.3 V without wordline boost and can operate down to 1.8 V with a single-stage boost. The other is to provide a scalable path to high-throughput embedded Flash.

It turns out that the key to high-speed, high-throughput, and power-efficient Flash is the ability to compact the array V_t distribution to no more than 0.5 V at the low V_t state. The low end of the V_t distribution is limited to no less than 0.8 V, below which the sum of the subthreshold leakage of all the bits tied to a bitline starts to affect the sensing of the selected bit. The high end of the V_t distribution coupled with the worst-case wordline voltage during read access defines the worst-case read current. If sufficient read current under the worst-case condition cannot be guaranteed, then higher wordline voltage must be supplied, either by raising supply voltage or by adding another stage of boost. Neither is desirable.

COMPACTION OF V_t DISTRIBUTION. It is well known that the bulk erase of 512-kbit blocks typically seen in ETOX arrays usually results in a V_t distribution of greater than 2 V. It will be difficult to bring that distribution to 0.5 V through compaction without creating the "overerase" problem. In order to practically compact the low V_t distribution to 0.5 V, the transition from high V_t states to low V_t states has to happen with a much smaller granularity, byte-wide or page-wide (<1 kbit). This means the low V_t state has to be designated as the programmed state because for Flash, by definition, the erased state is achieved by a bulk operation whereas the programmed state is achieved by a byte-wide or page-wide operation. This concept is illustrated in Figure 9.10.

The task of compaction is much more manageable if Fowler–Nordheim tunneling is used both for programming and erasure. Byte-wide or page-wide programming can be achieved with edge tunneling where a positive drain voltage and a negative control gate voltage can be employed to pull electrons out of the floating gates. Compaction can then be accomplished by performing a repetitive combina-

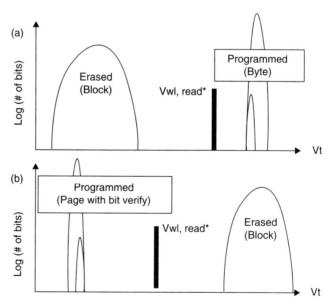

Figure 9.10. Comparison of V_t distribution. (a) ETOX-type V_t with low V_t being the block-erased state and (b) MoneT V_t distribution with low V_t being the page-programmed state. V wl, read is the minimum wordline voltage in the read mode that guarantees enough read current from the worste-case bit.

tion of program–verify–program–verify sequence. The V_t of the verified bits can be "frozen" by turning off the corresponding drain biases. The trade-off between the narrowness of the V_t distribution and the total programming time determines the optimal amount of incremental programming.

Bulk erasure can be achieved with uniform channel tunneling where a positive control gate voltage can be applied to inject electrons into the floating gates. When embedding into a deep submicron logic process, the amount of the positive control gate voltage is divided between a positive control gate voltage of lower magnitude and a negative well voltage. This requires the creation of an isolated P-well for the cell array.

PHYSICAL STRUCTURE OF CELL. The top view of the MoneT cell can be seen in Figure 9.11(a). The cell size is $2.7\,\mu m^2$ at $0.45\,\mu m$ technology, roughly 13 square features. Figure 9.11(b) shows the side-view cut along the floating-gate wordline. Figure 9.11(c) shows the side-view cut along the bitline.

CELL OPERATION. The electrical schematic of the MoneT cell is shown in Figure 9.12(a). Table 9.9 contains the typical bias conditions for read, program, and erase.

ARRAY ARCHITECTURE AND OPERATION. The array schematic is shown in Figure 9.12(b) in a 2 × 2 mini array fashion. Cell A is the selected bit to be programmed. Cell B shares the same bitline with cell A and is subject to drain disturb. Cell C shares the same wordline and is subject to control gate disturb. Cell D is undisturbed during programming.

Each array segment (erase block) can have up to 128 rows and 2048 columns. Every bit in one array segment has to be erased simultaneously as a block. Each row is divided into 8 pages with a page size of 256 bits. For a typical page programming time of 1 ms, the drain disturb time for cell B is up to 127 ms and the control

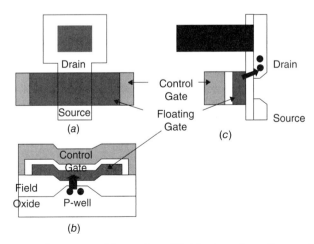

Figure 9.11. MoneT cell structure. (a) top view, (b) side view along the wordline, and (c) side view along the bitline. The arrow in (b) indicates the direction of electron flow through the channel during the erase operation. The arrow in (c) indicates the direction of electron flow into the drain region during the program operation.

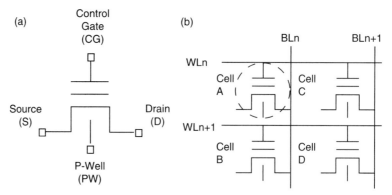

Figure 9.12. MoneT cell and array schematics. (a) Four-terminal single-cell schematic. (b) A 2 × 2 array schematic. During the program operation, cell A is the selected cell. Cell B is subject to drain stress and cell C is subject to gate stress. Cell D is not subject to program disturbs.

TABLE 9.9. Operating Conditions for the MoneT Cell

Operation	P-well	Source	Bitline	Wordline	Bitline + 1	Wordline + 1
Read	0 V	0 V	1 V	3.3 V	0 V	0 V
Program	0 V	Float	5 V	−9 V	0 V	0 V
Erase	−9 V	−9 V	Float	9 V	Float	9 V

gate disturb time for cell C is up to 7 ms. Erase block to erase block isolation is achieved with a block select transistor.

9.5 EMBEDDED FLASH MEMORY DESIGN

A Flash memory for embedded applications is typically designed as a module that contains the Flash memory core (FMC) and the bus interface unit (BIU). The FMC is the essential part of the embedded Flash memory module, which contains the Flash memory array (FMA) and the memory control block (MCB). The BIU provides the necessary interface between the Flash memory core and the host logic device. While FMC is designed to be portable, reconfigurable, and reusable for different host logic devices, the BIU is designed to meet specific interface requirements between the FMC and the particular bus of the host logic device to which the FMC is connected.

9.5.1 Special Requirements and Consideration

Design of an embedded Flash memory module, especially the Flash memory core (FMC), is similar to design of a stand-alone Flash memory. Any circuit techniques applicable to the stand-alone Flash memory are also applicable to the embedded Flash memory. However, there are significant differences between the embedded and stand-alone Flash memory. Special requirements and consideration for the embedded Flash memory are given below:

1. An embedded Flash memory is designed as a module that can readily be integrated into the host logic device. A stand-alone Flash memory is a self-contained chip that has I/O pads or pins to interface with other logic devices.
2. A Flash module for embedded applications must meet and be optimized for special requirements on array size, memory configuration, speed, and particular bus protocol for the host logic device and is considered a custom or semicustom design. Stand-alone Flash memory products typically have a common set of features and specs.
3. A Flash memory module designed for embedded applications must provide special test features, such as scan path or joint test action group (JTAG) interface capabilities called for by the host logic device. A stand-alone Flash memory may not have these requirements.
4. A Flash memory module for embedded applications is typically designed to support extensive reuse of intellectual property (IP) for multiple product families and multiple products within a product family. For this reason, a Flash memory module must be designed with a built-in memory generator or compiler to provide ease for designing new derivatives. A stand-alone Flash chip is typically designed and optimized for one specific array configuration.
5. Design of an embedded Flash memory can be rather logic intensive due to the required memory control block (MCB) and bus interface unit (BIU). For this reason, behavioral model, synthesis, and automated place-and-route design approaches are employed to shorten the design cycle time. A stand-alone Flash memory on the other hand may rely extensively on custom design to minimize silicon area and maximize speed.
6. The embedded Flash memory module typically shares the same tester with the host logic device, which costs more than a dedicated memory tester for stand-alone Flash memory because of high speed and high pin-count requirements for the host logic device. For this reason design for test time reduction is more critical for the embedded Flash memory than for the stand-alone Flash memory.
7. An embedded Flash module core once designed is to be reused by many different design groups from different companies, with varied degrees of knowledge and expertise in Flash memory technology and, therefore, good portability, good documentation, and ease of use are critical to the Flash memory core design. For the stand-alone Flash memory, there is typically one design for a given generation of technology and little reuse, if any.
8. An embedded Flash memory has less direct accessibility, observability, and controllability than stand-alone Flash memory due to lack of accessible I/O pins. Also, supply noise and substrate operating temperature may increase for an embedded Flash memory product due to integration of other modules on the same chip.
9. In order to serve as wide a range of product families as possible, the embedded Flash core (FMC) must be designed to meet the highest performance and density required among the product families to be served. Also, performance margins required for the embedded Flash memory have to be

higher than that for the stand-alone Flash memory because, unlike the stand-alone flash memory that can be specified for multiple speed grades, there is typically only one speed grade for the host logic device.

10. An embedded Flash memory module may involve special requirements of very high speed operation, very low voltage operation, or ease of design and manufacturing not available for a stand-alone Flash memory. For this reason, selection of the Flash memory cell, Flash memory process, and Flash array architecture for embedded application can be different from the stand-alone Flash memory and are optimized for the particular application space.

9.5.2 Flash Module Design for Embedded Applications

As previously stated, a Flash memory module for embedded applications consists of a Flash memory core (FMC) and a bus interface unit (BIU). The FMC comprises two key parts, a Flash memory array (FMA) and a memory control block (MCB). The FMA contains all the basic array elements that include cells, decoders, sense amplifiers, and program data buffers. The MCB contains all the essential control logic, register sets, state machines, high-voltage charge pumps, voltage regulators, level shifters, voltage switches, and I/O buffers required to support proper FMA operations. A simplified block diagram of the Flash module is shown in Figure 9.13, and Figure 9.14 shows a simplified FMA.

Flash memory core operations include array program, array erase, program/erase verify read, regular read, and any special test operations for device characterization or quality/reliability enhancement. The BIU, as the name implies, provides all interface needs between the Flash memory core and the host logic device, or specifically the bus of the host logic device such as MCU or DSP, to which the Flash memory module is connected. FMC is the essential portion of an embedded Flash memory module and contains all basic elements required for a stand-alone Flash memory. Figure 9.15 is a block diagram for the FMC showing key circuit blocks required for the MCB. The FMC is typically designed to be portable, reusable, and easily reconfigurable for other embedded Flash memory applications.

While the FMC can be a common design for many applications, the BIU is typically application or product family specific and must be designed to meet

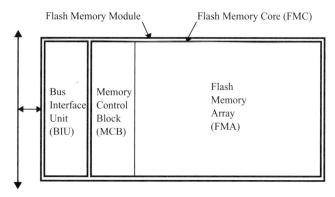

Figure 9.13. Embedded Flash memory module consists of bus interface unit (BIU) and Flash memory core (FMC) that contains Flash memory array (FMA) and memory control block (BCB).

EMBEDDED FLASH MEMORY DESIGN

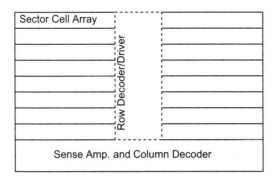

Figure 9.14. Typical Flash memory array (FMA) contains multiple sectors (8 shown) each of which includes cell and sector column decoder arrays.

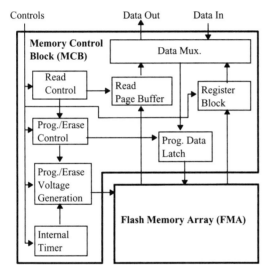

Figure 9.15. Flash memory core (FMC) consists of Flash memory array (FMA) and memory control block (MCB).

interface requirements between the Flash memory core and the specific bus system of a host logic device family.

Design of the FMC, which includes flash memory array (FMA) and memory control block (MCB), is very circuit intensive and highly cell, process, and device dependent. Since extensive knowledge in Flash cell, devices, and process as well as circuits is essential for carrying out an FMC design, the design work is typically done by qualified semiconductor manufacturers or foundries. The completed FMC design is made available from these design sources to users who can be owners of the host logic devices and who can design the BIU for the FMC to form a Flash memory module for chip integration. The FMC can also be offered by ASIC companies as a macrocell block for ASIC customers to integrate the Flash memory on chip.

The BIU typically contains control logic, state machines, and I/O buffers, along with the MCB is typically logic intensive and can be designed using behavioral model, synthesis, and auto-place-and-route techniques to reduce design cycle time.

Possible exceptions are portions in the critical data path, which require custom circuit design to optimize for speed performance. Design of the BIU for a Flash memory module is similar to design of a BIU for other peripheral modules. Knowledge of the Flash memory device and operation is helpful, but not essential, assuming that design source for the FMC provides all necessary design information and database including block definition, register transfer level (RTL) model, gate model, spice file, timing simulation file, as well as the design report.

9.5.3 Design Techniques for Embedded Flash Module

Design of a Flash memory core for embedded applications is similar to design of a stand-alone Flash memory. Circuit techniques applicable to stand-alone Flash memory design are, in general, applicable to embedded Flash module design. They are covered in the memory circuit technology chapter in this book and, therefore, will not be repeated here. Rather, this discussion of circuit design will concentrate only on those special techniques or subjects that are applicable, relevant, or related only to embedded Flash memory design. These special subjects include system architecture, array architecture, cell selection, process consideration, special circuit techniques, and low-voltage design.

Very high throughput and high-speed read access are often required for an embedded Flash memory to meet the high-performance needs of the host logic device. For example, the embedded Flash memory for DSP applications may require a relatively small array size but a fast sub-10-ns access time to support a near 200-Mhz DSP operation. Also, the embedded Flash memory for an MCU for automotive applications may require an array size of well over 8 million bits at an access time of less than 15 ns to support a near 150-MHz MCU operation.

Very high throughput and fast read access time is fundamental to a high-speed system and is often a key requirement for embedded Flash module design. To achieve a very fast read access time for an embedded Flash memory module, all factors that can contribute to high-speed performance must be considered and optimized. These factors include the Flash memory cell, process, array architecture, and circuit design.

9.5.3.1 System Architecture. A very high performance MCU or DSP operating at 150 MHz will require a very fast Flash memory with an access time of 10 ns less to support its operation. This is difficult to achieve for a relatively large cell array with single ended data signals from the Flash memory cells. To compensate for this Flash memory access time limitation, an optimized system or module architecture can be employed to meet the required high data throughput in order to maintain high-speed system operation. There are different techniques for achieving high data throughput. They include page mode read, burst mode read, buffered read registers, cache memory, and pipelined operation.

All these high data throughput techniques, except for pipelined operation, involve multiword parallel read of a page in the Flash memory array. The multiple bytes of data that make up the page are to be stored in buffers, registers, or SRAM array for fast subsequent read accesses. Cycle time for accessing a page from the cell array, or off-page access, can be more than one clock cycle, two, for example, and the subsequent access of the data from the buffers or registers, if available or "hit," is one cycle. A page contains multiple words, 8, for instance, and a word contains 4 bytes for a 32-bit core. An off-page read will result in 256 bits of data on the array to be accessed and sensed simultaneously by 256 on-module sense amplifiers. The sensed data is stored in a 256-bit register with the first data word appearing on the output data bus. For page mode read, all 8 words of data on the register can be accessed randomly for the on-page read at a rate of one word per system clock cycle. For burst mode read, the rest of the data (7 words) is clocked out sequentially, one word at a time per system clock cycle. For buffered read, multiple 256-bit registers can be randomly accessed one word at a time among registers and among the words in each register. For cache mode read, the off-page data is stored in a relatively large cache array of SRAM. This data is tagged and can be accessed in one clock cycle when hit.

9.5.3.2 Array Architecture. Array architecture plays a very important role for embedded Flash memory design because it has direct impact on all those critical performance factors important to the embedded Flash memory, including fast access time, flexible array size, test time, and cost reduction. Optimized array architecture for meeting specific performance needs for the host logic device is possible for the embedded Flash memory due to its custom or semicustom design flexibility, which is one of the important factors that differentiate the embedded from the stand-alone Flash memory. The most significant element for achieving a high-speed design through array architecture optimization is the reduction of the number of cells tied to a wordline or bitline at its lowest hierarchical level. This is accomplished by subdividing the wordline or bitline through a global/local hierarchy, subdividing the array in wordline or bitline direction, and/or subdividing the array into multiple separate arrays.

The architecture for implementing wordline subdivision is through the global/local wordline configuration shown in Figure 9.16. The wordline division technique is effective if the wordline delay is a speed-limiting factor in the critical read path, at the expense of an increased number of local wordline drivers and a larger silicon area. If bitline delay is the speed-limiting factor in the critical read path, bitline subdivision similar to global/local wordline subdivision can be implemented for speed enhancement, at the expense of an increased number of local bitline decoders and a larger silicon area.

The next level of array hierarchical improvement for speed enhancement is to subdivide the array into multiple sectors in the wordline direction as shown in Figure 9.17, at the expense of increased number of row decoders and drivers. The similar technique can be applied to subdivide the array in the bitline direction for speed enhancement, at the expense of an increased number of column decoders and sense amplifiers.

For some applications with a relatively large array size or very fast access time, sectoring of wordlines, bitlines, or array may not be sufficient. Further access time enhancement can be accomplished by subdividing the array into separate arrays of

Figure 9.16. Global/local wordline architecture for achieving reduced wordline delay.

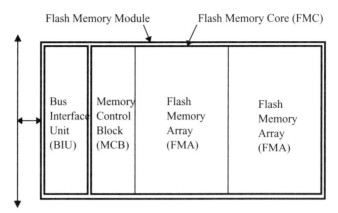

Figure 9.17. Block diagram for an embedded Flash memory module with split-array architecture for speed enhancement.

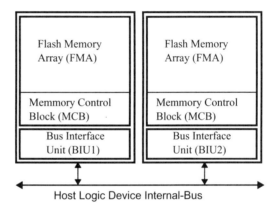

Figure 9.18. Multiple module architecture for speed enhancement at the expense of silicon area and power dissipation.

equal or nearly equal sizes as shown in Figure 9.18. Each of the subarrays forms a module and requires its own memory control block (MCB) and bus interface unit (BIU) and, therefore, is the least area efficient.

Additional design effort will be required if the array cannot be subdivided into equal sizes. The advantages of this approach, besides being able to provide the most speed improvement, is reduction of program time or test time by operating these modules in parallel on chip. When array division is introduced, decoding in the Z dimension for module selection is required.

9.5.3.3 Cell Selection. Cell selection is one of the most important steps in design and development of the embedded Flash memory. For embedded or stand-alone Flash application, there are commonly recognized attributes for a Flash memory cell. These attributes include small cell size, ease of design, high read cell current, low operating voltage, low program/erase current, low program disturb, high endurance, good manufacturability, good reliability, and good scalability. Ideal Flash memory cells do not exist; "best compromise" are the key words for cell selection to meet specific embedded Flash applications and to achieve a best balance in trade-offs among competing factors including cell size, cell array efficiency, process complexity, design complexity, experience, and facility compatibility. As stated earlier, the criteria for cell selection for embedded application is different from that for the stand-alone Flash memory. Different cells can be selected to optimize for specific application space, such as ease of design and good manufacturability, high density, and high speed, or very low voltage operation.

For ease of design and good manufacturability, split-gate or 1.5T cell such as SCSG [3] or SuperFlash cell [11] can be used. For very high density and high-speed design, single-transistor or 1T cell such as MoneT [7] cell or ETOX [23] cell can be utilized. For very low voltage of sub-1-V operation, two-transistor or 2T cell such as 2TS cell [4] can be utilized. Each of these cells possesses special attributes to meet special application requirements.

In general, high speed is the common requirements for all cells for embedded applications. Therefore, some common elements of cell structure are desirable for any cells for embedded applications. These elements include NOR-type cell structure, high cell current, low wordline capacitance, and low bitline capacitance.

9.5.3.4 Process Consideration.
The process for supporting a high-performance host logic device with an embedded Flash memory requires two different transistor types: low-voltage logic transistors and high-voltage transistors. Both transistors must be optimized for maximum performance or gain for a given maximum operating voltage by minimizing gate oxide thickness, channel length, and threshold voltage, while meeting the breakdown voltage and leakage current requirements. High-performance low-voltage logic transistors are required to meet high-speed operation of the logic core, such as CPU or DSP. The high-voltage transistors are used to support high voltages required for Flash memory program or erase operation. High-performance capability is also required for the high-voltage transistors because they can exist in the low-voltage critical read path, such as row or column select circuits, which contain nodes common to both low- and high-voltage circuits.

Performance of the high-voltage transistors can be maximized by minimizing the absolute maximum operating voltage required for programming and erasing of the Flash memory cell. The so-called split-voltage technique is one effective way for minimizing the absolute operating voltages. The technique splits a high voltage into the positive and negative halves and, therefore, reduces the absolute maximum operating voltage and gate oxide thickness nearly by half. The technique requires isolated well or triple-well process for high-voltage transistors at increased process complexity.

In addition to providing high-performance capability for both low- and high-voltage transistors, an embedded Flash memory process must support the required high-performance Flash memory cell. Silicided process, self-aligned source, local interconnect, and/or multilevel metal layers of three or more are often required for achieving high-density and high-speed design for the host logic device and the embedded Flash memory module.

9.5.3.5 High-Speed Circuit Techniques.
In general, all circuit techniques applicable to the stand-alone Flash memory are also applicable to the embedded Flash memory. Circuits for achieving fast access time design are no exceptions.

High-speed performance can be very critical to the embedded Flash memory, possibly more so than to the stand-alone Flash memory. For this reason, extra design effort to achieve fast access time can be employed to enhance speed of access time. For example:

1. *Boosted Wordline Voltage* For the single-ended Flash memory cell structure, higher cell current for read can provide higher speed operation. Higher cell current is one of the important goals for Flash memory cell development and can be further enhanced through circuit techniques by realizing a wordline voltage higher than V_{dd}. The higher wordline voltage can be accomplished by using a higher than V_{dd} wordline supply voltage, an on-chip high-voltage charge pump, or a synchronized wordline boosting technique. The high wordline supply voltage is most straightforward for design but requires an extra supply voltage that may not exist in the system. The charge pump technique achieves speed improvement without the need of an added supply voltage but requires extra circuits and standby current to avoid an access time penalty for the first access cycle from the standby state. The synchronized wordline boosting technique requires no standby current and makes possible "zero power" design but with reduced speed improvement possible compared with the other two techniques.

2. *Reduced High-Voltage Transistor Effect* The high-performance embedded Flash memory process contains both low- and high-voltage transistor types as stated earlier. High-voltage transistors have lower gains than the low-voltage transistors due to a thicker gate oxide and the longer channel length involved. Any use of high-voltage transistors in the critical read path can result in increased delay. For high-speed design, the speed impact from the high-voltage transistors must be minimized. The minimization can be accomplished by separating the high-voltage circuits from the low-voltage circuits and using the high-voltage transistors only as noncritical passing devices. For example, to achieve high-speed wordline access for read, there can be a low-voltage driver for read and a high-voltage driver for program or erase. To protect the low-voltage driver from being exposed to high voltage during program or erase operations, a high-voltage passing transistor is used to block the high voltage from reaching the low-voltage circuit. Since these high-voltage transistors in the passing gate are used mainly to block the high voltage from reaching the low-voltage circuits through the gate voltage's limiting effect, no critical switching of these high-voltage transistors is involved. The signal V_p applied to the gate of high-voltage pass transistor M can be a dc voltage of V_{dd} or a higher than V_{dd} voltage that will not cause the low-voltage transistors from being exposed to the breakdown voltage after a biased V_t drop. Figure 9.19 illustrates this technique.

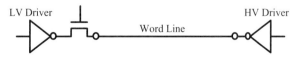

Figure 9.19. Separate low-voltage and high-voltage wordline drivers are used with a non-switching high-voltage pass transistor in the critical read path to minimize read delay.

9.5.3.6 Low-Voltage Design. In general, it is possible to achieve a lower voltage design for a logic device than for a Flash memory. This is because the threshold voltage for the low-logic state is higher for Flash memory cells than for logic transistors. In order to realize low power dissipation and to maximize battery life for handheld products, it is necessary that all devices in a system or all modules on a chip, including embedded Flash memory, operate at a lowest possible supply voltage. The lowest supply voltage can be the end-of-life single-cell battery voltage, which is typically specified at 0.9 V. To achieve a very low voltage operation of as low as 0.9 V for an embedded Flash memory, some special design approaches will be required. The approaches include the application of a special low-voltage Flash memory cell such as 2TS cell [4], which is discussed in Section 9.4.2.3, and special low-voltage circuit techniques.

The circuit techniques to achieve the low-voltage Flash memory read operation boost the selected wordline voltage above the V_{dd} and the cell threshold voltage to overcome the relatively high low-logic-state threshold voltage of Flash memory cells. There are two possible approaches for realizing a higher wordline voltage as discussed in Section 9.5.3.5. The magnitude of higher voltage required for the selected wordline is dependent on the type of Flash memory cell selected. The voltage will be higher for ETOX-type cells that use HCI (hot-carrier injection) for programming to increase the threshold voltage and lower for stack-gate-type cells that use Fowler–Nordheim tunneling for programming to lower the threshold voltage, as described in the embedded Flash memory cell section of this chapter.

Utilization of a specialized low-voltage cell such as 2TS cell [4] allows for the lowest wordline voltage possible for read operation, which translates to a minimum amount of wordline voltage increase required beyond V_{dd} to provide the required cell current for a reliable read operation. Since, in general, any circuit technique for enhancing low-voltage operation can be applied equally to all Flash memory cells; the cell selection may ultimately determine the lowest voltage operation possible.

REFERENCES

1. C. Kuo, M. Weidner, T. Toms, H. Choe, K.-M. Chang, A. Harwood, J. Jelemensky, and P. Smith, "A 512-kb Flash EEPROM Embedded in a 32-b Microcontroller," *IEEE J. Solid-State Circuits*, Vol. 27, No. 4, pp. 574–582, Apr. 1992.
2. L. Parker, R. Singh, C. N. Li, L. Walker, C. M. Hong, S. Liu, D. Farenc, K.-T. Chang, and P. Ingersoll, "Advanced Flash Devices Embedded in a 0.13 µm CMOS Process," paper presented at the 18th IEEE NVSM Workshop, Monterey, CA, pp. 59–61, Aug. 2001.
3. K.-M. Chang, S. Cheng, and C. Kuo, "A Modular Flash EEPROM Technology for 0.8 µm High Speed Logic Circuits," *CICC Proc.*13., pp. 18.7.1–18.7.4, May 1991.
4. W. Liu, K.-T. Chang, C. Cavins, B. Luderman, C. Swift, K.-M. Chang, B. Morton, G. Espinor, and S. Ledford, "A 2-Transistor Source-Select (2TS) flash EEPROM for 1.8V-Only Applications," paper presented at the 15th IEEE NVSM Workshop, Monterey, CA, pp. 4.1.1–4.1.3, Feb. 1997.
5. M. van Duuren, R. Van Schaijk, M. Slotboom, P. Tello, P. Goarin, N. Akil, F. Neuilly, Z. Rittersma, and A. Huerta, "Performance and Reliability of 2-Transistor FN/FN Flash Arrays with Hf Based High-k Interpoly Dielectrics for Embedded NVM," paper presented at the 21th IEEE NVSM Workshop, Monterey, CA, pp. 48–49, Feb. 2006.

6. K. Takahashi, H. Doi, N. Tamura, K. Mimuro, T. Hashizume, Y. Moriyama, and Y. Okuda, "A 0.9 V Operation 2-Transistor Flash Memory for Embedded Logic LSIs," paper presented at the Symp. VLSI Technology, Kyoto, Japan, pp. 21–22, June 1999.
7. C. Kuo, D. Chrudimsky, T. Jew, C. Gallun, J. Choy, B. Wang, S. Pessoney, H. Choe, C. Harrington, R. Eguchi, T. Strauss, E. Prinz, and C. Swift, "A 32-Bit RISC Microcontroller with 448K Bytes of Embedded Flash Memory", paper presented at the International NVM Technology Conference, Albuquerque, NM, pp. 28–33, June 1998.
8. D. Wellekens, J. Van Houdt, J. De Blauwe, L. Haspeslagh, L. Deferm, and H. Maes, "A Low Voltage, High Performance 0.35 μm Embedded Flash EEPROM Cell Technology," paper presented at the 16th IEEE NVSM Workshop, Monterey, CA, pp. 106–108, Aug. 1998.
9. J. Van Houdt, L. Haspeslagh, D. Wellekens, L. Deferm, G. Groeseneken, and H. E. Maes, "HIMOS—A High Efficiency Flash EEPROM Cell for Embedded Memory Applications," *IEEE Trans. Electron Devices*, Vol. ED-40, pp. 2255–2263, 1993.
10. J. Van Houdt, L. Haspeslagh, D. Wellekens, J. De Vos, P. Hendrickx, and J. Tsouhlarakis, "A Low-Cost Poly-Sidewall Erase HIMOS Technology for 130–90 nm Embedded Flash Memories," paper presented at the 20th IEEE NVSM Workshop, Monterey, CA, pp. 61–62, Aug. 2004.
11. S. Kianian, A. Levi, D. Lee, and Y.-W. Hu, "A Novel 3V-Only, Small Sector Erase, High Density Flash EEPROM," paper presented at the Symp. VLSI Technology, Kyoto, Japan, pp. 71–72, June 1999.
12. W.-M. Chen, C. Swift, D. Roberts, K. Forbes, J. Higman, B. Maiti, W. Paulson, and K.-T. Chang, "A Novel Flash Memory Device with Split Gate Source-Side Injection and ONO Charge Storage Stack (SPIN)," paper presented at the Symp. VLSI Technology, Kyoto, Japan, pp. 63–64, June 1997.
13. J. Bude, M. Mastrapasqua, M. Pinto, R. Gregor, P. Kelley, R. Kohler, C. Leung, Y. Ma, R. McPartland, P. Roy, and R. Singh, "Secondary Electron Flash—A High Performance, Low Power Flash Technology for 0.35 μm and Below," *IEDM Tech. Dig.*, pp. 279–282, Dec. 1997.
14. V. Sarin, A. Yu, V. Kowshik, R. Yang, T. Chang, N. Radjy, K. Huang, B. Ly, M. Chen, A. Kwok, A. Wang, K. Yoshikawa, J. Noda, L. Yeh, H Tseng, C. Hsu, and Y. Sheu, "A Low Voltage, Low Power p-Channel Memory for Embedded and System-on-a-Chip Applications," paper presented at the 16th IEEE NVSM Workshop, Monterey, CA, pp. 116–119, Aug. 1998.
15. J. Caywood, G. Batchelder, S. Murray, A. Parthasarathy, J. Pathak, and T. Turner, "A Simple p-Channel EEPROM," paper presented at the 17th IEEE NVSM Workshop, Monterey, CA, pp. 119–123, Feb. 2000.
16. C. N. Li, D. Farenc, R. Singh, J. Yater, S. Liu, C. L. Chang, S. Bagchi, K. Chen, P. Ingersoll, and K.-T. Chang, "A Novel Uniform-Channel Program-Erase (UCPE) Flash EEPROM Using an Isolated p-Well Structure," *IEDM Tech. Dig.*, pp. 779–782, Dec. 2000.
17. C. N. Li, D. Farenc, R. Singh, S. Liu, L. Parker, L. Walker, R. Rai, L. Duncan, K. Chen, P. Ingersoll, and K.-T. Chang, "A NAND/NOR Dual Architecture Flash EEPROM Using a Uniform-Channel Program-Erase (UCPE) Structure," paper presented at the 18th IEEE NVSM Workshop, Monterey, CA, pp. 50–54, Aug. 2001.
18. D. Shum, A. Tilke, L. Pescini, M. Stiftinger, R. Kakoschlke, K. J. Han, S. Kim, V. Hecht, N. Chan, A. Yang, and R. Broze, "Highly Scalable Flash Memory with Novel Deep Trench Isolation Embedded into High Performance CMOS for the 90 nm Node & Beyond," *IEDM Tech. Dig.*, pp. 344–347, Dec. 2005.
19. H. M. Lee, S. T. Woo, H. M. Chen, R. Shen, C. D. Wang, L. C. Hsia, and C. C.-H. Hsu, "Neo-Flash—True Logic Single Poly Flash Memory Technology," paper presented at the 21st IEEE NVSM Workshop, Monterey, CA, pp. 15–16, Feb. 2006.
20. R. Muralidhar, R. Steimle, M. Sadd, R. Rao, C. Swift, E. Prinz, J. Yater, L. Grieve, K. Harber, B. Hradsky, S. Straub, B. Acred, W. Paulson, W. Chen, L. Parker, S. G. H. Anderson, M. Rossow,

T. Merchant, M. Paransky, T. Huynh, D. Hadad, K.-M. Chang, and B. E. White Jr., "A 6V Embedded 90nm Silicon Nanocrystal Nonvolatile Memory," *IEDM Tech. Dig.*, pp. 601–604, Dec. 2003.

21. F. Ito, Y. Kawashima, T. Sakai, Y. Kanamaru, Y. Ishii, M. Mizuno, T. Hashimoto, T. Ishimaru, T. Mine, N. Matsuzaki, H. Kume, T. Tanaka, Y. Shinagawa, T. Toya, K. Okuyama, K. Kuroda, and K. Kubota, "A Novel MNOS Technology Using Gate Hole Injection in Erase Operation for Embedded Nonvolatile Memory Applications," paper presented at the Symp. VLSI Technology, Honolulu, pp. 80–81, June 2004.

22. C. Kuo, J. R. Yeargain, W. Downey, K. Ilgenstein, J. Jorvig, S. Smith, and A. Bormann, "An 80ns 32K EEPROM Using the FETMOS Cell," *IEEE J. Solid-State Circuits*, Vol. 17, No. 5, pp. 821–827, Oct. 1982.

23. S. Lai, "Flash Memories: Where We Were and Where We Are Going," *IEDM Tech. Dig.*, pp. 971–973, Dec. 1998.

10

TUNNEL DIELECTRICS FOR SCALED FLASH MEMORY CELLS

T. P. Ma

10.1 INTRODUCTION

Much like what has happened to microprocessors, dynamic random-access memory (DRAM), and static random-access memory (SRAM), the continued scaling of flash memory cells has caused a dramatic increase in memory capacity and reduction in cost per bit ever since its introduction. While in principle one wishes to scale down the tunnel oxide at the same rate as that for the gate oxide of complementary metal–oxide–semiconductor (CMOS) transistors, in practice the scaling rate of the former has been far slower than the latter in the 1990s [1] and is stagnant at best in the 2000s [2]. This is primarily because of the very stringent 10-year retention requirement, which places an upper limit on the leakage current through the tunnel oxide (on the order of 1 electron/day per cell [3]), and therefore a lower limit on the tunnel oxide thickness. This problem is accentuated due to the stress-induced leakage current (SILC) resulting from repeated program/erase (P/E) operations [3–10] and the possible read-disturb effect [3–10]. As a consequence, a tunnel oxide thickness of 8 nm has been set as a lower limit by some [3]. The International Technology Roadmap for Semiconductors [2] has put a scaling limit at 8 nm for NOR Flash and 6 to 7 nm for NAND Flash in recent years.

To overcome this scaling limit, it appears that any candidate tunnel dielectric must exhibit low leakage current and low SILC. In this context, this chapter will show that silicon nitride, when it is properly deposited, is an excellent candidate because of its significantly reduced gate current when compared to SiO_2 with the same equivalent oxide thickness (EOT) [11]. It has also been shown that the SILC for such nitride films can be much smaller than those typically observed for SiO_2

Nonvolatile Memory Technologies with Emphasis on Flash. Edited by J. E. Brewer and M. Gill
Copyright © 2008 the Institute of Electrical and Electronics Engineers, Inc.

films [12]. Therefore, there is a strong possibility that a silicon nitride film with a substantially thinner EOT than 8 nm can be used as the tunnel dielectric in scaled Flash memory cells. The lower electron and hole barrier heights of the nitride/Si interface (compared to those of the SiO_2/Si interface) can also be explored for more efficient P/E operations.

This chapter, therefore, focuses on silicon nitride as the tunnel dielectric for scaled Flash memory cells. It should be noted that a key reason for the lower gate current in silicon nitride than in silicon oxide of the same EOT is the higher dielectric constant of the former, which enables the use of a physically thicker gate dielectric. By the same token, dielectrics with even higher dielectric constants, such as HfO_2 and TiO_2, appear to be promising candidates as well, although leakage currents other than tunneling current may render them less viable. This aspect will also be discussed in this chapter.

10.2 SiO_2 AS TUNNEL DIELECTRIC—HISTORICAL PERSPECTIVE

First, to avoid confusion later in this chapter, let it be clear that by *tunnel dielectric* we mean generally the dielectric between the floating gate and the silicon substrate of a floating-gate memory transistor, regardless whether the P/E operations of this memory transistor actually uses the tunneling mechanism at all. For example, a Flash electrically programmable read-only memory (EPROM) cell that uses avalanche injection or substrate hot-electron injection mechanism for its P/E operation may still contain a "tunnel oxide" between the floating gate and the Si substrate by our definition, even though tunneling is not the dominant charge-transfer mechanism in this case.

The requirements for the tunnel dielectric are extremely stringent. It must not only satisfy all of the requirements for a CMOS gate dielectric (i.e., it must provide good transconductance and current drivability with excellent stability and integrity) but also allow extremely low leakage current and endure prolonged high-field high-current stress. It is therefore not surprising that an overwhelming majority of the tunnel dielectrics that have been reported in the literature are based on SiO_2 and its slight modifications, including those grown with a low thermal budget [13–15], textured oxide grown on polysilicon/Si substrate [16, 17], lower barrier oxide grown on As- or P-implanted Si substrate [18], and nitrided oxides [19–23].

These various approaches to produce modified oxides have achieved varying degrees of success. For example, the low-thermal-budget oxide has shown reduced SILC, improved effective electron mobility, and improved endurance [14]; the textured oxide offers higher electron injection rate and better reliability [16, 17]; the lower barrier oxide exhibits superior P/E characteristics and much improved endurance [18]; the nitrided oxides, in which the nitrogen is introduced either by NH_3 [19, 21], N_2O, or NO [20–23], have generally shown improved endurance compared to the conventional tunnel oxide. However, none of them have been able to scale below 8 nm because of the SILC problem and the very stringent retention requirement.

One significant variation from the above is the silicon-rich oxide (SRO) [24–26], which takes advantage of the high electron injection efficiency due to the enhanced local electric fields near the tiny silicon islands embedded in SiO_2. Like other SiO_2-based tunnel oxides, however, the SRO is not scalable below 8 nm for the same reasons discussed before.

The only alternative tunnel dielectric reported in the literature that is not SiO-based is silicon nitride [27, 28], and apparently only the Fujitsu group [27, 28] did significant work on it, first in the late 1970s [27] and continued through the mid-1980s.

10.3 EARLY WORK ON SILICON NITRIDE AS A TUNNEL DIELECTRIC

The aforementioned silicon nitride film was grown either by direct thermal nitridation of a silicon substrate [29, 30] or by plasma-assisted nitridation [30], and its most attractive feature was reported to be the lower barrier heights for both electrons and holes at the Si/nitride interface. The lower barrier heights enabled the use of lower voltages for the P/E operation [27, 28].

A 4-bit memory cell array in which a 9.5-nm-thick thermal nitride served as the tunnel dielectric in each cell was demonstrated by the Fujitsu group as early as in 1979 [27]. This cell used the avalanche injection mechanism for its P/E operations. Because of the nitride's relatively low barrier heights for both electrons and holes, relatively low voltages were sufficient, with 10 V, 10 V (gate, drain) for programming and −5 V, 10 V (gate, drain) for erase reported in the study. With these operating voltages, both writing and erase could be accomplished in 1 ms. Endurance more than 10^5 cycles and retention longer than 10 years at 125°C were projected.

Subsequently, a *graded bandgap* tunnel dielectric was reported by the Fujitsu group in which the tunnel dielectric was made of an oxidized thermal nitride film; that is, the tunnel dielectric was made by oxidizing a 6-nm thermal nitride film to approximately 10 nm in thickness [28]. The authors claimed that the resulting graded bandgap dielectric was almost entirely silicon nitride at the substrate interface, and almost entirely silicon oxide at the floating-gate interface, with a mixture of the two in between. A 2-kbit memory array with such a tunnel dielectric was demonstrated [28], with relatively low-voltage capability and good P/E endurance reported.

Despite these promising results, however, there has been no reported work on floating-gate memory devices using silicon nitride as the tunnel dielectric since the mid-1980s. It is not clear exactly why the thermal nitride approach was discontinued. It is very likely that the very high thermal budgets required to grow the silicon nitride film (e.g., over 2 h at 1200 to 1300°C for direct nitridation or 1050°C for plasma-enhanced nitridation) could be the main reason for its demise.

The reason for not using the more popular CVD (chemical vapor deposition) method to grow the silicon nitride tunnel barrier was because of the conventional CVD nitride's poor quality as a MOS gate dielectric in the past [31–41]; in particular, the high densities of electron and hole traps are most detrimental for this application. On the other hand, recent advances in rapid-thermal CVD (RTCVD)[42] and remote-plasma CVD (RPCVD) [43] have significantly improved the quality of CVD silicon nitride. More over, gate-quality silicon nitride films have recently been synthesized by a low-temperature jet-vapor deposition (JVD) method [11]. These JVD silicon nitride films exhibit low leakage current (lower than that for SiO_2 of the same EOT) [44] and high reliability [45], suggesting the strong possibility that a deposited silicon nitride film, if properly prepared, may be suitable as a tunnel dielectric in Flash memory cells. One of the more exciting results is the greatly suppressed SILC in MNS capacitors with JVD silicon nitride as the gate dielectric [45], which removes a major obstacle that prevents the scaling of tunnel oxide below the 8 nm.

In the next few sections, the properties of JVD silicon nitride films are presented as an example to illustrate what's achievable in the laboratory now. It is hopeful that similar properties (or even better ones) will soon be realized in silicon nitride films deposited by other techniques, such as CVD, as there is no reason why CVD technique cannot produce silicon nitride films with the same properties as those produced by JVD.

Since the JVD technique is not yet a commonly used technique, a brief introduction is given in the next section.

10.4 JET-VAPOR DEPOSITION SILICON NITRIDE DEPOSITION

Figure 10.1 shows the top view of a JVD system used to deposit the silicon nitride films reported here. One can see that the wafer is scanned with respect to the vapor jet to achieve film uniformity. A microwave cavity surrounds a portion of the jet nozzle and generates a gaseous discharge in the nozzle.

Figure 10.2 shows a closeup view of the coaxial dual-nozzle jet-vapor source used for silicon nitride deposition in which He is used as the carrier gas to rapidly transport the depositing species (silane in the inner nozzle and reactive nitrogen in the outer nozzle) to the substrate where they undergo reactions to form the silicon nitride film. The pressures in both nozzles are significantly higher than that in the chamber so that a supersonic flow occurs at the exit of both nozzles. It is believed that the high speed of the jet leads to two important consequences that contribute to the high quality of the deposited films: (1) the high impinging energy of the depositing species allowed the use of room temperature substrates, and (2) the short transit time of the depositing species minimizes the possibility of gas-phase nucleation, which is often a source of imperfection for CVD silicon nitride.

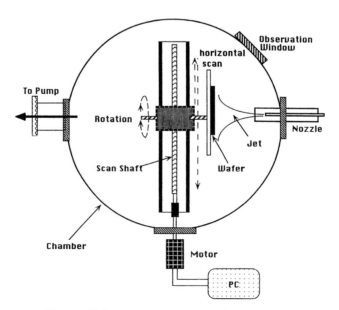

Figure 10.1. Schematic top view of a JVD system.

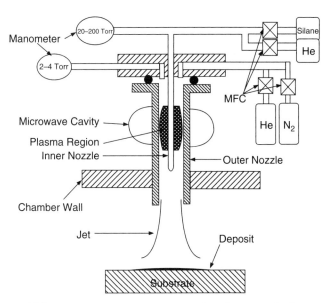

Figure 10.2. Dual-nozzle jet-vapor source for silicon nitride deposition.

The deposition rate is controlled by adjusting the flow rates of silane, nitrogen, and helium, the microwave power, and the wafer scan speed. For thin gate dielectric applications, a deposition rate lower than 0.5 nm/min is typically used, although it is possible to increase the rate significantly.

A postdeposition annealing (PDA) step normally follows the silicon nitride deposition to densify the film, and the typical PDA temperature is 800°C for 30 min in nitrogen. A postmetal anneal (PMA) step is also normally part of the processing sequence, and the PMA is typically performed in a forming gas ambient (5% H_2 + 95% N_2) at 400°C for 30 min to further improve the interface properties.

A more detailed description of the JVD process can be found in Ma [11].

10.5 PROPERTIES OF GATE-QUALITY JVD SILICON NITRIDE FILMS

Figure 10.3 shows the Auger depth profile of a JVD silicon nitride film deposited on Si. One can see the expected silicon and nitrogen signals. There is also an oxygen signal throughout the thickness of the film, although no oxygen gas was intentionally introduced into the deposition chamber. This unintentional oxygen is believed to come from the residual gas in the chamber, which is evacuated by mechanical pumps to a base pressure of ~1 mTorr. The effects of the oxygen on the film properties are not entirely clear, although it is known to lower the dielectric constant (it changes from 7.8 for pure nitride to 6.5 with 20% of oxygen) and thus increases the gate tunneling current for a given EOT [46]. Recent construction of an improved JVD system has resulted in lower residual oxygen pressure and consistently lower oxygen concentration in the silicon nitride films (<10%), and the electrical properties are similar to those reported here.

Figure 10.4 shows the Fourier transform infrared (FTIR) spectrum of a JVD silicon nitride film deposited on Si, which exhibits the expected SiN bonds. The

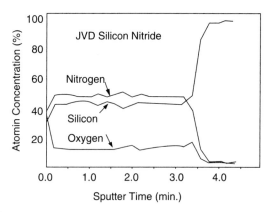

Figure 10.3. Auger depth profile of a JVD silicon nitride film deposited on Si.

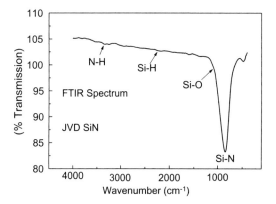

Figure 10.4. FTIR spectrum of a JVD silicon nitride film.

hydrogen concentration, as manifested by the NH and SiH signals, is relatively low as compared to the plasma-enhanced CVD silicon nitride. Furthermore, the NH and SiH signals do not change significantly after annealing at 800°C, suggesting that these bonds are much stronger compared to those in plasma enhanced CVD (PECVD) silicon nitride, which often dissociate at such a temperature and could act as electronic traps.

Next, some electrical properties of JVD silicon nitride films will be presented, with an emphasis on the properties of ultrathin (≤4nm of EOT) films, as the JVD silicon nitride was originally developed to be a potential replacement for SiO_2 as an advanced gate dielectric in scaled CMOS circuits.

Figure 10.5 shows a pair of high-frequency and quasi-static capacitance–voltage (C–V) curves for an MNS capacitor with a JVD nitride film of 4.2nm of equivalent oxide thickness (EOT). The coincidence of the two curves indicates very low density of interface traps. There is no discernible hysteresis in the high-frequency C–V curve when swept back and forth along the voltage axis, indicating negligible border traps. These results are comparable to quality MOS capacitors, and certainly much better than any other MNS capacitors made of conventional CVD silicon nitride reported in the literature.

Figure 10.6(a) depicts a histogram of the breakdown fields for a set of MNS capacitors with EOT of 4.2nm, showing an average breakdown field of over 13MV/

PROPERTIES OF GATE-QUALITY JVD SILICON NITRIDE FILMS 413

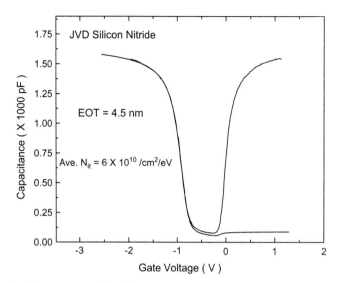

Figure 10.5. High-frequency and low-frequency C–V curves of an MNS capacitor with JVD silicon nitride as the gate dielectric.

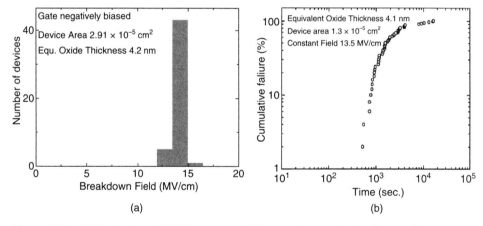

Figure 10.6. (a) Breakdown field histogram and (b) TDDB characteristics of a set of MNS capacitors made of JVD silicon nitride as the gate dielectric.

cm. Figure 10.6(b) shows the corresponding time-to-breakdown data under a very high field of 13.5 MV/cm. These data are comparable to high-quality thermal SiO_2.

Figure 10.7 shows the result from a constant-current stress experiment. Based on the saturation voltage shift (of order of 30 mV), one may estimate a trap density in the $10^{11}/cm^2$ range, which is comparable to the values for quality thermal SiO_2, but much lower than conventional CVD silicon nitrides, which typically contain at least one order of magnitude higher densities of traps [31–39].

The current–voltage (I–V) characteristics of the MNS capacitors fit the Fowler–Nordheim (FN) plot very well, as exemplified by the data in Figure 10.8, where J is the current density and E is the equivalent oxide field. Using an effective mass of 0.5 times the free electron mass, a barrier height of 2.1 eV is calculated from the

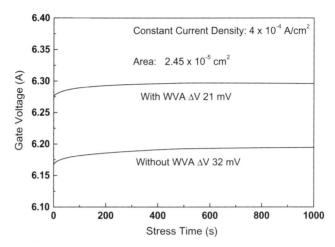

Figure 10.7. Gate voltage required to maintain a constant current. The sample for the upper curve received water vapor anneal (WVA) while the sample for the lower curve did not.

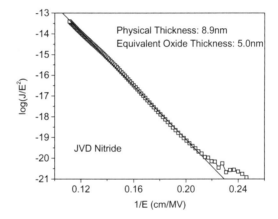

Figure 10.8. Fowler–Nordheim plot from the *I–V* characteristic of a JVD MNS capacitor.

slope of the straight line. Other MNS capacitors made with the JVD silicon nitride show similar barrier height values from their *I–V* characteristics.

Since for the conventional CVD silicon nitride the dominant current component is the Frenkel–Poole (FP) conduction mechanism, it is very interesting that the JVD nitride shows primarily FN tunneling mechanism. It is believed that the lack of traps in the JVD nitride is probably responsible for this, because the FP conduction requires a high density of traps.

The fact that the dominant conduction mechanism in the JVD nitride is tunneling is also confirmed by its relatively weak temperature dependence, as shown in Figure 10.9. As a comparison, the CVD nitride shows much stronger temperature dependence.

Theoretical calculations indicate that, if tunneling is the dominant current transport mechanism rather than Frenkel–Poole process, then the gate leakage current should be substantially lower for the nitride when compared with the thermal SiO_2

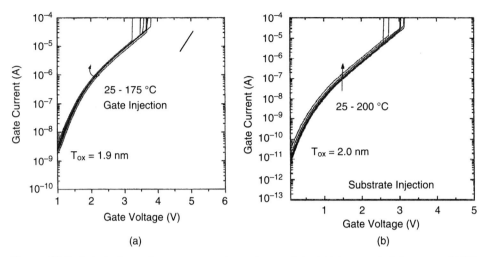

Figure 10.9. Relatively weak temperature dependence of the I–V characteristic for a set of MNS capacitors made of JVD silicon nitride as the gate dielectric: (a) electron injection from the gate electrode and (b) electron injection from the substrate.

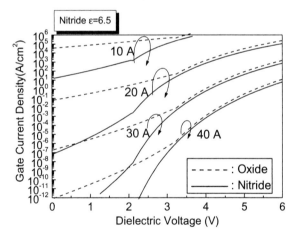

Figure 10.10. Theoretical gate currents as functions of the dielectric voltage for silicon oxides and silicon nitrides of various thicknesses.

of the same equivalent thickness, especially at low fields [11, 44]. Figure 10.10 shows an example where the dielectric constant of the nitride is assumed to be 6.5, which is a reasonable assumption based on the experimental results. Note that the kink in each curve corresponds to the transition from direct tunneling to Fowler–Nordheim tunneling as the dominant transport mechanism.

One can see that, for a given equivalent oxide thickness, the leakage current is substantially lower in the nitride, especially in the direct tunneling regime, where the difference is several orders of magnitude. By assuming a higher dielectric constant for the nitride, the calculation showed even lower nitride current [44].

Figure 10.11 shows the leakage currents measured in three JVD nitride films of 2.1, 2.9, and 3.9 nm of EOT, respectively. For comparison, the data for thermal SiO_2

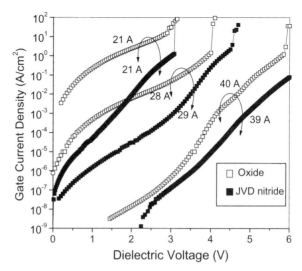

Figure 10.11. Comparison of measured gate currents between oxide and nitride samples.

of comparable thicknesses are also included. The data clearly indicate a lower leakage current through the nitride for a given oxide equivalent field, in qualitative agreement with the theoretical prediction. In the low-field regime, the leakage currents in the nitride films are not as low as what the tunneling calculations have predicted. It is suspected that the Frenkel–Poole conduction mechanism has not been entirely eliminated in these nitride films.

The reason that the tunneling current is much lower in silicon nitride for a given EOT, despite its lower barrier height than SiO_2, is because of its larger physical thickness (nearly twice as thick as SiO_2) which more than offsets the effect of its lower barrier height. Figures 10.12(a) and 10.12(b) show I–V characteristics for a set of ultrathin JVD silicon nitride films measured before and after high-field stress in both polarities, and the SILC values are very small.

Both N- and P-channel field-effect transistors (FETs) with JVD silicon nitride as the gate dielectric exhibit good quality. Figure 10.13(a) shows a family of drain current curves of an N-channel MNS transistor with EOT of 3.8 nm, while Figure 10.13(b) shows the corresponding transconductance characteristics. The data for a control MOSFET are also included for comparison. Note that both sets of data have been normalized with respect to the gate EOT in order to have a fair comparison. One can see that the low-field transconductance is lower for the nitride sample, but its high-field transconductance is higher than that of the oxide sample. This behavior may be attributed to carrier trapping by the slow interface traps (or border traps [47]) near the nitride/Si interface, as described elsewhere [48].

The data for the P-channel devices are shown in Figures 10.14(a) and 10.14(b), where the transconductance for the MNS device is also lower than that for the MOS device at low fields, but higher at high fields. The current drivability is comparable between the two dielectrics.

Superior transistor characteristics have also been realized for submicron MOSFETs made of JVD silicon nitride as the gate dielectric [49, 50].

DEPOSITED SILICON NITRIDE AS TUNNEL DIELECTRIC 417

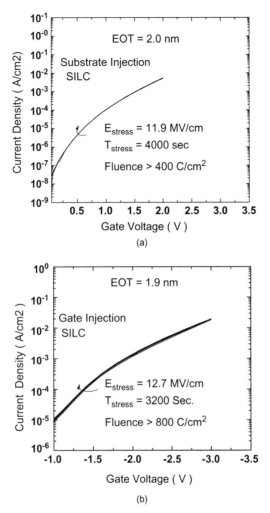

Figure 10.12. Stress-induced leakage current (SILC) is relatively small for MNS capacitors made with JVD silicon nitride as the gate dielectric: (a) electron injection from the silicon substrate and (b) electron injection from the gate electrode.

These properties are certainly sufficiently encouraging for someone to consider seriously the possibility of using deposited silicon nitride as the tunnel dielectric for scaled Flash memory devices.

10.6 DEPOSITED SILICON NITRIDE AS TUNNEL DIELECTRIC

To be acceptable as a tunnel dielectric, a key requirement is that the gate leakage current must be sufficiently low to satisfy the 10-year data retention requirement. The allowable gate leakage current has been estimated to be between 10^{-24} A/cell to 2×10^{-18} A/cell (for a cell area of 2×10^{-8} cm^2), depending on the duty cycle of the disturb voltage (with duty cycle ranging from 100% to 10^{-6}) [51]. Even if one

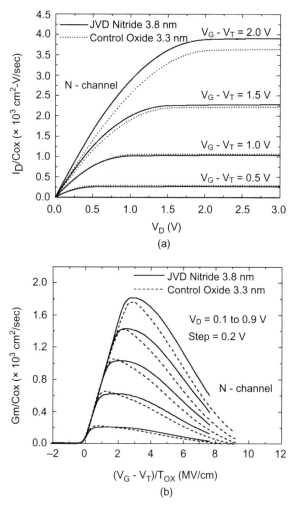

Figure 10.13. Comparison of N-channel transistors made of oxide gates with those made of JVD nitride gates: (a) family of I–V curves and (b) corresponding transconductance curves.

takes the most demanding value of 1×10^{-24} A/cell (or 1×10^{-16} A/cm^2) as the allowable limit, then in principle a 5-nm thermal SiO$_2$ layer could be used, if it weren't for the excessive SILC that is expected for such a thin oxide after repeated write/erase (W/E) cycles [51]. Because of SILC, the thinnest practical tunnel oxide has been estimated to be 7 to 8 nm [3, 51].

Figure 10.15 gives an example of the SILC data reported for a set of silicon oxide MOS devices [52], where it is obvious that the SILC problem becomes more severe as the oxide gets thinner in the range studied.

In view of the report that silicon nitride films deposited by the JVD method are almost free of SILC (see Fig. 10.12 and [45]), and that their gate leakage currents are significantly lower than their silicon oxide counterparts (see Fig. 10.16 and [44, 45]), one might wonder if a silicon nitride film of 4 to 5 nm of EOT could be used as a tunnel dielectric. Some preliminary experiments have been done, and the results are very encouraging, presented as follows.

DEPOSITED SILICON NITRIDE AS TUNNEL DIELECTRIC

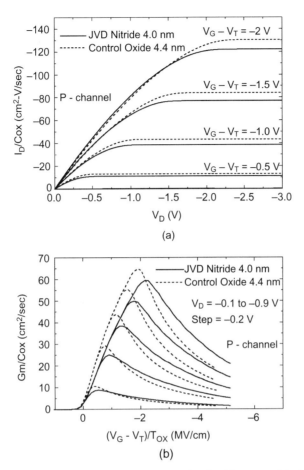

Figure 10.14. Comparison of P-channel transistors made of oxide gates with those made of JVD nitride gates: (a) family of I–V curves and (b) corresponding transconductance curves.

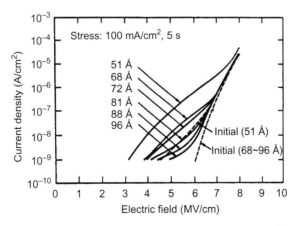

Figure 10.15. Stress-induced leakage current is more severe as gate oxide gets thinner (after Karamcheti et al. [50]).

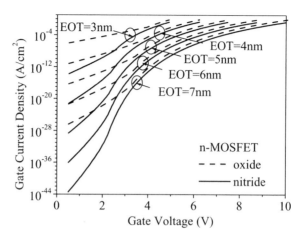

Figure 10.16. Comparison of theoretical gate tunneling currents between oxide and nitride samples.

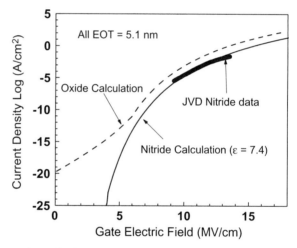

Figure 10.17. Comparison of calculated tunneling gate currents between oxide gate and nitride gate of 5.1 nm of EOT. The available experimental data for JVD nitride are also included.

Figure 10.17 shows the calculated current densities as functions of the gate dielectric field for a silicon nitride film and a silicon oxide film, both of 5.1 nm of EOT. Some experimental data for the nitride are also shown, which match the calculated results quite well. These results indicate that both the oxide and the nitride would satisfy the $\leq 10^{-16}\,\text{A/cm}^2$ leakage current requirement at a gate field lower than 3 MV/cm, with the silicon nitride showing much lower current.

The problem with the ~5-nm oxide, however, is that the SILC is unacceptably high, as exemplified by the data shown in Figure 10.18, where a 5.2-nm oxide exhibits more than an order of magnitude increase in the gate current after stressing at a field of 11 MV/cm for 100 s. In contrast, the silicon nitride sample of a comparable EOT shows nearly no SILC after 5 times the stress, as also shown in Figure 10.18.

In fact, even thinner JVD silicon nitride samples showed good SILC characteristics, as exemplified by the data in Figure 10.19 for a sample with EOT of 4.6 nm.

DEPOSITED SILICON NITRIDE AS TUNNEL DIELECTRIC

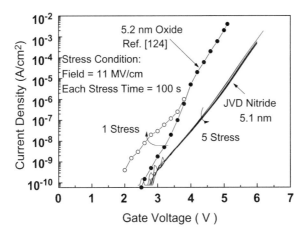

Figure 10.18. *I–V* characteristics for an oxide sample and a JVD nitride sample before and after high-field high current stress.

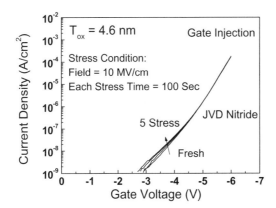

Figure 10.19. Stress-induced leakage current is relatively small in this JVD nitride sample with EOT = 4.6 nm.

It should be noted that not all silicon nitride samples exhibit low SILC. Figure 10.20 shows that, if not properly prepared, even a silicon nitride gate can be made to exhibit a significant amount of SILC.

Thus the results in Figures 10.16 through 10.19 strongly suggest that a properly synthesized silicon nitride film with an EOT of 4 to 5 nm could be used as a tunnel dielectric from the data retention point of view.

Consistent with the low SILC, the hot-carrier-induced threshold voltage change (ΔV_T) and transconductance degradation (ΔG_m) are also very small in MNSFETs made of silicon nitride as the gate dielectric, as shown in Figures 10.21 and 10.22, where the silicon nitride films were deposited in the same run as that shown in Figure 10.19. These results serve to demonstrate the robustness of deposited silicon nitride as a gate dielectric.

In the above, the substrate current for either the N-channel or the P-channel device arises from the impact ionization process near the drain caused by the channel hot carriers. This impact ionization process plays a very important role in

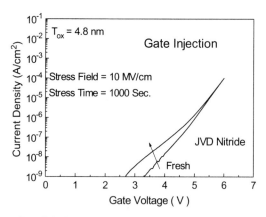

Figure 10.20. Stress-induced leakage current can be very significant even in silicon nitride samples if the silicon nitride is not properly prepared.

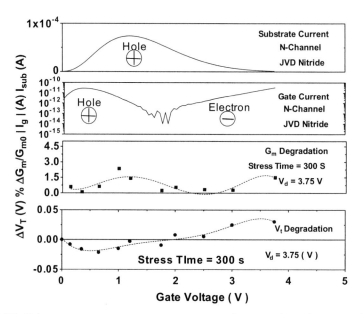

Figure 10.21. Substrate current, gate current, transconductance degradation, and threshold voltage degradation in a N-channel MNSFET resulting from channel-hot-carrier stress at $V_d = 3.75\,V$ and various V_g.

the nature of the gate current. For small gate voltages, the hole gate current in the N-channel device (see Fig. 10.21) and the electron gate current in the P-channel device (see Fig. 10.22) originate from the carriers generated by this impact ionization.

One major potential advantage of silicon nitride as a tunnel dielectric arises from its lower barrier heights for both electrons and holes compared to silicon oxide, as shown in Figure 10.23. As a consequence of the lower electron barrier, one would expect a much higher gate electron current injection efficiency for a given channel hot-electron (CHE) current, and thus faster programming efficiency for a nitride-

DEPOSITED SILICON NITRIDE AS TUNNEL DIELECTRIC 423

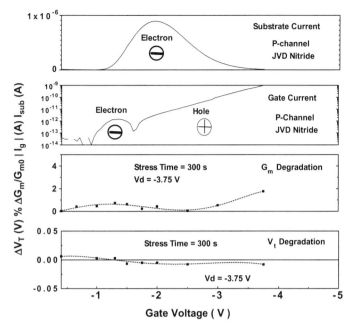

Figure 10.22. Substrate current, gate current, transconductance degradation, and threshold voltage degradation in a P-channel MNSFET resulting from channel-hot-carrier stress at $V_d = -3.75\,V$ but various V_g.

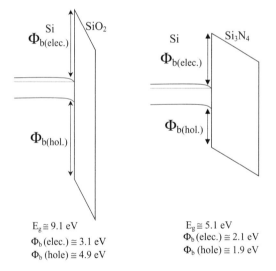

Figure 10.23. Electron energy band diagrams for metal insulator semiconductor (MIS) capacitors with silicon oxide (left) and silicon nitride (right) as the gate dielectric.

barrier Flash cell that utilizes CHE to program. Preliminary results are indeed consistent with the above expectation, as presented below.

Figure 10.24 compares the channel hot carrier (CHC)-induced gate current between a MOSFET with an oxide gate and a MNSFET with a nitride gate. Both

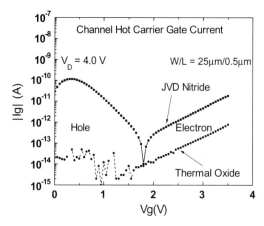

Figure 10.24. Gate currents as functions of gate voltage under a fixed drain voltage of 4V for two transistors with different gate dielectrics, thermal oxide and JVD nitride.

gate dielectrics have approximately the same EOT of 4.2 nm, and the gate dimensions are width of 25 μm and length of 0.5 μm. The drain voltage, V_D, is fixed at 4 V during the measurement. Note that these transistors are designed for logic circuits, and therefore are not optimized for CHC programming; in particular, they do not have abrupt drain junctions. Nonetheless, these transistors do serve to illustrate the qualitative trend when one switches from silicon oxide to silicon nitride as the tunnel dielectric.

One can see that, for V_g higher than about 2 V, the gate current is dominated by electron current for either the oxide-barrier device or the nitride-barrier device, but the electron current in the nitride-barrier device is 50 to 80 times higher than that in the oxide-barrier device. This is consistent with the expectation that the lower barrier height of the nitride allows a higher rate of electron injection into the gate. These results suggest that the use of silicon nitride as the tunnel dielectric may lead to a faster CHC programming speed by a factor of 50 to 80.

One striking feature of the data in Figure 10.24 is the hole current for V_g below 1.8 V. While the hole current in the oxide-barrier device is very small (of the same order as the noise in the measurement), it gets to be very significant in the nitride-barrier device, especially for low V_g. In fact, below $V_g = 1$ V the hole injection efficiency in the nitride-barrier device far exceeds the electron injection efficiency measured at $V_g = 3.5$ V.

Such a high hole injection efficiency arises primarily from the relatively low hole barrier at the nitride/Si interface (1.9 eV, as shown in Fig. 10.23), which is not only lower than the hole barrier at the oxide/Si interface (4.9 eV) but even slightly lower than the electron barrier (2.1 eV) at the nitride/Si interface.

The results in Figure 10.24 suggest strongly the possibility of using the CHC mechanism for both the write (via hot-electron injection) and the erase (via hot-hole injection) operations by applying the appropriate control gate voltages. For example, for the device shown in Figure 10.24 to induce hot-electron injection one may apply a relatively high control gate voltage such that the floating gate potential is at 3.5 V above the source, while to induce hot-hole injection a relatively low control gate voltage may be applied such that the floating-gate potential is at 1 V above the source. Note that the nitride-barrier device illustrated here is not designed for Flash applications, and it is expected that the operating voltages for a realistic nitride-barrier Flash device can be optimized by appropriate device engineering and circuit design.

The bit-erasable capability with the CHC mechanism is also desirable for some applications. The next two sections will show some preliminary results on floating-gate Flash memory devices with JVD silicon nitride as the tunnel dielectric.

10.7 N-CHANNEL FLOATING-GATE DEVICE WITH DEPOSITED SILICON NITRIDE TUNNEL DIELECTRIC

The N-channel floating-gate devices used in this study were based on Flash memory cells originally designed to use channel FN tunneling for both program and erase operations, and not optimized for CHE (channel hot-electron) or CHH (channel hot-hole) injection. The tunnel oxide in the original cells was replaced with 55 Å EOT JVD silicon nitride, with all other processing steps kept unchanged. The devices had a drawn channel width (W) and length (L) of 0.35 and 0.25 μm, respectively. The coupling ratio to the control gate was 0.60. Since the Si–SiN barrier height is about 2 eV for electrons and about 1.9 eV for holes, which are significantly less than those for the Si–SiO2, CHE and CHH injections were used for program and erase operations, respectively.

Figure 10.25 shows the gate current, I_g, as a function of gate voltage for various drain voltages, for a Flash memory cell with an electrical contact to the floating gate, indicating that both the electron current (for V_g greater than 2.5 V) and the hole current (for V_g less than 2 V) can be substantial at modest programming voltages. These currents are orders of magnitude higher than those observed in control devices with SiO$_2$ tunnel dielectric.

To assess the possible speed of the program operation, a full program/erase curve for the device was measured between 5 μs and 10 ms (Fig. 10.26). The erased device threshold voltage was about 0.5 V. As seen from Figure 10.26, a pulse width of 75 μs would program the cell to about 2.5 V. This pulse width was then used in the subsequent cycling experiment (Fig. 10.27). The rise/fall times of the pulses

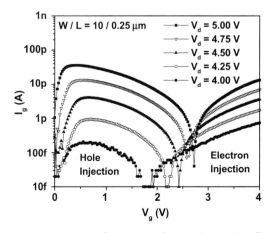

Figure 10.25. Hot carrier gate current for various drain voltages in a floating-gate contacted Flash memory cell.

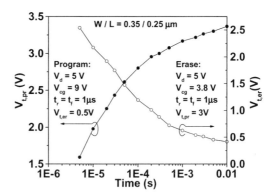

Figure 10.26. Typical program/erase curves for a Flash memory cell with 55-Å JVD silicon nitride tunnel dielectric. The erased device threshold voltage was about 0.5 V.

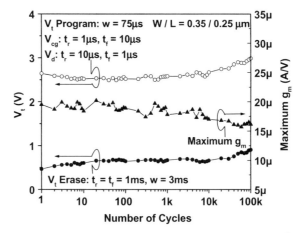

Figure 10.27. Endurance characteristics of a Flash memory cell with 55-Å JVD silicon nitride tunnel dielectric. Program: $V_d = 5\,V$, $V_{cg} = 9\,V$; erase: $V_d = 5\,V$, $V_{cg} = 3.8\,V$; $V_s = V_b = 0$.

applied to the drain and to the control gate were tailored with respect to each other in such a way that the voltage on the control gate is always high whenever the drain voltage is rising or falling. This is done in order to prevent accidental hole injection during programming, since a low control gate and a high drain voltage favor hole injection (Fig. 10.25).

From Figure 10.27 one can see that up to about 100 cycles a slight V_t window closing trend is observed. This can be explained by both electron and hole trapping, which lowers, respectively, the electron and hole injection efficiencies. After about 1000 cycles the V_t window shows a parallel upward shift, which is due to the electron trap-up and interface trap generation in the channel. Starting from the same point, the maximum transconductance, G_m, also shows some degradation due to the increased interface traps.

Figure 10.28 shows threshold voltage distribution data from 100 cycles of 10 cells after an initial 10,000 cycles, and after a 150°C bake for 2000 h followed by an additional 10,000 cycles. Both program and erase threshold voltage distributions are tight and show a very small postbake change. Threshold voltages of these 10

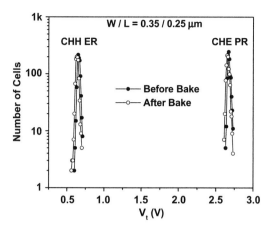

Figure 10.28. Program/erase threshold voltage distributions of 10 cells cycled 100 times each after initial 10,000 cycles, and after 150°C bake for 2000 h and additional 10,000 cycles.

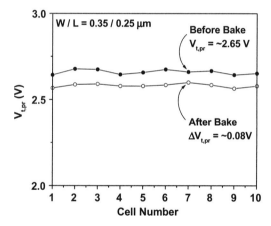

Figure 10.29. Threshold voltage of 10 programmed cells before and after 150°C bake for 2000 h.

programmed cells before and after the bake are shown in Figure 10.29. The average postbake V_t shift is only about 0.08 V.

To speed up the program operation and to lower the program voltage, we utilized the pulse agitated substrate hot electron injection (PASHEI) programming technique [53] (Fig. 10.30). When the PASHEI programming voltages are kept the same as during CHE program ($V_{dc} = 5\,\text{V}$, $V_{cg} = 9\,\text{V}$), it is possible to reduce the programming time to 36 μs (Fig. 10.31). If low-voltage operation is desired, the control gate voltage, V_{cg}, during PASHEI programming can be reduced down to 5 V, with a longer programming time of 500 μs (Fig. 10.32). Thus, the cell can be operated from a single polarity +5 V supply. It is worth noting that since the original cells modified for this study were not designed for CHC or PASHEI operation, the injection efficiencies for the cells with silicon nitride as tunnel dielectric are not as high as one would like. We believe, however, that by proper junction engineering the injection

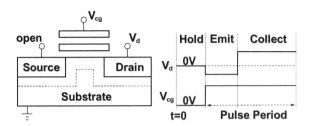

Figure 10.30. Principle of PASHEI programming. During one pulse applied to drain, the electrons are first emitted into the substrate by a forward-biased drain junction, then the bias on the drain is changed to high reverse, the electrons are accelerated in the depletion region, injected into the tunnel dielectric, and collected by the floating gate.

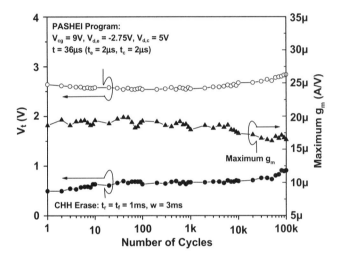

Figure 10.31. Endurance characteristics of a Flash memory cell with 55-Å JVD silicon nitride tunnel dielectric. Program: $V_{dc} = 5V$, $V_{cg} = 9V$; erase: $V_d = 5V$, $V_{cg} = 3.8V$, $V_s = V_b = 0$.

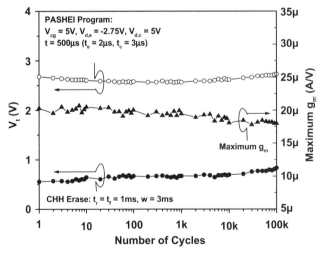

Figure 10.32. Endurance characteristics of a Flash memory cell with 55-Å JVD silicon nitride tunnel dielectric. Program: $V_{dc} = 5V$, $V_{cg} = 5V$; erase: $V_d = 5V$, $V_{cg} = 3.8V$, $V_s = V_b = 0$.

efficiencies can be significantly increased, which should result in much higher speeds of the program/erase operations and lower operating voltages.

P-channel floating-gate devices have also been fabricated, and the results are presented next. The P-channel Flash memory devices offer several advantages over their N-channel counterparts, including lower power, higher speed, and better reliability [52].

10.8 P-CHANNEL FLOATING-GATE DEVICE WITH DEPOSITED SILICON NITRIDE TUNNEL DIELECTRIC

The P-channel Flash memory devices used in this study were fabricated in a N-well that was formed by 150-keV phosphorus implantation. After active-area definition, 30-keV phosphorus channel implantation was introduced to adjust the threshold voltage. JVD nitride was then deposited at room temperature and annealed in N_2 at 800°C for 30 min. Afterward, the floating-gate layer was deposited and patterned. High-temperature oxide (HTO) and N^+ poly-Si were deposited for the interpoly dielectric and control gate, respectively. Standard back-end processing was used to complete the device fabrication. The intrinsic threshold voltage is close to −2.2 V, so a programmed device becomes a depletion mode transistor if the V_t shift exceeds 2.2 V. This issue can be avoided by increasing the channel doping concentration so that the device will always remain in enhancement mode.

In this study, P-channel devices are programmed by band-to-band tunneling-induced hot-electron (BBHE) injection for high efficiency [54]. Data are reported for devices with $W/L = 0.7\,\mu m/0.4\,\mu m$. For a device with a tunnel dielectric of 8 nm JVD nitride (~5 nm EOT), the threshold voltage shift reaches 2 V in 0.6 µs and 3 V in 1 µs when the control gate voltage (V_{cg}) is biased at only 6 V, as shown in Figure 10.33(a). Since BBHE programming consumes very little power, better than 2 ns/byte programming speed can be obtained if 512 bytes are programmed in parallel.

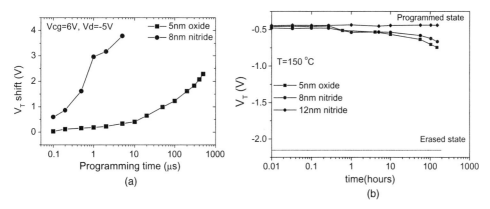

Figure 10.33. (a) P-channel JVD nitride Flash memory device can be programmed in 1 µs at low voltage. For 2-V V_t shift, it is almost 700 times faster than the oxide memory device. (b) Retention characteristics after 10^4 cycles. An 8-nm JVD nitride memory device has better retention than the 5-nm oxide control device. A 12-nm JVD nitride should meet the 10-year retention time requirement.

In contrast, for a device with a 5-nm SiO_2 tunnel dielectric, the V_t shift reaches 2 V at 400 μs, for the same programming voltage. This is almost 700 times slower than that of the device with a JVD nitride. The data on retention times in Figure 10.33(b) clearly show that the device with a JVD nitride has better retention than the one with an oxide, although 8 nm of JVD nitride (EOT ~4.5 nm) will not provide for 10-year retention time in this case, but 12 nm (EOT ~7 nm) of JVD nitride will meet the 10-year retention time requirement.

The P-channel JVD nitride Flash memory device can be erased by either of two mechanisms: conventional FN tunneling or hot-hole injection (Fig. 10.34). Erasing time of 60 μs is achieved by FN tunneling (−12 V, V_{cg}). Hot-hole injection can be used for even faster erase (<10 μs).

Endurance data are shown in Figure 10.35 for BBHE programming and hot-hole erasing. The device shows little degradation up to 10^5 cycles. The JVD nitride Flash memory device can alternatively be programmed using hot holes and erased using BBHE tunneling, although the data are not shown here.

The programming characteristics for a 12-nm JVD nitride device are shown in Figure 10.36. To achieve 3-V V_t shift in 1 μs, a control gate bias of 8 V is required,

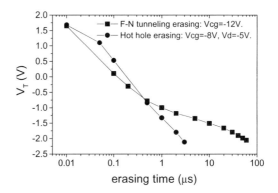

Figure 10.34. Erase characteristics for Flash memory device with 8 nm of JVD nitride tunnel dielectric. Either FN tunneling or hot-hole injection can be used to erase the device.

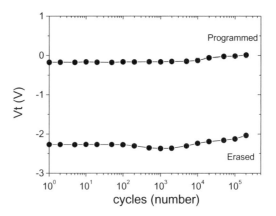

Figure 10.35. Endurance characteristics of a Flash memory device with 8 nm of JVD nitride tunnel dielectric for hot-electron programming and hot-hole erase.

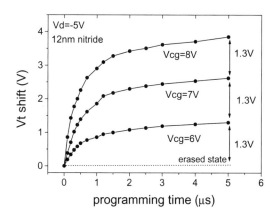

Figure 10.36. Programming characteristic of a Flash memory device with 12 nm of JVD nitride tunnel dielectric. It shows multilevel capability with low-voltage programming using BBHE.

which is 2 V more than that for 8-nm JVD nitride. This device possesses multilevel programming capability, with a uniform 1.3-V shift in V_t for each 1-V increment in V_{cg}. Thus, 2-bit storage per cell is possible for significant improvement in storage density. Similar multilevel programming was also observed for the 8-nm nitride device at correspondingly lower gate voltage. The self-convergent programming feature of BBHE is key to the multilevel storage property. It should be noted that the devices fabricated in this work are not optimized for hot-carrier injection. Reductions in programming voltage should be achievable by optimizing the dopant profiles in the channel and drain regions, as well as the JVD nitride thickness.

The lower electron barrier height of the nitride/Si interface should also significantly enhance the substrate-hot-electron (SHE) injection efficiency, as the SHE injection efficiency is an exponential function of the barrier height [55]. The same can also be said about the substrate-hot-hole (SHH) injection efficiency. This should greatly enhance the programming/erase efficiency for Flash devices that utilize the SHE and/or the SHH mechanisms for programming and/or erase [56, 57].

Figure 10.37(a) compares the calculated SHE injection probability between two N-channel devices, one with silicon nitride as the gate dielectric and the other with silicon oxide as the gate dielectric. The two devices are biased the same way, with $V_g = 3$ V, source and drain grounded, while the substrate is reverse biased with respect to the source/drain junctions. What is not shown but assumed is that excess electrons are being injected into the reverse-biased substrate from an external source (e.g., a forward-biased N^+-P junction in the vicinity of the transistor under test). The injection probability on the Y axis in Figure 10.37(a) is defined as the fraction of injected electrons that go through the gate dielectric of the transistor. The data in Figure 10.37(a) show clearly that the electron injection probability for the nitride-gate device is significantly higher than its oxide-gate counterpart. For example, at a reverse junction bias of 3 V, the nitride-gate device exhibits over 2 orders of magnitude higher injection probability.

The SHH injection probability in P-channel transistors is also substantially higher when silicon nitride is used in place of the silicon oxide as the gate dielectric, as shown in Figure 10.37(b). In fact, the hot-hole injection probability for the nitride-gate transistor is higher than 10^{-4} for a reverse-junction bias greater than 3 V, which is more than sufficient for most program/erase applications. The possibility of using

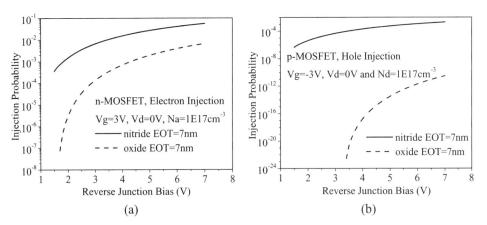

Figure 10.37. Comparison of substrate-hot-carrier (SHC) injection probability as a function of the substrate reverse bias between an oxide-gate and a nitride-gate transistor: (a) substrate-hot-electron injection in N-channel transistors and (b) substrate hot-hole injection in P-channel transistors.

hot-hole injection as a program or erase mechanism, which has not been practical with silicon oxide as the tunnel dielectric, should open new opportunities for silicon nitride as the tunnel dielectric in Flash memory devices and circuits.

10.9 RELIABILITY CONCERNS ASSOCIATED WITH HOT-HOLE INJECTION

Hole injection into or through a SiO_2 tunnel barrier has been found to cause reliability problems [58–60], and therefore the common strategy is to minimize hole injection, if possible. Thus the suggestion of using hole injection as a write or erase mechanism must be viewed with suspicion by many. Although it is a legitimate concern for SiO_2, hole injection through JVD silicon nitride does not seem to cause reliability problems. As shown in Figures 10.21 and 10.22, the threshold voltage shifts under hole injection conditions are very minor for both N- and P-channel devices, suggesting very little hole trapping in the silicon nitride. The transconductance degradation is also not very significant under hole injection conditions, suggesting that the damage to the interface due to hole injection is also not excessive. Although a lot more thorough investigation is needed to firmly establish the viability of hole injection as a write/erase mechanism, these results certainly do not suggest obvious show stoppers with hole injection.

10.10 TUNNEL DIELECTRIC FOR SONOS CELL

The special features of the JVD silicon nitride that improve the performance and reliability of floating-gate memory cells are also applicable to SONOS technology, where the use of JVD silicon nitride as the tunnel dielectric should give rise to the same advantages as those described above. In addition, one can also take advantage of the lower leakage current and higher dielectric constant of JVD silicon nitride to replace SiO_2 as the blocking layer on top of the tunnel dielectric. Thus an all-nitride dielectric stack has been proposed to succeed the conventional SONOS structure, named SNNNS [61]. As illustrated in Figure 10.38, SNNNS cells [61] may utilize CHE injection for programming and hot-hole injection for erase. Because of the substantially reduced barrier heights for both electrons and holes (Fig. 10.23), dramatically improved programming/erase (P/E) efficiency can be expected by using silicon nitride as the tunnel barrier [61].

Figure 10.39 shows the operation principle of tunneling-based P/E scheme, where a positive gate voltage programs the device by causing electron tunneling through the tunnel nitride (TN) from the Si substrate and trapped in the storage nitride (SN), while a negative gate voltage erases it by forcing the trapped electrons to tunnel back to the Si substrate; hole injection from the substrate also helps the erase operation.

Figure 10.40 demonstrates the memory effect of an SNNNS capacitor structure, where the C–V curves show a sizable memory window after applying program/erase voltages of 8V/–8V (V_g–V_{fb}, corresponding to 5.2 MV/cm of oxide field), which is close to state-of-the-art low-voltage SONOS programming voltage [62]. Charge retention test results, shown in Figure 10.41, indicate a charge loss rate of 50 mV/dec at room temperature, and 80 mV/dec at 85°C, which is comparable to the best retention performance of SONOS-type memories [63]. It should be noted that by further optimizing the nitride thicknesses and the trapping properties of the charge-storage

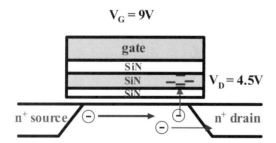

Figure 10.38. Channel-hot-electron programming of SNNNS nonvolatile memory transistor that contains a trap storage nitride sandwiched between a trap-free (TF) tunnel nitride at the bottom and another TF blocking nitride at the top.

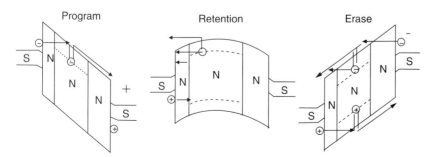

Figure 10.39. Operation principle of tunneling-based P/E scheme of a SNNNS nonvolatile memory structure.

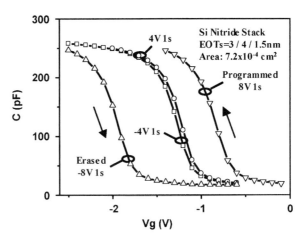

Figure 10.40. Significant memory window of an SNNNS structure is observed after 8V/−8V (V_g−V_{fb}) P/E.

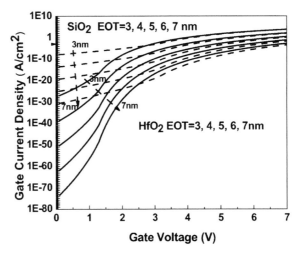

Figure 10.41. For a given EOT, the tunneling current is substantially lower for the sample with HfO_2 as the gate dielectric than the one with SiO_2 as the gate dielectric.

nitride layer, even better program/erase and charge retention characteristics could be expected.

10.11 PROSPECTS FOR HIGH-K DIELECTRICS

If silicon nitride can bring us significantly improved program/erase efficiency and longer retention time, due to its lower barrier height and larger dielectric constant than SiO_2, one might expect even more benefit with a tunnel dielectric having a still higher dielectric constant and lower barrier height. Let's examine this possibility for a representative high-K gate dielectric, HfO_2 [64, 65].

Since it has been shown that quality transistors with good transconductance and current drivability can be made with HfO_2 as the gate dielectric [64,65], what remain to be evaluated are the charge retention capability and long-term reliability issues associated with HfO_2 as a possible tunnel dielectric.

The HfO_2/Si barrier height for electrons has been reported to be approximately 1.2 eV [60], which is to compare with ~2.1 eV for silicon nitride and ~3.1 eV for silicon oxide. As a result of the reduced barrier height, the electron injection probability across the HfO_2/Si barrier is expected to be 70 to 80 times larger than that across the SiN/Si barrier, or 5000 to 6000 times larger than that across the SiO_2/Si barrier. Thus, the use of HfO_2 as the tunnel dielectric promises to improve the programming efficiency by many orders of magnitude if CHE or SHE mechanism is used.

Now, let's examine the retention aspect of the HfO_2/Si barrier. Figure 10.41 shows the theoretical tunneling $I-V$ characteristics for a set of HfO_2 films of various EOTs, ranging from 3 to 7 nm. A dielectric constant value of 20 is assumed for the HfO_2 films. Thus, in this case the physical thickness of a HfO_2 film with EOT = 3 nm is approximately 15 nm, and that with EOT = 7 nm is approximately 36 nm. One can see that, compared to SiO_2, the tunneling current is a lot lower in HfO_2 for a given EOT. Based on these results alone, one might jump to the optimistic conclusion that the use of HfO_2 may allow one to scale down the EOT of the tunnel dielectric considerably below 7 nm.

However, one must not overlook a detrimental problem for high-K dielectrics with barrier heights lower than 2 eV (which include TiO_2, Ta_2O_5, ZrO_2, HfO_2, and practically all other higher K dielectrics being studied, except for Al_2O_3, which has only a slightly higher K than Si_3N_4); that is, the very high thermionic current, which will make the retention time far shorter than 10 years. Figure 10.42 shows the calculated thermionic currents for the HfO_2/Si barrier at 27°C and 100°C, respectively, assuming a barrier height of 1.2 eV. One can see that the thermionic current is many orders of magnitude too high to meet the retention requirement even at room temperature.

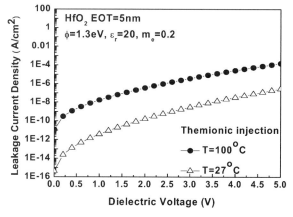

Figure 10.42. Calculated thermionic current density over HfO_2/Si barrier at 27°C and 100°C, respectively.

One effective way of reducing the thermionic current is to lower the operating temperature. For example, by cooling the device down to −50°C, one expects to reduce the thermionic current over a 1.2-eV barrier by 6 orders of magnitude, which will be low enough to satisfy the 10-year retention requirement. Thus there seems to be a possible niche application in low-temperature electronics for low-barrier tunnel dielectrics.

Besides giving rise to higher CHE injection efficiency, which is desirable for NOR Flash technology, the use of trapless silicon nitride as the tunnel dielectric is also advantageous for NAND technology that uses Fowler–Nordheim (FN) tunneling as the programming/erase mechanism, as discussed below.

Figure 10.43 compares the theoretical tunneling currents for a set of SiO_2 films and a set of Si_3N_4 films, with the latter 0.5 nm thinner for each film. One can see that, comparing 5 nm (EOT) of Si_3N_4 against 5.5 nm of SiO_2, the nitride shows higher current above 3 V and lower current below 3 V, suggesting higher FN programming efficiency and longer retention if one replaces 5.5 nm of SiO_2 by 5 nm (EOT) of Si_3N_4. The same can be said for the other three pairs.

Although the data shown in Figure 10.43 are based on a particular set of parameters for the silicon nitride (i.e., $k = 7.0$, $m = 0.5m_0$, and $\phi = 2.1\,\text{eV}$), the conclusion will be qualitatively the same for other reasonable sets of parameters.

Once again, the above holds true only if tunneling is the only conduction mechanism through the tunnel dielectric. This rules out a lot of high-K gate dielectrics that have shown to exhibit high densities of traps that give rise to trap-assisted conduction mechanism. In addition, most high-K dielectrics have low barrier heights, which give rise to intolerable thermionic currents that tend to ruin the retention, as discussed in the previous section.

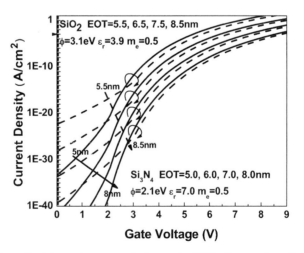

Figure 10.43. Theoretical tunneling currents for a set of SiO_2 films and a set of Si_3N_4 films that are 0.5 nm thinner for each film. One can see that, comparing 5 nm (EOT) of Si_3N_4 against 5.5 nm of SiO_2, the nitride shows higher current above 3 V and lower current below 3 V, suggesting higher FN programming efficiency and longer retention for the nitride. The same can be said for the other three pairs.

10.12 TUNNEL BARRIER ENGINEERING WITH MULTIPLE BARRIERS

The previous section shows that, the trapless silicon nitride exhibits sharper I–V than SiO_2, which is advantageous for NAND Flash technology that utilizes FN tunneling as the programming/erase mechanism. This section will show that sharper I–V may be realized by appropriate combinations of multiple barriers. Two examples will be illustrated: (1) crested barrier [66, 67] and (2) U-shaped barrier [68, 69].

10.12.1 Crested Barrier

Figures 10.44(*a*) and 10.44(*b*) serve to illustrate the advantage of the crested barrier for enhanced FN injection efficiency, as originally introduced by Likharev [66, 67]. For a single-layer tunnel barrier, as illustrated in Figure 10.44(*a*), the peak barrier height (at the left interface) is only slightly reduced with increasing applied gate voltage, due to the image force barrier lowering effect. Therefore, the tunneling current is a relatively slow function of the gate voltage, as exemplified by the two dashed curves in Figure 10.45 for SiO_2 barrier with thicknesses of 8 and 12 nm, respectively. In contrast, for a trilayer crested tunnel barrier shown in Figure 10.44(*b*), where the middle layer has the highest barrier, the applied gate voltage is able to pull the middle barrier *below* the left barrier, causing a much reduced peak barrier height and sharp rise of the tunneling current for high voltages, as shown by the solid curve in Figure 10.45, which represents the FN tunneling characteristic for a SiN/AlN/SiN trilayer crested barrier with 4 nm/5 nm/4 nm thickness combination.

Comparing the solid curve with the dashed curves in Figure 10.45, it is obvious that the solid curve has a much steeper slope. For a Flash cell that uses FN mechanism for programming/erase, the steeper slope translates to faster programming speed *and* longer retention (because the retention time is related primarily to leakage current at low fields).

As mentioned in the beginning of this section, the concept of the trilayer crested tunnel barrier was originally proposed by Likharev [66, 67], but convincing experimental demonstration is still lacking, primarily due to the difficulty of producing

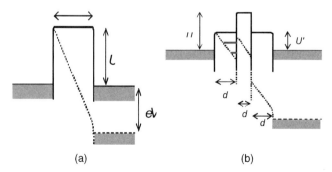

Figure 10.44. Conduction band edge profiles of two kinds of tunnel barriers (solid lines) and their deformation at a high applied voltage (dashed lines): (a) single-layer, uniform tunnel barrier and (b) trilayer crested barrier where the two thick horizontal lines show the position of electron subbands formed in the triangular quantum well at the interface between the first and the second layer (schematically). (Figures supplied by K. K. Likharev.)

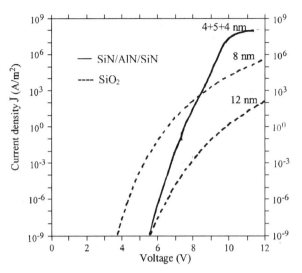

Figure 10.45. Fowler–Nordheim tunneling current density for a trilayer crested barrier (solid curve) with parameters corresponding to the N^+-Si/Si_3N_4/AlN/Si_3N_4/N^+-Si system and total barrier thickness $d' + d + d' = 4 + 5 + 4 = 13$ nm. Results for two SiO_2 barriers are also shown for comparison (dashed curves). (Figure supplied by K. K. Likharev.)

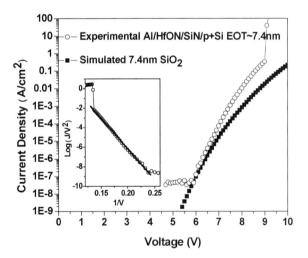

Figure 10.46. J-V curves of Al/HfON/Si_3N_4/P^+-Si stack (one-side crested barrier) and theoretical SiO_2. Inset shows the FN fitting of the J-V of crested barrier.

crested tunnel dielectrics that are trap free, both in each of the layers and at the interfaces.

Recently, our lab has developed nearly trap-free JVD HfON, as demonstrated by the lack of hysteresis in the C–V curves. The JVD HfON has a barrier height of ~1.3 eV on Si, which can be combined with Si_3N_4 to form a crested barrier.

Figure 10.46 shows the J–V characteristic of a crested barrier, consisting of HfON/Si_3N_4 made of JVD, where it exhibits sharper J–V than the SiO_2 control (taken from theoretically calculated tunneling current). One can see that it takes only 3 V for the crested barrier to achieve an 8-order increase in the current density but 4.2 V for SiO_2 with the same EOT. As shown in the inset of Figure 10.46, this J–V curve of the crested barrier fits the FN tunneling model over 7 orders of magnitude, which can only be realized with a nearly trapless $HfON/Si_3N_4$ stack as well as their interface.

10.12.2 U-Shaped Barrier

Another promising tunnel barrier engineering approach is the U-shaped barrier [68, 69], consisting of a relatively thick low barrier dielectric in the middle, sandwiched by two thinner, higher barrier dielectrics. It is also desirable that the middle barrier have higher dielectric constant than the two outer layers.

Figure 10.47 shows the energy band diagrams of an idealized symmetric U-shaped barrier. At low fields, the leakage current is low because the tunneling current is effectively damped by the thick high-K layer, while the thermionic emission current can be suppressed by the high barrier layers. At high gate voltages, on the other hand, FN tunneling current through the thin barrier is unimpeded by the other two barriers, causing a sharper I–V characteristic than it's SiO_2 counterpart of the same EOT. Again, it should be emphasized that the ability to produce trapless dielectric stacks is the key to success of the U-shaped barrier. Recent theoretical simulation and preliminary experimental results based on $SiO_2/Si_3N_4/SiO_2$ stacks have confirmed that it is a very promising approach [70].

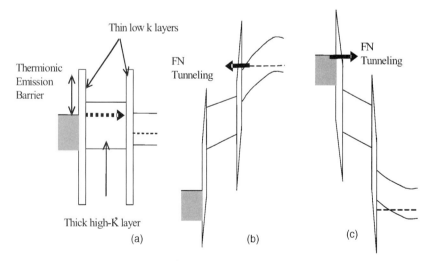

Figure 10.47. Energy band diagrams of an idealized symmetric U-shaped barrier. (a) At zero gate bias, which indicates that at a low field, the tunneling current is effectively damped by the thick high-K layer, whereas the thin high-barrier layers suppress themionic emission. (b) At a high negative gate voltage, or (c) at a high positive gate voltage, FN tunneling current through the thin barrier is unimpeded by the other two barriers.

10.13 SUMMARY

The tunnel dielectrics in commercially available Flash memory devices are all based on SiO_2 or its modifications, which have served the Flash technology very well for more than 2 decades. However, the SiO_2-based tunnel dielectric cannot be scaled below 7 to 8 nm in thickness because of problems with stress-induced leakage current that cannot meet the 10-year retention requirement.

An alternative tunnel dielectric, silicon nitride, has been examined, and the following attractive features have been identified: (1) much lower leakage currents at low fields compared to SiO_2 of the same EOT; (2) much reduced SILC compared to the case for SiO_2 as the tunnel dielectric; (3) sharper $I-V$ characteristics compared to SiO_2 of the same EOT; (4) lower SiN/Si barrier height (~2.1 eV, compared to ~3.2 eV for SiO_2/Si) gives rise to a much higher electron injection probability, or a much faster programming speed, for Flash devices that utilize the CHE or the SHE mechanism for write/erase operations; (5) relatively low hole barrier (~1.9 eV) makes hot-hole injection a possible write/erase mechanism. The first two features are desirable for both NOR and NAND Flash technology; the third is attractive for NAND Flash technology that uses FN tunneling as the programming mechanism; while the fourth is advantageous for NOR Flash technology that uses CHE or SHE for programming/erase operation; the last one is potentially useful for single-transistor electrically erasable programmable ROM (EEPROM) technology.

The potential of materials with even higher dielectric constants has also been examined. It has been concluded that a major problem that shortens the retention time is the high thermionic current for these high-K dielectrics because they all tend to have low barrier heights on Si. A potential niche application for the high-K tunnel dielectric is in low-temperature electronics, where the thermionic current is greatly suppressed and a 10-year retention time is possible.

Tunnel barrier engineering based on trilayer crested barrier and trilayer U-shaped barrier has been discussed. Provided tunneling is the only conduction mechanism through these barriers, sharper $I-V$ characteristics than SiO_2 of the same EOT can be expected, which are attractive for NAND Flash technology that utilizes FN tunneling mechanism for program/erase operation. The key to success of either approach is the ability to produce tunnel dielectric stacks that are practically trap free.

REFERENCES

1. K. Yoshikawa, S. Mori, E. Sakagami, N. Arai, Y. Kaneko, and Y. Ohshima, "Flash EEPROM Cell Scaling Based on Tunnel Oxide Thinning Limitations," *Symp. VLSI Technol. Dig. Tech. Papers*, pp. 79–80, 1991.
2. *International Technology Roadmap for Semiconductors*, Semiconductor Industry Association, 1997–2005 Editions, available online at www.itrs.net/reports.html.
3. K. Yoshikawa, "Flash Memory Technology and Reliability Issues Associated with SiO_2/Si Interface," Paper 8.1, Extended Abstract, paper presented at the 28th IEEE Semiconductor Interface Specialist Conference, Dec. 1997.
4. J. Maserjian and N. Zamani, "Behavior of the Si/SiO_2 Interface Observed by Fowler–Nordheim Tunneling," *J. Appl. Phys.*, Vol. 53, p. 559, 1982.

5. R. Rofan and C. Hu, "Stress-Induced Oxide Leakage," *IEEE Electron Device Lett.*, Vol. 12, p. 632, 1991.
6. D. J. Dumin and J. R. Maddux, "Correlation of Stress-Induced Leakage Current in Thin Oxides with Trap Generation Inside the Oxides," *IEEE Trans. Electron Devices*, Vol. ED-40, p. 986, 1993.
7. E. F. Runnion, S. M. Gladstone, IV, R. S. Scott, Jr., D. J. Dumin, L. Lie, and J. C. Mitros, "Thickness Dependence of Stress-Induced Leakage Current in Silicon Oxide," *IEEE Trans. Electron Devices*, Vol. 44, No. 6, pp. 993–1001, June 1997.
8. D. A. Baglee and M. C. Smayling, "The Effects of Write/Erase Cycling on Data Loss in EEPROMs," *IEDM Tech. Dig.*, pp. 624–626, 1985.
9. J. DeBlauwe, J. Van Houdt, D. Wellekens, G. Groeseneken, and H. Maes, "SILC-Related Effects in Flash E2PROM's—Part II: Prediction of Steady-State," *IEEE Trans. Electron Devices*, Vol. 45, No. 8, pp. 1751–1760, Aug. 1998.
10. *The National Technology Roadmap for Semiconductors*, Table 14, Semiconductor Industry Association, 1997 Edition, available online at www.itrs.net/reports.html.
11. T. P. Ma, "Making Silicon Nitride Film a Viable Gate Dielectric," *IEEE Trans. Electron Devices*, Vol. ED-45, No. 3, p. 680, 1998.
12. M. Khare, X. W. Wang, T. P. Ma, T. Tamagawa, and B. L. Halpern, "Highly Robust Ultra-Thin Gate Dielectric for Giga Scale Technology," Technical Digest, *Symp. VLSI Technol. Dig. Tech. Papers*, pp. 218–219, 1998.
13. H. Watanabe, S. Aritome, G. Hemink, T. Maruyama, and R. Shirota, "Scaling of Tunnel Oxide Thickness for Flash EEPROMs Realizing Stress-Induced Leakage current Reduction," *Symp. VLSI Technol. Dig. Tech. Papers*, pp. 47–48, 1994.
14. J. Kim and S. T. Ahn, "Improvement of the Tunnel Oxide Quality by a Low Thermal Budget Dual Oxidation for Flash Memories," *IEEE Electron Device Lett.*, Vol. 18, No. 8, pp. 385–387, Aug. 1997.
15. G. Ghidini, M. Tosi, and C. Clementi, "Feasibility of Steam Tunnel Oxide for Advanced Non-Volatile Memories," *Solid-State Electron.*, Vol. 41, No. 7, pp. 975–979, July 1997.
16. S. L. Wu, D. M. Chiao, C. L. Lee, and T. F. Lei, "Characterization of Thin Textured Tunnel Oxide Prepared by Thermal Oxidation of Thin Polysilicon Film on Silicon," *IEEE Trans. Electron Devices*, Vol. 43, No. 2, pp. 287–294, Feb. 1996.
17. K. M. Chang, C. H. Li, B. S. Sheih, J. Y. Yang, S. W. Wang, and T. H. Yeh, "New Simple and Reliable Method to Form a Textured Si Surface for the Fabrication of a Tunnel Oxide Film," *IEEE Electron Device Lett.*, Vol. 19, No. 5, pp. 145–147, May 1998.
18. H. Nozawa, N. Matsukawa, and S. Morita, "EEPROM Cell Using a Low Barrier Height Tunnel Oxide," *IEEE Trans. Electron Devices*, Vol. 33, No. 2, pp. 275–281, Feb. 1986.
19. C. S. Jeng, T. L. Chiu, B. Joshi, and J. Hu, "Properties of Thin Oxynitride Films Used as Floating-Gate Tunneling Dielectrics," *IEDM Tech. Dig.*, pp. 811–812, 1982.
20. H. Fukuda, M. Yasuda, T. Iwabuchi, and S. Ohno, "Novel N_2O-Oxynitridation Technology for Forming Highly Reliable EEPROM Tunnel Oxide Films," *IEEE Electron Device Lett.*, Vol. 12, No. 11, pp. 587–589, 1991.
21. T. Arakawa and H. Fukada, "Effect of NH_3 Nitridation on Time-Dependent Dielectric Breakdown Characteristics of Heavily Oxynitrided Tunnel Oxide Films," *Electron. Lett.*, Vol. 30, No. 4, pp. 361–362, Feb. 1994.
22. Y. Okada, P. J. Tobin, K. G. Reid, R. I. Hegde, B. Maiti, and S. A. Ajuria, "Furnace Grown Gate Oxynitride Using Nitric Oxide," *IEEE Trans. Electron Devices*, Vol. 41, No. 9, pp. 1608–1613, Sept. 1994.
23. J. DeBlauwe, D. Wellekens, J. Van Houdt, R. Degraeve, L. Haspeslagh, G. Groeseneken, and H. E. Maes, "Impact of Tunnel Oxide Nitridation on Endurance and Read-Disturb Charac-

teristics of Flash E2PROM Devices," *Microelectron. Eng.*, Vol. 36, Nos. 1–4, pp. 301–304, June 1997.

24. D. J. DiMaria, K. M. Demeyer, and D. W. Dong, "Electrically-Alterable Memory Using a Dual Electron Injection Structure," *IEEE Electron Device Lett.*, Vol. 9, pp. 179–181, Sept. 1980.

25. D. J. DiMaria, D. W. Dong, F. L. Presavento, C. Lam, and S. D. Brorson, "Enhanced Conduction and Minimized Charge Trapping in Electrically Alterable Read-Only Memories Using Off-Stoichiometric Silicon Dioxide Films," *J. Appl. Phys.*, Vol. 55, No. 8, pp. 3000–3019, April 1984.

26. L. Dori, A. Acovic, D. J. DiMaria, and C. H. Hsu, "Optimized Silicon Rich Oxide (SRO) Deposition Process for 5 V-Only Flash EEPROM Applications," *IEEE Electron Device Lett.*, Vol. 14, No. 6, pp. 283–285, 1993.

27. T. Ito, S. Hijiya, T. Nozaki, H. Arakawa, H. Ishikawa, and M. Shinoda, "Low-Voltage Alterable EAROM Cells with Nitride-Barrier Avalanche-Injection MIS (NAMIS)," *IEEE Trans. Electron Devices*, Vol. 26, No. 6, pp. 906–913, June 1979.

28. S. Hijiya, I. Takashi, T. Nakamura, H. Ishikawa, and H. Arakawa, "A Nitride-Barrier Avalanche-Injection EAROM," *IEEE J. Solid-State Circuits*, Vol. 17, No. 5, pp. 852–856, Oct. 1982.

29. T. Ito, T. Nozaki, H. Arakawa, M. Shinoda, and Y. Fukukawa, "Very Thin Silicon Nitride Films Grown by Direct Thermal Reaction with Nitrogen," *J. Electrochem. Soc.*, Vol. 125, No. 3, pp. 448–452, 1978.

30. T. Ito, T. Nozaki, H. Arakawa, and M. Shinoda, "Thermally Grown Silicon Nitride Films for High-Performance MNS Devices," *Appl. Phys. Lett.*, Vol. 32, No. 5, pp. 330–331, 1978.

31. T. Ito, I. Kato, T. Nozaki, T. Nakamura, and H. Ishikawa, "Plasma-Enhanced Thermal Nitridation of Silicon," *Appl. Phys. Lett.*, Vol. 38, No. 5, pp. 370–372, 1981.

32. B. E. Deal, E. L. MacKenna, and P. L. Castro, "Characteristics of Fast Surface States Associated with SiO_2-Si and Si_3N_4-SiO_2-Si Structures," *J. Electrochem. Soc.*, Vol. 116, pp. 997–1005, 1969.

33. A. S. Ginovker, V. A. Gristsenko, and S. P. Sinitsa, "Two-Band Conduction of Amorphous Silicon Nitride," *Phys. Stat. Solid A*, Vol. 26, No. 2, pp. 489–495, 1974.

34. V. J. Kapoor, R. S. Bailey, and H. J. Stein, "Hydrogen-Related Memory Traps in Thin Silicon Nitride Films," *J. Vac. Sci. Technol.*, Vol. 1, No. 2, pp. 600–603, 1983.

35. S. Fujita, T. Ohishi, T. Toyoshima, and A. Sasaki, "Electrical Properties of Silicon Nitride Films Plasma-Deposited from SiF_4, N_2, and H_2 Source Gases," *J. Appl. Phys.*, Vol. 57, No. 2, pp. 426–431, 1985.

36. M. Maeda and H. Nakamura, "Hydrogen Bonding Configurations in Silicon Nitride Films Prepared by Plasma-Enhanced Deposition," *J. Appl. Phys.*, Vol. 58, No. 1, pp. 484–489, 1985.

37. K. Alloert, A. Van Calster, H. Loos, and A. Lequesne, "A Comparison between Silicon Nitride Films Made by PCVD of N_2-SiH_4/Ar and N_2-SiH_4/He," *J. Electrochem. Soc.*, Vol. 132, No. 7, pp. 1763–1766, 1985.

38. W. R. Knolle and J. W. Osenbach, "The Structure of Plasma-Deposited Silicon Nitride Films Determined by Infrared Spectroscopy," *J. Appl. Phys.*, Vol. 58, No. 3, pp. 1248–1254, 1985.

39. D. V. Tsu, G. Lucovsky, and M. J. Mantini, "Local Atomic Structure in Thin Films of Silicon Nitride and Silicon Diimide Produced by Remote Plasma-Enhanced Chemical-Vapor Deposition," *Phys. Rev. B*, Vol. 33, No. 10, pp. 7069–7076, 1986.

40. T. J. Cotler and J. Chapple-Sokol, "High Quality Plasma-Enhanced Chemical Vapor Deposited Silicon Nitride Films," *J. Electrochem. Soc.*, Vol. 140, No. 7, pp. 2071–2075, 1993.

41. M. K. Mazumder, K. Kobayashi, J. Mitsuhashi, and H. Koyama, "Stress-Induced Current in Nitride and Oxidized Nitride Thin Films," *IEEE Trans. Electron Devices*, Vol. 41, No. 12, pp. 2417–2422, 1994.
42. S. Song, H. Luan, M. Gardner, J. Fulford, M. Allen, and D. Kwong, "Ultra Thin (<20 Å) CVD Si_3N_4 Gate Dielectric for Deep-Sub-Micron CMOS Devices," Paper No. 14.1, *IEDM Tech. Dig.*, Dec. 1998.
43. Z. Lu, M. J. Williams, P. F. Santos-Filho, and G. Lucovsky, "Fourier Transform Infrared Study of Rapid Thermal Annealing of a-Si:N:H(D) Films Prepared by Remote Plasma-Enhanced Chemical Vapor Deposition," *J. Vac. Sci. Technol. A*, Vol. 13, No. 3, pp. 607–613, 1995.
44. X. W. Wang, Y. Shi, T. P. Ma, G. J. Cui, T. Tamagawa, J. Golz, B. Halpern, and J. Schmitt, "Extending Gate Dielectric Scaling Limit by Use of Nitride or Oxynitride," *Symp. VLSI Technol. Dig. Tech. Papers*, pp. 109–110, 1995.
45. M. Khare, X. W. Wang, T. P. Ma, G. J. Cui, T. Tamagawa, B. Halpern, and J. Schmitt, "Highly Robust Ultra-Thin Gate Dielectric for Giga Scale Technology," *Symp. VLSI Technol. Dig. Tech. Papers*, pp. 218–219, 1998.
46. X. Guo and T. P. Ma, "Tunneling Leakage Current in Oxynitride: Dependence on Oxygen/Nitrogen Content," *IEEE Electron Device Lett.*, Vol. 19, No. 6, pp. 207–209, June 1998.
47. D. M. Fleetwood, "Border Traps in MOS Devices," *IEEE Trans. Nucleic Sci.*, Vol. 39, pp. 269–271, 1992.
48. M. Khare and T. P. Ma, "Transconductance in Nitride-Gate or Oxynitride-Gate Transistors," *IEEE Electron Device Lett.*, Vol. 20, No. 1, pp. 57–59, Jan. 1999.
49. H. H. Tseng et al, "Application of JVD Nitride Gate Dielectric to A 0.35 Micron CMOS Process for Reduction of Gate Leakage Current AND Boron Penetration," Paper No. 26.6.1, *IEDM Tech. Dig.*, pp. 647–650, Dec. 1997.
50. Karamcheti, V. H. C. Watt, T. Y. Luo, D. Brady, F. Shaapur, L. Vishnubhotla, G. Gale, H. R. Huff, M. D. Jackson, K. Torres, A. Diebold, J. Guan, M. C. Gilmer, G. A. Brown, G. Bersuker, P. Zeitzoff, X. Guo, X. W. Wang, and T. P. Ma, "Electrical and Physical Characterization of Ultrathin Silicon Oxynitride Gate Dielectric Films Formed by the Jet Vapor Deposition Technique," Paper No. T5.7, presented at the MRS Fall Meeting, Hynes Convention Center, Boston, Nov. 29–Dec. 3, 1999.
51. E. F. Runnion, S. M. Gladstone, IV, R. S. Scott, and D. J. Dumin, "Limitations on Oxide Thickness in FLASH EEPROM Applications," paper presented at the International Reliability Phys. Symp., p. 93, 1996.
52. K. Naruke, S. Taguchi, and M. Wada, "Stress-Induced Leakage Current Limiting to Scale Down EEPROM Tunnel Oxide Thickness," *IEDM Tech. Dig.*, pp. 424–427, 1988.
53. Z. Liu and T. P. Ma, "A New Programming Technique for Flash Memory Devices," Paper No. G21, *Proc. Tech. Pap. Int. Symp. VLSI-TSA*, pp. 191–194, 1999.
54. T. Ohnakado, H. Onoda, O. Sakamoto, K. Hayashi, N. Nishioka, H. Takada, K. Sugahara, N. Ajika, and S. Satoh, "Device Characteristics of 0.35 μm P-Channel DINOR Flash Memory Using Band-to-Band Tunneling-Induced Hot Electron (BBHE) Programming," *IEEE Trans. Electron Devices*, Vol. 46, No. 1, pp. 1866–1871, 1999.
55. T. H. Ning, C. M. Osburn, and H. H. Yu, "Emission Probability of Hot Electrons from Silicon into Silicon Dioxide," *J. Appl. Phys.*, Vol. 48, pp. 286–293, Jan. 1977.
56. B. Eitan, J. L. McCreary, D. Amrany, and J. Shappir, "Substrate Hot-Electron Injection EPROM," *IEEE Trans. Electron Devices*, Vol. 31, No. 7, pp. 934–942, 1984.
57. N. Tsuji, N. Ajika, K. Yuzuriha, Y. Kunori, M. Hatanaka, and H. Miyoshi, "New Erase Scheme for DINOR Flash Memory Enhancing Erase/Write Cycling Endurance Characteristics," *IEDM Tech. Dig.*, pp. 53–56, Dec. 1994.
58. K. Tamer San, C. Kaya, and T. P. Ma, "Effects of Erase Source Bias on Flash EPROM Device Reliability," *IEEE Trans. Electron Devices*, Vol. 42, No. 1, pp. 150–159, Jan. 1995.

59. T. Ong, A. Fazio, N. Mielke, S. Pan, N. Righos, G. Atwood, and S. Lai, "Erratic Erase in ETOX Flash Memory Array," *VLSI Technol. Symp. Tech. Dig.*, pp. 83–84, May. 1993.
60. Brand, K. Wu, S. Pan, and D. Chin, "Novel Read Disturb Failure Mechanism Induced by Flash Cycling," *Proc. Int. Reliability Phys. Symp.*, p. 127, March 1993.
61. F. Yeh, Y. Liu, X. Wang, and T. Ma, "SONOS-Type Non-Volatile Memory with All Silicon Nitride Dielectric," Paper No. 5.4, presented at IEEE/SISC, Washington, DC, Dec. 1–3, 2005.
62. B. Eitan, P. Pavan, I. Bloom, E. Aloni, A. Frommer, and D. Finzi, "NROM: A Novel Localized Trapping, 2-Bit Nonvolatile Memory Cell," *IEEE Electron Device Lett.*, Vol. 21, No. 11, pp. 543–545, Nov. 2000.
63. M. She, H. Takeuchi, and Tsu-Jae King, "Silicon-nitride as a tunnel dielectric for improved SONOS-type flash memory," *IEEE Electron Device Lett.*, Vol. 24, pp. 309–311, 2003.
64. G. D. Wilk, R. M. Wallace, and J. M. Anthony, "High-k Gate Dielectrics: Current Status and Materials Properties Considerations," *J. Appl. Phys.*, Vol. 89, No. 10, pp. 5243–5275, May 2001.
65. W. Zhu, T. P. Ma, T. Tamagawa, Y. Di, and J. Kim, "Current Transport in Metal/Hafnium Oxide/Silicon Structure," *IEEE Electron Device Lett.*, Vol. 23, No. 2, pp. 97–99, Feb. 2002.
66. K. K. Likharev, "Layered Tunnel Barriers for Nonvolatile Memory Devices," *Appl. Phys. Lett.*, Vol. 73, No. 15, pp. 2137–2139, Oct. 1998.
67. N. Korotkov and K. K. Likharev, "Resonant Fowler-Nordheim Tunneling and Its Possible Applications," *IEDM Dig. Tech. Papers*, pp. 223–226, Dec. 1999.
68. M. Specht, M. Städele, and F. Hofmann, "Simulation of High-K Tunnel Barriers for Non-volatile Floating Gate Memories," Proceedings of 32th European Solid-State Device Research Conference, pp. 599–602, September 2002.
69. Govoreanu, P. Blomme, M. Rosmeulen, J. Van Houdt, and K. De Meyer, "VARIOT: A Novel Multilayer Tunnel Barrier Concept for Low-Voltage Nonvolatile Memory Devices," *IEEE Electron Device Lett.*, Vol. 24, No. 2, pp. 99–101, Feb. 2003.
70. J.-Fei Zheng, private communications.

11

FLASH MEMORY RELIABILITY

Jian Justin Chen, Neal R. Mielke, and Chenming Calvin Hu

11.1 INTRODUCTION

Flash integrated circuits (ICs) pose a more interesting and difficult reliability challenge than do most other IC devices. At one level, perhaps, there is little about Flash reliability that is entirely unique to Flash technology. Flash ICs are composed of the same basic materials and processing steps as other types of ICs (silicon, thermal, and deposited oxides, AlCu metallization, and so forth), and the same physical degradation mechanisms apply. One may speak of oxide breakdown and hot-electron damage, of failure-in-time (FIT) rates and bathtub curves. But Flash ICs are nevertheless significantly different from other kinds of ICs, and these differences create reliability characteristics that are unique and challenging.

The most obvious difference, of course, is the Flash cell itself. The Flash cell has *unique functionality*. It must be able to program, erase, and retain charge on its floating gate. The Flash cell is also unique for subjecting the device to much *higher electric fields* than usual due to its reliance on oxide tunneling and hot-electron injection for the normal operation of the cell. These are the mechanisms that cause reliability *failures* in other IC devices. Reliability concerns and the fact that these mechanisms depend more on, for example, the 3.2-eV Si–SiO$_2$ barrier than on channel length or width have impeded the scaling of certain Flash cell dimensions and voltages. For example, Flash tunnel-oxides have remained about 100 Å thick during a time period when logic gate oxides have scaled from about 200 to 30 Å, nor has the voltage applied across this dielectric scaled substantially. These high-voltage and limited-scaling characteristics carry over from the cell to the peripheral circuits. Flash peripheral circuits must control high voltages of 10 V or more in lithography generations when logic ICs often operate at 2 V or less. High-voltage

---

Nonvolatile Memory Technologies with Emphasis on Flash. Edited by J. E. Brewer and M. Gill
Copyright © 2008 the Institute of Electrical and Electronics Engineers, Inc.

transistors with graded junctions and thick gate oxides must be built into the process technology. To achieve high-speed performance, the process technology may need to support separate low-voltage, high-speed transistors, which require separately processed gate oxides and implants.

A final observation about the Flash cell is that its operation is *distinctively analog*. Most ICs rely on rail-to-rail logic, the dynamic random-access memory (DRAM) cell being a notable exception. A Flash cell data stores data in the form of a changeable threshold voltage, which is an analog function of the charge stored on the floating gate. These analog characteristics carry over into the peripheral circuits. Fundamentally, a Flash memory product is a mixed-signal device. The digital information is stored in the analog mode of threshold voltage distribution, and the analog information is later read and translated to digital information. The state machine and control circuits are digital, but the rest of the circuits—such as the charge pumps, sense amplifiers, and voltage generation and regulation circuits—are analog.

The analog nature of the Flash memory chip is nowhere greater than in a state-of-the-art multibit/cell Flash memory. In such a multibit or multilevel cell (MLC), more than 1 bit of information is stored in one physical cell by precise placement and readout of the threshold voltages. In such chips, the precise control and uniformity of different voltages and sense amps can have major impacts on the product reliability.

All of these differences have implications to reliability. Reliability engineers must characterize and model the unique functionality of the Flash cell itself, with particular attention to analog measurements of V_t distributions, program/erase times, and so forth. A loss (or gain) of 10,000 electrons from a floating gate can cause data loss. If 10,000 electrons are not to leak from the floating gate in 10 years, the average charge loss rate must be less than one electron every 10 h. No other kind of integrated circuit is sensitive to such low levels—less than 10^{-23} A—of dielectric leakage. Avoiding unwanted charge loss and gain during storage as well as during write, erase, and read operations is an important goal of a Flash reliability program. Familiar degradation mechanisms such as hot-electron damage and dielectric breakdown can also occur in the Flash cell, but they operate at a different voltage regime than is common with other ICs, and their effects on the functions of the Flash cell are in turn unique, affecting such characteristics as program and erase speed. In the peripheral circuitry, the reliability engineer must also consider the unusually high voltages and analog requirements of the circuitry.

It is helpful to separately consider the Flash memory reliability in two different modes of operation: low-voltage operation (read or powered-down mode), and high-voltage operation (program/erase). Low-voltage operation is similar to that of a typical IC, whereas high-voltage operation is unique. Most voltage-accelerated failure mechanisms must be characterized in high-voltage operation. Ironically, the apparent disadvantage of needing to support high voltages becomes a reliability *advantage* during low-voltage operation, when the high-voltage Flash process has ample margin against failure. The failure rate in lifetest, the catch-all mainstay of the reliability characterization of many kinds of ICs, may be almost too small to measure for Flash memories.

In general, modern Flash memory products are extremely reliable and have endurance often far exceeding the requirements of their actual applications. For example, for code-storage Flash memory applications, in which the memory is used to store program code in a system such as the motherboard, disk drive, or a cellular phone, the memory content is written and erased no more than a few hundred times. Yet most modern Flash memory products from the major manufacturers guarantee 100,000 to 1 million times endurance. For consumer mass-storage applications, such as the memory cards used in digital cameras and audio players, a user rarely erases and reprograms the card more than a thousand times. For example, for a 64-MB (megabyte) memory card, it can store 60 high-resolution photos, or 60 min of near-CD quality music in an MP3 file, cycling 1000 times translates to 60,000 high-resolution photos and 60,000 minutes of music. In real use, the card will outlive the host or the actual appliance. In fact, if the user continuously erases and reprograms the card day and night, it would take months to complete such task. (For example, for a 64-MB card, assume a write/erase speed of 1 MB/s, it will take 64 s to complete one program and erase cycle on the whole card, or 1350 cycles/day. In order to get to 300,000 cycles, it will take 222 days to finish.).

In this chapter, we will first discuss the unique reliability concerns of the Flash cell itself. Next, we will describe the reliability considerations for the peripheral circuits in a Flash memory, as well as design and system methods for improving reliability. Finally, we will describe how a Flash product is tested and qualified.

11.2 CYCLING-INDUCED DEGRADATIONS IN FLASH MEMORIES

11.2.1 Overview of Cycling-Induced Degradations

Currently, there are mainly three methods to program a Flash memory: channel hot-electron injection (CHE), Fowler–Nordheim (FN) tunneling from the edge of the source or drain, and FN tunneling from the channel region. There are also three methods to erase a Flash memory: FN tunneling through the channel region, FN tunneling through the source/drain junction edge, and FN tunneling to a floating gate through poly-to-poly. The carriers can be injected into the floating gate either over an energy barrier (hot-carrier injection) or through a barrier (tunneling). FN tunneling requires high electric field in the oxide, and the high electric field stresses the oxide, creates traps, and results in stress-induced leakage current (SILC) as discussed in a later section. Channel hot-electron program involves hot carriers, and this can also degrade the oxide and results in electron trapping, interface-state generation, and reduced program speed. Erase by FN tunneling through the junction edge involves biasing the device in such a way that band-to-band tunneling is present and potentially can generate hot-hole injection that can severely degrade the oxide. In this section, we will use the ETOX memory cell as an example to discuss the mechanisms of the hot-carrier-induced degradation and the methods to prevent the degradation and to improve the memory device hot-carrier reliability. A general overview of the cycling-induced window closing, transconductance degradation, and SILC are described here.

Figure 11.1(a) shows typical Flash memory P/E (program/erase) cycling characteristics with fixed program and erase pulses, with no verify operation. The cell is programmed with a single 2-μs program pulse at $V_d = 5.5$ V, $V_g = 10$ V, and $V_s = V_{sub} = 0$ V, and erased with a single 10-ms negative gate erase pulse at $V_s = 5$ V, $V_g = -10$ V. The program and erase threshold voltage window closing is caused by different mechanisms. The program threshold V_{tp} closing is due to electron trapping

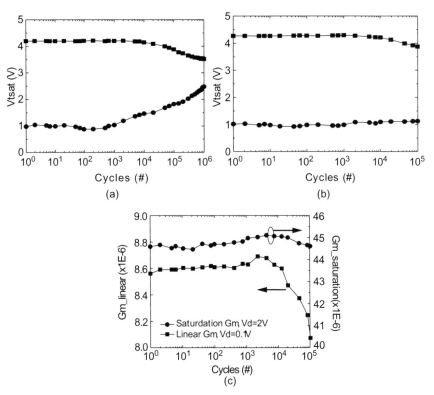

Figure 11.1. Typical stacked-gate Flash memory P/E cycling characteristics with fixed program and erase pulses. (Figures 1, 2, and 4 of [1].)

in the oxide near the drain, which suppresses the hot-electron injection (HEI) efficiency. The erase threshold V_{te} closing is due to electron trapping in the source–gate overlap region caused by FN tunneling, and the transconductance degradation is due to hot-carrier damage during programming. Repeated FN tunneling cycles at the source side causes electron trapping over the source–gate overlap region where the tunneling occurs. This reduces the tunneling current and therefore causes the V_{te} window closing. Transconductance degradation reduces the channel read current and causes the erased threshold voltage window closing.

In actual Flash memory products, program-verify and erase-verify operations are always performed. If after each program and erase pulse, correct margins are not reached, additional program and erase pulses are applied until a sufficient margin is obtained. Figure 11.1(b) shows the result of a Flash memory P/E endurance cycling test with the same conditions as in (Figure 11.1(a)), but with a variable number of program and erase pulses. The program V_{tp} and erase V_{te} window is maintained, and the degradation is reflected in the increased number of program and erase pulses required versus cycling. The memory cell linear transconductance ($V_d = 0.1$ V) and saturation transconductance ($V_d = 1.5$ V, similar to read condition) are measured as a function of cycling as shown in Figure 11.1(c). The more severe linear G_m degradation compared with the saturation G_m degradation suggests that the interface traps generated during programming are near the drain junction edge.

When $V_d = 0.1\,\text{V}$, they affect the linear transconductance easily. However, during read operation with $V_d = 1.5\,\text{V}$, much of the damage is shielded by the drain depletion region, and little saturation G_m degradation is shown.

11.2.2 Channel Hot-Electron Programming-Induced Oxide Degradation

11.2.2.1 Gate Current in the CHE Program Mode. The channel hot electron (CHE) has been widely used for programming erasable programmable read-only memory (EPROMs) and several types of Flash memories. Hot carrier means that carriers are not in equilibrium with the lattice. We will use the stacked-gate memory transistor as the principle example for the discussion in this section.

The typical CHE programming condition involves biasing the transistor in strong inversion with the control gate $V_{cg} \approx 10\,\text{V}$, the drain voltage $V_d \approx 5\,\text{V}$, and the source and the substrate grounded. Because $V_{fg} \geq V_d$, the direction of the oxide field near the drain favors electron injection from the channel to the floating gate; V_{fg} is the floating-gate voltage determined by V_{cg} and the charge stored on the floating gate. Channel hot-electron injection results from the presence of the "lucky electrons," which have gained sufficient energy to surmount the Si–SiO$_2$ barrier before suffering an energy-losing collision in the channel. The electric field responsible for accelerating the carriers is the lateral electric field. The barrier against electron injection into the oxide is 3.1 eV, and the barrier against hole injection into the oxide is 4.9 eV, which is higher than that for electrons. If the carriers acquire sufficient energy from the lateral electric field and become sufficiently "hot," they can surmount this barrier, jump into the oxide conduction band, and drift to the floating gate.

Figure 11.2 shows measured gate current characteristics for a device with $L_{eff} = 0.8\,\mu\text{m}$ and $T_{OX} = 10\,\text{nm}$ versus gate voltage V_g with the drain voltage V_d as the parameter. The channel hot-carrier gate current initially increases as V_g is increased,

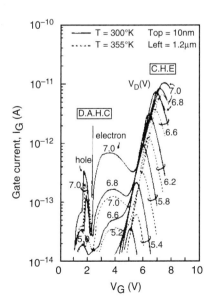

Figure 11.2. Measured gate current characteristics of a device with $L_{eff} = 0.8\,\mu\text{m}$ and $T_{ox} = 10\,\text{nm}$ for varied V_g and V_d. Two peaks in the $V_g < V_d$ region are observed besides the peak near $V_g \approx V_d$ resulting from CHE injection. These peaks corresponding to avalanche hot-hole and electron injection. At the balance point of positive hole current and negative electron current, the gate voltage is V_g^* and the gate current equals to zero. (Figure 1 of [2].)

due to the favorable vertical electric field at the drain and the increase in the number of carriers in the channel. The gate current peaks at the point where $V_g \approx V_d$. When V_g is increased beyond V_d, the metal–oxide–semiconductor field-effect transistor (MOSFET) is driven toward the linear region of operation and the lateral channel electric field is reduced. As the channel electric field decreases, fewer energetic electrons are created and hot-electron emission rate also declines. When drain bias V_d increases, hot-electron injection into the oxide increases due to increased lateral channel field. When V_g is smaller than V_d, the field in the oxide near the drain end of the channel is weak or repulsive, so that the electrons in the oxide conduction band at that location are driven or drift back toward the substrate. In this case, the electrons that jump over the energy barrier into the oxide simply return to the MOSFET channel. Since only the electrons that reach the gate contribute to the gate current, the gate current due to CHE will decrease as gate bias decreases even though the rate of electron emission at the Si–SiO$_2$ interface increases because of the higher electric field in the silicon.

At low V_g, the high electric field at the drain junction can produce additional carrier injection called drain avalanche hot-carrier (DAHC) injection. The electrons created by impact ionization are either collected by the drain or injected into the oxide. Holes are formed by the impact ionization process and give rise to the substrate current, providing an easily measurable current and a simple way of assessing the rate of impact ionization. Under conditions of very low gate bias, that is, high lateral electric field, some of these holes can also be injected into the oxide. This hot-hole injection is responsible for significant Flash memory oxide degradation during programming. When a Flash cell is biased in this condition, hole injection into the floating-gate current will result in a reduction of the memory threshold voltage. As the gate voltage is increased beyond the hole–hole injection regime, drain avalanche hot-electron injection is encountered. Simulation shows that a net electron gate current is observed here because the hole injection rate drops rapidly with increasing V_g due to a decrease in the lateral electric field and because of the very strong sensitivity of hole injection rate on the field. Further increase in gate voltage lowers the lateral electron field even more such that hot electrons generated by impact ionization no longer contribute significantly to the gate current. Gate current falls before rising again due to the channel hot-electron injection. At certain gate voltage V_g^* (~2 V in Fig. 11.2), the positive hot-hole current balances out the negative hot-electron current, and the gate current is zero. When the device is biased near this condition, there is little gate current observed, but the device still can be degraded due to hot-carrier injections.

11.2.2.2 Overerase Correction Programming. It is best to think of the horizontal axis of Figure 11.2 as $V_g - V_t$. The fact that the gate current decreases rapidly near 2 V as V_g decreases has a special application. In the development of high-density Flash electrically erasable programmable ROM (EEPROM), the broad distribution of threshold voltage after erasure is one of the most serious problems. Additionally, an "erratic erase" phenomenon occurs when a few random bits are severely overerased due to hole trapping in the oxide and the resultant higher tunneling current. The overerased cells can cause read error when other cells on the same bit line are read. So it is essential to control the distribution of the threshold voltages of erased cells to guarantee reliability and functionality. One way of achieving this is to program the overerased cells back to a higher threshold voltage by hot-carrier injection. For example, in a NOR-type Flash memory array, in order to detect overerase bits, read can be done after erase with the wordlines grounded. Ideally, the current should be zero when there are no overerased cells and all the cells have V_t larger than zero. However, if some of the cells are overerased with too low or negative threshold voltages, even zero gate voltage can not shut off the cells, and there is a column leakage current. When such column leakage current is detected, program drain voltage is applied to these columns with the gate grounded. Only the leaky cells are programmed to positive threshold voltages. It is important that the other cells on the same column are not disturbed such that they will fail erase verify, that is, have V_t's that are too high.

Yamada et al. [3] reported such a self-converging program-back scheme, using avalanche hot-carrier injection. Figure 11.3(b) shows the threshold voltages for the cells as a function of the program-back time, with different starting threshold voltages as parameters. The control gate and source are grounded and the drain is biased to 6 V. Except for the upper two curves, the threshold voltages reach a "steady state." This can be explained by the MOSFET gate current I_g characteristics shown in Figure 11.3(a). In the floating-gate memory cell, if the floating-gate potential $V_{fg} > V_g^*$, electrons are injected into the floating gate, which decreases V_{fg} gradually, and finally V_{fg} reaches the balance point V_g^*. Here V_g^* is the point at which avalanche hot-hole injection and hot-electron injection are in balance. If $V_g^* > V_{fg}$, holes are injected into the floating gate, which increases V_{fg}, and V_{fg} also reaches the balance point V_g^*. If V_{fg} is sufficiently low, the channel is off and neither hole nor electron injection takes place. Therefore, the starting threshold voltages lower than the ultraviolet-erased condition converge to a steady-state value, and those higher than certain value in Figure 11.3(b) do not shift as a result of the overerase correction program.

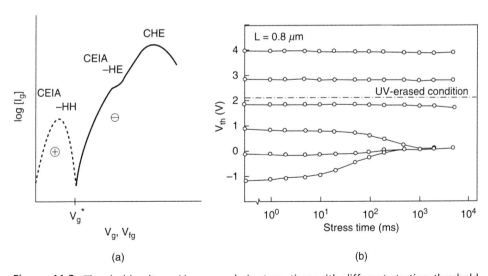

Figure 11.3. Threshold voltage V_{th} versus drain-stress time with different starting threshold voltage as the parameter. Threshold voltage lower than the UV-erased condition converge to a "steady-state" convergence threshold voltage V_{th}^*. The convergence point V_{th}^* can be controlled by many devices as well as operation parameters. (Figures 2 and 4 of [3].)

The value of the steady-state convergence threshold voltage V_{th}^* depends on the bias conditions such as the gate and drain voltages during the overerase correction program, as well as the hole and electron injection balance point V_g^*. The hole and electron injection balance point V_g^* is a function of various device parameters, including channel doping concentration and channel length. The understanding of V_{th}^* and V_g^* provides design guidelines and insights into device reliability. From measured V_{th}^*, we can determine the dependence of the avalanche hot-electron and hole injection balance point V_g^* on stress drain voltage as well as stress time. V_g^* is a constant for a memory device at certain V_d and V_s bias. For a floating-gate device whose threshold voltage converged to V_{th}, we can write the floating-gate voltage as:

$$V_{fg} = V_g^* = \alpha_d V_D + \alpha_s V_S + \alpha_g V_{cg} - \alpha_g (V_{th}^* - V_{t,UV}) \quad (11.1)$$

here α_d, α_s and α_g are the floating-gate to drain, source, and gate coupling ratios, V_{cg} the control gate voltage, and $V_{t,UV}$ the device ultraviolet (UV) threshold voltage. We have assumed that $V_{fg} \approx 0\,V$ after UV erasure. Rewrite the above equation for V_{th}^* as a function of various parameters:

$$v_{th}^* = v_{t,UV} + v_{cg} + \frac{1}{\alpha_g}(\alpha_d v_D + \alpha_s V_S - v_g^*) \quad (11.2)$$

From Eq. (11.2) we see that V_{th}^* is linearly proportional to $V_{t,UV}$ and V_{cg}, and can be shifted by changing $V_{t,UV}$ or applying different V_{cg}'s. Its value can also be easily controlled by changing the $V_{t,UV}$ through channel doping. In a Flash memory product, V_{th}^* can be designed and shifted to the desired value to achieve tighter erase V_t distribution and less hot-hole injection by optimizing the overerase correction program voltages [4]. The hot-electron and hot-hole injection balance point V_g^* can vary due to stress. Using the floating-gate technique, one can stress the device exactly at the V_g^* (or V_{th}^*) point where the same amount of electrons and holes are injected, and monitor V_g^* changes as a function of stress time and stress voltages.

11.2.2.3 Degradation Mechanism of Flash EEPROM Memory Programming. Hot-carrier-induced device degradation has been one of the major problems for submicron MOSFET device design reliability. In logic circuits, the use of LDD effectively controls the hot-carrier injection and related device degradation by suppressing the electric field in the channel. However, Flash memory devices intrinsically operate in the hot-carrier injection regime, and the hot-carrier mechanism is needed for the programming of the memory cell. High drain bias, shallow N⁺ drain junctions, and high-concentration channel doping are used to enhance the hot-carrier injection to achieve high program speed and efficiency. Electrons trapped in the oxide, holes trapped in the oxide, and interface states generated at the Si–SiO$_2$ interface are dominant causes for Flash memory device degradation. It is important to have high program speed, while minimizing program-induced device performance and reliability degradations.

After repeated CHE program and erase cycles, the main degradation mechanisms are electron trapping and interface states generation in the oxide near the drain junction, as shown in Figure 11.4. With the presence of trapped electrons in

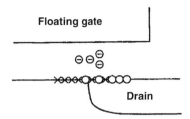

O : Interface state generated by $V_g \simeq V_d$ stress.
× : Interface state generated by $V_g \simeq V_d/2$ stress.
⊖ : Electrons trapped to accetor - like oxide traps generated by $V_g \simeq V_d/2$ stress.

Figure 11.4. Schematic representation of the location of the oxide traps and interface-states created by CHE programming. (Figure 7 of [5].)

the oxide near the drain, the transistor V_t is shifted higher, and less electron charge is needed on the floating gate for the memory to pass program verify. When these trapped electrons in the oxide are detrapped later due to various reasons, for example, high-temperature storage, the memory V_t is shifted lower and results in reduced program margin. The trapped electrons also retard the channel hot-electron injections and result in lower injection efficiency and slower program speed. This effect can be fairly significant in Flash memory cells after repeated P/E cycles as seen in Figure 11.1. That is, the cell program V_t decreases progressively versus P/E cycling when programming time duration is fixed, and the V_t versus cycling window closes.

As the cell is being programmed, the floating-gate potential rapidly collapses. The lateral channel electric field, as well as the average electron energy rises. As shown in Figure 11.2, hot-hole injection can be significant. Upon co-injection of hot electrons and hot holes, it is well known that interface state generation occurs [6]. This can lead to device transconductance (G_m) degradation as a result of mobility reduction. The reduced transconductance results in lowered read current and makes it harder for the device to pass erase verify. The cell has to be erased further to lower the threshold voltage to obtain enough read current to be read as an erased cell. One effect of this is increased erase time. More important is that other cells may be overerased and cause column leakage current. If enough cells are overerased, the overerase correction program might not be able to bring all cell V_t's to positive if the charge pump is not strong enough. And, of course, in the worst case, if the transconductance is so degraded, no matter how much the cell is erased, there is still not enough read current, and this results in erase failure. Therefore, the ending condition of cell programming operation also can have an important impact on device degradation.

Yamada et al. [5] have used charge pumping, G_m, and endurance measurements to study how oxide damage such as interface states and oxide charge leads to programming slowdown. They found that the reduction of electron injection into the floating gate is caused mainly by the interface traps in the drain overlap region rather than charge trapping in the oxide. During the initial phase of a program operation, when $V_{fg} \approx V_d$, interface traps are generated in the drain overlap region. This reduces electron injection into the floating gate but does not significantly reduce G_m or read current I_d current. During programming, these interface states would be filled with electrons that will retart electron injection to the floating gate. Later in the program operation, the gate becomes negatively charged with electrons,

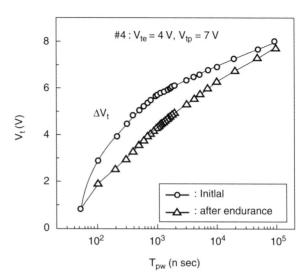

Figure 11.5. Program characteristics of a cell before and after 10^5 P/E endurance cycling. (Figure 9 of [5].)

and V_{fg} approaches $V_d/2$. In this regime, interface traps and acceptor-like oxide traps are created near the drain edge, which together can cause serious degradation of G_m and read I_d current. Figure 11.5 shows the program characteristics of a cell before and after 10^5 P/E endurance cycling.

11.2.2.4 Trap Reduction Through Nitridation.
The composition of the gate oxide can play a part in reducing the trap density in the tunnel oxide. Oxynitrides have drawn much attention as a candidate tunnel dielectric for nonvolatile memory due to its low charge-trapping rate and the low stress-induced leakage current. The interfacial nitrogen incorporation is desirable for thin tunnel oxide dielectric reliability. Several studies have shown that nitridation of tunnel oxide makes the tunnel oxide more resistant to electron trap generation [7–12]. Joshi et al. [7] reported that neutral electron trap generation could be significantly suppressed by nitridation and reoxidation.

Fukuda et al. [8] showed that both electron and hole trap densities are greatly reduced in heavily nitrided tunnel oxide films [RTONO, (rapid thermal ONO)]. As shown in Figure 11.6, three devices with 11-nm-thick tunnel oxide formed by RTO(1), RTON(2), and RTOO(3) were fabricated. The RTO(1) film is pure SiO_2, and the RTON(2) and RTONO(3) films are oxynitrided SiO_2 films. The Flash memory cells are programmed with channel hot-electron injection and erased with source-side FN erase. Figure 11.6 shows the typical endurance characteristics of the devices. After 5×10^4 cycles, the RTO cell shows 42% window closing, whereas the narrowing improved to 37% in the RTON cell. By contrast, dramatic improvement to less than 6% is achieved in the cell with RTONO tunnel oxide. The key is to reduce the trap density in the tunnel oxide. From secondary-ion mass spectrometry (SIMS) analysis, it was found that a large amount of N atoms ($>10^{21}$ atoms/cm^3) pile up at the SiO_2–Si interface in the RTON(2) and RTONO(3) films. It was also found that the number of H atoms, which are the origin of electrons traps [10], decreases in

CYCLING-INDUCED DEGRADATIONS IN FLASH MEMORIES

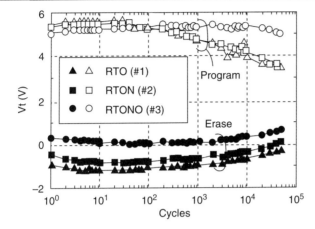

Figure 11.6. Endurance characteristics of the Flash memory cells with different tunnel oxide. (Figure 3 of [8].)

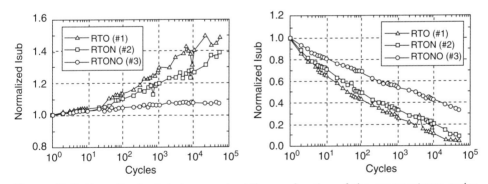

Figure 11.7. Normalized substrate currents shift as a function of the program/erase cycles. Heavily nitrided tunnel oxide films (RTONO) showed reduced electron and hole trap densities. (Figure 7 of [8].)

the RTONO films compared with the RTO(1) and RTON(2) films. To investigate the hot-carrier trapping behavior, substrate currents in both forward and reverse bias conditions as a function of program/erase cycles were studied. As anticipated, shown in Figure 11.7, the RTO(1) cells show an increase of I_{sub}, whereas almost no change of I_{sub} is observed for the RTONO(3) cell. The data indicate that in the RTONO tunnel oxide, both electron and hole trap densities are greatly reduced.

11.2.3 Tunnel-Erase-Induced Oxide Degradation

During the tunnel-erase operation, the tunnel oxide is subjected to a very high electric field (above 10 MV/cm) and substantial oxide current. The literature is rich with studies of mechanisms by which oxides can be damaged by high fields and by tunneling currents: trapping, hydrogen-release effects, hot-carrier generation within the oxide, polarization-induced damage, and so forth. In addition to high fields and currents *within* the oxide, hot carriers can be generated in the anode from high fields induced there and from impact ionization that occurs when energetic tunneling electrons enter the anode. It should not be surprising, therefore, that both intrinsic and defect-related degradation mechanisms occur as a result of repeated erase steps during program/erase cycling. This section will review the various erase schemes used in Flash memories and the resulting degradation mechanisms.

11.2.3.1 Erase Methods. Figure 11.8 illustrates the three common erase bias schemes used in NOR Flash memories: high-voltage source with grounded gate erase (HSE), negative gate with positive source erase (NGSE), and negative gate channel erase (NGCE). With any scheme, the basic requirement is to couple enough voltage across the tunnel oxide to generate the tunneling current required to erase the cell in the specified time.

Older Flash memories usually used the HSE scheme, which is the simplest and has the least impact to peripheral circuits. The main disadvantage of the HSE scheme is that the high voltage placed on the source junction can cause high band-

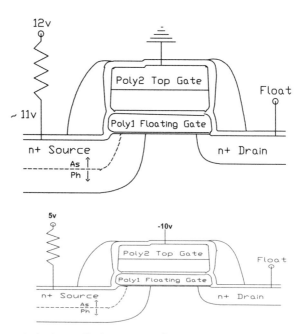

Figure 11.8. Three stacked-gate Flash memory cell erase operation conditions: (a) high-voltage source erase with grounded gate (HSE), (b) negative gate with positive source erase (NGSE), and (c) negative gate channel erase (NGCE). Note that the NGCE scheme does not require the source junction to be graded.

to-band tunneling current and generate hot-carrier damage. This damage is discussed in greater detail below. Also, the source current during erase could be substantial because in the block-erase scheme, the sources of all 512,000 cells are tied together. Even if the current from each cell is only 10 nA, the total current could be 5 mA. In order to reduce the band-to-band tunneling and to reduce the hot-carrier effects, a phosphorus implanted junction is needed to create a deep and graded source junction. The biggest drawback of this is that it increases the cell size and is a scaling limit as cell channel length becomes smaller.

Some of these problems can be reduced by the NGSE scheme, which couples the necessary tunnel voltage with a lower source bias through the help of a negative gate bias. The main drawback of this scheme is that generating and switching the negative gate voltage requires additional complexity in peripheral circuits. For example, in order to limit the peak voltage to around 10 V, a triple-well process is required. In the NGSE scheme, a phosphorus implanted junction is still needed to form the graded source junction in order to reduce the band-to-band current and to reduce hot-carrier degradation. But compared with the HSE scheme, where 11 to 12 V are applied to the source, in the NGSE scheme, the source voltage is only 5 V or less. The double source junction can be made less deep in this case and shorter cell channel length can be achieved. One advantage of both the HSE and NGSE schemes is that the program and erase paths are totally separated, so both the source and drain junction can be optimized. Since the erase is through the source and gate overlap region, the channel is not stressed, and there is no channel mobility degradation due to the erase operation.

Finally, with triple-well processes the channel region in addition to the source can be biased positively; this eliminates the band-to-band tunneling because there is no bias across the source junction. In this scheme, the FN tunneling process takes place through the entire channel, and the benefit is that the phosphorus implanted graded source junction is no longer needed. As device channel length becomes shorter with the advances of technology scaling, the HSE and NGSE schemes become more difficult to implement. The channel tunneling erase scheme becomes a logical choice for sub-quarter-micron technology. The absence of the large band-to-band tunneling current also makes channel erase more suitable for low-voltage applications. As compared with the HSE and NGSE methods, one drawback of NGCE method is that the channel mobility might be degraded due to FN erase operation, and there might be further degradation due to interaction between program and erase.

The relationship of gate and source voltages of the three erase methods is summarized in Figure 11.9. Erase voltages chosen give the same initial electric field in the tunnel oxide for the three different erase methods. As source voltage decreases, a larger negative gate voltage is required. In fact one can chose arbitrary V_g and V_s combinations on these lines. The key parameters used in this calculation are tunnel oxide and ONO layer thickness, the gate, drain, and source coupling ratios.

Yoshikawa et al. [13] studied the effects of process and device parameter fluctuations of the Flash cells on the erase V_t distribution for the three erase methods described above. Since a huge number of cells are simultaneously erased, the variations of different device parameters can lead to coupling ratio and electric field variations among the memory cells. When the same voltages are applied, cells with different parameters will have different electric fields and results in different erase speed. In fact, the different erase methods respond to the device parameter

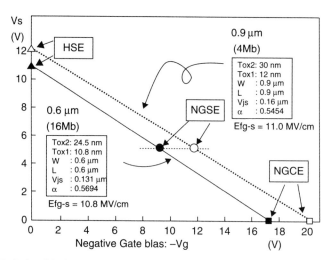

Figure 11.9. Relationship between the negative gate bias and the positive source bias. Voltage combinations on the line give the same initial electric erase field. (Figure 1 of [13].)

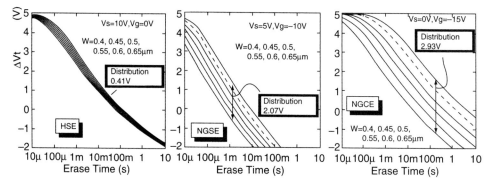

Figure 11.10. Simulated erase curves with different gate widths for the three erase methods. The HSE method gives the tightest erase V_t distribution. (Figure 6 of [13].)

variations differently, some are more sensitive and some are less sensitive to the process variations. It is beneficial that the erase scheme is more tolerant to the ever-present process variations.

Figure 11.10 shows the simulated erase curves with different gate widths of 0.45 to 0.65 μm for the three erase methods. In HSE, the erased V_t distribution width shrinks rapidly when the cells have completed erase, but it remains nearly the same for NGSE and NGCE. The erase method with higher V_s and lower V_g is less sensitive to the cell width variations. This is because for the HSE method, the tunnel field is determined by the source voltage and source coupling ratios and is not sensitive to the variations in the gate coupling ratio, which is strongly dependent on the cell width. On the contrary for the NGCE method, the erase field is determined by the gate voltage and gate coupling ratio, therefore it is extremely sensitive to the variations of cell widths. The NGSE method splits the voltage between the gate and

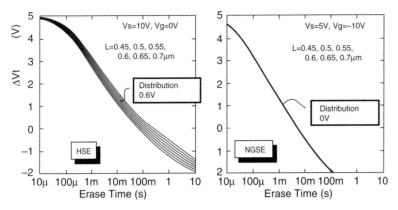

Figure 11.11. Simulated erase curves with different gate lengths for the HSE and NGSE methods. Little variation is observed for NGSE. (Figure 7 of [13].)

source, and its sensitivity to cell width variation lies between the case for HSE and NGCE.

Similarly, the erased V_t distribution sensitivity to the channel length variations can also be simulated. Simulated erase curves with different gate lengths for HSE and NGSE methods are shown in Figure 11.11. HSE is more sensitive to the channel length variations than NGSE and NGCE methods, while NGSE and NGCE are almost independent of channel length variations.

In a recent study of channel Fowler–Nordheim tunneling in Flash memory cells [14], Chan et al. observed enhanced erase behavior when a high P-well voltage is used, with the source and drain junctions of the cell floating. Their study indicates that the floating source and drain take on a high junction voltage during the P-well voltage transient. This causes transient band-to-band tunneling current, and in some cases, junction avalanche breakdown to occur in the source and drain junctions. Because the floating gate is negatively charged, hot-hole injection into the floating gate occurs to create the enhanced erase phenomenon.

11.2.3.2 Band-to-Band Tunneling Current. The band-to-band tunneling (BBT) current is due to electron–hole pairs generated in the gated-diode source junction during erase, as shown in Figures 11.12 and 11.13. When high voltages are applied to the source junction with the control gate grounded or negative biased, a deep depletion region is formed underneath the gate-to-source overlap region. The tunneling of valence band electrons into the conduction band generates electron–hole pairs. The electrons flow to the source and the holes to the substrate. The generated holes gain energy because of the large lateral potential gradient created by the source-to-substrate voltage. While the majority of these generated holes flow into the substrate, some of them gain sufficient energy to surmount the Si–SiO$_2$ barrier and are injected into the gate while some are trapped in the oxide [15–17].

As shown in Figure 11.13, avalanche breakdown can also occur during erase if the source voltage is high and source junction abrupt. In an arsenic N$^+$ to P-substrate abrupt junction, the avalanche breakdown current can be particularly detrimental to the cell, and leads to serious cell transconductance degradation [17–20]. The

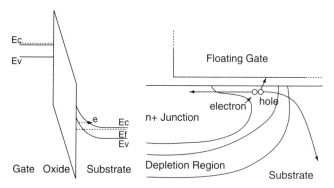

Figure 11.12. (a) Energy band diagram for a gated diode at the point of the highest band-to-band tunneling with the source (or drain) junction reverse biased. Valence band electrons tunnel to the conduction band, leaving a valance band hole. (b) The generation of electron–hole pairs and their trajectory. The holes flow to the substrate and the electrons flow to the source (or drain).

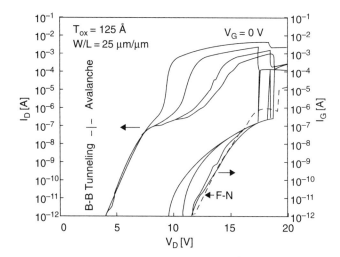

Figure 11.13. Measured drain and gate current for the 125-Å oxide device with channel surface concentration of 10, 5, 2, and $1 \times 10^{16}/cm^3$. The dashed curve represents the normalized FN gate current [16].

band-to-band tunneling current and the avalanche breakdown can be reduced by using a graded source junction. By introducing an additional phosphorous diffusion, not only is the BBT current shifted to a higher source (or drain) voltage, but the avalanche breakdown is also softened up as shown in Figure 11.14. The optimization of the double diffused N^- region is a trade-off between the source junction breakdown voltage and the capacitive-coupling ratio between the source junction and the floating gate.

11.2.3.3 Band-to-Band Current-Induced Degradation and Charge Trapping.
The reliability implications of the band-to-band tunneling effect depend

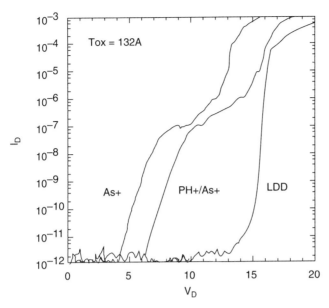

Figure 11.14. Comparison of BBT leakage currents in devices made with As⁺, P⁺/As⁺, and LDD junction [16].

upon the bias scheme used. With channel erase, the region of the source diffusion that experiences enough band bending to initiate tunneling actually inverts, forming a surface layer of holes. Once inversion has occurred, further band-to-band tunneling ceases. One would, therefore, expect relatively small reliability concerns with the channel-erase method, except for the band-to-band current due to the P-well voltage transient as discussed above.

In the other schemes, generated holes are swept out of the source by the lateral voltage drop between source and channel, and band-to-band tunneling continues throughout the erase operation. The band-to-band tunneling itself is not a reliability hazard because the generated electron–hole pairs are not energetic and therefore can do no damage. But the holes that are swept toward the channel pass through a potential drop of several volts and may become highly energetic. In the case of HSE, avalanche can occur, as demonstrated by a nearly vertical break in the I–V characteristic viewed from the source diffusion and the fact that this break point shifts to higher voltage at higher temperature. In the case of NGSE, avalanche is not observed because of the lower source voltage, but a voltage drop of several volts across a depletion region less than 0.1 μm wide is still sufficient to energize many of the holes. Additionally, since the floating gate is negative with respect to the source, some of the energetic holes may be able to pass through the tunnel oxide as gate current.

At the simplest level, energetic holes impinging on the oxide or passing through it may cause some holes to be trapped. It is well known that trapped holes will enhance tunneling, by virtue of the attractive field they generate to electrons. This current enhancement is important to the reliability issue of *disturbs*, which are discussed in detail in Section 11.4.

It is also well known that energetic holes can damage oxide interfaces, and if holes are injected through the oxide, it is possible to create bulk defects or defects

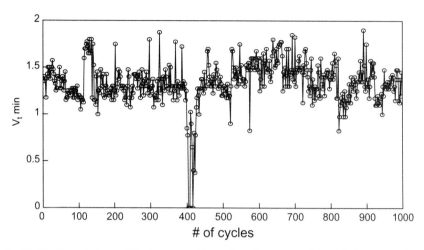

Figure 11.15. Erased V_t of a bit changes with cycling. The erase of this bit is erratic, that is, it develops a severe overerase problem, then suddenly recovers a few cycles later.

at the interface with the floating gate. Such defects can contribute to degradation of the transconductance of the cell and to some of the other defect-related degradation mechanisms described in this chapter, such as SILC-related data loss.

Reliable Flash memories have been constructed with all three of the erase schemes described above. The channel-erase scheme eliminates the band-to-band tunneling effects but requires the largest overhead in processing (triple well) and design (well bias during erase). For the HSE and NGSE schemes, developing a reliable memory involves optimized engineering of the graded source profile and the erase source bias.

11.2.4 Erratic Erase

In Section 11.2.2.2, it was noted that with a NOR architecture the Flash erase distribution must be controlled against overerase. If even a single bit erases to a negative V_t, the leakage current through that cell will cause an entire column to read improperly. With a block-erase scheme, the top of the erase V_t distribution is fixed (e.g., at 3 V) by the erase algorithm, which stops erasing when the highest V_t bit reaches the desired V_t. But this algorithm does not control the V_t of the *fastest* erasing bit in the block. Figure 11.15 illustrates the fact that the lowest V_t in the erase distribution is erratic with respect to cycles; indeed, as occurred around 400 cycles in the case illustrated, a block initially with a perfectly fine V_t distribution may suddenly develop a severe overerase problem, only to suddenly recover a few cycles later. This *erratic erase* behavior is an intrinsic property of tunnel-erased Flash memory [21].

The erratic-erase mechanism typically affects only a single bit at a time; for example, the overerase observed at about 400 cycles in Figure 11.15 was caused by a single bit. Figure 11.16 traces such a bit in the cycles immediately after the cycle in which it became overerased. The erased V_t erratically fluctuates between two distinct states in a manner similar to that of random telegraph noise. The bimodality

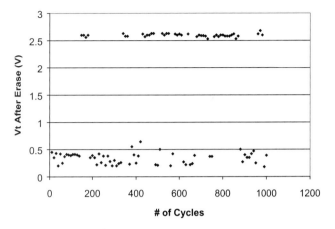

Figure 11.16. A bit that has been erased can oscillate between "stable" and "unstable" many times, behaving "erratically."

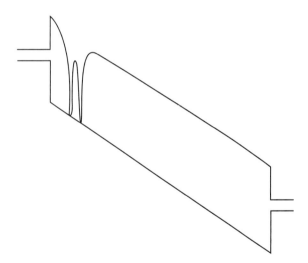

Figure 11.17. Band diagram of the tunnel oxide during memory cell erase operation. Here two positive electronic charges are placed in the oxide at distances of 10 and 15 Å from the cathode.

of the effect suggests a fundamentally quantum-mechanical mechanism, with a "normal" state switching to an "erratic" state with the change of a single quantum state.

Erratic erase has been modeled as enhanced tunneling caused by a small number of holes trapped at optimal positions in the oxide. The model is illustrated schematically in Figure 11.17, which shows the band diagram of the tunnel oxide during tunnel erase. In the figure, two positive electronic charges are placed in the oxide at distances of 10 and 15 Å from the cathode. The overlapping Coulomb potentials greatly reduce the tunneling distance.

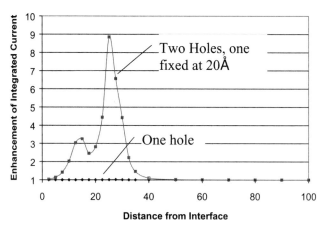

Figure 11.18. Increase of tunneling current with different number of trapped holes near the top interface. The charge location is measured from the top interface. (After [21].)

Quantitatively, the effect has been evaluated using the Wentzel–Kramers–Brillouin (WKB) approximation [21]. Figure 11.18 shows the enhancement of the tunneling current from one or two trapped holes, as a function of trap distance from the cathode interface. In this calculation, the enhancement is the increase in tunneling current, integrated over the entire tunneling area in a Flash cell. In the case of two holes, both are taken to be trapped at the same x-y location but at different distances from the interface; one is assumed fixed at 20 Å from the interface, while the distance to the second hole is varied. As shown, a single hole, regardless of position, has a negligible effect on the tunneling current. The *local* current enhancement, to the current density at the hole's location, may be high, but the integrated effect is small because the current is enhanced only in the cross-sectional area of the hole. With two holes, however, the current enhancement can be much larger, approaching an order of magnitude. As shown in Figure 11.19, the transition from normal to erratic behavior corresponds to a roughly 10 times increase in erase time, or about one order of magnitude in tunnel current. Larger effects observed on some bits can be simulated by adding a third trapped hole. The trapping or detrapping of one of the trapped holes is responsible for the quantum switching between the erratic and normal states.

Erratic erase is therefore viewed as caused by happenstance variations in the hole trapping that intrinsically occurs in tunnel oxides at high fields. Erratic cells are not so much *defective* as they are *unlucky*. The effect is not measurable in capacitor measurements, where the measured current is integrated over a very large area and individual regions of current enhancement are lost in the noise. But in microscopic Flash cells, with tunnel areas on the order of $10^{-10}\,\text{cm}^2$, these statistical variations become apparent. It should be noted that the tunneling enhancement is very large on a local scale. The observed increase in total tunneling current in a Flash cell may be one or two orders of magnitude, but for this to occur the current density at the location of the trapped holes must be enhanced by several orders of magnitude. In a Flash cell the charge passage (and energy dissipated) is limited by the small capacitance of the floating gate (on the order of 1 fF), such a drastic increase in current density does not destroy the oxide, as evidenced by the fact that erratic

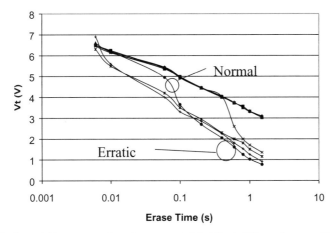

Figure 11.19. Erased V_t vs. erase time for an unstable bit at different cycles. The abrupt drop in the V_t curves exemplify the effect of hole trapping and supports the hole trapping and detrapping as the cause of erratic erase. (After [21].)

bits invariably recover with additional cycling. But one can speculate that the same kinds of statistical "defects" responsible for erratic erase could be a precursor to destructive oxide breakdown in capacitor studies, where energy and current are not so limited.

Experiments show that switching from the HSE to the NGSE scheme does not significantly reduce the erratic-erase effect, suggesting that the trapped holes are not caused by the band-to-band tunneling effects described earlier. This is not surprising, for the holes generated by band-to-band tunneling do not become energetic enough to cause damage until they have fallen through a substantial voltage drop as they are swept toward the channel. In contrast, the erase current in a Flash cell primarily occurs in the most highly doped regions of the source (near the gate edge), where there is little voltage drop within the source diffusion and thus the voltage across the tunnel oxide is at its greatest. It is likely, instead, that the trapped holes are generated by the high field in the oxide or by impact ionization of the tunneling electrons, either in the oxide or from the anode-hole injection mechanism once they enter the source silicon.

The erratic-erase effect is large enough that the failure rate would be unacceptable in high-density memories without some form of error management. The effect is not screenable during production testing because of its erratic and unpredictable nature. Instead, manufacturers use posterase repair methods to move an erratic bit's V_t back to the acceptable range. A self-convergence scheme for doing so has already been described. Such methods depend on the gate current characteristics of the Flash cell and can be difficult to control because of variability with cycling, process parameters such as channel length, and temperature. Alternatively, it is possible to reposition a cell's V_t using an active algorithmic approach [22]. With this scheme, overerased cells are detected using special sensing circuits and then softly programmed to a positive threshold voltage successively higher gate voltages embedded in an iterative loop. The resulting V_t is determined by the exit condition of the iterative loop rather than the characteristics of the cell, allowing

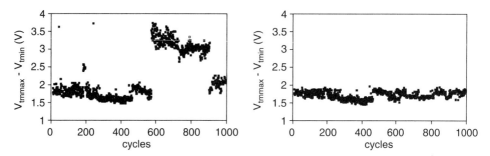

Figure 11.20. Width of erase distribution as a function of cycles, before (left) and after (right) posterase repair.

more complete control. Figure 11.20 shows the effect of this algorithm on a Flash array.

11.3 FLASH MEMORY DATA RETENTION

Data retention is the ability for the memory to retain the correct data over a prolonged period of time under storage conditions. In floating-gate memories, the charge is stored in the floating gate surrounded by insulting oxide layers, which have enough barrier height to prevent the stored charge from leaking out. The stored charges can leak away from the floating gate through the gate oxide or through the interpoly dielectric. The stored charge could be both negative and positive charge, and the symptom is a decrease or increase in the memory transistor threshold voltage. The "charge loss" and "charge gain" could be caused by many different reasons, such as mobile ions, oxide defects, detrapping of electrons or holes from the oxide traps, trap-to-trap hopping within the ONO film, or the generation of positive interface charge [23–32].

For UV light erasable EPROMs, the write states are negatively charged with electrons, and different charge loss mechanisms have been described in the literature [26, 27]. We will discuss these in detail in the following sections. Because of the long history of EPROMs, these mechanisms have been carefully studied and well understood. There are few data retention problems for modern EPROMs. However, Flash EEPROMs are electrically programmed and erased repeatedly, which subjects the oxide in the memory transistor to high electric field stress. For a state-of-the-art high-density Flash EEPROMs, 100,000 to 1 million program/erase cycles can be required. Even though in many applications such as cell phones and personal computer (PC) BIOS, the memory is not cycled more than a handful of times; however, due to competitive pressure, almost all manufacturers are forced to offer Flash memories with 100,000 cycles or more. Such stress can cause many reliability issues. For example, it has been reported that a major endurance failure mode of early FLOating-gate Tunneling OXide (FLOTOX) EEPROM is tunnel-oxide breakdown [33]. But more subtle reliability and performance impact is the slowing down of the program and erase speed due to electron trapping in the oxide [31, 32].

An even more subtle, that is, hard to detect, endurance failure mode is the loss of data retention ability after endurance stress. Apparently, oxide can become more conductive after electrical stress, resulting in stress-induced leakage current (SILC). Typically failures occur at a small number of random bits [34]. The SILC mechanism is described in detail in Section 11.5. Another subtle effect is that after high-field FN stress and hot-electron stress, an oxide will trap charges, and the charge detrapping can result in threshold shifts and thus in data retention failure. This is a more serious problem for high-density multilevel cell technology, as the margins between states are much smaller.

Long-term (>10 years) data retention test is usually accelerated by temperature to a period of days or weeks. This is done for charge-loss mechanisms with known activation energy. It has been discovered that for advanced Flash EPROM devices with very thin tunnel oxide (~100 Å), the charge-loss activation energy could be lower after endurance cycling, and this makes it even harder to test for data retention with temperature acceleration. In such cases, it is possible to augment temperature acceleration with electric field acceleration.

In this section, we will first discuss the concept of activation energy and the method to measure the charge-loss activation energy. Electric field acceleration mechanisms are then discussed. Then we will discuss the various charge-loss/gain mechanisms in a floating-gate memory device, such as intrinsic charge loss, mobile ions, and charge hoping in the ONO film. We will further discuss in detail data retention concerns related to modern Flash memories, especially data retention issues caused by repeated write/erase cycling.

11.3.1 Activation Energy and Accelerated Data Retention Bake Tests

11.3.1.1 Data Retention Bake. During the reliability development and verification of new nonvolatile technologies, all of the associated process and design rules need to be verified. While many reliability issues are generic among the various semiconductor technologies, there are a few failure mechanisms that are unique to Flash memories. One of them is data retention, especially after write/erase endurance cycling.

The data retention duration is verified by a data retention bake after the specified number of endurance cycles as part of the development characterization and product qualification tests. This stress exposes the devices to unbiased storage at an elevated temperature, normally 150°C for plastic packages or 250°C for hermetic packages. These are the highest practical temperatures the applicable package can sustain to accelerate the loss of charge off the floating gate. For plastic packages, 160°C is at the maximum safe storage temperature because the glass transition temperature of most epoxies is below 165°C. Above 165°C, the mechanical and chemical stability of the plastic is uncertain, thus prolonged exposure can create failure mechanisms that would otherwise not be observed. For solder seal hermetic packages, 260°C is the maximum temperature before damaging the solder seal. For glass to seal hermetic packages, 300°C is the maximum prolonged storage temperature before introducing unpredictable effects in the packaging materials. Often temperatures above 175°C are not used because of oxidation of the lead finish.

11.3.1.2 Activation Energy and Temperature-Accelerated Test. Temperature-accelerated life tests are widely used as characterization methods for charge retention capability. The industry standard methodology to extrapolate data retention lifetime is to use the Arrhenius diagram to correlate failure rate obtained at high temperature to lower operation temperatures. It assumes lifetime distribution to be lognormal. Thermal activation temperature indicates the effect that the increased temperature has on frequency of the failures, the higher the activation energy the greater the effect. The classical temperature-activated Arrhenius law is shown below with the failure rate R determined by the activation energy:

$$R = A\exp\left(-\frac{E_A}{kT}\right) \quad (11.3)$$

where $K = 8.623 \times 10^{-5}\,eV/K$, Boltzmann constant
A = proportionality constant
R = mean rate to failure
E_A = activation energy
T = temperature, in Kelvin

And the mean time to failure t_R is expressed as follows:

$$t_R = \frac{1}{A}\exp\left(\frac{E_A}{kT}\right) \quad (11.4)$$

The acceleration factor (AF) is calculated as:

$$\text{AF} = \frac{R_2}{R_1} = \exp\left[\frac{E_A}{k}\left(\frac{1}{T_1} - \frac{1}{T_2}\right)\right] \quad (11.5)$$

Often the data retention failure mode is not known, and one must conduct experiments to find the correct activation energy for the specific device and failure. The activation energy can be obtained by baking the device at different temperatures, and plotting the charge-loss/gain rate as a function of temperatures, as shown in Figure 11.21. The obtained activation energy can be a good clue as to what is the cause of the failures and what acceleration factor should be used in design of data retention bake tests.

In nonvolatile memory (NVM), the data retention failure can be caused by different mechanisms with different activation energies. We will discuss the different failure mechanisms and typical activation energies associated with the different mechanism in the following sections. It is necessary to test for data retention at higher temperature (250°C, e.g.) to catch the failures caused by mechanism with higher activation energy (e.g., contamination's) in an accelerated test time. Some examples of acceleration factor (AF) for charge retention bake experiments are shown here. Using the AF equation (11.3), a typical commercial Flash memory product that has a data retention specification of 10 years at 85°C can be calculated as shown below.

Assume an $E_a = 1.0\,eV$, the AF values for 85°C are:

If baked at 150°C, AF = 145, for 10 years at 85°C, need to bake 600 h.
If baked at 250°C, AF = 27,515, for 10 years at 85°C, need to bake 3 h.

FLASH MEMORY DATA RETENTION

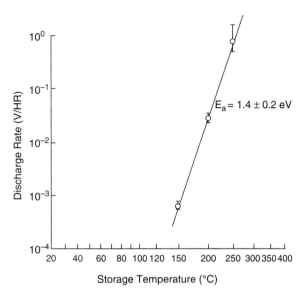

Figure 11.21. Plots of initial discharge rate versus storage temperature. (Figure 6 of [27].)

Assume an $E_a = 0.6\,\text{eV}$, the AF values for 85°C are:

If baked at 150°C, AF = 19.8, for 10 years at 85°C, need to bake 4416 h, or 26 weeks.

If baked at 250°C, AF = 461, for 10 years at 85°C, need to bake 190 h, or 8 days.

Assume an $E_a = 0.3\,\text{eV}$, the AF values for 85°C are:

If baked at 150°C, AF = 2.6, for 10 years at 85°C, need to bake 3.8 years.

If baked at 250°C, AF = 21, for 10 years at 85°C, need to bake 173 days, or 5.7 months.

11.3.1.3 Voltage-Accelerated Data Retention Test. For the cases when the thermal activation energy is small, such as 0.3 eV, the thermal acceleration factor is small. As the example above, in order to test for data retention of 10 years at 85°C, the baking times are 3.8 years at 150°C and 5.7 month at 250°C. Obviously, this is not practical and other means of acceleration must be used. Many of the charge-loss/gain mechanisms are related to electric field and can be accelerated by manipulating voltage. For example, after endurance cycling the charge-loss mechanism for Flash memory is SILC in the tunnel oxide. This mechanism is trap-assisted tunneling, with low thermal activation energy, and can be accelerated by voltage. Similar to oxide breakdown test and screening, both temperature and voltage acceleration are often utilized to accelerate the data retention failures.

The simplest voltage acceleration technique is to measure the threshold voltage of a Flash cell as a function of time and to extrapolate the curve to the failure point.

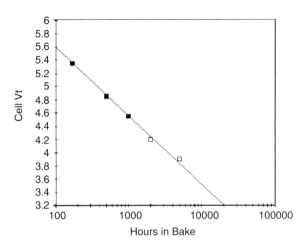

Figure 11.22. V_t loss versus time in bake, for a defective Flash cell after cycling. The line is fitted to the data taken within the first 1000 h (solid squares) and is shown to accurately predict the data out to 5000 h (open squares).

Charge loss is gradual; a cell's V_t decreases along a well-behaved curve that can be extrapolated. For example, Figure 11.22 shows the V_t as a function of time in bake for a bit with SILC-related leakage. The data fit a log-linear relationship that can be extrapolated. As shown in the Figure 11.22, the fit to data obtained in the first 1000 h accurately predict later data taken through 5000 h. Extrapolating even beyond 5000 h, the fit implies an ultimate time to failure of 10,000 h if the trip point of the sense amplifier is 3.5 V. The effective 10× acceleration of this technique (10,000 vs. 1000 h), when combined with the modest but nonzero temperature acceleration of the mechanism, can allow years of operation to be simulated in a practical amount of time.

How well one trusts these extrapolations depends on how well one trusts the model to which the data are fit—such as a log-linear model as shown in Figure 11.22. Charge loss in a nonvolatile memory can be modeled as the discharge of a capacitor through an impedance, which represents the dielectric leakage. If one knows the I-V characteristic of the impedance, one can solve for the V_t versus time characteristic. If the impedance is a simple resistance, the V_t will decrease according to a standard RC decay curve. If the impedance is exponential, then the V_t will follow a log-linear curve as shown in the figure. This conclusion can be derived as follows.

Let a floating-gate capacitance C start with some initial voltage V_0 and discharge over time through a leakage current $I = I_0 e^{KV}$. The circuit solution is simple[1]:

$$I = I_0 e^{KV} = C \frac{dV}{dt} \quad (11.6)$$

---

[1] For simplicity, voltages are taken to be positive, even though the voltage of the floating gate is negative.

Integrating,

$$t = \frac{C}{I_0 K}(e^{-KV} - e^{-KV_0}) = \frac{C}{I_0 K} e^{-KV}(1 - e^{-K(V_0 - V)}) \qquad (11.7)$$

This equation is plotted in Figure 11.23 for the case in which $K = 3$. Two curves are shown, for starting voltages (V_0) of 2.5 and 3.5 V. As shown, the curves asymptotically approach a straight line on a log-linear plot. In the formula, this asymptote occurs because the V_0 term becomes insignificant once enough charge loss has occurred for $K(V_0 - V)$ to be large. The measured quantity in retention bake, cell V_t, is linearly related to the floating-gate voltage V and will therefore follow the same shape of curve.

Figure 11.23 illustrates two points. The first point is the validity, in the case of an exponential I-V characteristic, of using a log-linear plot (similar to Fig. 11.22) for extrapolation purposes. The second point is that another way to accelerate time to data is to set memory cells up to higher initial V_t's. In Figure 11.23, the cell loses charge from 3.5 to 3.0 V significantly faster than from 2.5 to 2.0 V. In fact [see Eq. (11.7)] raising V_0 by some increment ΔV accelerates by a factor of $e^{K \Delta V}$ the time required to lose a given amount of charge.

The question then becomes how to determine the I-V characteristic of the charge-loss leakage so that a model for the V_t versus time behavior can be generated and extrapolated to a time to failure. There are a limited number of common conduction mechanisms: ohmic, Frenkel–Poole hopping conduction, Schottky emission, tunneling, and so forth. Each has a well-known I-V characteristic, and empirical charge-loss studies can determine which characteristic best fits the charge-loss behavior. (A complication is that the Fowler–Nordheim tunneling equation applies only to tunneling through a triangular barrier; tunneling involved in Flash charge loss typically occurs at low oxide fields where the tunnel barrier is trapezoidal.) In practice, the quasi-exponential I-V characteristics applied to Frenkel–Poole, Schottky, and tunneling mechanisms are indistinguishable from each other, and from a pure exponential characteristic, except over a wide range in electric field.

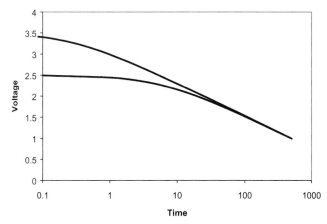

Figure 11.23. Plot of charge loss with exponential leakage I–V characteristic. The curve shown has $K = 3$. The voltage and time scales are arbitrarily scaled.

For extrapolating charge-loss characteristics over a limited range of V_t, an exponential will suffice for any of these mechanisms.

11.3.1.4 New Extrapolation Law for Data Retention Lifetime of Nonvolatile Memories.
There are some limited recent studies that dispute the accuracy of this industry standard model [35–37]. De Salvo et al. [35] reported that the retention time, namely $\log(t_R)$, varies linearly with temperature T rather than with $1/T$ as commonly assumed, yielding reduction in the extrapolated time to failure. Using stacked-gate EPROM cells with 11-nm dry thermal oxide and 11-nm equivalent low pressure CVD (LPCVD) ONO interpoly dielectric, the intrinsic charge-loss rate as a function of baking temperature is shown in Figure 11.24.

The Arrhenius plot of retention time characteristics for all baking temperatures is shown in Figure 11.25. It shows that the experimental data do not fit a straight line, as the $1/T$ model would require. However, a good fitting is instead obtained with a new T model, where mean time to failure t_R expressed in the form

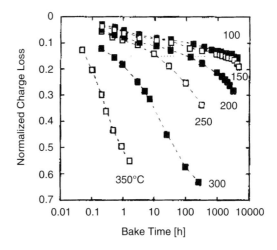

Figure 11.24. Normalized charge loss (equal to the ratio of the current threshold voltage shift to the initial programmed one: $\Delta V_{th}(t)/\Delta V_{th0}$) vs. bake time, corresponding to data experiments performed at different baking temperatures. (Figure 1 of [37].)

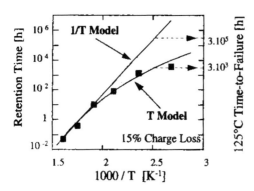

Figure 11.25. Arrhenius plot of retention time characteristics for all baking temperatures. Time to failure using the $1/T$ model or the T model are extracted at 125°C. (Figure 2 of [37].)

$$t_R = T_0 \exp\left(-\frac{T}{T_0}\right) \tag{11.8}$$

where T_0 is the characteristic temperature of data retention. The important implication of this result is that the nominal operation time to failure extracted at 125°C is significantly overestimated by the $1/T$ model (3×10^5 hours instead of 3×10^3 hours).

It should be noted that intrinsic charge loss is not the limiting reliability issue in terms of data retention for Flash memories. Rather, it is the other issues such as stress-induced leakage current (SILC), and program/erase cycling-induced oxide degradation and charge trapping that are the main issues for data retention reliability. For this isolated study, other charge-loss mechanisms such as Fowler–Nordheim and direct tunneling that do not follow the Arrhenius law might be responsible for the charge loss. Currently, Arrhenius law is still the standard technique of temperature acceleration for nonvolatile memory failures, and the technique has served the industry well.

11.3.2 Charge-Loss and Gain Mechanisms in EPROMs and Flash EPROMs

In this section, we will discuss the different charge-loss/gain mechanisms for EPROM, and Flash EPROM-related charge-loss/gain mechanism are discussed in the next section.

11.3.2.1 Intrinsic Charge Loss. For UV-light-erasable EPROM, the write state has always been negatively charged with electrons, and different charge-loss mechanisms have been described in the literature [26, 27]. When charge is stored in the floating gate, electric field exists across the barrier oxides, which will tend to pull the charge out of the floating gate. This is the intrinsic charge loss. The device is designed such that it can retain the charge with the worst operation and storage conditions. This mechanism is not dominant since the thermal emission barrier height of 1.4 eV is considerably higher than the activation energies for the other charge-loss mechanisms.

Detail quantitative study of long-term charge-loss characteristics in a floating-gate EPROM cell with ONO (oxide–nitride–oxide) interpoly stacked dielectric were reported by Pan et al. [38, 39]. It was shown that the major cause of the long-term shift charge loss is due to electrons leaking out of the top oxide of the ONO stacks. Although Fowler–Nordheim injection usually dominates the carrier conduction across silicon dioxide, it is not applicable to the EPROM charge-loss case because the electric field is too small, less than 3 MV/cm. The long-term charge-loss rate is exponentially dependent on electric field.

11.3.2.2 Charge Loss Due to Oxide Defect. The second mechanism is related to defects in the oxide and results in both charge loss and gain, depending on the biasing conditions. It is typically of single bits, and these bits can be screened out in sort and burn-in testing. The activation energy for this kind of charge loss is about 0.6 eV as shown in Figure 11.26 [27].

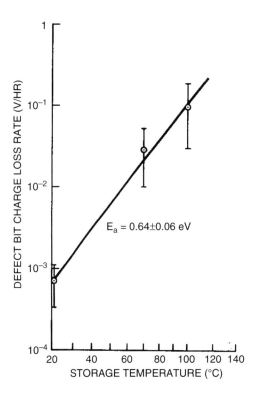

Figure 11.26. Plots of initial discharge rate vs. storage temperature with oxide defects. (Figure 7 of [27].)

11.3.2.3 Charge Loss Due to Contamination. The third mechanism is related to contamination. Positive ions entering the memory cell may compensate a part of the negative charge stored on the floating gate. These positive ions can come from photoresists that were used in the process if the floating gate is not probably sealed by oxide and interlayer dielectrics. Due to the nature of this mechanism, it typically shows in a cluster. It was found that the radius of the circular zone grows with the square root of the time. The activation energy for charge loss by contamination is about 1.2 eV and can be detected with high-temperature data retention bake. Figure 11.27 is a bit map display of a programmed device with contamination entering the array causing charge loss. Ultraviolet light brought the cell thresholds down to normal erased levels, but a subsequent bake caused a threshold increase again. It was reported that the presence of carbon in oxidizing ambient could also lead to EPROM charge loss [40].

11.3.2.4 Charge Loss through ONO. There are two oxides in a floating-gate memory device: the thermal oxide between the substrate and the floating gate, often called the first gate oxide, and the oxide (dielectric) between the floating gate and the control gate, often called interpoly oxide. Charge leakage is more likely to occur through the interpoly oxide than the gate oxide because the asperities at the rough floating polysilicon gate surface create field enhancement at the tips of the asperities. This reduces the average electric field or voltage that is required to produce a certain Fowler–Nordheim tunneling current by a factor of 2 to 5. Scaling down interpoly dielectric thickness is one of the most important subjects for Flash memory

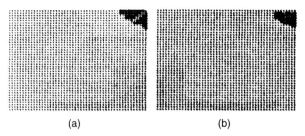

Figure 11.27. Bit-map display of (a) a programmed chip after it failed a 250°C retention bake and (b) the same device after UV erase and another 250°C bake. Failing cells here loss charge in (a) and gained charge in (b). (Figure 11 of [28].)

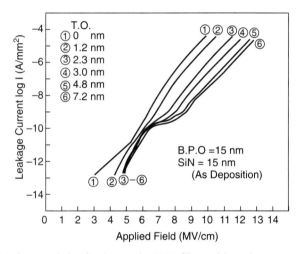

Figure 11.28. E–I characteristics for interpoly ONO films with various top oxide thicknesses. (Figure 18 of [30].)

scaling. For nonvolatile memory applications, the interpoly dielectric films must have low defect density and long mean time to failure, as well as good charge retention capability. ONO interpoly dielectrics are currently used for many nonvolatile memories due to its lower defect density and lower leakage current. Mori et al. [30, 32] and Wu et al. [31] investigated the ONO interpoly dielectric thickness scaling effect on charge retention characteristics in nonvolatile memories.

Figure 11.28 shows the $E–I$ characteristics for interpoly ONO film with various top-oxide thicknesses. Thicker anode side oxide results in smaller leakage current in the relatively high electric field region for ONO films formed on Si substrate and on poly-Si. This was attributed to the top-oxide blocking effects against hole injection. Figure 11.29 illustrates the energy band diagram for ONO structure and possible charge loss mechanisms. The $E–I$ characteristics for the ONO structure have three parts. In the low electric field region, Poole–Frenkel (PF) current of holes and

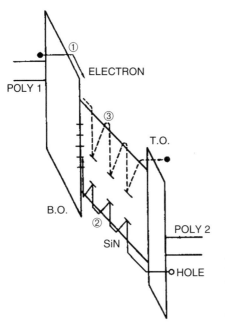

Figure 11.29. Possible leakage current through a ONO film. Component 1 is FN tunneling current through the bottom polyoxide. Components 2 and 3 are Poole–Frenkel (PF) hole and electron currents in the SiN film. (Figure 19 of [30].)

electrons in SiN (component 2 and 3 in the figure) is larger than the FN current in the bottom polyoxide (component 1). The hole conduction in SiN (component 2) is much larger than electron conduction (component) when the top oxide is thin and holes are easily injected from an electrode by direct tunneling. Holes injected by component 2 accumulate at the bottom polyoxide/SiN interface and enhance the electron conduction in the bottom polyoxide. The scaling and composition of ONO have to be defined carefully, considering both charge retention and dielectric reliability. A thicker anode-side top oxide can significantly reduce hole injection component 2 and should be kept above a certain critical value (~3 nm). SiN thickness scaling can improve charge retention and thinning down to 5 nm may be possible. However, SiN scaling leads to degradation in time-dependent dielectric breakdown (TDDB) characteristics, and too much thinning sometimes leads to degradation in charge retention capability. Thinning of bottom oxide down to 10 nm does not lead to degradation if bottom oxide quality can be well-controlled.

Wu et al. [31] also studied the charge retention characteristics of an EPROM cell with a stacked ONO film used as interpoly dielectric at elevated temperatures. It was found that there are three dominating charge-loss mechanisms through the ONO film. The first phase is an initial fast threshold voltage shift, which increases with nitride thickness. This phase is likely due to an intrinsic nitride property. The physical origin of the fast shift could be carrier movement in the nitride or a nitride polarization effect. The second phase is caused by electron movements within the nitride film. The majority of trapped electrons in the nitride or at the nitride/bottom oxide interface are introduced during programming. This electron transport inside the nitride follows a linear ohmic-like conduction. The activation energy for this phase has been found to be 0.35 eV. This is about the energy level difference of trap levels in nitride. It was believed that trap-to-trap hopping could be the transport mechanism of this second phase. The third phase is a long-term charge loss due to electrons leaking through the top oxide. This charge loss can be minimized with a thick top oxide.

11.3.2.5 Charge Loss Due to Charge Detrapping.
The reliability of channel oxide is very important for Flash EPROMs because these devices are programmed and erased with high electric fields. The current through the oxide degrades the quality of the oxide and builds up both positive and negative charges in the oxide. For Flash EPROMs with channel hot electron for programming, a typical programming condition is $V_{cg} = 12\,\text{V}$, $V_d = 5.5\,\text{V}$, and $V_s = V_{sub} = 0\,\text{V}$. The cell is biased in the so called $V_{fg} > V_d$ regime so that the direction of the oxide field favors electron injection. This causes oxide degradation such as electron trapping (N_{ox}) in the nearby oxide, and electron trap generation. Liang et al. [41] observed that preexisting electron traps in the oxide would trap-up electrons to a saturation level independent of the oxide field when it is below 5 MV/cm. At an oxide field larger than 5 MV/cm, electron trap generation, and the amount of electron, trapping can increase dramatically.

In a floating-gate memory, the presence of trapped electrons in the oxide near the drain edge can repel the channel hot-electron injection and reduce the injection efficiency, and thereby reduce the programming speed. This can be fairly significant in Flash memory cells where programmability degrades substantially after repeated program/erase (P/E) cycles. As a result of the electron trapping in the oxide, some of the charge that separates the different states of data are not stored in the floating gate but rather in the oxide traps. They can be detrapped and result in charge loss and data retention problems. Because of the low activation energy of these trapped electrons, detrapping happens even at room temperature. The only way to deal with them is to put enough charge on the floating gate so that even after all the trapped electrons are detrapped enough margin for the stored data is still maintained. This can be particularly problematic for Flash memories that store multiple bits in a single cell since the margin requirement is much more stringent. The other method to lessen this problem is to try to generate as few traps as possible the first place, such as reduced stress voltages, and trap reduction through nitridation [7–12].

11.3.2.6 Charge Loss Due to Cycling-Induced Oxide Damage.
One of the most challenging task for Flash memory is the ability to retain data and margins after the cells have been cycling hundreds and even million of times. Such a problem is typically related to Flash memories that use thin tunneling oxide near 100 Å. The oxide can be degraded due to hole generation and trapping in the oxide caused by high field during program or erase [21, 42]. Such a charge-loss mechanism could have activation energy as low as 0.25 eV. This poses a particularly difficult reliability problem due to the extremely long time needed to study, characterize, and to screen out such failures. We will discuss Flash memory cycling-induced data retention in more detail in the next section. The different-charge loss mechanisms and their activation are summarized in Table 11.1.

11.3.3 Flash EEPROM Cycling-Induced Data Retention Issues

For advanced Flash EPROMs, typically an endurance cycle of 100,000 to 1 million is required. The memory cells are either programmed with channel hot-electron

TABLE 11.1. Activation Energy of Different Charge-Loss Mechanisms

Failure Mode	Activation Energy (eV)	Reference
Intrinsic charge loss	1.4	27
Oxide defect	0.6	27
Ionic contamination	1.2	27
Tunnel oxide breakdown	0.3	33
ONO	0.35	31
Cycling induced charge loss	0.25–1.1	23

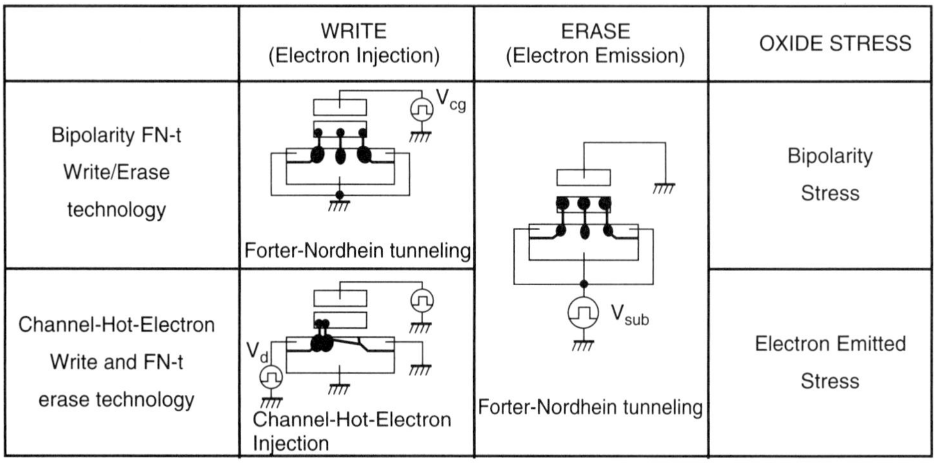

	WRITE (Electron Injection)	ERASE (Electron Emission)	OXIDE STRESS
Bipolarity FN-t Write/Erase technology	V_{cg} — Forter-Nordhein tunneling	Forter-Nordhein tunneling	Bipolarity Stress
Channel-Hot-Electron Write and FN-t erase technology	V_d — Channel-Hot-Electron Injection	V_{sub} — Forter-Nordhein tunneling	Electron Emitted Stress

Figure 11.30. Comparison between (a) bipolarity FN tunneling write/erase technology and (b) channel hot-electron (CHE) write and FN erase technology. (Figure 22 of [42].)

injection or FN tunneling, and erase with FN tunneling, either through source side, channel, or poly-to-poly. All these repeated program and erase subjects the oxide in the memory transistor to high electric field stress. Oxide can become more conductive after electrical stress results in SILC, electrons and holes can be trapped in the oxide, and then subsequently detrapped, and trapped holes can lower the oxide barrier and lower the activation energy. All these could affect the data retention characteristics of the Flash memory cells.

Aritome et al. [42] compared the data retention characteristics of Flash memory cells programmed by two different write/erase (W/E) technologies. Figure 11.30 illustrates these two W/E techniques. The first is the bipolarity W/E technology, which is a uniform write and erase technology. During the write operation, a high voltage (V_{cg}) is applied to the control gate, with the substrate and source–drain regions grounded. Electrons are injected from the substrate to the floating gate over the whole channel area of memory cells. In the erase operation, a high voltage (V_{sub}) is applied to the substrate and source–drain regions and the control gate grounded. Electrons are then emitted from the floating gate to the substrate. In this write/erase method, the high field stress of the thin oxide corresponds to bipolarity stress.

The other write/erase method utilizes channel hot-electron (CHE) injection for the write and FN tunneling for the erase. This technology is a nonuniform write and uniform erase technology. The erase operation is the same as that in the bipolarity W/E technology. However, during the write operation, high voltages are applied to the control gate and the drain. Thus, channel hot electrons are generated by the lateral electric field, and electrons are injected from the substrate to the floating gate. In this case, the high field stress of the thin oxide is electron injection stress. The data retention will be different for these two W/E technologies because the thin oxide leakage current is different for bipolarity and electron-emitted stress of thin oxide. Moreover, in the conventional Flash memory erasing method, a high voltage is applied to the source. However, in this experiment, a high voltage is applied to the substrate, as well as the source–drain region, in order to prevent the degradation of the thin oxide due to hole injection caused by the band-to-band tunneling.

Data retention characteristics are measured under various gate voltage conditions in order to accelerate the retention test. The memory cells are subjected to 10^5 program/erase cycles. In the case of the CHE write and FN tunneling erase technology, the stored positive charge rapidly decays as a function of time. As a result, the threshold voltage window decreases as shown Figure 11.31. However, in the case of the bipolarity FN tunneling W/E technology, data loss of the stored positive charge is significantly reduced. This phenomenon can be explained by the fact that the thin oxide leakage current is reduced by the bipolarity FN tunneling stress [43–45].

Figure 11.32 shows the data retention time after write/erase cycling as a function of the tunnel oxide thickness. The data retention time is defined by the time it takes V_{th} to reach -1.0 V during the gate voltage stress. In devices with a 7.5-nm

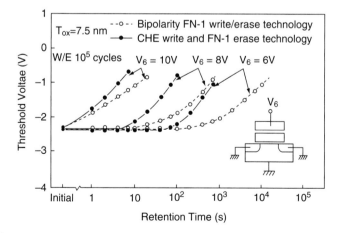

Figure 11.31. Data retention characteristics under gate voltage stress for bipolarity FN tunneling write/erase technology and channel hot-electron (CHE) write and FN tunneling erase technology. (Figure 24 of [42].)

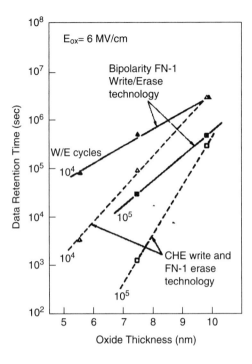

Figure 11.32. Data retention time of Flash memory cell after write and erase cycling as a function of tunnel oxide thickness. The data retention time is defined by the time at which V_{th} reaches -1.0 V during the applied gate voltage stress. (Figure 25 of [42].)

tunnel oxide, the data retention time obtained for the bipolarity FN tunneling write/erase technology is 50 times longer than that of CHE write and FN tunneling erase technology after 10^5 write/erase cycles. However, in devices with a 90-Å tunnel oxide, the data retention time is almost the same for both technologies. For very thin (<90 Å) tunnel oxides the bipolarity write/erase technology offers improved data retention times in comparison with the CHE write and FN tunneling erase technology. Therefore, this technology may facilitate the downscaling of tunneling oxides. Reducing the tunnel oxide thickness results in lower programming voltages and in faster read operations because the read current is increased.

Flash memory cell data retention lifetimes are typically determined by accelerated high-temperature data retention bake and calculated with the known activation energy. Such tests involve cycling a large number of devices to the specific number of write/erase cycles, then store the data at high temperature for extended period of time, and periodically check for the lost margin. Such tests are very time consuming and costly, and the activation energy value used is extremely critical for correct lifetime prediction. It was also observed that for some advanced Flash memory technologies that use source-side FN erase (such as ETOX), after extended write/erase endurance cycling, some single bits lose charge. The activation energy for such charge loss is about 0.25 eV, much less than the 0.5 to 0.6 eV used throughout the industry when classifying Flash EPROMs charge-loss failures. This single-bit charge loss resulted from trapped holes in the tunnel oxide of the affected cell due to extended write/erase cycling. After extended storage or read, the cell may lose charge, resulting in a random single-bit failure. It was believed [46] that the trapped holes generated during the source-side FN erase were responsible for such charge loss.

11.3.4 Data Retention Characteristics Related to Tunnel Oxide and Floating-Gate Poly Texture

For the ETOX stacked Flash memory, the overerase phenomenon has always been a serious reliability problem. Overerase is the phenomenon where the V_{th} of one or more cells becomes negative after the erase operation. As a result, the cell current can not be cut off during read even when the cell is not selected. This causes error in read as well as in program. Overerase can be caused by too wide a distribution of the memory cell V_{th} in the erased state or by erratic bits that release charge from the floating gate during erase at an excessively high rate. Almost all the Flash memory manufactures devote significant amount of efforts to investigate how to obtain a tight erase V_{th} distribution in the erased state, from all direction such as device, circuit, process, and materials.

In one study [47], Muramatsu et al. reported that overerase was strongly influenced by the grain size of the floating-gate poly-Si. The floating gate with smaller grain size realizes a narrower erase distribution. In their model, shown in Figure 11.33, an "oxide valley" region contains a high concentration of phosphorus and as a result is a high-conductivity site where overerase may occur. Small-grain poly-Si ensures more uniform erase and a tighter erase distribution because it creates a larger number of smaller oxide valleys. In manufacturing, small-grain poly-Si can be achieved by lowering the phosphorous doping to about $1 \times 10^{20} \text{cm}^{-3}$, which suppresses grain growth during annealing.

In their study of amorphous Si (a-Si) as a memory floating gate [48], Kubota et al. also found that by lowering the phosphorous doping concentration in the floating gate, stress-induced leakage current (SILC) is suppressed and data retention prolonged. In addition, they found that it is advantageous to deposit an a-Si film as the floating gate (FG). The a-Si turns into small-grain poly-Si films during annealing and results in improved SILC and longer data retention. The a-Si FG has a lower concentration of phosphorus at the FG–tunnel oxide interface than does the poly-Si FG, so the a-Si results in the same benefit as does lowering the phosphorous con-

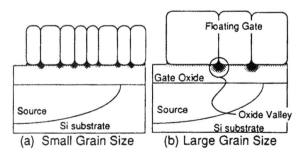

Figure 11.33. Schematic diagram of an oxide valley. A large number of small oxide valleys in one cell results in uniform erase speed. A smaller number of oxide valleys results in wide variation in the erase speed: (a) small grain size and (b) large grain size. (Figure 6 of [47].)

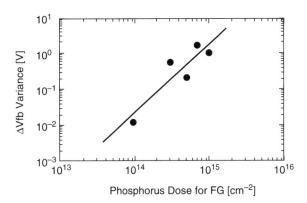

Figure 11.34. Cell data retention characteristics after 125°C 400h bake. ΔV_{th} variance is calculated from ΔV_{th} distribution of 30 programmed parallel cell arrays. The cells used in the study are conventional self-aligned stacked-gate structures with 11-nm gate oxide. (Figure 3 of [48].)

centration. Using arrays of 128,000 parallel cells, they varied the floating-gate phosphorous dose from 10^{14} to 10^{15} cm^{-2}. Figure 11.34 shows that lowering the FG doping density improves the data retention, reducing the V_{th} shift from the initial programmed state.

Using a 16-Mbit NAND Flash memory, Arai et al. [34] studied the effect of Si surface microdefects on the data retention characteristics. The cells have a stacked-gate structure with 90 Å, 0.32-μm^2 tunnel oxide. The write operation is executed by applying a programming pulse of 22 V, 3 ms to the control gate while grounding the source, drain, and P-well. The erase operation is executed by applying an erasing pulse of 22 V, 5 ms to the source, drain, and P-well while the gate is grounded. After 10^4 to 10^6 W/E cycles, the memory cells are programmed within 2.3 to 3.2 V, and within 3.9 to 4.7 V in two tests. The chips are stored at room temperature or baked at 125 °C, and then the device threshold voltage V_t distributions are measured.

Figures 11.35 and 11.36 show the results of 1000h of storage at room temperature after 10^5 and 10^6 W/E cycles for cells programmed to a window of 2.3 to 3.2 V and 3.9 to 4.7 V. The main population showed almost no shift, but some cells (called tail bits) experienced large charge loss through the tunnel oxide. The charge loss increases significantly when the electric field is larger than 1.4 MV/cm. This is why the V_t distribution of the tail bit is smaller with a lower programmed V_t. The number of tail bits increases with increasing endurance cycles.

Further experiments showed that some of the tail bits reappear after reprogramming and one more retention bake, but some behave as normal cells and do not exhibit overerasure. Apparently, the cells that exhibit anomalous leakage have two states: in one state a leakage current flows, and in the other state the mechanism of the leakage current vanishes or is inactive. By tracking the bits individually, Arai et al. [34] found that some of the bits are suddenly transformed into normal cells during retention test at room temperature. These observations indicate that the anomalous leakage current of a tail bit flows through one or a few spots where conduction is enhanced by hole trapping or trap-assisted tunneling. When the tun-

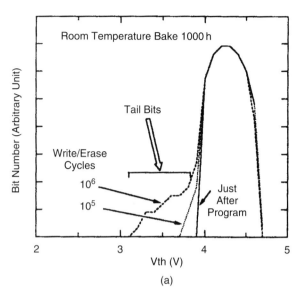

Figure 11.35. Measured V_t distribution before and after bake. Solid line indicates V_t just after programming. Dotted line and dashed line are V_t distribution after 1000 h bake at room temperature. (Figure 3(a) of [34].)

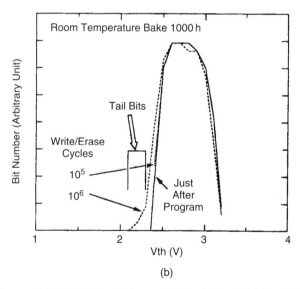

Figure 11.36. Measured V_t distribution before and after bake. Solid line indicates V_t just after programming. Dotted line and dashed line are V_t distribution after 1000 h bake at room temperature. (Figure 3(b) of [34].)

neling path is closed due to charge trapping or detrapping, the excess leakage stops. These findings are similar to what Ong et al. first reported [21] in their study of erratic erase.

Another important finding by Arai et al. [34] is that the same cycled device after 1000 h of bake at 125°C showed no tail bits. As shown in Figure 11.37, only an intrinsic shift of 0.3 V was observed due to intrinsic charge loss. This suggests that

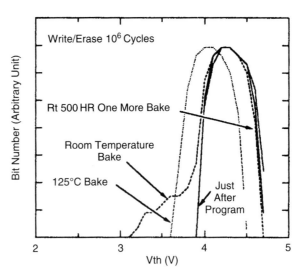

Figure 11.37. V_t distribution after 25°C and 125°C bake. After a 1000-h bake at 125°C, V_t shifted 0.3 V, but no tail bits were observed. No tail bits were seen either after an additional program operation and a 500-h RT storage. (Figure 9 of [34].)

the tail bit charge-loss mechanism was removed by the 125°C bake, perhaps by annealing out the traps or trapped holes. Reprogramming and a room temperature data retention test were done again, and this time no tail bits appeared. This is shown in the Figure 11.37. The V_t distribution after 500 h of room temperature storage almost overlaps the distribution just after programming. Clearly, the tail bits disappear after a 125 °C bake independent of the initial V_t, and the bake can inactivate or eliminate the leakage paths. Same results were also obtained after a 25 h of bake at 125 °C; however, a 4-h bake did not completely remove the tail bits.

The preparation and cleaning of the Si surface before the growing of the tunnel oxide, as well as the oxidation temperature are extremely important for good oxide quality and good data retention. For example, process damage of the Si substrate may induce weak points in the tunnel oxide, and if the oxidation temperature is not high enough, surface microdefects may be built in the tunnel oxide. Arai et al. [34] also investigated the effect of process damage on the tail bits. They adjusted the fabrication process to intentionally vary the Si surface microdefect density. As shown in Figure 11.38, after 10^6 W/E cycles, the tail bit generation rate strongly depends on the surface microdefect density. This dependence suggests a direction for process optimization of a more highly reliable tunnel oxide.

11.3.5 Soft Errors

Although Flash memories are sensitive to a variety of unique data-loss mechanisms discussed in this section, they are remarkably *insensitive* to radiation-induced soft errors. This insensitivity is important because soft errors in DRAMs and static random-access memory (SRAMs) can produce failure rates on the order of

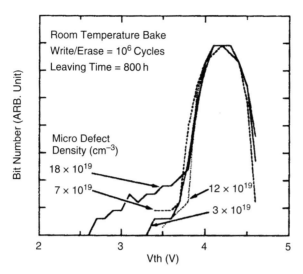

Figure 11.38. Number of tail bits is correlated with the Si surface microdefects. The surface microdefects are caused by impurity implantation, RIE damage, and stress. (Figure 12 of [34].)

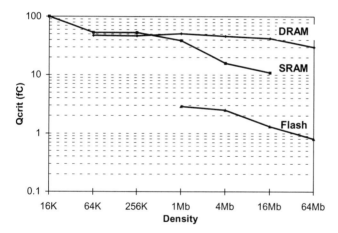

Figure 11.39. Critical stored charge in a single memory cell as a function of process generation. DRAM and SRAM data from [49]; Flash data previously unpublished.

1000 FITs, corresponding to a time-to-1%-fail of only about one year [56]—whereas soft errors have been undetectable in Flash and other nonvolatile memories [50, 51].

As shown in Figure 11.39, a current-generation Flash cell stores substantially less charge than does a current SRAM or DRAM cell. Since the amount of stored charge is the dominant determiner of soft-error rates in the volatile memories, it is clear that the Flash cell must lose very little charge when struck by a charged particle.

Figure 11.40 schematically illustrates a charge track caused by an incident charged particle, either an alpha particle or a recoil nucleus from a cosmic-ray event. A high-energy charged particle creates a charge track of electron–hole pairs in silicon by direct interaction of the charged particle with the electrons in the Si valence band. The generated carriers rapidly become thermalized as high-energy electrons and holes dissipate any excess energy in a cascade of follow-on pair creation.

In the case of a diffused node, in an SRAM or DRAM cell (left half of Fig. 11.40), minority carriers on either side of the junction are swept to the opposite side of the junction through a combination of drift (from the built-in field and formation of a "funnel" region) and diffusion. This flow of current is impeded by no energy barrier, and the fact that the carriers are thermalized does not inhibit the process. Depending upon the doping densities and the time constant of the memory cell (long for DRAMs, short for SRAMs), a diffused node will efficiently collect charge generated within tenths of a micron to as much as several microns below the junction.

The situation for a Flash cell is far different (right half of Fig. 11.40). Carriers generated in the floating gate or substrate silicon must overcome the 3.1-eV barrier between Si and SiO_2 if they are to change the amount of charge stored on the floating gate. Since the vast majority of generated carriers are thermalized, very few carriers can overcome this barrier. It is also possible for charge generated in the dielectrics (tunnel oxide or interpoly dielectric) to drift to or from the floating gate, but this process is inefficient because of the following:

1. The dielectrics are thin, about 10 to 20 nm, and therefore intercept a much smaller fraction of an incident particle's charge track than does the charge-collection regime below a diffused junction (typically 100 to 10000 nm).
2. Fewer pairs are generated per unit length of the charge track in dielectrics because the energy required to generate a pair is much lower in Si than in dielectrics (1 pair per 18 eV deposited in SiO_2, versus one pair per 3.6 eV in Si).

As a result of these effects, single-event upsets have not been observed in Flash or other floating-gate nonvolatile memories. In 1983 Caywood and Prickett [50] reported no upsets in an equivalent 2 million years of alpha exposure, and in 1998 Eto et al. [51] reported no upsets in an equivalent 20,000 years of cosmic-ray neutron exposure.

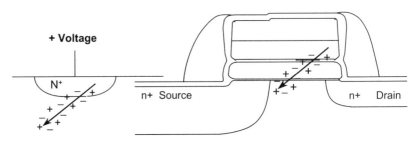

Figure 11.40. Charge generation caused by an incident charged particle. Left side: SRAM or DRAM node; right side: Flash cell.

11.4 FLASH MEMORY DISTURBS

Disturb is a phenomenon intrinsic to all memory arrays. Disturb occurs when reading, programming, or erasing some memory cells causes unwanted alteration to the data or margins of the memory cells at these or different locations. Write/erase cycling degrades the oxide and can make the Flash memory devices more vulnerable to disturb. Flash memory requires special considerations in design and processing to minimize the possible disturb problems. All possible disturb conditions have to be studied and taken into consideration for proper circuit and system designs.

11.4.1 Read Disturb and the Effects of Cycling

Read disturb is the unintentional changing of the content in a cell while reading that cell or another. This occurs because the voltage required to read the cells may alter the charges in the memory cells. Shown in Figure 11.41 is a portion of a NOR-type Flash memory sector organized as a 512 bitline × 1024 wordline array. During read, bitline voltage of about 1.0 V, and wordline voltage of 5 V are applied to the cell being read. The cell itself is stressed with the gate voltage of 5 V and drain voltages of about 1.0 V. All cells that share the same bitline as this cell are stressed with the read drain voltage but their gate voltage remains zero. All cells that share the same wordline as this cell are stressed with the read gate voltage of 5 V. The read disturb lifetime has to be longer than the de facto standard of 10 years.

If the cell that is being read is an erased cell, the channel is strongly on with 5 V on the gate. Even with only about 1.0 V on the drain, a small amount of charge may be injected into the gate, and the cell maybe disturbed by channel hot electron injection. The read disturb lifetime can be extrapolated by stressing the device at

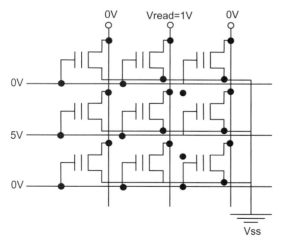

Figure 11.41. During the read operation of a typical NOR-type Flash memory array, drain voltage of about 1.0 V is applied to the selected bitline and gate voltage of 5 V is applied to the selected wordline. Many cells are stressed with the read drain voltage and many other cells are stressed with the read gate voltage. Finally, the cell being read is stressed with the gate voltage of 5 V and drain voltages of about 1.0 V.

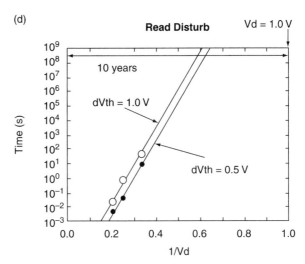

Figure 11.42. Read disturb lifetime is extrapolated by stressing the erased cell at higher drain voltages and monitoring the threshold voltage shifts. (Figure 7 of [52].)

higher drain voltages and $V_g = 5\,\text{V}$ and monitoring the cell threshold voltage shift. One example is shown in Figure 11.42. If the acceptable read disturb threshold shift is $\Delta V_{th} = 0.5\,\text{V}$. Then in order to have a read disturb lifetime larger than 10 years, the drain voltage during read has to be smaller than $(1/0.62)\,\text{V}$, or $1.6\,\text{V}$. In this example, the bitline voltage during read is $1.0\,\text{V}$, and the read disturb lifetime is much larger than 10 years. For the cells that share the same bit line, the read disturb is negligible with $1.0\,\text{V}$ on the drain and gate voltage $V_g = 0\,\text{V}$.

The cells that share the same wordline as the cell being read are stressed with the read gate voltage of $5\,\text{V}$ (V_{cc}). On a high-density Flash memory, many cells are put into this low-field stress condition when one byte is read. As has been discussed in Section 11.3, thin oxides exposed to high-field stress and high levels of charge injection may conduct via stress-induced leakage-current (SILC) at low field [53–57]. Usually the low field encountered in read disturb is not large enough to cause charge flow onto or off the floating gate. However, at the microscopic level SILC is not entirely intrinsic—the level of SILC varies widely from cell to cell, and cells falling into the upper probability tail of the SILC leakage can gain charge under a 5-V gate stress. Under prolonged or direct current (dc) read conditions, these tail cells can become programmed. As program/erase cycling requirements increase, and the Flash memory tunnel oxide thickness is decreased, the EEPROM becomes more susceptible to tunnel oxide leakage. It has been shown that the charge gain takes place by electron tunneling through a corrupted oxide barrier [53]. The barrier reduction is caused by positive charge trapping at the tunnel oxide in the vicinity of the source to tunnel oxide junction. The charge trapping is caused by hole generation during FN erase.

A typical test methodology is described in Brand et al. [53]. The read disturb tests are conducted on a 1-Mbit test chip. In the normal operation of Flash memory read, only one wordline is selected out of 1024 wordlines. The test chip provides a special mode in which the read disturb stress voltage was applied to every wordline

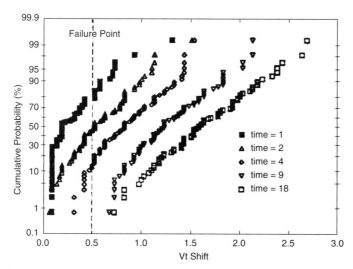

Figure 11.43. V_t shift distribution of the worst bit in arrays that have been cycled for 30,000 times. The read disturb stress voltage was 6.5 V at temperature of 125°C for different stress times. (Figure 2 of [53].)

in the array, so an acceleration factor of 1024 was achieved. For each stress condition evaluated, a group of 48 to 75 arrays were program/erase cycled, erased, and then stressed. The cycling is conducted at worst-case operation conditions of high temperature and high program and erase voltages. After cycling, each array is erased, and the array V_t was measured. The arrays are then stressed at the specified control gate voltage in an oven with elevated temperature. Periodically the arrays are removed to measure the array V_t.

The array read disturb failure rate is tracked by the cumulative probability of reaching a particular array V_t value. Figure 11.43 shows a plot of cumulative probability versus ΔV_t for arrays that were cycled 30,000 times and stressed at 6.5 V at 125°C. The plot shows that the distribution of the worst bits became normal. Then the entire distribution continues to shift toward higher V_t, at a rate that is logarithmic with time. The failure criterion is an increase of $\Delta V_t = 0.5$ V, and the time to a particular probability of failure can be interpolated from the cumulative probability plot. Since the probability of failure is lognormal with time, most failures occurs at a time on the order of the MTTF (mean time to failure). The time before the bulk of the failures occur can be thought of as a latency period. As long as the read disturb stress is applied to a cell for less than this latency time, there will be tremendously lower chance of a failure than if the stress time reaches the MTTF. The time to a single bit failure under constant stress conditions is reduced due to cycling as shown in Figure 11.44. The MTTF decreases by half with each decade of cycling, and there does not appear to be any saturation in the reduction of failure time to cycling. The reduced time to failure shows that cycling is responsible for the generation of the defects that caused read disturb charge leakage. There is a power-law relationship between time to failure and number of cycles.

As is typical for tunneling, there is a higher activation energy at low fields than at high fields. An empirical relation between failure time and stress tempera-

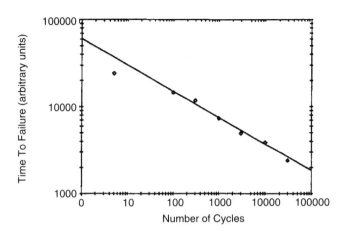

Figure 11.44. Time-to-failure vs. number of program/erase cycles. Cycling leads to generation of the defects responsible for read disturb charge leakage. The relationship between time to failure and number of cycles is a power law. (Figure 4 of [53].)

ture, stress voltages, and number of cycles is expressed in the following equation [53]:

$$t_m = t_0 \exp\left(\frac{-0.42}{kT}\right) \exp\left(\frac{\frac{3.6}{kT} - 65}{V}\right) (\text{cyc\#})^{-0.30} \qquad (11.9)$$

Here k is the Boltzmann constant, T is the temperature in Kevin, V is the stress voltage, and cyc# is the number of program/erase cycles the memory has received. This relation predicts $E_A = 0.30$ eV at 5.0 V, and $E_A = 0.09$ eV at 7.0 V. A low activation energy is consistent with the expectation that the leakage mechanism is tunneling.

There are several factors that determine the read disturb failure rate. One is the number of cycles, and the others are the stress time, voltage, and temperature. The number of cycles determines the probability of having created a leaky bit, and the stress time determines whether a leaking bit is able to reach the failure point. The actual failure rate of a memory part varies significantly depending on the distribution of cycles over the lifetime of the part and the total read stress time experienced by each row. For example, read disturb is minimized if the program/erase cycles are evenly distributed through the unit's lifetime, that is, different data are written before the previous data had enough chance to be disturbed, and there is very short stress time between cycles. For most applications, the Flash memory read stress time is less than 1 μs per read, which is extremely short, and Flash memory failure rate due to read is low, which is not a major concern. A well-designed memory has read disturb lifetime well over 10 years, while in actual applications the same cell is never read continuously for 10 years. It is fairly easy to accelerate the read disturb failures by operating the memory at high read voltage and dc stress mode where all rows are stressed at the same time.

11.4.2 Program Disturb

Program disturb is the unintentional changing of the content in an unselected location as a result of programming another location, such as a bit on the same bitline or wordline. This occurs because the high voltage required to program is not isolated from the unselected bits. Program disturb can be minimized by proper circuit and device design. Again, take the 512 columns by 1024 rows, 64-kbyte NOR-type Flash memory array as an example. Shown in Figure 11.45, the cells are programmed with channel hot-electron injection, and erased with negative gate erase between the gate and the common source junction.

During program operation, a drain voltage of about 6V is applied to the selected bitline, and a gate voltage of about 11 V is applied to the selected wordline. All cells that share this bitline are stressed with the program drain voltages with their gate voltages remaining at zero. This disturb condition is sometimes referred to as drain disturb. Some of the programmed cells might lose charge due to FN tunneling, and this causes loss of stored data or margin as shown in Figure 11.46. In the meantime some of the erased cells on the same bitline might gain some charge and this results in loss of erase margins due to hot-electron injection. Fortunately, the memory array is divided into 64-kbyte sectors so that the program disturb time is quite small.

The total program drain disturb time any bit will see before the whole sector is programmed can be calculated this way. Assume the typical program pulse width is 10 μs, the typical total program drain disturb time is 10 μs × 1024 rows, or about 10 ms. Even if after 100,000 program/erase cycling, program time increases from 10 to 50 ms, the new total program drain disturb time is only 50 μs × 1024 rows, or about 50 ms. This time is short, and typically drain disturb is not a serious problem for Flash memories operating in sector mode. It is important that different sectors do not share the same bitline, and the whole sector is erased before any rows in it are programmed a second time.

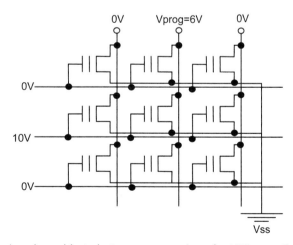

Figure 11.45. During channel hot electron programming of a NOR-type Flash memory, drain voltage of about 6V are applied to the selected bitline and gate voltage of about 10V are applied to the selected wordline. All cells that share this bitline are stressed with the program drain voltage. All cells that share this wordline are stressed with the gate voltage of 11V.

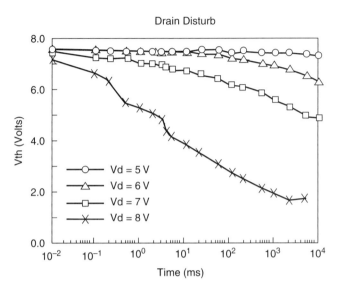

Figure 11.46. Drain disturb characteristics of a programmed cell versus drain stress voltages. This disturb is due to FN tunneling between the drain and the negatively charged floating gate. (Figure 7(a) of [52].)

Substrate-injection-induced program disturb was also observed [58, 59]. This disturb is caused by substrate injection of thermally generated electrons, and it disturbs an erased cell during programming of others on the same bitline. The memory drain junction is typically very shallow and heavily doped while the channel has a lower doping concentration. The possible source of substrate currents are band-to-band tunneling current in the N^+P junction, thermal generation, and the N^+P junction breakdown if drain voltage is close enough to the junction breakdown voltage. In an erased cell, with some positive voltage coupled to the floating gate by the drain, the electric field can accelerate some substrate hot electrons to the floating gate. The substrate-injection-induced program disturb is found to be accelerated by high temperature and has an activation energy of about 1.0 eV [59]. It was believed that the temperature dependence of the program disturb identifies the thermal generation as responsible for the source of the substrate current at least in that case. Different temperature acceleration has been reported in the literature. For example, Dunn et al. [58] reported a smaller temperature activation energy of 0.4 eV.

Similarly, during program, cells that share the same wordline as the program cell are stressed with the programming gate voltage of about 11 V, a stress sometimes called gate disturb. Two possible disturb phenomena exist. First, electrons may pass through the interpoly ONO dielectric layer to the control gate from the floating gate, which results in a loss of charge in the programmed cell. Second, electrons may pass through the tunnel oxide from the channel to the floating gate, which results in a gain of charge in an erased cell. For the Flash memory sector in this example, there are 512 bitlines. If the product is operated in the so-called ×8 mode, that is, 1 byte, or 8 bits are programmed at the same time, then it takes 64 program operations to finish programming one row of the memory. Assuming the

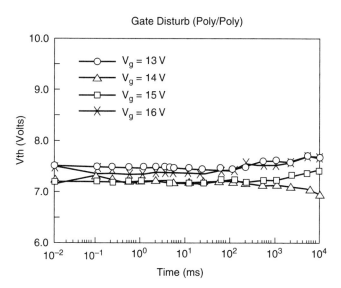

Figure 11.47. Program gate disturb characteristics due to poly1 to control gate leakage through the ONO dielectric for a programmed cell vs. different gate stress voltages. (Figure 7(b) of [52].)

typical program pulse width of 10 µs, then the total gate disturb time for any bit before the whole sector is erased can be calculated as 10 µs × 64 passes, or about 0.64 ms. Even if after 100,000 program/erase cycling, program time increased from 10 to 50 ms, the total gate disturb time is 50 µs × 64 passes, or about 3.2 ms. The gate disturb characteristics due to poly-to-poly leakage for a programmed cell are shown in Figure 11.47 as a function of different gate voltages. The gate disturb characteristics due to tunnel oxide leakage for a erased cell versus different stress voltages are shown in Figure 11.48. In both these examples, the amount of disturb is small for the voltage and time and is acceptable for actual device operations. In practice, in well-designed Flash memories, these disturb mechanisms are negligible for normal cells. The issue with these disturbs is a statistical one: What small fraction of cells are affected by oxide or ONO defects? Such defects will primarily cause failures at time zero—a quality DPM (defects per million) issue—but there can be some level of reliability fallout after cycling if the program time, and thus the disturb time, increases with cycling. If the DPM impact is significant (greater than, say, 100 DPM), then care must be taken in the test flow for the product to screen these defects through disturb stresses. This disturb stressing should anticipate worst-case conditions that may occur in field use, such as longer program times after cycling and at high use temperatures.

A more direct reliability issue is the degradation in the time to failure induced by cycling for the tunnel-oxide gate-disturb mechanism if the erase step injects holes into the tunnel oxide. Figure 11.49 schematically illustrates the hole injection mechanism in a Flash cell erased wholly with a positive source voltage (top gate grounded). Band-to-band tunneling from the high vertical field in the gate–source overlap region generates low-energy electron–hole pairs. The large horizontal field, from the

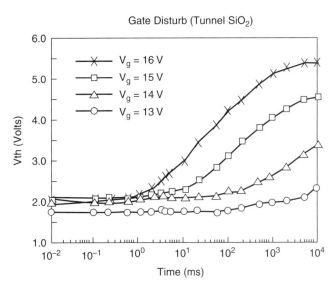

Figure 11.48. Program gate-disturb characteristics due to tunnel leakage for a erased cell vs. different gate stress voltages. (Figure 7(c) of [52].)

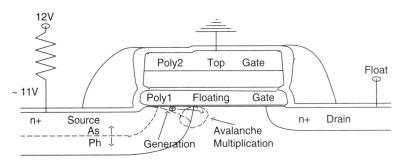

Figure 11.49. Generation of hot holes during Flash erase.

12-V source bias, accelerates the holes, resulting in hot holes at the oxide interface near the junction edge. Since the floating gate is negative at the beginning of erase, there is an attractive vertical field that promotes the injection of these hot holes into the tunnel oxide where many become trapped. As is well known, trapped holes reduce the voltage required for electrons to tunnel through an oxide because of the internal field generated by the holes. As a result, charge gain during gate disturb is increased.

Figure 11.50 shows how cycling can increase the amount of charge gain from this disturb mechanism. A small (4000-cell) array was put through cycling, which was periodically interrupted so that the array could be erased to about 3.2 V and then disturbed at accelerated voltage. Figure 11.50 shows the postdisturb V_t versus cycle count. The disturb shift increases dramatically in the first 100 to 1000 cycles and then saturates. Such saturation behavior is typical with hole-trapping mechanisms, such as the "window widening" effects observed in earlier generations of

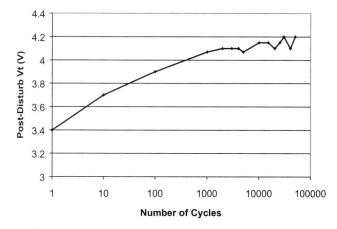

Figure 11.50. Postdisturb cell V_t as a function of cycles.

EEPROMs. The cycling degradation of about 0.8 V in Figure 11.50 corresponds to about an order-of-magnitude reduction in time to fail; this degradation needs to be built into product design and screening.

Most current-generation Flash products use negative-gate erase, which dramatically reduces the source voltage during erase (from ~12 to ~5 V). As one would expect, the lower source voltage dramatically reduces the amount of hole injection, and the degradation effect described above becomes much less of an issue.

11.4.3 Erase Disturb

Erase disturb is the unintentional changes of the content in a cell while erasing another cell. This occurs because the high voltage required to erase is not isolated from the unselected cell. This is usually not as serious a problem as program disturb. Flash memories are erased in sectors or blocks so that there is no erase disturb within a sector. In actual Flash memories, the erase voltages are always isolated between the different erase sectors or blocks. During erasure, the different sectors and blocks are very much independent of each other, and no disturb occurs.

11.4.4 Block-to-Block Disturbs

Depending upon the architecture of the Flash array, different blocks may or may not be entirely isolated from each other during program/erase cycling. If they are not isolated, disturbs can be a major concern. Figure 11.51 shows an example in which two adjacent blocks share wordlines. When one cell is programmed, cells on the same wordline see a gate-disturb stress, *even if those cells are in another block*. If in a field application the first block gets cycled 10,000 times, and the second block never gets cycled, then the second block is subjected to a gate-disturb stress 10,000 times longer than the times calculated earlier in this section. If blocks are to share any program or erase voltages (gate, drain, or source), then very careful attention must be paid to this issue. Because of this issue, it is good practice in Flash product qualifications to cycle each block to its maximum allowed number of cycles (i.e., 100,000 cycles) and to check after each block is cycled in this way that none of the

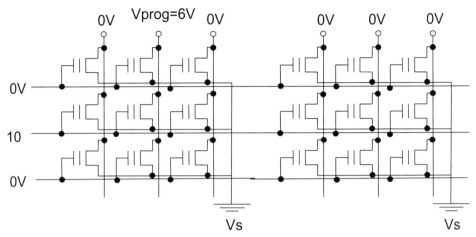

Figure 11.51. Block-to-block disturb.

other blocks were disturbed. If one performs cycling by programming and erasing the entire array, one will be blind to possible block-to-block disturbs.

11.5 STRESS-INDUCED TUNNEL OXIDE LEAKAGE CURRENT

As Flash memory devices are scaled down, it is advantageous to scale down the thickness of the tunnel oxide to reduce the internal programming and erase voltages. This would improve the shrinkability of the memory cell and the periphery high-voltage circuitry. Most commercially available Flash memories are programmed and/or erased by high-field (Fowler–Nordheim) injection of electrons through a very thin tunnel oxide to charge or discharge the floating gates. The ETOX stacked-gate Flash memory is erased by source-side FN tunneling. The DiNOR and AND-type stacked-gate Flash memory are programmed by drain-side FN tunneling and erased by channel FN tunneling. For the NAND-type Flash memory, both program and erase are also accomplished through channel FN tunneling. The few exceptions are SanDisk's and SST's (Silicon Storage Technology) Flash memory cells that use field-enhanced poly-to-poly erase and employ thick oxides.

It is well known that high-field stressing of the gate oxides increases the low-field leakage current known as stress-induced leakage current (SILC) [25, 54, 56, 57, 60]. SILC can cause serious limitations on data retention and read disturb in Flash memories, which become more severe with tunnel oxide thickness scaling [25, 61]. SILC has been extensively studied over the past several years [9, 25, 42–45, 54, 56–58, 60–90]. SILC has been modeled as a trap-assisted tunneling process caused by the generation of oxide electron traps from high-field stressing. Trap-assisted SILC has been linked to soft breakdown in thin oxides. The stress-induced leakage current depends on many factors, such as the stress field, time, waveform, polarity, oxide thickness, and oxide fabrication processes. SILC remains the major barrier to tunnel oxide scaling. Great effort has gone into reducing tunnel oxide trap generation through methods such as N_2O nitridation [9, 12, 83, 91–95].

In this section, we will first discuss the experimental results related to SILC in Flash memories, SILC as a function of tunnel oxide thickness, stress conditions, and temperatures, focusing on oxide thickness in the practical range of 70 to 130 Å for Flash memory. The trap-assisted tunneling model will be presented. We will then discuss SILC under unipolar and bipolar stress, corresponding to the operations of different types of Flash memories, the microscopic and statistical SILC results, and the process factors that affect the tunnel oxide SILC characteristics such as thermal budget, polysilicon gate concentration, and N_2O oxynitridation process.

11.5.1 Uniform SILC in Thin Oxide

The current flow through a thin oxide degrades the quality of the oxide and causes wearout of the tunnel oxide. The reliability of thin oxide has been investigated extensively by many authors. It has been found that high-field stress induces a low-field leakage current in thin oxides with thickness less than 100 Å [25, 54, 56, 57]. This may or may not be a problem for logic circuits, but it is most certainly a problem for Flash memory because it leads to read disturb and data retention failures.

Figure 11.52 shows the oxide current density versus electrical field before and after stress. The measurements were performed on capacitors having 51 to 96 Å thickness oxides. It can be seen that the oxide leakage at low electric field, induced by the charge injection stress increases as the oxide thickness decreases. A stress of 100 mA/cm^2 and time of 5 s results in a fluence of 0.5 C/cm^2. Assuming a Flash memory cell having a gate coupling ratio of 0.5, and a program/erase V_t window of 6 V, the amount of charge passing through the oxide per program/erase cycle is 10^{-6} C/cm^2. The fluence of 0.5 C/cm^2 in this study represents the stress of 500,000 program/erase cycles.

Moazzami and Hu [57] studied SILC for oxide thicknesses between 60 and 130 Å and attributed the leakage to trap-assisted tunneling of electrons. SILC is easily detectable in very thin oxides (<80 Å), but in this study SILC was observed even in oxides thicker than 100 Å. Figure 11.53 shows the I–V characteristics sub-

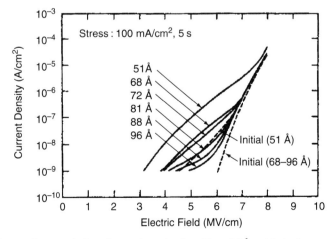

Figure 11.52. J–E characteristics of capacitors having 51 to 91 Å oxide before and after charge injection stress. (Figure 1 of [56].)

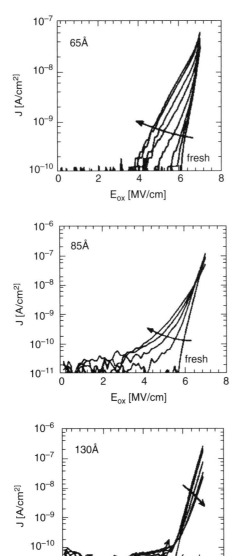

Figure 11.53. Conduction characteristics following FN stress of (a) 65 Å, (b) 85 Å, and (c) 130 Å oxide. The capacitors are stressed with a constant field at 9.5 MV/cm for increasing time. The initial current observed below 1 MV/cm after stress in (c) is attributed to filling of slow interface states generated during stress. (Figure 1 of [57].)

sequent to FN stress. There are two distinct effects that occur as a result of the stress. At low fields (≤ 1 MV/cm), there is a low-field leakage that is attributed to the filling of slow interface states generated during the stress. This current appears to be caused by charging of the interfacial traps that are observed to discharge when the bias is removed. Stress-induced current is also observed at fields near the onset of tunneling (3 to 5 MV/cm). Both current components are very reproducible even after repeated voltage scans.

The enhanced conduction in the thicker oxides is not a steady-state current but follows a t^{-1} decay over long time periods as shown in Figure 11.54(b). As more

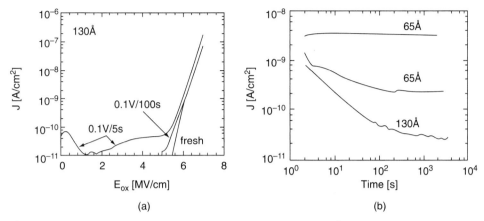

Figure 11.54. (a) Stress-induced current in oxides greater than 75 Å is a function of the voltage ramp rate suggesting that it is not a steady-state current. (b) As the oxide thickness increases, stress-induced current is transformed from a steady-state current to a transient current. The current is measured at a field of 5.4 MV/cm following stress at 9.5 MV/cm for 20 to 30 C/cm^2. (Figure 2 of [57].)

electrons are injected during the stress, the magnitude of stress-induced current increases. However, after a sufficient electron fluence (typically several C/cm^2), the current saturates at a level that is dependent on both the stress field and the field at which the stress-induced current is monitored. The dc stress experiments provide reasonable estimates for the actual magnitude of stress-induced current, since very little frequency or duty cycle dependence was found under pulse dc conditions. The generation mechanism of the stress-induced current does not exhibit large temperature acceleration. The activation energy for stress-induced current following stressing at different temperatures is only 0.14 eV, while the current–voltage characteristics following stress at 150°C have an activation energy of only 0.11 eV. The stress-induced current is also observed to decrease after high temperature anneals as shown in Figure 11.55. Almost identical stress-induced current was observed in both polarities following stress in either polarity, suggesting a symmetric conduction mechanism.

There have been many theories to explain SILC. Early work on thin SiO$_2$ films postulated charge-assisted tunneling via positive-charge traps [86] or thermally assisted tunneling at stress-induced weak spots with lowered barrier heights at the Si/SiO$_2$ interface [54]. The most convincing model is sequential trap-assisted tunneling via neutral electron traps proposed in Moazzani and Hu [57], and supported by many others [67, 68, 75, 79].

From experimental observations over a wide range of oxide thicknesses, Moazzami and Hu [57] concluded that the conduction mechanism is trap-assisted tunneling. As the density of trap generated within the oxide increases with stress duration, there is a higher probability for direct electron tunneling into trap sites near the cathode. For thin oxides, there is also a high probability of direct tunneling out of trap sites into the anode, and steady-state current flows when there is equilibrium between the trap filling and emptying processes. For the 65-Å case in Figure 11.54, the traps are extremely efficient for electron transport through the oxide and no

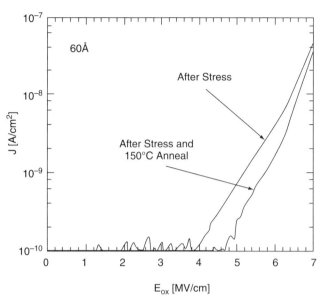

Figure 11.55. A 150°C anneal for 15 min is sufficient to reduce the stress-induced current created by room temperature stress. (Figure 8 of [57].)

decay is observed in the stress-induced current. As the oxide thickness increases, the initial stress-induced current is higher than the steady-state value (e.g., the 85-Å case), since a large fraction of the initial current is due to electron flowing from the cathode into the traps, that is, due to trap filling. For even thicker oxides, the probability for field emission out of the traps is extremely low. Therefore, the stress-induced current gradually decays as the trap-filling process is completed. By integrating the stress-induced current for a thick oxide, it is possible to estimate the trap density within a tunneling distance from the cathode. Whenever an applied bias is removed, a reverse current flows as the traps are discharged. Both the stress-induced current transient and the trap discharge current follow the t^{-1} dependence predicted by the trap-assisted tunneling model [87].

Studies by Runnion et al. [63] also reveal similar thickness dependence of stress-induced leakage current in thin oxide. They have shown that SILC consists of two components, transient and dc. Both components were due to trap-assisted tunneling charging and discharging of the stress-generated traps near the two interfaces. The dc component is caused by trap-assisted tunneling through the oxide. For oxide thicknesses most relevant to Flash memory applications, that is, around 10-nm-thick oxides, the majority of the SILC is the transient current as shown in Figure 11.56. Transient SILC has been observed to cause a short-term threshold voltage decay after programming in EEPROMs [89].

The model describing the trap-assisted tunneling processes as a function of the oxide thickness is shown in Figure 11.57. During the high-voltage stress, traps were generated relatively uniformly throughout the oxide [90]. In two oxides, one thin and one thick with equal trap densities, the dc component of the SILC would be higher in the thinner oxide. It would take only one tunnel transition for an electron to traverse from the cathode to the anode in a thin oxide. Both the dc components

STRESS-INDUCED TUNNEL OXIDE LEAKAGE CURRENT

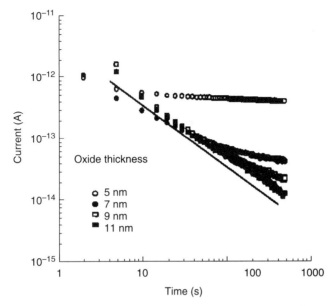

Figure 11.56. Time-dependent SILC measured after stressing at 1.4C/cm² for oxide thickness range from 5 to 11 nm. The stress field was 10 MV/cm, and the SILC was measured at 4 MV/cm. (Figure 9 of [63].)

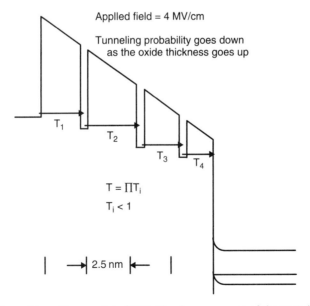

Figure 11.57. Tunnel transition model of SILC. The dc component of the SILC depends on oxide thickness and the trap density. (Figure 15 of [63].)

of the SILC and the transient $1/t$ components of SILC are proportional to the trap densities in the oxides. In the thinner oxides, the electrons have a high probability of tunneling completely through the oxide. In the thicker oxides, the tunnel charging and discharging of the traps near the interface dominates the conduction process.

11.5.2 SILC in Thin Oxide after Bipolarity Stress

In the previous section we discussed SILC in thin oxides after unipolar stress, which is similar to the stress present during the erase of DiNOR and AND, as well as the ETOX-type cells when channel erase is used. However, for the NAND-type Flash memory, both program and erase are accomplished through channel FN tunneling. Electrons are injected into and removed from the floating gate through the same path, and the tunnel oxides are subjected to bipolar stress.

It has been reported that thin oxides have longer time-dependent dielectric breakdown (TDDB) lifetime under bipolar stress than under unipolar stress [43–45]. The breakdown time under unipolar stress does not depend on frequency. However, as shown in Figure 11.58, the breakdown time under bipolar stress strongly depends on the frequency. The bipolar stress lifetime in the range of 10 khz to megahertz is about 40 times higher than the unipolar stress lifetime. The Flash memory FN tunneling W/E occurs in this range, as typical Flash memory FN tunnel W/E pulses are in the range of a few microseconds to 10 ms. It is suggested [43] that the improvement in TDDB and SILC under bipolar stress is due to the confinement of the holes near the anode under the alternating field. For the same reason, the interface trap generation is enhanced under the bipolar stress conditions because hole trapping at or near the interface leads to the generation of interface traps [44].

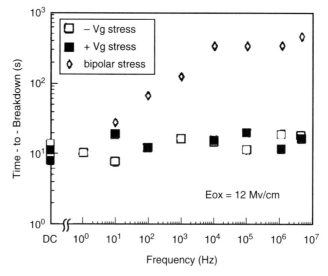

Figure 11.58. Thin oxide t_{BD} under bipolar and unipolar stress. The lifetimes under dc and unipolar dynamic stress conditions are similar but t_{BD} is longer and is frequency dependent under bipolar stress. T_{ox} = 110 Å. (Figure 2 of [45].)

Aritome et al. [42, 88] studied the influence of the waveform and polarities of the stress voltage on the degradation of the tunnel oxide. Three types of high-field dynamic stress were used to study the thin oxide leakage currents. Figure 11.59 shows the applied dynamic stress waveforms. In the case of bipolarity stress, positive high voltage is alternately applied to the gate and to the substrate–source–drain regions. FN tunneling occurs from the substrate to the gate and from the gate to the substrate alternately. The applied gate voltage is comparable to the floating-gate voltage of a memory cell during the write/erase operation. The substrate voltage is chosen in such a way that the tunnel currents in the two directions are the same.

The I_g–V_g characteristics before and after dynamic stress for a 7.8-nm oxide thickness are shown in Figure 11.60. The SILC of thin gate oxide subjected to three types of simulated W/E stressing are compared. It is observed that the thin oxide leakage current induced by bipolarity stress is about one order of magnitude smaller than that reduced by both the electron-emitted stress and the electron-injected stress. This result shows that the origin of the thin oxide leakage current can be suppressed by a reverse high-field stress. According to Rosenbaum et al. [43], bipolarity stress confines the hole trapping to regions near the interfaces and reduces the generation of traps near the center of the oxide. Continuous SILC depends on traps near the center of the oxide.

Most SILC investigation used capacitors as test devices. However, when capacitors are used, it is difficult to apply the same stress as that actually experienced by the tunnel oxide of a memory device. In the studies using capacitors as test devices, typically a constant current stress is used. However, in actual Flash memory operations, the stress is not a constant current stress. Also, large area capacitor characteristics may not represent the actual memory device with extremely small channel length and width with complications introduced by different isolation technologies. Satoh et al. [61] directly calculated the SILC from the read disturb characteristics of the Flash memory cell and investigated the degradation mechanism of the read

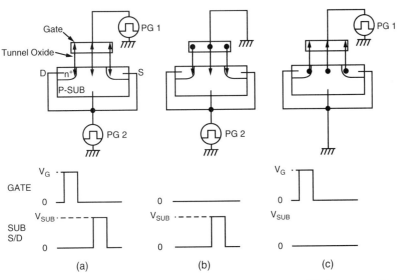

Figure 11.59. Setup and stressing waveform for (a) bipolarity stress, (b) electron-emitted stress, and (c) electron-injected stress. (Figure 1 of [88].)

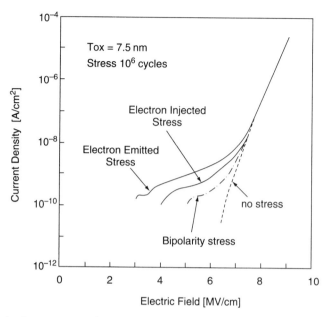

Figure 11.60. Leakage current of tunnel oxide of 7.5 nm oxide after bipolarity stress, electron emission stress, and electron injection stress. In the case of bipolarity stress, oxide leakage current is the smallest. (Figure 2(b) of [88].)

disturb characteristics over a wide range of tunnel oxide thickness under bipolar W/E stress. The W/E conditions corresponded to the operation of an NAND memory. They concluded that the read disturb characteristics is determined by the steady-state region of the SILC and that a high-temperature program/erase and read operation increases the stress-induced leakage for an oxide thickness of 5.7 to 10.6 nm.

The SILC can be directly calculated from the threshold voltage shift (ΔV_{th}) of the memory cell during a read operation. Figure 11.61(a) shows the read disturb characteristics after 10^6 W/E cycles. Using this data, the SILC can be calculated by

$$I_{Leak} = C_{ONO} \frac{\Delta V_{th}}{\Delta t} \qquad (11.10)$$

where I_{Leak} is the SILC, C_{ONO} the capacitance between the control gate and the floating gate, V_{th} the threshold voltage measured on the control gate, and t the read disturb time during the read stress.

The threshold voltage (V_{th}) is determined by the floating-gate charge (Q_{fg}) and the oxide trapped charge (Q_{ot}), which consists of trapped electrons and/or holes. During the read disturb, the Q_{fg} change is caused by the electron injection from the inversion layer to the floating gate through SILC. The Q_{ot} change is caused by the trapping or detrapping of carriers in the tunnel oxide. Therefore the SILC calculated has two terms. One is the differential of the Q_{fg}, which is described as the steady-state leakage current; the other is the differential of the Q_{ot}, which

STRESS-INDUCED TUNNEL OXIDE LEAKAGE CURRENT

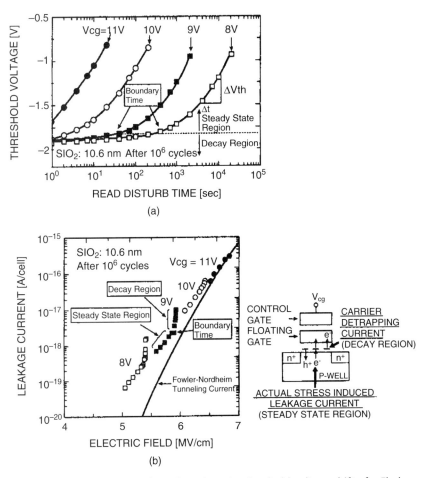

Figure 11.61. Characterization of SILC based on the threshold voltage shift of a Flash memory cell under the read disturb condition. (Figure 2 of [61].)

is described as the time decay leakage current. Therefore, the SILC can also be written as

$$I_{Leak} = -\frac{dQ_{fg}}{dt} + \frac{C_{ONO} + C_{OX}}{C_{OX}} \frac{dQ_{OT}}{dt} \qquad (11.11)$$

where C_{OX} is the capacitance of the tunnel oxide. It is assumed here that Q_{ot} is localized near the Si–SiO$_2$ interface. The electric field over the tunnel oxide is a function of the floating-gate voltage (V_{fg}) and is given by

$$E_{OX} = \frac{V_{fg} + \phi_f - \phi_s}{T_{OX}} \qquad (11.12)$$

where ϕ_f is the Fermi potential of the floating gate, ϕ_s the surface potential of the P-well, and T_{ox} the tunnel oxide thickness. The floating-gate voltage is given by

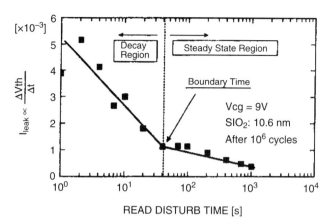

Figure 11.62. Plot of the differential of the threshold voltage vs. the read disturb time. A decay region and a steady-state region were observed. (Figure 3 of [61].)

$$V_{fg} = \frac{C_{ONO}}{C_{ONO} + C_{OX}}(V_{cg} - V_{th}) + V_{fgth} \qquad (11.13)$$

where V_{cg} is the control gate voltage, and V_{fgth} the threshold voltage as measured at the floating gate of the memory.

As shown in Figure 11.61(b), the leakage current has two regions, a decay region and a steady-state region. To further illustrate this, the differential $\Delta V_{th}/\Delta t$ is plotted in Figure 11.62. At the beginning of the read stress after the W/E cycling, $\Delta V_{th}/\Delta t$ quickly decays. In this region, the leakage is caused by both the fast decay of the SILC and the decay of the number of trapped carriers in the tunnel oxide, which is generated during W/E cycling. After this decay region of $\Delta V_{th}/\Delta t$, a steady-state region is reached in which $\Delta V_{th}/\Delta t$ gradually decreases with the read disturb time. In this steady-state region, the gradual decrease of $\Delta V_{th}/\Delta t$ is caused by the oxide electric-field dependence of the SILC. Very low level leakage current (10^{-20} A) can be evaluated by this method.

Figure 11.63 shows the stress-induced leakage current after 10^6 W/E cycles for oxide thicknesses ranging from 5.7 to 10.6 nm. The SILC increases greatly as the tunnel oxide thickness decreases. As shown in Figure 11.64, the threshold voltage shift of the decay region after 10^6 W/E cycles is independent of both the control gate voltage during the read disturb condition and the tunnel oxide thickness for the range of 5.7 to 10.6 nm. Since the initial threshold voltage shift is only about 0.1 V after 10^6 W/E cycles, the read disturb lifetime is not determined by the decay region but mainly by the steady-state region. Therefore, with respect to the read disturb lifetime, it is important to reduce the saturated leakage current, rather than the time-dependent leakage current.

The high-temperature (125°C) write/erase and read operations degrade the read disturb characteristics further as compared to room temperature operation. This is shown in Figure 11.65. The steady-state leakage current measured at 125°C after 125°C W/E cycling operation increases about three times in comparison with a device cycled and read at room temperature. This is true for oxide thickness ranging from 6 to 11 nm. Therefore, a high-temperature accelerated test can be used for W/E-cycling-related read disturb tests.

STRESS-INDUCED TUNNEL OXIDE LEAKAGE CURRENT

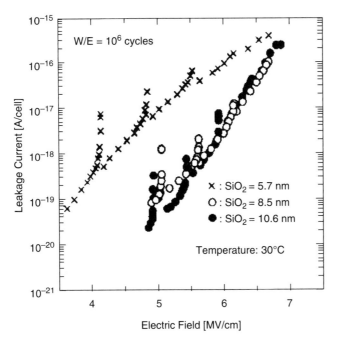

Figure 11.63. The SILC after 10^6 W/E cycles in the tunnel oxide with a thickness from 5.7 to 10.6 nm. (Figure 5 of [61].)

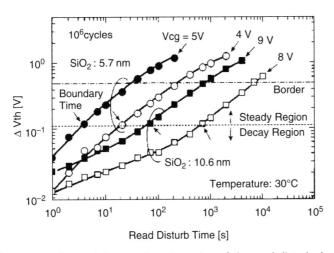

Figure 11.64. Decay region and the steady-state region of the read disturb characteristics for 5.7 and 10.6 nm oxide. The ΔV_{th} in the decay region is about 0.1 V, so the read disturb time is determined by the steady-state region. (Figure 6 of [61].)

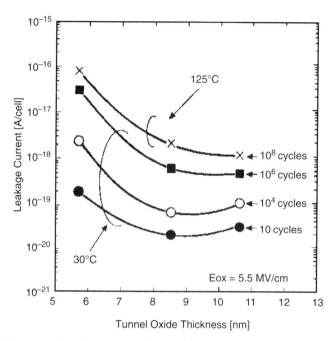

Figure 11.65. SILC as a function of tunnel oxide thickness and W/E and read temperatures. At 125°C, the steady-state leakage current increases about 3 times as compared with room temperature. (Figure 7 of [61].)

11.5.3 Microscopic Characteristics of Stress-Induced Leakage Current (mSILC)

All the above studies have been on large flat capacitors or single memory cells. However, to ensure data retention in a Flash memory product, the characteristics of billions of cells have to be studied. Statistical variation of the cell characteristics have to be accounted for in the design, process development, and product testing. Characteristics of stress-induced leakage current at the microscopic cell level (mSILC) have been studied by analyzing data on a large numbers of memory cells [66, 70]. It is found that the SILC characteristics fluctuate in the microscopic regions but follow the Fowler–Nordheim field dependence. The α's and β's obtained by measuring the mSILCs are widely distributed and have a strong exponential relation.

By measuring 28,800 stacked-gate transistors, microscopic investigation of 28,800 mSILC characteristics are studied for each 0.15-μm² area. The transistor has a width of 0.3 μm, length of 0.5 μm, tunnel oxide thickness of 10 nm, and poly-to-poly dielectric thickness of 18 nm. Current stresses are applied to all transistors by FN W/E operation of 10,000 cycles. Afterward, V_t shifts of each transistor is monitored at the resolution of 5 mV up to 300 h and microscopic E–J characteristics are calculated as described in the previous section. The 28,800 mSILCs show a range of SILC characteristics, and most of them fit the FN formula well as shown in Figure 11.66. The 28,800 mSILCs yield 28,800 pairs of α's and β's of the following FN equation:

STRESS-INDUCED TUNNEL OXIDE LEAKAGE CURRENT

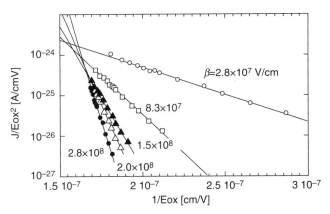

Figure 11.66. Typical FN plots and fitted lines of mSILC characteristics obtained. They fit the FN formula well, though the values of β vary greatly. (Figure 2 of [66].)

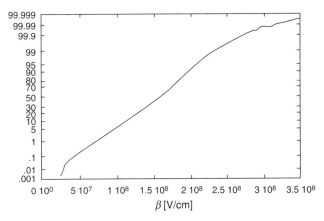

Figure 11.67. β distribution obtained by analyzing 28,800 cells. β is distributed widely from the intrinsic SiO$_2$ value to 2×10^7 V/cm. (Figure 1 of [66]

$$J = \alpha E_{OX}^2 \exp\left(-\frac{\beta}{E_{OX}}\right) \quad (11.14)$$

The β value ranges from 2.5×10^8 V/cm, which is almost the intrinsic β value of SiO$_2$ films, to 2×10^7 V/cm as shown in Figure 11.67. The continuous distribution of β indicates that the mSILC is not caused by singular defects in SiO$_2$, but by statistically distributed defects. In other words, the characteristics of SILC in a large flat capacitor only show the average characteristics of microscopic SiO$_2$ defects. It was also found that there is a strong exponential relation between the α's and β's as in the following equation, though both the α's and β's fluctuate widely:

$$\alpha = 9.77 \times 10^{-25} \exp(-1.96 \times 10^{-7} \beta) \quad (11.15)$$

As described above, SILC is due to trap-assisted electron tunneling. The β distribution of mSILC is caused by the fluctuation in trap-trap distance between the critical trap pairs. The exponential relation between the α's and β's can be qualitatively explained by the electron tunneling process between the current-limiting traps, that is, the traps with the longest distance.

11.5.4 Stress-Induced Leakage Current in Oxynitride

Many factors affect the tunnel oxide SILC. The quality of the tunnel oxide is a function of the quality of the starting substrate, the oxidation temperature, the incorporation of nitrogen, the cleaning process prior to the oxidation, and the floating-gate doping concentration [12, 84, 85, 91, 92]. For example, in Figure 11.52, the 96-Å oxide showed considerable SILC after 100-mA/cm² stress for 5 s. However, in Kimura and Koyama [75] a 92-Å capacitor was stressed with 100-mA/cm² constant current. Even after 100-s stress, with a fluence of 1×10^1 C/cm², no SILCs were observed. The differences of these results may reflect the progress being made in Flash memory tunnel oxide fabrication technology.

It has been observed by many that N_2O nitridation of SiO_2 reduces electron trap generation and suppresses SILC [9, 12, 83, 91–95]. It is a promising approach to facilitate scaling of the tunnel oxide thickness. Fukuda et al. [9] reported that rapid thermal growth of the thin oxide in an N_2O ambient yielded a thin oxynitride that exhibited excellent SILC. After a standard cleaning procedure, SiO_xN_y films were formed in N_2O (99.998%) at 1200°C for 5 to 20 s in a rapid thermal processing (RTP) chamber. This nitridation greatly reduced bulk electron traps, as shown in

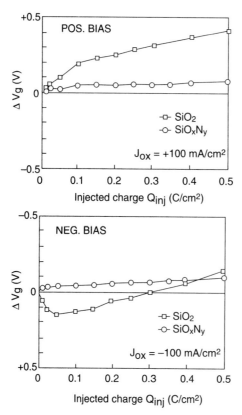

Figure 11.68. Gate voltage shifts vs. injected charge for 80 Å pure SiO_2 and N_2O-oxynitridation oxide under both positive and negative bias stress. The oxide with N_2O-oxynitridation exhibits less charge trapping. (Figure 2 of [9].)

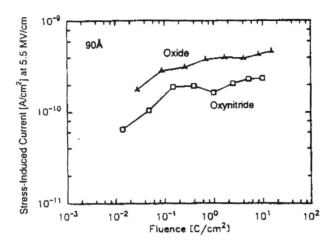

Figure 11.69. SILCs measured at 5 MV/cm for 90 Å pure oxide and oxynitride films. (Figure 8 of [110].)

Figure 11.68. It is believed that the Si–N bond in bulk SiO_2 suppresses the electron-trap generation during high-field stressing.

Most importantly, Fukuda et al. [9] observed that the oxynitride showed less SILC than did pure SiO_2 after high-field stress. Figure 11.69 shows example data as reported by Sharma et al. [91]. This result holds for dc, unipolar, and bipolar stress [79]. Hole trapping and interface trap generation are also suppressed by N_2O annealing [94].

11.5.5 Stress-Induced Leakage Current as the Limiting Factor for Tunnel Oxide Scaling

The tunnel oxide SILC is one of the most challenging reliability issues for Flash EPROMs. It is of particular concern to read disturb and data retention. There has been no particularly good solution to this problem, and SILC remains one of the main limitations to Flash memory tunnel oxide and high-voltage periphery transistor scaling. In order to maintain high data integrity, manufactures have taken the conservative approach and kept the tunnel oxide thickness at 80 Å or above. While logic technology now routinely uses oxide thickness well below 50 Å, the tunnel oxide thickness for Flash memory has not changed very much over the past 20 years, spanning many generations of technology. This posts an added complexity to process integration. A process technology must support very thick oxides for high-voltage transistors (400 Å oxide for 25-V peak voltage, or 150-Å oxide for 10-V peak voltage), thin oxides (50 Å) for high-speed low-voltage logic circuits, as well as the tunnel oxide thickness at a third value of around 100 Å. New breakthroughs in material technology and extensive use of system approaches (error management) are necessary for the further scaling of Flash memory tunnel oxide thickness.

11.6 SPECIAL RELIABILITY ISSUES FOR POLY-TO-POLY ERASE AND SOURCE-SIDE INJECTION PROGRAM

For Flash memory technology that uses thin oxide FN tunneling for program and erase, the method to reduce the device operation voltages is to reduce the tunnel oxide thickness. However, as discussed in Section 11.5, this method faces the limitations of the stress-induced leakage current (SILC) and data retention as well as thin oxide defect density. To overcome this limitation, the other methods of erasing the memory is the geometrically or field-enhanced poly-to-poly erase. There are some highly successful cells that utilize this mechanism for memory erasure; the best examples being the SanDisk triple-poly cell and the Silicon Storage Technology (SST) cell both of which have been in mass production for many years.

The main disadvantage of the conventional channel hot-electron injection mechanism for programming is its low injection efficiency, and consequently, its high power consumption. To overcome this problem, the source-side injection (SSI) program method has been proposed that offers more than 1000 times more efficient hot-electron programming. The best examples of cells that utilize this mechanism for programming are the SST cell and the High Injection MOS (HIMOS) cell. There are some unique benefits to these approaches as well as some special reliability features and concerns that are discussed in detail in this section.

11.6.1 Poly-to-Poly Erase and Its Reliability Issues

11.6.1.1 Poly-to-Poly Erase and the Formation of the Injectors. One way of obtaining high field in Si–SiO$_2$ structure that does not require the use of thin oxide layers is the use of the textured polysilicon gate Flash EPROM technology. The memory is erased through charge transfer from poly-to-poly by geometrically or field-enhanced Fowler–Nordheim tunneling. The interpoly oxide through which the tunneling takes place can be significantly thicker (40 to 100 nm) than those for the conventional Flash memory since the electric field is enhanced by geometrical effects of the injector at the corner or surface of the polysilicon structures. The most current examples of such technologies in commercial mass production are the SanDisk and SST cells. Because no thin tunnel oxides are used in such technologies, poly-to-poly erase Flash memory devices exhibit excellent reliability in failures due to random oxide defects.

The nonuniformity of the field oxide combined with the surface roughness of the polysilicon surface increased the localized field strength near the surface of the floating gate so that FN tunneling started at lower voltages than would have been expected for such a thick oxide. Figure 11.70 shows the band diagram of the conventional ultrathin tunnel oxide and the geometrically or field-enhanced thick oxide tunneling. The electron tunneling is encouraged by sharp needlelike protrusions extending from the surface of the polysilicon into the thermal oxide. These asperities concentrate fields at their tips and support enhanced localized conduction, which is as much as an order of magnitude greater than on the normal oxide. The poly-to-poly erase currents are determined primarily by the shape and size of asperities and less by the oxide thickness.

There are different ways to create the geometrically or field enhanced tunneling features. The horn-shaped tunneling injector on the floating gate is formed by oxidation of the polysilicon, similar to the formation of the field oxide "bird's beak"

SPECIAL RELIABILITY ISSUES

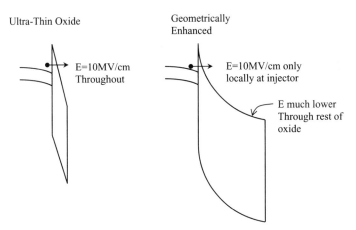

Figure 11.70. Band diagram of conventional ultrathin tunnel oxide (~100Å) and geometrically enhanced thick oxide. (After [96].)

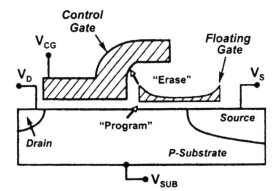

Figure 11.71. Schematic diagram of the split-gate SST Flash memory cell with a horn-shaped floating gate for poly-to-poly erasing, and split-gate for source-side injection programming.

on single-crystal silicon. After forming a 15-nm first gate oxide, a 170-nm polysilicon layer was then deposited, followed by the deposition of a 150-nm silicon nitride layer. The silicon nitride layer was then selectively removed with a reverse-gate mask, so as to expose the floating-gate region. Wafers were then oxidized to form a LOCOS-type polyoxide on top of the exposed polysilicon layer. The remaining nitride layer was then stripped by hot phosphoric acid, exposing the underlying polysilicon layer. The exposed polysilicon layer was subsequently etched away, while the remaining polysilicon layer underlying the LOCOS-type polyoxide was preserved during the polysilicon etch as the LOCOS-type polyoxide serves as a hard mask during the etch, thus leaving the horn-shaped floating gate underneath the LOCOS-type polyoxide as shown in Figure 11.71.

The asperities also occur because oxidation progresses faster along some crystal directions. Since crystal orientation is random in deposited polysilicon, there are points on the surface where oxide growth is enhanced. It has long been observed that the oxide growth rate is greatly reduced. In a unique study by Kuo et al. [97],

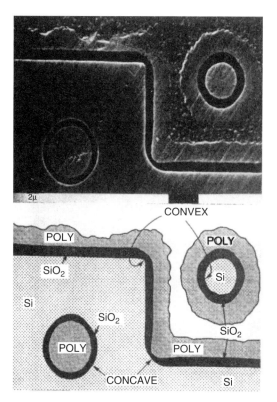

Figure 11.72. Difference of oxide thickness' among a convex Si surface, a straight surface, and a concave surface. The SEM is shown at the top, and the schematic at the bottom. The oxide on the concave surface is thinner than the oxide on the convex surface. The oxide on the straight surface is thicker that the oxide at the corners, resulting in sharp corners. (Figure 4 of [97].)

they showed that oxidation of curved Si surface is retarded by the stress associated with the nonplanar viscous deformation of the oxide, and that the retardation is more severe on concave structures than in convex structures. This effect is more pronounced at lower temperatures and less severe at higher temperatures. The oxide thickness is a strong function of the radius of the curvature of the Si surface—the smaller the radius or sharper the corner, the thinner the oxide, as shown in Figure 11.72. In other words, polysilicon with a $90°$ corner after oxidation will have a corner of less than $90°$ and hence forms a sharp corner.

One advantage of the poly-to-poly erase technique is that it offers improved erase performance and reliability by confining the high erase voltages within the memory array to the erase polysilicon gate only, rather than to diffusion regions within the substrate. This eliminates the high erase currents and device degradation associated with high junction field and leakage. Scaling of the device is also made simpler as the device junctions do not see the high erase voltages.

11.6.1.2 Poly-to-Poly Erase Charge Trap-up.
One often mentioned issue with polysilicon-to-polysilicon erase is electron trapping, or simply referred to as trap-up, which can result in memory cycling window closing. The trapped charge in

SPECIAL RELIABILITY ISSUES

the oxide reduces the effective injection field at the tunneling interface and decreasing the tunneling current. The reason is that since the tunneling happens at the very few field-enhanced injection points the current density at these injection points is extremely high and results in enhanced charge trap-up. An early study compared the trap-up of thin oxide in FLOTOX and textured-poly oxide [98] as shown in Figure 11.73. More modern data typically shows less trap-up for both normal thin oxide erase and poly-to-poly erase.

A good example of the more current poly-to-poly erase cycling trap-up rate is given in the literature [99–101] in which the source-side injection program and poly-to-poly erase cell is used in a multilevel analog voice storage application. Test array structures with access to poly-1 floating gates are used to characterize the trap-up as shown in Figure 11.74. The tunnel voltage increases as a function of time and stress current due to electron trap-up in the tunnel oxide, and the amount of shift is larger than conventional thin tunneling oxide. Figure 11.75 shows that as the number of write erase cycles increases, higher erase voltages are needed. It also shows that there is a relatively wide range of erase voltages from about 10 to 13.5 V before any cycling. This spread in erase voltage is due to the nature of the varying detail shapes of the polysilicon erase injectors. After 50,000 program/erase cycles, the erase voltages increased from about 10 to 13.5 to 12.5 to 14.5 V. The behavior is nearly logarithmic. In the study reported in Kordesch et al. [99], the maximum erase voltage, therefore the program/erase cycling endurance, is limited by the reliability consideration of the peripheral circuits. In order to achieve better cycling endurance, a necessary condition is that the peripheral circuits have to be able to support voltages higher than the highest erase voltage after program/erase cycling.

There are several simple and effective solutions to the charge trap-up problem as described below. Currently, Flash memory products with poly-to-poly erase are in high volume production with excellent reliability and endurance.

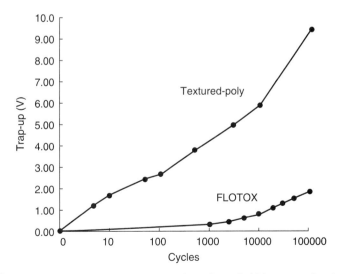

Figure 11.73. Comparison of normal thin tunnel oxide and thick textured polyoxide charge trap-up. Trap-up is measured as the increase in programming/erasing voltages required to program/erase the cell as a result of cycling. (After [98].)

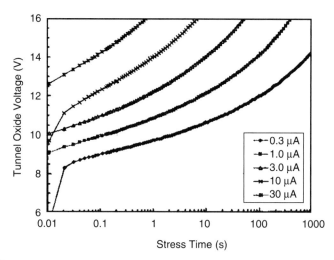

Figure 11.74. Tunnel voltage versus time curves for the poly-2/poly-1 oxide at five different current stress levels. The test structures with accessible poly-1 floating gate have 5000 cells in parallel. (Figure 5 of [99].)

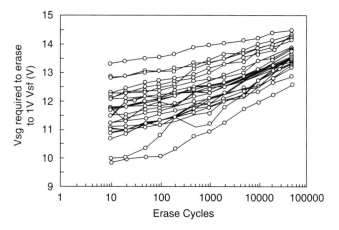

Figure 11.75. Select gate (poly-2) voltage required for erase as a function of the number of program/erase cycles. Note the initial wide spread of the erase voltages due to the different shapes of the erase injectors. (Figure 7 of [99].)

The first approach takes the advantages of the split-channel feature of the poly-to-poly erase cell to overcome the overerase problem. The split-cell structure has one part of the memory channel controlled by the control gate; therefore, it enables the cell to remain in the enhancement mode even if the floating-gate transistor is overerased and becomes a depletion mode device after erase. As suggested in Ono et al. [102], one simple way to counter the poly-to-poly erase trap-up problem is simply to overerase the memory at the beginning. With program/erase cycling as the poly-to-poly erase traps-up, even after the desired number of cycles, the cell is less erased but is still erased enough since at time zero it was overerased.

Another more optimized method is to adjust the erase voltage automatically to compensate for charge trapped in the erase tunnel oxide during expended cycling. An erase verify follows each erase operation. With this method there is no erase performance degradation as the product ages. The cells do not need to be heavily overerased and high program speed can therefore be achieved [97].

Recently, it was also reported that by adding an N_2O annealing after interpoly oxide formation, improved cycling endurance was achieved due to reduced electron trap-up during poly-to-poly erase [103, 104]. One of the studies was done on the SST type of cell, which uses the horn-shaped floating gate to control gate poly-to-poly tunneling for erasure.

11.6.2 Source-Side Injection and Its Reliability Issues

The main disadvantage of the conventional channel hot-electron injection mechanism for programming is its low injection efficiency and, consequently, its high power consumption. To overcome this problem, a source-side injection (SSI) program method has been first proposed by Wu et al. [105], and there are a variety of SSI cells that have been realized [106–109]. In most cases, the MOS channel between the source and drain regions is split into two "subchannels" controlled by two different gates. The gate on the source side of the channel is biased at the condition for maximum hot-electron generation, that is, very close to the threshold voltage of the channel. The gate at the drain side, which is the floating gate of the cell, is coupled to a high potential in order to favor hot-electron injection to the floating gate. This can be accomplished either by implementing an additional gate with high coupling ratio to the floating gate [109], or by using a high drain-coupling ratio [106]. As a result, most of the drain to source voltage is dropped across the narrow gap between the two subchannels. The high lateral field peak and therefore the hot-electron injection point is obtained in the gap between the two subchannels as shown in Figure 11.71. Because of the high floating-gate potential, the vertical field at the injection point is favorable for electrons, and high injection efficiency (on the order of 10^{-3} to 10^{-2}) can be achieved. This is almost three orders of magnitude higher than the conventional channel hot-electron injection as shown in Figure 11.76 where gate current of both source-side injection and channel hot-electron injection were measured.

However, there are some unique reliability issues related to this approach in terms of the split-gate "gap region" formation and gap region charge trapping and interface generating, as we shall discuss in the following.

11.6.2.1 Reliability Issues of the Gap Region Formation.
For the SST cells shown in Figure 11.71, one of the scaling and process and reliability challenges is the engineering of the gap region between the floating gate and the select gate [111]. During the thermal oxidation to grow the interpoly oxide, the two-dimensional nature of oxide growth produces a sharp reentrant corner between the select gate and the floating gate and also forms a bird's beak under the floating gate. These mechanisms can degrade both the reliability and performance of the split-gate source-side injection device. By using two-dimensional simulations and experimental data, Bhattacharya et al. [111] proposed a composite interpoly

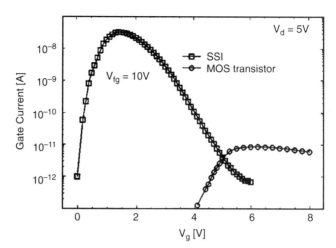

Figure 11.76. Comparison of injected gate current for source-side injection (SSI) and channel hot-electron injection, both measured at drain voltage of 5 V. The SSI gate current is about 10^3 times higher than that of channel hot-electron injection. (After [110].)

dielectric consisting of thermally oxidized polysilicon and deposited TEOS (tetra-ethoxy-silane) oxide for improved hot-electron injection efficiency and reliability. This analysis is described here.

In typical split-gate formation, after defining the floating gate, a 250- to 300-Å oxidation forms the interpoly oxide, the oxide in the gap region, and the select gate oxide simultaneously. Figure 11.77 shows the simulated results; showing a sharp reentrant corner between the select gate and floating gate. Subsequent conformal deposition of the select-gate polysilicon makes the sharp corner a high-field electrode that poses a reliability concern in terms of program disturb in the memory array. This is confirmed in the XTEM (cross-sectional TEM) photograph shown in Figure 11.78. The bird's beak increases the effective oxide thickness under the floating gate in the vicinity of the gap region. The simulated vertical and lateral fields in Figure 11.77 show that even though the applied voltage is dropped entirely across the gate, the bird's beak effectively separates the physical location of the peak lateral field from the peak vertical field. Good coupling of the lateral and vertical field is crucial to the effectiveness of source-side injection. The bird's beak therefore reduces the hot-electron injection efficiency and degrades the device program performance.

This problem can be avoided by using a thin grown thermal oxide (<100 Å) followed by CVD TEOS and densification to produce a high-quality composite interpoly oxide. Due to the growth of only a thin thermal oxide, the bird's beak under the floating gate is significantly reduced, and the conformal deposition of TEOS eliminates the reentrant corner. This is simulated as shown in Figure 11.79 and confirmed in the XTEM photograph in Figure 11.80. Device simulation data in Figure 11.79 shows higher vertical and lateral electric fields in the channel under the gap region for the composite interpoly oxide. Due to the significant reduction of the bird's beak, there is excellent coupling between the vertical and lateral fields in the channel under the gap region and results in improved hot-electron injection efficiency.

In their study, Bhattacharya et al. [111] found that the Flash cells with composite thermal/TEOS oxide have better than three times the program speed compared to

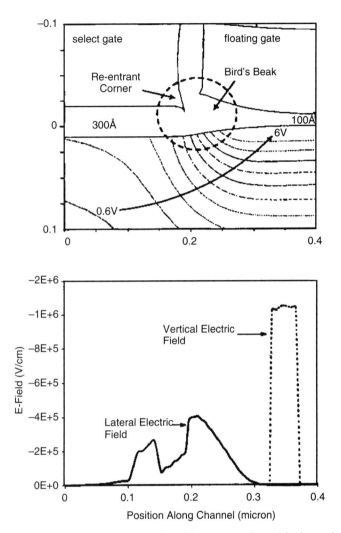

Figure 11.77. Simulation results in the vicinity of the gap region with thermal oxide. Equal-potential contours are in 0.6-V steps. Vdrain = 6V, Vsource = Vsubstrate = 0V, Vcontrol_gate = 1.5V, and Vfloating_gate = 10V. The vertical and lateral electric fields under the Si–SiO$_2$ interface are shown as a function of the position along the channel. (Figure 2 of [111].)

those with thermally grown oxide, confirming the benefits of suppression of the bird's beak. It was also observed that the composite oxide and the thermal have comparable quality in terms of breakdown and electron trapping in the gap region. Therefore, the composite oxide is more suitable for optimized split-gate source-side injection Flash memory technology.

11.6.2.2 Charge Trapping in the Gap Region for Source-Side Injection Cells. Source-side injection is concentrated near the split-gate narrow gap region; a small amount of electron may get trapped in the spacer like oxide between the floating-gate edge and the control gate. This can change the local electric field,

Figure 11.78. XTEM photograph of the gap region for a thermally grown interpoly oxide showing the reentrant corner and the bird's beak between the select gate and floating gate. (Figure 5a of [111].)

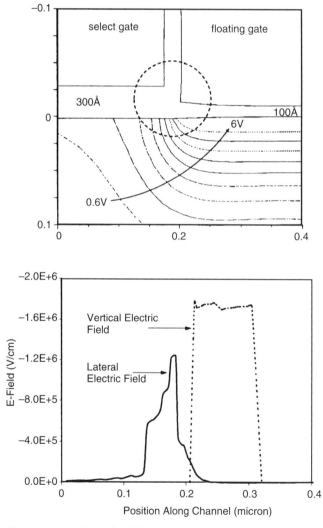

Figure 11.79. Simulation results in the vicinity of the gap region with composite 70 Å thermal and 230 Å TEOS interpoly oxide. All other conditions are the same as in Figure 11.77. The vertical and lateral electric fields under the Si–SiO$_2$ interface are shown as a function of the position along the channel. (Figure 4 of [111].)

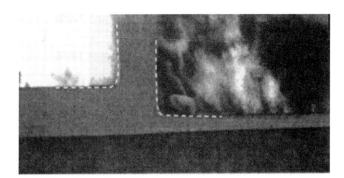

Figure 11.80. XTEM photograph of the gap region for the thermal/CVD-TEOS composite interpoly dielectric. Both the reentrant corner and the bird's beak are removed. (Figure 5b of [111].)

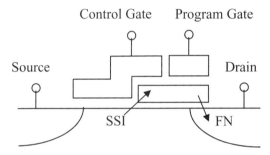

Figure 11.81. Schematic representation of the HIMOS cell. The cell is programmed with source-side injection (SSI), and erase with drain-side FN tunneling. (Figure 1 of [113].)

reduce the read current, degrade the program performance, and limit the endurance of these structures. Due to the de-trapping of these trapped charges, it is also a data retention concern. Wellekens et al. conducted detailed studies on the degradation and reliability issues of source-side injection Flash memory cells using the higher injection MOS (HIMOS) cell [109, 112–114] as shown in Figure 11.81.

The detailed operations of the HIMOS cell are described in Van Houdt et al. [114]. The typical operation conditions for the different operation modes are summarized in Table 11.2. The oxide under the floating gate is 85 Å, and the interpoly dielectric layer is 300 Å of thermally grown polyoxide. A special large area cell was used for this particular study. Figure 11.82 presents the endurance characteristics of the cell, showing large window closing due to decrease in the readout current at the erased state. This is largely due to the electrons trapped at the gap region, and the G_m degradation due to interface states creation.

In order to separate the different causes for the window closing, the memory cell I_d-V_{pg} curves after ultraviolet (UV) light erasure are measured as a function of different P/E cycles. In this way the I–V characteristics only reveal any possible insulator damage that has occurred within the transistor channel. From Figure 11.83, one can see that there are negative charges and/or acceptor-type interface traps in

TABLE 11.2. Typical Operation Voltages of the HIMOS Cell in Different Operating Modes

	Source (V)	Drain (V)	CG (V)	PG (V)
Write	0	5	1.5	12
Erase	—	5	0	−12
Read	0	2	3	0

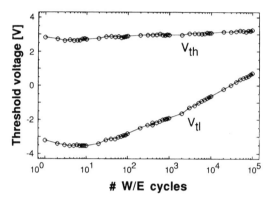

Figure 11.82. Measured endurance characteristics of the HIMOS cell, showing large window closing due to local charge trapping. (Figure 2 of [109].)

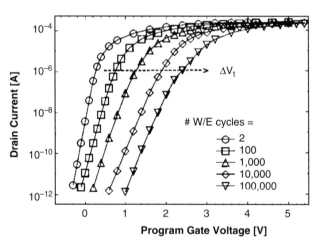

Figure 11.83. Memory cell drain current characteristics at UV states as a function of endurance cycles. After 2, 10^2, 10^3, 10^4, and 10^5 cycles, the cell is UV erased, and the drain current is then measured. It shows significant V_t increases due to charge trapping and G_m degradation due to interface states generation. (Figure 3 of [109].)

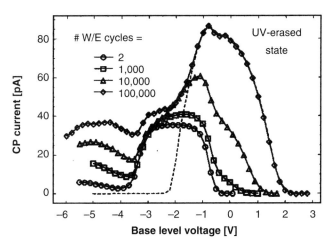

Figure 11.84. Charge pumping currents after different numbers of W/E cycles and after UV erasure. The dashed line is for the case when the drain is left floating after 10^5 W/E cycles. (Figure 5 of [109].)

the gate oxide. The amount of shift in V_t is significantly larger than that of the channel hot-electron injection programmed Flash memory cells. After P/E cycling, a severe degradation of the subthreshold slope is also observed caused by interface trap generation. There is also some degradation due to the FN erasing charge trapping, and the mechanism is the same as the ETOX cell erasure slow down as discussed in Section 11.2.

Charge pumping (CP) measurements were also performed after a different number of P/E cycles as shown in Figure 11.84. The variable base-level pulses with a constant amplitude of 4 V were applied at a frequency of 1 MHz. The pulse rise and fall times were 100 ns, and a small reverse bias of 200 mV was applied to source and drain. Only the program gate (PG) was pumped since the damages are located underneath the floating gate. During the CP measurement, the control gate (CG) was kept at 1.5 V and the select gate channel is inverted. A few conclusions can be drawn from the CP curves.

The P/E cycling induces a large positive shift in the falling edge of the CP curves after UV erasure, giving rise to a so-called shoulder in the CP characteristics. This is indicative of localized electron trapping in the floating-gate channel since it is caused by interface traps that are pumped at higher voltage levels. Consequently, the increase of this with P/E cycling is indicative of the generation of additional interface traps within the degraded part of the channel. The dashed line in Figure 11.84 is the CP characteristics measured with the drain contact left floating. The shift of the rising edge to higher base level proves that negative charge is present at the source side of the floating-gate channel. The difference between the rising edge of the CP curves with source and drain connected and the CP curves with the drain floating is a direct measure of the effect caused by hot-electron damage at the split point.

An additional feature of the CP characteristics is the current tail at low base levels, which increases with the number of P/E cycles. This tail is indicative of positive oxide charge and is related to the erase operation, which may generate positive charges by band-to-band tunneling current. As shown in Figure 11.85, the origin of

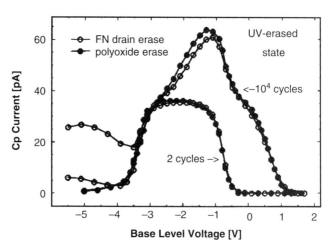

Figure 11.85. Comparison between the charge pumping characteristics of devices erased with drain-side FN tunneling erase and devices with the floating to select gate poly-to-poly erase. The low-voltage current tail disappears in the case of poly-to-poly erasure, when no positive charge was generated in the channel due to band-to-band tunneling current. (Figure 6 of [109].)

this tail has been identified by comparing CP results on a cell erased with the conventional drain side FN tunneling erase scheme to results of a device erased through the poly-to-poly erase between FG and SG. It is clear that both initial and degraded CP curves are essentially identical, apart from the low-voltage tail, which completely disappears when the device is erase through the polyoxide. This proves that the CP current at low base levels is a consequence of oxide damages generated during drain-side FN erasure.

From both the I–V measurements after UV erasure, and from the charge pumping measurements, it is clear that the program/erase degradation of the SSI device is mainly caused by charge trapping and interface traps creation at the injection point. The damages are highly localized near the split point where the injection occurs. It is estimated that the damaged region is less than 100 Å, but it has profound effects on the reliability of the source-side injection cells. The generated interface states degrade the G_m and the cell read current, the trapped charge degrades the program performance, and the de-trapping of the trapped charge at high temperature can affect the data retention behavior.

De Blauwe et al. [113] studied the impact of high-temperature operation on the reliability of the HIMOS cell. Two devices were first cycled for 10^4 P/E cycles; they were then UV erased and baked at room temperature and at 150°C. The results are shown in Figure 11.86 with the measured threshold voltages as a function of the baking time. When the retention test is performed at room temperature, the V_t shift after 5×10^4 s, or 14 h, is only about 10 mV. However, when the baking temperature is 150°C, the total V_t shift is about 200 mV after 14 h. Note there are two effects that have been taken into consideration here: first the V_t measured at 150°C is about 220 mV smaller as compared to V_t measured at 20°C, and second, the temperature ramp-up from room temperature to 150°C took about 200 s. Therefore, an important portion of the Q_{ip} is already lost during the temperature ramp-up. When both

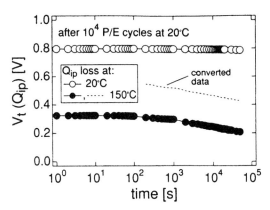

Figure 11.86. Threshold voltage (V_t) shift during a retention test due to the detrapping of injection-point charge (Q_{ip}). Q_{ip} has been built up during preceding P/E cycling at room temperature. At 150°C the initial V_t is smaller due to Q_{ip} emission during the temperature ramp-up (≈200 s), and the influence of the temperature on V_t. Taking these two effects into account, the dashed line shows the converted data, showing a V_t shift of about 200 mV after 5 × 10^4 s, or 14 h of bake. (Figure 4 of [113].)

effects are adjusted, the dashed line shows a pure V_t shift of about 200 mV from time 0 to 14 h. It is expected that if the P/E cycles are higher, the amount of Q_{ip} will increase, and the amount of charge de-trapping will also increase. This could pose a long-term data retention concern for this source-side injection cell and make it more complicated to guarantee long-term nonvolatility after cycling.

11.7 PROCESS IMPACTS ON FLASH MEMORY RELIABILITY

There are many process-related issues that can have significant impacts on the reliability of Flash memory. Some of these effects are common to the complementary metal–oxide–semiconductor (CMOS) transistors, but some are extremely subtle and indirect, and their long-term impact difficult to characterize and understand. For example, floating-gate polysilicon grain size and doping concentration, the junction doping profiles, and the stack gate etch conditions can affect the Flash memory erased V_t distribution. The starting substrate can affect the defect density and the tunnel oxide quality and trapping rate. The tunnel oxide may have local oxide thinning and oxide trapping problems. Interpoly dielectric (ONO) can affect the charge leakage and data retention. The intermetal and passivation layers can affect the device hot-carrier reliability, cycling endurance, moisture, contamination, and data retention. During the plasma etching process, in-line charging can degrade and damage oxides.

Different manufacturers have different specific process sensitivities; many of the findings are proprietary and well guarded. However, some of these findings and process improvements have been published in the form of papers or patents. A great deal can be learned by carefully studying these documents. In this section, we select a few topics for detailed discussion.

11.7.1 Tunnel Oxide Process and Nitrogen Incorporation

Flash memory gate oxide or tunnel oxide process is arguably the most important process step for Flash memory. It can have major impact on the oxide charge injection and charge trapping property. The quality of the oxide affects the Flash memory product program/erase endurance and reliability. Many factors and subtle process details affect the oxide and Flash memory reliability. The starting substrate, the cleaning process prior to the oxidation, the oxidation temperature, the incorporation of nitrogen, the floating-gate doping methods and doping concentration, the etching of the staked gate, and the plasma damage during dry can all affect the oxide quality and reliability [7–9, 12, 34, 47, 48, 83–85, 92–95, 115, 116].

The composition of the gate oxide plays an important part in reducing the trap density in the tunnel oxide. Many have found that the interfacial nitrogen incorporation is desirable for the tunnel oxide reliability since it reduces both electron and hole trap densities [7–9, 12, 83, 92–95, 115]. As discussed in detail in Section 11.5, N_2O nitridation of SiO_2 improves the resistance to electron-trap generation and hot-carrier degradation and suppresses stress-induced leakage current (SILC), because the Si–N bond in bulk SiO_2 suppresses the electron-trap generation during high-field stressing.

The preparation and cleaning of Si surface before the growing of the tunnel oxide, as well as the oxidation temperature are extremely important for good oxide quality and good data retention. For example, process damage of the Si substrate may induce weak points in the tunnel oxide, and if the oxidation temperature is not high enough, surface microdefects may be built in the tunnel oxide. Arai et al. [34] investigated the influence of the process damage on the data retention and charge-loss bits. They intentionally varied the Si surface microdefect density by adjusting the fabrication process. They found that after 10^6 W/E cycles, charge-loss bit generation rate is a strong function of the surface microdefect density.

11.7.2 Effects of Floating-Gate Process and Morphology

It is reported [117, 118] that local oxide thinning and oxide surface roughness at the polyoxide interface is greatly influenced by post-gate-oxide-growth temperature

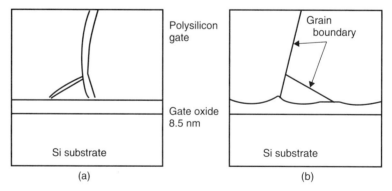

Figure 11.87. Cross-sectional view of the polysilicon-gate tunneling oxide interface (a) before and (b) after the annealing. (Figure 6 of [117].)

treatment, as shown in Figure 11.87. These are direct consequences of the grain growth of the polysilicon gate and the viscous flow of the oxide, which are enhanced with increasing annealing temperature and time. In addition, the dopant level and species of the polysilicon floating can play a major role in the local variation in tunneling current. Phosphorous atoms segregated at the grain boundaries of the poly gate were found to diffuse to the polyoxide interface and react with the oxide to form a phosphorous-rich SiO_2 or so-called oxide ridges. The phosphorous-doped SiO_2, due to the change in stoichiometry, exhibits a barrier-height lowering effect, thereby causing the FN tunneling current to vary, depending on the amount of dopant incorporation. Post-gate-oxide-process optimization can play a major role in the control of the erased V_t distribution.

Observed by many [47, 48], as discussed in Section 11.3, erased V_t distribution is strongly influenced by the grain size of the floating-gate poly-Si. The floating gate with smaller grain size realizes a tighter erased V_t distribution. Such small-grain poly-Si can be manufactured by lowering the phosphorous doping concentration to around $1 \times 10^{20} \text{cm}^{-3}$ as high phosphorous concentration enhances grain growth during annealing. It was also found that by lowering the phosphorous doping concentration in the floating gate, SILC, is suppressed and data retention prolonged.

For example, in the study by Nkansah and Hatalis [119], they found good correlation among the fast-erase bit density in a 2-Mbit array, the FN tunneling currents, and the floating-gate deposition and doping process. Three different floating-gate polysilicon deposition and doping techniques were examined: in situ doped, implanted, and undoped + in situ doped stacked floating-gate of 1500 Å thickness. The different floating gates and the measured grain size are summarized in the Table 11.3. From atomic force microscope (AFM) measurement, the undoped plus in situ doped stack for the floating-gate poly exhibits the smoothest morphology. This is further confirmed with TEM grain size characterization as shown in Table 11.3. The I–V characteristic shows that the undoped plus in situ doped stack has reduced tunneling current for a given field compared to other floating-gate stacks. The tunnel oxide is 95 Å nitrided oxide.

In Nkansah and Hatalis [119], the memory V_t distribution is measured after erase with 5 V on the drain and −8 V on the control gate (this process is called "program" in their study). They consistently observed that the fast bits are always fast and did not see evidence of the erased V_t toggling between fast and normal V_t states, while normal bits are always normal. It was concluded that locally enhanced

TABLE 11.3. Summary of Floating-Gate Polysilicon Grain Size Characterization

Grain Size Process Types	Process A Phosphorous Implantation	Process B In Situ Doped Phosphorus	Process C Undoped + In Situ Doped Phosphorus Stack
Median grain size	2540.3 Å	2219.1 Å	1643.4 Å
Grain size range	1900–4564 Å	1530–3360 Å	962–1731 Å
Number measured points	53	53	53
Standard deviation	805.6 Å	560.1 Å	352.7 Å
Variance	78.1 Å	37.6 Å	17.3 Å

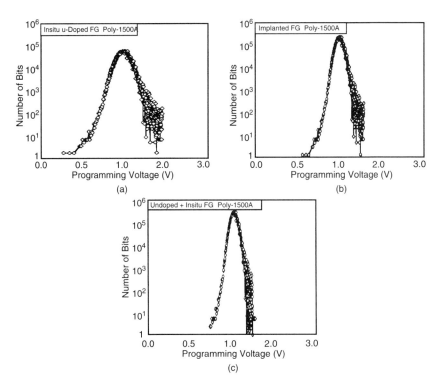

Figure 11.88. A 2-Mbit Flash array are erased (called "programmed" in this study) with control gate held at −9 V and drain junction at 5 V. The measured V_t distribution as a function of different floating gate doping process: (a) in situ phosphorous doped, (b) phosphorous implanted, and (c) undoped plus in situ phosphorous doped stacked floating gate. (Figure 10 of [119].)

tunneling field results from the floating-gate polysilicon grain structure is responsible for the fast bits. The erase V_t distribution of the 2-Mbit array as a function of the poly gate deposition process is shown in Figure 11.88. It shows that the V_t distribution is narrowed for the floating-gate stack that showed reduced FN tunneling current and smaller grain sizes. The floating-gate polysilicon with smaller grain size shows more uniform tunneling and tight erase V_t distribution.

11.7.3 Stacked Gate SAS (Self-Aligned Source) Etch Process and Erase Distribution

The etching of the stacked floating-gate structure is another critical process step that can greatly affect the erased V_t and erase-speed distribution of Flash memory. This process step is also complicated and many subtle factors, such as the etch process sequence, etch machine power, and etch gas chemistry, as well as the junction doping profile and thermal cycles can affect the reliability of the tunnel oxide.

A common method to achieve higher density in an ETOX memory array is to use the so-called self-aligned source (SAS) etch described in Tang and Lu [120]. The patent discloses a method for forming a source region that is self-aligned with the poly 2 wordline. The end edges of the field oxide regions are vertically aligned with the polysilicon wordline with no bird's beak encroachment and corner rounding effect remaining in what will become the source region. The source region, formed between the ends of the field oxide regions of neighboring cells, is thus self-aligned with both the field oxide regions and the polysilicon gate wordlines. High-selectivity oxide etch is used to remove the field oxide at the source side of the device. This oxide etch, using the gate as a protective barrier, creates a source diffusion junction that is self-aligned to the gate. As reported [120], for their specific machine, the high-selectivity oxide etch is a plasma etch process with a radio frequency (RF) power of 1200 W, a chamber pressure of 50 mTorr, and a CHF3 to O_2 gas flow rate of 75:5 standard cubic centimeters per minute (SCCM). This self-alignment of the source and gate allows closer placement of polysilicon wordlines without any decrease in source width and thus requires less physical separation between one memory transistor to the next memory transistor. Reduced cell size and greater array density is thus achieved.

As shown in Figure 11.89, the SAS etch is used after a stacked gate etch as a way to reduce overall cell size in a ETOX Flash memory process. However, during

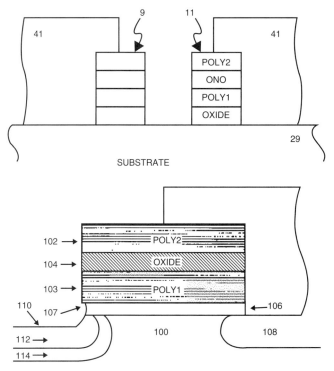

Figure 11.89. (a) Flash memory cell cross-sectional view. (b) The source profile after the SAS etch, when the source is implanted after the self-aligned source (SAS) etch. Note that since the source/drain implants are performed after the SAS etch their profiles are sensitive to silicon gouge and damage by the SAS etch.

the SAS etch, the stacked gate edge is exposed to the SAS etch. This has a significant negative impact on the tunnel oxide integrity. Electron and hole traps could be created in the tunnel oxide edge due to exposure to plasma etch. In addition, the building implants, which consist mainly of the source diffusion implant, are done after the SAS etch. Since the SAS etch has a tendency to etch away or gouge away silicon in the source region, the implant profiles might not be uniform at the source and may change the profile of the surface source that overlaps below the stacked gate. In that case, the erase integrity and erase distribution of the Flash memory array may be significantly degraded. Also if the gate–source overlap area is too great, then the source coupling may be too large and interferes with the erase operation. If it's too small, then there may not be enough overlap area for FN tunneling erasure to happen. Typically, due to the severity of SAS over-etch, source edge FN erasure is significantly impeded due to the lack of sufficient dose underneath the source–gate overlap region.

To solve this problem, an improved SAS etching method was disclosed [121], which maintains the small cell size but protects the gate edge with oxide spacer during the SAS etch step. In this new method, the sequence of the SAS etch step and source implant step is reversed so that the source implant is performed prior to the SAS step to ensure a uniform source doping profile. In addition, an oxide or polysilicon spacer was formed on the stacked gate edge prior to self-aligned source etch to protect the tunnel oxide from the gauging that occurs during the SAS etch. Any damage that occurs due to the SAS etch takes place away from the stacked gate edge. Thus the tunnel oxide integrity and the uniformity of the source junction doping profile are preserved.

In the new method, the source implant is performed along with the drain implant. It is this implant and its resultant high surface concentration under the source–gate overlap region that facilitates the proper erasure of the Flash memory. The spacer is then deposited and the cross-sectional view of the Flash memory after the spacer-etch step is shown in Figure 11.90(*a*). Next, the N^+ source/drain (S/D) implant and a source implant are performed. The N^+ implant is also used to connect the source line from the actual source region and field region that is SAS etched in conventional process. The resulting cross-sectional view of the Flash is shown in Figure 11.90(*b*). In this cell the tunnel oxide integrity is improved and there is a uniform source doping concentration under the source–gate overlap area. Uniform erasure of the memory array can then be realized and the number of overerase bits can be minimized.

11.7.4 In-Line Plasma Charging Damage

Plasma-induced oxide damage has been a critical issue in CMOS device scaling and reliability. Gate oxide, tunnel dielectrics, and interpoly dielectric (ONO) can be damaged by plasma etching processes [122, 123]. The undesirable plasma-induced in-line charging and stress damage in nonvolatile memories can cause severe instability in UV-erased threshold voltage and transconductance degradation. This can lead to serious memory circuit malfunction and deteriorate product yield and performance [124]. The nonuniformity in the plasma across the wafer surface can cause local charge buildup, which can damage the tunnel dielectric. It can cause charge traps in the interpoly ONO dielectric films too.

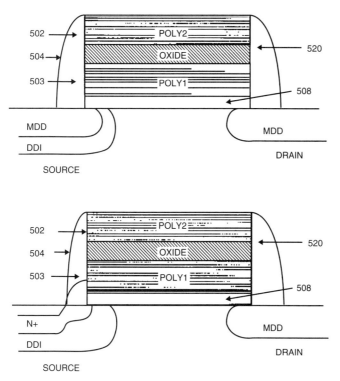

Figure 11.90. (a) Flash memory cell cross-sectional view before SAS etch and (b) after the SAS etch. Since the SAS etch is done after the source drain implant and is also protected by the spacer, the source doping profile is not sensitive to silicon gouge and damage. Uniformity is then realized.

Fang et al. [124] investigated plasma-induced charging in the ONO dielectric film and tunnel oxide damage in Flash memories by using a 0.5-μm, double-level metal Flash memory process. Eight different kinds of test structures with different antenna ratios were designed for monitoring the in-line plasma charging and stress damage on the Flash memory cells. The detail of each structure is given in Table 11.4. The poly-1 flash memory devices have contacts to the poly-1 gate, and they are useful in separating the charge trapping in the bottom tunnel oxide and the ONO film. The antenna ratio is defined as the ratio of gate pad area to device active area. The P^+ and N^+ protection diodes are connected to the control gates of Flash cells through the metal-1 layer.

Figure 11.91 shows the measured transconductance characteristics for various types of Flash devices. Significant transconductance degradation was observed on test structures with large antenna ratios. In addition, the transconductance curve for the unprotected device was shifted by +0.8 V from the one for the P^+ diode protected. The corresponding threshold voltages as a function of the antenna ratio and device structures are displayed in Figure 11.92. For both Flash devices and poly-1 transistors, UV-erased threshold voltage increased with increasing antenna ratio. Electron trapping in tunnel oxides during the plasma etch gives rise to the $V_{ts,uv}$ increases in medium and large antenna ratio poly-1 transistors.

TABLE 11.4. Flash and Poly-1 Device Structures with Estimated Antenna Ratios

Name	Test Structure	Antenna Ratio[a]
FL	Standard Flash device	2300
FM	Standard Flash device	2000
FS	Standard Flash device	50
FP	Flash device with P$^+$ diode connected to the gate	50
FN	Flash device with N$^+$ diode connected to the gate	50
PL	Standard poly-1 transistor	2300
PM	Standard poly-1 transistor	2000
PS	Standard poly-1 transistor	50

[a] The antenna ratio is defined as the ratio of gate pad area to thin oxide area.

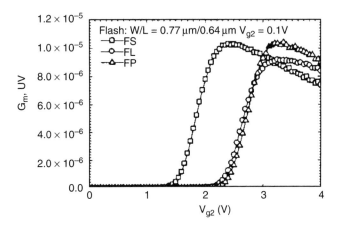

Figure 11.91. Measured transconductance G_m characteristics for standard Flash device with antenna ratio of 50, antenna ratio of 2300, and Flash device with P$^+$ diode connected to the gate. (Figure 3 of [124].)

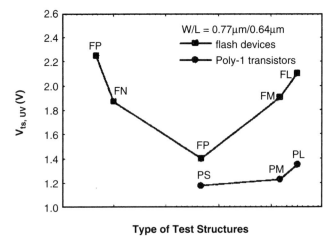

Figure 11.92. Measured $V_{ts,uv}$ for various poly-1 and Flash test structures as described in Table 11.4. (Figure 4 of [124].)

The pronounced difference in $V_{ts,uv}$ between floating-gate Flash and poly-1 devices manifests the charge trapping effect in the ONO dielectric. Meanwhile, a Flash device with its gate connected to a P^+ diode had the closest $V_{ts,uv}$ to the designed value with no G_m degradation. This structure provides the best protection against in-line plasma charging induced damage. The device with the gate connected to the N^+ diode did not provide the same level of protection. A positive charge buildup on the gate during plasma processing is responsible for the degradation and damage. This is confirmed by simulating the plasma stress with positive gate voltage stress. The plasma-induced high-field stress during the dry etch processing is the main cause for the observed $V_{ts,uv}$ increase and G_m degradation in large antenna ratio Flash devices. The V_{ts} turn-around behavior in the small antenna ratio Flash device stemmed from an initial positive charge trapping in the ONO dielectric film at low gate voltage stress followed by a negative charge trapping at high gate voltage stress. This study found that the impacts of the ONO charge trapping on device program and erase characteristics are relatively small because the trapped charge mainly just shifts the $V_{ts,uv}$, which can be compensated by the floating-gate charges.

Methods of reducing the oxide damage by the dry etching processes include modification to the reactor design and operation, and to minimize charging damage by appropriate antenna design rules and protection devices. A recent study describes the use of an electron cyclotron resonance (ECR) plasma etching technique for reducing gate charging during polysilicon gate patterning [125]. The 10-μs pulse time modulated ECR plasma makes it possible to realize extremely high selectivity, high etch rate, high anisotropy, and notching-free etching in simple Cl_2 gas.

11.7.5 Impacts of Intermetal Dielectric and Passivation Films on Flash Memory Reliability

With the scaling of the memory device, it becomes difficult to fill the gap between the first metal lines without void formation and realize planarized surface topology using conventional silane-based plasma CVD SiO_2 (P-SiO_2). To solve this problem, plasma CVD using tetra-ethoxy-silane (P-TEOS) has been used because of its better step coverage. However, such dielectric layers with good step coverage usually absorb a large amount of water-related species and degrades the hot-carrier immunity of the transistors. The water-related species contained in intermetal dielectric layers can have adverse effects on the reliability of floating-gate devices. They can accelerate the memory cell charge loss and hot-carrier-induced degradation of peripheral N-channel MOSFETs during the high-temperature bake [126–129].

The influence of the intermetal dielectric layer on nonvolatile memory reliability is evaluated in detail in Sakagami et al. [128]. The schematic cross section of the cell structure used in the study is illustrated in Figure 11.93. Various kinds of dielectric films were deposited on metal 1 and the reliability impacts of these films on memory cell charge retention and N-channel device hot-carrier (HC) degradation were evaluated. The dielectric films were atmospheric pressure CVD (APCVD) phosphosilicate glass (PSG), plasma CVD SiO_2 by the decomposition of tetra-ethoxy-silane (P-TEOS), and silane-based plasma CVD SiO_2 (P-SiO_2) as summarized in Table 11.5.

Figure 11.94 shows the hot-carrier lifetime of N-MOSFET and memory cell charge-loss characteristics with various P-TEOS passivation structures. The P-SiO_2

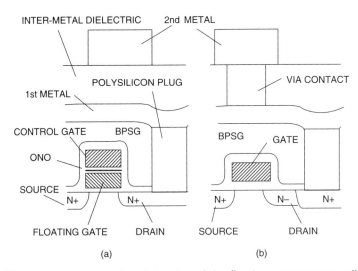

Figure 11.93. Schematic cross-sectional drawing of the floating-gate memory cell and the N-channel transistor in the peripheral circuit. (Figure 1 of [128].)

TABLE 11.5. Structure of Different Passivation Layers

Sample	Bottom P-SiO$_2$ (nm)	P-TEOS (nm)	Cap P-SiO$_2$ (nm)
A	0	100	1100
B	200	900	1100
C	400	700	1100
D	1100	0	0
E	400	0	0
F	400	1000	0

layer is used as the top passivation layer to block contamination by mobile ions and the absorbing of water into the P-TEOS layer. Sample A with a bottom P-TEOS showed the worst HC lifetime degradation and charge retention characteristics after the bake. In the case of samples B and C that have bottom P-SiO$_2$ layer, charge retention capability and HC lifetime degradation of N-MOSFET are dramatically improved. Charge-loss rate of the 200-nm P-SiO$_2$ (sample B) is higher than that of the 400-nm P-SiO$_2$ sample; thus the bottom P-SiO$_2$ layer should be thicker than 400 nm to suppress the charge loss.

The cell charge loss and HC lifetime degradation are enhanced by a bottom CVD PSG or P-TEOS layer. It is known that CVD PSG and P-TEOS tend to absorb water. The water in these layers passes through the borophosphosilicate glass (BPSG) layer and reacts with the weak bonds in the gate oxide to form wafer-related traps. In the case of water-rich intermetal dielectric layers, the high-temperature bake test causes the water to diffuse from the intermetal layer to the BPSG layer. A water-blocking layer such as P-SiO$_2$ under the water-rich intermetal dielectric layer can avoid the degradation of HC lifetime. The memory cell charge loss is related to the amount of water-related species in intermetal dielectric layer, through

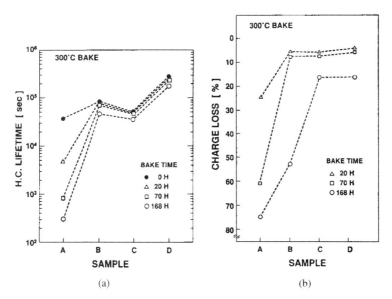

Figure 11.94. (a) HC lifetime of NMOSFET and (b) charge loss of the cells with various P-TEOS passivation layers. (Figures 4 and 5 of [128].)

a "two-step" process. The first step is the diffusion of water from intermetal dielectric layer into the BPSG layer under metal 1. Second step is the release of positive ions (sodium) gettered in the interlevel BPSG layer due to the increase in the amount of water. These released mobile ions drift and neutralize the negative charges on the floating gate. Secondary ion mass spectrometry (SIMS) profiles have shown a pile up of sodium ions in the BPSG layer. It is suggested that water can mobilize or activate the normally immobile sodium ions. The best solution to this problem is to form a thin P-SiO$_2$ layer under the P-TEOS layer. The thin P-SiO$_2$ layer blocks the water species from reaching the gate and tunnel oxide.

Plasma silicon nitride (P-SiN) and plasma silicon oxynitride (P-SiON) have been widely used as the top passivation film because they're excellent barriers to water penetration. The impacts of the passivation film deposition, compositions, and structures on the Flash memory tunnel oxide reliability are reported by Shuto et al. [164]. They found that the enhancement of the tunnel oxide degradation strongly depends on the refractive index of the P-SiON passivation film, that is, films with higher refractive index (higher nitrogen content) shows higher tunnel oxide degradation. It is explained that water diffusion toward the tunnel oxide results in the tunnel oxide degradation, and water diffusion to the outside results in the recovery of degradation. If the passivation film is water permeable (low refractive index), water in the BPSG film diffuses out due to the thermal budget during passivation film deposition and results in lower degradation. However, if the passivation film is hardly water permeable (high refractive index), then water does not diffuse out and most of it is confined n the BPSG film. Parts of this water diffuse into tunnel oxide and results in oxide degradation by creating water-related electron traps. Control of many process factors such as the amount of water contained before the passivation film deposition, the refractive index, or water permeability of the passivation film, and the thermal budget after passivation film deposition are critical to enhance the reliability of the Flash memory.

11.8 HIGH-VOLTAGE PERIPHERY TRANSISTOR RELIABILITY

11.8.1 High-Voltage Transistor Technology

The high voltage (HV) required to operate Flash memories today ranges from a low of 10 V for certain NOR Flash memory to nearly 25 V for NAND and poly-to-poly tunnel-erase Flash memories. These voltages are generated on-chip with internal charge pumps. These high program and erase voltages are not expected to be scaled much below the current levels. Flash memories employ FN tunneling for erase and perhaps program; the tunnel oxide thickness is not expected to be scaled much below the 80 Å level due to stress-induced leakage current (SILC) as discussed in Section 11.5. Therefore the voltages required for FN tunneling erase or program are not expected to scale much below the current levels. The voltage required for hot-carrier program cannot be reduced much either if the gate oxide thickness can not be reduced.

The circuits that control the memory array use low voltage for Read and high voltage for Write and Erase operations. The low voltages may be 5 V, 3.3 V, or lower, and the high voltages may be 10 to 25 V. To provide these two types of transistors, dual-gate oxide technologies are almost always used—one to handle high voltages, with oxide thickness ranging from 450 to 150 Å; and the other to provide high-performance low-voltage transistors for high-speed logic, with oxide thickness ranges from 120 to 70 Å or lower. The exact oxide thickness and voltages vary depending on the type of Flash memory device and the technology generations. In the future, the low-voltage transistors in embedded Flash memory will use even thinner oxides.

The high-voltage devices must withstand the Write and Erase voltages without suffering punchthrough, junction breakdown, gate oxide breakdown, well breakdown, and long-term reliability degradations. In order to tolerate these voltages, the field oxide isolation process has to provide a sufficiently high-field threshold voltage and junction breakdown voltage. It is also important that the field isolation be compact and not take too much space in high-density memory chips. The standard approach is to modify the LOCOS process to use a much thicker field oxide than a low-voltage logic technology of a comparable generation. Trench isolation technology is also used today.

To obtain a high junction breakdown voltage, source and drain junctions of HV transistors are typically formed by double diffusion or with well diffusions. The source and drain N^+ diffusions of the N-channel HV transistor are surrounded by deep N^- diffusions or N-wells. There are many different ways to realize the high-voltage transistors. For example, in Fang [130], the 512-kb embedded Flash EEPROM has a 0.8-µm twin-well CMOS double-metal technology as the baseline process for the logic circuits with a gate oxide thickness of 150 Å. The high-voltage process module for the high-voltage NMOS and PMOS transistors is integrated into the baseline process. The high-voltage transistors use poly 1 gates and thicker gate oxide of 350 Å. The same lightly doped drain (LDD) process, optimized for the high-speed baseline process, is used for high-voltage transistors without modifications. Four additional photoresist steps are required over the baseline process for the high-voltage process module.

Conventional LOCOS isolation with ~8500-Å thick field oxide and modified LOCOS process has been used in 4-Mbit memories [131, 132]. Effects like bird's beak, punch-through, and field oxide thinning limit the field oxide thickness (to about 6000 Å) and minimum geometry (to about 0.8 µm) achievable with LOCOS-based technology. Trench and oxide refill technologies are also being used [168, 169]. Channel stop boron ions are implanted into the trench sidewall to increase the field threshold voltage.

11.8.2 Reliability of HV Transistors in Flash Memory Products

One subject not often discussed is the reliability of high-voltage transistors and circuits in the Flash memory products. Often the Flash memory product failures are not due to the Flash memory cells but are due to reliability problems stemming from high-voltage periphery devices.

A conventional row decoder circuit for a 5-V-only stacked-gate Flash memory product using negative-gate erase is described in Umezawa et al. [135] and shown in Figure 11.95. During the erase operation, the negative voltage generated by a negative charge pump is applied to the wordline. The N^+/P^- substrate diode would be forward biased if a negative voltage is applied to the N^+ region with the P^- substrate at ground level. P-MOS transfer gates have been used in the first generation of Flash memories using negative-gate erase as one way to overcome this problem. In this scheme, the P-MOS transfer gate isolates the wordline driver N-MOS transistors from the negative voltage during the erase operation as shown in Figure 11.95. In this row decoder, P-MOS gates are inserted between each wordline and an N-MOS driver, and a gate–drain-connected P-MOS transistor is utilized as a diode between each wordline and the negative charge pump. A high positive voltage is applied to the P-MOS gate during the erase operation to isolate the negative voltage. The P-MOS gate is negatively biased during other operations.

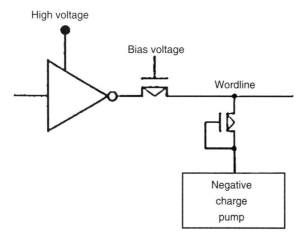

Figure 11.95. Common row decoder circuit for NOR-type negative gate erase Flash memory. (Figure 2 of [135].)

There are several problems with this scheme. First, a high-stress voltage is applied to the P-MOS gate. Due to the low hole mobility and the small device width that is available, a large negative voltage has to be applied to the gate of the P-MOS transistor in order to minimize its resistance during the read and program operations. The P-MOS gate voltage has to be negative (perhaps -3 V) in order to supply 0 V to a nonselected wordline during the read and program operations. During programming, a selected wordline needs to be raised to perhaps 10 V, creating a 13-V bias across the gate oxide of the P-MOS transistor. This bias can limit gate oxide scaling. Second, the negative voltage required for the output of the negative charge pump is larger because the negative voltage applied to the wordlines is reduced by the diodelike P-MOS transistor. Finally, this scheme requires a standby current because a negative voltage, generated by a clocked pump, is needed during standby operation, which is not desirable for portable systems.

To overcome these problems, a row decoder utilizing a triple-well process technology was developed [135] and is shown in Figure 11.96. In this circuit, N-MOS transistors used in the row decoder are in the P-wells, which are isolated from the P-substrate by a deep N-well. During the erase operation, a negative voltage of -10 V is applied to the P-well, and all wordlines are set to -10 V by the N-MOS transistors in the P-well. During other operations, the P-well is set to ground level. This scheme prevents the forward bias problem. The row decoder circuit consists of a NAND gate, an inverter, and transfer gates. It is designed to minimize the electric field on the gate oxide. The maximum stress voltage on the gate oxide is reduced to 10 V, and thinner oxide can be used.

The transistors in the X-decoder circuit subjected to the maximum stress during memory program and erase are shown in Figure 11.97. In order to ensure the reliability of the X-decoder circuit over the lifetime of the product, proper oxide thickness and transistor channel length have to be used. The main reliability concerns here are gate oxide reliability and defect density, and hot-carrier-induced dc and alternating current (ac) stress at about 10 V. The total stress time can be estimated from the device operation conditions, such as the required number of program/erase cycles, the program and erase pulse widths, and the average number of pulses.

In a NAND-type Flash memory, the program and erase of the memory is through uniform channel FN tunneling. As shown in the literature [137, 138], the NAND array is positioned in a P-well region of an N-type substrate wafer. The peripheral CMOS circuitry and the NAND memory array are located in different P-wells: peripheral circuit inside P-well 1 and NAND memory array inside P-well 2. The two P-wells are electrically isolated from each other with a well depth of about 12 µm. The CMOS logic circuits use 160-Å gate oxide, whereas transistors for generating and transferring the high program/erase voltage use 500-Å gate oxide. In the erase operation, P-well 2 and N-substrate are raised to 20 V, the source and the bitlines are floating, and P-well 1 is always grounded. In the program operation, P-well 2 is grounded, 20 V is applied to the selected control gate, and 10 V is applied to the unselected control gate.

The typical hot-carrier injection (HCI) stress conduction during program and erase operation for X-decoder transistors in a NAND-type Flash memory are shown in Figure 11.98. The junction breakdown and punch-through voltages of these transistors have to be higher than 20 V to withstand the stress. The main reliability concerns for these high-voltage transistors are ac hot-carrier stress, high-voltage oxide damage, high-field junction damage, ac lifetime, latch-up, and snapback.

HIGH-VOLTAGE PERIPHERY TRANSISTOR RELIABILITY

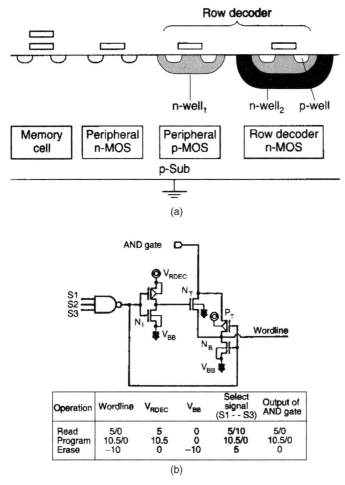

Figure 11.96. Row decoder circuit utilizing triple-well process technology for negative gate erase Flash memory. (a) Row decoder transistors cross-sectional view and (b) circuits and operation. (Figure 2 of [135].)

Figure 11.99 shows the drain, source, and substrate current of a high-voltage transistor before and after stress. The high-voltage transistor reliabilities have to be examined carefully to ensure the product reliability.

11.8.3 Process Defects: The Role of Cycling and Burn-in

The reliability failure rate of a well-designed integrated circuit often is determined by random defects that produce the infant mortality and random-life components of a standard reliability "bathtub curve." Typically, whether the product is an SRAM, DRAM, or logic device, the reliability engineer turns to accelerated burn-in and life test in order to estimate the defect-induced failure rate. Flash products often perform very well in burn-in and life test, resulting in outstandingly low failure rate predictions—and in many field applications the observed failure rate should indeed be very low. But burn-in and life test stimulate only the *low-voltage* operation of a

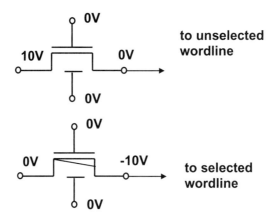

Figure 11.97. Worst hot-carrier injection stress during program and erase operation for X-decoder transistors in NOR-type Flash memory that utilize negative erase and triple-well technology [136].

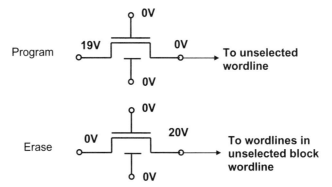

Figure 11.98. Typical HCI stress during program and erase operation for X-decoder transistors in a NAND-type Flash memory [136].

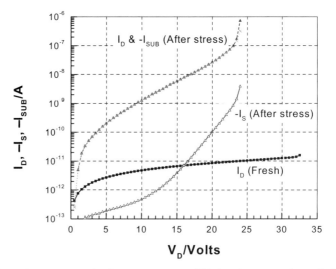

Figure 11.99. Drain current and substrate currents of high-voltage transistors before and after stress of $V_d = 21.5\,V$ for 100 mins. The other nodes are grounded [136].

Flash memory: the read operation and standby mode. To fully understand the defect-related reliability of a Flash memory, one needs to perform cycling evaluations to stimulate the *high-voltage* operation of the product. In cycling, too, there will be a "bathtub curve," with characteristic infant mortality, random life, and wearout regimes. If a field application is cycling intensive, the failure rate estimated from the cycling evaluation could be more important than that estimated by the burn-in/life test evaluation.

It is worthwhile to pause to consider why burn-in and life test are so effective for evaluating the reliability performance of other kinds of integrated circuits (ICs). Figure 11.100 illustrates a common defect type: a conductive particle between two metal lines. During operation, if one metal line is at 5 V and the other at ground, then the dielectric separating the two lines is subject to premature breakdown because of physical thinning, field enhancement from the particle itself, and mechanical or chemical damage to the dielectric from the presence of the particle. As with defects in gate oxides, defects between interconnects cause time-dependent breakdown that is accelerated by voltage and temperature. During burn-in, life test, and field application, gates and interconnect nodes are toggled between supply voltage and ground, resulting in voltage stress applied across such defects. Significantly, these voltage stresses typically have high duty cycles. For example, in an SRAM cell the two internal nodes will always be in opposite voltage states, so the space in between them will be stressed 100% of the time. In a logic device, nodes have a chaotic mix of high and low states, so that a typical defect experiences a voltage stress for 50% of the overall stress time.

There are three main reasons for the good performance of Flash memories in burn-in and lifetest—and, indeed, during low-voltage operation in field applications. First, Flash processes are designed for high-voltage operation and simply are not stressed close to failure by operation at low voltage. Gate oxides are thick, and junctions are graded. Electric fields, which drive so many reliability failure mechanisms, are low during read and standby. If, despite this intrinsic margin, there were a defect that had a short time to failure at read/standby voltage levels, then that defect is unlikely to survive even one program/erase cycle—and Flash manufacturers typically perform several such cycles during internal testing. The result is that most of the defects that might potentially cause burn-in and life test failures, or failures in field application during read/standby operation, get screened by the manufacturer during normal Flash-related testing. The astute reader will object that these arguments do not apply to every circuit. A Flash product may contain both high-voltage and low-voltage transistors, and the low-voltage transistors and

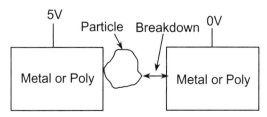

Figure 11.100. Schematic illustration of burn-in defect, showing two interconnect lines in cross section.

associated circuitry will not necessarily have the advantages noted above. But the argument certainly applies to the entire Flash memory array itself, which makes up the largest area of the IC, and to many large high-voltage circuits such as charge pumps and the final stages of decoder circuits. And the arguments largely apply to an important class of low-cost Flash products in which high-voltage transistors are used for all circuits to avoid the extra processing cost of supporting dual-gate oxides. Finally, even in a process with dual-gate oxides it will be common for the low-voltage transistors to share (for cost reasons) some of the characteristics of their high-voltage counterparts: for example, the same graded source–drain profiles.

Second, the Flash array itself sees very little stress during read and standby modes. Unlike, for example, an SRAM array, a Flash array has no direct connection to the supply voltage. During standby, all wordlines and bitlines are grounded. During read, only one row and one group (byte or word) of columns is selected—and the columns are typically raised to only 1 V rather than the full supply voltage. Since there are thousands of wordlines and bitlines in a Flash memory, defects in a Flash memory get stressed for only a tiny fraction of the total operation time, resulting in correspondingly smaller failure rates.

Third, Flash technologies often have fewer layers of metal than logic technologies do, and in practice many of the defects responsible for failures in burn-in, life test, and field usage are found in those layers. For example, Flash technologies in the 0.25-μm generation typically have two layers of metal, compared to five for logic ICs. A NOR-type Flash array fundamentally requires a single layer of metal, for the bitlines, with the wordlines carried in poly; a second metal layer is typically added for speed, to bypass the wordline resistance, and for packing density in the peripheral support circuits. Logic and SRAM designs benefit greatly, in packing density and speed, from additional metal layers.

All three of these effects are largest for those products that emphasize basic Flash capability over performance or enhanced features. A basic, low-cost, low-performance Flash product will consist mostly of the Flash array itself, and the peripheral circuits will contain high-voltage transistors exclusively and few layers of metal. At the other extreme, a high-performance embedded-Flash product with high-speed demands and substantial on-chip logic will have a large area devoted to high-performance low-voltage transistors with multiple layers of interconnect.

How big are these effects? As a rough guide, one may suppose that half of the die area is the Flash array itself, which is nearly immune to failures in low-voltage operation, which would result in a 2× reduction in the failure rate. Of the remaining peripheral circuit, perhaps half consists of high-voltage circuits with similar immunity, resulting in another factor of 2. The remaining low-voltage area may contain only two layers of metal, compared to five for a logic device, resulting in a factor of 2.5 reduction in failure rate if all layers contribute equally to the overall defect density. One would therefore expect an improvement of about one order of magnitude in the failure rate ($2 \times 2 \times 2.5 = 10$). Table 11.6 summarizes the burn-in results

TABLE 11.6. Comparison of Flash to Logic Burn-In Results

Product	Burn-in Results	Die Size	Defect Density
Logic	18/25219	0.781	9×10^{-4}
Flash	0/13777	0.343	$<2 \times 10^{-4}$

from a Flash product and a logic product built on closely related 0.25-µm processes that are run in the same fabrication facility on the same tool set. In each case, the burn-in conditions were run at highly accelerated conditions, through elevated voltage and temperature. The defect density measured on the Flash product was one-fifth, perhaps less, than that of the logic product. In fact, the Flash product was run for a somewhat longer burn-in time and at a more elevated voltage level than the logic product, suggesting that the difference is even larger than indicated by the raw numbers. This improvement factor is consistent with the simple order-of-magnitude estimate made above.

During program/erase cycling, of course, the Flash array and support circuits are indeed subjected to high voltages, and it is during cycling that many of the more conventional IC failure mechanisms can come into play. Infant mortality and random-life failure rates need to be measured with respect to the number of cycles performed, in addition to conventional burn-in hours. Failure mechanisms that can occur in peripheral circuits as a result of cycling include:

1. Gate oxide breakdown in high-voltage support circuits
2. Shorting between high-voltage metal lines because of random defects
3. Hot-electron degradation
4. Transistor damage from snapback or latchup

11.9 DESIGN AND SYSTEM IMPACTS ON FLASH MEMORY RELIABILITY

There are three major forms of Flash memory depending on their applications. For Flash memories that are used in code storage applications such as PC BIOS, cellular phone and the like, they are typically sold in single-chip form and are sometimes called *socket Flash* because often they are socket exchangeable among different venders. For Flash memories that are used for mass-storage applications, such as PCMCIA card or CompactFlash (CF) card, the memories are sold as a system, and these Flash memories are called *system Flash*. For Flash memories that are embedded in some special chips such as a microcontroller to store code or data, they are called *embedded Flash*. However, regardless of what they are called, all modern Flash memory chips are indeed Flash memory systems, with highly intelligent on-board controller or state machine to execute complex operations and sequences. Reliable Flash memory products are only possible through the intelligent system approach.

Reliability and program/erase wear out of Flash memory cells depend on many factors. For example, to verify program and to minimize program overstress, intelligent embedded program algorithms are used. In order to obtain a tight erased V_t distribution and avoid overerase, intelligent erase and overerase correction algorithms are also employed. To achieve the highest possible performance and longer endurance, wear-leveling techniques are used to ensure that in Flash memory systems, different blocks are cycled at the same rate. Fault-tolerant techniques are also effective for realizing high reliability. At present, two major techniques are used in Flash memory chips, redundancy and error correction methods. Replacing failure cells (initial) or mapping of defects by redundant cells drastically increases chip yield. The error correction method can solve some of the soft failures such as loss of data margins due to charge gain/loss/detrapping caused by endurance cycling.

The design for the memory reliability concept is most important. By studying the modes and mechanisms of the degradations and failures, rules, models, and algorithms can be developed. Reliable devices/circuits/systems can be designed by applying these rules, models, and algorithms. With the understanding of different reliability limitations and failure mechanisms, many design and system techniques can be employed to achieve highly reliable, high endurance, fault- and process-tolerant Flash memory systems. We shall discuss some of the techniques used to improve Flash memory reliability or to achieve functionality.

11.9.1 Embedded Erase and Program Algorithm

For ETOX-type stacked Flash memory, one of the toughest challenges is to control the erase threshold voltage distribution. The memories are erased in blocks or sectors by applying a source voltage to the common source of the sector and a negative gate voltage to all the wordlines while leaving the drain floating. Due to many different causes such as cell-to-cell variation in coupling ratio, tunneling oxide thickness, and "erratic bits," not all cells exhibit identical erase characteristics. Some cells are slow to erase and other cells are fast to erase. The fastest erasing cell must be exposed to the same amount of erase pulses as the slowest erase cell. Some of the fast erase cells may be overerased to a negative threshold voltage when the last cell in the sector has passed erase verify. This could lead to a leakage column during read and results in data error. In the worst case, the conduction cell can draw so much current that it will collapse the on-chip charge pump and make it impossible to program other cells in the same column. When this happens, the whole device fails. There are generally two approaches to solving this problem, and combinations of both are used in actual Flash memory products. The first approach is to control the placement of the erase cell V_t distribution, and the second approach is to tighten the erase V_t distribution by checking and correcting the overerased bits.

Optimized device structure, tight process control, and product testing can lead to tighter after-erase V_t distribution and less column leakage. Preprogram was also included as a part of the erase algorithm to ensure all cells are erased from the same starting programmed state. During the erase operation, if there are memory cells that are never programmed, the cells will be erased repeatedly over many cycles. Given enough cycles, these cells will eventually be overerased and become leaky cells. In order to prevent this from happening, all cells need to be preprogrammed before erase. The preprogram resets all cells to a programmed charge state and assures good threshold control over a large number of program/erase cycles. But due to the complex device structure and hard to control events such as erratic erase and oxide degradation caused by cycling, overerase can still happen.

The best way to guarantee that no overerase column leakage error happens is through intelligent system approach to check and correct the overerased cells, as used by many later generations of Flash memory products. The after-erase V_{th} distribution can be tightened by a self-convergence-erasing scheme [3]. This method employs drain or source posterase stress to inject channel hot electrons only to the overerased cells. As previously discussed in Section 11.2, Figure 11.3 presents typical characteristics of the channel hot-carrier-induced gate current versus V_{fg} of a typical cell in which V_g^* corresponding to the saddle point where channel hot-hole injection is in balance with the channel hot-electron injection. Using this device phenomenon, cells starting with different erased V_t's will converge to a common threshold voltage V_{th}^*. Figure 11.101 demonstrates the tightening of erased threshold voltage distribution before and after the convergence.

The complete erase algorithm is described here and summarized in Figure 11.102. First, all cells of the sector are initially preconditioned by preprogramming. After the preconditioning takes place, the state machine initiates the erase operation by applying pulses to the source and gate of each Flash memory cell. After each erase pulse, the state machine performs an erase margin read against the erase reference current and compares to check if all the cells within the associated block are erased. Data address selection is controlled by an internal address counter. The state machine halts this internal margin–sense/data–compare function when a byte location does not verify. In this instance, another erase pulse is applied. The location

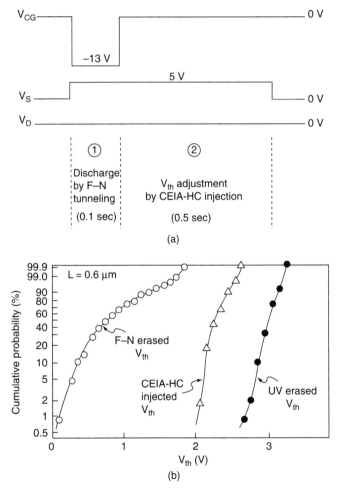

Figure 11.101. (a) Timing diagram used in experiments on the new erase sequence and (b) threshold voltage distribution before and after convergence by channel hot-electron injection. (Figures 8 and 9 of [3].)

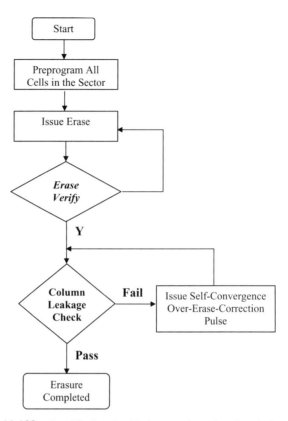

Figure 11.102. Simplified embedded erase algorithm for Flash memory.

last checked is then retested, and if successful, address sequencing and verify testing recommences; if not, another erase pulse is applied. After the last byte of the data is checked for erase verify, overerase bits or leaky columns are checked. This is accomplished by reading the memory while grounding all the control gate wordlines. If there are no overerased bits and therefore no leaky column, it passes the leaky column check and the erasure is complete. If some cells are overerased and conducting current even with grounded gate, then the leaky columns are detected. Bitline voltages are applied to these identified addresses and soft-program pulse is applied. The array is checked again for leaky columns until the last column has passed. For a properly engineered Flash memory product, if after erase there is no leaky column, the overerase correction may never be invoked. But the algorithm provides a last line of defense against any possibility of data error and rare cases of overerased bits.

The embedded program algorithm for an industry standard ETOX Flash is shown in Figure 11.103. The memory is configured as ×8 (by 8), and programmed 1 byte (8 bits) at a time. These 8 bits are located on the same wordline within a sector. Programming is done by applying high voltages to the wordline and their respective bitlines. After each program pulse, the memory is read and checked for program verify. If the cell is programmed, then the bitline voltage is turned off to avoid further stress. If the cell is not programmed, then more program pulses are applied until all the bits in the byte are verified.

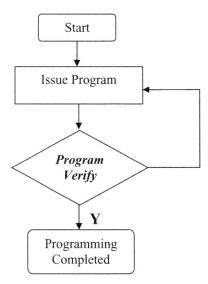

Figure 11.103. Simplified embedded program algorithm for Flash memory.

The first-generation Flash devices required extensive off-chip firmware to implement the above algorithms and to simply program and erase the device. The subtle nature of the algorithms often led to software implementation errors that resulted in costly failures and degraded reliability. To address this issue, embedded program and erase algorithms were developed by including the software in the silicon. The embedded algorithms reduce the Flash firmware requirements to a simple program or erase command sequence. All the firmware and algorithm described above are built on the chip and are implemented by an on-chip state machine.

11.9.2 Redundancy and Defect Mapping

The failures and reliability issues in Flash memories can be classified into the following categories: the ones that are common to all VLSI technologies and the ones that are specific to nonvolatile Flash memories. The former include the initial or short-term hard failures, such as MOSFET gate short, wiring open/short due to particles, and long-term failures such as MOSFET degradation due to hot carriers, capacitor short due to time-dependent dielectric breakdown (TDDB), and wiring resistance degradation due to electromigration, for example. The failures and reliability issues that are specific to Flash memories are data retention, data margin loss due to disturbs, charge detrapping, and endurance cycling, for example. To overcome these failures and to improve product reliability, a combination of redundancy, defect mapping, error correction techniques, and wear leveling have been used.

The use of redundant elements on a Flash memory chip to replace defective elements can result in higher yields and earlier introduction of new products on existing wafer fabrication lines or in new process technology. Several methods have been developed for effecting replacement of redundant elements. These include laser blow fuses, laser-annealed diffusions, electrically blown links, EEPROM, EPROM cells, and metal fuses, for example. One option that has been used on EPROM is a polysilicon fuse associated with a content-addressable memory (CAM) circuit element. CAMs are used to program the redundant row decoders to respond to the addresses of any row. This technique has the benefit that implementing the redundancy can be done either at electrical wafer sort or after packaging at final test. Redundancy on EEPROM and Flash memory has also been implemented with the use of an EEPROM cell and a CAM [139]. Each redundant row has one CAM associated for each row select address. In addition, it has a CAM that can be programmed to select that row for use at the address encoded in the CAMs. During testing the number of defective rows and their addresses are remembered. If this is equal or less than the number of redundant rows available, then the redundant rows are tested and programmed in.

Physical defects in memory devices give rise to hard errors. Data becomes corrupted whenever it is stored in the defective cells. In conventional memory devices such as RAMs and hard disks, any physical defects arising from the manufacturing process are corrected at the factory. In RAMs, spare redundant memory cells on chip may be patched on, in place of the defective cells. In the traditional disk drive, the medium is imperfect and susceptible to defects. To overcome this problem manufacturers have devised various methods of operating with these defects present, the most usual being defect mapping of sectors. In a normal disk system the media is divided into cylinders and sectors. The sector is the basic unit in which data is stored. When a system is partitioned into the various sectors, the sectors containing the defects are identified and are marked as bad and not to be used by the system. This is done in several ways. A defect map table is stored on a particular portion of the disk to be used by the interfacing controller. In addition, the bad sectors are marked as bad by special ID and flag markers. When the defect is addressed, the data that would normally be stored there is placed in an alternative location. The requirement for alternative sectors makes the system assign spare sectors at some specific interval or location. This reduces the amount of memory capacity and is a performance issue in how the alternative sectors are located.

Harari et al. [140, 141] invented a Flash memory system that allows defect mapping at cell level in which a defective cell is replaced by a substitute cell from the same sector. The defect pointer that connects the address of the defective cell to that of the substitute cell is stored in a defect map. Every time the defective cell is accessed, its bad data is replaced by the good data from the substitute cell. It also allows defect mapping at the sector level. When the number of defective cells in a sector exceeds a predetermined number, the sector containing the defective cells is replaced by a substitute sector. The defective cells or defective sectors can be remapped as soon as they are detected, thereby enabling error correction codes to adequately rectify the relatively few errors that may crop up in the system. Error correction code (ECC) is employed at all times to correct for soft errors as well as any hard errors that may arise. As soon as a hard error is detected, defect mapping is used to replace the defective cell with a spare cell in the same sector block. Only when the number of defective cells in a sector exceeds the defect mapping's capacity for that specific sector will the whole sector be replaced as in a conventional disk system. This scheme minimized wastage without compromising reliability.

11.9.3 Error Correction Concepts and Techniques

For high-density Flash memory applications, single-bit endurance failures have been the limiting reliability constraint. To improve endurance reliability and to obtain high performance, error correction circuits (ECC) are sometimes utilized [142, 143]. An on-chip ECC technique can detect and correct intermittent operational errors, such as α-particle-induced soft errors in DRAMs and SRAMs, and single-bit charge gain/loss/detrapping in Flash memories. Although ECCs are widely used in memory board systems, it is not as widely used in standard code-storage Flash memories, except in certain mass-storage Flash memory chips [144, 145]. However, with high-density Flash memory, especially multilevel Flash memory, the use of ECC techniques is extremely beneficial and can drastically increase the Flash memory reliability.

The nature of the Flash EEPROM device predicates a higher rate of cell failure especially with increasing program/erase cycling. The hard errors that accumulate with use would eventually overwhelm the ECC and render the device unusable. One solution to this problem [140, 141] is where the system is able to correct for hard errors whenever they occur. Defective cells are detected by their failure to program or erase correctly. Also during read operation, defective cells are detected and located by the ECC. As soon as a defective cell is identified, the controller will apply defect mapping to replace the defective cell with a spare cell located usually within the same sector. This dynamic correction of hard errors, in addition to conventional error correction schemes, significantly prolongs the life of the device.

Error correction involves two steps: One is detecting the error and the other is correcting it. Error detection and correction techniques are based on redundancy; this can be redundancy in the hardware of the system or redundancy in the way the data is represented, called coding. Depending on the amount of redundancy and the way it is used, error detection or error detection combined with correction is possible.

Two forms of hardware redundancy can be distinguished: replication and reconfiguration. Replication relies on identical copies of memory to mask errors. For example, a memory module is replicated three times each containing the same data, the output data is passed to a voter that produces as results the data that two or three memory modules agree with, or else an error is produced. Reconfiguration is based on the use of an error detection mechanism and the availability of spares. Reconfiguration is widely used by manufacturers for yield improvement by using spare rows and columns in the memory array, and this is often called redundancy. Reconfiguration can also be used to improve reliability and make the system more fault tolerant. The most widely used form of error detection and correction is based on coding because it offers high fault coverage with a minimal amount of extra hardware, and because of the availability of custom chips to perform the error detection and correction operations.

Figure 11.104 [146] shows a block diagram of a memory system using error detection and correction. The information stored in the memory array consists of words that contain B data bits and C check bits used for error detection and correction. The error detection and correction logic generates the check bits upon a write operation and uses the check bits upon a read operation for error detection or detection and correction purposes. In case of error correction, fault masking takes place such that correct information will be produced in the presence of one or more errors in the bits of the data word.

Parity check codes are very widely used error detection codes [145, 147]. They append one check bit to every word of data and are able to detect all odd number of faults in a word. The hardware required for encoding and decoding parity is

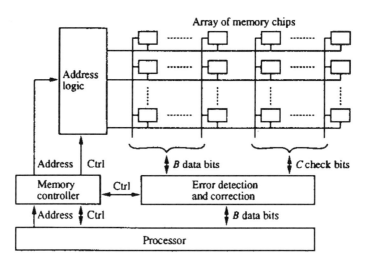

Figure 11.104. Block diagram of a memory system with error detection and correction. (Figure 12.36 of [146].)

simple and fast; it consists of a tree of XOR gates. They are used at the board level as well as at the chip level. This form of parity, applied to a memory word or byte, is also referred to as horizontal parity and it is one dimensional. Parity check codes can also be used two dimensionally, such that in addition to a horizontal parity bit per word, a vertical parity bit per column is used. The vertical parity bits are stored in a special word, called parity word, which has to be updated with each write operation.

Hamming codes are well-known simple codes capable of correcting single-bit errors. In order to be able to correct a single error, the code word has to contain an addition to the B data bits C check bits. The C check bits have to be able to specify the cases: no error or an error in one of the B + C bits of the code word. The number of check bits therefore has to satisfy the equation:

$$2^C \geq B + C + 1$$

A Hamming code capable of correcting single-bit errors in a 4-bit word requires 3 check bits and results in a 7-bit code word. For an 8-bit word, the number of check bits is 4 and results in a 12-bit code word [142]. For a 1985 1-Mb ROM chip, error correction is performed on 64 data bits using 7 parity bits for a total data field of 71 bits. For a $2^{10} = 1024$-bit word, the number of check bits is $10 + 1 = 11$ and results in 1035 code word. The extra space is relatively small.

Ting et al. [142] reported an example of the error correction and detection method for nonvolatile memory. In order to improve the endurance and achieve good data retention and reliability, a modified Hamming code for single-bit error detection and correction (SEDC) was designed for the 256 kbit EEPROM. Each byte contains 8 data bits and 4 parity bits for SEDC. During write, the parity generator generates 4 parity bits according to a modified Hamming code. These 4 parity bits, along with the 8 data bits, are latched into the column latches before they are written into the memory array. During read, the 12-bit data are sensed out through the sense amplifier. The error correction unit corrects any single-bit errors and provides 8 bits to the output buffers. The area penalty for the SEDC is about 21%.

Figure 11.105 illustrates a mathematical model of the endurance improvement with the SEDC technique based on a random defect generation model in tunnel dielectric. For example, to achieve a 1% failure probability, one is allowed to have 0.01 defects in the 256 kbit array without any error correction. But with the error correction design, one is allowed to have 15 defects in the array, which represents an improvement of over 3 orders of magnitude.

Here Hamming code was selected according to the principle of optimizing the speed and minimizing the hardware implementation. Each parity bit is generated as an EXCLUSIVE-OR of five input data. Each check bit is an EXCLUSIVE-OR of a parity output and five data sense outputs. The parity-bit generator and the check-bit generator share the same EXCLUSIVE-OR circuits, and a multiplexer is used to swap the proper inputs to the EXCLUSIVE-OR gates according to the read-write status.

The drawbacks of error correction techniques are the added circuit and system complexity and die size increase, as well as impact on read performance. That is why it is less used in code-storage-type socket Flash memories that require fast random reads. However, for high-density mass-storage-type system Flash where the reading of data is serial, the random access speed is not as important, the ECC technique is more suitable. For the Flash memory ATA card [166, 170], a Flash controller including ATA interface is used to manage data written in or read out from Flash chips. The ECC is mounted only on the controller and is commonly used for many Flash chips. Tanzawa et al. [148] reported a compact on-chip ECC for single-chip applications without a controller that cost only 2% die size penalty. The on-chip ECC, employing 10 check bits for 512 data bits, was implemented on a 64-Mbit NAND

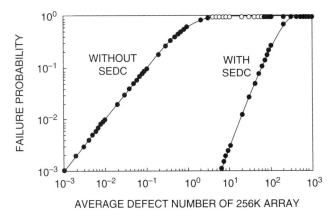

Figure 11.105. Failure probability plotted against average number of defects in a 256 kbit EEPROM array using Hamming code type error correction and without error correction. The endurance of the EEPROM is improved with the use of single-bit error detection and correction (SEDC). (Figure 12 of [143].)

Flash memory. The cumulative sector error rate was improved from 10^{-4} to 10^{-10}, by 6 orders of magnitude. The random-access time has increased by 1.5 times due to ECC.

11.9.4 Wear Leveling

Many of the Flash memory reliability issues are the result of extended program/erase endurance cycling; the errors that accumulate with use would eventually overwhelm the ECC and render the device unusable. Another solution to this problem and to lengthen the system lifetime is the "wear-leveling" technique [140, 141] where the Flash system cycles all erase blocks at an equal rate.

Studies have shown that in an application environment, certain areas of a Flash memory array such as those blocks in which application programs are stored are subject to very infrequent rewriting. On the other hand, blocks where data is accumulated for a particular operation by an application program, change and must be rewritten very frequently. Consequently, a particular block may be subject to much more rapid failure than other blocks if something is not done to equalize the amount of switching over the entire array.

In the Flash system an algorithm can be implemented that prevents any portion of the mass storage from being cycled a substantially larger number of times than any other portion [140]. The system operates so that a metering record is kept of the usage of each block of Flash memory, and the usage list is continually sorted in order to provide a free list, or priority list, of blocks according to the least amount of usage. From this least-amount-of usage selection, the microprocessor in this system can select the next least-used block of Flash memory as the next memory area to be utilized. This prevents any one block of the mass storage from failing and becoming unusable earlier than any other block, thereby extending the life of the entire mass storage. The processor uses algorithmic software functions to sort the usage value of each block of Flash memory so that the system will select the least-used memory block for the next cycle of memory usage. Some overhead header information is required for each Flash memory block to store the usage value.

11.10 FLASH MEMORY RELIABILITY SCREENING AND QUALIFICATION

11.10.1 Introduction to Reliability Testing and Screening

In the highly competitive Flash memory and semiconductor memory industry, the requirement for manufacturers to ship highly reliable products has resulted in an overall philosophy of valuing reliability and quality. To ensure these standards, care is taken to design in reliability during the product development, and all products go through vigorous reliability screening and qualifications. Ongoing reliability tests on samples from finished goods are routinely performed in order to monitor device performance and to accumulate statistical data on failure rate and understanding of failure mechanisms for further improvements. The results of these tests provide the basis for production decisions. The test methods and the proper interpretations of these data are essential in projecting the correct overall reliability of the product with confidence.

Reliability is the probability that the product remains effective in performing its designated function as in the operating environment for a specific period of time. At the component level, reliability is typically expressed as failures in time (FIT), or one part per million after 1000 h of operation.

$$1 \text{ FIT} = 1 \text{ failure}/10^9 \text{ device hours}$$

The purpose of reliability testing is to quantify the expected failure rate of a device at various points in its life cycle. Reliability screens and accelerated aging tests rely on the rapidly declining failure rate of infant mortality failure to reduce the early life failure rate. Burn-in, which is conducted at elevated temperature and voltage, is an example of reliability screens. The failure rate of memory chips due to manufacturing defects can be improved significantly with burn-in. The failure rate for semiconductor devices normally follows a predictable curve called the bathtub curve, which is shown in Figure 11.106. This curve can be divided into three different regions for reliability analysis.

1. *Infant mortality:* Early life failures showing a high initial value. The failures are usually related to manufacturing defects. Screening, testing, and burn-in procedures are the primary means to weed out such failures.
2. *Normal:* This region shows a relatively stable failure rate. Proper design increases the longevity of this period and decreases the rate of random failures in this region.
3. *Wear out:* After the normal life period the failure rate increases rapidly due to wear out.

The infant mortality period may be a few weeks, while the stable or normal life period could be more than 20 years or so. Most vendors have subjected their product to a short period of accelerated burn-in stress prior to shipment in an effort to reduce the infant mortality. Many accelerated stress tests have been developed to accelerate various failure mechanisms as discussed in the following sections. To produce reliable Flash memory products, it is necessary to know the failure modes

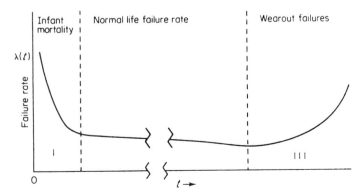

Figure 11.106. Empirical bathtub curve depicting the typical time progression of failure rate of memory devices.

that may afflict the device, and tests are then implemented to weed out the affected parts.

Besides the normal semiconductor chip failure modes, the Flash memory reliability is complicated by data retention, disturbs, high voltage, and program/erase cycling endurance. Flash memories have failure mode characteristics that are not applicable to other types of semiconductor products. In this section, we will describe various commonly used accelerated tests, some general Flash memory wafer-sort, product qualification, and burn-in procedures and considerations. Finally we will describe how to calculate the failure rates from burn-in data.

11.10.2 Classification of Flash Memory Reliability Tests

There are two categories of degradation mechanisms affecting reliability: electrical reliability issues and environmental reliability issues. Different accelerated life tests have been developed to effectively accelerate and screen these failures. Failures can also be classified as being intrinsic or defect induced. Intrinsic failure mechanisms such as hot electron, endurance, program/erase speed degradation, and the like can be characterized by accelerated testing on discrete test structures and their effects translated to product failure rates through some acceleration factors. Failure rates due to random defects can be estimated from accelerated product life test data that are translated to nominal operation life once the failure mechanisms are understood. By subjecting a device to extreme operation conditions for a shorter time, one can with some specified confidence level predict device lifetime under normal operation conditions. Many of these tests are described in detail in the Joint Electron Device Engineering Council (JEDEC) publication JESD22 and can be found at the website www.eia.org. Some of the most commonly used tests are described below.

11.10.2.1 High-Temperature Operating Life Test (HTOL). This test is sometimes called high-temperature dynamic life test (HTDL), or simply as dynamic life test (DLT). The device is operated at high temperature and/or a power supply voltage above the nominal value. The objective is to accelerate the latent failures such as contamination, mechanical stress, circuit sensitivity, and MOS transistor failures to monitor and quantify the early life reliability. This early life reliability monitor measures the effectiveness of screens and tests employed to identify and eliminate the infant mortality failure modes. During the HTOL test, the memory is sequentially addressed and the outputs are exercised but not monitored. A checkerboard data pattern is sometimes used to simulate random patterns expected during actual use. The typical time for this test is 2000h at an ambient temperature of 125°C, in a N_2 purged chamber.

11.10.2.2 Low-Temperature Operation Life Test (LTOL). This test is similar to HTOL except at low temperature to detect the effects of hot-electron injection into the gate oxide as well as package-related failures. The temperature is typically at the low end of the specified product operation temperature such as −10°C or −45°C.

11.10.2.3 Data Retention Storage Life (DRSL). This test is used to accelerate charge loss/gain at the floating gate. This stress exposes the devices to unbiased storage at an elevated temperature, usually 150°C for plastic packages or 250°C for wafers or hermetic packages. These are the highest practical temperatures the package can sustain to accelerate the retention failures. For Flash memory, this test should be performed after the data-sheet-specified endurance cycling.

11.10.2.4 Endurance Cycling. This test consists of repeated programming and erasing the device between the all-ones pattern and the all-zeros pattern. It should replicate the user's writing conditions for the device; otherwise failures in peripheral circuits that are bypassed by endurance test modes may not be assessed. Worst-case voltage levels are used to maximize the stress. Endurance cycling is used to detect failures due to oxide leakage and rupture or charge trapping of the charge-transfer dielectric of the memory cell or failures in peripheral transistors. The data-sheet-specified number of cycles must be performed before HTOL or DRSL.

11.10.2.5 Electrostatic Discharge. Electrostatic discharge (ESD) testing is done to measure a device's tolerance to damage caused by electrostatic discharge due to machine or human handling. All products incorporate on-chip ESD protection networks at the appropriate input/output pads. Human handling is modeled in specification Mil-Std-883 method 3015.7 with a 100-pf capacitor and a 1.5K series resistor. The machine model uses a 200-pf capacitor with no resistor. Three devices are stressed at each input/output pin and are zapped a minimum of three times for each polarity. A failure is defined as any failure to meet data sheet specifications. Typically, the following pin combinations are tested.

- Each I/O pin with respect to all other I/O pins tied together and grounded. The supply pins are left floating.
- Each pin with respect to electrically discrete groups of V_{CC} pins. All other pins are left floating.
- Each pin with respect to electrically discrete groups of ground pins. All other pins are left floating.
- All supply pins with respect to each electrically discrete supply. All other pins left floating.

11.10.2.6 Latch-up. Latch-up is caused by parasitic transistor turn-on, creating a current path from power to ground. This condition is characterized by a sudden increase in supply current to the device. During latch-up testing, the V_{CC} voltage is monitored for a sudden drop due to loading. For example, a stepped voltage may be applied to each pin of the device with all other pins open, while V_{CC} and V_s pins are powered up. The tests are performed at 0, 25, and 125°C. Latch-up may result in a temporary malfunction or permanent damage, and the vulnerability to latch-up is a function of process and layout design.

11.10.2.7 Temperature Humidity Bias (THB) Test. This accelerated temperature and humidity bias stress is performed at 85°C and 85% relative humidity, and is sometimes called the 85/85 test. The objective is to determine the performance of the device under extreme environmental conditions with an applied bias. The high temperature accelerates the effect of humidity on corrosion of aluminum lines under electrical bias. In general, the worst-case bias condition is the one that minimizes the device power dissipation and at a larger applied voltage. Higher power dissipation tends to lower the humidity level at the chip surface and lessen the corrosion susceptibility. Typical test duration is 2000 h.

11.10.2.8 Pressure Pot Operating Test (PPOT).

This test sometimes is also called autoclave. It applies high pressure in addition to high humidity and high temperature but without bias to accelerate chemical corrosion failure mechanisms. The objective of the test is to accelerate the problems found in very moist environments. The stress exposes the devices to saturated steam (100% humidity) at an elevated temperature and pressure. Example conditions are two atmospheres of pressure and temperature of 121°C. The steam stress accelerates moisture penetration through the plastic packaging to the surface of the die. The minimum test duration for surface mount packages is 336 h.

11.10.2.9 Highly Accelerated Stress Test (HAST).

The HAST test combines the worst-case characteristics of THB and the high-temperature, high-pressure characteristics of PPOT stressing. The stress condition is 140°C and 85% relative humidity in accordance with JESD-22, Test Method A110. HAST is often used in process control as a rapid test for moisture-related reliability assessment. Optimum bias conditions are the same as used for THB. A typical duration is 96 h for surface mount packages.

11.10.2.10 Temperature Cycle Test.

The device is cycled between the specified upper and lower temperature without power in an air or nitrogen environment. The common temperature extremes are −65°C to +150°C. The units must be transferred between temperatures within 1 min and must reach the specified temperature within 15 min, and kept at that temperature for a minimum of 10 min. Heating and cooling are done by convection. The typical number of cycles is from 200 to 1000.

This test is performed to evaluate the mechanical integrity of the device when exposed to temperature extremes. Mechanical failure mechanisms such as package cracking, die cracking, thin-film cracking, bond wire lifting, and die attach problems are accelerated by this stress. Temperature cycling also accelerates changes in electrical characteristics due to mechanical displacement or rupture of conductors and insulating materials.

11.10.2.11 Thermal Shock.

This is a more severe version of temperature cycling. Liquid is used rather than air.

11.10.2.12 Solderability Testing.

This test is designed to determine the solderability of the device leads using a standardized soldering procedure after a specified preconditioning (steam aging). Rejection criteria are based on physical appearance of the soldered leads (porosity, pinholes, nonwetting, dewetting, foreign material, etc.).

11.10.2.13 Lead Integrity.

This method tests the leads of a device by bending them and rejecting the device if the bending results in a broken or loosened lead or damage to the device hermeticity.

11.10.3 Acceleration Models of the Reliability Tests

For the different accelerated life tests, the main acceleration factors are temperature, voltage, humidity, and temperature cycles. Often combinations of these factors are used. For example, in Flash memory data retention test after endurance cycling, sometimes due to the very low value of the activation energy and low acceleration factor, very long test times are required. In order to accelerate such cycling-induced data retention failures, both temperature and voltage are needed to accelerate the charge-loss/gain mechanisms. Failure analysis techniques have been developed that allow determination of the acceleration factors through identification of the specific failure mechanisms. The different types of accelerated life tests and the acceleration factors are described below. These tests accelerate both the chip and the package failure mechanisms, and each test has separate acceleration models.

11.10.3.1 Temperature Acceleration Model. High temperature is often used as a stress factor, and high-temperature operation life test (HTOL) is typically used as the main temperature acceleration test. The commonly accepted high-temperature acceleration model is the Arrhenius model. The acceleration is determined from the Arrhenius relation:

$$\text{AF}_{\text{Temperature}} = \exp\left[\frac{E_A}{k}\left(\frac{1}{T_1} - \frac{1}{T_2}\right)\right]$$

where $\text{AF}_{\text{Temperature}}$ is the acceleration factor, E_A is the activation energy, T_2 is the stress junction temperature in Kelvin, and T_1 is the operation junction temperature in Kelvin, k is the Boltzmann constant. The junction temperature is determined from the equation

$$T_J = T_A + P_D/\theta_{\text{ja}}$$

where T_J is the junction temperature, T_A is the ambient temperature, P_D is the power dissipation, and θ_{ja} is the junction ambient thermal resistance.

The key parameter and only unknown in the model is the activation energy, and it can have a large effect on the acceleration factor. The activation energy can either be determined from lifetime tests at multiple temperatures or by adopting an activation energy that is commonly assigned to a particular type of failure mechanism. Table 11.7 shows some commonly cited activation energy values for several failure mechanisms.

11.10.3.2 Voltage Acceleration Model. An accurate time-dependent dielectric breakdown (TDDB) model is of great importance to the evaluation of oxide reliability under normal operation conditions through accelerated breakdown tests and extrapolation. The model for the voltage acceleration factor is another area where error in extrapolations can occur. There are two principal models used in the industry today, commonly referred to as the thermochemical E [149] model and the hot-induced breakdown $1/E$ [150] model. Both are used to predict oxide lifetime and to calculate the acceleration factors with reasonable results, specifically

TABLE 11.7. Activation Energy of Different Failure Mechanisms

Failure Mechanism	Activation Energy (eV)
Oxide defect	0.3
Silicon defect	0.3
Na ionic contamination	1.0
Electromigration	0.4–0.6
Charge injection	1.3
Metal corrosion	0.7
Gold–aluminum interface	0.8

with the E model more accurate at low field and the $1/E$ model more accurate at the high stress field.

The E model explains the oxide breakdown in terms of field-enhanced bond breakage, and this model predicts that the logarithm of lifetime is in linear relation with the electrical field. The E model predicts that the field acceleration factor remains a constant under different stress fields. The $1/E$ model explains the breakdown based on the process of electron injection into oxide and subsequent anode hole generation, back injection, and trapping, which forms a positive feedback that leads to dielectric breakdown. The $1/E$ model predicts that the logarithm of lifetime is proportional to the reciprocal of the electric field, and the field acceleration factor is inversely proportional to E_{Ox}^2. Recently, a model has been proposed that unifies the E model and $1/E$ model [151]. It assumes that the field-enhanced bond breakage and the hole generation and trapping processes coexist, both contribute to the degradation of the oxide. The total degradation rate is the sum of the two individual rates. The unified model also accounts for the effect of gross defects which limit the oxide reliability in real ICs.

As found experimentally by many, the E model is more accurate at low field, and $1/E$ model more accurate at the high-stress field. For modern VLSI oxide technology, this transition point between low field and high field occurs around 6 to 8 MV/cm. For most properly designed Flash memory circuits, the oxide fields are in the "low field" regime. For example, in the typical ETOX NOR memory, the highest voltage is 10 V and the transistor oxide thickness is around 150 Å. This means the peak operation field is about 6.6 MV/cm. For the 5 V logic circuit, the electric field across the oxide is even smaller. If the technology uses a single oxide thickness, then the operation field at 5.5 V is only 3.67 MV/cm. For some advanced technology that uses dual-gate oxide for high-speed operation, the 5 V (or 3 V) logic uses gate oxide as thin as 80 Å with a peak electric field of 6.8 MV/cm. An example of the E model is:

$$\text{AF}_{\text{Voltage}} = 10^{0.4(E_2-E_1)\exp\left(\frac{0.07}{kT}\right)}$$

where E_1 and E_2 are the operation and stressing fields. Sometime a simplified version is also used by some as shown below.

$$\text{AF}_{\text{Voltage}} = 10^{7(E_2-E_1)}$$

11.10.3.3 Temperature and Humidity Acceleration Model. The temperature and humidity (T&H) tests are used to evaluate the chip and package susceptibility to moisture. The model used for T&H acceleration factor has the same dependence on temperature as in the Arrhenius model. The acceleration factor due to humidity for a nonhermetic package is determined as follows:

$$A_{\text{Humidity}} = e^{0.08(\text{RH}_{\text{Stress}} - \text{RH}_D)}$$

where A_{Humidity} is the acceleration factor due to humidity, RH_D is the relative humidity at device surface for use condition (%), and $\text{RH}_{\text{Stress}}$ is the relative humidity for stress condition (%). If the device surface relative humidity during operating is 40%, and it is 85% during the accelerated test, then the acceleration factor due to humidity is 37.

11.10.3.4 Temperature Cycling Acceleration Model. The temperature cycling test is used to simulate the power on/off cycling experienced by the devices in field applications. The goal is to determine the capability of the device to withstand the thermal stresses created by changes in temperature. These stress could affect the die bond, wire bonds, or metallization of the device. The Coffin–Manson equation is used as a model for the acceleration factor between the stress condition and the field conditions:

$$A_{\text{Cycle}} = \left(\frac{\Delta T_{\text{Stress}}}{\Delta T_{\text{use}}}\right)^n$$

where A_{Cycle} is the acceleration factor due to temperature cycling, ΔT_{Stress} is the stress junction temperature range, and ΔT_{Stress} is the use junction temperature range. A typical value for n is 2, although some studies have found the value of n to be in the range of 4 to 7, dependent on the type of failure mechanism [152, 153].

11.10.4 Flash Memory Sort and Reliability Test Flow

Flash memory wafer sort and screen tests are performed to guarantee performance within specifications and screen those dies that do not meet the specifications and to provide feedback for yield improvement. The ever-increasing complexity and capacity of Flash memory has been matched by similar developments in test and reliability philosophies and practices. Large capital investments are required to establish a high-performance test that balances high quality, high yield, high throughput, and provides meaningful feedback to the wafer fabs. With the increase in device complexity and performance, it was necessary to introduce functional testing (such as timing specifications) as well as parametric testing.

For high-density memory, a 100% test for all possible operating combinations is almost impossible due to the extremely long test time required. Algorithm functional testing has been developed that involves the use of patterns of an algorithmic nature. Algorithm testing performs test combinations that are sufficient to weed out failures and still minimize test time. Expensive test equipment, sophisticated test software, and skilled engineering expertise are necessary for successful Flash memory sort and screen test. Stress tests at wafer sort are also used to weed out marginal devices before they go through the costly assembly and final test. The data retention tests are reliability screens unique to nonvolatile memory. They can be done both at the wafer level as well as the package level, with or without voltage bias.

Table 11.8 shows a generic production test flow for Flash memory [155]. Depending on the specific memory type, the maturity of the product, and the manufacturer, some tests might be removed and other tests added. For example, in this case in order to screen the programmed-state charge loss and erased-state charge gain, the wafer has to be probed and sorted three times. This can be very expensive due to the time involved, and probing the same bond pads three times might introduce bonding problems. If the product is mature and the manufacturer has collected enough statistical data, it may be possible to take out some of the data retention bake test and burn-in test without sacrificing quality and reliability.

The reliability monitor program provides a continuous, timely measure of the memory reliability and allows the manufacturers to detect potential problems early and initiate corrective actions. Certain qualities of devices in the final stages of testing are randomly sampled and subjected to a battery of reliability tests. After

TABLE 11.8. Generic Flash Memory Production Test Flow

Steps	Remarks
Wafer Processing	
Initial wafer sort	Parametric testing, check for shorts/opens. Test for input/output leakage currents. Check for static and dynamic standby currents. Setting reference cell. Test the array for erase, program operations. Stress and disturb tests. Repair bad columns/rows/bits with redundancy. Condition the array for programmed state data retention bake.
Bake at 250°C for 24 h	Unbiased bake at wafer level to look for charge loss.
Second wafer sort	Check device for charge-loss bits. Repair the leaky bits with redundancy. Erase the device for erased state data retention bake.
Bake at 250°C for 24 h	Unbiased bake at wafer level to look for charge gain.
Third wafer sort	Check device for charge gain. Repair the leaky bits with redundancy.
Assembly	Package Devices
Initial class test @25°C	Check device parameterics prior to static burn-in. Check gross functionality.
Static burn-in @125°C for 24 h	Look for charge gain under bias, such as gate voltage stress mode.
Second class test @75°C	Check for charge gain. Program the CKBD pattern for dynamic burn-in.
Dynamic burn-in @125°C for 48 h	Look for biased charge gain and charge loss. Look for adjacent bits charge transfer.
Third class test @75°C	Check for charge loss/gain. Check for speed degrades, or parametric shift. Final speed test and speed selection. Erase the array before shipping.
Q.A. sample test	Ongoing reliability monitor program.
Shipping	

production test flow, these devices are retested and undergo a 48-h 125°C life test. A percentage of the remaining devices are stressed for 1000 h of operating life. Readouts are taken at 168, 500, and 1000 h to determine time dependence of failures. One of the most important aspects of the reliability monitor program is the in-depth failure analysis performed on each failed device. The failure analysis data are used to identify problems and to determine corrective actions.

11.10.5 Flash Memory Product Qualification Flow

Before releasing product to market manufacturers always undertake many levels of qualification activities. All the incoming parts and materials must meet the dimensional, physical, electrical, and reliability characteristics defined in the procurement specifications. The process control data, yield, and reliability are examined in the trial manufacturing cycle. Further qualification tests are carried out when changes are made in any of the following: design, process, package, and materials. During this phase, prototype samples for performance checks are released to selected customers. Further lots of the products are run through the pilot run phase in order to gather statistically significant manufacturability and reliability data. In this phase, nonqualified parts and materials, process and process flows are no longer permitted, and the quality control system has to be in place. Only then can commercial samples shipments be made [154].

A qualification plan and schedule should be developed. The plan includes activities that need to be completed to qualify the technology. Data are gathered from wafer level stress testing and packaged test chip stress testing to verify the technology reliability. Some qualification tests are performed with chips in packages representative of those to be used in the actual applications, while others are performed in special packages such as ceramic packages that allow higher temperature as well as UV light erase.

For Flash memory product that is shipped as a system, such as a memory card, a second-level packaging qualification must be conducted. The definition of second-level packaging is the attachment of memory modules to a printed circuit carrier. The process of attaching surface mount packages is critical as the size of the pitch and pads decrease. Different tests at the package level and the assembly level are needed to ensure the reliability of the product, and some of these tests are:

- Ionic contamination testing on card assemblies after assembly operations to verify cleanliness.
- Biased insulation resistance test on assembled cards using 50°C/80% RH with 15-V bias for 500 h. Insulation resistance measurements and visual inspections are used to check for corrosion.
- Accelerated thermal cycling testing on card assemblies to verify that solder joint reliability will meet the thermal cycling requirement of the applications. Card assemblies are cycled from 0 to 100°C for up to 1000 cycles, and solder joint resistance values are monitored to detect if degradation is occurring.

Figure 11.107 shows a typical reliability qualification test plan and illustrates the significance of program/erase endurance cycling and data retention tests [155]. Some qualification plans and qualification data can be found at the websites

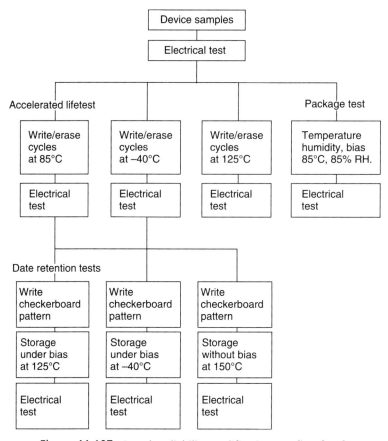

Figure 11.107. Sample reliability qualification test flow [155].

of various Flash memory manufacturers. Example of a qualification database for an industry-standard 0.35-µm socket Flash memory product is given below [156].

Am29F400B Product Qualification Database

Table of Contents
- I. Introduction
- II. Block Diagram
- III. Die Photograph
- IV. Die Processing Summary
  - Key Features of CS-39 Process
- V. Memory Address Bit Map
  - Physical Layout
  - Single Sector Enlargement
- VI. Assembly Packaging Summary
- VII. Assembly Bonding Diagrams

FLASH MEMORY RELIABILITY SCREENING AND QUALIFICATION

- VIII. Electrical Test Summary
  - Test Equipment
  - Test Coverage
  - Test Correlation and Guard Banding
  - Test Flow
- IX. Product Test Flow
- X. Reliability Data
  - High Temperature Operating Life Test
  - ESD
  - Pre-Conditioning
  - Temperature Cycle
  - Steam Pressure Pot
- XI. Burn-in Schematics
- XII. Write/Erase Endurance
- XIII. CMOS Latch-up Immunity
- XIV. Device Characterization Data
  - Tacc vs. Temperature
  - Tce vs. Temperature
  - Toe vs. Temperature
  - Pre-Program and Erase Times at $-40°C$
  - Pre-Program and Erase Times at $25°C$
  - Pre-Program and Erase Times at $90°C$
  - Icc Active @ 5 MHz vs. Temperature
  - Icc Active vs. Frequency
  - Icc Standby vs. Temperature
  - Icc Standby (CMOS) vs. Temperature
  - Vil vs. Temperature
  - Vih vs. Temperature
  - Vol vs. Temperature
  - Voh vs. Temperature
  - Tacc vs. Vcc
  - Tce vs. Vcc
  - Toe vs. Vcc
  - Vih (AC) vs. Vcc
  - Vil (AC) vs. Vcc
- XV. Quality Assurance Guidelines
  - Excelsior Quality Monitor Program
  - Advance Change Notification
  - Customer Corrective Action Request

XVI. Reliability Monitor Program
   • Reliability Monitor Stress Conditions
XVII. Statistical Process Management
   • Highlights of the current SPC programs

11.10.6 Burn-In and Reliability Monitoring Program

The requirement for low early life failure rate generally necessitates burn-in. Burn-in is typically planned for and developed during the qualification phase. During the design of the chip, different test modes are designed in so that burn-in can be performed efficiently. Burn-in is intended to cause devices with reliability defects to fail, effectively removing most of the devices that fail in the infant mortality (early life) failure region of the bathtub curve. Failing while undergoing burn-in is preferable to having the devices fail after they are shipped. By excluding such weak parts, the reliability of the product can be significantly improved.

Flash memory failure modes, which can be accelerated by a combined elevated temperature and higher voltage stress, are threshold voltage shifts (due to slow charge trapping/detrapping, contamination with mobile ions), oxide TDDB, and data retention degradation (charge loss and charge gain). Once the temperature and voltage acceleration factors are known, the burn-in temperature and voltages can be adjusted to reach the desired value. The number of infant mortality failures is generally kept in the production line database along with the number of initially nonfunctional parts detected at initial functional test. A yield figure for functional parts after burn-in is the true production yield.

Burn-in is accomplished using a dynamic test with temperature and voltage acceleration. The duration of the burn-in can be adjusted during the qualification phase based on the fallout observed at various readout intervals. The objective is to have enough time to cover the early life period considering the voltage and temperature acceleration values. Once the production phase is started, fallout is monitored and, when data indicate burn-in is not needed, burn-in can be eliminated.

The instantaneous failure rate of parts that fail by a mechanism that exhibits a lognormal distribution decreases monotonically in time. Early failure oxides have been shown to exhibit such lognormal behavior. Thus the device failure rate decreases as the stress time increases, and this is the logical basis of the burn-in screen tests. Figure 11.108 shows an example of how such a burn-in procedure reduces the failure rate of parts under normal operating conditions. The instantaneous failure rate of two groups of oxides is shown. In one group (black dots), a continuous stress of 2 MV/cm (which we assume to be the normal operating electric field stress) is applied, while in the other group (open circles), 2.5 MV/cm is applied for the first second of the test. Then the stress is reduced to 2 MV/cm. During this first second, the failure rate in the second group is increased. Upon dropping the stress to its normal operating value (2 MV/cm), the failure rate drops to a value well below that of the first group. That is, it falls to a value that is not attained by the first group until a much later time.

While longer burn-in screen time lowers the failure rate, it also increases the yield loss. It is necessary to determine the optimal set of burn-in conditions that minimizes the burn-in time, yet adequately lowers the failure rate with minimal yield loss. Moazzami and Hu [158] developed a model that predicts the effects of various

FLASH MEMORY RELIABILITY SCREENING AND QUALIFICATION   565

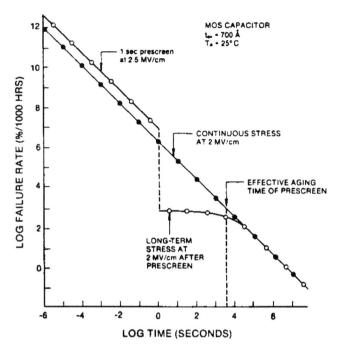

Figure 11.108. Instantaneous failure rate data showing the effect of burn-in aging stress at elevated field [157].

burn-in conditions and enables the optimization of burn-in conditions. In order to determine the appropriate stress time, field, and temperature, an accurate quantitative mode of predicting the failure rate and TDDB for these oxides with certain area is required. However, in many cases such an accurate model is not available, and the early failure can be due to mechanisms other than TDDB oxide breakdown. Often the burn-in duration is set arbitrarily or determined by experimental results that are applicable to only certain products and technology. When faced with a raw population of Flash memory chips whose reliability is to be improved by screening, the first step is a thorough reliability life testing, followed by detailed analysis of the failures. Once the dominant failure mechanisms have been identified and a failure rate computed, the type and conditions of the screens can be chosen rationally.

Screens and burn-ins can be very expensive because of the time and equipment needed to implement them. Overkill is a measure of the number of parts that fail a screen but that would never have failed in the field. A good screen has little overkill yet still lowers the failure rate. A screen may be put into production to combat a new failure mode with little regard to overkill or complexity. Once the failure rate is under control, the screen can be modified to optimize its cost with respect to its effectiveness in improving reliability.

11.10.7 Failure Rate Calculations

Accelerated life testing has been used widely to verify that a chip manufacturing line is meeting its reliability failure rate goals. Failure rates due to random defects

TABLE 11.9. Effect of Failure Rate Goal on Minimum Sample Size Requirements Assuming Poisson Statistics

Failure Rate Goal (FIT)	No. of Devices 90% UCL		No. of Devices 60% UCL	
	0 Fails	2 Fails	0 Fails	2 Fails
1000	355	835	143	463
100	3550	8350	1437	4630
10	35500	83500	14370	46300

Source: After Cook [159].

are estimated on the base of accelerated product life test data and failure acceleration models. The most effective and common method of assessing parts reliability is to run accelerated life tests on a randomly selected set of parts from one or more wafer lots. The sample size is dictated by the desired FIT rate and the test result confidence limits level. If we assume that the failure distribution is Poisson, then the number of samples required for a life test is shown in Table 11.9.

The actual estimation of the device normal operating life failure rate is the sum of the endurance, data retention, and operating life failure rates. This does not include any contribution to the failure rate from extended mechanical stresses, for example, temperature cycling, loose bond wires, and so on. Endurance preconditioning, to the data-sheet-specified number of cycles, must be performed prior to data retention and life test in order to assure the total failure rate is accurately assessed. Full parametric, functional, and timing testing at the worst-case conditions (voltage and temperature) must be performed after reliability stresses. Here we present a detailed example of how to calculate the failure rate. It consists of seven steps as described below.

EXAMPLE. Failure rate calculations for 60% upper confidence level for a 125°C burn-in/life test of 2 lots with 1200 samples.

Step 1. Accumulate data from burn-in through lifetime test of each qualification lot as shown below with a V_{cc} = 5.5 V. Three failures occurred at 168, 500, and 1000 h.

In this example, the lot size decreases from one test to another without a corresponding number of identified failures. This may be due to a variety of factors. Many tests require smaller sample sizes, and as a result all parts from a previous test do not necessarily flow through to a succeeding test. In addition, various parts are pulled from a sample lot when mechanical or handler failures or human errors occur. These "failures" are not a result of the test just completed but are nonetheless removed from the sample lot size and are not included in any failure rate calculation.

Step 2. Determine the failure rate mechanism for each failure and assign an activation energy (E_A) corresponding to each failure mechanism. Please refer to the activation energy and data retention discussion in Section 11.10.3.

FLASH MEMORY RELIABILITY SCREENING AND QUALIFICATION

TABLE 11.10. Reliability Data from 125°C Life Test @V_{CC}= 5.5 V for Example Calculation

	48 h	168 h	500 h	1000 h	2000 h
Lot 1	0/1000	1/1000	0/999	0/998	0/935
Lot 2	0/221	0/201	0/201	0/100	0/100
Totals	0/1221	1/1201 Failure A	1/1200 Failure B	1/1098 Failure C	0/1035

Note: Here 0/1000 = #failures/total # devices. Failure analysis is performed for the three failure samples.

The failure mechanisms of the above three failures are identified and activation energies of 0.6 eV for failure sample B and C, and 1.0 eV for failure sample A are assigned.

Step 3. Calculate the total number of device hours from 48 h of burn-in through the lifetime test.

EXAMPLE. A 125°C burn-in/lifetime test and a two-lot sample as shown in Table 11.10.

$$\text{Device hours} = (\text{number of devices}) \times (\text{number of hours})$$

$$\begin{aligned}\text{Total device hours} &= 1221 \times 48 + 1201 \times (168\,\text{h} - 48\,\text{h}) + 1200 \times (500\,\text{h} - 168\,\text{h}) \\ &\quad + 1098 \times (1000\,\text{h} - 500\,\text{h}) + 1035 \times (2000\,\text{h} - 1000\,\text{h}) \\ &= 2.185 \times 10^6 \text{ Device hours}\end{aligned}$$

Step 4. Calculate the rise in junction temperature T_J due to the thermal resistance of the package (θ_{JA}). T_J must be added to the desired and actual burn-in/life test temperatures.

$$T_{\text{TEST}} = T_J + T_A = \theta_{JA}(IV @ T_A) + T_A$$

Assume in this example the I_{CC} active current is 57 mA at $T_A = 125°C$, and I_{CC} active current is 60 mA at $T_A = 55°C$. Also assume that $\theta_{JA} = 35°C/W$, then

$$T_2 = (35°\text{C/W}) \times (57\,\text{mA}) \times (5\,\text{V}) + 125°\text{C} \approx 135°\text{C} = 408\,\text{K}$$

$$T_1 = (35°\text{C/W}) \times (60\,\text{mA}) \times (5\,\text{V}) + 55°\text{C} \approx 65°\text{C} = 338\,\text{K}$$

Step 5. Use the Arrhenius relation to find the equivalent device hours at a desired temperature for each activation energy (failure mechanism). The mean time to failure t is

$$t = \frac{1}{A}\exp\left(\frac{E_A}{kT}\right)$$

The acceleration factor is

$$\text{AF}_{\text{Tempeature}} = \frac{t_1}{t_2} = \exp\left[\frac{E_A}{k}\left(\frac{1}{T_1} - \frac{1}{T_2}\right)\right]$$

For example, for $E_A = 0.6\,\text{eV}$, $T_2 = 398\,\text{K}$, $T_1 = 328\,\text{K}$, then

$$\text{AF}_{\text{Temperature}} = 41.7$$

Therefore, 1 h at 125°C is the equivalent of 41.7 h at 55°C for a failure mechanism having an activation energy $E_A = 0.6\,\text{eV}$. And 41.7 is the thermal acceleration factor.

Step 6. Organize the burn-in/life test data by E_A, total device hours at the burn-in/life test temperature T_2, thermal acceleration factors for each failure mechanism, number of failures for each failure mechanism, and the calculated equivalent total device hours at the desired operation temperature T_1. The failure rate is calculated by dividing the number of failures by the equivalent device hours and is expressed as a %/1000 h.

To arrive at a confidence level associated failure rate, the failure rate is adjusted by a factor related to the number of device hours using a chi-square distribution. A conservative estimate of the failure rate is obtained by including zero failures at 0.3 eV even though no failures occurred at such an activation energy. The failure rates for each failure mechanism and the total combined failure rate can be calculated using the data in Table 11.10 and the following formula:

$$\%\text{Fail}/1000\,\text{h} = [\chi^2(n,\alpha)/2T](10^5)$$

where $\chi(n,\alpha)$ is the value of the chi-squared distribution for n degrees of freedom and confidence level of α; T is the total equivalent device hours at T_1. The total combined rate is just the sum of the individual failure rates for each failure mechanism.

For a 60% upper confidence level (UCL), the above formula converts to the following:

Number of Failures	% Fail/1000 h 60% UCL
0	$0.915 \times 10^5 / T$
1	$2.02 \times 10^5 / T$
2	$3.105 \times 10^5 / T$
3	$4.17 \times 10^5 / T$
3 < # < 15	$\dfrac{1.049(\#\,\text{Failures for a particular}\,E_A)+1.0305}{\text{Equivalent hour}@T_1} = 10^5$
>15	$\dfrac{(0.2533+\sqrt{(4\times\#\,\text{failed})+3})^2}{4T} 10^5$

The failure rate for our example of three failures is summarized in Table 11.11.

Step 7. In addition to the standard life test at $V_{CC} = 5.5\,\text{V}$, tests were also performed at higher V_{CC} to accelerate the oxide breakdown failures. Since the oxide breakdown has a activation energy of 0.3 eV, this data plus the above standard dynamic life test data are used to calculate the 0.3-eV failure rate. Assume that an additional lot of 800 Flash devices is burned in at $V_{CC} = 6.5\,\text{V}$. Using Table 11.12, a voltage acceleration factor of 55 results from a 20% voltage overstress (5.5 to 6.5 V).

$$\text{Device hours} = 800 \times (48\,\text{h} - 0\,\text{h}) + 800 \times (168\,\text{h} - 48\,\text{h}) + 799 \times (500\,\text{h} - 168\,\text{h})$$
$$= 3.997 \times 10^5\,\text{h}$$

The voltage acceleration factor (VAF) is calculated using the following formula:

$$\text{VAF} = 10^{0.4(E_2 - E_1)\exp(0.07/kT)}$$

where E_1 and E_2 are the operation and stressing fields. Assume the 5-V logic oxide thickness in this Flash device is 150 Å. Then the 5.5-V operation electric field is 3.67 MV/cm, and the stressing field is 4.33 MV/cm. The voltage acceleration factor calculated from the above equation of 46,416 results from a 20% voltage overstress (5.5 to 6.5 V).

Since this voltage accelerated stress is used to predict an oxide breakdown failure rate, the total device hours for the 5.5-V burn-in life test for $E_A = 0.3\,\text{eV}$ and the 6.5-V burn-in life test for 55°C equivalent hours are listed in Table 11.13.

TABLE 11.11. Summary of Failure Rate Calculation Example Data

E_A (eV)	Actual Device Hours @125°C	Acceleration for 135°C to 65°C	Equivalent Hours @55°C	Number of Fails	55°C %Fail/ 1000 h
0.3	2.185×10^6	5.85	1.287×10^7	0	0.0081
0.6	2.185×10^6	34.18	7.4×10^7	2	0.0042
1.0	2.185×10^6	359.93	7.86×10^8	1	0.0003
		Total combined failure rate = 0.0126 = 126 FIT			

TABLE 11.12. Test Results from Burn-in of an Additional Lot of 800 Devices

	48 h	168 h	500 h
Lot 3	0/800	1/800 Failure C	0/799

TABLE 11.13. Equivalent 55°C Hours Derived from 125°C Life Tests

125°C Burn-in/ Life Test	E_A (eV)	Actual Device Hours @125°C	Acceleration Factors for 135°C to 55°C	Equivalent Hours @55°C
5.5 V	0.3	2.185×10^6	5.85	1.278×10^7
6.5 V	0.3	3.997×10^5	(5.85×46416)	1.085×10^{11}
Total equivalent $E_A = 0.3$ eV Device Hours = 1.085×10^{11}				

TABLE 11.14. Summary of Failure Rate Prediction Data

E_A (eV)	Actual Device Hours @125°C	Acceleration for 135°C to 65°C	Equivalent Hours @55°C	Number of Fails	55°C %Fail/ 1000 h
0.3	2.185×10^6	5.85	—	—	—
0.3 + 46416[a]	3.997×10^5	(5.85×46416)	1.085×10^{11}	1	0.000002
0.6	2.185×10^6	34.18	7.4×10^7	2	0.0042
1.0	2.185×10^6	359.93	7.86×10^8	1	0.0003
Total combined failure rate = 0.0045 = 45 FIT					

[a]The notation 0.3 + 46416 is used to show that 6.5 and 5.5 V burn-in life test equivalent hours have been combined.

The failure rate predictions shown in Table 11.14 include the total equivalent 55°C, $E_A = 0.3$ eV device hours found here.

11.11 FOR FURTHER STUDY

11.11.1 Introduction

This chapter has reviewed the important concepts that have guided the optimization of Flash memory reliability from its conception in the 1980s to the position it achieved at the turn of the millennium—a nearly indispensable form of digital storage. Understanding of the physical mechanisms involved in Flash reliability has deepened with time. Since early 1999 when the first draft of this chapter was written, there has been significant growth and advancement in the Flash memory industry and technologies, while the fundamentals covered in this chapter remain the same. This section summarizes some recent works as a guide for those wishing to pursue this topic in greater detail.

11.11.2 Erratic Erase

A series of papers by Chimenton, Irrera, Modelli, Olivo, and Pellati [160–169] investigated the erase tail and erratic erase in particular. The work upheld the basic picture outlined earlier, that erratic erase is due to statistical variations in oxide-trapped charges, in particular clusters of trapped holes. Detailed analysis has indicated that the erase tail consists primarily of erratic bits and that the number of such bits can be strongly modulated by high-temperature bakes and exposure to ultraviolet light, both of which effects are ascribed to detrapping of the oxide charge. The severity of the erratic-erase effect has been shown to be a strong function of the stress electric field; reduced peak E-field and/or erasure with short erase pulses separated by delays have been shown to significantly reduce the number of erratic bits. The E-field dependence, and the finding that thinner oxides are less susceptible to erratic erase, have been shown to be consistent with anode hole injection as the source for the trapped holes leading to the effect.

11.11.3 Stress-Induced-Leakage-Current Related Retention Effects

Stress-induced-leakage-current related retention effects have been an area of intensive study. A number of studies [170–174] have focused on statistical accelerated reliability models for the effect. Such models attempt to generate practical reliability failure rate predictions that incorporate the important determining factors: cycle count, retention time, retention temperature, cycling voltages and other conditions, and oxide thickness. It has been found that individual leaky cells vary widely in behavior and often have characteristics that are difficult to model, such as erratic leakage currents that occasionally stop and later restart. The statistical distributions of cells, however, are well behaved. Reliability modeling activity has focused on fitting such statistical distributions. This technique is sometimes called the equivalent-cell method, in which one models, for example, the behavior of the 1-ppm point of a distribution of cells, as if that 1-ppm point represented a specific cell. Using this approach, a number of generally consistent models have been generated that have been shown to fit retention shifts occurring over multiple years of retention bake. It has been found that the SILC-related tail of the statistical distribution has an extreme-value shape and that the threshold voltage shifts with the logarithm of time (which is consistent with an exponential or quasi-exponential I–V characteristic for the leakage). The number of tail cells has been found to scale approximately as a power-law in cycle count, though deviations from a pure power-law relationship have been reported. Retention temperature acceleration is low, with reported activation energies ranging from near zero to about 0.25 eV.

Progress has been made in separating the effects of program and erase in generating the dielectric damage leading to SILC retention loss, for NOR-type memories in which programming is by channel-hot-electron injection. It has been reported for one manufacturer's technology [173] that evidence pointed to the damage being overwhelmingly caused by the erase step: changing erase voltages during cycling strongly modulated the resulting SILC, but changing the programming did not. In particular, raising the source bias used with the negative gate–source erase scheme strongly increased the number of tail bits, giving evidence for hot-hole injection being an important source of damage. For another manufacturer's technology [175–177], it was shown that SILC-related damage had a large component due to drain-disturb-induced hole injection during CHE programming of other cells on the same bitline. Evidence for this included electrical bias experiments that localized the leakage near the drain and correlation work, which showed that the cells with the most drain disturb exposure had the most SILC. The mechanics of such hole injection has been investigated [178]. Finally, for a third manufacturer [179] it was found that about 99% of the SILC defects were over the source, where erase occurred, but about 1% were near the drain and were attributed to

CHE programming itself. These apparently contradictory results can be understood when one recognizes the widely varying degree to which band-to-band tunneling resulting in hot holes exists in each of the technologies. Whenever BBT is present, to generate a large supply of holes, and sufficient voltage drop exists to make them energetic, then SILC is generated near the point of hole injection. When BBT of holes is eliminated during erase, BBT of holes during drain disturb can be dominant. When all sources of BBT are eliminated, it remains possible that anode hole injection (AHI) generated holes play a crucial role.

Progress has been made in explaining the behavior of individual SILC-affected cells in terms of quantum-mechanical models for the trap-assisted tunneling mechanism [180–185]. Consensus has emerged that the SILC responsible for retention loss in nonvolatile memory cells is the result of trap-assisted tunneling through percolation paths consisting of two traps in the most common case. Across a wide range of tunnel-oxide thicknesses evidence of single-trap or multiple-trap conduction can sometimes be found. Modeling of this trap-assisted tunneling has moved beyond the simple one-dimensional WKB-based models to comprehend three-dimensional effects and enhancement from phonon-assisted tunneling. Nonexponential characteristics of the tunneling current, though not apparent in the statistical modeling mentioned above, have been detected in individual memory cells, most notably in cutoff behavior in which leakage abruptly drops below a certain electric field. Such cutoff behavior have been attributed to forbidden transitions that arise when trap energy no longer lines up with available states in the anode or the cathode. These single-cell models have been extended to predict statistical distributions of cells in Ielmini et al. [186, 187].

As with erratic erase, it has been shown that SILC effects may be reduced by erasing with a larger number of shorter erase pulses, with delays in between [169, 188].

11.11.4 Detrapping-Related Retention Effects

With the increasing importance of multilevel-cell Flash memories and the dominance of channel erase in both NOR and NAND memories, retention shifts from charge detrapping (see Section 11.3.2.5) have become more important [189–196]. Attention has focused both on the emptying and refilling of traps and on the thermal annihilation of the traps themselves. Trap emptying has been attributed both to direct tunneling and to thermal emission. Trap annihilation has been attributed to both interface and bulk traps. The detrapping effects have been found to become more significant with narrower geometries, which has been attributed to mechanical stress generated by trench isolation edges and deposited nitride films. Detrapping attributed to annihilation has been found to be highly thermally accelerated, with an activation energy of 1.1 to 1.2 eV. For this mechanism it has been found that detrapping between cycles can be important: Any detrapping that occurs between cycles represents charge that cannot detrap during a retention period. It has been found that the detrapping between cycles is also thermally accelerated, with an activation energy of about 1.1 eV. As a result, cycling conditions can be designed to match expected use by performing the cycling over a specific number of days and at a specific temperature chosen so that the temperature-extrapolated time matches the expected use condition.

11.11.5 Qualification Methods

A number of changes have occurred to Flash memories that have rendered certain past qualification practices incomplete or misleading. The most important changes are:

1. Dominance in retention of the SILC and detrapping mechanisms rather than oxide defects
2. Growth in memory density, especially multiblock memories
3. Growing importance of multilevel-cell memories
4. Increasing cell-to-cell interactions

Standard retention tests were designed around the defect-related retention mechanisms (see Section 11.3.2.2), with an activation energy of 0.6 eV. With such an activation energy, a 1000-h 125°C retention test would be equivalent to about 5 years of retention at a use temperature of 55°C, which is a reasonable requirement. Flash memories today, however, are dominated by the SILC mechanism ($E_A \approx 0$) and the detrapping mechanism ($E_A \approx 1.1$ eV). That same life test then amounts to only 1000 h of use for the former and 100 years for the latter, neither of which is reasonable. In addition, for the detrapping mechanism it has been established that the cycling rate and temperature are important parameters, neither of which was addressed in published standards.

Growth in memory density has posed a simple but intractable problem: Established qualification standards required all cells of a memory device to be cycled to the maximum endurance specification of the device (often 100,000 cycles), but in a large multiblock memory this could require each device to be cycled continuously for several years. This would be impossible in a qualification flow and highly unrealistic for actual use.

Although multilevel-cell (MLC) memories experience the same physical mechanisms as single-level Flash memories, logistically they must be handled differently. Published standards required cycling every cell between the program state and the erased state, but MLC devices have multiple program states. It is often supposed that the worst-case condition is to cycle a cell from its lowest V_T state to its highest, ensuring the maximum oxide flux. That is certainly true for some mechanisms, but it is not for others. Programming an MLC device to an intermediate V_T state involves a careful, multiple-pulse charge-placement algorithm that can involve more stress time than programming to the highest level. In addition, the probability of the programming operation failing may be higher for the intermediate levels because of the smaller margins for those levels. As a result, cycling to extreme levels neither simulates actual use nor guarantees a worst-case result.

The increase in cell-to-cell interactions can take many forms. In all floating-gate memories, as lateral dimensions have been reduced the capacitive coupling between adjacent floating gates has increased. As a result, programming one cell affects the V_T of the adjacent ones. In NAND Flash memory, V_T interactions also occur because the V_T of a cell depends on the series resistance created by the other cells in the same string. These particular interactions tend to increase programmed threshold levels, which reinforces data integrity for the programmed cells in a single-level NOR memory but can detract from data integrity in an inner state of a multilevel NOR memory or in a NAND memory (which can suffer a string failure if a cell programs too high). Other cell-to-cell interaction effects, such as hot-carriers generated in one cell affecting the threshold of another, can be critically dependent on the nature of the data in the adjacent memory cells, often in hard-to-predict ways. Established qualification standards did not consider such effects. As a practical matter, how such interactions occur in a qualification flow depends upon the data pattern used. Established qualification standards generally specified simple data patterns that would create certain cell-to-cell interactions but not others.

As a result of these and other changes, published qualification standards often did not emulate reasonable use, and manufacturers generally developed customized qualification flows that did. There has been a recent effort to update the JEDEC standards to better reflect current practice and emulate realistic use [197, 198], and there has been increasing emphasis in the industry on establishing accepted methods for generating customized qualification flows [199]. The new focus has addressed the issues above through:

1. Retention requirements based upon the acceleration factors for the dominant retention mechanisms
2. Procedure for selecting the cycling rate and temperature in a qualification
3. Limitation on the number of cells cycled, for high-density memories
4. Selection of cycling and retention data patterns appropriate to multilevel-cell memories and sensitive to cell-to-cell interaction effelcts

11.11.6 Flash Memory Floating-Gate to Floating-Gate Coupling

With the aggressive scaling of the Flash memory, the distance between floating gates is becoming increasingly smaller, new physical phenomena and scaling limitations and reliability issues arise. One such issue that occurred in the scaled floating-gate Flash memory is the coupling between adjacent floating gates, first observed during the development of NOR MLC Flash technology in early 1997 [200]. Due to the Coulomb field by the adjacent cells, the threshold voltage of one, the data of one bit, or the data margin of one bit can be affected by the subsequent writes of the adjacent cells. The amount of threshold voltage shift or margin lost is a function of the magnitude of the voltage swing of the subsequent write of the adjacent cell. This eats into already tight threshold voltage margins of MLC operation, and negatively impacts gate coupling ratio, data retention margins, overprogram margins, and write speed. As illustrated in Figure 11.109 for the NAND Flash memory [201], this FG-to-FG coupling, or parasitic capacitance effects in NAND Flash become very important as one of the scaling limitations. The amount of coupling between two adjacent wordline cells becomes inversely proportional to their distance.

In order to counter this problem, several methods have been proposed. One obvious method is the reduction of the floating-gate height so the FG-to-FG capacitor area is reduced. But this has its limitations and makes it difficult to achieve proper gate coupling ratio. One process method is the shielding of the adjacent floating gate from the coupling [202], but for the case of NAND, it is only effective for the bitline directions, while not possible for the cells from adjacent wordlines. Another process method is the reduction of the capacitance by using lower K dielectric materials [200, 202]. The air spacer method was proposed [203, 204] that

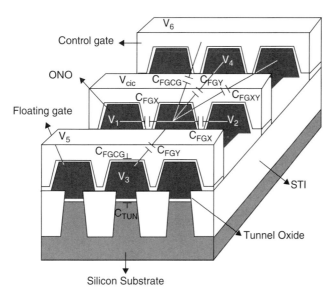

Figure 11.109. Adjacent floating-gate to floating-gate coupling by Coulomb field.

has the lowest dielectric constant. By applying the air spacer technology on poly-Si/Wsix stack gate of 90-nm design rule NAND Flash device, the coupling was reduced. Other methods are also possible such as by device operation algorithms and circuit design techniques. Charge-trapping devices have also been proposed as having little or no adjacent coupling effect [205].

11.11.7 New Program Disturb Phenomenon in NAND Flash Memory

Program disturb is one of the most critical scaling and reliability issues for NAND Flash memory. Especially these concerns are more significant for multilevel cells (MLCs) where much higher program verify and program voltage levels are needed, while margins between states are significantly reduced. Various program inhibit boosting schemes have been studied and used by different vendors [206].

As the NAND device is further scaled down, due to high boosted voltages in the various junctions, and the different floating-gate potentials depending on it's data state (high V_t's or low V_t's), a highly reverse-biased junction under the negatively charged gate can cause gate-induced drain leakage (GIDL) current [207]. GIDL involves electrons leaking into the boosted channel. It occurs with a large bias in the junction and a low or negative gate voltage, which is precisely the case when the source-side neighbor cell is programmed and the drain junction is boosted. GIDL will cause the boosted voltage to leak away prematurely, resulting in programming error. GIDL is more severe with the abruptly and highly doped junctions, which is required as cell dimensions are scaled. If the leakage current is high enough, the boosting potential in the channel region will go down and can result in program disturb. Careful optimization of the cell junction profiles and optimized boosting schemes are required to minimize such problems.

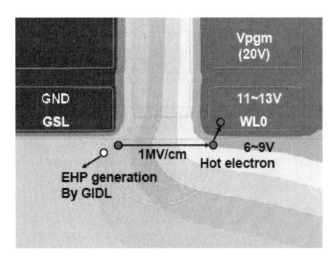

Figure 11.110. Simulation of the hot-carrier disturbance phenomenon.

Recently, a new programming disturbance phenomenon in the NAND Flash cell operation is reported and modeled as "source/drain hot-carrier injection disturbance" [208]. The source/drain hot electrons are supplied from GIDL current at the GSL (ground-select line) transistor gate edge and are accelerated in the source/drain region of GSL–WL0 (wordline 0) cell, then are injected to the WL0 cell. The simulation results (Fig. 11.110) show that the space of GSL–WL0 cell should be kept over 110 nm to suppress the hot-carrier disturbance, which is also verified by experiment. This obviously has significant consequence to the NAND product design and technology scaling.

To overcome this problem, Park et al. [209] proposed a WL shielding scheme by using dummy cells in NAND Flash memory. At the two ends of the NAND string, two dummy wordlines were inserted to eliminate the edge WL effects, which cause degradations of program disturbance and V_{th} distribution. This is shown in Figure 11.111.

11.11.8 Impacts of Random Telegraph Signals and Few-Electron Phenomena on the Scaling of Flash Memories

The random telegraph signal (RTS) has been a major concern for analog and radio-frequency (RF) applications in advanced CMOS technologies and will also become a serious issue in digital logic circuits and memories as device dimensions shrink. Especially in MLC Flash memory, some authors believe RTS will start to become an issue for the following reasons [205]. The threshold voltage amplitude due to RTS in Flash memory can be estimated by the simple formula $q/L_{eff}W_{eff}C_{ox}R_c$, where q is the elementary charge, L_{eff} is the effective channel length, W_{eff} is the effective channel width, C_{ox} is the capacitance of the gate dielectric per unit area, and R_c is the coupling coefficient of the control gate to the floating gate. The stack-gate structure in Flash memory amplifies the threshold voltage by the coupling coefficient (R_c). Besides, C_{ox} cannot be enlarged because it is difficult to decrease the gate oxide

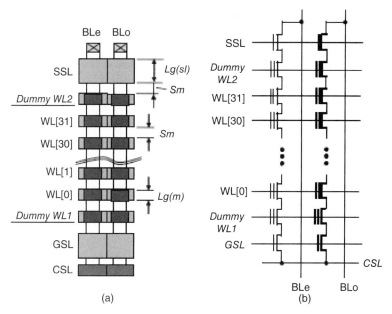

Figure 11.111. Proposed dummy wordline scheme [207].

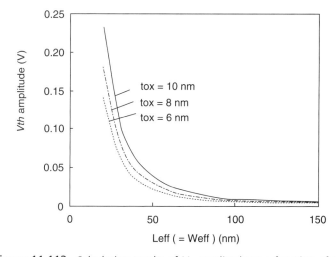

Figure 11.112. Calculation results of V_{th} amplitude as a function of L_{eff}.

thickness (t_{ox}) due to stress-induced leakage current. The V_{th} amplitude due to RTS increases abruptly as the device scaling advances, and it could be in excess of 100 mV at the 45-nm node, as shown in Figure 11.112. It will become a significant concern especially for the MLC operation. In Kurata et al. [210], threshold voltage fluctuation due to RTS was observed and confirmed the existence of the tail bits generated by RTS, as shown in Figure 11.113. Further, with write/erase cycling, additional interface traps are generated, and this results in increased RTS and hence the increased V_{th} due to RTS. In the experiment reported [210], the V_{th} amplitude more than doubled after 10,000 W/E cycles. (Fig. 11.114)

As the dimensions of Flash memories are scaled down, the number of electrons representing one bit dramatically reduces while the use of multilevel cell memory technologies will result in an even more reduced number of electrons per bit. For example, a sub-40-nm floating-gate MLC device will have less than 100 electrons to differentiate between two states. For MLC devices operating with more than 2 bits/cell, the number of electrons per state is further reduced by half. A quantitative evaluation of the intrinsic reliability limits of floating-gate (FG) memories is pre-

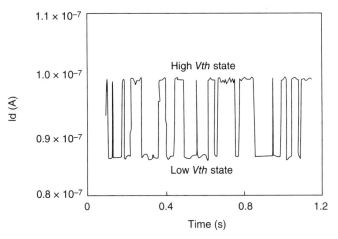

Figure 11.113. Example of time-series change in drain current in 90-nm-node Flash memory.

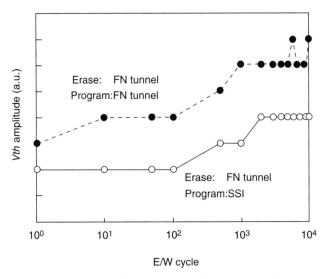

Figure 11.114. V_{th} amplitude due to RTS increases as E/W cycles.

sented in [211], predicting the wider dispersion of the memory intrinsic retention time and program distribution. This is understandable that when the difference between states is only tens of electrons, each program pulse adds only a few electrons.

So far there has been little study and data on the real impacts of RTS and few and fewer electron phenomena on the Flash memories. With the rapid scaling of NAND Flash memory, devices will enter those regimes in the next few generations; more study and solutions will be emerging from the industry. Flash memory reliability remains as interesting and challenging as ever, and just like Flash memory as a whole, full of opportunities for new ideas and innovations.

REFERENCES

1. J. Chen, N. Radjy, S. Cagnina, and J. Lien, "Degradation Mechanism of Flash EPROM Program/Erase Endurance," paper presented at the 13th Annual IEEE Nonvolatile Semiconductor Memory Workshop, Monterey, CA, Feb. 1994.
2. E. Takeda, N. Suzuki, and T. Hagiwara, "Device Performance Degradation due to Hot-Carrier Injection at Energies below the Si-SiO$_2$ Energy Barrier," *IEEE IEDM Tech. Dig.*, p. 396, 1983.
3. S. Yamada, T. Suzuki, E. Obi, M. Oshikiri, K. Naruke, and M. Wada, "A Self-Convergence Erasing Scheme for a Simple Stacked Gate Flash EEPROM," *IEEE IEDM Tech. Dig.*, p. 307, 1991.
4. J. Chen, L. Cleveland, S. Hollmer, M. Kwan, D. Liu, and N. Radjy, "Adjustable Threshold Voltage Conversion Circuit," U.S. Patent 5,521,867, May 28, 1996.
5. S. Yamada, Y. Hiura, T. Yamane, K. Amemiya, Y. Ohshima, and K. Yoshikawa, "Degradation Mechanism of Flash EEPROM Programming After Program/Erase Cycles," *IEEE IEDM Tech. Dig.*, p. 23, 1993.
6. B. Doyle, M. Bourcerie, J. C. Marchetaux, and A. Boudou, "Interface State Creation and Charge Trapping in the Medium-to-High Gate Voltage Range ($V_d/2 < V_g < V_d$) during hot-carrier stressing of n-MOS," *IEEE Trans. Electron Devices*, Vol. 37, p. 744, 1990.
7. B. Joshi and D. L. Kwong, "Comparison of Neutral Electron Trap Generation by Holt-Carrier Stress in N-MOSFET's with Oxide and Oxynitride Gate Dielectrics," *IEEE Electron Device Lett.*, Vol. 13, p. 260, 1992.
8. H. Fukuda, A. Uchiyama, T. Kuramochi, T. Hayashi, T. Iwabuchi, T. Ono, and T. Takayashiki, "High-Performance Scaled Flash-Type EEPROMs with Heavily Oxynitrided Tunnel Oxide," *IEEE IEDM Tech. Dig.*, p. 465, 1992.
9. H. Fukuda, M. Yasuda, T. Iwabuchi, and S. Ohno, "Novel N$_2$O-Oxynitridation Technology for Forming Highly Reliable EEPROM Tunnel Oxide," *IEEE Electron Device Lett.*, Vol. 12, No. 11, p. 587, 1991.
10. S. K. Lai, J. Lee, and V. K. Dham, "Electrical Properties of Nitride-Oxide Systems for Use in Gate Dielectric and EEPROM," *IEEE IEDM Tech. Dig.*, p. 190, 1983.
11. H. Fukuda, A. Arakawa, and S. Ohno, "Thin-Gate SiO$_2$ Films Formed by Insitu Multiple Rapid Thermal Processing," *IEEE Trans. Electron Devices*, Vol. 39, No. 1, pp. 127–133, Jan. 1992.
12. J. Kim, J. Choi, W. Shin, D. Kim, H. Kim, K. Mang, S. Ahn, and O. Kwon, "Scaling Down of Tunnel Oxynitride in NAND Flash Memory: Oxynitride Selection and Reliabilities," *Proc. of International Reliability Physics Symp.*, p. 12, 1997.

13. K. Yoshikawa, S. Yamada, J. Miyamoto, T. Suzuki, M. Oshikiri, E. Obi, Y. Yamada, Y. Ohshima, and S. Atsumi, "Comparison of Current Flash EEPROM Erasing Methods: Stability and How to Control," *IEEE IEDM Tech. Dig.*, p. 595, 1992.
14. V. H. Chan and D. K. Y. Liu, "An Enhanced Erase Mechanism During Channel Fowler-Nordheim Tunneling in Flash EPROM Memory Devices," *IEEE Electron Device Lett.*, Vol. 20, No. 3, p. 140, 1999.
15. J. Chen, T. Y. Chan, I. C. Chen, P. K. Ko, and C. Hu, "Subbreakdown Drain Leakage Current in MOSFET," *IEEE Electron Device Lett.*, Vol. 8, p. 515, 1987.
16. C. Chang and J. Lien, "Corner-field induced drain leakage in thin oxide MOSFET's," *IEEE IEDM Tech. Dig.*, p. 714, 1987.
17. T. Y. Chan, J. Chen, P. K. Ko, and C. Hu, "GIDL and Its Impact on Scaling," *IEEE IEDM Tech. Dig.*, p. 718, 1987.
18. S. Haddad, C. Chang, B. Swaminathan, and J. Lien, "Degradations Due to Hole Trapping in Flash Memory," *IEEE Electron Device Lett.*, Vol. 10, p. 117, 1989.
19. Y. Inura, H. Matsuoka, and E. Takeda, "New Device Degradation Due to 'Cold' Carriers Created by Band-to-Band Tunneling," *IEEE Electron Device Lett.*, Vol. 10, p. 227, 1989.
20. K. Yoshikawa, S. Mori, E. Sakagami, Y. Ohshima, Y. Kaneko, and N. Arai, "Lucky-Hole Injection Induced by Band-to-Band Tunneling Leakage in Stacked Gate Transistors," *IEEE IEDM Tech. Dig.*, p. 577, 1990.
21. T. C. Ong, A. Fazio, N. Mielke, S. Pan, N. Righos, G. Atwood, and S. Lai, "Erratic Erase in ETOX Flash Memory Array," *IEEE Symposium on VLSI Technology*, p. 83, May 1993.
22. N. Mielke, G. Atwood, and A. Merchant, "Method of Repairing Overerased Cells in a Flash Memory," U.S. Patent 5,237,535, Aug. 17, 1993.
23. G. Verma and N. Mielke, "Reliability Performance of ETOX Based Flash Memories," *IEEE International Reliability Physics Symp.*, pp. 158–166, 1988.
24. S. Aritome, K. Sakui, K. Ohuchi, and F. Masuoka, "Extended Data Retention Characteristics after More Than 10^4 Write and Erase Cycles in EEPROMs," *Proc. IEEE International Reliability Physics Symp.*, pp. 158–166, 1990.
25. D. A. Baglee and M. Smayling, "The Effects of Write/Erase Cycling on Data Loss in EEPROMs," *IEEE IEDM Tech. Dig.*, pp. 624–626, 1985.
26. K. Naruke, S. Taguchi, and M. Wada, "Stress Induced Leakage Current Limiting to Scale Down EEPROM Tunnel Oxide Thickness," *IEEE IEDM Tech. Dig.*, pp. 424–427, 1988.
27. R. E. Shiner, J. Caywood, and B. Euzent, "Data Retention in EPROMs," *Proc. International Reliability Physics Symp.*, pp. 238–243, 1980.
28. R. N. Mielke, "New EPROM Data Loss Mechanisms," *Proc. International Reliability Physics Symp.*, pp. 106–113, 1983.
29. J. S. Witters, G. Groeseneken, and H. E. Maes, "Degradation of Tunnel-Oxide Floating Gate EEPROM Devices and the Correlation with High Field Current Induced Degradation of Thin Gate Oxide," *IEEE Trans. Electron Devices*, Vol. 36, pp. 1663–1682, Sept. 1989.
30. S. Mori, Y. Kaeko, N. Arai, Y. Ohshima, H. Araki, K. Narita, E. Sakagami, and K. Yoshikawa, "Reliability Study of Thin Interpoly Dielectrics for Nonvolatile Memory Application," *Proc. International Reliability Physics Symp.*, pp. 132–144, 1990.
31. K. Wu, C. S. Pan, J. J. Shaw, P. Freiberger, and G. Seryet, "A Model for EPROM Intrinsic Charge Loss Through Oxide-Nitride-Oxide (ONO) Interpoly Dielectric," *Proc. International Reliability Physics Symp.*, pp. 145–149, 1990.
32. S. Mori, E. Sakagami, H. Araki, Y. Kaneko, K. Narita, Y. Ohshima, N. Arai, and K. Yoshikawa, "ONO Inter-Poly Dielectric Scaling for Nonvolatile Memory Applications," *IEEE Trans. Electron Devices*, Vol. 38, pp. 386–391, Feb. 1991.

33. D. A. Baglee, "Characteristics and Reliability of 100A Oxides," *IEEE International Reliability Physics Symp.*, p. 152, 1984.
34. F. Arai, T. Maruyama, and R. Shirota, "Extended Data Retention Process Technology for High Reliability Flash EEPROMs of 10^6 to 10^7 W/E Cycles," *IEEE International Reliability Physics Symp.*, pp. 378–382, 1998.
35. B. De Salvo, G. Ghibaudo, G. Pananakakis, G. Guillaumot, and P. Chandelier, "Experiment and Modeling of Data-Retention in NVM's," *Proc. IEEE Nonvolatile Semiconductor Memory Workshop*, Monterey, CA, p. 76, 1998.
36. B. De Salvo, G. Ghibaudo, G. Pananakakis, G. Guillaumot, P. Chandelier, and G. Reimbold, "A New Physical Model for NVM Data-Retention Time-to-Failure," *Proc. IEEE International Reliability Physics Symp.*, p. 19, 1999.
37. B. De Salvo, G. Ghibaudo, G. Pananakakis, G. Guillaumot, P. Chandelier, and G. Reimbold, "A New Extrapolation Law for Data-Retention Time-to-Failure of Nonvolatile Memories," *IEEE Electron Device Lett.*, Vol. 20, p. 197, May 1999.
38. C. S. Pan, K. Wu, and G. Sery, "Physical Origin of Long Term Charge Loss in Floating Gate EPROM with an Interpoly ONO Stacked Dielectric," *IEEE Electron Device Lett.*, Vol. 12, No. 2, pp. 51–53, Feb. 1991.
39. C. S. Pan, K. Wu, D. Chin, G. Sery, and J. Kiely, "High Temperature Charge Loss Mechanism in a Floating Gate EPROM with an ONO Interpoly Stacked Dielectric," *IEEE Electron Device Lett.*, Vol. 12, No. 9, pp. 503–509, Sept. 1991.
40. S. A. Barker, "Effects of Carbon on Charge Loss in EPROM Structures," *IEEE International Reliability Physics Symp.*, p. 171, 1991.
41. M. S. Liang, C. Chang, W. Yang, C. Hu, and R. W. Brodersen, "Hot Carrier-Induced Degradation by Substrate Hot Electron Injection," *IEEE IEDM Tech. Dig.*, p. 186, 1983.
42. S. Aritome, R. Shirota, G. Hemink, T. Endoh, and F. Masuoka, "Reliability Issues of Flash Memory Cells," *Proc. IEEE*, Vol. 81, No. 5, pp. 776–787, May 1993.
43. E. Rosenbaum, Z. Liu, and C. Hu, "Silicon Dioxide Breakdown Lifetime Enhancement Under Bipolar Bias Conditions," *IEEE Trans. Electron Devices*, Vol. 40, p. 2287, 1993.
44. E. Rosenbaum and C. Hu, "High Frequency Time-Dependent Breakdown of SiO_2," *IEEE Electron Device Lett.*, Vol. 12, p. 267, 1991.
45. E. Rosenbaum, Z. Liu, and C. Hu, "The Effect of Oxide Stress Waveform on MOFET Performance," *IEEE IEDM Tech. Dig.*, p. 719, 1991.
46. K. Tamer San and T. P. Ma, "Determination of Trapped Oxide Charge in Flash EPROM's and MOSFET's with Thin Oxides," *IEEE Electron Device Lett.*, Vol. 13, p. 439, 1992.
47. S. Muramatsu, T. Kubota, N. Nishio, H. Shirai, M. Matsuo, N. Kodama, M. Horikawa, S. Ssaito, K. Arai, and T. Okazawa, "The Solution of Over-Erase Problem Controlling Poly-Si Grain Size-Modified Scaling Principles for Flash Memory," *IEEE IEDM Tech. Dig.*, pp. 847–850, 1994.
48. T. Kubota, K. Ando, and S. Muramatsu, "The Effect of the Floating Gate/Tunnel SiO_2 Interface on Flash Memory Data Retention Reliability," *IEEE International Reliability Physics Symp.*, pp. 12–16, 1996.
49. C. Lage, D. Burnett, T. McNelly, K. Baker, A. Bormann, D. Dreier, and V. Soorholtz, "Soft Error Rate and Stored Charge Requirements in Advanced High-Density SRAMs," *IEEE IEDM Tech. Dig.*, p. 821, 1993.
50. J. Caywood and B. Prickett, "Radiation-Induced Soft Errors and Floating Gate Memories," *IEEE International Reliability Physics Symp.*, p. 167, 1983.
51. Eto, M. Hidaka, Y. Okuyama, K. Kimura, and M. Hosono, "Impact of Neutron Flux on Soft Errors in MOS Memories," *IEEE IEDM Tech. Dig.*, p. 367, 1998.

52. N. Aijka, M. Ohi, H. Arima, T. Matsukawa, and N. Tsubouchi, "A 5 Volt Only 16 M Bit Flash EEPROM Cell with A Simple Stacked Gate Structure," *IEEE IEDM Tech. Dig.*, p. 115, 1990.
53. Brand, K. Wu, S. Pan, and D. Chin, "Novel Read Disturb Failure Mechanism Induced by Flash Cycling," *IEEE International Reliability Physics Symp.*, p. 127, 1993.
54. P. Olivio, T. N. Nguyen, and B. Ricco, "High Field Induced Degradation in Ultra-Thin SiO_2 Films," *IEEE Trans. Electron Devices*, Vol. 35, p. 2259, Dec. 1988.
55. G. Verma and N. Mielke, "Reliability Performance of ETOX Based Flash Memories," *Proc. International Reliability Physics Symp.*, p. 158, 1988.
56. K. Naruke, S. Taguchi, and M. Wada, "SILC Limiting to Scale Down EEPROM Tunnel Oxide Thickness," *IEEE IEDM Tech. Dig.*, pp. 424–427, 1988.
57. R. Moazzami and C. Hu, "Stress-Induced Current in Thin Silicon Dioxide Films," *IEEE IEDM Tech. Dig.*, p. 139, 1992.
58. C. Dunn, C. Kaya, T. Lewis, T. Strauss, J. Schreck, P. Hefley, M. Middenford, and T. San, "Flash Memory Disturb Mechanisms," *Proc. International Reliability Physics Symp.*, p. 299, 1994.
59. Roy, R. Kazerounian, A. Kablanian, and B. Eitan, "Substrate Injection Induced Program Disturb- A New Reliability Consideration for Flash-EPROM Arrays," *IEEE International Reliability Physics Symp.*, p. 68, 1992.
60. Hu, "Thin Oxide Reliability," *IEEE IEDM Tech. Dig.*, p. 368, 1985.
61. S. Satoh, G. Hermink, K. Hatakeyama, and S. Aritome, "Stress-Induced Leakage Current of Tunnel Oxide Derived from Flash Memory Read Disturb Characteristics," *IEEE Trans. Electron Devices*, Vol. 45, No. 2, pp. 482–486, 1998.
62. Y. H. Lin, C. L. Lee, and T. F. Lei, "Correlation of Stress-Induced Leakage Current with Generated Positive Trapped Charges for Ultrathin Gate Oxide," *IEEE Trans. Electron Devices*, Vol. 45, No. 2, pp. 567–570, 1998.
63. E. F. Runnion, S. M. Gladstone IV, R. S. Scott Jr., D. J. Dumin, L. Lie, and K. C. Mitros, "Thickness Dependence of Stress-Induced Leakage Currents in Silicon Oxide," *IEEE Trans. Electron Devices*, Vol. 44, No. 6, pp. 993–1001, 1997.
64. K. Sakakibara, N. Ajika, M. Hatanaka, H. Miyoshi, and A. Yasuoka, "Identification of Stress-Induced Leakage Current Components and the Corresponding Trap Models in SiO_2 Films," *IEEE Trans. Electron Devices*, Vol. 44, No. 6, pp. 986–992, 1997.
65. K. Sakakibara, N. Ajika, K. Eikyu, K. Ishikawa, and H. Miyoshi, "A Quantitative Analysis of Time-Decay Reproducible Stress-Induced Leakage Current in SiO_2 Films," *IEEE Trans. Electron Devices*, Vol. 44, No. 6, pp. 1002–1007, 1997.
66. N. Tsuji, K. Sakakibara, N. Ajika, and H. Miyoshi, "Microscopic and Statistical Approach to SILC Characteristics," *IEEE Symposium on VLSI Technology*, pp. 196–197, 1998.
67. Young-Bog Park and Dieter K. Schroder, "Degradation of Thin Tunnel Gate Oxide Under Constant Fowler-Nordheim Current Stress for a Flash EEPROM," *IEEE Trans. Electron Devices*, Vol. 45, No. 6, pp. 1361–1368, 1998.
68. E. Rosenbaum and L. F. Register, "Mechanism of Stress-Induced Leakage Current in MOS Capacitor," *IEEE Trans. Electron Devices*, Vol. 44, No. 2, pp. 317–323, Feb. 1997.
69. S. Kamohara, D. Park, and C. Hu, "Deep-Trap SILC (Stress-Induced Leakage Current) Model for Nominal and Weak Oxides," *Proc. of International Reliability Physics Symp.*, pp. 57–61, 1998.
70. S. Yamada, K. Amemiya, T. Yamane, H. Hazama, and K. Hashimoto, "Non-Uniform Current Flow through Thin Oxide after Fowler–Nordheim Current Stress," *Proc. of International Reliability Physics Symp.*, pp. 108–112, 1996.

71. A. Teramoto, K. Kobayashi, Y. Matsui, M. Hirayama, and A. Yasuoka, "Excess Currents Induced by Hot-Hole Injection and F-N Stress in Thin SiO_2 Films," *Proc. of International Reliability Physics Symp.*, pp. 113–116, 1996.
72. K. Sakakibara, N. Ajika, M. Hatanaka, and M. Miyoshi, "A Quantitative Analysis Stress-Induced Excess Current in SiO_2 Films," *Proc. of International Reliability Physics Symp.*, p. 100, 1996.
73. K. Shimizu, T. Endoh, and H. Iizuka, "Mechanism of AC-Stress-induced Leakage Current in EEPROM Tunnel Oxides," *Proc. of International Reliability Physics Symp.*, p. 56, 1995.
74. N. Matsukawa, S. Yamada, K. Amemiya, and H. Hazama, "A Hot Carrier Induced Low Level Leakage Current in Thin Silicon Dioxide Films," *Proc. of International Reliability Physics Symp.*, p. 162, 1995.
75. M. Kimura and H. Koyama, "Stress Induced Low Level Leakage Mechanism in Ultrathin Silicon Dioxide Films Caused by Neutral Oxide Trap Generation," *Proc. of International Reliability Physics Symp.*, p. 167, 1994.
76. K. Kobayashi, A. Teramoto, and M. Hirayama, "Electron Traps and Excess Current Induced by Hot Hole Injection into Thin SiO_2 Films," *Proc. of International Reliability Physics Symp.*, p. 168, 1995.
77. N. Matsukawa, S. Yamada, K. Amemiya, and H. Hazama, "A Hot Hole-Induced Low-Level Leakage Current in Thin Silicon Dioxide Films," *IEEE Trans. Electron Devices*, Vol. 43, p. 1924, 1996.
78. J. De Blauwe, J. Van Houdt, D. Wellekens, R. Degraeve, P. Roussel, L. Haspeslagh, L. Deform, G. Groeseneken, and H. E. Maes, "A New Quantitative Model to Predict SILC Related Disturb Characteristics in Flash EEPROM Devices," *IEEE IEDM Tech. Dig.*, pp. 343–346, Dec. 1996.
79. D. J. Dumin and J. R. Maddux, "Correlation of Stress-Induced Leakage Current in Thin Oxide with Trap Generation inside the Oxides," *IEEE Trans. Electron Devices*, Vol. 40, p. 986, 1993.
80. R. Rofan and C. Hu, "Stress Induced Oxide Leakage," *IEEE Trans. Electron Devices Lett.*, Vol. 12, No. 6, p. 632, 1991.
81. D. J. DiMaria and E. Cartier, "Mechanism for SILCs in Thin Silicon Dioxide Films," *J. Appl. Phys.*, Vol. 78, p. 3883, 1995.
82. H. Satake and A. Toriumi, "Common Origin for Stress-Induced Leakage Current and Electron Trap Generation in SiO_2," *J. Appl. Phys. Lett*, Vol. 67, p. 3489, 1995.
83. G. W. Yoon, A. B. Joshi, J. Kim, and D. L. Kwong, "High Field Induced Leakage in Ultrathin N_2O Oxides," *IEEE Electron Devices Lett.*, Vol. 14, p. 231, 1993.
84. D. Wristers, H. H. Wang, I. D. Wolf, L. K. Han, D. L. Kwong, and J. Fulford, "Ultra Thin Oxide Reliability: Effects of Gate Doping Concentration and Poly-Si/SiO_2 Interface Stress Relaxation," *Proc. of International Reliability Physics Symp.*, p. 77, 1996.
85. T. Kaneoka, M. Anma, H. Itoh, and M. Hirayama, "Impact of An In-Situ Nitrogen and Phosphorus Co-Doped Amorphous Silicon as a Scalable and Reliable Floating Gate of Flash Memory," *IEEE IEDM Tech. Dig.*, p. 511, 1996.
86. J. Maserjian and N. Zamani, "Observation of Positively Charged State Generation near the Si/SiO_2 Interface During Fowler-Nordheim Tunneling," *J. Vac. Sci. Tech.*, Vol. 20, No. 3, p. 743, 1982.
87. Lundstrom, S. Christensson, and C. Svensson, "Carrier Trapping Hysteresis in MOS Transistors," *Phys. Stat. Sol. (a)*, Vol. 1, p. 395, 1970.
88. S. Aritome, S. Shirota, R. Kirisawa, T. Endoh, N. Nakayama, K. Sakui, and F. Masioka, "A Reliable Bi-Polarity Write/Erase Technology in Flash EEPROMs," *IEEE IEDM Tech. Dig.*, p. 111, 1990.

89. M. Kato, N. Miyamoto, H. Kume, A. Satoh, T. Adachi, M. Ushiyama, and K. Kimura, "Read-Disturb Degradation Mechanism Due to Electron Trapping in the Tunnel Oxide for Low-Voltage Flash Memories," *IEEE IEDM Tech. Dig.*, p. 45, 1994.
90. R. S. Scott Jr., E. F. Runnion, T. W. Hughes, D. J. Dumin, and B. T. Moore, "Properties of High-Voltage Stress Generated Traps in Thin Silicon Oxides," *Proc. of International Reliability Physics Symp.*, p. 131, 1995.
91. U. Sharma, R. Moazzami, P. Tobin, Y. Okada, S. K. Cheng, and J. Yeargain, "Vertically Scaled, Highly Reliability EEPROM Devices with Ultra-Thin Oxynitride Films Prepared by RTP in N_2O/O_2 Ambient," *IEEE IEDM Tech. Dig.*, pp. 461–464, 1992.
92. J. Kim and S. Ahn, "Improvement of the Tunnel Oxide Quality by a Low Thermal Budget Dual Oxidation for Flash Memories," *IEEE Electron Device Lett.*, Vol. 18, No. 8, p. 385, 1997.
93. J. C. Chen, Z. Liu, J. T. Krick, P. K. Ko, and C. Hu, "Degradation of N_2O Annealed MOSFET Characteristics in Response to Dynamic Oxide Stressing," *IEEE Electron Device Lett.*, Vol. 14, No. 5, p. 225, 1993.
94. J. Ahn, A. Joshi, G. Q. Lo, and D. L. Kwong, "Time-Dependent Dielectric Breakdown Characteristics of N_2O Oxide under Dynamic Stressing," *IEEE Electron Device Lett.*, Vol. 13, No. 10, p. 513, 1992.
95. Z. Liu, H. J. Wann, P. K. Ko, C. Hu, and Y. C. Cheng, "Improvement of Charge Characteristics of N_2O-Annealed and Reoxidized N_2O-Annealed Thin," *IEEE Electron Device Lett.*, Vol. 13, No. 10, p. 519, 1992.
96. D. Guterman, "Non-Volatile Flash Memory Based Solid State Mass Storage Technologies," *IEEE IEDM Tech. Dig.*, Short Course, 1995.
97. D. B. Kuo, J. P. McVittie, W. D. Nix, and K. Saraswat, "Two-Dimensional Silicon Oxidation Experiments and Theory," *IEEE IEDM Tech. Dig.*, p. 388, 1985.
98. N. Mielke, A. Fazio, and H. C. Liou, "Reliability Comparison of FLOTOX and Textured-Poly Silicon E2PROM's," *Proc. IEEE International Reliability Physics Symp.*, p. 85, 1987.
99. A. V. Kordesch, S. Awsare, J. Brenan, Jr., P. Gao, M. Hemming, M. Herman, P. Holzmann, E. Ng, C. M. Liu, K. Su, C. H. Wang, and M. Wu, "The Reliability of Multilevel Analog Memory in a Voice Storage and Playback System Using Source-Side Injection," *International Symposium on VLSI Technology, Systems and Applications*, pp. 266–269, June 1999.
100. C. M. Liu, J. Brenan, Jr, P. Guo, P. Holzmann, P. Klinger, A. V. Kordesch, M. Kwan, I. S. Liu, K. Su, C. H. Wang, and S. Yoon, "A Multilevel Analogy Storage Memory Using Source-Side Injection Flash Array," *Proc. of International Symposium on VLSI Technology, Systems and Applications*, p. 187, June 1999.
101. C. M. Liu, M. Hemming, P. Klinger, A. V. Kordesch, C. M. Liu, and K. Su, "On the Cell Misalignment for Multilevel Storage Flash E2PROM," *Proc. of VLSI Technology, Systems and Applications*, p. 191, June 1999.
102. T. Ono, T. Mori, T. Ajioka, and T. Takyashiki, "Studies of Thin Poly Si Oxides for E and E2PROM," *IEEE IEDM Tech. Dig.*, p. 380, 1985.
103. F. C. Jong, T. Y. Huang, T. S. Chao, H. C. Lin, L. Y. Leu, K. Young, C. H. Lin, and K. Y. Chiu, "Improved Flash Cell Performance by N_2O Annealing of Interpoly Oxide," *IEEE Electron Device Lett.*, Vol. 18, p. 343, 1997.
104. J. H. Klootwijk, H. Van Kranenburg, P. H. Woerlee, and H. Wallinga, "Deposited Inter-Polysilicon Dielectrics for Nonvolatile Memories," *IEEE Trans. Electron Devices*, Vol. 46, p. 1435, 1999.
105. Wu, T. Y. Chan, P. K. Ko, and C. Hu, "A Novel High-Speed, 5 V Programming EPROM Structure with Source-Side Injection," *IEEE IEDM Tech. Dig.*, p. 58, 1986.

106. S. Kianian, A. Levi, D. Lee, and Y. W. Hu, "A Sovel 3 Volts-Only, Small Sector Erase, High Density Flash E2PROM," *IEEE Symposium on VLSI Technology*, p. 71, 1994.

107. Y. Ma, C. S. Pang, J. Pathak, S. C. Tsao, C. F. Chang, Y. Yamauchi, and M. Yoshimi, "A Novel High Density Contactless Flash Memory Array Using Split-Gate Source-Side-Injection for 5V-Only Applications," *IEEE Symposium on VLSI Technology*, p. 49, 1994.

108. Y. Ma, C. S. Pang, K. T. Chang, S. C. Tsao, J. E. Frayer, T. Kim, K. Jo, J. Kim, I. Choi, and H. Park, "A Dual-Bit Split-Gate EEPROM (DSG) Cell in Contactless Array for Single Vcc High Density Flash Memories," *IEEE IEDM Tech. Dig.*, p. 57, 1994.

109. D. Wellekens, J. F. Van Houdt, L. Faraone, G. Groeseneken, and H. E. Maes, "Write/Erase Degradation in Source Side Injection Flash EEPROM's: Characterization Techniques and Wearout Mechanisms," *IEEE Trans. Electron Devices*, Vol. 42, p. 1992, 1995.

110. W. Brown and J. Brewer (Eds.), "Nonvolatile Semiconductor Memory Technology," *IEEE Press Series on Microelectronic Systems*, p. 590, Oct. 1997.

111. S. Bhattacharya, K. Lai, K. Fox, P. Chan, E. Worley, U. Sharma, L. Hwang, and G. P. Li, "Improved Performance and Reliability of Split Gate Source-Side Injected Flash Memory Cells," *IEEE IEDM Tech. Dig.*, p. 339, 1996.

112. J. F. Van Houdt, D. Wellekens, G. Groeseneken, and H. E. Maes, "Investigation of the Soft-Write Mechanism in Source-Side Injection Flash EEPROM Devices," *IEEE Electron Device Lett.*, Vol. 16, p. 181, 1995.

113. J. De Blauwe, D. Welleken, G. Groeseneken, L. Haspeslagh, J. Van Houdt, L. Deferm, and H. E. Maes, "Read-Disturb and Endurance of SSI-Flash E2PROM Devices at High Operating Temperatures," *IEEE Trans. Electron Devices*, Vol. 45, p. 2466, 1998.

114. J. F. Van Houdt, L. Haspeslagh, D. Wellekens, L. Deferm, G. Groeseneken, and H. E. Maes, "HIMOS-A High Efficiency Flash EEPROM Cell for Embedded Memory," *IEEE Trans. Electron Devices*, Vol. 40, p. 2255, 1993.

115. G. Q. Lo, J. Ahn, D. L. Kwong, and K. K. Young, "Dependence of Hot-Carrier Immunity on Channel Length and Channel Width in MOSFET's with N_2O-Grown Gate Oxides," *IEEE Electron Device Lett.*, Vol. 13, no. 12, p. 651, 1992.

116. C. Dunn, P. Hefley, S. Pope, T. Lewis, D. Chong, S. Desai, P. Patel, D. Baker, D. Dolby, T. Strauss, S. Gunturi, P. Wright, and P. Sudak, "Process Reliability Development for Nonvolatile Memories," *Proc. of International Reliability Physics Symp.*, p. 133, 1993.

117. K. Yoneda, Y. Fukuzaki, K. Satoh, Y. Hata, Y. Todokoro, and M. Inoue, "Reliability Degradation Mechanism of the Ultra-Thin Tunneling Oxide by Post-Annealing," *IEEE Symposium on VLSI Technology*, p. 121, 1990.

118. M. Ushiyama, Y. Ohji, T. Nishimoto, K. Komori, H. Murakoshi, H. Kume, and S. Tachi, "Two Dimensionally Inhomogeneous Structure at Gate Electrode/Gate Insulator Interface Causing Fowler-Nordheim Current Deviation in Nonvolatile Memories," *Proc. of International Reliability Physics Symp.*, p. 331, 1991.

119. F. D. Nkansah and M. Hatalis, "Effects of Flash EEPROM Floating Gate Morphology on Electrical Behavior of Fast Programming Bits," *IEEE Trans. Electron Devices*, Vol. 46, p. 1355, 1999.

120. D. N. Tang and W. J. Lu, "Process for Self Aligning a Source region with a Field Oxide Region and a Polysilicon Gate," U.S. Patent 5,120,671, June 1992.

121. D. Liu, C. Chang, and Y. Sun, "Method Protecting a Stacked Gate Edge in a Semiconductor Device from Self Aligned Source (SAS) Etch," U.S. Patent 5,693,972, Dec. 1997.

122. H. Shin, C. C. King, and C. Hu, "Thin Oxide Damage by Plasma Etching and Ashing Processes," *Proc. of International Reliability Physics Symp.*, p. 37, 1992.

123. S. Fang and J. McVittie, "A Mechanism for Gate Oxide Damage in Nonuniform Plasmas," *Proc. of International Reliability Physics Symp.*, p. 13, 1993.

124. H. Fang, S. Haddad, C. Chang, and J. Lien, "Plasma-Induced In-Line Charging and Damage in Non-Volatile Memory Devices," *IEEE IEDM Tech. Dig.*, p. 467, 1994.
125. S. Samukawa and K. Terada, "Pulse Time Modulated ECR Plasma Etching for Highly Selective, Highly Anisotropic and Less-Charging Poly-Si Gate Patterning," *Proc. of International Reliability Physics Symp.*, p. 13, 1993.
126. D. A. Baglee, L Nannemann, and C. Huang, "Building Reliability into EPROM," *Proc. of International Reliability Physics Symp.*, p. 12, 1990.
127. N. Shimoyama, K. Machida, K. Murase, and T. Tuchiya, "Enhanced Hot-Carrier Degradation due to Water in TEOS/O_3-Oxide and Water Blocking Effect of ECR-SiO_2," *IEEE Symposium on VLSI Technology*, p. 92, 1992.
128. E. Sakagami, N. Arai, H. Tsunoda, H. Egawa, Y. Yamaguchi, E. Kamiya, M. Takebuchi, K. Yamada, K. Yoshikawa, and S. Mori, "The Impact of Intermetal Dielectric Layer and High Temperature Bake Test on the Reliability of Nonvolatile Memory Devices," *Proc. of International Reliability Physics Symp.*, p. 359, 1994.
129. S. Shuto, M. Tanaka, M. Sonoda, T. Idaka, K. Sasaki, and S. Mori, "Impact of Passivation Film Deposition and Post-Annealing on the Reliability of Flash Memories," *Proc. of International Reliability Physics Symp.*, p. 17, 1997.
130. H. Fang, "Reliability for Non-Volatile Memories," *IEEE IEDM Tech. Dig., Short Course*, 1998.
131. H. Arima, et al., "A Novel Process Technology and Cell Structure for Megabit EEPROM," *IEEE IEDM Tech. Dig.*, p. 472, 1988.
132. N. Ajika, M. Ohi, T. Futatsuya, H. Arima, T. Matsukawa, and N. Tsubouchi, "A Novel Cell Structure for 4 Mbit full Feature EEPROM and Beyond," *IEEE IEDM Tech. Dig.*, p. 295, 1991.
133. P. C. Fazan and V. K. Mathews, "A Highly Manufacturable Trench Isolation Process for Deep Submicron DRAMs," *IEEE IEDM Tech. Dig.*, p. 257, 1990.
134. T. Hori, et al., "A 0.1 mm CMOS Technology with Tilt-Implanted Punchthrough Stopper (TIPS)," *IEEE IEDM Tech. Dig.*, p. 75, 1994.
135. Umezawa, S. Atsumi, M. Kuriyama, H. Banba, K. Imamyia, K. Naruke, S. Yamada, E. Obi, M. Oshikiri, T. Suzuki, and S. Tanaka, "A 5 V-Only Operation 0.6 um Flash EEPROM with Row Decoder Scheme in Triple-Well Structure," *IEEE J. of Solid-State Circuits*, Vol. 27, No. 11, p. 1540, 1992.
136. J. Pathak, "Non-Volatile Memory Design," *IEEE International Solid State Circuits Conference Short Course*, 1996.
137. T. Tanaka, M. Momodomi, Y. Iwata, Y. Tanaka, H. Oodaira, Y. Itoh, R. Shirota, K. Ohuchi, and F. Masuoka, "A 4-Mbi NAND-EEPROM with Tight Programmed Vt Distribution," *IEEE Symposium on VLSI Circuits*, p. 105, 1990.
138. R. Kirisawa, S. Arilome, R. Nakayama, T. Endoh, R. Shirota, and F. Masuoka, "A NAND Structured Cell with a New Programming Technology for Highly Reliable 5-V Only Flash EEPROM," *IEEE 1990 Symposium on VLSI Technology*, p. 130, 1990.
139. G. S. Gongwer and K. H. Gudger, "A 16 K EEPROM Using EE Element Redundancy," *IEEE Journal of Solid-State Circuits*, Vol. 18, No. 5, p. 550, 1983.
140. E. Harari, R. Norman, and S. Mehrotra, "Non-Volatile Memory System Card with Flash Erasable Sectors of EEPROM Cells Including a Mechanism for Substituting Defective Cells," U.S. Patent 5,535,328, July 1996.
141. E. Harari, "Flash EEPROM System with Defect Handling," U.S. Patent 5,671,229, Sept. 1997.
142. S. Mehrotra, T. C. Wu, T. L. Chiu, and G. Perlegos, "A 64 Kb CMOS EEPROM with On-Chip ECC," *IEEE International Solid State Circuits Conference*, p. 142, 1984.

143. T. K. Ting, T. Chang, T. Lin, C. S. Jenq, and K. L. C. Naiff, "A 50-ns CMOS 256 K EEPROM," *IEEE Journal of Solid State Circuits*, Vol. 23, No. 5, p. 1164, 1988.

144. S. Mehrotra, J. H. Yuan, R. A. Cernea, W. Y. Chien, D. C. Guterman, G. Samachisa, R. D. Norman, M. Mofidi, W. Lee, Y. Fong, A. Mihnea, and E. Harari, "Serial 9 Mb Flash EEPROM for Solid State Disk Applications," *IEEE Symposium on VLSI Circuits*, p. 24, 1992.

145. D. Lee, R. A. Cernea, M. Mofidi,. S. Mehrotra, E. Y. Chang, W. Y. Chien, L. Goh, J. H. Yuan, A. Mihnea, G. Samachisa, Y Fong, D. Guterman, and R. D. Norman, "An 18 MB Serial Flash EEPROM for Solid-State Disk Applications," *IEEE Symposium on VLSI Circuits*, pp. 59–60, June 1994.

146. A. J. van de Goor, *Testing Semiconductor Memories: Theory and Practice*, John Wiley and Sons, Chichester, 1991.

147. B. W. Johnson, *Design and Analysis of Fault-Tolerant Digital Systems*, Addison-Wesley, Reading, MA, 1989.

148. T. Tanzawa, T. Tanaka, K. Takeuchi, R. Shirota, S. Aritome, H. Wattanabe, G. Hemink, K. Shimizu, S. Sato, Y. Takeuchi, and K. Ohuchi, "A Compact On-Chip ECC for Low Cost Flash Memories," *IEEE Journal of Solid-State Circuits*, Vol. 32, No. 5, p. 662, 1997.

149. J. W. McPherson and D. A. Baglee, "Acceleration Factors for Thin Gate Oxide Stressing," *Proc. International Reliability Physics Symp.*, p. 1, 1985.

150. J. Lee, I. C. Chen, and C. Hu, "Statistical Modeling of Silicon Dioxide Reliability," *Proc. IEEE International Reliability Physics Symp.*, p. 131, 1988.

151. C. Hu and Q. Lu, "A Unified Gate Oxide Reliability Model," *Proc. International Reliability Physics Symp.*, p. 47, 1999.

152. C. C. Harry and C. H. Mathiowetz, "ASIC Reliability and Qualification: A User's Perspective," *Proceedings of the IEEE*, Vol. 81, No. 5, p. 759, 1993.

153. C. F. Dunn and J. W. McPherson, "Temperature-Cycling Acceleration Factors for Aluminum Metallization Failure in VLSI Applications," *Proc. International Reliability Physics Symp.*, p. 114, 1989.

154. E. Takeda, K, Ikuzaki, H. Katto, Y. Ohji, K. Hinode, A. Hamada, T. Sakuta, T. Funabiki, and T. Sasaki, "VLSI Reliability Challenges: From Device Physics to Wafer Scale Systems," *Proceedings of the IEEE*, Vol. 81, No. 5, p. 653, 1993.

155. B. Prince, *Semiconductor Memories: A Handbook of Design, Manufacture, and Application*, 2nd Edition, John Wiley & Sons, Chichester, 1991.

156. Am29F400B Product Qualification Database, http//www.amd.com

157. D. L. Crook, "Method of Determining Reliability Screens for Time Dependent Dielectric Breakdown," *Proc. International Reliability Physics Symp.*, pp. 1–7, April 1979.

158. R. Moazzami and C. Hu, "Projecting Gate Oxide Reliability and Optimizing Reliability Screens," *IEEE Trans. Electron Devices*, Vol. 37, p. 1643, 1990.

159. R. L. Cook, "Evolution of VLSI Reliability Engineering," *Proc. International Reliability Physics Symp.*, pp. 2–11, March 1990.

160. P. Pellati, A. Chimenton, P. Olivo, and A. Modelli, "Dynamics of Fast-Erasing Bits in Flash Memories," *Proc. 30th European Solid-State Device Res. Conf.*, pp. 288–291, Sept. 2000.

161. Chimenton, P. Pellati, and P. Olivo, "Analysis of Erratic Bits in Flash Memories," *Proc. 39th IEEE Reliability Physics Symposium*, pp. 17–22, 30 April 30–May 2001.

162. Chimenton, Pellati and P. Olivo, "Analysis of Erratic Bits in Flash Memories," *IEEE Trans. Dev. and Mat. Reliability*, Vol. 1, No. 4, pp. 179–184, Dec. 2001.

163. Chimenton and P. Olivo, "Impact of Tunnel Oxide Thickness on Erratic Erase in Flash Memories," *Proc. 32nd European Solid-State Device Res. Conf.*, pp. 363–366, September 2002.

164. Chimenton and P. Olivo, "Erratic Erase in Flash Memories—Part I: Basic Experimental and Statistical Characterization," *IEEE Trans. Elect. Dev.*, Vol. 50, No. 4, pp. 1009–1014, April 2003.
165. Chimenton and P. Olivo, "Erratic Erase in Flash Memories—Part II: Dependence on Operating Conditions," *IEEE Trans. Elect. Dev.*, Vol. 50, No. 4, pp. 1015–1021, April 2003.
166. Chimenton, P. Pellati, and P. Olivo, "Overerase Phenomena: An Insight into Flash Memory Reliability," *Proc. IEEE*, Vol. 91, No. 4, pp. 617–626, April 2003.
167. Chimenton and P. Olivo, "Reliability of Flash Memory Erasing Operation under High Tunneling Electric Fields," *Proc. 42nd IEEE Reliability Physics Symposium*, pp. 216–221, April 2004.
168. Chimenton and P. Olivo, "Impact of High Tunneling Electric Fields on Erasing Instabilities in NOR Flash Memories," *IEEE Trans. Electron Devices*, Vol. 53, No. 1, pp. 97–102, Jan. 2006.
169. Chimenton, F. Irrera, and P. Olivo, "Ultra-Short Pulses Improving Performance and Reliability in Flash Memories," *IEEE Nonvolatile Semiconductor Memory Workshop, Monterey*, CA, pp. 46–47, Feb. 2006.
170. Scarpa, G. Tao, J. Dijkstra, and F. Kuper, "Fast-Bit-Limited Modeling of Advanced Floating Gate Non-Volatile Memories," Integrated Reliability Workshop Final Report (2000), pp. 24–28.
171. P. J. Kuhn, A. Hoefler, T. Harp, B. Hornung, R. Paulsen, D. Burnett, and J. Higman, "A Reliability Methodology For Low Temperature Data Retention In Floating Gate Non-Volatile Memories," *Proc. International Reliability Physics Symp.*, pp. 73–80, 2001.
172. D. Ielmini, A. Spinelli, A. Lacaita, A. Modelli, "Equivalent Cell Approach for Extraction of the SILC Distribution in Flash EEPROM Cells," *IEEE Electron Device Lett.*, Vol. 23, No. 1, pp. 40–42, Jan. 2002.
173. H. Belgal, N. Righos, I. Kalstirsky, J. Peterson, R. Shiner, N. Mielke, "A New Reliability Model for Post-Cycling Charge Retention of Flash Memories," *Proc. IEEE International Reliability Physics Symp.*, pp. 7–20, 2002.
174. Hoefler, J. Higman, T. Harp, and P. Kuhn, "Statistical Modeling of the Program/Erase Cycling Acceleration of Low Temperature Data Retention in Floating-Gate Nonvolatile Memories," *Proc. International Reliability Physics Symp.*, pp. 21–25, 2002.
175. D. Ielmini, A. S. Spinelli, A. L. Lacaita, R. Leone, and A. Visconti, "Localization of SILC in Flash Memories after Program/Erase Cycling," *Proc. IEEE International Reliability Physics Symp.*, pp. 1–6, 2002.
176. D. Ielmini, Alessandro S. Spinelli, Andrea L. Lacaita, and Angelo Visconti, "Statistical Profiling of SILC Spot in Flash Memories," *IEEE Trans. El. Dev.*, Vol. 49, No. 10, pp. 1723–1728, Oct. 2002.
177. Chimenton, A. S. Spinelli, D. Ielmini, A. L. Lacaita, A. Visconti, and P. Olivo, "Drain-Accelerated Degradation of Tunnel Oxides in Flash Memories," *IEEE IEDM Tech. Dig.*, pp. 167–170, 2002.
178. D. Ielmini, A. Ghetti, A. Spinelli, and A. Visconti, "A Study of Hot-Hole Injection during Programming Drain Disturb in Flash Memories," *IEEE Trans. Elect. Dev.*, Vol. 53, No. 4, pp. 668–676, April 2006.
179. M. Suhail, T. Harp, J. Bridwell, and P. J. Kuhn, "Effects of Fowler Nordheim Tunneling Stress vs. Channel Hot Electron Stress on Data Retention Characteristics of Floating Gate Non-Volatile Memories," *Proc. International Reliability Physics Symp.*, pp. 439–440, 2002.
180. F. Schuler, R. Degraeve, P. Hendricks, and D. Wellekins, "Physical Description of Anomalous Charge Loss in Floating Gate Based NVM's and Identification of Its Dominant Parameter," *Proc. International Reliability Physics Symp.*, pp. 26–33, 2002.

181. D. Ielmini, A. S. Spinelli, A. L. Lacaita, and A. Modelli, "Modeling of Anomalous SILC in Flash Memories Based on Tunneling at Multiple Defects," *Solid-State Electronics* Vol. 46, Issue 11, pp. 1749–1756, Nov. 2002.

182. D. Ielmini, A. S. Spinelli, A. L. Lacaita, and A. Modelli, "A Statistical Model for SILC in Flash Memories," *IEEE Trans. El. Dev.*, Vol. 49, No. 11, pp. 1955–1961, Nov. 2002.

183. F. Schuler, R. Degraeve, P. Hendricks, and D. Wellekens, "Physical Charge Transport Models for Anomalous Leakage Current in Floating Gate-Based Nonvolatile Memory Cells," *IEEE Trans. Dev. and Mat. Reliability*, Vol. 2, No. 4, pp. 80–88, Dec. 2002.

184. Gehring and S. Selberherr, "Modeling of Tunneling Current and Gate Dielectric Reliability for Nonvolatile Memory Devices," *IEEE Trans. Dev. and Mat. Reliability*, Vol. 4, No. 3, p. 319, Sept. 2004.

185. R. Degraeve, F. Schuler, B. Kaczer, M. Lorenzini, D. Wellekens, P. Hendricks, M. van Duuren, G. J. M. Dormans, J. Van Houdt, L. Haspeslagh, G. Groeseneken, and G. Tempel, "Analytical Percolation Model for Predicting Anomalous Charge Loss in Flash Memories," *IEEE Trans. Elect. Dev.*, Vol. 51, No. 9, pp. 1392–1400, Sept. 2004.

186. D. Ielmini, A. S. Spinelli, A. L. Lacaita, and M. J. van Duuren, "Correlated Defect Generation in Thin Oxides and Its Impact on Flash Reliability," *Proc. IEEE IEDM Tech. Dig.*, pp. 143–146, 2002.

187. D. Ielmini, A. Spinelli, A. Lacaita, and M. van Duuren, "Impact of Correlated Generation of Oxide Defects on SILC and Breakdown Distributions," *IEEE Trans. Elect. Dev.*, Vol. 51, No. 8, pp. 1281–1287, Aug. 2004.

188. F. Irrera and B. Ricco, "Pulsed Tunnel Programming of Nonvolatile Memories," *IEEE Trans. Elect. Dev.*, Vol. 50, No. 12, pp. 2474–2480, Dec. 2003.

189. R. Yamada, Y. Mori, Y. Okuyama, J. Yugami, T. Nishimoto, and H. Kume, "Analysis of Detrap Current Due to Oxide Traps to Improve Flash Memory Retention," *Proc. International Reliability Physics Symp.*, pp. 200–204, 2000.

190. R. Yamada, T. Sekiguchi, Y. Okuyama, J. Yugami, and H. Kume, "A Novel Analysis Method of Threshold Voltage Shift Due to Detrap in a Multi-Level Flash Memory," *IEEE Symposium on VLSI Technology*, pp. 115–116, 2001.

191. J. Lee, J. Choi, D. Park, and K. Kim, "Degradation of Tunnel Oxide by FN Current Stress and Its Effects on Data Retention Characteristics of 90-nm NAND Flash Memory," *IEEE International Reliability Physics Symp.*, p. 497, 2003.

192. J. Lee, J. Choi, D. Park, and K. Kim, "Data Retention Characteristics of Sub-100nm NAND Flash Memory Cells," *IEEE Electron Device Lett.*, Vol. 24, No. 12, pp. 748–750, Dec. 2003.

193. J. Lee, J. Choi, D. Park, and K. Kim, "Effects of Interface Trap Generation and Annihilation on the Data Retention Characteristics of Flash Memory Cells," *IEEE Trans. Dev. and Mat. Reliability*, Vol. 4, No. 1, pp. 110–117, March 2004.

194. J. Om, E. Choi, S. Kim, H. Lee, Y. Kim, H. Chang, S. Park, and G. Bae, "The Effect of Mechanical Stress From Stopping Nitride to the Reliability of Tunnel Oxide and Data Retention Characteristics of NAND FLASH Memory," *Proc. IEEE International Reliability Physics Symp.*, pp. 257–259, April 2005.

195. N. Mielke, H. Belgal, I. Kalastirsky, P. Kalavade, A. Kurtz, Q. Meng, N. Righos, and J. Wu, "Flash EEPROM Threshold Instabilities due to Charge Trapping During Program/Erase Cycling," *IEEE Trans. Dev. and Mat. Reliability*, Vol. 2, No. 3, pp. 335–244, 2004.

196. N. Mielke, H. Belgal, A. Fazio, Q. Meng, and N. Righos, "Recovery Effects in the Distributed Cycling of Flash Memories," *Proc. IEEE Reliability Physics Symposium*, pp. 29–35, March 2006.

197. JEDEC Standard JESD22-A117A, "Electrically Erasable Programmable ROM (EEPROM) Program/Erase Endurance and Data Retention Stress Test," JEDEC Solid State Technology Association, March 2006.
198. JEDEC Standard JESD47E, "Stress Driven Qualification of Integrated Circuits," JEDEC Solid State Technology Association, 2006.
199. JEDEC Publication JEP148, "Reliability Qualification of Semiconductor Devices Based on Physics of Failure Risk and Opportunity Assessment," JEDEC Solid State Technology Association, April 2004.
200. J. Chen and Y. Fong, "High Density Non-Volatile Flash Memory without Adverse Effects of Electric Field Coupling between Adjacent Floating Gates," U.S. Patent 5,867,429, Feb. 1999.
201. J. D. Lee, S. H. Hur, and J. D. Choi, "Effects of Floating-Gate Interference on NAND Flash Memory Cell Operation," *IEEE Electron Device Lett.*, Vol. 23, No. 5, pp. 264–266, May 2002.
202. H. Chien and Y. Fong, "Deep Wordline Trench to Shield Cross Coupling between Adjacent Cells for Scaled NAND," U.S. Patent 6,894,930, June 2002.
203. D. Kang, H. Shin, S. Chang, J. An, K. Lee, and J. Kim, "The Air Spacer Technology for Improving the Cell Distribution in 1 Giga Bit NAND Flash Memory," *IEEE Nonvolatile Semiconductor Memory Workshop*, Monterey, CA, pp. 36–37, Feb. 2006.
204. J. Chen and M. Higashitani, "Use of Voids Between Elements in Semiconductor Structures for Isolation," U.S. Patent 7,045,849, May 2006.
205. C. H. Lee, J. Choi, C. Kang, Y. Shin, J. S. Lee, J. Sel, J. Sim, S. Jeon, B. I. Choe, D. Bael, K. Park, and K. Kim, "Multi-Level NAND Flash Memory with 63 nm-Node TANOS (Si-Oxide-SiN-Al$_2$O$_3$-TaN) Cell Structure," *IEEE Symposium on VLSI Technology*, pp. 24–25, 2006.
206. S. H. Hur, J. D. Lee, M. C. Park, J. D. Choi, K. C. Park, K. T. Kim, and K. N. Kim, "Effective Program Inhibition beyond 90 nm NAND Flash Memories," *IEEE Non-Volatile Semiconductor Memory Workshop, Monterey*, CA, pp. 44–45, Feb. 2004.
207. J. W. Lutze, J. Chen, Y. Li, and M. Higashitani, "Source Side Self Boosting Technique for Non-Volatile Memory," U.S. Patent 6,859,397, Feb. 2005.
208. J. D. Lee, C. K. Lee, M. W. Lee, H. S. Kim, K. C. Park, and W. S. Lee, "A New Programming Disturbance Phenomenon in NAND Flash Memory by Source/Drain Hot-Electrons Generated by GIDL Current," *IEEE Nonvolatile Semiconductor Memory Workshop*, p. 31, Feb. 2006.
209. K. T. Park, S. C. Lee, J. Sel, J. Choi, and K. Kim, "Scalable Wordline Shielding Scheme using Dummy Cell beyond 40 nm NAND Flash Memory for Eliminating Abnormal Disturb of Edge Memory Cell," *International Conference on Solid State Devices and Materials, Yokohama*, pp. 298–299, 2006.
210. H. Kurata, K. Otsuga, A. Kotabe, S. Kajiyama, T. Osabe, Y. Sasago, S. Narumi, K. Tokami, S. Kamohara, and O. Tsuchiya, "The Impact of Random Telegraph Signals on the Scaling of Multilevel Flash Memories," *IEEE Symposium on VLSI Circuits*, pp. 112–113, 2006.
211. G. Molas, D. Deieruyelie, B. DeSalvo, G. Ghibaudo, M. Gely, S. Jacob, D. Lafond, and S. Deieonibu, "Impact of Few Electron Phenomena on Floating-Gate Memory Reliability," *IEEE IEDM Tech. Dig.*, pp. 877–880, 2004.

12

MULTILEVEL CELL DIGITAL MEMORIES

Albert Fazio and Mark Bauer

12.1 INTRODUCTION

Multilevel cell (MLC) technology represents a cost breakthrough for Flash memory devices by enabling the storage of 2 bits of data in a single Flash memory transistor. This chapter will discuss the evolution of the 2-bit/cell technology from conception to production, using the Intel StrataFlash memory as an example of one such implementation of an MLC technology.

The Flash memory business has grown from about $50 million in 1987 to multibillion dollars in 2002 due to its unique mix of functionality and cost. Flash memory devices are now found in virtually every personal computer (PC) and cellular phone and are one of the key components of the emerging digital imaging and audio markets. Cost per bit reduction of Flash memory devices has been traditionally achieved by aggressive scaling of the memory cell transistor using silicon process-scaling techniques such as photolithography line width reduction. In an attempt to accelerate the rate of cost reduction beyond that achieved by process scaling, a research program was started in 1992 to develop methods for the reliable storage of multiple bits of data in a single Flash memory cell. By storing 2 bits in a single memory transistor, the memory cell area is effectively cut in half, allowing the storage of twice as much data in the same area as the standard single-bit-per-cell technology. This chapter discusses the evolution of the 2-bit/cell capability from conception to production and the challenges that were successfully overcome to produce a high-quality product compatible with the standard single-bit/cell devices.

The concept of MLC is ideally suited to the Flash memory cell. The cell operation is governed by electron charge storage on an electrically isolated floating gate.

Nonvolatile Memory Technologies with Emphasis on Flash. Edited by J. E. Brewer and M. Gill
Copyright © 2008 the Institute of Electrical and Electronics Engineers, Inc.

The amount of charge stored modulates the Flash cell's transistor characteristic. MLC requires three basic elements: (1) accurate control of the amount of charge stored, or placed, on the floating gate such that multiple charge levels, or multiple bits, can be stored within each cell, an operation called placement; (2) accurate measurement of the transistor characteristics to determine which charge level, or data bit, is stored, an operation called sensing; and (3) accurate charge storage, such that the charge level, or data bit, remains intact over time, an operation called retention. These elements are achieved by exploiting stable device operation regions and by the direct cell access of the ETOX Flash memory array.

12.2 PURSUIT OF LOW-COST MEMORY

History has shown that as the price of memory drops and the density increases, the application usage and demand for that memory will increase. The cost for semiconductor memories [i.e., dynamic random-access memory (DRAM), static random-access memory (SRAM), read-only memory (ROM), and Flash] is largely determined by the amount of silicon area it takes to store a data bit of information. As with other semiconductor memories, Flash memory, which retains its data even when the power is removed, achieves higher density and lower cost through traditional silicon process scaling techniques such as feature size reduction. To build on process scaling, a concept called multilevel cell (MLC) technology is introduced. This technology lowers the cost by enabling the storage of multiple bits of data per memory cell, thereby reducing the consumption of silicon area. The 2-bit/cell MLC technology provides a cost structure equivalent to the next generation of process technology while using the current generation of process technology equipment. Figure 12.1 illustrates the substantial acceleration of the rate of cost reduction possible with MLC.

A discussion of the MLC technology first requires a brief overview of the standard ETOX Flash memory technology and its use. A more detailed discussion occurs elsewhere in this book. Flash memory is a member of the nonvolatile class of memory devices, storage devices that maintain their data in the absence of applied power. The ETOX technology is the predominate Flash technology, representing over 70% of Flash memory shipments. Data is entered into the Flash

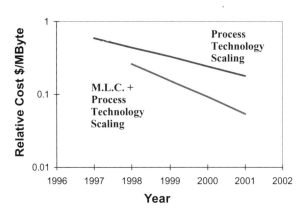

Figure 12.1. Accelerated cost reduction using MLC.

memory on a bit, byte, word, or page boundary through an operation called programming. Once data is entered into the device, it will remain, regardless of the presence or absence of power. Data is cleared from the Flash memory with an erase operation. The contents of the Flash memory are erased on a block boundary, where a block size can be anywhere from 8k bits to 1Mbits depending on the product design.

The ETOX Flash memory storage element, or memory cell, shown in Figure 12.2, is a single transistor with the addition of an electrically isolated polysilicon floating gate capable of storing charge (electrons). The amount of stored charge modifies the behavior of the memory cell transistor. This change in transistor behavior is translated into stored data: The presence of charge is interpreted as data 0; the absence of charge is interpreted as data 1. The single transistor memory cell results in a small cell size, and thus a small amount of silicon area is consumed for the storage of one bit of data, resulting in low cost.

The growth of the Flash memory market has been driven by a continual increase in density and reduction in cost, enabling new applications to emerge and further fuel the demand for more Flash. Figure 12.3 illustrates the rapid increase in the Flash market size driven by the reduction in memory price. As the price of memory was reduced, new applications for Flash memory emerged (some examples are shown in Fig. 12.3) fueling further market growth.

Traditionally, cost reduction and density increase for Flash memory has been driven by process scaling in the same way as other semiconductor memory devices

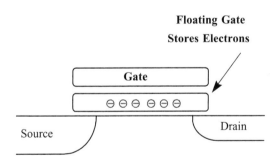

Figure 12.2. Single-transistor Flash memory cell.

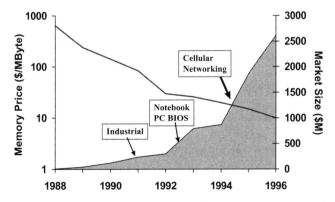

Figure 12.3. Flash memory price and market size. (Courtesy Semico Research, May 1997).

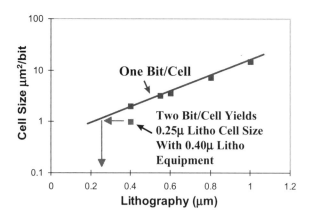

Figure 12.4. Cell area as a function of lithography.

such as DRAMs and SRAMs. As the ability of the semiconductor manufacturing process equipment improves, smaller features can be resolved on the silicon wafer resulting in a smaller memory cell and thus more bits in a given amount of silicon area. More bits in a given silicon area result in higher density memories and lower cost per bit. Using the technique of process technology scaling, the Flash memory cell size has been reduced by over 200× since the inception of Flash memory. The reduced cell area combined with increases in the size of the memory product (brought about by improved manufacturing techniques and yields) has resulted in a product density increase of over 1000× in the same time period.

The Flash memory cell is a single transistor; one bit of data is stored in one transistor. By comparison, an SRAM memory cell requires six transistors (or four transistors and two resistors), a DRAM memory cell requires one transistor and one capacitor, and an electrically erasable programmable read-only memory (E^2PROM) cell requires two transistors.

A single transistor has been generally considered the smallest practical unit for the storage of a bit of data. In 1992, an Intel Flash development team began a research effort to reduce the amount of silicon required to store a bit of data to a fraction of a transistor through the storage of more than one bit in a single memory cell transistor. The MLC memory technology provides the cost structure of the next-generation process technology while using the current generation process technology equipment (Fig. 12.4).

12.3 MULTIBIT STORAGE BREAKTHROUGH

12.3.1 Intel StrataFlash Technology

As discussed earlier, the Flash memory device is a single transistor that includes an isolated floating gate. The floating gate is capable of storing electrons. The behavior of the transistor is altered depending on the amount of charge stored on the floating gate. Charge is placed on the floating gate through a technique called programming. The programming operation generates hot electrons in the channel region of the memory cell transistor. A fraction of these hot electrons gain enough energy to surmount the 3.2-eV barrier of the Si–SiO_2 interface and become trapped on the floating gate. For single-bit/cell devices, the transistor either has little charge (<5000 electrons) on the floating gate and thus stores a 1 or it has a lot of charge (>30,000 electrons) on the floating gate and thus stores a 0. When the memory cell is read, the presence or absence of charge is determined by sensing the change in the behavior of the memory transistor due to the stored charge. The stored charge is manifested as a change in the threshold voltage of the memory cell transistor. Figure 12.5 illustrates the threshold voltage distributions for a half-million cell ($\frac{1}{2}$-Mc) array block. After erasure or programming, the threshold voltage of every memory cell transistor in the $\frac{1}{2}$-Mc block is measured, and a histogram of the results is presented. Erased cells (data 1) have threshold voltages less than 3.1 V, while programmed cells (data 0) have threshold voltages greater than 5 V.

The charge storage ability of the Flash memory cell is a key to the storage of multiple bits in a single cell. The Flash cell is an analog storage device not a digital storage device. It stores charge (quantized at a single electron) not bits. By using a controlled programming technique, it is possible to place a precise amount of charge on the floating gate. If charge can be accurately placed to one of four charge states (or ranges), then the cell can be said to store 2 bits. Each of the four charge states is associated with a 2-bit data pattern. Figure 12.6 illustrates the threshold voltage distributions for a $\frac{1}{2}$-Mc block for 2-bit-per-cell storage. After erasure or precise

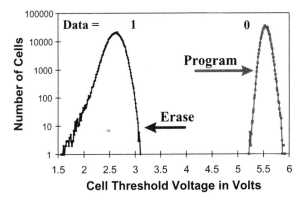

Figure 12.5. Single-bit/cell array threshold voltage histogram.

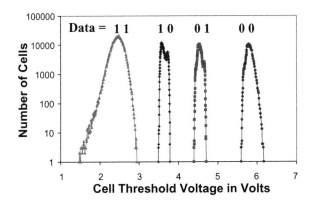

Figure 12.6. Two-bit/cell array threshold voltage histogram.

programming to one of three program states, the threshold of each of the $\frac{1}{2}$-Mc blocks is measured and plotted as a histogram. Note the precise control of the center two states, each approximately 0.3 V (or 3000) electrons in width.

Higher bit/cell densities are possible by even more precise charge placement control. Three bits per cell requires eight distinct charge states; 4 bits per cell requires 16 distinct charge states. In general, the number of states required is equal to $2N$ where N is the desired number of bits.

The ability to precisely place charge on the floating gate and at some later time sense the amount of charge that was stored has required substantial innovations and extensive characterization and understanding of cell device physics, memory design, and memory test. The aspects are discussed later in this chapter.

12.3.2 Evolution of MLC Memory Technology Development

This section will outline the development of the Intel StrataFlash memory technology from conception in 1992 to production in 1997, highlighting the key innovations along the way. The 64-Mbit product initially introduced in 1997 differs markedly from the 1992 view of what a 2-bit/cell product might look like. Today, MLC products have become the mainstream Flash memory technology. The learning that has occurred over the years has enabled the development of a 2-bit/cell memory device that functionally looks identical to a 1-bit/cell device, far exceeding the capability that was considered possible when development began.

12.3.3 Multilevel Cell Concept

Storage of analog data in a floating-gate memory device is not a new concept. It was suggested as early as 1971 for erasable programmable ROM (EPROM) devices [1] and was implemented on E²PROM devices for use in neural networks, voice recorders, and toys as early as 1982. These analog storage applications can tolerate a high error rate and thus do not place stringent requirements on the memory reliability or accuracy. Neural networks are, by their nature, fault tolerant. Voice storage and simple talking toys can tolerate a few lost bits without any audible impact. These high error rate lossy memories are generally not usable for mainstream digital storage and thus have had limited acceptance. The goal of the MLC program was to produce a 2-bit/cell digital storage technology capable of penetrating the larger nonvolatile memory market, enabling the growth of new digital Flash memory applications.

12.3.3.1 The 1992 View of MLC. In the early 1990s, Flash memory was considered as a potential replacement for hard disks at lower densities for applications that require small, rugged, and low-power storage. One of the main issues for use of Flash in this application was the high cost of the Flash memory as compared to that of magnetic storage. A lower cost Flash memory was required. The hard disk requirements are much relaxed over silicon memory due to the inclusion of error correction in the hard disk subsystem, the block transfer of data (no byte access), and the relative low read performance. Multilevel cell technology appeared to be an ideal solution for the solid-state disk, addressing the lower cost through 2-bit-per-cell (and later 3 or 4 bits/cell) technology. The use of error correction and the large block transfer of data in the solid-state disk would address any reliability issues with multilevel storage. The Intel MLC program was thus started with a goal of a high-density, low-cost, solid-state disk.

The basic techniques for accurate charge placement and sensing were developed in the lab and implemented into a 32-Mbit silicon test chip. During this time frame, the three major challenges for multibit storage were identified:

- *Precise Charge Placement*: The Flash memory cell programming must be very accurately controlled, requiring a detailed understanding of the physics of programming as well as the control and timing of the voltages applied to the cell.
- *Precise Charge Sensing*: The read operation of an MLC memory is basically an analog to digital conversion of the analog charge stored in the memory cell to digital data, a concept new to memory devices.
- *Stable Charge Storage*: Meeting the data retention goals would require the stored charge to be stable with a leakage rate of less than one electron per day.

The 32-Mbit test chip clearly demonstrated the ability to store multiple bits in a single cell. Based on the functionality of this device, the MLC technology was announced in 1994.

12.3.3.2 The First MLC "Product." With the knowledge gained from the 32-Mbit test chip, the first attempt at a 2-bit/cell storage product was started. This device was aimed at the solid-state disk goal. The solid-state disk system would include error correction and would generate nonstandard voltages to interface to the 2-bit/cell memory device. A special dc-to-dc (direct current) voltage converter was commissioned that would generate $12\,\text{V} \pm 1\%$ and $5.5 \pm 1\%$. The MLC part required these precise supply voltages to perform the accurate program and read operations. Also designed to be integrated with the other control logic of the solid-state disk was an error corrector. A paper based on this 32-Mbit MLC memory was presented at the prestigious International Solid State Circuits Conference (ISSCC) in 1995 [2], winning the best paper of the conference award. The 32-Mbit chip became the workhorse for the MLC technology development effort, demonstrating the ability of MLC to meet stringent reliability requirements and to produce yield equivalent to single bits per cell Flash memories. It was also used to develop the MLC testing and to debug the manufacturing process for test and packaging.

12.3.3.3 Question of Reliability. The primary concern for MLC was the reliability of the storage of the multiple charge states. Charge states would be separated by a few thousand electrons in an MLC device, and a loss of one electron per day from the floating gate could result in a bit error after 10 years of storage. To understand the detailed physics of charge storage, a large experiment was started to monitor the charge storage behavior of 200 billion cells (2×10^{11} cells). This massive experiment could resolve changes in the stored charge of as small as 100 electrons on all of the cells under evaluation. The rate of charge loss was accelerated through the use of elevated temperatures. The knowledge gained and models developed based on this experiment have resulted in changes to the design of both the product and the process, allowing removal of the error correction requirement for 2 bits per cells [3]. This data fundamentally changed the direction of the multibit storage program.

12.3.3.4 Removing the Constraints. Toward the end of 1995, the MLC project had grown from a small research effort to a full-blown program. Almost 2 years worth of reliability data was showing excellent performance, indicating that the error corrector was not required. The 32-Mbit device had demonstrated the viability of the circuit techniques and the device physics used for the precision program and read operations. Moreover, the yield was looking excellent, and the manufacturing issues were understood. Test circuits had demonstrated the ability to provide the required voltages and voltage regulation on the memory chip, eliminating the need for the external dc-to-dc converter. It became clear that the project could accomplish much more than the initial vision of a solid-state disk. The team believed that it was possible to remove the two major requirements initially envisioned for MLC: error correction and precision external power supplies. The solid-state disk market, while developing, had not reached the desired volume levels. The decision was made to not take the 32-Mbit device to production and focus on the design of an MLC 2-bit/cell part with functionality substantially equivalent to the standard 1-bit/cell products.

12.3.3.5 The 1997 View of MLC. The first 2-bit/cell Intel StrataFlash memory device was introduced in September of 1997, a 64-Mbit device. This device has functionality that is largely equivalent to the standard 1-bit/cell Flash products. A highlight comparison of the Intel StrataFlash memory features to an Intel 16-Mbit single-bit/cell product is shown in Table 12.1.

Read performance was in line with expectations for memories of 32- and 64-Mbit densities with about a 20% increase in read access time for a doubling of memory density. Two bits/cell doubles the erase block size as compared to 1 bit/cell since each cell now stores twice as much data. The power supply was maintained at the 5 V industry standard used at that time. The 2 bits/cell write performance was maintained equivalent to 1 bit/cell, even with the more complex (and slower) precision write algorithm, through the use of an 8-byte write buffer and a higher write bandwidth into the array. The 10,000 erase/write endurance specification was more than acceptable for virtually all Flash applications and easily justified by the reduced cost.

TABLE 12.1. Comparison of 1-Bit/Cell and 2-Bit/Cell Product Features

	1-Bit/CellFlash Memory	Intel StrataFlash 2-Bit/Cell Memory	
Density	16 Mbits	32 Mbits	64 Mbits
Read speed	100 ns	120 ns	150 ns
Block size	64 kbytes	128 kbytes	
Architecture	×8	×8 / ×16	
V_{cc} power supply (±10%)	5 V	5 V	
V_{pp} (program/erase voltage)	5 V or 12 V	5 V	
Effective write speed	6 µs/byte	6 µs/byte	
I_{ccr} (read current)	35 mA	55 mA	
$I_{ppw} + I_{ccw}$ (write current)	75 mA	90 mA	
Endurance	100,000 cycles	10,000 cycles	
Operating temperature	Extended	Commercial	

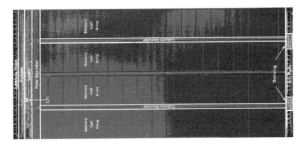

Figure 12.7. Intel StrataFlash 64-Mbit memory.

The 64-Mbit device integrated all of the knowledge gained from the two previous test vehicles and advanced beyond them with the introduction of precision internal voltage regulation and internal test capability. The 64-Mbit 2 bit/cell Intel StrataFlash memory was less than 5% larger than the 32-Mbit 1 bit/cell device on the 0.4 µm ETOX Flash memory process, delivering on the promise of 2× the bits in 1× the space and setting a new cost paradigm for Flash memory devices. A photomicrograph of the 64-Mbit Intel StrataFlash memory is shown in Figure 12.7.

12.4 VIEW OF MLC TODAY

Today, further advances in MLC capabilities have resulted in the functionality of MLC products to be indistinguishable from their single-bit-per-cell counterparts. Random byte read access times below 90 nS, V_{cc} of 1.8 V, and flexible read-while-write features on state-of-the-art lithography are the mainstream for both products [4].

12.4.1 Multilevel Cell Key Features

The concept of MLC is ideally suited to the Flash memory cell. The cell operation is governed by electron charge storage on an electrically isolated floating gate. The amount of charge stored modulates the Flash cell's transistor characteristic. MLC requires three basic elements: (1) accurate control of the amount of charge stored, or placed, on the floating gate such that multiple charge levels, or multiple bits, can be stored within each cell, an operation called placement; (2) accurate measurement of the transistor characteristics to determine which charge level, or data bit, is stored, an operation called sensing; and (3) accurate charge storage, such that the charge level, or data bit, remains intact over time, an operation called retention. These elements are achieved by exploiting stable device operation regions and by the direct cell access of the ETOX Flash memory array.

12.4.2 Flash Cell Structure and Operation

An explanation of MLC first requires a review of the Flash memory cell. Key points relevant to MLC operation are covered here. The ETOX Flash memory cell and products [5] have a long manufacturing history, having evolved in the late 1980s from EPROMs, which had been an industry standard from the early 1970s.

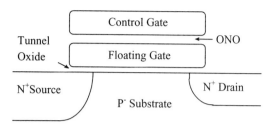

Figure 12.8. ETOX Flash memory cell cross section.

12.4.2.1 Cell Structure.
Figure 12.8 shows a cross-sectional view of a Flash cell. It consists of an N-channel transistor with the addition of an electrically isolated polysilicon floating gate. Electrical access to the floating gate is only through a capacitor network of surrounding SiO_2 layers and source, drain, transistor channel, and polysilicon control gate terminals. Any charge present on the floating gate is retained due to the inherent Si–SiO_2 energy barrier height, leading to the nonvolatile nature of the memory cell. Characteristic of the structure is a thin tunneling oxide (~100 Å), an abrupt drain junction, a graded source junction, ONO (oxide–nitride–oxide) interpoly oxide and a short electrical channel length (~0.3 μm).

Because the only electrical connection to the floating gate is through capacitors, the Flash cell can be thought of as a linear capacitor network with an N-channel transistor attached. The total capacitance of the cell (C_{TOT}) is equal to the additive capacitance of the network. For convenience, coupling ratio terms, which are defined as the ratio of terminal voltage coupled to the floating gate, can be defined as follows:

GCR = control gate coupling ratio
DCR = drain coupling ratio
SCR = source coupling ratio

Therefore, a change in control gate voltage will result in a change in the floating-gate voltage, $\Delta V_{FG} = \Delta V_{CG} \times$ GCR. The basic equation for the capacitor network is

$$V_{FG} = Q_{FG}/C_{TOT} + GCR \times V_{CG} + SCR \times V_{SRC} + DCR \times V_{DRN} \quad (12.1)$$

where Q_{FG} is the charge stored on the floating gate.

A simple first-order transistor equation of drain current says

$$I_D = G_M(V_{FG} - V_{CG} - V_{DRN}/2)V_{DRN} \quad (12.2)$$

where G_M is $q\mu e C_{OX} Z_E/L_E$.

This equation is very inexact for the small geometry of the Flash cell, but nevertheless the conclusions derived hold. Substituting V_{FG} of the basic coupling ratio Eq. (12.1) into the basic transistor I–V Eq. (12.2) leads to the conclusions that the transconductance of the transistor (and also the prethreshold slope) degrades by GCR, while the threshold voltage, V_T, depends upon Q_{FG}, the charge stored on the floating gate. Therefore, the V_T depends upon Q_{FG}, while the I–V shape does not.

VIEW OF MLC TODAY

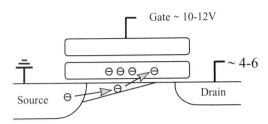

Figure 12.9. Cell bias conditions during programming.

Very simply, the Flash cell can be thought of as a capacitor, which is charged and discharged, the charge value being determined by the amplification of the transistor I–V. To give an idea of the amount of charge, every volt of cell threshold corresponds to approximately 10,000 electrons of floating-gate charge.

12.4.2.2 Cell Operation: Programming.
Figure 12.9 shows the cell bias conditions during program operation. A high drain-to-source bias voltage is applied, along with a high control gate voltage. The gate voltage inverts the channel, while the drain bias accelerates electrons toward the drain.

Programming a Flash cell means that charge, or electrons, are added to the floating gate. Programming a Flash cell, by channel hot electrons, can be understood by use of the lucky electron model [6]. In the lucky electron model, an electron crosses the channel without collision, thereby gaining 5.5 to 6.0 eV of kinetic energy, more than sufficient to surmount the 3.2-eV Si–SiO$_2$ energy barrier. However, the electron is traveling in the wrong direction. Its momentum is directed toward the drain. Prior to entering the drain and being swept away, this lucky electron experiences a collision with the silicon lattice and is redirected toward the Si–SiO$_2$ interface, with the aid of the gate field. It has sufficient energy to surmount the barrier. However, an electron does not have to be completely lucky. It can be "somewhat lucky" or "barely lucky," making the process of programming efficient. We can observe from this model that the lateral field, determined by bias voltage, junction profiles, electrical channel length, and channel doping are important to the effectiveness of generating energetic electrons and are therefore key to the MLC placement operation. Hence the abrupt drain junction and short channel length of the cell structure. After programming is completed, electrons have been added to the floating gate, increasing the cell's threshold voltage. Programming is a selective operation, uniquely occurring on each individual cell.

12.4.2.3 Cell Operation: Erase.
The distinguishing feature between EPROM and Flash memory is the erase operation. EPROM removes electrons from the floating gate by exposure to ultraviolet (UV) light. A photon of this light source has high enough energy that if transferred to an electron on the floating gate, that electron will have enough energy to surmount the Si–SiO$_2$ energy barrier and be removed from the floating gate. This is a rather cumbersome operation requiring a UV-transmissive package and a light source. It is also rather slow and costly, often requiring the removal of the memory from the system. In Flash, the contents of the memory, or charge, are removed by means of applying electrical voltages, hence to be erased in a *Flash*, with the memory remaining in the system.

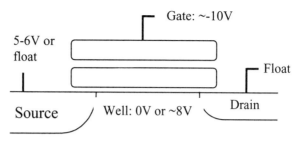

Figure 12.10. Cell bias conditions during erase.

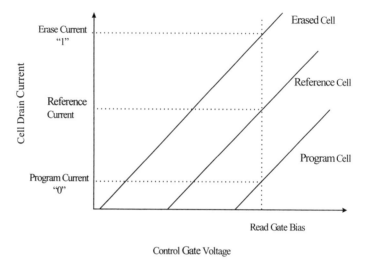

Figure 12.11. Erase, program, and reference cell I–V.

The electrical erase of Flash is achieved by the quantum-mechanical effect of Fowler–Nordheim tunneling [7], for which the bias conditions are shown in Figure 12.10. Under these conditions, a high field (8 to 10 MV/cm) is present between the floating gate and the source for source side erase, or between the floating gate and the channel for channel erase. Electrons tunneling through the first ~30 Å of the SiO_2 are then swept into the source. After erase has been completed, electrons have been removed from the floating gate, reducing the cell threshold. While programming is selective to each individual cell, erase is not, with many cells (typically 64 kbytes) being erased simultaneously.

12.4.2.4 Cell Operation: Read. The read operation of the cell should now be apparent. Storing electrons (programming) on the floating gate ($Q_{FG} < 0$), increases the cell V_t. By applying a control gate voltage and monitoring the drain current, the difference between a cell with charge and a cell without charge on their floating gates can be determined (Fig. 12.11). A sense amplifier compares the cell drain current with that of a reference cell (typically a Flash cell that is programmed to the reference level during manufacturing test). An erased cell has more cell current than the reference cell and therefore is a logical 1, while a programmed cell

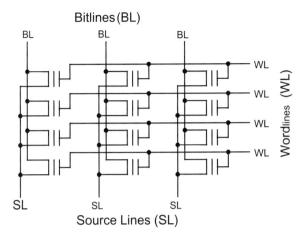

Figure 12.12. Array configuration.

draws less current than the reference cell and is a logical 0. The floating-gate charge difference between these two states is roughly 30,000 electrons.

12.4.2.5 Array Configuration. Figure 12.12 shows a schematic drawing of the Flash memory cells in a NOR array configuration. In this configuration, cells on the same wordline, or row, share common control gates. Cells on common bitlines, or columns, share common drains, which are connected via low-resistance metallization, providing direct access to each cell's drain junction. The sources for cells in the array are common. They are connected locally via common degenerately doped silicon and globally via low-resistance metallization. Decoders are linked to the control gate wordlines and drain bitlines to uniquely select cells at the cross-point location. The direct access to the cell in this configuration versus alternative array architectures that have parasitic resistance or devices, ensure that accurate voltages can be applied to the cell and infrared (IR) drops are minimized. This is a key aspect to achieving MLC operation.

12.4.3 Multilevel Cell Operation

We have reviewed thus far how a 1-bit-per-cell (1B/C) Flash memory operates. As can be inferred from the previous discussion, MLC is simply a means by which charge on the floating gate is modulated and detected to levels less than the 30,000 electrons described above, such that intermediate charge levels, or states, can be extracted from the cell. These states can now represent not just the simple 1B/C 1 and 0, but rather an MLC representation with four distinct charge states: 11, 10, 01, and 00, or 2 bits in one cell (2B/C). These four distinct levels are illustrated in the *I–V* curve of Figure 12.13. The key aspects of achieving these intermediate states, or levels, are precise charge placement, precise charge sensing, and precise charge retention.

12.4.3.1 Precise Charge Placement. A comparison of Figures 12.11 and 12.13 shows that MLC requires a means to control how much programming occurs within a cell. For a 1-bit/cell product all that is necessary is to have enough program-

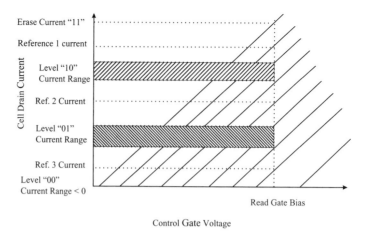

Figure 12.13. Cell and reference *I–V* curves of 4-level 2B/C.

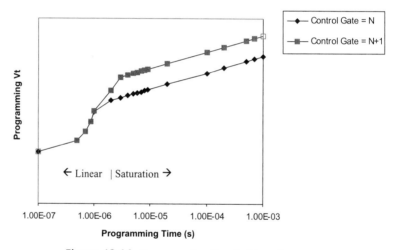

Figure 12.14. Programming threshold vs. time curve.

ming to change a 1 into a 0. Overprogramming a cell to much higher V_t's (adding more floating-gate charge) would be fine. This is not the case for MLC, where too much programming would cause an intermediate level to overshoot onto the next level. For instance, if a 10 was desired, but a cell was overprogrammed, a 01 might occur, leading to erroneous data. Therefore, a method of controlling precisely how much charge is transferred to the floating gate is required. Enough charge is needed to reach a state level without overshooting the desired level.

To gain insight into how such precise control can be obtained, let's take a deeper look into the Flash cell's programming characteristics. Figure 12.14 shows how the Flash cell's V_t changes as a function of log time, at two different bias conditions. Two regions of operation are shown: linear and saturation, so-called because the linear region is linear when plotted in linear time and the saturation region is where the cell V_t changes little with time, analogous to a metal–oxide–semiconductor

(MOS) transistor I–V curve. Note also that in the linear region, the control gate voltage has little influence on the rate of programming, while in the saturation region, the control gate voltage has a strong dependence upon the saturated V_t. A characteristic from Figure 12.14 is that the Flash cell programming slows as more charge is added to the floating gate.

In the linear region, energetic electrons, near the drain, are attracted to the floating gate. As programming progresses, the floating gate [that is coupled to the control gate and drain biases as governed by Eq. (12.1)] becomes charged more negatively, until it eventually reaches the same bias potential as the drain voltage. At this point, the energetic electrons become repelled by the floating-gate charge. Programming slows, as near-drain electrons must tunnel through the SiO_2 barrier, or less energetic midchannel electrons "jump" over the barrier. The strong gate dependence results from the vertical field limitation in this region.

One can also see from Figure 12.14 that the saturated V_t increases in a one-for-one fashion with an increase in the programming control gate voltage. This is a simple result of the coupling equation (12.1).

Given this characteristic curve, one could devise several possible methods of controlling the charge transfer to the floating gate. These methods would have to pass the criteria of being reliable (no overshoot), controllable (simple to implement), and fast (to ensure compatibility with standard Flash memory product features). Programming in the linear region, while being fast, is not controllable. In this region, programming V_t is exponentially dependent upon time and the electron energy distribution (as determined by drain bias, channel length, doping profiles, etc.). Small variations will lead to large changes in the cell threshold and therefore overshoot of the desired state, thereby having a high likelihood of being unreliable. Minimization of these variations would be also difficult to implement. In the saturated region, the cell V_t simply depends upon the applied control gate voltage. Control in this region is more achievable. With ease of control, design optimization practices can be employed to achieve fast programming. This will be shown later. Therefore, to achieve speed and control, a placement algorithm that employed programming in the saturated region was developed.

This leaves us with reliability. Unlike Fowler–Nordheim tunneling, used for programming in addition to erase in some versions of Flash memories and subject to erratic programming due to the presence or absence of as few as one or two holes trapped in the oxide [8], channel hot-electron programming has no erratic programming mechanism. The programming threshold in saturation is simply a linear function of the applied control gate voltage. Programming in this region can be forced into an unstable operating point, known as impact-ionization-induced latch-up [9]. This is the point where an excess of holes in the silicon substrate, created by the collisions of the energetic electrons with the silicon lattice, build up to the point where the parasitic NPN transistor in the silicon substrate turns on. Proper architectural design of the silicon process flow and Flash cell architecture can easily prevent this from happening.

Therefore, exploiting stable device operation regions, namely programming in saturation, satisfies the three success criteria. Now, a simple placement algorithm was chosen for implementation (outlined in the flowchart in Fig. 12.15). The algorithm consists of the simple loop of programming in saturation, checking the cell V_t to determine if the desired state has been reached, stopping if the desired state is reached, or if not, incrementing the control gate voltage and providing an additional

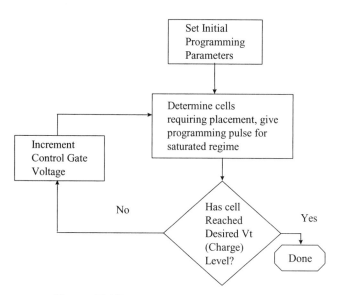

Figure 12.15. Placement algorithm flowchart.

programming pulse and continuing in this fashion until the desired V_t has been achieved. In a typical MLC memory 2-bit/cell device, each programming pulse within the placement algorithm will transfer roughly 3000 electrons of charge to the floating gate.

12.4.3.2 Precise Charge Sensing. As can be seen from the flowchart in Figure 12.15, integral to the placement algorithm is a means of detecting whether or not the desired cell V_t has been achieved. Without a precise means of sensing the floating-gate charge, precise charge placement would not be possible. A look back at Eq. (12.2), the cell drain current–voltage relationship, gives some insight into what is required to achieve precise charge sensing.

Control gate and drain voltage control and process, L_{eff}, Z_{eff}, mobility, and oxide capacitance control are important aspects of precise charge sensing. Drain voltage control is facilitated by direct access to the cell drain junction (bypassing any resistive IR drops) allowable in the ETOX NOR Flash memory array architecture; and by applying a high enough drain voltage to operate in the saturated mode (normal MOS device saturated I–V, not programming saturation as previously discussed) where drain bias variations have minimal current impact. Process control is important since the multimega cells within one product represent a >10 sigma variation, and is achieved by proper process architecture and manufacturing process control, derived from more than a decade of manufacturing experience with Flash memories. Control gate voltage control is achieved by an on-chip read regulation circuit, which is fully explained in a later section.

Flash memory has a unique feature associated with its nonvolatility: The data write (placement) can occur under one condition of ambient temperature and system power supply, while the readout of data (sensing) can occur at a later date, at different ambient temperature and system power supply. Being fundamentally a

MOS transistor, the Flash cell's drain current is a function of these ambient conditions. As such, the precise charge sensing is required to span wide ranges of operation. To facilitate this needed precision, the reference levels that separate charge state levels are generated by reference Flash cells contained on-chip. These reference cells, whose V_t levels are precisely placed at manufacturing test under a controlled environment, will have the same tracking with temperature and power supply as the array Flash cells. This contrasts to reference levels generated by other transistor types (i.e., NMOS or PMOS), which have different temperature, voltage, and process tracking than the Flash memory cell. This lessens the necessary constraints on the read regulation circuitry.

12.4.3.3 Precise Charge Retention. Due to the nonvolatility requirement of Flash memory, it is important that any charge placed on the floating gate remain intact for extended periods of time, typically many years. This translates to a requirement of not losing more than one electron per day from the floating gate. If electron loss occurs from even one memory cell in an array of millions, the data will be corrupted. The inherent storage capability exists due to the Si–SiO$_2$ energy barrier that traps electrons on the floating gate. The inter-polysilicon oxide (ONO film mentioned in the cell structure) is processed to maximize charge storage capabilities [10]. Under normal circumstances, the energy barrier allows charge storage for hundreds of years. There are conditions of trapped oxide charge, known as intrinsic charge loss [11], which can cause one-time shifts in threshold. These shifts are rather small and are compensated for during manufacturing test. Random defects in the insulating oxides that can lead to charge loss are less of an issue with low-defect high-yielding process technologies, but if still present are screened out by the manufacturing tests. These defects are driven to low enough levels on ETOX Flash memories where error-correcting codes (ECC) are not needed. The remaining concern for charge retention is any degradation to the insulating oxides that occurs as a result of the stresses of device operation.

During normal operation, high fields are applied to the Flash cell. The presence of the high fields over time can degrade the charge storage capabilities of the device, effectively by lowering the energy barrier, or by providing traps sites in the oxide that can act as intermediate tunneling locations. The benefit of channel hot-electron programming, compared to tunneling for programming, is that fast programming can occur at lower internal fields thereby lessening the probability of oxide damage. Nevertheless, occurrence of damage needed to be understood to ensure the stability of the MLC charge. Consequently, the charge retention ability of the insulating oxides under various process and bias field conditions were studied in great detail. Over the course of the 4-year MLC development period, in excess of 200 billion (2×10^{11}) Flash cells were studied for charge retention, each to a resolution of floating-gate charge of ~100 electrons. This exhaustive study provided more physical insight into the oxide damage mechanisms and has enabled us to build large-scale empirical models for charge retention. The net result of this study was the ability to optimize process recipes and operating bias fields to maximize charge retention. A detailed review of the reliability aspects of Flash memory cells charge retention can be found in Belgal et al. [3]. Additionally, a more detailed discussion of the constraints of precise charge storage, sensing, and retention can be found in Modelli et al. [12].

12.4.4 Mixed Signal Design Implementation

The circuit design of precise charge placement and precise charge sensing will now be described through the implementation that was done on the original Intel StrataFlash memory. The implementation of the described charge placement algorithm and charge sensing operation required a mixed signal circuit design of both digital and precision analog voltage generation, regulation, and control circuits. The placement algorithm is executed by utilizing an on-board control engine, or the Flash algorithmic control engine (FACE). FACE runs the placement algorithm by sequencing through the programming and sensing loops. During a read operation (sensing of data at a later time), the user has random access to the memory array. A read operation performs a precision-sensing operation and invokes circuitry controlling the precise cell bias voltages.

12.4.4.1 Placement Algorithm Implementation.

The placement algorithm executed by FACE is stored in a small on-chip programmable Flash array. The programmable microcode allows for flexibility in algorithm changes. FACE, illustrated in Figure 12.16, consists of the microcode storage array, program counter (PC), arithmetic logic unit (ALU), instruction decoder, clock generator, register files, and input/output circuitry. FACE uses 6000 transistors for logic and 32 kbits of Flash memory for algorithm storage.

To describe the implementation of the placement algorithm, let us assume that a group of cells (i.e., a double-word, or 32 logical bits, 16 physical cells) is to be placed and is initially in the erased state (lowest floating-gate charge state). Any cells not to remain in the erased state (representing logical data 11) will receive a programming pulse. FACE will look up the drain and initial control gate voltage

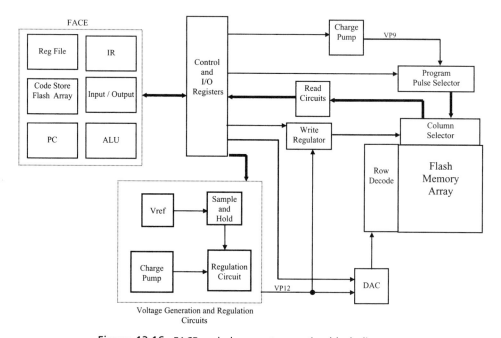

Figure 12.16. FACE and placement operation block diagram.

stored in a permanent read-only register located on-chip. FACE will then set the control gate voltage through the digital-to-analog converter (DAC). The DAC circuit receives the FACE digital input and divides the on-chip generated ~12-V power supply (VP12) to achieve the desired control gate voltage for that particular programming pulse. The drain voltage, used during the programming pulse, is generated from a regulation circuit, which sets the gate voltage on a source follower. FACE will continue to supply the programming voltages for the predetermined amount of time sufficient to reach the saturation region. When the programming pulse is complete, FACE will reconfigure the circuits to perform the sensing portion of the algorithm, an operation called verification. The drain and control gate voltages are now set to the same values as used in a user read access to ensure common mode between verification and read. FACE will take the result of the verification and determine which cells have reached their destination charge level and which have not. Those that have not will require an additional programming pulse with an increased control gate voltage. A cell that no longer requires additional programming pulses will have the drain voltage disabled by the program pulse selector circuit. This sequence of events continues until all cells in the double word have completed programming.

12.4.4.2 Analog Circuit Blocks for Precise Charge Placement.

Placement requires precision voltages covering a range of 4 to 12-V, while the chip V_{cc} (user-supplied voltage) is kept at a typical value of 1.8 to 5 V, depending upon the system design. The voltages applied to the memory array need to be internally generated and precisely regulated. On-chip voltage generation is achieved by use of charge pumps, in which switched capacitors boost the user-supplied V_{cc} to higher values. Voltages are controlled using a precision voltage reference circuit and voltage regulation circuits (Fig. 12.17).

During a programming pulse, two charge pumps are used. One charge pump generates the internal 12 V supply (VP12). This is used to supply a precision control

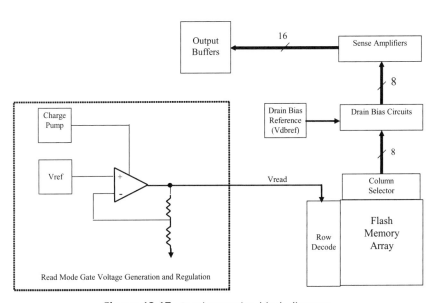

Figure 12.17. Read operation block diagram.

gate voltage to the Flash cells, through the DAC circuit VP12 also serves to generate the precision Flash drain voltage through the write regulation circuit (WRC). The WRC generates a voltage that is applied to an NMOS transistor configured as a source follower. This transistor is in the bitline (or drain) path of the Flash cell. The Flash cell drain current is supplied through a second pump that generates the signal VP9. This pump is required to supply the programming current for up to 32 Flash cells at a time.

During the placement algorithm, voltage stability is critical to precise charge storage. Any variations in the reference circuit voltages will be seen as variations in the Flash control gate voltage, to which the programming-saturated V_t is directly related. To achieve this absolute stability in the voltage reference circuit, a sample and hold circuit is employed. At the start of the placement algorithm, the sample and hold circuit samples the reference voltage and holds the value on a capacitor during the running of the entire algorithm. This guarantees the control gate voltage varies from pulse to pulse by only the desired step value and not by any additional components.

12.4.4.3 Circuit Blocks for Precise Charge Sensing. When the device is in the read mode of operation, FACE is disabled and the user has control to access the memory array. A read operation consists of sensing 16 bits worth of data from a random location in the memory array. With MLC, 8 Flash cells are used to obtain 16 bits of data. During the read operation (Fig. 12.17), the Flash cell control gate voltage is controlled through a read regulator circuit (RRC). Minimizing this voltage variation will minimize the variations in cell current [Eq. (12.2)]. This allows for more precise measurement of the charge level stored on the floating gate. Drain voltage stability is also important to ensure that the Flash cell being sensed has a high enough drain voltage to keep the memory transistor operating in the saturated region of the MOS I–V.

Due to fluctuations in user-supplied V_{cc} and a lower value than may be needed during read, an internal voltage charge pump is used during a read operation to generate the internal voltage to supply the Flash cell control gate. The RRC uses the same voltage reference circuit that is used for voltage regulation during a placement operation, as mentioned above. However, in the case of a read operation not as much voltage stability is required so the sample and hold circuitry is not used.

12.4.4.4 Parallel Charge Sensing. High-speed random access and precise charge sensing can be accomplished through a parallel charge-sensing scheme. Through direct connections to each memory cell, the data read operation determines the level of each memory cell quickly, accurately, and reliably. The data read operation senses which of the four levels the memory cell falls within based on the threshold voltages of three reference cells. This is done simultaneously with three sense amplifiers (Fig. 12.18), where each sense amplifier compares the Flash cell current being sensed to the current of the Flash reference cells.

The memory cell and the reference cells are biased in such a way that each conducts a current (I_{cell} and I_{ref}) proportional to their respective threshold voltage (V_t and $V_{t,ref}$). During a read operation, V_{read} is placed on the control gates of the memory and reference cells, the source terminals are grounded, and the drain voltages are set through a bias circuit that utilizes a precision voltage reference circuit.

LOW-COST DESIGN IMPLEMENTATION

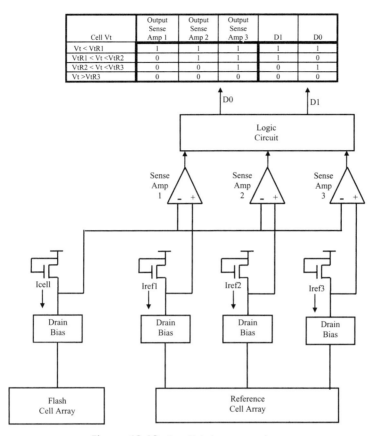

Figure 12.18. Parallel charge sensing.

The current for the memory cell being sensed is compared to the current of the three reference cells. The memory cell and reference cell current is converted to a voltage through an active load transistor. Three sense amplifiers compare the resultant voltages. A sense amplifier is associated with each of the three reference cells. Each sense amplifier also has an input from the Flash cell being sensed. If the current of the cell being sensed is greater than the current of the reference cell ($I_{cell} > I_{ref}$ or $V_t < V_{t,ref}$), the sense amplifier output is a logic 1. If the current of the cell being sensed is less than the current of the reference cell, the sense amplifier output is logic 0. The outputs of the three sense amplifiers are connected to a logic circuit, which interprets the two data bits in parallel. In recent years, subsequent circuit development and design has shown that a serial sense approach to MLC sensing can offer even more precise charge sensing at no performance penalty [4].

12.5 LOW-COST DESIGN IMPLEMENTATION

Traditionally a storage element in a memory corresponds to one bit of information. To double the amount of memory, the memory array or memory storage elements would need to be doubled. In addition to doubling the number of memory elements in the array, certain memory interface circuits must also be doubled. In particular, the memory array needs to be decoded requiring wordline and bitline decoders. In a typical single transistor nonvolatile memory device (Flash, EPROM), approximately 20% of the silicon area used is due to these interface circuits required to access the array. These interface circuits typically do not scale with process technology at the same rate as the memory array because they have high-voltage and analog requirements.

Intel StrataFlash memory doubles the storage capacity of a memory device without doubling the memory array and the associated interface decoding circuitry. Additional circuitry is required to achieve the multiple bits per cell, but takes up a relatively small additional area. The additional overhead for circuitry is due mostly to the additional sense amplifiers, reference circuitry, and circuitry for voltage generation or charge pumps. The additional silicon area required for this circuitry represents less than additional 5% over what is necessary for a 1-bit/cell device. Implementations of MLC that require externally supplied components (i.e., microcontroller, ECC, voltage regulators) have the cost savings of MLC diminished by these peripheral overheads. MLC memories achieve 2× the density at very close to 1× the area.

12.6 LOW-COST PROCESS MANUFACTURING

ETOX Flash memory has a long manufacturing history. As such, it was necessary that any implementation of MLC not disrupt that history by having unique process requirements, which would cause a slow yield learning period or poor manufacturing throughput. First and foremost for MLC to be successful, it must be able to ride on a technology that produces error-free 1 bit/cell Flash memory. This requirement throughout Intel's ETOX NOR Flash memory's history has resulted in tight manufacturing margins and the learning necessary for achieving such margins. Memories that rely on ECC for even 1-bit/cell operation have little margin built into the basic technology. Throughout the previous discussions mention has been made of process manufacturing attributes for MLC. These attributes have been achieved by utilizing the same process flow as the standard 1-bit/cell Flash memory. This approach has maintained shared learning and has led to lower costs. In other words, low-cost process manufacturing was achieved through an understanding of MLC requirements up-front in the design of the basic process architecture at the generation where MLC is introduced. The tight manufacturing margins required for MLC are a natural extension of the learning from manufacturing of error-free 1-bit/cell Flash memory and are well within the manufacturing, equipment, and process module capability.

12.7 STANDARD PRODUCT FEATURE SET

One of the main challenges in implementing MLC is maintaining product performance, usability, and reliability at the same levels as standard Flash memories. If the implementation of MLC resulted in a product that did not satisfy these goals, it would be relegated to a niche in the marketplace. Key features for a nonvolatile

STANDARD PRODUCT FEATURE SET 613

memory are programming speed, read speed, power supply requirements, and reliability. This chapter shows how the implementation of MLC achieves these features. Before finishing, however, let us briefly discuss each one of them.

12.7.1 Programming Speed

Programming speed is achieved by choosing a placement algorithm that exploits stable device operating points to enable circuit performance optimization to occur, with little limitations of Flash device operation. Parallel cell programming (32 cells, or 64 bits) at a time also amortizes the placement algorithm run time. The choice of charge sensing approaches also affects programming speed, as it is integral to the placement algorithm. Sensing approaches other than those described in this chapter can be used. An example would be a sensing scheme that varies control gate voltage to detect the threshold voltage directly. Such a scheme, while a more direct measure of floating-gate charge, does not exploit the current drive capability of the Flash cell, the drive used for sensing speed performance. To sum up, the choices of algorithms, optimizations, and architecture are what allow MLC programming to be as good as or better than 1 bit/cell Flash memories.

12.7.2 Read Speed

As mentioned above, the choice of fixed control gate sensing and utilization of the Flash cell's current drive capability allows fast read operation. In addition, parallel charge sensing allows for fast decode of the logic level, with little circuit overhead. As such, the read speed of Intel's StrataFlash memory is consistent with that of 1-bit/cell Flash memories of comparable bit density.

12.7.3 Power Supply

As also discussed, the on-chip voltage generation and regulation is key to the implementation of MLC One could specify an MLC product that uses externally supplied precision voltages, but such a product would be more costly to the user, who would have to pay for the power supply, memory, and board space. Having the voltages generated and regulated on-chip allows for the MLC memory to plug directly into existing Flash memory applications. Today's MLC products operate with standard single 1.8-V V_{cc} power supplies.

12.7.4 Reliability

Starting with high-yielding, low-defect memory, exhaustive cell studies and process and bias optimizations allow for an implementation of MLC that has the potential to achieve nonvolatility and high reliability without requiring on-chip or system ECC. Thus the user can interface to the device with random memory location access, without latency for correction.

These standard Flash memory product features, coupled with low-cost circuit design and manufacturing process implementation, allow users to benefit from the low cost of MLC without having to sacrifice needed features or performance.

12.8 FURTHER READING: MULTILEVEL FLASH MEMORY AND TECHNOLOGY SCALING

As process technology scaling follows Moore's law, continued evolution and innovations in MLC techniques are required. The fundamentals of MLC development presented in this chapter will continue to hold for many generations to come, however, new circuit techniques and semiconductor device types and materials will be necessary to mitigate some of the undesirable electrical degradations associated with scaling.

The principles of precise charge sensing for advanced MLC Flash remain, but evolutions in circuit design have been required to ensure adequate signal-to-noise margin to mitigate electrical scaling effects when performing a sensing operation on a Flash memory array. Constant current, variable gate voltage sensing schemes have been developed to minimize memory cell current effects in a memory array and random transistor variability across a device [13, 14].

What were considered to be second- and third-order effects on earlier technology generations become first-order effects as technologies continue to scale. For example, as a Flash memory becomes smaller, Flash memory transistors interact during charge placement where the floating-gate storage nodes capacitive couple. In addition, random device noise in the memory cell becomes problematic during charge placement due to the degraded signal noise during the verification operation. Circuit, chip architectural solutions such as ECC, and algorithmic are required to overcome some of the device physics limitations [15–18].

In addition to solutions to overcome scaling degrading the memory's electrical properties, system applications using multilevel Flash memory require improvements in device performance. Many architectural developments and algorithms are required to improve device performance by adding more parallelism and faster sensing and programming techniques [18–20].

12.9 CONCLUSION

It has been shown how MLC memories achieve multiple bits per cell, coupled with traditional process scaling, to provide an advance in memory cost reduction. The MLC requirements of precise charge placement, precise charge sensing, and precise charge retention are achieved by exploiting stable device-operating points and direct access to the memory cell, employing mixed signal digital and analog design. Non-cell-related costs are held low by riding on the tight manufacturing margins developed for error-free 1 bit/cell Flash memories. A standard product feature set ensures that the cost advantages of MLC are available to the mainstream Flash memory market.

REFERENCES

1. D. Frohman-Bentchowsky, "Floating Gate Solid State Storage Device and Methodology for Charging and Discharging Same," U.S. Patent No. 3,755,721, Aug. 28, 1973.
2. M. Bauer, R. Alexis, G. Atwood, B. Baltar, A. Fazio, K. Frary, M. Hensel, M. Ishac, J. Javanifard, M. Landgraf, D. Leak, K. Loe, D. Mills, P. Ruby, R. Rozman, S. Sweha, S. Talreja, and K.

REFERENCES

Wojciechowski, "A Multilevel-Cell 32 Mb Flash Memory," *Tech. Dig. IEEE Int. Solid State Circuits Conf.*, pp. 132–133, 1995.

3. H. P. Belgal, N. Righos, I. Kalastirsky, J. J. Peterson, R. Shiner, and N. Mielke, "A New Reliability Model for Post-Cycling Charge Retention of Flash Memories," *Proc. 40th IEEE Int. Reliability Phys. Symp.*, p. 7, 2002.

4. D. Elmhurst, R. Bains, T. Bressie, C. Bueb, E. Carrieri, B. Chauhan, N. Chrisman, M. Dayley, R. De Luna, K. Fan, M. Goldman, P. Govindu, A. Huq, M. Khandaker, J. Kreifels, S. Krishnamachari, P. Lavapie, K. Loe, T. Ly, F. Marvin, R. Melcher, S. Monasa, Q. Nguyen, B. Pathak, A. Proescholdt, T. Rahman, B. Srinivasan, R. Sundaram, P. Walimbe, D. Ward, D. R. Zeng, and H. Zhang, "A 1.8 V 128 Mb 125 MHz Multi-Level Flash Memory with Flexible Read While Write," *Tech. Dig. IEEE Int. Solid State Circuits Conf.*, Vol. 1, pp. 286–287, 2003.

5. V. N. Kynett, A. Baker, M. L. Fandrich, G. P. Hoekstra, O. Jungroth, J. A. Kreifels, S. Wells, and M. D. Winston, "An In-System Reprogrammable 256 K CMOS Flash Memory," *Tech. Dig. IEEE Int. Solid State Circuits Conf.*, Vol. 23, No. 5, pp. 132–133, 1988.

6. S. Tam, P. K. Ko, and C. Hu, "Lucky-Electron Model of Channel Hot Electron Injection in MOSFET's," *IEEE Trans. Electron Devices*, Vol. 31, No. 9, pp. 1116–1125, Sept. 1984.

7. M. Lenzlinger and E. H. Snow, "Fowler-Nordheim Tunneling into Thermally Grown SiO_2," *J. Appl. Phys.*, Vol. 40, No. 1, pp. 278–283, Jan. 1967.

8. T. C. Ong, A. Fazio, N. Mielke, S. Pan, N. Righos, G. Atwood, and S. Lai, "*Erratic Erase in ETOX™ Flash Memory Array*," paper presented at the IEEE VLSI Symposium, May 1993, pp. 83–84, pp. 145.

9. B. Eitan and D. Frohman-Bentchkowsky, "Surface Conduction in Short-Channel MOS Devices as a Limitation to VLSI Scaling," *IEEE Trans. Electron Devices*, Vol. ED-29, No. 2, pp. 254–266, Feb. 1982.

10. K. Wu, C.-S. Pan, J. J. Shaw, P. Freiberger, and G. Sery, "A Model for EPROM Intrinsic Charge Loss through Oxide-Nitride-Oxide (ONO) Interpoly Dielectric," *Proc. 28th IEEE Int. Reliability Phys. Symp.*, pp. 145–149, March 1990.

11. N. Mielke, "New EPROM Data-Loss Mechanisms," *Proc. 21st IEEE Int. Reliability Phys. Symp.*, pp. 106–111, April 1983.

12. A. Modelli, A. Manstretta, and G. Torelli, "Basic Feasibility Constraints for Multilevel CHE-Programmed Flash Memories," *IEEE Trans. Electron Devices*, Vol. 48, No. 9, pp. 2032–2042, Sept. 2001.

13. M. Bauer, and K. Tedrow, "A Scalable Stepped Gate Sensing Scheme for Sub-100 nm Multilevel Flash Memory" paper presented at the International Conference on Integrated Circuit Design and Technology, 2005, pp. 23–26.

14. C. Villa, D. Vimercati, S. Schippers, E. Confalonieri, M. Sforzin, S. Polizzi, M. La Placa, C. Lisi, A. Magnavacca, E. Bolandrina, A. Martinelli, V. Dima, A. Scavuzzo, B. Calandrino, N. Del Gatto, M. Scardaci, F. Mastroianni, M. Pisasale, A. Geraci, M. Gaitiotti, and M. Sali, "A 125 MHz Burst-Mode Flexible Read-While-Write 256 Mb NOR Flash Memory," *ISSCC Dig. Tech. Papers*, Vol. 1, pp. 52–53, Feb. 2005.

15. S. Lee, Y.-T. Lee, W.-K. Han, D.-H. Kim, M.-S. Kim, S.-H. Moon, H. C. Cho, J.-W. Lee, D.-S. Byeon, Y.-H. Lim, H.-S. Kim, S.-H. Hur, and K.-D. Suh, "A 3.3 V 4 Gb Four-Level NAND Flash Memory with 90 nm CMOS Technology," *ISSCC Dig. Tech. Papers*, Vol. 1, pp. 53–53, Feb. 2004.

16. R. Micheloni, R. Ravasio, A. Marelli, E. Alice, V. Altieri, A. Bovino, L. Crippa, E. Di Martino, L. D'Onofrio, A. Gambardella, E. Grillea, G. Guerra, D. Kim, C. Missiroli, I. Motta, A. Prisco, G. Ragone, M. Romano, M. Sangalli, P. Sauro, M. Scotti, and S. Won, "A 4 Gb 2 b/Cell NAND Flash Memory with Embedded 5b BCH ECC for 36 MB/s System Read Throughput," *ISSCC Dig. Tech. Papers*, pp. 142–143, Feb. 2006.

17. H. Kurata, K. Otsuga, A. Kotabe, S. Kajiyama, T. Osabe, Y. Sasago, S. Narumi, K. Tokami, S. Kamohara, and O. Tsuchiya, "The Impact of Random Telegraph Signals on the Scaling of Multilevel Flash Memories," *Symp. VLSI Circuits Dig. Tech. Papers*, pp. 1140–1141, June 2006.
18. M. Taub, R. Bains, G. Barkley, H. Castro, G. Christensen, S. Eilert, R. Fackenthal, H. Giduturi, M. Goldman, C. Haid, R. Haque, K. Parat, S. Peterson, A. Proescholdt, K. Ramamurthi, P. Ruby, B. Sivakumar, A. Smidt, B. Srinivasan, M. Szwarc, K. Tedrow, and D. Young, "A 90 nm 512 Mb 166 Mhz Multi-Level Cell Flash with 1.5 MB/s Programming," *ISSCC Dig. Tech. Papers*, Vol. 1, pp. 54–55, Feb. 2005.
19. K. Imamiya, H. Nakamura, T. Himeno, T. Yarnamura, T. Ikehashi, K. Takeuchi, K. Kanda, K. Hosono, T. Futatsuyama, K. Kawai, R. Shirota, N. Arai, F. Arai, K. Hatakeyama, H. Hazama, M. Saito, H. Meguro, K. Conley, K. Quader, and J. J. Chen, "A 125-mm^2 1-Gb NAND Flash Memory with 10-MByte/s Program Speed," *IEEE J. Solid-State Grants*, Vol. 37, No. 11, pp. 1493–1499, Nov. 2002.
20. K. Takeuchi, Y. Kameda, S. Fujimura, H. Otake, K. Hosono, H. Shiga, Y. Watanabe, T. Futatsuyama, Y. Shindo, M. Kojima, M. Iwai, M. Shirakawa, M. Ichige, K. Hatakeyama, S. Tanaka, T. Kamei, J.-Y. Fu, A. Cernea, Y. Li, M. Higashitani, G. Hemink, S. Sato, K. Oowada, S.-C. Lee, N. Hayashida, J. Wan, J. Lutze, S. Tsao, M. Mofidi, K. Sakurai, N. Tokiwa, H. Waki, Y. Nozawa, K. Kanazawa, and S. Ohshima, "A 56 nm CMOS 99 mm^2 8 Gb Multi-Level NAND Flash Memory with 10 MB/s Program Throughput," *IEEE J. Solid-State Grants*, Vol. 42, No. 1, pp. 219–232, Jan. 2007.

13

ALTERNATIVE MEMORY TECHNOLOGIES

Gary F. Derbenwick and Joe E. Brewer

13.1 INTRODUCTION

Flash memory is the major high-density, reprogrammable nonvolatile semiconductor memory technology today. Its small memory cell size provides a low cost per bit in a commodity memory business that is driven by cost, and its memory architecture has overcome many of the scaling and reliability problems of earlier electrically reprogrammable nonvolatile semiconductor memories. However, Flash memory may reach fundamental scaling limits because the minimum tunnel oxide thickness required to achieve 10-year retention necessitates high internal voltages for programming charge onto the floating gates. Isolation of these high internal voltages requires relatively large spaces, and hot-carrier injection dynamics require larger transistor sizes, limiting scaling of the memory cell. Attempts to reduce the programming voltages have not yet been successful.

Are there any alternative semiconductor memory technologies that might replace Flash memory technology, as Flash memory may approach its scaling limits? Memory technologies that use low internal programming voltages, such as ferroelectric, magnetic, chalcogenide, and resistive (crossbar) semiconductor memories, are being developed. These memory types are revolutionary approaches, but major challenges must be over come to achieve small memory cell sizes and aggressive technology design rule nodes competitive with Flash memory.

Other alternative technologies are more evolutionary in nature. Nitrided-ROM (NROM) combines features of silicon-oxide-nitride-oxide-silicon (SONOS) memo-

---

Nonvolatile Memory Technologies with Emphasis on Flash. Edited by J. E. Brewer and M. Gill
Copyright © 2008 the Institute of Electrical and Electronics Engineers, Inc.

ries and Flash memories. NROM memories use a similar programming method to that of Flash memory, but charge is stored within an insulating silicon nitride layer instead of on a conductive floating gate, similar to SONOS technology. Like Flash memories, NROM memories use small single-transistor memory cells, but in addition they can store two physical bits per memory cell. Other alternative memories, such as NOVORAMs and single-electron memories also evolve from Flash memory architectures.

As editors of this chapter, we have a word of caution: The semiconductor industry has consistently broken through anticipated scaling barriers, such as use of optical photolithography, use of oxide-nitride-oxide dielectrics for dynamic random-access memory (DRAM) capacitors, and overcoming short-channel effects. Flash memories can be scaled for more technology generations before anticipated fundamental scaling issues become critical, and the industry may well find solutions for these scaling barriers. Ballistic effects of the carriers in Flash memory, both for programming and sensing, provide a possibility for reducing the internal programming voltage and delaying the impact of the scaling limitations of Flash memory. Therefore, replacement of commodity Flash memory with alternative nonvolatile memory technologies will not occur quickly and may not occur at all.

The purposes of this chapter are to review briefly the scaling barriers of Flash memory and to survey a limited number of emerging nonvolatile semiconductor memory technologies, some in initial production and some far out. Even if some of these alternative memory technologies turn out to scale to smaller design rules than Flash memory, each memory type has its own issues that must be confronted and solved.

For an alternate memory technology to be competitive with commodity Flash memory, it must be of equal or lower cost. Customers are willing to tolerate reduced functionality and more complex operation of commodity memory chips in order to achieve reduced cost. For example, DRAM memories require constant refresh, and low-cost Flash memories can require complex erase cycles to prevent overerase conditions that would render the memory inoperable. Provided the amount of memory required on a printed circuit board is sufficiently large, the additional cost of controlling a low-cost commodity memory is warranted. Whether any alternative nonvolatile memory architectures can meet the low-cost requirements is unknown today.

Alternative memory technologies may find application in smaller markets where unique performance characteristics are needed. Ferroelectric, magnetic, and chalcogenide memories have faster programming speeds, lower programming voltages, and higher endurance than Flash memories. NROM, memories have similar programming speeds, programming voltages, and endurance to those of Flash memories. Ferroelectric, magnetic, chalcogenide, NROM, and resistive memories are expected to have superior radiation hardness characteristics compared to Flash memories.

The semiconductor memory markets are segmented because no semiconductor memory today offers all features for all market applications, and there are significant cost/feature trade-offs depending on the underlying memory technology. Should a semiconductor memory be developed that provides all features and provides the lowest cost per bit, then such a memory could be a universal memory that replaces all existing semiconductor memories. Table 13.1 lists some of the characteristics of an ideal "universal" memory.

TABLE 13.1. Characteristics of an Ideal Universal Memory

Feature	Universal Memory Characteristic
Read access	Random and word
Write access	Random and word
Erase access	Random, word and block
Read speed	Fast, comparable to SRAM
Write speed	Fast, comparable to SRAM
Erase speed	Fast, comparable to SRAM
Endurance	Infinite, comparable to SRAM and DRAM
Power	Low
Volatility	Nonvolatile, 10-year data retention, comparable to Flash

13.2 LIMITATIONS OF FLASH MEMORY

Stefan Lai and Gary F. Derbenwick

13.2.1 Introduction

As compared to the ideal attributes for a universal memory listed in Table 13.1, Flash memory lacks several features. These include low internal programming voltage, fast programming time, unlimited endurance, and random erase. However, Flash memory is reliable and relatively inexpensive, so the limitation in features is not a significant issue for most applications.

The high internal programming voltage required for Flash memory ultimately limits the scaling of the memory cell and certain peripheral circuitry, and, therefore, the ultimate density of Flash memory. Design for the high internal programming voltage also requires several extra mask levels, affecting the cost of embedded Flash memory.

13.2.2 Programming Voltage

Flash memory has had a history of dramatically increasing density over short time frames, similar to the history of increased density in DRAMs. However, the isolation of nodes in the Flash memory cell for high programming voltages and scaling of the storage transistor with hot-carrier injection limits the size of the memory cell. A typical cross section and a typical layout of a Flash memory cell are shown in Figures 13.1 and 13.2, respectively. As the technology progresses to smaller and smaller design rules, it is clear that simple shrinking of the cell size is not viable. More and more increases in density will be contingent on breakthroughs in technology or design.

The minimum programming voltage in a Flash memory is set by the minimum thickness of the tunnel dielectric. When the tunnel dielectric is too thin, charge tunnels or leaks off the floating gate through the tunnel dielectric and 10-year data retention cannot be met. Most Flash memories use hot-carrier injection through a thin dielectric onto the floating gate for programming. Approximately 20,000 electrons are stored on the floating gate for the 0.25-μm technology node. If the margin of the sense amplifier allows the loss of 20% of the electrons over 10 years, then

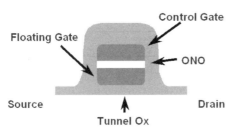

Figure 13.1. Typical cross section of flash memory cell.

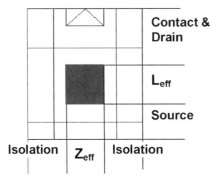

Figure 13.2. Typical layout of flash memory cell.

the allowable leakage current through the tunnel dielectric is approximately 10^{-15} A/cm^2. This leakage corresponds to approximately one electron per day, a convenient number to remember.

The Flash memory leakage current requirement is far more stringent than that for DRAMs. A typical DRAM can tolerate a loss of approximately 10^8 electrons per second. Therefore, the leakage current requirement on a Flash memory tunnel dielectric is approximately 13 orders of magnitude more stringent than that for a DRAM!

Despite years of research and development, the minimum tunnel dielectric thickness for floating-gate memory cells using SiO_2 as the tunnel dielectric continues to be limited to approximately 8 nm. This limitation is due to Fowler–Nordheim tunneling current on the intrinsic main population of memory cells, not freak leakage of defective memory cells. In fact, optimizing the fabrication process has all but eliminated the occurrence of freak memory cells in today's Flash memories. The thickness of the tunnel oxide for the memory cells is larger than that for logic transistors by almost a factor of 2 at 0.18-μm design rules.

If the amount of charge stored on the floating gate corresponds to a potential of 2 V, limiting the tunnel current to less than 10^{-15} A/cm^2 requires tunnel oxides greater than 6 nm. However, the tunnel current increases as a result of charge injection stress due to repeated programming, as shown in Figure 13.3 [1]. Calculations and experimental data (Fig. 13.4) show that a minimum tunnel oxide thickness of approximately 8 nm is required to obtain a tunnel current of less than 10^{-15} A/cm^2 after typical charge injection stress in a practical Flash memory transistor.

For a tunnel oxide of 8 nm and a coupling ratio of 0.6, a programming voltage of at least 10 V on the control gate is required to charge the floating gate of a stacked-gate Flash memory cell in a programming time of a few milliseconds or less.

LIMITATIONS OF FLASH MEMORY

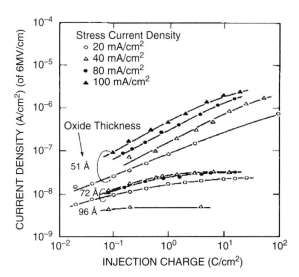

Figure 13.3. Stress-induced leakage current at 6 MV/cm as a function of injection charge passed through oxides of three different thicknesses [1].

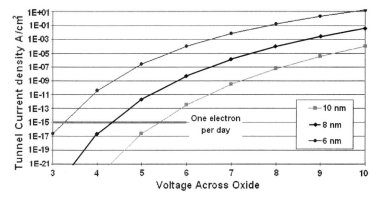

Figure 13.4. Fowler–Nordheim current for three different oxide thicknesses.

The programming voltage must be applied to the memory cell, decoders, and any other circuitry carrying the programming voltage. Sufficient voltage is required on the drain to generate hot electrons with energy greater than 3.2 eV to inject over the Si/SiO_2 barrier. In a typical case, only approximately one in 10^6 electrons in the channel achieves sufficient energy to be injected, but this is sufficient to program the floating gate.

As a result of improvements in technology and the resulting control of the write/erase characteristics of Flash memory to provide tighter transistor threshold voltage distributions, it has become possible to store more than one bit of information in each Flash memory cell. This is referred to as multilevel cell (MLC) technology. For example, if 4 discrete threshold voltage levels can reproducibly be sensed, then 2 bits of information can be stored in a single memory cell, as shown in Figure 13.5. Figures 13.6 and 13.7 show the charge placement for a Flash memory using 3 and 4 bits per memory cell, respectively. Sensing the required 8 voltage levels for 3 bits per cell

622 13. ALTERNATIVE MEMORY TECHNOLOGIES

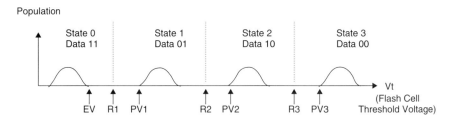

Figure 13.5. Sensing of four discrete threshold voltage levels to store two logical bits per Flash memory cell.

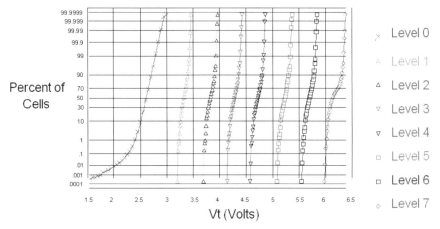

Figure 13.6. Sensing of eight discrete threshold voltage levels to store three logical bits per Flash memory cell.

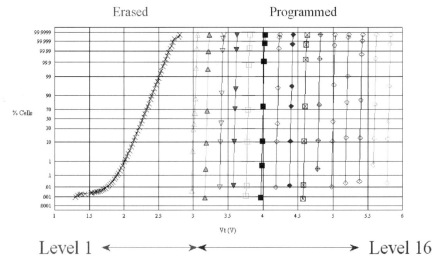

Figure 13.7. Sensing of 16 discrete voltage levels to store 4 logical bits per Flash memory cell.

requires a voltage resolution better than 500 mV. Sensing 16 discrete levels for 4 bits per cell requires a threshold voltage distribution control of less than 100 mV, a difficult technical challenge. With 4 logic bits per physical memory cell, Flash memory chips may well be capable of reaching hundreds of gigabits in density.

13.2.3 Programming Speed

A disadvantage of Flash memory as compared to volatile memory is the slow programming speed. The write and erase speeds of a Flash memory are controlled by the dynamics of charge injection or removal from the floating gate. For typical Flash memories, the fastest write speeds obtained are of the order of 10 μs to 2 ms. These are very slow speeds compared to volatile memories such as static random-access memory (SRAM) and DRAM where write speeds are shorter than 100 ns and, in the case of SRAMs, can be as short as a few nanoseconds.

The slow erase speeds in Flash memories are mitigated by erasing entire blocks of the memory at once. This is one origin of the word "Flash" since the memories erase in a flash compared to bit-by-bit or byte-by-byte erasure. The block or sector erase feature of Flash memories limits the architecture of the system in which the Flash memory is used, but the cost per bit is low.

13.2.4 Endurance

Endurance of Flash memories is limited by intrinsic charging of the tunnel oxides and damage to the tunnel oxides due to repeated charge injection through them. Over the last 30 years, endurance levels have improved from less than 10^3 write/erase cycles to 10^5 to 10^6 write/erase cycles with failure rates acceptable to the customer. This has been accomplished by improved process technology for fabricating the tunnel oxide and surrounding structures. However, in the last several years, little additional improvement has been made in the endurance capability of Flash memories.

At write/erase cycles higher than those specified, either poor programming (and hence, limited retention) due to charge trapping in the tunnel dielectric or unacceptable failure rates due to rupture of the tunnel dielectric makes operation unreliable. Neither of these mechanisms can be screened to very low failure levels using a nondestructive electrical test at the beginning of the life of the memory.

Because there is a correlation between endurance level and failure rate, the endurance characteristics also limit the reliability of Flash memories. Therefore, the failure rate of Flash memories may be higher than that for other commodity memories, such as DRAM and SRAM.

13.2.5 Scaling

Cross sections of several generations of Flash memory cells produced by Intel are shown in Figure 13.8. Scaling projections for Flash memory cells as a function of the minimum lithographic dimension are shown in Figure 13.9. Because of the high programming voltages, the memory cell active area has only scaled slowly and is typically not limited by lithography. The effective channel length is limited by hot-electron operation, punch-through voltage, and leakage issues. The effective channel width is limited by read current requirements and isolation encroachment of the

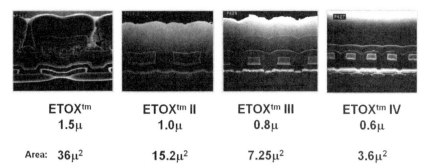

ETOX™ 1.5μ   ETOX™ II 1.0μ   ETOX™ III 0.8μ   ETOX™ IV 0.6μ

Area: 36μ²   15.2μ²   7.25μ²   3.6μ²

Figure 13.8. SEM cross sections of Flash memory cells for four generations of Flash technology.

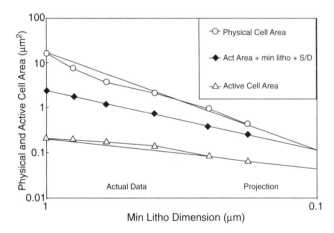

Figure 13.9. Physical and active flash memory cell area as a function of lithography.

field oxide. The inactive area of the cell is a high percentage of the cell area and may be reduced by improvements in the lithographic resolution and alignment. The inactive area can also be reduced by process innovations such as trench isolation and self-aligned sources. Design innovations, such as virtual ground arrays, can also reduce the inactive area by eliminating the need for contact windows in the memory array.

Based on historical trends, the physical memory cell size of a Flash memory is expected to be $10F^2$ where F is the technology node minimum feature size. At the 2007 65-nm node, the NOR cell would be about $0.042\,\mu m^2$.

13.3 NROM MEMORIES

Eli Lusky, Assaf Shappir, Guy Cohen, Ilan Bloom, Meir Janai, Eduardo Maayan, Oleg Dadashev, Yoram Betser, Yan Polansky, Shai Eisen, Yair Sofer, and Boaz Eitan

13.3.1 Introduction

The nonvolatile memory industry is continuously striving to achieve an improved memory cell in terms of size, memory density, performance, reliability, and cost. The industry standard cell is based on charge storage in a conductive floating gate of a metal–oxide–semiconductor (MOS) transistor. Though this basic concept has gone through many optimization stages in the course of its evolution, no single floating-gate-based cell technology supports the full range of applications, namely data (NAND), code (NOR), and embedded Flash.

NAND applications are conceived as the most appealing segment with an overwhelming yearly growth over the past few years. Recently, DRAM has been beaten to production at the leading-edge lithography feature size by the NAND technology. Nonetheless, it is speculated that scaling NAND technology beyond the 50-nm node is very difficult and charge-trapping technology is considered as the most probable alternative solution [2, 3].

Saifun NROM is a unique localized charge-trapping-based nonvolatile memory technology that employs two separate physical charge packets, realizing 2 and 4 bits per cell [4, 5]. NROM technology is able to provide data, code, and embedded Flash with a single low-cost fabrication process and minor architectural adjustments [6–9]. Scaling is projected to follow complementary MOS (CMOS) planar scaling path with no foreseen roadblock as a 20-nm channel length node was recently demonstrated [10].

Saifun NROM technology was the first to realize 4 bits per cell in a product [5], addressing the ever-growing demand for high-density and low-cost data applications. In a virtual ground array, the cell size is ~$5F^2$ or ~$1.25 F^2$/bit in a 4-bit product. Using a bit size of $0.021\,\mu m^2$ in 130-nm technology, an 85-mm^2 1-Gb data Flash product was realized [8]. Usage of short program pulses (150 ns), erase pulses (100 μs), and advanced algorithms, enable NROM to meet the very demanding low power consumption and high write rate requirements of data Flash products.

The NROM storage medium consists of cladding oxide layers and a nitride layer, forming an oxide–nitride-oxide (ONO) stack. Programming is performed by channel-hot-electron (CHE) injection and erase by band-to-band tunneling (BTBT) induced hot-hole injection (HHI). Read is performed by interchanging the role of the source and the drain [4]. Due to the localized charge storage property, the technology is immune to oxide defects and single bit failures, thus allowing bottom-oxide scaling down to the direct tunneling limits, ~3 nm; significantly thinner than the projected ultimate scaling limit of ~6 nm in FG technology [11]. The equivalent coupling ratio is 1, which is significantly larger than the counterpart FG technology. This improves CHE programming performance to 100 ns and helps in reducing the cell size. Due to the single poly process, WL–WL coupling limitations associated with floating-gate technology are removed in NROM technology.

13.3.2 Memory Cell and Array; Structure and Operation

13.3.2.1 Cell Structure and Operation.
As illustrated in Figure 13.10 the NROM cell is an NMOS field-effect transistor (FET) with ONO stacked dielectrics with a typical oxide equivalent thickness of ~16 to 20 nm used as the gate dielectric. Bottom-oxide thickness is ~3 to 7 nm, nitride is ~3 to 6 nm, and the top-oxide is ~8 to 12 nm. Poly gate and buried N$^+$ diffusion form wordlines (WLs) and bitlines

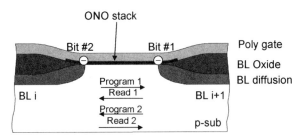

Figure 13.10. NROM cell: cross section along a wordline. The electron storage regions are indicated by arrows and will be referred to as bit 1 and bit 2.

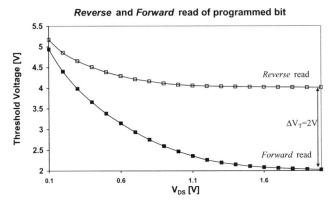

Figure 13.11. Read characteristics under "reverse" and "forward" read modes.

(BLs), respectively. To isolate between the WLs and the BLs various schemes are used, for example, continuous ONO, thermal oxide, and low-temperature deposited oxide.

Programming is performed by CHE and erased by BTBT-induced HHI. Read is performed in the "reverse" direction by applying biases to the gate (~4 to 5 V), drain (1.2 to 1.6 V), and grounding the source and the substrate. Note that when programming bit 1 as illustrated in Figure 13.10, BL $i + 1$ is the drain and BL i is the source. During the read operation, BL i is the drain and BL $i + 1$ is the source. The localization of the trapped charge next to the N$^+$ junctions is critical to separate the two physical bits. One bit does not affect the information of the other bit due to the narrow storage region [12]. It allows the circuit to "read-through" the trapped charge region, even at low V_{DS} voltages.

A demonstration of the unique read concept is shown in Figure 13.11 for a case where single bit is programmed to $\Delta V_T = 2$ V under $V_{DS} = 1.6$ V reverse read conditions. In the figure, the threshold voltage of the programmed bit is shown both in the "reverse" and "forward" read modes as a function of the drain to source voltage, V_{DS}. Under reverse read operation, ΔV_T reduces as a function of the V_{DS} level due to "localized" drain-induced barrier lowering effect [13]. For $V_{DS} > 1.5$ V, the barrier lowering effect saturates and the threshold voltage becomes constant.

Under forward read conditions where the role of the drain and the source is identical to the programming operation, the effect of the drain-to-source voltage is crucial. Only $V_{DS} > 1.3$ V conditions assure that the programmed bit does not affect the unprogrammed bit. Note, for low V_{DS} conditions (<1 V), decoupling between bit 1 and 2 is degraded.

Programming is carried out using gate voltage (V_G) ~7 to 9 V, drain voltage (V_D) ~4 to 5 V and source and substrate grounded. The accelerated electrons are injected into the ONO, self-aligned with the junction edge. The programming time is <150 ns, and the programming current is ~90 µA in 90-nm and ~60 µA in 63-nm technologies. Typical programming characteristics are shown in Figure 13.12 where the ΔV_T is shown for consecutive programming pulses, each of 150 ns where the programming voltages are $V_G = 9$ V and the drain voltage increases by 0.2-V steps, each pulse.

The combination of fast programming characteristics, multiple-bit parallel programming, and smart algorithms in NROM data flash [6] results in an 8-MB/s programming rate, hence, addressing the very demanding programming rate requirements in data Flash applications. The program algorithm relies on the fast and uniform program characteristics and enables completion of the program operation within 2 pulses of <150 ns each. Programming of the NROM cells is performed with a stable wordline and with a voltage pulse applied to the bitline. The pulse shape at each drain terminal being programmed is required to have no overshoots, to have very fast rise and fall times (<50 ns both), and to be insensitive to the drain pump operating conditions.

Erase involves BTBT-induced hot hole-injection [14–16]. Hole flow is generated by BTBT in the deep depletion layer inside the N^+ junction [Fig. 13.13(a)]. The holes accelerate laterally and high tail energy holes have high probability for injection [Fig. 13.13 (b)]. The tunneling of hot holes through the potential barrier due to the relatively high vertical field improves the erase speed. Typical erase operation requires a positive drain voltage and a negative gate voltage. Both erase and program operations are self-aligned with the junction edge.

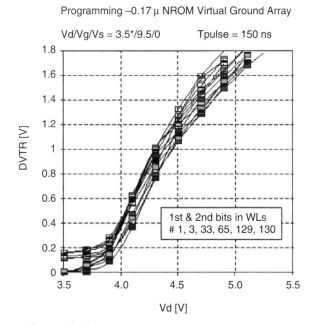

Figure 13.12. NROM programming characteristics [6].

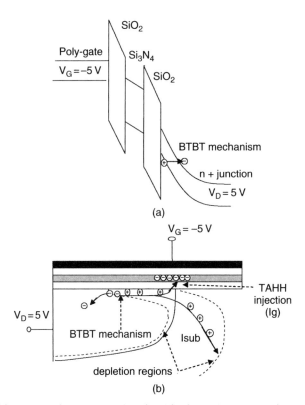

Figure 13.13. (a) BTBT mechanism is initiated inside the N⁺ junction under typical erase conditions of $V_D = 5\,V$, $V_G = -5\,V$, $V_B = 0\,V$, $V_S = F$. (b) The BTBT-generated holes are accelerated toward the channel. The most energetic holes are injected to the nitride, self-aligned to the trapped electrons.

Typical program and erase characteristics of bit 1 and bit 2 are shown in Figure 13.14. The measurements indicate that bit 1 and bit 2 are completely decoupled in all states; when bit 1 is programmed to ΔV_T of ~2 V, no effect is observed on bit 2. The same apply to the bit 1 programmed state when bit 2 is being programmed. Similarly to the program operation, when bit 1 is being erased, no effect is observed on the programmed bit 2, which maintains its high V_T level.

13.3.2.2 Array Architecture. The array architecture is a symmetric high-density virtual ground (VG) array as demonstrated in Figure 13.15 with a WL pitch of $2F$ and BL pitch of 2 to $3F$ depending on the technology. For example, at the 90-nm technology node the cell size can be as small as ~$4.4\,F^2$/cell while in 63-nm technology it is ~$5.6\,F^2$/cell. The VG array is segmented along the bitlines by select (SEL) transistors, creating isolated physical sectors. Based on bitline access load capacitance, voltage drops, sector size, and disturb time considerations, the physical sectors are partitioned. The segment size between two selects is typically 1000 WLs in a data application. A multiple SEL per global metal bitline (GBL) scheme is used to solve the mismatch between the GBLs pitch and the tighter local bitlines pitch

NROM MEMORIES   629

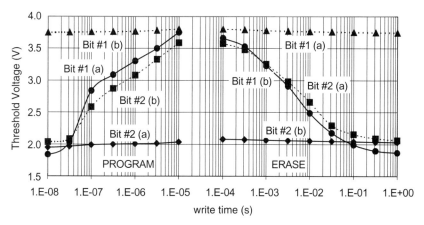

Figure 13.14. Programming curves for bit 1 and bit 2 in a $W/L_{eff} = 0.35\,\mu m/0.32\,\mu m$ device. Curves (a) solid: only bit 1 is programmed, while bit 2 is left in the erased state; biases are $V_{WL} = 9\,V$, $V_{BL1} = 4.5\,V$, $V_{BL2} = 0\,V$. Curves (b)—dashed: bit 1 is left in the programmed state, while bit 2 is programmed; biases are $V_{WL} = 9\,V$, $V_{BL1} = 0\,V$, $V_{BL2} = 4.5\,V$. Erase curves. Curves (a)—dashed: only bit 2 is erased, while bit 1 is left in the programmed state; biases are $V_{WL} = -5\,V$, $V_{BL1} = 5\,V$. Curves (b)—solid: bit 2 is left in the erased state, while bit 1 is erased; biases are $V_{WL} = -5\,V$, $V_{BL2} = 5\,V$.

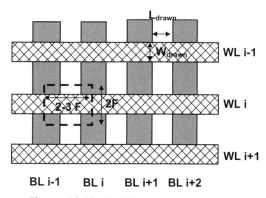

Figure 13.15. NROM virtual ground array.

(Fig. 13.16). The multiple SELs scheme allows reducing the GBLs density, improving yield, and enabling the GBLs width-space trade-off taking into account cross talk and resistance considerations.

One of the important advantages of the VG array is that a BL contact is only needed every 16 or 32 WLs. This relives the process from a very elaborate and complicated contact development, simplifies the process, and improves the yield.

Either source-side, close to ground, or drain-side sensing schemes are used in VG arrays [6, 8], enabling fast, accurate, and power-efficient read access. Sensing is performed with a stable wordline voltage (~4 to 5 V) where the array is initially precharged to ground. Then, ~1.2 to 1.6 V are driven on to the cell bitline(s), and the sensing is performed on the cell source or drain side. Simple latch sense amplifiers are used to compare the signal developed by an array cell to that of a programmed reference cell (REF).

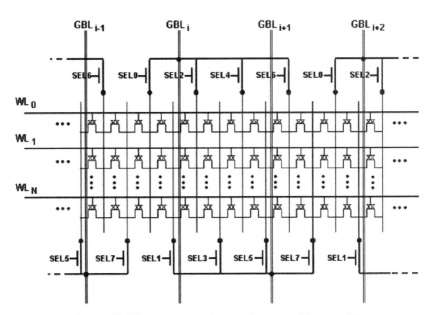

Figure 13.16. NROM virtual ground array architecture [6].

A REF cell program algorithm verifies the REF V_T placement using the existing sense amplifier (SA) in the product. Using a REF per SA compensates for the SA offset, hence, the array cells operate with the same program-erase window regardless of which SA is being used. There is no need to cycle or readjust the REF cells V_T over the product lifetime.

13.3.2.3 Array Disturbs. As a typical NROM memory array is composed of a dense crisscross of BLs and WLs, the minimal cell population that is erased simultaneously may not include all of the WLs and the BLs of the specific physical sector. Hence, continuous programming and erasure of subgroups may cause a shift in the threshold voltages of cells in adjacent groups, up to the point of data content alteration, referred to as BL disturb or gate disturb. Gate disturb can be suppressed significantly by using thinner bottom oxide [17, 18].

The main BL disturb mechanism is unintentional erasure of cells due to a low rate of hole creation at the surface of a highly biased drain junction in a cell with an unbiased gate [19]. This disturb is easily prevented by biasing the gate of cells not intended to be erased to a low positive voltage (wordline inhibit), thus suppressing hole creation. Wordline inhibit can eliminate the first-order BL disturb mechanism. Nonetheless, higher order disturbs may appear, which require additional measures.

Slightly biasing the wordline allows the NROM cell to start conducting current in the subthreshold region. The high erase drain voltage accelerates the small electron current and creates high-energy electrons. These electrons can be injected into the nitride and disturb the erased cell. An impact ionization event of these hot electrons creates hot holes near the drain that can disturb a programmed bit. Subsequently, erased bits may be slowly programmed and programmed bits may be slowly erased—the polarity of the injected carrier will be dependent on the local vertical field near the drain junction.

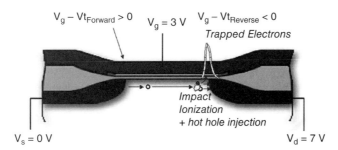

Figure 13.17. Schematic illustration of BL disturb in inhibited erase sectors.

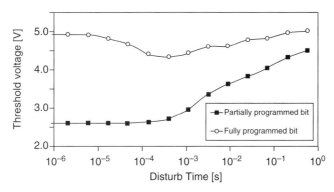

Figure 13.18. BL disturb in inhibited erase sectors in programmed and erased cells [19].

Such a mechanism is schematically illustrated in Figure 13.17, which shows the case of a single-sided programmed cell under a low gate voltage and a high drain voltage. Most of the channel is inverted and allows the conduction of electrons. The high drain voltage punches through the trapped electron-induced potential barrier, allowing current to flow through the cell.

Nevertheless, the ONO field at the drain end of the channel is repulsive to electrons and attractive to holes, causing unintentional erasure. Figure 13.18 demonstrates this disturb mechanism on a programmed NROM cell under direct current (dc) stress conditions. The bit above the drain of the cell was initially programmed, while the adjacent bit has been partially erased. Biasing the gate to a low voltage, 2.7 V, and the drain to a high voltage, 5.5 V, causes the erased bit to start programming after 1 ms of stress. The programmed bit is initially erased due to hot-hole injection, yet changes direction due to the programming of the adjacent bit. The wide electron injection, under these bias conditions, severely hinders the 2-bit separation.

This disturb mechanism can be eliminated by process optimizations, architectural adjustments, and design improvements. A straightforward approach is based on cutting the subthreshold leakage path within the cell, hence eliminating the disturb source. This is readily achievable by applying voltages in a sequence where the source is increased before the erase drain voltage reaches its final high value. This is a very effective method in NROM. Since it is a non-floating-gate technology, there is no source or drain potential coupling into a floating gate that may further influence the device and result in a subthreshold current that could serve as a source for a disturb. Figure 13.19 shows test results from NROM-based memory products

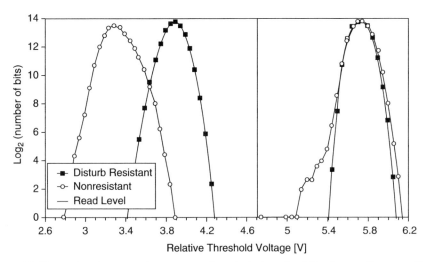

Figure 13.19. V_T distributions of two NROM-based Flash units following cycling on adjacent subgroups along the same bitlines. The unit that does not incorporate bitline resistance optimizations shows a tail of bits extending from the program distribution down to the read level [19].

that demonstrate the ability to withstand program and erase operations on adjacent subgroups along the same bitlines.

13.3.3 Storage Mechanism

Charge-trapping-based nonvolatile memory devices exhibit threshold voltage shifts with time (retention loss) due to the discharge of the storage medium. Such shifts are commonly attributed to direct or damage-assisted tunneling of carriers back to the Si substrate. The charge loss may alter the state of the device and corrupt the stored data.

An attempt has been made to explain threshold voltage drifts in the NROM cell by vertical charge transport [20, 21]. While such models may be attributed to discrete storage node-based nonvolatile memories featuring a tunnel oxide, the relatively thick oxide layers incorporated in NROM devices, >3 nm, inhibit vertical charge transport.

Gate stress experiments [18] have further revealed that thinning down the bottom oxide layer thickness, significantly reduces the threshold voltage shift associated with gate stress and improves the retention loss [20]—a striking contradiction to the projected outcome based on a vertical charge-loss transport.

The unique NROM structure and operating conditions govern the storage characteristics and discharge mechanisms in NROM technology. The relatively thick bottom oxide prevents electron discharge via the bottom oxide layer, and the desired spatial localization of stored charge is impacted by mechanisms that tend to laterally redistribute the charge.

In the following, the nitride trapping levels will be reviewed first [22, 23] followed by addressing the spatial charge distribution in NROM [12, 24–30]. Finally, the retention loss after cycling characteristics will be reviewed [31, 32].

13.3.3.1 Nitride Trapping Characteristics.

A spectroscopy method for analysis of the nitride layer using the NROM cell and the gate-induced drain leakage (GIDL) measurement was proposed recently [22, 23]. The proposed method allows probing of both electron and hole traps in the entire bandgap with almost no fitting parameters. The energy levels of occupied charge traps are extracted following a thermionic emission model.

It was found that the peak energy distribution of the electron traps is located ~2.2 eV below the nitride conduction band with a full-width at half-maximum (FWHM) of 0.16 eV, while the peak energy distribution of the hole traps is located ~1.5 eV above the nitride valence band with an FWHM of 0.64 eV.

The extraction of the energy levels follows a thermionic charge emission that is responsible for the lateral redistribution of the localized trapped charge. Under the influence of the internal field, the charges drift away and are retrapped in remote sites. The change in GIDL threshold voltage following the emission process $[\Delta V'_{BTB}(t, T, \Delta V_{BTB})]$ is assumed to reflect the charge concentration decay at the center of the spatial distribution. The drift-diffusion conductance in the nitride layer is not considered as the rate limiter for the charge redistribution process. The emission time is assumed to follow Boltzmann statistics:

$$\tau = \tau_0 \exp\left(\frac{\phi_\tau}{kT}\right) \qquad (13.1)$$

where ϕ_τ is the energy level of the trap, τ_0 is the reciprocal of the "attempt-to-escape" frequency (~10^{-13} s), and T is the temperature. Associating $\Delta V'_{BTB}(t, T, \Delta V_{BTB})$ with the corresponding energy level according to Eq. (13.1) allows one to characterize the occupied traps; $\Delta V'_{BTB}(t, T, \Delta V_{BTB})/\Delta V_{BTB}$ is the normalized retention loss, a measure of the relative portion of the filled traps that are depleted during the measurement time. This portion, $f(\phi_t)$, is a function of the trap's energy level. It is defined as the trap occupancy function:

$$f(\phi_t) = \frac{1}{N(t=0)} \int_0^{\phi_t} n_0(\phi'_t) \, d\phi'_t \qquad (13.2)$$

where $N(t = 0)$ is the total occupied trap concentration at time $t = 0$ and $n_0(\phi'_t) d\phi'_t$ is the initially occupied trap concentration that is characterized by energy levels between ϕ'_t and $\phi'_t + d\phi'_t$. This mechanism is depicted in Figure 13.20, where the electrons case is illustrated.

Figure 13.21 summarizes the nitride trap distributions as derived by the methods described above. The peak electron distribution is centered at ~2.2 eV below the conduction band with FWHM of 0.16 eV while the peak hole distribution is centered at ~1.51 eV above the valence band with FWHM of 0.64 eV.

The outcome of this work provides a qualitative explanation of the nature of the retention loss characteristics in an NROM cell after extensive injection and trapping of holes and electrons. The relatively shallower trapping centers for holes indicates that hole redistribution in the nitride layer determines the retention loss characteristics.

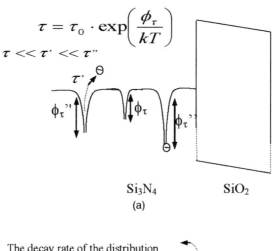

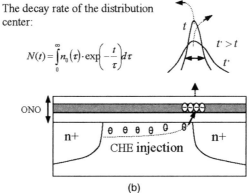

Figure 13.20. (a) The electron traps are distributed over the nitride bandgap. The emission time constants, $\tau \ll \tau' \ll \tau''$ exponentially depend on the various energy levels, $\phi_\tau < \phi'_\tau < \phi''_\tau$. (b) The decay of the trapped charge at the distribution center is thermal emission controlled [22].

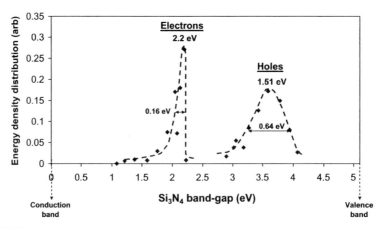

Figure 13.21. Energy trapping levels distribution within the nitride layer for electrons and holes [23].

13.3.3.2 Spatial Distribution of Localized Charge.

Due to the trapping characteristics of the ONO stacked dielectrics, the charge is mostly trapped in the exact physical location of injection. The narrow spatial distribution of the charge is very crucial for the two-bit capability of the NROM device. Narrow distribution width enables decoupling between the two-programmed bits and allows use of the reverse read method as demonstrated in the Figure 13.22 cell concept B approach where negligible cross talk is maintained down to L_{eff} ~100 nm [32]. According to the work presented in the literature [12, 24–30], the charge distribution is narrow, ~10 to 15 nm over the channel region.

An estimation of the distribution width is achieved by fitting two-dimensional simulated I–V's to the measurements. Subthreshold and GIDL characteristics are mostly used. Both methods are sensitive to localized charge above the N^+ junction and the channel region next to it. The fringing fields associated with the localized trapped charge govern the subthreshold and the GIDL I–V shift and slope.

Following the above procedure, the electron and hole distributions that account for the program and erased states are extracted as outlined in Figure 13.23(a). The

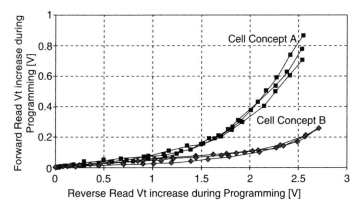

Figure 13.22. Increase of the forward and reverse threshold voltages during a single sided programming of NROM cells from two cell concepts, showing the 2-bit cross talk. Cell concept B is clearly superior [5].

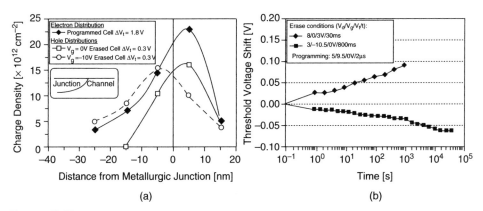

Figure 13.23. (a) Extracted trapped electron and hole distributions following charge injection. (b) Threshold voltage drifts with time as a function of two bias conditions.

number of trapped electrons in 0.35-um technology is estimated as ~1700, while in 63-nm node technology the number is scaled to ~200 to 300.

The electron and hole distributions described above account for the erase-state threshold voltage shift as being a manifestation of lateral charge redistribution within the nitride layer [24]. The erase-state threshold voltage drifts that correspond to Figure 13.22(a) are shown in Figure 13.22(b) for two different biasing conditions (same program erase window). The spatial location of the holes relative to the trapped electrons determines the rate and the polarity of the lateral hole redistribution process and the associated threshold voltage drift—an unexplained phenomena in terms of the vertical model.

13.3.3.3 Retention Loss after Cycling. As shown in Figure 13.21 the electron trapping levels are significantly deeper than those of the holes, and as a result they are relatively immobile at elevated temperatures [22, 23]; see Figure 13.24.

Hence, NROM cell retention loss after cycling is governed by hole redistribution within the nitride layer through a lateral Poole–Frenkel conduction mechanism [24]. This model has been shown to agree well with measurements and to explain findings unaccounted for by vertical tunneling models.

Following extensive program and erase cycling, holes may accumulate above the N⁺ junction of the NROM cell. This is reflected in Figure 13.25, which shows that while the threshold voltage operating window is kept constant, the gate voltage (V_{BTB}) necessary to generate 40-pA GIDL current shifts to lower values as cycling progresses. This indicates that a large amount of holes have been trapped above the N⁺ junction, while electrons may still reside above the transistor channel. Charge recombination, following lateral hole redistribution, will cause the transistor threshold voltage to decrease.

The proposed origin of this phenomenon is the increasing spatial misalignment between electrons and holes injected into the dielectric stack during the program and erase steps, respectively. Charge redistribution may proceed via trap-to-trap

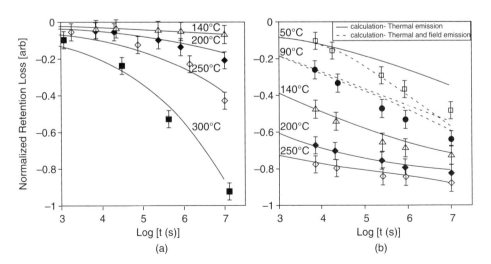

Figure 13.24. Normalized retention loss [$\Delta V'_{BTB}(t, T, \Delta V_{BTB})/\Delta V_{BTB}$] of a electrons (a) and holes (b) [23].

NROM MEMORIES

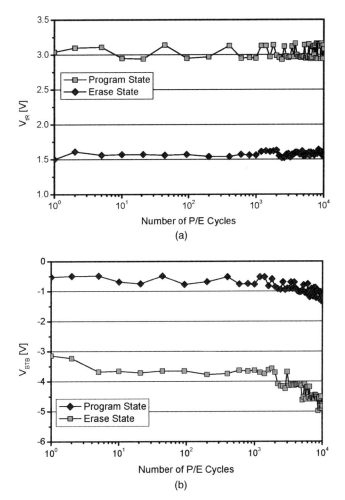

Figure 13.25. (a) Threshold voltage-operating window and (b) the gate voltage (V_{BTB}) necessary to generate a 40-pA GIDL current as a function of cycling [31].

tunneling (hopping conductance), trap-to-band tunneling (Fowler–Nordheim tunneling), or thermionic emission (Poole–Frenkel conduction). At large electric fields and high temperatures the latter dominates, hence the retention loss may be approximated by:

$$\Delta V_T(t, T) \approx \int C \int J_{PF}[E(t'), \phi, T] \, d\phi \, dt' \qquad (13.3)$$

Numerical simulations were performed based on the above model. Figure 13.26 shows the good agreement between the simulations and the measurements. Redistribution and recombination of the trapped carriers procure a decay of the electric field. Therefore, the retention curves are stretched and exhibit a fixed saturation level. The single saturation retention loss that all retention results are converging to is key to the NROM reliability. Note that similar fitting to product results can be achieved at elevated temperature (>150°C) by using only hole energy

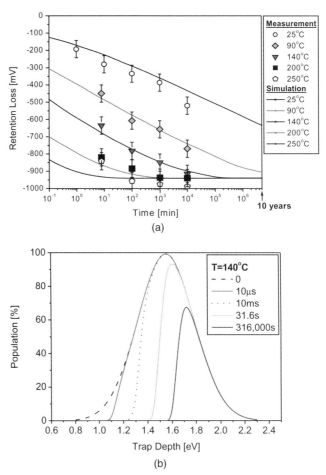

Figure 13.26. (a) Simulations vs. measurements of the retention loss as a function of time at different temperatures. (b) Associated simulation of hole traps population [32].

distribution and associated thermionic emission time constants, regardless of the Poole–Frenkel mechanism [31].

13.3.4 Reliability

NROM technology has an inherent immunity to failure due to point defects in the gate dielectric layer. Electrons (or holes) are trapped individually and loss of one charge through a defect does not affect other trapped charges that are not directly over the point defect. This contrasts to floating-gate technology where the charge is stored in a conductor over the channel, and any point defect in the floating-gate insulator may cause draining of the entire charge of the cell. NROM product reliability is achieved via an integrated solution including technology optimization, special programming and erase algorithms, a sensing scheme that incorporates error detection and correction, and a reliability model to accurately predict the lifetime of the product. Endurance limitations will be reviewed first followed by product retention loss characteristics. Then, the concept of window sensing is discussed and the path for optimization is outlined. Finally, a reliability model for lifetime prediction is presented.

13.3.4.1 Endurance. Endurance and retention in NROM technology are coupled. While no real limitation exists on the endurance capability below 10^5 cycling, the retention loss is strongly correlated to the number of cycles. Due to the inherent oxide defects immunity, the endurance is limited mainly by the deterioration of the erase voltages during cycling.

Endurance is not limited by material degradation but rather by the capability to inject the required amount of charge into the nitride layer. Though the effect of hot-hole injection has been widely reported in the literature to degrade the oxide integrity in floating-gate devices [30], the susceptibility to this degradation in NROM technology is not as severe due to the trapping characteristics of the nitride layer. Accordingly, cells can be cycled very easily to more than 10^5 cycles [33].

Typical cycling results are demonstrated in Figures 13.27 through 13.29 utilizing 0.35-µm EEPROM (electrically erasable programmable read-only memory) products. The operating window (Fig. 13.27) as a function of cycle number is essentially a constant. Figure 13.28 shows the program and erase voltages before and after cycling, respectively. The programming voltages (metal BL voltages) decrease during the 10^5 cycling by ~0.6 V while the erase voltages increase by ~1 V. The increase of the required erase voltage is the endurance limiter in NROM technology.

The physical mechanism responsible for the above characteristics is the misalignment between electron and hole distributions. This mismatch is increased over cycling and results in reduced programming voltages and increased erase voltages. The accumulation of uncanceled electrons over the channel induces enhanced erase voltages, and an associated number of holes are required to be injected over the N^+ junction region to compensate for the larger number of electrons over the channel. Subsequently, program voltage also decreases. The dynamic of charge injection during cycling determines the erase degradation rate, which ultimately establishes the endurance capability.

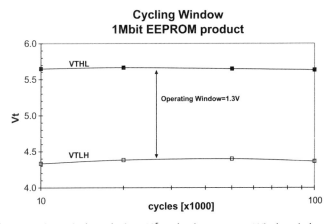

Figure 13.27. Operating window during 10^5 cycles is constant. Window is kept fixed at 1.3 V during cycling by the program and erase algorithms.

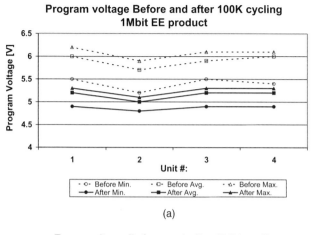

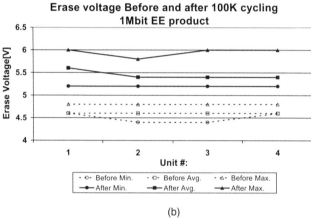

Figure 13.28. (a) Program voltages before and after 100,000 cycles in 0.35-µm EEPROM product. (b) Erase voltage before and after 100,000 cycles in 1-Mbit EEPROM product.

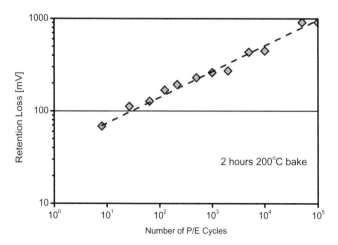

Figure 13.29. Retention loss of NROM-based EEPROM products, following a 200°C bake, as a function of the number of P/E cycles [32].

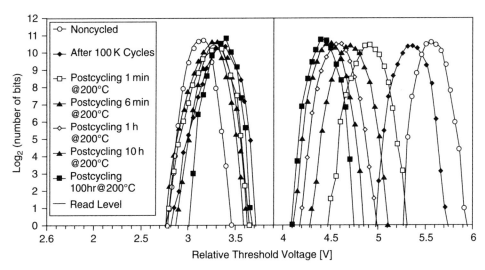

Figure 13.30. Threshold voltage distribution of a typical NROM-based EEPROM unit, as a function of the unit history. The unit was cycled 10 times at 105°C and subsequently administered to a 100-h bake at 200°C [19].

13.3.4.2 Retention Loss. The retention characteristics are determined by both the trapping levels in the nitride layer and by the interaction between electrons and holes as reflected in the correlation between retention loss and the number of cycles shown in Figure 13.29.

Figure 13.30 shows threshold voltage distributions relative to a fixed reference voltage level of a typical NROM EEPROM during 200°C retention bake, following 10^5 cycles at 105°C. The lowest bits in the programmed distribution appear to have reached their saturation level. Hence, data retention is ensured under any practical conditions and timescales. The distribution of the cycled cells is shown to shift rigidly. No irregular bits appear at the tail of the distribution, neither after cycling nor under the thermal stress.

A wide margin remains between the program and erased distributions, demonstrating the high reliability and robustness of the technology. The saturation of the programmed distribution is slightly faster for the low tail than the high tail. For the erased bits, we see two trends; first a slight reduction followed by a slow increase in the V_T of the erased state. In the checkerboard pattern, V_T movement has two sources. The first trend is downward, due to the retention loss of the programmed neighbor (in the same cell). The second trend is upward, due to its own erased state retention hole redistribution. This trend can be eliminated simply by thinning down the bottom oxide thickness [18].

13.3.4.3 Window Sensing. NROM products can achieve excellent reliability when implementing a fixed margin scheme, that is, predefined margins between the read level and the program verify and erase verify levels. The erase margin—the distance between the read and erase verify level—is commonly set to 500 mV to compensate for the second bit effect (Fig. 13.22), any residual erase-state threshold

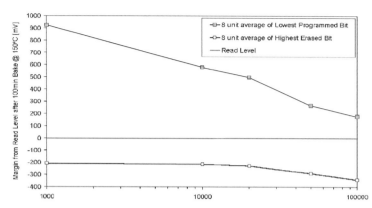

Figure 13.31. Retention loss following 100-min 150°C bake, as a function of the number of cycles. The upper line denotes the lowest bit in the programmed distribution, while the lower line denotes the highest bit in the erased distribution [19].

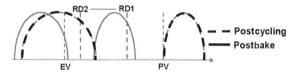

Figure 13.32. Schematic descriptions of the adjustable read level approach.

voltage drift and for design deficiencies. The program margin—the distance between the read and program verify level—is commonly set above 1100 mV, mostly to compensate for retention after cycling.

It is clear that such a fixed margin concept is not optimal, as a large retention loss is only expected near the products' end of life. Figure 13.31 shows the retention loss of both the program and erase state as a function of the number of program and erase cycles performed on the NROM EEPROM device.

However, once the retention loss of the programmed bits becomes excessive and their margin to the read level becomes small, the erased bits also show a downward shift and a wide differentiating window remains. By incorporating an internal error detection mechanism, the read level can be set in real time to the optimal location between the programmed and erased cell distributions [5, 8, 34], as demonstrated schematically in Figure 13.32. Hence sufficient margins from the read level can be maintained even when substantial threshold voltage shifts occur, guaranteeing data integrity.

The single most important feature of the technology that enables the implementation of the above-mentioned scheme is the absence of single-bit failures. In Figure 13.33, a V_T distribution from a 256-Mb product is shown after cycling and retention. The entire distribution moves rigidly, with no single-bit failures. This enables the use of a simple error detection scheme with a moving reference to fix any retention loss, as long as a window remains. This single-bit failure-less error detection scheme translates to an on-chip simple error correction scheme for any number of failures.

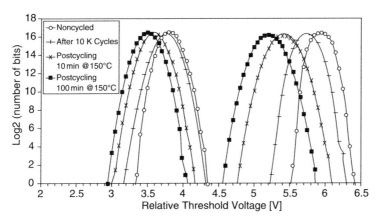

Figure 13.33. Threshold voltage distribution of a typical NROM-based 256-Mb data Flash unit, following 10^4 cycles and a subsequent bake at 150°C. The distribution shifts rigidly and a 500-mV differentiating window remains between the programmed and erased states [34].

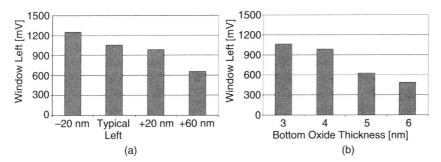

Figure 13.34. Window left following 10^4 cycles and bake at 200°C for 2h with the following parameters: (a) L_{eff} and (b) bottom oxide thickness.

It is possible to optimize the cell for data retention so that any available window is exploited. Suppression of the electrical cross talk between the two sides of the NROM cell (second bit effect) is a key topic, as shown in Figure 13.22 in 100-nm cell length regime. In Figure 13.34(a) the window left (margin between an erased bit and a programmed bit in the same cell) as a function of L_{eff} is shown. By scaling L_{eff}, the window left after cycling and bake is only reduced by 20% (from 1500 to 1200 mV). The effect of the bottom oxide on the window left after cycling and bake is shown in Figure 13.34(b). Thinning down the bottom oxide thickness improves the cell's reliability and fits well with scaling trends.

13.3.4.4 Reliability Model. In Section 13.3.3 the physical mechanisms governing the NROM cell characteristics were presented. Modeling of the erase-state and program-state retention loss was shown to account for the measurements with excellent fitting. However, such a modeling is numerically based and is not suitable for product lifetime prediction.

A more simplified and practical approach to model the program state retention loss utilizes a stretched exponent function to fit the retention curves [35]. The degradation rate of the threshold voltage of cycled cells is shown to be a multiplication of three functions: (1) bit density, (2) endurance, and (3) storage time and temperature. The retention loss is interpreted in terms of thermally activated lateral migration of trapped holes in the ONO layer. Saturation of the retention loss is demonstrated at threshold voltage levels well above the neutral state of the device. From the retention loss function a time-to-failure formula is derived. The retention loss is modeled as:

$$R(n,t,T,c) = -Z(n)Y(t,T)F(c) \tag{13.4}$$

where Z is a statistical function depending on the number of bits n, Y is a function of time and T temperature that varies from 0 to 1 as t varies from 0 to infinity, and F is a function of the number of cycles c endured by the product prior to storage bake. For a given value of n, F determines the saturation value of the retention loss function as $t \to \infty$.

The V_T distribution of noncycled bits is highly stable, fixed at the level of the initial programming. In contrast, the distribution of the cycled bits shifts nearly rigidly with bake time. The shape of the V_T distribution as demonstrated in Figures 13.30 and 13.33 indicates a nearly perfect normal distribution.

Examination of a variety of units, for different products and for different bit counts, cycle counts, and bake durations showed that during the process of retention loss the distribution expands linearly as retention loss increases, as demonstrated schematically by Figure 13.35. It shows the standard deviation of the V_T distribution after bake, as a function of the peak retention V_TPK. It can be seen that for all products and technologies, σ_f increases linearly with the decrease of V_TPK according to

$$\sigma_f = \sigma_i - b\,\Delta V_T \text{PK} \tag{13.5}$$

where σ_i is the initial standard deviation and b is a technology-dependent coefficient.

The retention loss can be presented as:

$$R = \Delta V_T \text{PK}(t,T,c)Z(n) \tag{13.6}$$

where Z is a function of the bit ensemble, being a Gaussian distribution. ΔV_TPK is a function of t, T, and c. It can be represented as follows:

$$\Delta V_T \text{PK} = -F(c)Y(t,T) \tag{13.7}$$

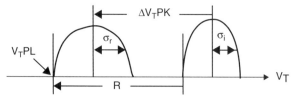

Figure 13.35. Schematic illustration of the relation between peak and tail retention loss and the change of the standard deviation of the V_T distribution with retention [35].

NROM MEMORIES

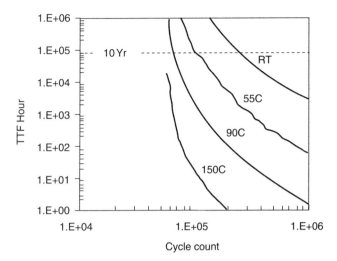

Figure 13.36. Time to failure of product as a function of cycle count for different storage temperatures for allowed retention loss of 1.1 V [35].

where Y is a stretched exponent, depending on time and temperature:

$$Y(t,T) = 1 - \exp[-(t/\tau)^\beta] \qquad (13.8)$$

and F is a power function where A and γ are fitting parameters:

$$F(c) = Ac^\gamma \qquad (13.9)$$

This leads to a simple equation for calculating the time-to-failure (TTF) of NROM devices:

$$R = -Z(n)F(c)(1 - \exp[-(t/\tau)^\beta]) \qquad (13.10)$$

The time to failure can be extracted then:

$$t = \tau(-\ln\{1 + R/[Z(n)F(c)]\})^{1/\beta} \qquad (13.11)$$

Equation (13.11) gives the time to reach retention loss R of an ensemble n of bits after c cycles. Note that R is a negative number. The temperature T is embedded in the parameter τ [35]. Extracted TTF curves are plotted in Figure 13.36.

13.3.5 Quad NROM Technology

The realization of a 4-bit NROM cell [5] is facilitated by having two physically separated bits on each side of the cell. Key features of a 4-bit product are optimized technology, accurate and fast programming algorithm (1 MB/s write speed), no single-bit failures, and window sensing using a moving reference as an error detection and correction scheme.

A schematic NROM cell cross section for 2- and 4-bit cells is shown in Figure 13.37. The inherent advantage of the 4-bit NROM is that only 4 V_T levels on each side of the cell are required. In the floating-gate approach 16 different V_T levels are required for a 4-bit cell, which is practically impossible for current technology.

The foremost requirement of a 4-bit per cell product is the ability to achieve distinguishable threshold voltage levels. The programming algorithm has to incorporate two factors, the cross talk between the two sides of the NROM cell (second bit effect) and high accuracy for multilevel implementation. To overcome this cross talk, the programming algorithm must implement simultaneous 4-bit programming on every cell. The cross talk between the first bit and the second bit was shown in Figure 13.22. The increase in V_T of the second bit must be incorporated into the programming algorithm. A floating-gate multilevel programming concept would not work under the constraint of the second bit requirement. A typical result from a 1-Gb product is shown in Figure 13.38.

The key programming performance concept has two phases. The first phase provides a fast programming rate using drain stepping to reach a lower V_T than the final target. This is done to all levels on a given page simultaneously. The second phase achieves accurate programming by using gate stepping and an adjusted drain level. This two-phase programming approach is illustrated in Figure 13.39 where the gate stepping method for each of the programmed levels can achieve a 20× improvement in accuracy.

The key to a successful implementation of the 4-bit programming is in the detailed sequence tailored to eliminate any cross talk between the two sides of the cell. The final optimization of this algorithm sequence and parameters is possible only by the flexibility of the embedded microcontroller programmable code.

The erase sequence is very important for performance and placing all the cells into a uniform initial state. The same erase algorithm implemented in 2-bit products

Figure 13.37. Two and 4 bits per NROM cells.

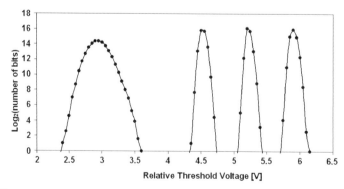

Figure 13.38. Example of measured distributions in a 1-Gb NROM based 4-bit per cell product (130-nm technology generation) [5].

NROM MEMORIES

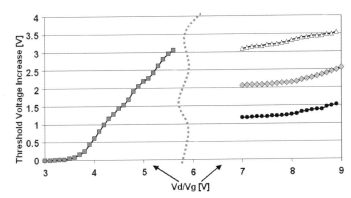

Figure 13.39. Demonstration of the two-phase multilevel programming approach, The cell is first programmed by drain voltage and then switched to gate step programming for improved accuracy [5].

is used in the 4-bit products with minor modifications. A programming speed of 3.5 MB/s is achieved with a page size of 2 kB. The write speed is 3 MB/s and is mainly limited by the erase operation overhead.

An obvious requirement for a 4-bit technology is to have an inherent good retention net window. The improvement in the final window after cycling is demonstrated in Figures 13.33 and 13.34 with shorter L_{eff} and thinner bottom oxide.

The design plays a center role in the reliability, performance, and functionality of the 4-bit product. Sensing scheme accuracy and speed, programming algorithm flexibility, and error detection and correction in page segmentation are main features.

Accurate, error free, read is a major requirement for any memory product. A drain-side alternating current (ac) sensing scheme is utilized, which eliminates the leakage current that may flow through the virtual ground array during the read operation. Current sensing with a very large signal in the SA input contributes to fast and accurate sensing. The cell read signal is then compared to a global dc level by a sense amplifier having an offset cancellation scheme. A schematic diagram of this system is shown in Figure 13.40 [5, 8]. A set of three reference cells is used to readout in parallel the 2 bits stored in each accessed storage node (2 per cell)—see Figure 13.41 [8]. A time-domain sensing scheme is adopted to minimize the number of reference cells.

An error detection block of 128 bytes is implemented to reduce the bit statistics effect on reliability. Each of the three programmed levels employs a zero counter to determine the correct placement of the read level. The zero counters are realized with extra bits, operated in 2-bit mode to enhance their reliability. Each of the sensing levels can be adjusted separately to match its window location.

Working with small bit ensembles, as demonstrated in Figure 13.42, guarantees that a sensing window will exist throughout the life span of a multi-giga-bit product. The die size penalty for this simple error detection and correction is 4%. The window sensing with no single-bit failures guarantees that any number of errors can be corrected with no extra overhead. In standard error correction algorithms, more than two corrections per page are not practical! An embedded microcontroller is implemented with its own programmable microcode array. This enables full flexibility in the programming algorithm optimization.

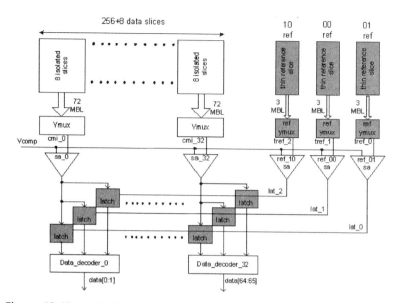

Figure 13.40. Block diagram of the time-domain multilevel sensing scheme [8].

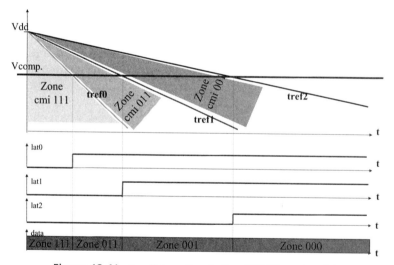

Figure 13.41. Parallel reading in QuadNROM product [8].

The 1-Gb 4-bit/cell product characteristics are in excellent agreement with the above-described work. Figure 13.43 shows histograms of the sensing windows between levels in 64 pages of a typical unit. The programmed pattern of a random population of the 4 levels (bits) is constructed. Large sensing windows remain even after 10^4 program and erase cycles. The apparent reduction in the margins after cycling is a manifestation of the retention loss mechanism that occurs at short time scale.

The importance of the adjustable read level in Quad NROM is demonstrated on a first-generation Quad product in Table 13.2. Following 100 cycles and subsequent bake at 200°C for 2 h, execution of the first read causes complete corruption

NROM MEMORIES

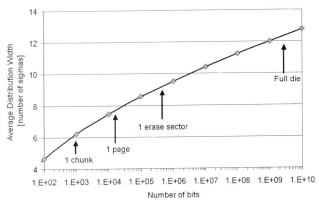

Figure 13.42. Average V_T distribution width as a function of the bit ensemble.

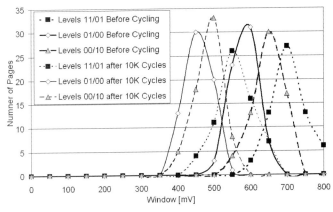

Figure 13.43. Histograms of the sensing windows between levels in 64 pages of a typical 1-Gb 4-bit/cell unit, before and after 10^4 program and erase cycles [5].

TABLE 13.2. Number of Failures as Function of Number of Read Level Adjustments in Different Units

Unit Number	Number of Failures 0 Moves	Number of Failures 3 + 1 Moves	Number of Failures 3 + 3 Moves
1	>100,000	0	0
2	>200,000	2	0
3	>100,000	0	0
4	>1,000,000	10	0
5	>400,000	3	0
6	>800,000	8	0

of the data retention. Adjustment of the read level following the first read operation, significantly reduced the number of failures with a few remaining in each unit. Following two additional reads and associated read-level adjustments fixed all read failures completely.

Figure 13.44 shows the average sensing windows between levels, per page, associated with Table 13.2 results. Additional process and algorithm improvements are underway to achieve reliable data storage after 10^3 program and erase cycles, which will guarantee system-level compliance with all real-world commercial applications.

The Saifun NROM was the first 4-bit/cell product to emerge. The 4-bit reliability concepts are a combination of the inherent physics of the NROM cell and very advanced design concepts. A reliability predictor developed for the 2-bit products fully supports the reliability screening of 4-bit products. The approach is compatible with small cell sizes, and small die sizes for high bit capacity chips are realizable. Certainly product densities of 8 and 16 Gb are practical using 90- and 65-nm technologies.

13.3.6 Fabrication

The NROM technology uses standard silicon process modules without any exotic materials. For most applications, only two or three extra masks are required. The basic process flow is described in Table 13.3. It starts with a conventional self-aligned twin well and LOCOS/STI isolation depending on the technology in use. The ONO stacked dielectric is formed afterwards by thermal oxidation growth of the bottom oxide, CVD nitride deposition, and wet oxidation step of the nitride layer to form the top oxide layer. Following the ONO formation only two extra masks are required:

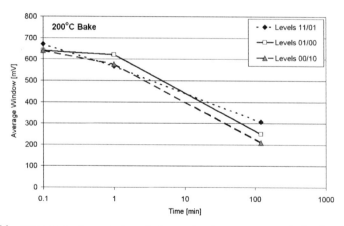

Figure 13.44. Average sensing windows between levels, per page, vs. bake time (200°C) after 100 program and erase cycles, for a typical 1-Gb 4-bit/cell unit. 200°C 2 h represents 10 years of storage at 55°C [5].

TABLE 13.3. NROM Process Flow Outline

Twin-well and STI (LOCOS) isolation
ONO formation
ONO removal from the CMOS region
Gate oxide
BL formation and BL oxide deposition
Poly WL patterning

array protect mask that removes the ONO from the CMOS zone to form the high-voltage (HV) gate oxide followed by bitline mask (poly hard mask) for the buried junctions. Basically, from this stage to the final step of the process the CMOS and array are processed simultaneously with no specific array-oriented extra steps; the BL oxide (~50 nm) [Fig. 13.45(a)] is formed by low-temperature oxide deposition. WL poly [Fig. 13.45 (b)] is then deposited and self-aligned patterned to the poly BL mask.

The formation of the ONO stacked dielectric involves a few issues. Wet oxidation of the nitride layer process results with ammonia that diffuses through the nitride and the bottom oxide and reacts with the Si–SiO$_2$ interface [33]. As a result, the quality of the gate oxide may be affected and should be carefully treated. The wet oxidation of the nitride layer is performed at a high temperature to enable the decomposition and morphology that are required to provide good trapping characteristics [36]. The ONO formation results in a well-defined nitride layer. The top oxide can also be formed by high temperature oxide (HTO) deposition combined with short high thermal treatment.

The above description refers to the basic process flow required to form the NROM array. However, there are various extra steps and masks that may be integrated in order to optimize the cell performance or to align with the product application. An example of such a process flow with extra steps is shown in Figure 13.46. In Figure 13.46(a) the cell implant mask is added to improve the punch-through immunity, and the STI isolation is used both for the CMOS region and the BL-to-BL isolation at the BL contact region, typically implemented every 32 WLs. To enable negative gate voltage conditions in erase, deep N-Well is also implemented. In Figure 13.46(b) and (c) gate oxide and BL implant are implemented sequentially, hence minimizing the BL thermal drive.

In the above example, only high-voltage gate oxide was considered. However, to realize stand-alone applications with $V_{CC} = 1.8\,V$ as well as embedded Flash applications, this requirement is essential. In such a case, a few more additional process steps have to be considered, namely low and medium voltage gate oxides, adjustment V_T implants, and extra NLDD and PLDD implants. Nevertheless, NROM technology is the most adequate technology to realize these applications due to its compatibility with the CMOS process.

To summarize, the simplicity of the manufacturing process results from the following:

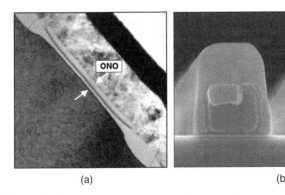

Figure 13.45. NROM SEM cross sections: (a) along WLs and (b) between BLs [6].

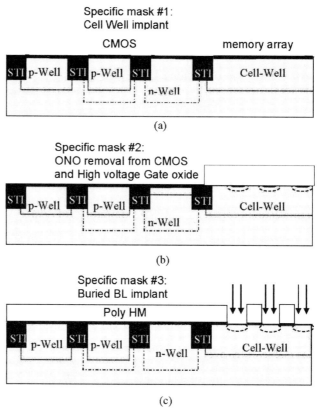

Figure 13.46. NROM process flow: (a) To improve punch-through immunity, cell-well implant is added to the memory array, (b) ONO removal from the CMOS region and gate oxide is formed, and (c) buried BL are defined in the memory array.

1. Only three simple masks are needed to generate the array; fewer masks means an inherently lower defect density.
2. NROM technology relies on the nitride as the charge retaining material, enabling immunity from point defect in the tunnel oxide and scalability down to ~3 nm.
3. No exotic materials are required to relax WL–WL coupling issue in sub-50-nm technology nodes as in floating-gate technology.

13.3.7 Scaling

As the technology generations advance toward smaller feature size, scaling the NROM cell seems relatively simple. Following the standard CMOS scaling methodology, the channel length, width, and doping, junction depth and ONO thickness are scaled down.

The industry state-of-the-art floating-gate technology suffers from severe scaling limitations in the sub-50-nm technology node [2]. WL–WL coupling interference imposes the most important limitation—see Figure 13.47. To maintain acceptable

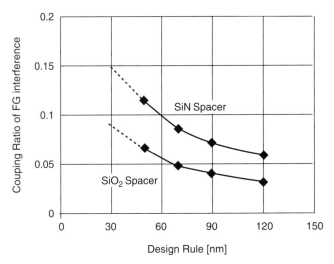

Figure 13.47. Floating-gate interference in oxide and nitride spacers [2].

TABLE 13.4. NROM Scaling Issues

Horizontal Scaling	Voltage Scaling
X direction: bitline width and space	Voltages during program
Y direction: wordline width and space lithography	Voltages during erase
	Voltages during read
	Initial V_t
Vertical Scaling	**Electrical**
Bottom oxide	Punch-through
Storage medium (nitride)	Junction breakdown
Top oxide	Dielectric breakdown
Wordline	Write speeds
Bitline oxide	Program current
	Two-bit separation

interference, the floating-gate poly thickness is scaled causing a reduced control gate (CG) to floating-gate coupling reflected in high operating voltages and associated reliability concerns. Since the ultimate floating-gate tunnel oxide thickness is 6 to 8 nm, scaling of the operating voltage faces a roadblock. The reduced CG–FG coupling ratio and the smaller cell size also imply that the number of stored electrons in sub-50-nm node technology is only ~100; hence the tolerance is a loss of a few electrons only.

In an NROM cell, WL–WL interference is not an issue as the poly WL stack is relatively thin and the WL-to-Si coupling ratio is 1. Furthermore, tunnel oxide can be scaled down to ~3 nm, hence assuring operating voltage scalability. The number of electrons is scaled approximately linearly with the channel width rather than with the cell area and the CG–FG coupling.

The main issues for NROM scaling are described in Table 13.4. They can be broken down into horizontal and vertical two-dimensional scaling, operating voltages, and electrical characteristics. These quantities and constraints determine the actual cell size with a theoretical minimum of $4F^2$. Further scaling may be considered either by using three-dimensional cell structures, three-dimensional array, or alternative denser planar cell approaches.

The motivation to minimize the feature size in the X and Y directions and to reduce the operating voltages, imposes strict limitations upon the electrical characteristics and the vertical dimensions. Special emphasis is put on the punch-through and the drain-induced barrier lowering (DIBL) effects due to the high voltages applied during program and erase operations. As a result, the channel and bulk doping concentration are relatively high, thus reducing the junction breakdown voltages and enforcing the scaling of the operating voltages as well.

The scaling of the operational voltages is also required because of the reduced dielectric breakdown when scaling the ONO thickness. Yet, the physical characteristics of the Si–SiO$_2$ interface, ~3.2 eV for electrons and 3.8 to 4.8 eV for holes, imply that a reduction of the operating voltages is limited due to the channel hot-electron and BTBT-induced hot-hole injection mechanisms. Further scaling of the floating-gate device will require either modification of the physical mechanisms or replacing the bottom oxide with alternative material with reduced potential barriers.

Use of stacked ONO allows scaling of the operating voltages and improvement of the punch-through immunity. The top and bottom oxide layers are limited to ~3 nm to avoid direct tunneling from the nitride layer. The nitride thickness should be determined in accordance with the required retaining characteristics, ~2 to 3 nm. Bottom oxide scalability to less than 3 nm is demonstrated in Figure 13.48 that shows the window left after 10^4 cycles followed by bake at 200°C for 1 h.

To be compatible with the fast programming rate required for data Flash, lower programming current is also a major issue. Linear scaling is achieved as the channel width scales with the technology node, typically being 1 F, the minimum feature size. The program voltage as a function of channel length and the associated improved retention after cycling are shown in Figure 13.49—the shorter effective channel length reduces retention loss after cycling.

Another unique aspect that has to be considered when discussing NROM scaling is the separation between the two physical bits. It has been demonstrated [12, 25–28] that the typical width of the charge distribution above the channel is ~10 to 20 nm. Theoretically, this limits the conventional scaling to L_{eff} ~30 to 60 nm.

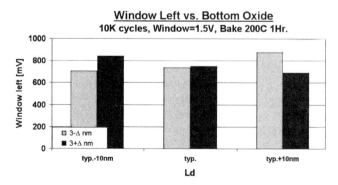

Figure 13.48. Effect of bottom oxide thickness on retention loss after 10^4 cycles followed by 1 h 200°C bake.

NROM MEMORIES

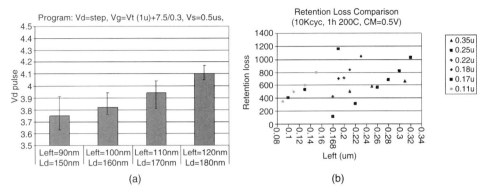

Figure 13.49. (a) Drain program voltage and (b) retention loss after 10^4 cycles after bake as a function of effective channel length in different technologies.

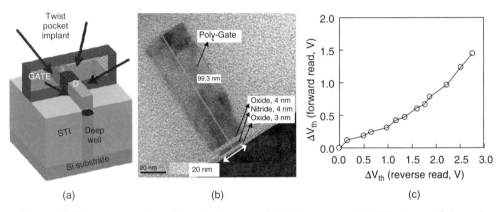

Figure 13.50. A 20-nm channel length FinFET NROM: (a) structure, (b) SEM picture of channel length, and (c) second bit margin loss [10].

Recently reported Fin-FET NROM cell with channel length 20 nm showed reasonable electrical characteristics [10] as demonstrated in Figure 13.50.

In a standard planar cell the second bit as a function of effective channel length is demonstrated in Figure 13.51—good decoupling between both bits is obtained down to 50 nm effective channel length.

The operating voltage of the reverse read conditions is yet another consideration that NROM technology scaling should address. In Figure 13.52 the second bit margin loss as a function of the reverse read voltage is plotted for 80 nm effective channel length—acceptable margin loss is obtained down to V_{DS}~1 V.

13.3.8 Products

NROM technology-based products have been penetrating the nonvolatile market for the past few years. All Flash segments are addressed with minor architecture modifications and simple integrated process. Data Flash—2- and 4-bit/cell [6, 8], code [7] Flash, and embedded Flash are available in the market with the relevant reliability specifications as presented in Table 13.5. The most interesting is the Quad NROM data Flash product, the world's first 4-bit/cell product [5, 8].

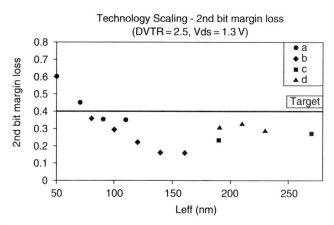

Figure 13.51. Second bit margin loss in planar cell as a function of the effective channel length.

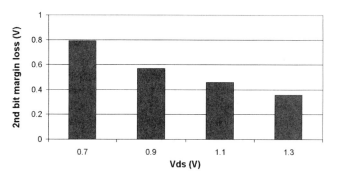

Figure 13.52. Second bit margin loss in planar cell as a function of the reverse read voltage for L_{eff} = 80 nm.

TABLE 13.5. NROM Technology Segments and Reliability Specifications

Endurance/ 10 years at 55°C	Product Data	Quad Data	Code	Serial	Embedded	EE
10,000 cycles/1,000 cycles		√				
100,000 cycles/10,000 cycles	√		√	√		
100,000 cycles/100,000 cycles					√	√

In a virtual ground array the cell size is 4 to $5F^2$ or 1 to $1.25 F^2$/bit in a 4-bit product. Using a bit size of $0.021\,\mu m^2$ in 130-nm technology, a 85-mm² 1-Gb data Flash with 70% array efficiency was realized—see Figure 13.53 [8]. Usage of extremely short program (150 ns) and erase (100 μs) pulses and advanced algorithms enables meeting the very demanding low power consumption and high write rate requirements of data Flash products. The product uses a single 3-V voltage supply. It is typically accessed in blocks of 2 kB for program operation and 128 bytes for read operations with dedicated error detection and correction on chip. First byte

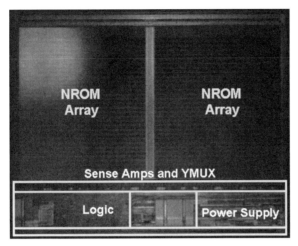

Figure 13.53. A 1-Gb Quad NROM product 85-mm² die size with 70% array efficiency in 130-nm technology node [8].

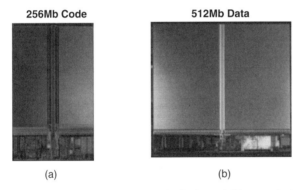

Figure 13.54. Two-bit per cell (a) 256-Mb code flash, and (b) 512-Mb data flash [6, 7].

TABLE 13.6. Quad NROM Technology Roadmap

Technology	2-bit Product	4-bit Product
90 nm	4 Gb	8 Gb
70 nm	8 Gb	16 Gb
50 nm	16 Gb	32 Gb

read access latency is 25 µs and burst read is 20 MB/s. The chip includes an embedded microcontroller for flexible algorithm implementation.

First-generation 130-nm Quad NROM products write at a rate of 1 MB/s, while the second generation will write at more than 4 MB/s and will support both 3- and 1.8-V operation. The Quad NROM technology roadmap is presented in Table 13.6 where the objective is to provide the highest densities available in the data Flash market.

First-generation 2-bit/cell 256-Mb code Flash and 512-Mb data Flash produced in 0.17-µm technology node are presented in Figure 13.54(a) and (b), respectively.

Figure 13.55. Second-generation 2-bit/cell code Flash for wireless application introduced by SPANSION Inc. under the trademark Mirrorbit.

The 256-Mb code Flash has a 55-mm^2 die size and a 90-ns random-access read [7]. The 512-Mb data Flash has a 78-mm^2 die size and an 8-MB/s program rate [6].

The second-generation 2-bit/cell 512-Mb code Flash for wireless applications is available in the market, under the trademark Mirrorbit produced by SPANSION Inc., a licensee of the NROM technology (Fig. 13.55).

13.3.9 Summary

As conventional nonvolatile memories approach a scaling roadblock in the sub-50-nm technology node, charge storage alternatives are most appealing due to their inherent immunity to floating-gate technology scaling limitations. NROM technology features many advantages as a possible candidate to replace the floating-gate technology; being scalable down to at least ~20-nm node, enabling the first (and currently the only) 4-bit/cell products, addressing all nonvolatile memory market segments and fully compatible with standard CMOS processes. In addition, NROM requires no novel materials such as high and low K dielectrics.

To realize state-of-the-art NROM products, an integrative approach incorporating design, device, and process optimization were presented and a reliability model to predict the product lifetime was explained.

13.4 FERROELECTRIC MEMORIES

Alan D. DeVilbiss and Gary F. Derbenwick

13.4.1 Introduction

Ferroelectric memories have low internal programming voltages, virtually infinite endurance, fast programming times, and greater than 10-year retention, providing performance advantages over Flash memory. However, the cost per bit remains high, and the memory capacities are small compared to Flash memories.

Early attempts to develop cross-point ferroelectric memories using ferroelectric capacitors were unsuccessful because disturb voltages were applied to deselected memory cells [37]. The first ferroelectric memory to overcome this problem was invented in 1959 by Anderson [38] as shown in Figure 13.56. Each memory cell consisted of a ferroelectric capacitor in series with an access device. The access devices prevented disturb voltages from being applied to deselected memory cells because the deselected memory cells could not turn on at one-half of the programming voltage, the voltage applied to deselected memory cells. Only the selected memory cell had the full programming voltage applied. Present-day switched-plate capacitor ferroelectric memory cells are similar to those invented by Anderson except that transistors are used in place of the back-to-back diodes for the access devices.

Semiconductor products containing ferroelectric capacitor memories are in the market. The long development time for ferroelectric memories was related to the difficulty in finding an acceptable ferroelectric material and electrode combination, and the difficulty of integrating the ferroelectric process module with conventional silicon processing. Integration difficulties have been incompatibility problems between the ferroelectric material and those materials used in conventional silicon processing, developing electrodes for the ferroelectric high-permittivity dielectrics, developing barriers to prevent degradation of the ferroelectric by hydrogen exposure, integrating the ferroelectric devices with multilevel metal processes, silicides, and salicides, and developing high-density stacked memory cells using conductive barrier layers that can withstand the high-temperature processing required by some ferroelectric materials. Most of these problems have been solved, enabling volume commercial production to commence.

Many ferroelectric materials have been evaluated for use in ferroelectric semiconductor memories. Lead zirconate titanate (PZT) has been a popular material because the annealing temperatures are sufficiently low to be compatible with conventional silicon processing and common barrier layers. PZT allows denser memory cells to be fabricated, such as those utilizing a ferroelectric memory capacitor stacked on top of the access transistor. PZT can provide for fast programming times

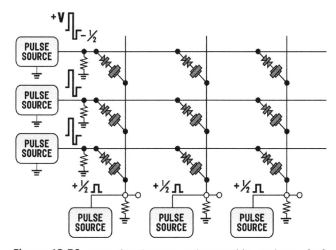

Figure 13.56. Ferroelectric memory invented by Anderson [38].

(under 50 ns), low programming voltages (under 2 V), and very high endurance (greater than 10^{12} to 10^{16} write/erase cycles). Ferroelectric memories are specified for 10-year retention at 85°C, but not yet at 125°C. Use of electrodes other than platinum, such as iridium oxide (IrOx) or lanthanum strontium cobalt oxide (LSCO) significantly improve the nonvolatile characteristics of the PZT ferroelectric storage devices.

The introduction of strontium bismuth tantalate (SBT), a layered perovskite ferroelectric material, has made it possible to produce ferroelectric memories with platinum electrodes that have excellent nonvolatile memory properties [39]. However, SBT requires a higher annealing temperature than PZT to form the ferroelectric phase and SBT has therefore proven more difficult to integrate with conventional silicon processes at small memory cell sizes.

Recent efforts have been directed at developing a ferroelectric memory that uses a ferroelectric field-effect transistor (FET) for the nonvolatile storage device instead of a ferroelectric capacitor [40–45]. The ferroelectric layer is incorporated in the gate dielectric of the FET and can provide a small memory cell size competitive with those of Flash memories. However, ferroelectric FETs using conventional ferroelectric materials with high dielectric permittivities have generally had poor retention. New classes of ferroelectric materials with dielectric permittivities under 50 are being investigated for use in ferroelectric FETs. Low dielectric permittivity ferroelectric materials typically have less polarization and switched charge than ferroelectrics used in ferroelectric capacitors, and therefore have been studied less. However, ferroelectric FET memories require only of the order of 1 to 10% of the polarization and switched charge to operate as compared with ferroelectric capacitor memories. The main issues with ferroelectric FET storage devices are achieving adequate programming field across the ferroelectric layer and achieving adequate retention [46]. Retention is limited by depolarization fields and charge injection and trapping that offsets the effect of the electric field due to polarization.

13.4.2 Storage Mechanism

Above a critical temperature, known as the Curie temperature, ferroelectric materials exhibit paraelectric behavior. This means that they act as conventional insulating dielectrics with no polarization effect. However, for temperatures below the Curie temperature, the lattice distorts along one axis and the material exhibits ferroelectric behavior. This means that an ion in the unit cell has two stable states on either side of the center of the unit cell along an elongated axis. Since these ions are charged, the movement of the ions creates a displacement of the charge centroid within the dielectric that results in an image charge forming at the electrodes that can be sensed.

A common ferroelectric structure, that of PZT, is the perovskite unit cell structure shown in Figure 13.57 [47]. Below the Curie temperature, the perovskite structure distorts to orthorhombic, rhombohedral, or tetragonal structures, depending on whether the elongated axis is along a face diagonal, body diagonal, or side of the cube, respectively. Layered perovskite ferroelectrics have the additional property that they spontaneously organize into layers at the molecular level. This is shown in Figure 13.58 for SBT.

For example, SBT forms two unit cells of perovskite strontium tantalate, one layer of bismuth oxide, two more unit cells of strontium tantalate, and so forth. It is

FERROELECTRIC MEMORIES

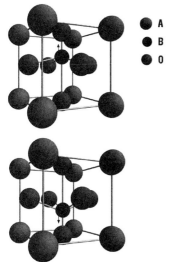

Cubic Perovskite

Figure 13.57. Perovskite unit cell. The ion labeled "B" has two stable states along a stretched axis of the unit cell.

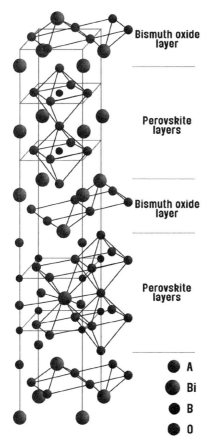

Figure 13.58. Layered superlattice ferroelectric structure.

believed that this molecular layering is the reason SBT has virtually fatigue-free behavior with platinum electrodes independent of the number of write–erase cycles, although this theory has not been proven definitively. The layering within SBT grains is shown in the transmission electron microscopy (TEM) shot in Figure 13.59.

The displacement of the ions in the ferroelectric material as a function of applied electric field causes the nonvolatile memory effect. This is often displayed as a hysteresis curve, where the polarization of the ferroelectric layer is plotted against the applied electric field, as shown in Figure 13.60. The current that flows between the plates of the ferroelectric capacitor as the polarization of the ferroelectric dielectric is switched can be used to detect the state of the memory. Referring to Figure 13.60, as the electric field increases positively from its most negative value in a counterclockwise direction along the hysteresis curve, the polarization changes

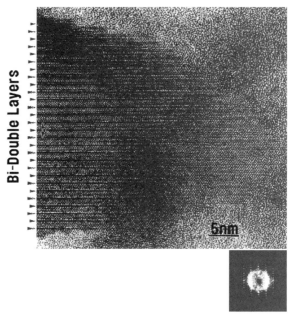

Figure 13.59. TEM showing superlattice layering in SBT. (*Courtesy of Infineon Technologies.*)

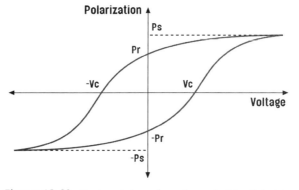

Figure 13.60. Hysteresis loop for a ferroelectric dielectric.

to the opposite state P_s and a switching current can be sensed. As the external electric field returns to zero, the remnant polarization labeled P_r remains. If the electric field is then reversed to the negative direction and subsequently returned to zero, negative switching current results and the polarization returns to the remnant polarization labeled $-P_r$.

The value of the voltage at which the hysteresis loop is midway between the maximum positive and negative polarizations is referred to as the coercive voltage, labeled V_c in Figure 13.60. Caution should be exercised when referring to the coercive voltage because the hysteresis curve may not be centered along the x axis, and the coercive voltage is a function of the frequency at which the hysteresis loop is obtained [48, 49]. For example, if a hysteresis loop is obtained by cycling the applied voltage at 10 kHz, the coercive voltage may be up to three times higher when the hysteresis loop is obtained by cycling the applied voltage at 20 MHz. Also note that when the applied voltage has increased to the coercive voltage, one-half of the polarized charge has switched. Therefore, voltages considerably less than the coercive voltage can disturb the memory state of a ferroelectric capacitor.

If a polarization state has been set by application of an electric field, the immediate application of another electric field of the same polarity does not result in further displacement of the B cations, and there is no external switching current. If there is no switching current, only the displacement current due to the linear capacitance flows through the circuit, as shown in Figure 13.61.

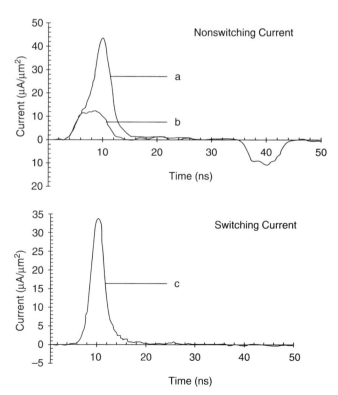

Figure 13.61. Fast pulse measurements on a ferroelectric capacitor. Curve *a* is the switched plus linear displacement current. Curve *b* is the linear displacement current. Curve *c* is the difference between curves *a* and *b*, the switched current.

Curve *a* corresponds to switching the polarization plus the linear displacement current, and curve *b* shows the linear displacement current with no ferroelectric switching.

In a ferroelectric capacitor memory, the state of a logical 1 or 0 can be determined by sensing if there is a switching current or not when application of an electric field of a given polarity is applied to the ferroelectric capacitor. In a ferroelectric FET memory, the state of a logical 1 or 0 can be determined by sensing the conductivity of the channel for a given gate voltage. The threshold voltage of the ferroelectric FET depends on the state of the polarization in the gate dielectric.

13.4.3 Memory Cells and Arrays

There are two basic memory cell structures for ferroelectric memories. One structure uses a single-transistor FET where the ferroelectric material is incorporated within the gate dielectric. This ferroelectric FET memory cell uses a nondestructive read, similar to that of Flash memory. The second structure is similar to that of a DRAM and uses a ferroelectric capacitor in series with a pass transistor. This ferroelectric capacitor memory cell uses a destructive read, similar to that of DRAM memory. The only ferroelectric memories in production today use a ferroelectric capacitor for the storage device.

13.4.3.1 Ferroelectric Transistor Memory Cell. A single-transistor ferroelectric memory device incorporates the ferroelectric material as a layer in the gate dielectric of an FET [40]. Inclusion of the ferroelectric material into the gate dielectric has been a challenge. Normally an insulator is placed between the ferroelectric material and the silicon in the FET to avoid direct contact between the ferroelectric material and the silicon. Earlier approaches attempted to use a high dielectric insulator because of the very high dielectric permittivities of typical ferroelectric materials. However, the dielectric permittivity of conventional ferroelectric materials is usually at least an order of magnitude higher than the highest dielectric constant of conventional insulators. Because of the high permittivity of the ferroelectric, any insulator with low dielectric constant in series with the ferroelectric material prevents adequate programming voltage from being applied to the ferroelectric material because the applied voltage drops across the layers inversely proportional to their dielectric permittivities. A more promising approach is to use a low dielectric permittivity ferroelectric so that conventional low dielectric constant insulators can be used between the silicon and the ferroelectric material.

Because a uniform electric field typically is applied across the entire area of the ferroelectric device for writing and erasing, it is normal to include a pass transistor in series with a simple ferroelectric FET to isolate the ferroelectric device from the bitline to accomplish random programming of memory cells. Alternatively, researchers have stacked a ferroelectric capacitor on top of the gate electrode of a conventional MOS transistor to form a more complex ferroelectric FET, and then placed another gate on top of the ferroelectric layer. In this case the MOS transistor gate electrode becomes a type of floating gate. The behavior of this structure is not well understood and will not be discussed here.

Two transistor ferroelectric memory cells with a pass transistor are too large to be cost competitive with Flash memory cells. True single-transistor ferroelectric memory cells may be developed in the future to compete with Flash memory.

FERROELECTRIC MEMORIES

However, developing a memory array architecture that allows for a small memory cell size, decoding the programming and read operations, avoiding disturb problems on deselected memory cells during programming and reading, and avoiding materials incompatibility problems are significant issues remaining to be solved.

13.4.3.2 Ferroelectric Capacitor Memory Cell. A small ferroelectric memory cell can be obtained by connecting a ferroelectric capacitor in series with a pass transistor. The cell structure is similar to that of a stacked DRAM, except the plate of the capacitor not connected to the pass transistor is decoded for the ferroelectric memory cell. This is called the plate line, so the one-transistor, one-capacitor (1T/1C) ferroelectric memory cell has a bitline, a wordline, and a plate line, as shown in Figure 13.62. These memory cells use a destructive read followed by a restore cycle, just like DRAMs. The plate line is switched to sense the state of the memory cell. Alternatively, the plate line can be continuously held at a voltage midway between the power supply voltage and ground. However, care must be taken to avoid data disturb problems with the latter nonswitched plate line architecture.

Ferroelectric memories of the DRAM-type structure are in volume production today and are the most popular ferroelectric memory cell architecture today. If the ferroelectric capacitor can be stacked on top of the access transistor, these ferroelectric memories will have memory cell sizes comparable to those of stacked-cell DRAMs depending on the etching characteristics of the ferroelectric device.

A cross-sectional TEM of a 1T/1C memory cell is shown in Figure 13.63. The ferroelectric capacitor is placed above the field oxide next to the pass transistor. This memory cell is often called an offset cell or planar cell. Up to one-half of the memory cell area may be due to the etch requirements for the ferroelectric capacitor structure. To reduce these cell sizes requires improvements in the etches and design rules associated with the ferroelectric layer and its electrodes.

To be competitive in cell size with NOR Flash memory, the ferroelectric capacitor must be fabricated above the pass transistor. This is referred to as a stacked cell. To maximize the ferroelectric capacitance and signal to be sensed, the bottom electrode of the ferroelectric capacitor can be made taller than it is wide. For example,

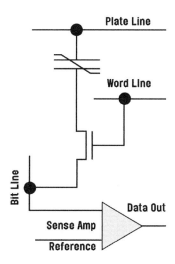

Figure 13.62. Schematic of a one-transistor, one-capacitor ferroelectric memory cell.

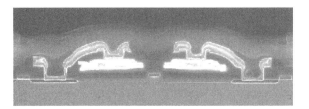

Figure 13.63. Cross section of a 1T1C ferroelectric memory cell. Two adjacent memory cells are shown. The drains and gates of the two pass transistors are at the left and right edges of the SEM. (*Courtesy of Hynix Corporation.*)

if the height is two times the base, the capacitance will be nine times that of a simple planar structure for the same footprint. For a stacked memory cell, the cell size may be of the order of $11\,F^2$, where F^2 is the design rule feature dimension squared. For example, for a 0.3 µm design rule, the cell area is of the order of 1 µm². This is larger than a NOR Flash memory cell, which has a cell size of the order of $8\,F^2$. NAND Flash memory cells can have cell sizes as small as 4 to $5\,F^2$, but they are not true random-access memories.

The operation of a 1T/1C ferroelectric memory cell is best understood by examining the switching and nonswitching current characteristics of the ferroelectric capacitor shown earlier in Figure 13.61 along with the schematic in Figure 13.62. During the rise time of an applied voltage pulse, the linear displacement current, $i = C(dV/dt)$, flows and is labeled b in Figure 13.61. A similar negative linear displacement current flows during the fall time of the applied voltage pulse. If the ferroelectric capacitor does not switch and the capacitor has a small leakage current, then only the linear displacement current flows to the bitline. However, if the state of the ferroelectric capacitor switches, an image charge flows between the plates of the capacitor as the charge centroid in the ferroelectric shifts, labeled c in Figure 13.61. A switching capacitor delivers more charge to the bitline than a nonswitching capacitor, allowing the state of the memory cell to be sensed.

Typical read and write operation of a 1T/1C ferroelectric memory cell is shown in Figure 13.64. First, the bitlines are precharged to 0 V and the plate line is high. The wordline is pulsed to connect the ferroelectric capacitors on the selected wordline to their respective bitlines through the pass transistors, causing possible switching of ferroelectric capacitors, depending on their programmed state. A current due to the linear capacitance of the capacitor flows and starts charging the bitline. If the ferroelectric capacitor switches, then an additional switched charge is placed on the bitline and the bitline charges faster. This difference in charge rate enables a "1" state to be differentiated from a "0" state, shown by the two bitline curves. At a predetermined time, the sense amplifier is powered or enabled. As the bitlines are driven to the power supply rail voltages, and the plate line is pulsed low, the relevant ferroelectric capacitors are automatically switched or restored back to their original states. This completes the ferroelectric memory cell read operation. Because of the fast switching time of the ferroelectric capacitor, normally enough charge is stored on the chip that the restore operation occurs completely even if the power supply voltage is removed during the read/restore operation.

A programming operation is similar to the end of a read operation because the read cycle applies a restore pulse after the destructive readout. In the case of a

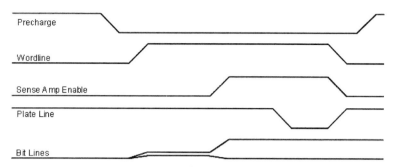

Figure 13.64. Operation of a 1T/1C ferroelectric memory cell.

TABLE 13.7. Applied Voltages for Arrayed Memory Cells Shown in Figure 13.65

Line	Signal
WL2, PL2	Low
WL1, PL1, BL1, BL2	Per the waveform drawing shown in the Figure 13.65

programming operation, the appropriate voltages are applied to the bitlines and the restore sequence occurs.

An alternative operating scheme to that shown in Figure 13.64 can be obtained by turning off the access device before the plate line voltage is returned to its initial state [38]. This allows charge to be trapped on the isolated ferroelectric capacitor nodes. Effectively, a longer rewrite time is obtained until the trapped charge leaks off the isolated node. Care must be taken to assure that the longer effective restore time does not cause unacceptable time-dependent dielectric breakdown to occur.

13.4.3.3 Ferroelectric Capacitor Memory Cell Array Architecture.

Ferroelectric memories capable of Flash memory replacement may have array architecture similar to that of a DRAM, except the plate line (electrode of the cell capacitor not connected to the pass transistor) is decoded, as mention previously. The plate line is normally run parallel to the wordline. When run parallel to the wordline, smaller disturb voltages appear on the deselected cells providing an advantage if disturb is a problem. Because the plate line is decoded, the size of the array may be larger than that of a DRAM, depending on the layout. Four 1T/1C memory cells in an array are shown in Figure 13.65. For illustrative purposes, assume the selected memory cell is in the upper left corner with plate line 1, word line 1, and bit line 1. The other three memory cells are deselected. Applied signals for the operation of the array are described in Table 13.7. All bits along a selected wordline are read.

It is important to consider disturb voltages on the deselected memory cells. Small repetitive voltages on a deselected memory cell can eventually destroy the data in the deselected cell because ferroelectric programming voltages are small.

13.4.3.4 Programming Voltage.

Ferroelectric materials can be polarized and switched at low voltages. Figure 13.66 shows typical polarization for a ferroelectric material versus applied voltage for two different ferroelectric thicknesses.

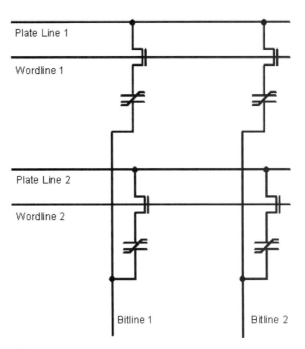

Figure 13.65. Arrayed 1T1C ferroelectric memory cells.

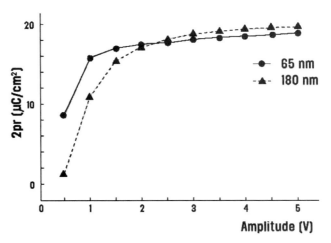

Figure 13.66. Polarization vs. voltage for a typical ferroelectric capacitor.

PZT with IrOx electrodes and SBT with Pt electrodes give similar results. It can be seen that the programming voltage scales with the ferroelectric thickness, and that the programming voltages are low compared to those of floating-gate memories. Charge pumps to increase programming voltages are not required for ferroelectric memories but wordline boost may be required. However, programming voltage, ferroelectric film thickness, and programming time are not independent. For write times of the order of 20 ns on 200-nm films, voltages above 3 V may be required for complete switching. For write times of the order of 10 ns on 40-nm films, voltages under 1.5 V may be sufficient for complete switching.

Because of the low voltage programming of ferroelectric devices, caution must be exercised in the circuit design of ferroelectric memories to prevent disturb of deselected memory cells during reading and programming. Caution must also be exercised during power-up and power-down operations because there are no charge pumps to be disabled to assure data integrity.

13.4.3.5 Programming Speed. The ferroelectric material itself switches its polarization state in less than 1 ns [50]. Accurate measurements of the switching speed have been difficult to measure due to the constraints of high-frequency instrumentation. In a practical circuit with pass transistors and bitlines, read access and read cycle times comparable to those of DRAMs can be obtained. The primary speed limitation in a circuit is switching the plate lines that have high capacitance.

The switching time of typical ferroelectric capacitors at room temperature is shown in Figure 13.67 as a function of programming pulse width for several

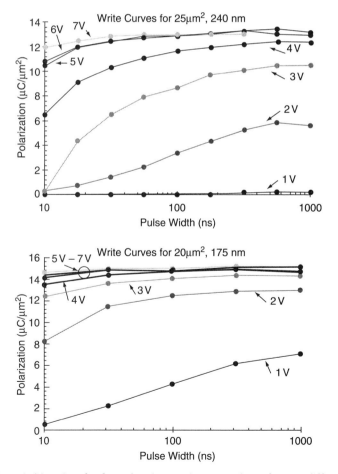

Figure 13.67. Switching time for ferroelectric capacitors vs. voltage for two different ferroelectric thicknesses.

programming voltages. Complete switching of the capacitor occurs for programming times as short as 10 ns at the higher voltages. It is seen that if thickness of the ferroelectric film is scaled, these same switching times can be obtained for lower programming voltages so long as the electric field in the ferroelectric material remains comparable. As a rule of thumb, for very fast switching times, voltages at least in excess of three times the low-frequency coercive voltage should be used.

The switching time for a given ferroelectric capacitor is also a function of the temperature, with slower switching occurring at lower temperatures. This temperature dependence places a limitation on programming times in ferroelectric memories that must operate over a wide temperature range.

13.4.4 Fabrication

For the stacked ferroelectric cell, the ferroelectric materials are isolated from the silicon by dielectric layers and conductive plugs. Fabrication of the ferroelectric capacitor occurs after the underlying CMOS is completed, except for the metallization. For SBT, the ferroelectric capacitor is formed before the metallization layers are applied, but for PZT, the ferroelectric capacitor may also be formed after the metallization layers are applied because the anneal temperatures for PZT can be kept below 450°C. Both approaches allow the CMOS circuitry to be formed in a conventional manner. Since the ferroelectric programming voltages are at or below the power supply voltage, separate high- and low-voltage P- and N-wells and P- and N-channel transistors are not required as they are for Flash memories. This reduces the mask count of ferroelectric memory compared to that of Flash memories.

Since platinum electrodes with a thin layer of titanium "glue" underneath are conventionally used for the bottom electrodes to a ferroelectric capacitor, a barrier metal must be used to isolate the polysilicon plug from the bottom electrode in a stacked memory cell. Since the lowest annealing temperature in oxygen for SBT to form the ferroelectric phase is at or above 720°C, conventional titanium nitride barriers cannot be used because they oxidize and lose their effectiveness at these temperatures. Therefore, special high-temperature barrier materials have to be developed for SBT processing. Alternatively, lower temperature processing could be developed for SBT if the nonferroelectric fluorite phase that forms at lower temperatures can be avoided. The latter is preferable because the high-temperature processing required for SBT is also incompatible with conventional CMOS scaling.

The lower annealing temperature required by PZT allows conventional barrier metals, such as titanium nitride, to be used. Use of IrO_x for the top electrode and doping the PZT with La, Ca, and Sr or using IrO_x for both electrodes can have dramatic effects on improving the nonvolatile properties of PZT.

A typical process flow for a ferroelectric capacitor added to a CMOS process prior to the metallization steps is shown in Figure 13.68. In Figure 13.68, a forms the bitline, b forms the gate of the pass transistor, e forms the ferroelectric capacitor bottom electrode, d forms the ferroelectric layer, and c forms the ferroelectric capacitor top electrode.

Ferroelectric films are sensitive to hydrogen, so degradation in the switching characteristics and breakdown characteristics may be observed after hydrogen exposure. Care must be taken to protect the ferroelectric film from hydrogen

FERROELECTRIC MEMORIES

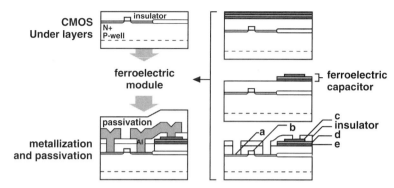

Figure 13.68. Typical integration of a ferroelectric capacitor into a CMOS process.

exposure caused by interlevel dielectric deposition, metal plug formation, and forming gas alloy anneals. This can be accomplished by engineering the film to be less sensitive to hydrogen or by using dielectric layers impervious to hydrogen to encapsulate the ferroelectric capacitor. Alumina and silicon nitride have been used to encapsulate the ferroelectric device.

Both PZT and SBT ferroelectric films tend toward a chemically reduced state when heated at high processing temperatures. Therefore, it is typical to use oxygen anneals in the process to prevent chemical reduction. Care must be taken to assure that the oxygen anneals do not cause degradation of other materials, such as titanium silicide, titanium salicide, or the resistivity of barrier materials. SBT contains so much oxygen due to the bismuth oxide layering that even in inert ambients, such as nitrogen and argon, enough oxygen can exit the film as to cause oxidation of surrounding materials.

13.4.5 Nonvolatile Characteristics

13.4.5.1 Endurance. High endurance is critical for one transistor–one capacitor ferroelectric memory cells because a destructive read operation is used. This means that a restore operation occurs for every read, and unlike Flash memory, the endurance specification applies to read operations as well as write operations. Continuously reading a bit at a 100-ns cycle time for 10 years corresponds to approximately 3×10^{15} write cycles. While this scenario is unlikely to occur during operation in a real system, 3×10^{15} endurance is a good target for ferroelectric memories that use destructive read operations and require high read or write endurance.

Figures 13.69 and 13.70 show the endurance of typical PZT and SBT capacitors, respectively. In both cases, the polarization remains near its initial value over the entire measurement range and both ferroelectric materials show vastly superior endurance as compared to Flash memory. Both PZT and SBT can be engineered so there is virtually no endurance degradation for practical lifetimes of semiconductor memories. SBT exhibits such behavior with Pt electrodes, and PZT exhibits such behavior with IrO_x electrodes. Extrapolated data for PZT may show endurance levels higher than 10^{16} write/erase cycles [51].

One technique for measuring endurance is to measure the lowest power supply voltage at which the memory will operate. If even only a single bit starts showing

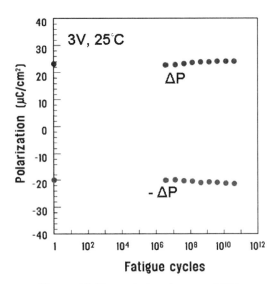

Figure 13.69. Typical endurance of PZT.

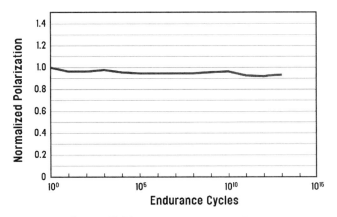

Figure 13.70. Typical endurance of SBT.

fatigue, then the lowest power supply voltage for which the memory operates will increase. Therefore, the endurance of the worst bit in the memory array can be inferred by when the lowest operating voltage begins to increase [52].

13.4.5.2 Retention. The switched charge obtained from a polarized ferroelectric capacitor is observed to decay linearly with the logarithm of time, as shown in Figure 13.71. The decay rate increases with higher temperature and after imprint. Therefore, practical ferroelectric devices must be guaranteed to have adequate retention at the worst-case high temperature and imprint, when programmed at the worst-case lowest temperature.

A typical test sequence for retention is to cycle the devices with unipolar pulses to the same polarization state at the highest device temperature. Then the devices are written to the opposite state at low temperature and the retention decay rate is measured at high temperature after baking the device at elevated temperatures.

FERROELECTRIC MEMORIES

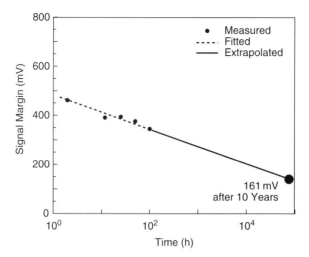

Figure 13.71. Polarization decay as a function of the logarithm of time. (*Courtesy of Hynix Corporation.*)

The writing at low temperatures can be simulated at room temperature by using shorter write times provided the temperature/write time characteristic is known. This testing can be accomplished on a sample basis for a given product to assure reliability. Certain design schemes allow this testing to be accomplished on every bit of every memory prior to shipment to a customer using margin testing. Use of such margin testing can result in ferroelectric memories with very high reliability levels.

Ferroelectric capacitors also exhibit a behavior known as imprint [53]. If a ferroelectric capacitor is programmed repeatedly to the same state, that state becomes preferred or reinforced. If the capacitor is then programmed to the opposite state, the retention in the opposite state is poorer than for a capacitor that has not been imprinted. The imprint effect may be stronger for higher temperatures. The physical mechanisms for imprint are not completely understood, but must be taken into account when projecting the nonvolatile performance of a ferroelectric memory.

13.4.6 Scaling

Ferroelectric memories may be scaled to the limit of the thinnest ferroelectric films that can be fabricated. As ferroelectric films are scaled thinner, the switched charge due to polarization reversal may increase. Hence the amount of signal available may increase, contrary to intuition.

Aggressive CMOS scaling requires lower annealing temperatures than those used today. PZT, with annealing temperatures as low as 375°C, is most compatible with CMOS scaling. The much higher annealing temperatures of SBT may prevent the smallest memory cells from being fabricated using SBT, unless a way is found to form the ferroelectric phase of SBT at lower temperatures.

Since the ferroelectric films are polycrystalline and the grain size is governed by the highest anneal temperature (grains grow larger at higher temperatures), there is probably an ultimate limitation to the minimum film thickness. This limitation will most likely manifest itself by increased leakage current through the film and degraded time-dependent dielectric breakdown characteristics. It is generally accepted that the grain size should be small compared to the area of the ferroelectric device to provide uniform ferroelectric properties across all memory cells. This also means that the anneal temperatures should be kept as low as possible.

Ferroelectric film thicknesses of the order of 50 nm have been fabricated with acceptable electrical characteristics. As metal organic CVD (MOCVD) deposition of the films improves, it should be possible to use even thinner films. The control of the thickness of the films, both on a single wafer and wafer-to-wafer, is similar to that of an low presscure CVD (LPCVD) film, typically less than 1.5% variation across a wafer.

As a possible scaling scenario, it can be assumed that the electric field for polarizing the ferroelectric film remains constant. Therefore, the programming voltage decreases linearly with ferroelectric film thickness. A voltage of 5 V on a 200-nm film provides adequate polarization. It may be projected that the grain size may limit the thickness of the ferroelectric film to approximately 15 nm. Assuming a smooth scaling of the ferroelectric film thickness from 70 to 15 nm, and assuming a stacked memory cell with a bottom electrode aspect ratio of 2:1, ferroelectric memory cell sizes can be projected. The projected memory cell size changes from approximately $11 F^2$ to $8 F^2$ as the design rules decrease from 0.5 to 0.1 µm. The memory cell size for a ferroelectric FET can be projected to be 4 to $5 F^2$, provided that the disturb issues on deselected memory cells can be solved. These projections assume that progress is made in reducing the design rules associated with etching the ferroelectric layer and the electrodes.

Ideally, the ferroelectric capacitor would be formed in the contact window of the pass transistor for the smallest memory cell size. This would give rise to scaling scenarios superior to those just discussed. For this to occur, process integration solutions and increases in switching current per unit area are needed.

Ferroelectric memories in production are several technology generations behind Flash memory. Typical ferroelectric process design rules in 2006 were 0.18 to 0.5 µm. Should current development efforts be successful, ferroelectric memories may approach similar design rules to those for Flash memories. Because of the similarity of the ferroelectric process to that of a stacked capacitor DRAM process, ferroelectric memories may take advantage of the learning curve for DRAMs.

This scaling analysis assumes that the ferroelectric technology will not encounter any freak memory cell defect problems that are not solvable or that are not in such high concentration as to reduce yield significantly. Freak ferroelectric memory cells may be those that exhibit poorer retention than the main population. It is probably possible to screen memories containing these freak memory cells electrically with nondestructive testing. Alternatively, error detection and correction may be used at the system level.

13.4.7 Reliability

Reliability of PZT production ferroelectric memories has been comparable to or better than that observed for other semiconductor memories. The reliability of small 1000 bit ferroelectric memories has been compared to the reliability of small floating-gate EEPROM memories [54]. At room temperature, the ferroelectric memory

FERROELECTRIC MEMORIES

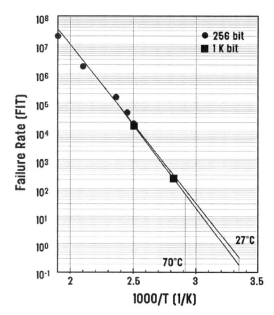

Figure 13.72. FIT failure rates of ferroelectric memories.

failure rate is found to be less than that for the EEPROM memories. At higher temperatures, the failure rate becomes comparable to that of EEPROM memories. As larger density ferroelectric memories are manufactured in volume production, the statistics of failure will become better known for ferroelectric memories.

Failure rates for SBT-embedded ferroelectric memories have been characterized. Figure 13.72 shows the FIT failure rates of both 288-bit and 1120-bit SBT embedded memories [55]. One FIT is one device failure per 10^9 device operating hours. The failure rate for both memories is observed to be under 100 FITs at 70°C and is also observed to be independent of the memory density. Since the design margins for these two memory designs are different, measuring the same failure rates for memories with different bit densities might not be unexpected.

Of concern for any memory technology is the failure rate for freak single-bit cells, not the failure rate for nominal memory cells. Production ferroelectric memories have shown no significant occurrence of freak single-bit failures in the field.

13.4.8 Die and Test Cost

A general comparison between ferroelectric memory and Flash memory with respect to die and test cost can be made. While a ferroelectric memory cell is larger than a NOR Flash memory cell, the cost of processing should be lower because extra reticles are not required in the ferroelectric process to provide transistors able to withstand high programming voltages. Therefore, comparable silicon costs might be possible provided that the die yields are comparable.

Because of the fast programming times for the ferroelectric memory, the test time for a ferroelectric memory is significantly less than that for a Flash memory. Typical programming times for ferroelectric memory are under 100 ns, whereas typical programming times for Flash memory are greater than 100 µs and typically around 2 ms. With programming times 1000 to 50,000 times shorter for ferroelectric memory, test costs are dramatically reduced. Therefore, the tested die cost for ferroelectric memory manufactured at the same design rules as Flash memory is projected to be comparable to or less than that of Flash memory.

13.4.9 Ferroelectric Products

Many ferroelectric products are in production, including stand-alone and embedded memories. Figures 13.73 through 13.75 show examples of a few production ferroelectric products.

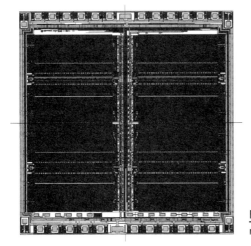

Figure 13.73. A 256-kbit FeRAM using 2T/2C memory cell. (*Courtesy Ramtron Corporation.*)

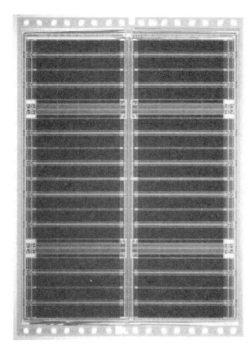

Figure 13.74. A 1-Mbit FeRAM using 1T/1C memory cell. (*Courtesy Fujitsu Corporation.*)

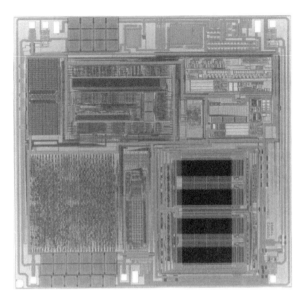

Figure 13.75. Smart Card LSI with 4-kbyte ferroelectric embedded memory (lower left quadrant). (*Courtesy Fujitsu Corporation.*)

13.4.10 Ferroelectric Memory Summary

Ferroelectric memories are in volume production today. Stand-alone ferroelectric memories of 1-Mbit density have been sold in volume, and higher densities have been prototyped. Ferroelectric memory, with its low-voltage programming and fast programming time, is ideal for embedded memory used in smart cards and radio frequency identification (RFID) tags. The largest production volumes of ferroelectric devices have been in this latter category.

It will be difficult for ferroelectric memory to compete directly with commodity Flash memory because of the larger memory cell size for ferroelectric memory. In addition, process technologies for ferroelectric memory are several generations behind those of Flash memory. Ferroelectric memory must take advantage of the similarities in processing stacked-cell DRAMs in order to catch up with the technology generations of Flash memory, and this is unlikely to occur in the near future. In addition, freak memory cell occurrences in ferroelectric memory at the higher memory densities must be sufficiently rare that comparable yields are obtainable. Ultimately lower wafer fabrication costs and lower test costs for ferroelectric memories may help to offset the disadvantage of the larger memory cell size.

Ferroelectric memories have several performance advantages over Flash memory, including low programming voltages, fast programming times, and virtually unlimited endurance. These performance features must be obtained at little or no additional cost in order for ferroelectric memory to sell into broad markets.

The cost of integration of both DRAM and Flash memory on a single chip for embedded memory applications is expensive. It is argued that a single memory technology, such as ferroelectric technology, can replace both of these types of embedded memory as well as embedded EEPROM at lower cost. Therefore, ferroelectric memory is being considered as a more ideal nonvolatile memory for embedded applications.

13.5 MAGNETIC MEMORIES

Romney R. Katti

13.5.1 Introduction

Early random-access memories used magnetic cores [56] to store nonvolatile data in a row-and-column array architecture. Plated wire memories [56] were also developed based on magnetic materials to provide nonvolatile memory. Both core and plated wire memories had the advantages of inherent nonvolatility and infinite endurance, and the disadvantages by current standards of relatively low sense currents, large size, and high cost. Developments in magnetic materials and devices, such as in magnetoresistive devices used in hard disk drives [57–60], demonstrated that magnetic materials and devices with ever-increasing capabilities and performance could be manufactured to increase storage density and capacity at sustainable compound growth rates.

Magnetoresistive devices emerged that could support higher density and higher capacity magnetic random-access memories. The magnetoresistive devices emerged with higher signal levels to support readback and lower switching thresholds and higher switching margins to support the write process. These magnetic devices could be integrated with semiconductor technology to make integrated, monolithic magnetic random-access memories.

A perspective on the advancement of magnetic devices for nonvolatile magnetic random-access memory (MRAM) applications is presented in Figure 13.76. In particular, Figure 13.76 shows the increase in device magnetoresistance (MR) coefficient as a function of time, which is important to the development of manufacturable MRAMs. The ability for magnetic devices to continue to provide higher read signals

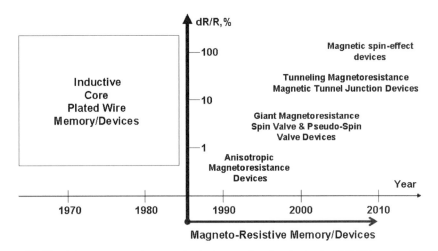

Figure 13.76. Advancement of magnetic devices and technologies for MRAM applications.

[56–90] supported device requirements in semiconductor memory and manufacturing environments and applications. Advancements in semiconductor and materials processing, thin-film magnetic materials, and the ability to integrate magnetic and semiconductor devices and processes [56, 79, 80, 82] enabled the creation of monolithic nonvolatile semiconductor magnetic memories and MRAM products.

A few different approaches have been developed to integrate thin-film magnetic storage elements into a semiconductor process. Two of the more prominent and timely approaches have been MRAMs that use the giant magnetoresistance (GMR) effect and the tunneling magnetoresistance (TMR) effect. Magnetic spin-valve-type devices [61–70, 74, 75, 82–84, 87] have been developed that exhibit the GMR effect. Magnetic tunnel junction (MTJ) devices have been developed [56, 79–95] that exhibit the TMR effect. Both types of MRAMs involve two sets of magnetic thin films that are separated by a nonmagnetic layer. GMR memories sense the change in resistance in the plane of the magnetic layers, whereas MTJ TMR memories sense the change in resistance perpendicular to the plane of the magnetic layers. This change in resistance results when the magnetic directions of the two magnetic layers on opposing sides of the nonmagnetic spacer layer are switched between the parallel and antiparallel states. The change is induced by changes in the relative spin polarization, the coupling and scatlering of electrons in the magnetic structure. The magnetic thin-film layers can be integrated with standard semiconductor processing, providing a fully integrated, monolithic memory. Since the switching properties of the magnetic layers never wear out, high levels of endurance are obtained.

Spin-valve-based memories use the GMR effect. The GMR effect is the change in resistance produced in a magnetic multilayer structure that depends on the direction of magnetization in two individual magnetic layers within the multilayer structure. The read, or sense, process is nondestructive since it involves reversible switching of one of the magnetic layers, a magnetic sense layer, whereas the other magnetic layer, the magnetic storage layer, is not switched during the sense operation.

Giant magnetoresistance technology [61–70, 74, 75, 87] provides from 2 to 10 times as large a change in resistance than that produced by the anisotropic magnetoresistance (AMR) effect. This is the origin of the use of "giant" in the GMR name. The GMR effect is a current-in-plane effect involving different scattering rates between the two electron polarization states. Obtaining adequate signal is critical for successful design of magnetic semiconductor memories.

The magnetic tunnel junction (MTJ) memory with the TMR effect has a still larger change in resistance for read sensing than the GMR memory. MTJ devices, using the TMR effects, in turn, offer more than 10 times the change in resistance than GMR devices. The TMR effect is a current-perpendicular-to-plane effect involving spin-polarized electron tunneling between two ferromagnetic metals.

13.5.2 Magnetic Random-Access Memory with Giant Magnetoresistive Devices

13.5.2.1 Giant Magnetoresistive MRAM Bit Storage and Read Mechanisms. In a GMR-based magnetic random-access memory, a magnetic multilayer structure modulates the resistance of a sensing magnetic layer depending on the direction of magnetization of two magnetic layers. Sensing the state of the memory cell is achieved using the magnetoresistance effect, as induced through the GMR effect. A sensing layer in the magnetic multilayer, which supports the GMR effect for example, allows determination of the magnetic state, and therefore the logical state, of the storage layer without disturbing the state of the storage layer.

Magnetic RAM cells can use magnetic pseudo-spin valve (PSV) devices as the magnetic multilayer to form the basic memory cell. A PSV device, as shown in Figure 13.77, is a magnetic, multilayered structure composed of three main layers. The resistance of one of the magnetic layers is modulated depending on the directions of magnetization of the two magnetic layers.

A NiFeCo layer, which can be composed of a NiFe/CoFe bilayer, can serve as the storage layer. The direction of magnetization in the storage layer determines the logical state of the stored data. A thinner NiFeCo layer, which can also be composed of a NiFe/CoFe bilayer, can serve as the sensing layer. The sensing layer can be switched reversibly, without switching the magnetization of the thicker layer, to allow sensing the state of the thicker layer using the GMR effect. As shown in Figure 13.78, when the magnetization between the thick and thin layers is antiparallel, the resistance in the sensing layer is maximized. When the magnetization between the thick and thin layers is parallel, the resistance in the sensing layer is minimized. The ability to change and to sense, using the GMR effect, the magnetization in the thin layer reversibly relative to the storage layer allows the state of the cell to be read.

The third layer in the PSV is a thin, nonmagnetic, metallic spacer, such as Cu, placed between the thick and thin magnetic layers. The spacer reduces the coupling between the two magnetic layers and improves the switching properties. The spacer also enhances the GMR effect by creating interfaces that support spin-dependent scattering and provide magnetizationdependent resistance.

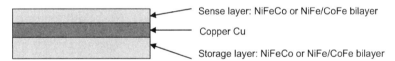

Figure 13.77. Cross section of a pseudo-spin valve device.

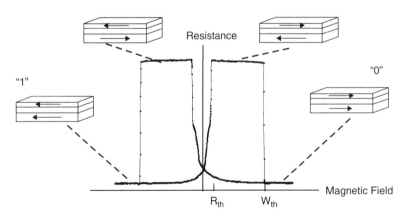

Figure 13.78. GMR effect. Shown is the effect for a pseudo-spin valve (PSV) magnetic multilayer.

MAGNETIC MEMORIES

13.5.2.2 Giant MR Cell Structure and Operation. As shown in Figure 13.79, the basic GMR cell in a PSV-based MRAM is connected to a sense line and placed in proximity to a wordline and an optional bias line. Using a specific sequence of currents applied to the various lines allows writing to a selected cell without writing to cells that are not selected.

To program the state of a GMR cell, a sense current is applied that lowers the field required for magnetic switching of the storage layer. Applying a sense current reduces the word current needed to program the cell, thereby creating the property of cell selectivity, which is the ability to write to a specific cell in the array without disturbing deselected cells. A current can be applied to the optional bias line if additional reduction is desired in the switching field. Optionally, additional bias can be added from a metal line that runs parallel to the cell. Coherent switching and/or coupled switching of the magnetization [71–73, 76–78] are used to make the switching abrupt between the two data states.

Once the sense current is applied, the word current is applied. The direction of the word current sets the direction of the magnetization in the storage layer of the magnetic multilayer and therefore sets the logical state of the cell. As shown in Figure 13.78, W_{th} is the magnitude of the word current needed to write a selected cell. An architecture that shows the placement of cells in an array is shown in Figure 13.80. Figure 13.81 shows how currents produce magnetic fields that affect the

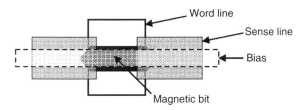

Figure 13.79. GMR cell layout. The PSV cell is shown with a wordline, sense line, and optional bias line.

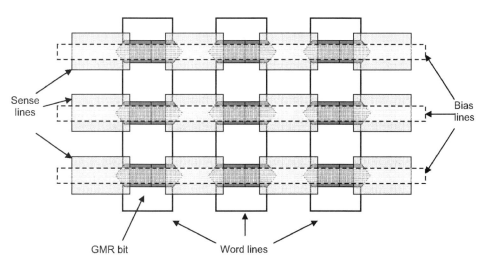

Figure 13.80 Array architecture of a GMR-based magnetic memory.

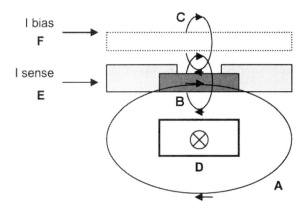

Figure 13.81. Cross section of a GMR PSV cell. View of (A) word magnetic field, (B) sense magnetic field, and (C) bias magnetic field induced by (D) word current, (E) sense current, and (F) bias current.

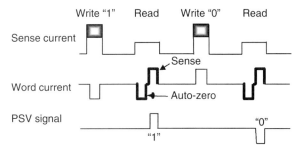

Figure 13.82. Write and read current timing in a GMR magnetic memory.

magnetization of layers within the PSV cell. The timing diagram for writing to a GMR cell is shown in Figure 13.82.

To sense the state of a PSV cell, current flows in the plane of the PSV that produces a signal dependent on the state of the storage layer through the GMR effect. As shown in Figure 13.78, if the magnetization directions of the thicker and thinner layers are parallel, the resistance is minimized. If the magnetization directions of the thicker and thinner layers are antiparallel, the resistance is maximized. If the magnetization directions of the thicker and thinner layers are at an angle, ø, between π and $-\pi$ radians, the total resistance, R_{total}, is given by:

$$R_{total} = R_{min} + \Delta R \sin^2(|0.5\phi|) \tag{13.12}$$

where R_{min} is the minimum resistance of the PSV, ΔR is the change in resistance of the PSV, and $\Delta R/R_{min}$ is the magnetoresistance coefficient. The resistance and change in resistance values are generally given in ohms. The magnetoresistance coefficient is typically quoted as a percentage.

To read the state of a PSV cell, a sense current is applied to allow sensing the magnetization direction of the sense layer relative to the storage layer within the PSV device. The bias line is not used. As shown in Figure 13.78, a word current is applied with a magnitude of R_{th} that is less than W_{th}. Under this condition, the state of the sensing layer can be switched so that the state of the PSV device is read without changing the state of the storage layer. This process allows reading to occur without overwriting the cell. As shown in Figure 13.82, a bipolar word current is used to allow switching the sense layer of the PSV device reversibly so that the magnetic state of the PSV device before and after sensing is unchanged. The bipolar word current therefore supports a nondestructive read operation. Because the polarity of the read signal produced by the GMR effect depends on the magnetization direction of the storage layer, the bipolar current allows determining the relative magnetization between the sense layer and the storage layer in the PSV device and reading the logical state of the GMR memory cell.

13.5.2.3 Giant MR Magnetic Memory Array Architecture. As shown in Figure 13.80, a GMR magnetic memory architecture is generated when PSV device cells are placed in a matrix. Wordlines run in columns perpendicular to the long axes of cells. Wordlines allow selecting a set of cells in a particular word. The wordline currents are generated by current sources and sent to a particular wordline by decoding logic and pass transistors. Column addressing is used to gate the word current to particular columns. The word currents produce magnetic fields that are parallel to the long axes of the cells.

Sense lines and optional bias lines run in rows parallel to the long axes of cells. Sense lines are string connected to PSV devices using contacts that connect the sense metal with the ends of PSV devices. Sense lines intercept a string of cells. Sense currents run through each cell in a selected row and produce fields that cause the magnetization in the thin and thick layers to become perpendicular to the long axis of the PSV. Bias lines can be run above the PSV devices, separated from the PSV devices by a dielectric. The sense currents are generated by peripheral circuitry. Row addressing is used to gate the sense current to particular rows. A digital line that runs parallel to the sense line and perpendicular to the wordlines can also be used to increase selectivity. During programming, the combined word and sense fields set the magnetization of the thick magnetic layer in the selected cell and cause that cell to be written. During sensing, a selected cell is affected by both the read current and the word current. The row in which the selected cell resides is decoded and connected to a sense amplifier to provide gain and to enable signal detection as shown in Figure 13.83.

Figure 13.81 shows the currents and fields needed to write and to read a PSV cell in a magnetic memory. Writing to a selected cell is achieved by setting the state of its storage layer and is accomplished by applying a magnetic field from the sense

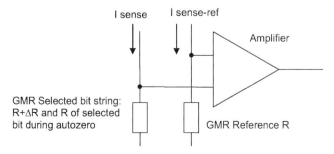

Figure 13.83. GMR magnetic memory read architecture.

current and the word current. Applying both currents allows writing to the selected cell and inhibits changing the state of the other cells that see only one of the currents or neither current. The direction of the word current for a selected cell dictates the state (i.e., 1 or 0) of the written bit. An optional current can be applied along with the sense current to provide additional bias to help writing to a selected bit if desired.

Referring again to Figure 13.82, sensing is achieved by providing a word current in the presence of a sense current. The word current is used to switch the sense layer and then switch it back to its original state, thereby accomplishing nondestructive read because the magnetic storage layer is never switched during the sense operation. The magnetoresistance is sensed when the thin layer is magnetized in each direction. The polarity of the read signal determines which logical state has been read.

13.5.3 Magnetic Random-Access Memory with Magnetic Tunnel Junction Devices

As depicted in Figure 13.84, a magnetic tunnel junction [56, 79–86, 88–95] consists of two magnetic layers separated by an insulating tunnel barrier layer. One of the magnetic layers is the storage layer, in which the direction of magnetization is used to store the digital memory state for the cell. The other magnetic storage layer is a pinned layer, which is magnetically fixed in orientation using ferromagnetic–antiferromagnetic coupling between a ferromagnetic pinned layer and an antiferromagnetic layer. During readback, a tunneling current flows from one magnetic layer to the other magnetic layer through the insulating barrier.

Figure 13.85 presents a conceptual magnetoresistance hysteresis curve for an MTJ where magnetoresistance is shown as a function of applied field. If the mag-

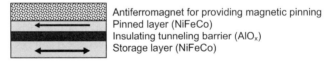

Figure 13.84. Cross section of a magnetic tunnel junction (MTJ).

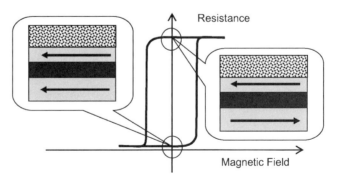

Figure 13.85. Magnetoresistance as a function of applied field for a magnetic tunnel junction (MTJ) device. The magnetization directions for the two storage layer resistance states are shown in the two insets.

MAGNETIC MEMORIES

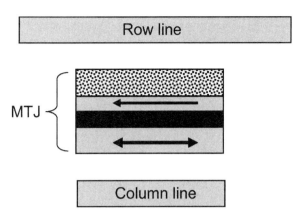

Figure 13.86. Write row-and-column addressing and read row-and-column addressing for a magnetic tunnel junction (MTJ) device.

netization direction of the storage layer is in the same direction as the magnetization in the pinned layer, electron scattering is reduced because of the tunneling magnetoresistance (TMR) effect that is a consequence of the spin polarization of the electrons. The MTJ is therefore in the low resistance state. If the magnetization direction of the storage layer is in the opposite direction of, that is, antiparallel to, the magnetization in the pinned layer, electron scattering is increased and the MTJ is in the high resistance state. The difference in resistance is used to discriminate the two information states during readback. Readback is nondestructive since the written information state is not changed during the read process.

Shown in Figure 13.86 is an MTJ device embedded within a row line and column line architecture for performing writing and readback. In a memory array, during the write process, the MTJ that is selected experiences magnetic fields from both row line and column line currents. The magnetization in the selected MTJ's storage layer switches in the desired information state direction. During the read process, a read tunneling current flows through the selected MTJ through the row line or column line conductors to ground. The magnitude of the read tunneling current depends on the magnetoresistance and therefore information state of the MTJ. The MTJ device produces a read signal that is amplified to facilitate readback.

Shown in Figure 13.87 is an MTJ bit [90, 91, 95] contained in an MRAM bit cell. The digit line and bitline is shown in the bit cell. Also shown is a transistor in the cell. When the transistor is on, a tunneling current flows to allow sensing the state of the bit cell. When the transistor is off, the bit cell is not in the read mode and is either in a standby mode, off mode, or the write mode. Parallel magnetization states of the bit supports low resistance for readback. Antiparallel magnetization states of the bit supports high resistance for readback.

13.5.4 Programming Characteristics

13.5.4.1 Programming Voltage. Programming of magnetic memory cells depends on current magnitudes and therefore does not require high programming voltages. Magnetic memories operate at 5 and 3.3 V and appear scalable to 2.5 V, 1.8 V, and lower power supply voltage standards. As supply voltages and

686 13. ALTERNATIVE MEMORY TECHNOLOGIES

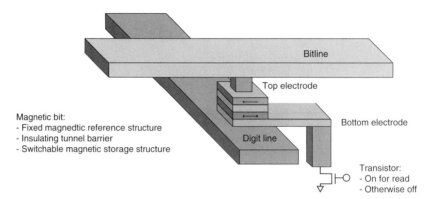

Figure 13.87. Bit cell for an MRAM based on an MTJ device.

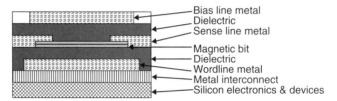

Figure 13.88. Cross section of a magnetic memory element.

lithographic feature sizes are expected to continue to scale to finer dimensions, bitlines and digit lines are expected to scale accordingly. In addition, switching fields are expected to reduce as switching dimensions and variability continue to reduce. On-chip voltage regulation can be implemented if needed.

13.5.4.2 Programming Speed. The switching times for the magnetic materials used in magnetic memories are in the nanosecond to subnanosecond time regimes. The programming speed is therefore dominated by the times needed to switch currents and to perform address selection and decoding. Programming times are typically a few tens of nanoseconds per word. The word length is set by the memory architecture. Word lengths have been 8 bits and have proceeded to 16-bit words.

13.5.5 Fabrication

Magnetic memory processing technology is compatible with semiconductor processes for both stand-alone memory and embedded memory applications, including application-specific integrated circuits, digital signal processing circuits, and encryption electronics. As indicated in Figure 13.88, a magnetic device becomes a memory element when integrated and processed with interconnection metal layers on top of silicon underlayers. A metal layer beneath the magnetic device is used to provide the word current and to provide interconnection to silicon electronics. A metal layer is used to interconnect rows of magnetic devices and to provide the sense current. A metal layer can be placed above the magnetic bit, if necessary, to provide additional bias that augments the sense current's magnetic field. The electronics underlayer generally consists of the passive and active semiconductor devices and metallization interconnect and power routing layers. One or more metal layers can be used in the electronic underlayer, between the silicon processing and magnetic layers.

13.5.6 Nonvolatile Characteristics

13.5.6.1 Endurance. As in the case of other magnetic memories, such as plated wire memories and magnetic core memories, the magnetic device cells in a magnetic memory have no wear out mechanisms and no imprint. Virtually unlimited endurance can be achieved. Trillions of bit cycles have been performed with no signs of degradation, as expected.

Magnetic memories use currents for programming and sensing. As long as proper design techniques are used and followed in circuit design, fatigue mechanisms in the peripheral circuitry, such as electromigration in metal lines, can be avoided. Copper lines can also be used instead of aluminum lines to reduce the resistive and electromigration effects. The endurance of magnetic memories is ultimately expected to be limited by mechanisms not directly related to the magnetic storage element.

13.5.6.2 Retention. The magnetic materials used in magnetic memories are ferromagnetic and have Curie temperatures that approach or exceed 700°C, generally in the range of 400 to 1200°C [72, 76–78, 96]. For temperatures at typical memory operating and storage ranges, changes in magnetization and anisotropy coefficients are small. Activation energies for magnetic retention are relatively high, on the order of a few electron volts. Temperature coefficients of magnetic switching fields are also characterizable and can be matched to circuit capabilities. Magnetic memories can readily meet retention specifications for high-reliability applications.

13.5.7 Scaling

13.5.7.1 Magnetic GMR Cell Scaling. A magnetic device cell is scalable. Considered here will be a GMR PSV device, noting that the characteristics apply to other magnetic devices. As the length and width of a PSV device are scaled in proportion, the resistance is constant because the number of squares of resistance is kept constant. Therefore, for a constant GMR coefficient, signal amplitude can be kept constant as the area is decreased. As the thickness of the PSV device is decreased, the switching fields also decrease proportionately. This reduces switching currents and switching power. As the thickness of the PSV device is reduced, the sheet resistance increases so that the read signal also increases proportionally as a result of the product of the magnetoresistance coefficient and the resistance of the cell. As the width of the PSV device is decreased, the switching fields and selectivity increase. Switching fields and selectivity can be optimized by the choice of width and thickness. These combined dependencies allow maintaining or increasing operating margins as the technology is scaled down to smaller design rules and layer film thicknesses.

13.5.7.2 Magnetic Tunnel Junction Scaling. An MTJ cell is also scalable. The resistance and change in resistance of the MTJ increase as the area of an MTJ cell is scaled to smaller dimensions. For a constant TMR coefficient, signal amplitude increases as the area is decreased. As the thickness of the MTJ is decreased, the switching fields also decrease proportionately. This reduces switching currents and switching power. As the thickness of the MTJ device is reduced in proportion to the area, the resistance can be kept constant so that the read signal can also be kept constant as a result of the product of the magnetoresistance coefficient and the resistance of the cell. As the width of the bit is decreased, the switching fields and selectivity increase. Switching fields and selectivity can also be optimized by how the width and thickness are chosen. These combined dependencies allow maintaining or increasing operating margins as the technology is scaled down to smaller design rules and layer film thicknesses.

13.5.8 Reliability

Magnetic memory technology involves the integration of magnetic technology with semiconductor technology. As such, the reliability and FIT failure rates of magnetic memory technology [97] are comparable to those achieved in semiconductor memory devices and magnetic transducers. It is expected that magnetic memory reliability is dominated by the reliability of semiconductor devices and metal layers, which are subject to known effects and characteristics such as hot-carrier effects, stress, leakage, and electromigration [97]. If desired, error detection and correction may also be applied to lower bit error rates.

13.5.9 Die and Test Cost

A general comparison between magnetic memory and Flash memory with respect to die and test cost can be made. Since Flash memory is a high-volume production technology, Flash memory is currently expected to have a cost advantage. From a technological perspective, magnetic memories use standard semiconductor processing for the underlayers. The underlayers can include standard silicon technologies such as bulk CMOS or silicon-on-insulator CMOS. The magnetic multilayer film deposition and processing processes are compatible with standard integrated circuit processing techniques and equipment. Since magnetic memory cell processing occurs after semiconductor silicon underlayer and metallization processing, standard semiconductor processing costs apply to the CMOS portion.

The memory cell area of a magnetic memory is comparable to that of Flash memory. The number of masks used is expected to be fewer than those used for Flash memory. High-volume die cost of magnetic memory may be comparable to that of Flash memory provided that the yields are comparable.

Because of the fast programming times for magnetic memory, the test time for magnetic memory may be less than that for a Flash memory. Typical programming times for magnetic memories are well under 1 µs, whereas typical erasing and programming times for Flash memories can be greater than 100 µs. With programming times at least 100 times shorter for magnetic memories, test costs may be reduced. Therefore, the tested die cost for magnetic memories manufactured at the same design rules as Flash memory may be comparable to or less than that of Flash memory.

Because the tested die cost of both magnetic memories and Flash memories fabricated at the same design rules may be comparable, magnetic memories may have the potential to compete with Flash memory. However, magnetic memory technology emerging [86] into the marketplace, are several generations behind those of Flash memory. Power and supporting on-chip circuitry considerations and trade-offs also need to be made. Since magnetic processing occurs late in the wafer fabrication process of a magnetic memory and the memory cells are small, magnetic memories may be able to make progress to higher densities and capacities, depending on market development.

13.5.10 Magnetic Memory Summary

Magnetic memories have historically served as random-access memories, ranging from the middle to late twentieth century with the use of core and plated wire memories. Advancements in magnetic materials and devices and the ability to integrate these devices with more advanced semiconductor technologies has been realized in the late twentieth century and early twenty-first century. In particular, GMR devices and MTJ devices with higher signal levels have been developed, which have also found applications in high-density hard disk drive storage.

Magnetic random-access memories (MRAMs) are offering performance advantages as a desirable nonvolatile memory. MRAMs offer nonvolatility, low programming voltages, including operation at standard chip power supply voltages, fast programming times and fast read times, in the nanosecond regime, and virtually unlimited endurance with excellent reliability characteristics. Magnetic memory may be desirable in a variety of stand-alone and embedded applications, with desirability enhanced as costs are further reduced. GMR-based MRAMs have been developed and demonstrated. MTJ-based MRAMs have been developed and demonstrated and have entered the marketplace as a production technology.

13.6 SINGLE-ELECTRON AND FEW-ELECTRON MEMORIES

Konstantin K. Likharev

13.6.1 Introduction

The continuing trend of scaling down integrated circuit components [11], in particular floating gates of nonvolatile memories, is rapidly leading us to the point where the gate will be charged with just a few electrons. Figure 13.89 shows the number of electrons that should charge a gate to produce a voltage of 1 V, as calculated using a simple but reasonable electrostatic model. One can see that at the gate size of 25 nm (predicted to be attainable by the mid-2010s), the number is reduced to some 20 electrons. At that stage some questions arise: What are the major consequences of the discreteness of the electron charge, and would this discreteness lead to additional problems or present new opportunities for floating-gate memories? Fortunately, due to substantial solid-state physics research since the 1950s (and especially since the invention of the single-electron transistor in 1985), answers to most of these questions are already on hand—see, for example, the reviews [98–101].

13.6.2 Electric Charge Quantization in Solids

Probably, the most important result of single-electron physics research has been the discovery of the following *impedance restriction* on electron discreteness: The

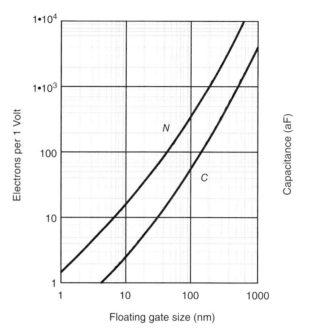

Figure 13.89. Floating-gate capacitance C and the number n of electrons necessary to charge the gate to 1 V, as calculated for a simple model. In this model the gate is a cube embedded into a dielectric matrix with the dielectric constant $\varepsilon = 4$, with two sides occupied by tunnel junctions with a barrier thickness of 10 nm.

number N of electrons in any conducting "island" (e.g., a floating gate) is a well-defined integer only if the full conductance $G \equiv \text{Re } Z^{-1}(\omega)$ of the circuit elements connecting this island to the conducting environment (e.g., the bit and wordlines) is low enough:

$$G \ll 1/R_K \qquad (13.13)$$

where R_K is the so-called quantum unit of resistance:

$$R_K = h/e^2 \approx 25 \text{ k}\Omega \qquad (13.14)$$

More specifically, Eq. (13.13) should be satisfied at frequencies ω of the order of E_C/h, where E_C is the single-electron charging energy:

$$E_C = e^2/2C \qquad (13.15)$$

with C being the island capacitance. (In cases of practical interest, E_C/h is in the range between 10^{13} and 10^{15} Hz).

If the condition expressed by Eq. (13.13) is not satisfied, the laws of quantum mechanics cause the electron density to be spread around, and their number N be undetermined at any fixed instant, even if on the average N is small. The physical interpretation of this fact is that in systems with high conductance the electron wavefunctions are well extended, and electrons cannot be treated as classical, localized particles. One very important example: A small-size MOSFET does not exhibit any immediate single-electron effects even if the average number of carriers in its channel is smaller than one since, in these devices the channel-contact resistance is typically much lower than R_K. On the other hand, in nonvolatile memories the active conductance of the tunnel barrier separating the floating gate from other memory cell components is very low, and the number of electrons in the gate is a well-defined "c-number," though for present-day implementations this number is very large (~10^5), and its discreteness is not readily noticeable.

Another source of uncertainty of N is classical, thermal fluctuations. These fluctuations may be important if their energy scale $k_B T$ becomes comparable to the charging energy E_C defined by Eq. (13.15). In fact, the root-mean-square (rms) thermal fluctuation of N is

$$\delta N \approx (k_B T / E_C)^{1/2} \qquad (13.16)$$

so that it is negligibly small only at

$$k_B T \leq E_C / \ln^{1/2}(1/p) \qquad (13.17)$$

where $p \ll 1$ is the probability of getting an undesirable value of N. Notice, however, that for floating-gate memories Eq. (13.16) describes the statistics of a large ensemble of write/erase cycles; within the time intervals between these operations the gate charge is fixed and the number of electrons does not fluctuate.

13.6.3 Single-Electron Effects in Memory Cells

When applied to floating-gate memories, the above statements mean that as soon as the gate is decreased to a few tens of nanometers, the quantization of the gate charge should become visible. This effect has been indeed repeatedly demonstrated experimentally, first in cryogenic environment where Eq. (13.17) may be more easily satisfied. For example, Figure 13.90(a) shows a scanning electron microscopy (SEM) picture of the so-called single-electron trap, where a 100-nm-scale metallic floating gate is separated from the bias line by a sort of one-dimensional array of small tunnel junctions [102, 103]. (Such an array acts similarly to a tunnel layer, creating a potential barrier that may be suppressed by an external electric field—see Likharev [101] for details.) The trap state may be read out by a sensitive electrometer (in this particular case, a single-electron transistor). Figure 13.90(b) presents a typical hysteretic charging curve showing a bistability caused by trapping of a single additional electron. In these experiments, an electron retention time above 12 h has been demonstrated [103].

Later, such experiments have been extended to room temperature [104–106]. For example, Figure 13.91 illustrates a nanoscale floating-gate memory cell (~7-nm floating gate, ~10-nm MOSFET) using the ordinary Fowler–Nordheim tunneling through a silicon oxide barrier, and the results of testing it at room temperature [104]. Figure 13.91(b) shows that the addition of a single electron to the gate changes the transistor current considerably. Single-electron charging effects have been also observed in numerous experiments (pioneered by an IBM group [107–109]) with floating gates consisting of many ($M \neq 1$) nanocrystals.

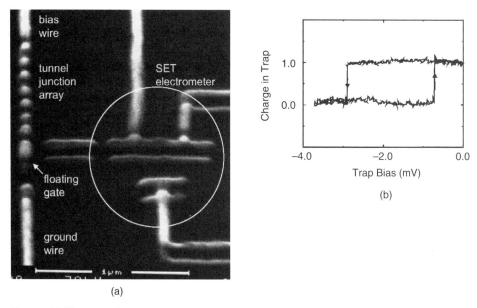

Figure 13.90. One of the single-electron traps explored at Stony Brook: (a) scanning electron microscope picture, and (b) the result of testing at low temperatures (below 1 K). (*Picture courtesy by J. Lukens Stony Brook University.*)

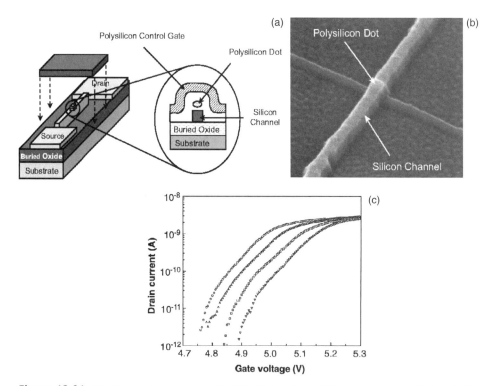

Figure 13.91. Floating-gate memory cell of the University of Minnesota group: (a) the device scheme, (b) SEM picture taken before the control gate deposition, and (c) the MOSFET subthreshold curves for several numbers of electrons added to the floating gate. (After [104], with permission.)

13.6.4 Single-Electron Memories

One could infer that the floating-gate charge quantization could be useful for practical nonvolatile memories, since it might help to discriminate between discrete output signals on a noisy background, and possibly use this effect for multivalue storage using several (rather than 2) values of N. Unfortunately, these opportunities run into two problems.

First, in order to make N reproducible cycle-to-cycle, Eq. (13.17) has to be satisfied. For $T = 300\,\text{K}$, and a reasonable bit error rate p (say, 10^{-10}), this requires very small islands (below ~3 nm) that cannot be fabricated with acceptable precision of a few percent by any currently available, or even reasonably envisioned, industrial-scale patterning technology [11].

Second, all single-electron devices (and in fact virtually all sub-10-nm electron devices [110]) are affected by the infamous effect of random background charge; see Fig. 13.92). The dielectric environment of a floating gate may contain randomly located charge impurities. Even a single impurity located at a distance of the order of the gate size changes the electric potential of the gate by a random amount of the order of $E_C/e = e/2C$, that is, of the difference between the quantized values corresponding to the neighboring values of N. Using an optimistic estimate of $10^{11}\,\text{cm}^{-2}$ for the minimum possible surface/interface charged trap concentration, and assuming an island size of 3 nm (i.e., footprint ~$10^{-13}\,\text{cm}^2$), we see that a substantial fraction ($\varepsilon \sim 1\%$) of the cells will have a considerable background charge fluctuation of the quantized voltage levels.

Such a small fraction of bad bits may be readily tolerated using either the replacement of the worst memory matrix lines for spares, or error correction codes, or a combination of these two techniques [111], but if ε is substantially larger, such exclusion may lead to unacceptable losses of chip area.

13.6.5 Few-Electron Memories

Both problems outlined above may be relieved if we operate with a small but finite number of electrons, just ignoring the discreteness of the charging states even if it is visible. Several experiments of this kind have been reported. (Unfortunately,

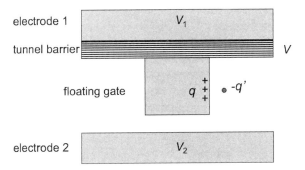

Figure 13.92. Effect of a single charged impurity on the nanoscale floating gate of a nonvolatile memory cell (schematically). If the gate capacitance is small, the image charge $q \sim e$ induced by the impurity changes the voltage drop across the tunnel barrier significantly, even if the voltages $V_{1,2}$ applied to the external electrodes and the total charge $Q = ne$ of the gate are fixed.

some authors speak of these devices as "single-electron memories," creating a certain terminological confusion.) Probably, the largest publicity has been created by a demonstration, by a Hitachi group, of the first "128-Mb single-electron memory chip" [112]. Figure 13.93 shows the basic idea of operation of the cells used in this memory [113, 114]. MOSFETs with a granular silicon channel show memory effects due to the capture of single electrons in certain stand-alone grains of the channel. After the capture, the electron changes the conductivity of the more conductive parts of this channel. The main problem of such a memory cell is that the exact geometry of the grains, and hence the memory cell switching thresholds, are inherently irreproducible. This drawback could not be eliminated even when the Hitachi group started to use a few electrons for coding each bit [112–114]. I believe that this problem does not leave any hope for the implementation of practical memories based on this principle.

It seems much more realistic to operate with a few electrons using either the usual Flash memory structures, or the cells suggested for NOVORAM (see Fig. 13.103), or similar cells with the readout MOSFET replaced with a single-electron transistor (SET)—see Figure 13.94(b).

Such a transistor [117] (which is probably the most important of all single-electron devices [98–102]) is generally similar to the FET, but the channel is replaced by a small conducting island separated from the source and drain by tunnel barriers with small transparency—see Eq. (13.13). Since tunneling of a single electron from the source to the island requires a charging energy of the order of E_C [see Eq. (13.15), where C is now the total capacitance of the transistor island], at sufficiently low temperatures ($k_B T \ll E_C$) the source–drain current is blocked until the source–drain voltage V reaches a certain threshold value $V_t \sim e/C$. This value can be controlled with the voltage V_g applied to the gate (in this particular case, the floating gate), which is capacitively coupled to the device island since the voltage changes the initial electrostatic potential of the island. As a result, the gate voltage effectively controls the source–drain current at $V \geq V_t$. In the SET, this control differs from that in the MOSFET in two important ways:

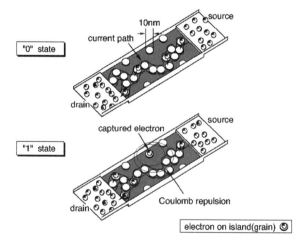

Figure 13.93. Graphic showing the operation principle of Hitachi's memory cell. A single electron captured in a silicon grain may block the transfer of electrons along a percolation current path. [*Courtesy by T. Ishii (Hitachi)*].

SINGLE-ELECTRON AND FEW-ELECTRON MEMORIES

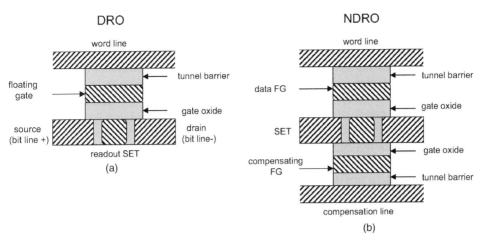

Figure 13.94. Possible structure of few-electron memory cells with (a) destructive [115] and (b) nondestructive [116] readout.

1. The SET sensitivity is very high: The addition of just one electron to the gate charge Q changes the current considerably, and changes of Q as small as $10^{-5}e$ may be detected.
2. The current is a periodic function of the gate voltage V_g, so that the transconductance sign alternates with the change of V_g.

These features make the SET a valuable device for several applications, including ultrasensitive electrometry [98–101]. Unfortunately, the device suffers from the effect of the background charge randomness discussed above, which hinders its direct use in digital circuits [101]. In order to overcome this handicap, two methods have been suggested.

The first, the destructive readout (DRO) option [Fig. 13.94(a)] [115], uses the fact that during the write process, the electrons tunneling into the floating gate ramp up its voltage V_g. This ramp up causes a few oscillations of the SET current. These oscillations may be picked up, amplified, and rectified by a sense amplifier (one MOSFET per line seems sufficient for this purpose), with the resulting signal V_{out} serving as the output. The main idea behind this proposal is that the random background charge will cause only an unpredictable shift of the initial phase of the current oscillations, which does not affect the rectified signal. This concept has been verified in experiments [116] with a low-temperature prototype of the memory cell.

An attractive feature of this design is the relatively mild fabrication requirements. Room temperature operation is possible with an electron addition energy of about 250 meV. This requires a minimum feature size of about 3 nm, that is, much larger than that required for purely single-electron digital circuits. (The reason for this relief is that in this hybrid memory the single-electron transistor is used in essentially an analog mode, as a sense preamplifier/modulator, and can tolerate a substantial rate of thermally activated tunneling events.) The estimated density of the memory at this fabrication level is close to 100 Gb/cm². Further scaling down the device runs into the limitation imposed by the discreetness of the floating-gate charge, which has been described above. As a result, the floating-gate potential and hence read/write operation thresholds become sensitive to the effect of random charged impurities. On the negative side, the destructive readout (combined with write 1) requires the subsequent restoration of the initial contents of the cell, but this operation is not much more complex than refresh in ordinary DRAM. The second, relatively minor, drawback is the need for a sense amplifier/rectifier. Estimates show that since the signals are preamplified with the single-electron transistor with its very low noise, one FET amplifier may serve up to 100 memory cells and hence the associated chip real estate overhead per bit is minor.

The second way, which allows nondestructive readout, has been proposed in Korotkov [118]. In this design [Fig. 13.94(b)]; the random background charge of the island is compensated by weak capacitive coupling with the additional ("compensating") floating gate. The necessary few-electron charge of this gate may be inserted from an additional ("compensation") wordline before the beginning of memory operation, when special peripheral CMOS circuits measure the background charge of each single-electron transistor (this may be done in parallel for all bits on a given wordline) and develops an adequate combination of bitline signals. During the actual memory operation, the compensating charge is constant but may be adjusted periodically if necessary. Due to the compensation, the SET may be biased reliably at the steep part of its control characteristic and hence used for nondestructive readout of the cell state. Moreover, the signal charge swing may be relatively small, $\Delta Q_0 \sim (k_B T/E_a)e$, conveniently decreasing with the island size. This may allow the single-electron island to be scaled down to ~1 nm, with the ~3-nm floating gate storing just 2 to 3 electrons.

For both designs, the estimated power density (~3 W/cm^2, mostly in the sense amplifiers) also seems quite acceptable.

Of course, the 3-nm SET islands necessary for the implementation of the few-electron memories discussed above cannot be fabricated reproducibly using the current, and even rationally envisioned, patterning technologies. This is why right now the prospects for such memories are overshadowed by those for resistive memories, and their extension to hybrid CMOS/nanodevice memories, which are described in the next section. Note, however, that deep (sub-10-nm) scaling of such memories may also require single-electron devices, though in a different, molecular-electronic form.

13.7 RESISTIVE AND HYBRID CMOS/NANODEVICE MEMORIES

Konstantin K. Likharev

13.7.1 Introduction

All memories described above require at least one field-effect (or single-electron) transistor per cell. This imposes heavy limitations on memory cell scaling, virtually forbidding its extension into the sub-10-nm range [112]. There is, however, a class of memories that may work without a transistor. Figure 13.95 shows the generic structure of such "resistive" (a.k.a. crossbar, or 0T, or 0T/1R) memories.

In such memory, information bits are stored in two-terminal bistable nanodevices (called either "programmable diodes" or "latching switches"), which are

RESISTIVE AND HYBRID CMOS/NANODEVICE MEMORIES

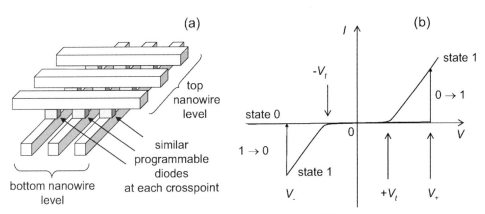

Figure 13.95. Resistive memory: (a) crossbar array structure and (b) I–V curve of a single nanodevice (schematically).

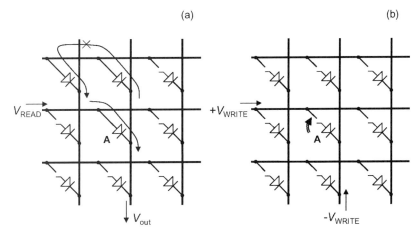

Figure 13.96. Equivalent circuits of the array showing (a) read and (b) write operations for one of the cells (marked A). On (a), the arrow by cell A shows the useful readout current. The looping arrow shows a potential parasitic current going to the wrong output wire.

formed at each crosspoint of a crossbar structures [Fig. 13.95(a)]. The device I–V curve [Fig. 13.95(b)] has two branches corresponding to its two possible internal states. (In the equivalent circuits shown in Figure 13.96, this bistability is represented by a switch. Note that a quantitative analysis of memory performance requires more accurate equivalent circuits.) In the low-resistive state presenting binary 1, the nanodevice is essentially a diode, so that the application of voltage $V_t < V_{\text{READ}} < V_+$ to one (say, horizontal) wire leading to the memory cell gives substantial current injection into the second wire [Fig. 13.96(a)]. This current pulls up voltage V_{out}, which can now be read out by a sense amplifier. [The diode's property to have low current at voltages above $-V_t$ prevents parasitic currents that might be induced in other state-1 cells by the output voltage—see the looping arrow in Fig. 13.96(a) and a quantitative analysis in Amsinck et al. [119]. It also reduces shot noise in the output line.]

In state 0 (which presents binary zero) the crosspoint current is very small, giving a nominally negligible contribution to output signals at readout. In order to switch it to state 1 (i.e., write binary 1 into the cell), the two wires leading to the device are fed by voltages $\pm V_{\text{WRITE}}$ [Fig. 13.96(b)], with $V_{\text{WRITE}} < V_+ < 2V_{\text{WRITE}}$. (The left inequality ensures that this operation does not disturb the state of "semiselected" devices contacting just one of the biased wires.) The write 0 operation is performed similarly using the reciprocal switching with threshold V_- [Fig. 13.95(b)]. It is evident from Figure 13.96 that the read and write operations may be performed simultaneously with all cells of one row. (Actually, only one of the "write 0" and "write 1" operations can be performed simultaneously with all cells. Because of the opposite polarity of the necessary voltages across nanodevices for these two operations, the complete write may be implemented in two steps, e.g., first writing 0's and then writing 1's.)

Thanks to the no-transistor structure of the resistive cell, its area may be very close to $4F^2$, even without pursuing any three-dimensional integration options. So, at fixed design rules the resistive memories may potentially have the highest ("ultimate") cell density. Another significant advantage of the resistive memories over other prospective memory technologies is that the two-terminal devices with necessary characteristics have only one [in Fig. 13.95(a), vertical] critical dimension. This dimension may be defined by the thickness of deposited film(s) and does not require lithographic definition, opening a way toward deep scaling of the devices and memory cells.

13.7.2 Programmable Diode Technologies

Two-terminal devices with the functionality of programmable diodes [Fig. 13.95(b)] have been experimentally demonstrated using several structures, notably including amorphous metal–oxide films [120–127], relatively thick organic films (both with [128–132] and without [133–135] embedded metallic clusters), self-assembled molecular monolayers [136–141], and thin chalcogenide layers [141, 142]. Despite the very early stage of development, the reproducibility of some of these devices is already very impressive (Fig. 13.97 [127]).

For most programmable diode types, the device bistability mechanism is not yet clear, but for the currently most reproducible metal–oxide devices it is probably due to electron trapping in localized states [143]. The basic drawback of the diodes based on this mechanism is that in order to feature reasonable current density, the distance between active localized states cannot be much smaller than ~3 nm. In order to be statistically reproducible, the device should have a large number of the states. This is why the extension of the excellent reproducibility demonstrated for crosspoint devices with $F_{\text{nano}} > 100$ nm (as in [127]) to cells with $F_{\text{nano}} < 10$ nm may present a major challenge. (A similar conclusion may be made concerning the programmable diodes based on organic materials, where reversible formation of conductive filaments is the leading candidate for the interpretation of experimental results.)

This problem may be addressed using uniform self-assembled monolayers of specially designed molecules (Fig. 13.98 [144]) implementing single-electron latching switches—see Section 13.6, in particular Figures 13.90 and 13.94 and their discussion. So far molecular implementations have been only demonstrated [145–149] for the main components of these devices, single-electron transistors. A major challenge

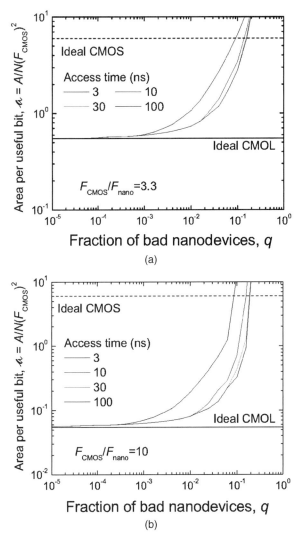

Figure 13.97. Copper-oxide programmable diode developed by Spansion LLC: (a) microphotograph of the device, (b) its I–V curve, and (c) ON and OFF current statistics. (From [127], with permission.)

here is the reproducibility of the interface between the monolayer and the second (top) metallic electrode, because of the trend of the metallic atoms to diffuse inside the monolayer during the electrode deposition [150]. Recent very encouraging results toward the solution of this problem have been obtained using an intermediate layer of a conducting polymer [151]. Another opportunity is to terminate the synthesized molecules, before their self-assembly, with special metallic clusters or large acceptor groups that would play the role of "floating electrodes" [152]. The recently reported experimental results [153] may be considered as the first step toward the practical implementation of this opportunity.

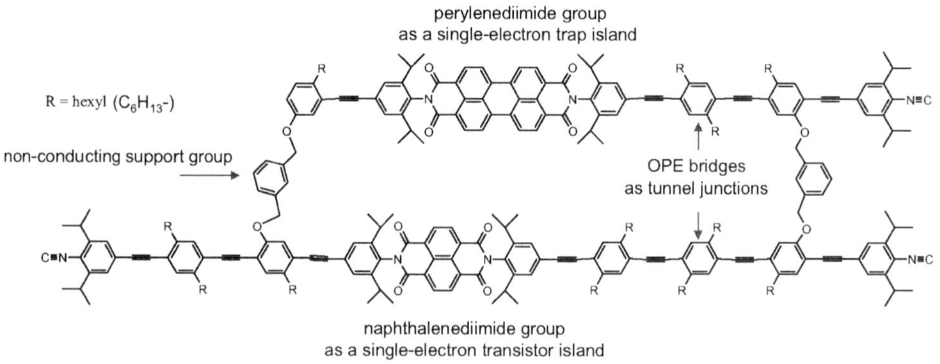

Figure 13.98. Proposed molecular implementation of a single-electron trap.

13.7.3 Hybrid CMOS/Nanodevice Resistive Memories

Besides scaling the programmable diodes below 10 nm, the implementation of resistive memories of this range would require similar scaling of the crossbar wiring [Fig. 13.95(a)]. This may be done by several advanced patterning technologies, such as nanoimprint [154, 155] or interference lithography [156, 157], which may combine acceptable fabrication speed with very high resolution, at acceptable cost. Indeed, nanoimprint technology has already allowed a crossbar memory prototype with $F = 30$ nm [140] and a nanowire crossbar with $F = 17$ nm [158] to be experimentally demonstrated, and there are good prospects for the half-pitch reduction to 3 nm or so within the next decade [154].

Unfortunately, these high-resolution patterning technologies do not offer equally accurate layer alignment and cannot be used for the fabrication of the necessary peripheral circuits including decoders, line drivers, sense amplifiers, and the like. This is the main reason for the recent intensive work on hybrid CMOS/nanodevice circuits, which focuses mostly on augmenting a nanowire/nanodevice crossbar [Fig. 13.95(a)] with an appropriate CMOS subsystem—see the literature [152, 159–162] for recent reviews.

The main conceptual problem here is interfacing a crossbar with its nanoscale half-pitch F_{nano}, with the much cruder CMOS wiring. Several initial suggestions for forming such interfaces at the crossbar array periphery do not seem very practicable [153]. Recently, our group suggested [163–165] (see also [152]) a new approach, dubbed CMOL (CMOS molecular circuits), in which the interface is provided all over the chip surface, using pins with the CMOS pitch but nanoscale-sharp tips (Fig. 13.99). The main feature of the CMOL topology is that it provides a unique access from the CMOS subsystem to each crossbar nanowire (whether in the bottom or the top layer) with the theoretical 100% fabrication yield even in the absence of *any* alignment between the CMOS and crossbar subsystems. (Note that the last statement is only true for the recently introduced [162] CMOL version in which each pin, going to the upper nanowire level, intentionally interrupts a lower level wire—see Fig. 13.99.)

It is important that CMOL circuits (with appropriate modifications of their CMOS subsystems) may be used not only for resistive memories but also for high-performance logic circuits [165] and mixed-signal neuromorphic networks [166] for advanced signal processing (e.g., ultrafast optical image recognition [167]). These

RESISTIVE AND HYBRID CMOS/NANODEVICE MEMORIES

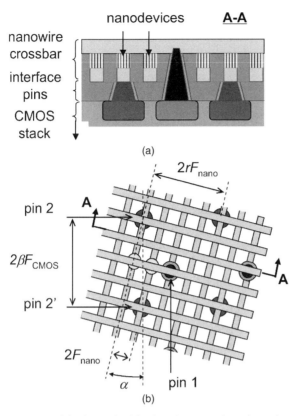

Figure 13.99. CMOL circuit: (*a*) schematic side view (cross section along the A-A line) and (*b*) a top view. For clarity, the last panel shows only two adjacent crosspoint devices that may be addressed via pin pairs {1, 2} and {1, 2'}. The figure shows that the CMOS system has a unique access to each nanowire (and hence to each nanodevice) if the nanowire crossbar is rotated relative the interface pin array by a specific angle $\alpha = \arctan(1/r) = \arcsin(F_{nano}/\beta F_{CMOS}) \ll 1$. Here r is an integer, and β is the distance between the adjacent pins (leading to the same crossbar level), expressed in terms of the CMOS pitch $2F_{CMOS}$. (For a more detailed discussion of the CMOL topology, see, e.g., [152].)

additional potential applications may leverage the resistive memory development costs.

13.7.4 Expected Performance

A recent thorough analysis [168] has shown that a synergy of bad line exclusion and advanced error correction codes may enable the hybrid CMOS/nanodevice resistive memories to combine high defect tolerance (which may be necessary at the initial stage of crosspoint nanodevice development) with extremely high density and access speed. Figure 13.100 illustrates the main result of this work. Even at the initial stage of the CMOL technology development (when the ratio F_{CMOS}/F_{nano} is of the order of 3), the hybrid memories may become denser than the purely semiconductor memories for a fraction of bad devices as high as ~15%, and at the (quite realistic) value $q = 5\%$ provide a nearly fivefold density edge. For $F_{nano} = 15\,\text{nm}$ this would mean a useful density of ~30 Gbits/cm², that is, the level that the semiconductor

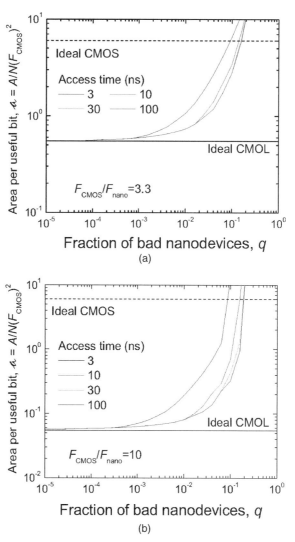

Figure 13.100. Normalized total chip area per one useful memory cell, calculated as a function of the bad bit fraction q, for several values of the memory access time and two typical values of the F_{CMOS}/F_{nano} ratio. The horizontal lines indicate the cell area for the hybrid and "ideal" semiconductor memories. In the latter case, this line corresponds to the results for negligible q, while for the former case we use the ITRS forecast for Flash memories.

technology may be able to reach in the very end of scaling, if ever. [Note that the cell area for semiconductor memories is assumed to equal $(2F_{CMOS})^2$, so that the comparison is rather conservative.]

Moreover, as the hybrid circuit technology matures, and the F_{CMOS}/F_{nano} ratio approaches an order of magnitude (say, F_{CMOS} = 32 nm, F_{nano} = 3 nm), the hybrid memory superiority may become quite spectacular. Indeed, as Figure 13.100(b) shows, for the defect fraction q = 2% (which looks quite plausible), the cell area factor may become as low as 0.1, implying the dimensional density as high as 1 Tbit/cm^2, far beyond the most optimistic projections for other memory technologies.

13.7.5 Resistive Memory Summary

The resistive memories are in the very beginning of their development, and much still should be done to confirm (or deny) the optimistic theoretical predictions. However, given the recent high rate of new outstanding experimental results, one should not be surprised if these memories, especially in their hybrid CMOS/nanodevice version, will fast become the leading candidates for the "ideal" digital memory.

13.8 NOVORAM/FGRAM CELL AND ARCHITECTURE

Konstantin K. Likharev

13.8.1 Introduction

A recent proposal [169, 170] for a substantial improvement of floating-gate memories is based on using layered (so-called crested) tunnel barriers. Theory shows that these barriers, based on available and silicon-compatible materials, may allow the standard 10-year retention time to be combined with very short, sub-10-ns write–erase time. Simultaneously, the maximum electric field applied to the barrier may be decreased below 10 MV/cm, apparently enabling a substantial increase in cell endurance, possibly well beyond 1 billion cycles. As a result, such memories may become one of implementations of the "ideal" memories discussed in the beginning of this chapter—see Table 13.1. On the other hand, even under optimized crested barriers may be used to substantially improve the Flash memory performance including operation at lower voltages and hence higher endurance.

13.8.2 Crested Tunnel Barriers

Let us compare the properties of the standard, uniform tunnel barrier [Fig. 13.101(a)] and an idealized "crested" barrier of symmetric, triangular shape [Fig. 13.101(b)]. Straightforward calculations show [169] that if at small voltages the barriers have similar transparency (and hence may provide similar retention time), at higher electric fields the current through the crested barrier changes much faster. The reason for this improvement is that while in the uniform barrier the electric field tilts the potential profile, it has virtually no effect on its highest point U_{max} [Fig. 13.101(a)]. On the contrary, in the crested barrier the highest part (in the middle) is pulled down by the electric field very quickly [Fig. 13.101(b)]:

$$U_{max}(V) = U_{max}(0) - eV/2 \qquad (13.18)$$

thus increasing the barrier transparency.

The implementation of triangular crested barriers is straightforward in composite semiconductors, where the barrier shaping may be achieved with either a gradual change in the layer composition during its epitaxial growth or by modulation doping. However, the maximum barrier height (conduction band offset) available in these materials is too small to provide sufficient retention time at room temperature. For most prospective wideband materials (Al_2O_3, SiO_2, Si_3N_4, etc.) both these approaches

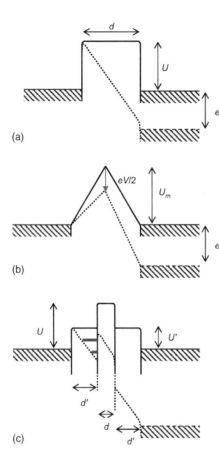

Figure 13.101. Conduction band edge diagrams of various tunnel barriers without (solid lines) and with (dashed lines) applied electric field: (a) a uniform barrier, (b) idealized triangular crested barrier, and (c) layered crested barrier. Thick dashes in (c) show the resonant energy levels (subbands for the full electron energy) formed at the interface of layered barriers in strong electric field.

run into fabrication problems; for example, suitable dopants with shallow levels, necessary for modulation doping, have not yet been found for these materials.

Fortunately, there is another option that seems much more practical [169, 170].[1] A perfect triangular barrier [Fig. 13.101(b)] may be reasonably well approximated by the "staircase" potential pattern formed in a trilayer barrier [Fig. 13.101(c)]. Several theoretical calculations of electron tunneling through such structures have been carried out [169–176]. Figure 13.102 shows a typical result of calculation of the function $j(V)$ for a good trilayer barrier candidate, in comparison with that for the usual uniform barrier. The calculation results show that with the appropriate choice of layer thicknesses, the voltage V_W corresponding to the floating gate recharging in sub-10-nanosecond-range may be just a factor of 2 or so above the voltage V_R corresponding to the standard retention time of 10 years.

---

[1] Actually, several suggestions to use tunneling through layered barriers in floating-gate memories were made long ago—see, for example, [171, 172]. However, the authors of those works considered asymmetric barriers. Though the injection characteristics of such barriers may be even better than those of their symmetric counterparts, this is only true for one current direction (say, for electron injection into the floating gate, i.e., for the "write 1" operation). The speed of the reciprocal process ("write 0" or "erase") is low, thus excluding the possibility of bit-addressable applications. Of course, this opportunity may be restored by connecting in parallel two barriers with reciprocal barrier profiles, but this option seems too complex for practical applications.

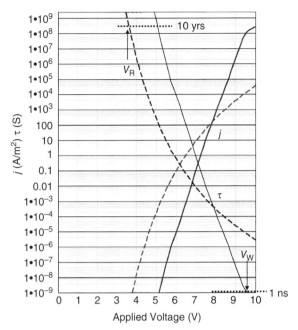

Figure 13.102. Tunneling current density j (in A/m², dashed lines) and the floating-gate recharging time scale $\tau = C_0 V/j$ (in seconds, solid lines) for the barriers shown in Figure 13.101(a) and (c), as functions of applied voltage V, calculated using the WKB approximation. Dashed lines: a uniform SiO$_2$ barrier ($U = 3.2$ eV, $m = 0.3 m_0$, $d = 8$ nm). Solid lines: trilayer crested barrier PGAO/Al$_2$O$_3$/PGAO, where PGAO is plasma-grown aluminum oxide ($U' = 2.0$ eV, $m' = 0.5 m_0$, $\kappa' = 5$, $d' = 2.5$ nm), while for the 5-nm γ-phase Al$_2$O$_3$ layer the bulk values ($U = 3.6$ eV, $m = 0.5 m_0$, $\kappa = 8.5$) have been accepted. C_0 is the specific capacitance of the barrier.

The effect of the fast increase of the crested barrier transparency with the applied electric field may be enhanced by an appropriate relation of the layer dielectric constants κ. Ideally, κ of the middle layer (providing a higher barrier) should be higher than that of the side layers, so that the electric field $E = D/\kappa\varepsilon_0$ (where D is constant for all the layers) in the former layer were as low as possible. (In this case, the top of the conducting band edge profile remains almost flat, ensuring a faster increase of the tunneling current in the moment it is aligned with the Fermi level of the electron source.) Moreover, such difference of κ alone may provide I–V curves that are steeper than those of the uniform barriers [175], though for realistic values of parameters this effect it weaker that that of the barrier height difference.

So far, promising initial experimental results have been obtained for Si$_3$N$_4$/SiO$_2$/Si$_3$N$_4$ trilayers [177, 178] and also bilayers SiO$_2$/ZrO$_2$ [179], HfON/Si$_3$N$_4$ [180], and SiO$_2$/HfO$_2$ [181], though the range of steep current change is still smaller (in static fields, up to 9 orders of magnitude) than that promised by the theory (Fig. 13.102). Evidently, more work on layer materials and deposition is necessary. However, the results obtained in the course of this work [177–183] already may be used to improve Flash and other floating-gate memories. For example, the demonstrated [182, 183] plasma-grown, postannealed aluminum oxide layers with charge to breakdown as high as 10^5 C/cm² at electric field $E \cong 10$ MV/cm (providing write/erase time below

10 ns) may be used for fast, high-endurance RAMs (FGRAM) requiring refresh at reasonable time intervals of the order of 0.1 s [183].

13.8.3 NOVORAM/FGRAM Cell and Architecture

Generally, steeper I–V curves provided by crested barriers are useful for virtually any floating-gate memory architecture. However, it is more notional to use them with a very simple, single-transistor memory cell structure [Fig. 13.103(a)] and the usual NOR memory architecture (Fig. 13.104) [169, 170, 182]. In this cell, digital 1 and 0 are stored as, correspondingly, $+Q$ and $-Q$ charge of the floating gate, with Q somewhat lower than CV_R, where C is the total capacitance of the gate. In the storage mode, electric potentials of all the lines (relative to the conducting substrate) are zero, so that the voltage across the crested barrier is below V_R, guaranteeing a large retention time.

In order to write binary 1 or 0 (while erasing the prior cell contents), the word line potential is raised to high positive or negative value, leading to Fowler-Nordheim tunneling of electrons to or from the floating gate. Note that in contrast to the usual non-volatile memories, the tunneling is performed at the back of the floating gate, while the gate oxide is kept thick enough to suppress tunneling through it at all times. In this geometry, with a metal-oxide tunnel barrier, the floating gate may be metallic, and its thickness reduced to a few nanometers. This may dramatically reduce the electrostatic crosstalk of the adjacent cells and make the memory scalable to at least 10 nm.

The only difference between two versions of the memory, nonvolatile RAM (NOVORAM) and floating-gate RAM (FGRAM), is that the latter version, enabled by the already demonstrated uniform AlOx barriers [182], requires periodic refresh similar to that used in the usual DRAM. In the hypothetical nonvolatile version of the memory, such refresh would not be necessary because of the very long retention time provided by the crested barrier.

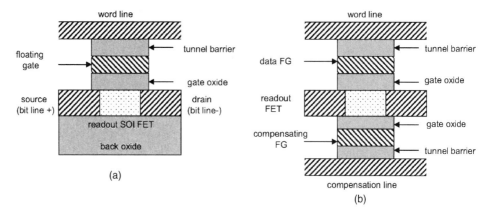

Figure 13.103. NOVORAM cells (*a*) without and (*b*) with the compensating floating gate.

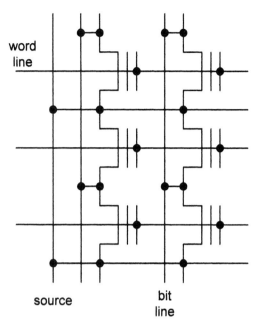

Figure 13.104. NOVORAM/FGRAM architecture.

13.8.4 NOVORAM/FGRAM Summary

Because of its high density (at comparable speed) NOVORAM/FGRAM may present a serious challenge for DRAM and SRAM even at the present technological level, since its nonvolatility is very important for many electronics applications. However, the most important feature of such memories is high scalability, limited mostly by the exponentially growing sensitivity of readout MOSFETs to fabrication spreads. The use of advanced double-gate SOI MOSFETs may push this limit to physical gate length ~5 nm. Apparently, an extension of this limit is possible using individual compensation of the threshold gate voltage of each readout MOSFET using an additional floating gate [Fig. 13.103(b)].

13.9 PHASE CHANGE MEMORIES

Joe E. Brewer, Greg Atwood, and Roberto Bez

13.9.1 Introduction

Phase change memory (PCM) technology is currently (2006) the subject of intense research and development by multiple companies and is a promising future alternative to Flash. In terms of cell size, die size, and cost it is approximately on a par with

Flash. It offers fast random read performance ~50 ns, fast write performance ~100 ns, and good data retention >10 years. PCM is a "direct write" technology. There is no need to erase a block of data and then program it. The technology can be made to fit into a CMOS fabrication process. It exceeds Flash in bit granularity and in endurance. In the long term it appears to have good scalability, and it has multilevel potential.

Phase change memory is certainly not a new concept. For example, the September 28, 1970, issue of *Electronics Magazine* [184] contained the die picture shown in Figure 13.105. This was a 1970 chip with a 256-bit capacity that measured 122 × 131 mils. Reset was accomplished by <200 mA at <25 V for 5 µs. Set required 5 mA at ~25 V for 10 ms. The read mode current was 2.5 mA with a supply <5 V. The history of the technology can be traced back to multiple patents filed by Stanford R. Ovshinsky beginning in the 1960s.

This primitive circa 1970 chip did not prove to be producible or competitive with the emerging floating-gate and MNOS technologies. Soon interest waned and the approach was abandoned. Later phase change technology experienced a rebirth in a new context—optical memory. Chalcogenide films were found to offer advantages when used as the memory medium for rotating disks. The rewritable phase change compact disk (CD-RW), the rewritable digital versatile disk (DVD-RW), the random-access memory digital versatile disk (DVD-RAM), and the blue-ray disk are commercially successful examples.

The optical disk in many ways is a simpler platform for investigation of the phase change phenomena. The active bit storage regions are not individually contacted. It was not necessary to first develop compatible electrode contacts and devise suitable heating elements. The medium is programmed by a high-power laser beam and interrogated by a lower power laser combined with an optical sensor. The primary issue of fast reliable and reversible phase conversion could be examined without the distraction of dealing with these other issues. The disk business provided

Figure 13.105. Photo titled "First Amorphous Semiconductor Memory." (From [184].)

a sound financial incentive, and research on the materials properties progressed rapidly. The basic concepts and materials have been refined, and now the technology may become an important competitive approach. A major part of the intellectual property that makes up PCM technology is controlled by Ovonyx, Inc. (www.ovonyx.com), and the company has teamed with and licensed the technology to many large device manufacturers.

13.9.2 Storage Mechanism

The name "phase change" accurately describes the storage mechanism. Certain materials can be made to change from a polycrystalline state to an amorphous state in a reversible manner. In the crystalline state (called the SET state) the material exhibits a low resistance while the amorphous state (called the RESET state) has a high resistance. Transitions are induced by heating a small volume of the material. When this local region is heated to the melting temperature, it assumes a glassy amorphous form. If it is rapidly cooled after reaching the melting point, the material remains in the amorphous state. To cause amorphous material to change to the microcrystalline form, it is again heated to a temperature below the melt temperature or cooled slowly from the melt state. Under this condition the crystallization process proceeds and low-resistivity polycrystals form. This lower level heating must be maintained until the conversion to the crystallized state is completed.

Ovshinsky [185] once described phase change memory as being based on a material that can exist in two separate structural states in a stable fashion. In order to change the state, an energy barrier must be overcome. This barrier provides the stability of the two different structures. The energy to cause a transition can be applied in different ways. If the energy exceeds a threshold value the material will be excited into a high mobility state where it is possible to rapidly rearrange bond lengths and angles by slight movement of the individual atoms.

13.9.3 GST Phase Change Material

Chalcogenide alloys such as $Ge_2Sb_2Te_5$ and related materials can be made to undergo an electrically initiated, reversible phase change between a glassy amorphous state and a microcrystalline state. The chalcogen elements are oxygen, sulfur, selenium, and tellurium, and their compounds are called chalcogenides. These are nonmetals and semimetals in group 16 of the periodic table (by the old numbering scheme column VI of the periodic table). Most current PCM work is focused on $Ge_2Sb_2Te_5$, but studies have been reported [186] on GeTeAsSi, GeTe, GeSbTe, GeTeBi, GeSb(Cu,Ag), GeTeAs, InTe, AsSbTe, SeSbTe, and PbGeSb, for example.

The notation GST is commonly used to refer to germanium antimony tellurium alloys (Ge–Sb–Te). Compositions along the pseudo-binary tie line $GeTe$–Sb_2Te_3 as typified by $Ge_2Sb_2Te_5$ are of special interest. Figure 13.106 shows the phase diagram and the tie line.

Compositions along this tie line have the unique property of being able to rapidly and reversibly go through the phase change sequence of events. $Ge_2Sb_2Te_5$

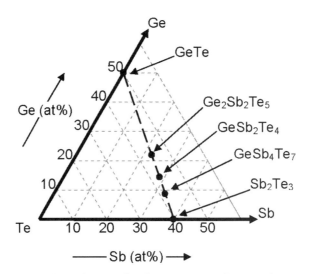

Figure 13.106. Ternary phase diagram for the GST system showing the GeTe–Sb_2Te_3 pseudo-binary tie line and several stoichiometric compositions that lie on the tie line.

TABLE 13.8. Germanium, Antimony, and Tellurium Elemental Properties

Symbol	Atomic Number	Atomic Weight	Ionic Radius	Electronic Shell Configuration											
				$1s$	$2s$	$2p$	$3s$	$3p$	$3d$	$4s$	$4p$	$4d$	$4f$	$5s$	$5p$
Ge	32	72.59	0.53	2	2	6	2	6	10	2	2	—	—	—	—
Sb	51	121.75	0.62	2	2	6	2	6	10	2	6	10	—	2	3
Te	52	127.6	2.11	2	2	6	2	6	10	2	—	10	—	2	4

is often referred to using just the general designation GST, or more specifically as GST225. Crystalline and amorphous GST is referenced as c-GST and a-GST, respectively. Some basic properties of the elements Ge, Sb, and Te are listed in Table 13.8.

Studies conducted in the context of the optical disk application explored the performance and detailed structure of the GST compositions in the vicinity of the GeTe–Sb_2Te_3 pseudo-binary line and found that the crystallization temperature increased and the switching speed degraded as the compositions moved away from the line [187]. The optical disk normal operating scenario is to radiate the material with a high-power narrow pulse width laser beam and impart sufficient thermal energy to the spot until it melts. The resultant thermal agitation destroys any crystalline order in the material structure. When the beam is removed, the spot cools rapidly. For practical disk structures cooling rates $>1 \times 10^{10}\,°C/s$ have been mentioned in the literature [187]. The material then solidifies in an amorphous state. It is unclear what very short range structure may exist in the a-GST region. Figure 13.107 shows the difference in surface appearance between the a-GST and c-GST states.

PHASE CHANGE MEMORIES

The amorphous region can be converted to a metastable NaCl-type crystalline structure by the application of a short lower power laser pulse that heats the material to a point below the melting temperature followed by rapidly cooling it to lock in the metastable state. Figure 13.108 illustrates the metastable crystal form as presented in Yamada and Matsunagea [188]. This structure has been described as being a distorted rocksalt cell. Te atoms occupy the lattice sites of one face-centered-cubic (fcc) sublattice, and Ge and Sb randomly fit into the sites of a second sublattice. A number of the sites are vacant. The lattice parameter is 6.02 Å.

If the metastable c-GST is subjected to further heating at about 225°C or more, it will shift into a stable crystal state composed of planar layers that are stacked to achieve close packing. Atomic diffusion processes rearrange GST225 into a cyclic ABC-type nine-layer structure [187]. Figure 13.109 shows the crystal cell stack as reported in the literature [187]. The individual A, B, C layers fit together to achieve a minimum energy, smallest volume, configuration.

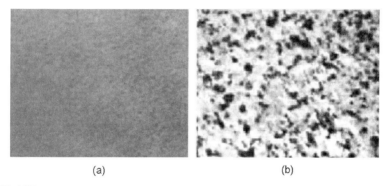

Figure 13.107 Microphotograph of GST film surface (a) in the amorphous state and (b) in the crystalline state.

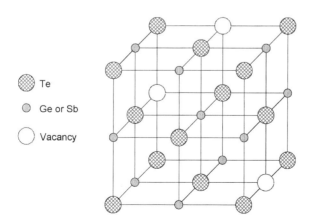

Figure 13.108 Metastable GST crystalline form formed by rapidly heating and cooling amorphous phase.

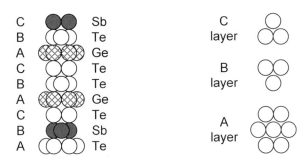

Figure 13.109. Stacking model of GST225 as reported in [187].

Studies of GST225 oriented toward the PCM application, of course, made extensive use of the optical disk base of knowledge but must account for the unique features of PCM. Zhang et al. [186] summarized the overall knowledge of the GST225 system and evaluated samples to obtain data in the context of PCM usage of the material. The observed crystallization temperature was about 175°C and the melting temperature was about 610°C. The GST material changed from a-GST to an fcc crystal (metastable NaCl-type crystal) at about 200°C, and then changed into a hexagonal close-packed (hcp) structure (the cyclic nine–layer structure) at about 400°C. (It is a common practice in the PCM literature to refer to the metastable crystal form as being fcc and the stable form as hcp. This is obviously incorrect in that the cells are more complex than those simple forms, but it does serve as a convenient shorthand notation.) The PCM application requires passing current through the GST material to both sense resistance and to accomplish heating that induces phase changes. The GST material must be interfaced in such a manner as to achieve stable electrical contacts that can withstand the severe thermal cycling of normal device operation.

13.9.4 Memory Cell

13.9.4.1 Basic Structure. The graphic shown in Figure 13.110 presents the essential structural features of a PCM cell. The memory cell core is a series-resistive circuit where currents are forced to flow though the top electrode, the chalcogenide layer and the local chalcogenide volume where the phase change occurs, the heater resistor material, and the bottom electrode material. In this simple model the heater and active volume are assumed to be cylindrical. In addition to the structure shown, some device such as a transistor must be present to act as a switch that connects the cell to power.

Figure 13.111 summarizes the basic PCM cell series string of effective resistances. Only the variable resistance R_V serves as the data storage element. The difference between $R_{V\text{-reset}}$ and the sum of $R_{V\text{-set}} + R_C + R_0 + R_H$ is the change in resistance that must be detected in order to distinguish between a stored ONE and a stored ZERO.

13.9.4.2 Cell Operation Concepts. Both thermal and electrical issues must be treated in the cell design. To achieve the phase changes, the temperature of the variable volume must meet certain minimum criteria. Figure 13.112(a) summarizes

PHASE CHANGE MEMORIES

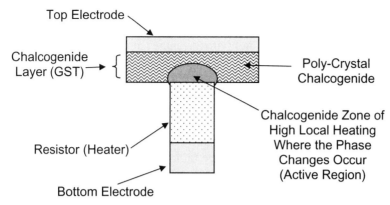

Figure 13.110. Simplified graphic showing basic structural features of a PCM cell.

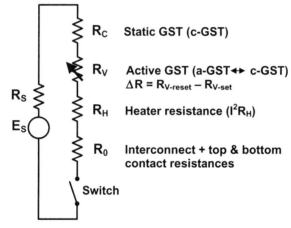

Figure 13.111. Simplified equivalent circuit for a PCM cell.

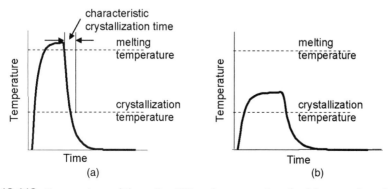

Figure 13.112. Temperature of the active GST region versus time for (a) conversion of c-GST to a-GST (RESET), and (b) conversion of a-GST to c-GST (SET).

the essential thermal requirements for a RESET operation. In order to convert from crystalline to amorphous, the material must be heated to its melting temperature. This is accomplished by applying a current pulse, and $I^2(R_H + R_V)$ heating at the active GST region results in a rapid localized temperature rise to a value above the amorphization temperature (melting temperature). While the figure shows temperature versus time, the current pulse that causes the heating has a similar looking waveshape. The duration of the heating period should be long enough to assure that the entire active region becomes molten. Typically, a current pulse width of ~10 ns is sufficient to convert the c-GST to a-GST [189]. Once the active region has been converted to a-GST, the current must be rapidly reduced and the thermal environment must allow the active region to cool faster than the GST crystallization time constant—typically a few nanoseconds.

Figure 13.112(b) shows the temperature versus time requirement for crystallization of the active region. The temperature must be raised to a level above the crystallization temperature but below the melting temperature. The crystallization process begins with nucleation. Nucleation sites can be anywhere in or on the boundary of the active region, but since the active region interfaces to the rest of the crystallized GST film, it is likely to occur at multiple sites on that interface. Once nucleation has taken place the crystal growth proceeds. The current pulse must be long enough to accommodate both the nucleation and growth periods.

Figure 13.113 shows an alternate approach to accomplishing the SET process. This temperature versus time arrangement has a leading edge identical to the RESET pulse but ramps down smoothly with a fall time longer than the characteristic crystallization time. Once the temperature falls to the point where no significant crystal growth will continue, the pulse is terminated. This approach is called SET-sweep, and it is a more reliable way to achieve a low-resistance SET state [190]. It allows all portions of the active volume to be subjected to an optimal crystallization temperature for an optimal duration. It tends to compensate for temperature gradients within individual cells and for variations among all the cells in an array [191].

13.9.4.3 Electrical Characteristics and Programming Waveforms. The RESET and SET operations require current pulses to establish the I^2R_H heating. To

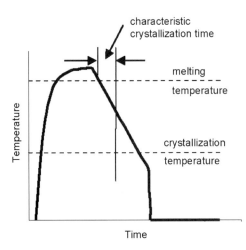

Figure 13.113. SET-sweep programming technique.

READ the cell a much lower current is used that has essentially no heating effect. The READ operation must distinguish between the high-resistance (amorphous) and low-resistance (crystallized) states. Figure 13.114 shows a typical voltage–current (V–I) characteristic for a cell and labels the regions for RESET, SET, and READ. The current axis is normalized to the magnitude of the RESET current.

The sketch in Figure 13.114 shows the V–I curve for the low-resistance SET state as a solid line, and the curve for the high-resistance RESET state as a dashed line. When the cell is in the high-resistance RESET state, an applied voltage must rise to the threshold voltage (V_{th} labeled in the figure) before switching will occur. A negative resistance is encountered, and the material changes from the high-resistance state to the low-resistance SET state. Once the current exceeds a minimum level, the two V–I characteristics become identical independent of the initial state.

A close examination of the V–I characteristic for the RESET state [194] shows that from 0 to about 0.2 V the curve is ohmic. From about 0.2 V up to about 0.7 V it shows an exponential increase in current with an increase in voltage. As the voltage continues to rise and approach V_{th} it becomes superexponential.

Both c-GST and a-GST are semiconductors and the large difference in conductivity of the two phases is due to the presence of electronic defect states. Pirovano et al. [194] developed the bandgap model of c-GST and a-GST shown in Figure 13.115. That model, together with the concurrent action of impact ionization and trapping of free carriers in the amorphous material, was able to account for both the dc and transient behavior of the material. The essential features of the Pirovano model are:

1. A valence band discontinuity between the crystalline and the amorphous GST
2. The presence of vacancies in the crystal leading to acceptorlike traps [188, 195]
3. The pronounced band tail of the valence band for the amorphous state due to localized tellurium *lone pairs* (C_2^0) [195–197].
4. Structural defects along the Te–Te chains in the amorphous GST giving a high density of donor/acceptor defect pairs ($C_3^+ - C_1^-$) [198, 199]

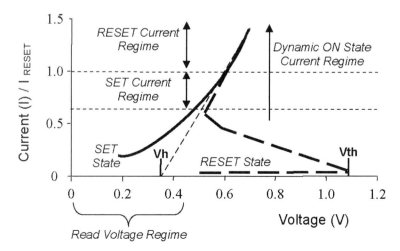

Figure 13.114. Voltage–current characteristic for a PCM cell. (After [192, 193].)

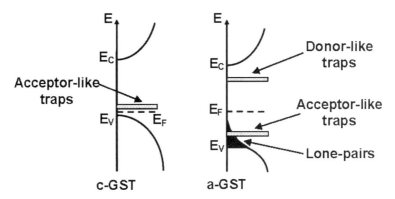

Figure 13.115. Band diagrams of c-GST and a-GST proposed by [194].

Comparatively speaking, the c-GST state offers relatively high mobility and high carrier concentration while the a-GST state has relatively low mobility and low carrier concentration. In the c-GST state lattice defects are present in concentrations as high as 15% [189]. The defects act as shallow acceptorlike traps causing the material to be a defect semiconductor that exhibits degenerate P-type conductivity with a bandlike mobility [194]. The optical bandgap of the c-GST is 0.5 eV [189].

When the c-GST is melted, the lattice and the lattice defects disappear, and the resulting structural disorder produces different kinds of electronic states. The most numerous are spatially localized C_2^0 lone-pair valence band tail states that cause a very low trap-limited hole mobility. In addition there are valence alternation pair (VAP) states believed to be caused by special defect bonding configurations between the nonbonding orbitals on the chalcogen atoms [189]. These states are the donor-like and acceptorlike traps shown in Figure 13.115. The donorlike traps are described as being C_3^+ states, and the acceptorlike traps are C_1^- states [194].

Pirovano et al. [194] describes the conduction processes for a-GST in the following way. When the voltage is low, the conduction is ohmic. As the electric field increases, impact ionization takes place, and the current rises exponentially due to secondary hole generation. Initially, most of the secondary electrons get trapped in the donorlike traps. Further increase in the voltage begins to fill up the traps and the free-electron density increases. The electron quasi-Fermi level is forced to move closer to E_C. Just beyond V_{th} impact ionization dominates over carrier recombination, the traps are filled and a voltage snap-back occurs. Now higher current can flow at a lower voltage. Impact ionization still takes place, but at a reduced multiplication rate. A much higher free-carrier density now sustains the current.

In Flash and many other memory technologies it is necessary to perform two steps to establish the state of the individual cells. Typically, an erase operation is followed by a program operation. In contrast, PCM is a direct write technology. No matter what the original state of the cell is, the final state of the cell can be established by the magnitude of current passed through the cell. For PCM it is appropriate to speak of a "read" operation and a "write" operation, but no doubt the legacy of Flash will lead to maintaining the terminology of "programming" as opposed to "write."

Figure 13.116 provides an alternative view of the writing process. Here resistance for a particular cell is plotted against the programming current. The procedure

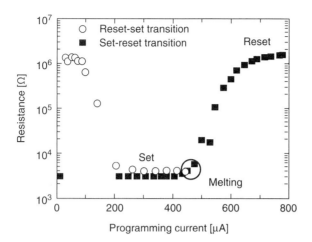

Figure 13.116. Resistance characteristic plot for SET–RESET and RESET–SET transitions [193].

followed was to initialize the cell in a particular state and then apply a current pulse and record the resulting resistance. Both the transition from a-GST to c-GST and the transition from c-GST to a-GST are shown in the figure. The applied current pulses had a rapid falling edge to ensure that the state of the material at the trailing edge was locked in. That is, the crystallization or amorphization process was terminated by rapid cooling that froze the material phase state.

For the RESET–SET transition the resistance decreases rapidly for currents higher than 100 µA because the crystallization process was initiated. For the opposite transition where the initial state was low resistance, the resistance remains unchanged until the level increases to about 450 µA where the amorphization process is initiated. Because the applied current pulses had a rapid falling edge, incomplete state conversions are locked in, and the resistance curve shows clearly the intermediate resistance values between the binary SET and RESET levels.

The function of the SET and RESET currents is to heat the active volume of GST. For efficiency reasons it is desired that the current produce the heat only in the active volume, which implies that the volume should be thermally insulated from the rest of the integrated circuit structure. At the same time it is desired that the active volume cool rapidly after removal of the current. This implies that the active volume should have a good thermal conduction path to the ambient. These conflicting demands, and the limitations of the thermal properties of the materials, point out the need for reasonable computer models that account for both thermal and electrical aspects of PCM.

The PCM cell is a string of resistances, and heat injection at specific points in the string depends on the particular resistance values. Heat loss depends on the thermal characteristics of the surrounding materials and the path to an ambient temperature. The heater resistor and the GST should be the points where the most heat is developed. Heat flow from the active volume occurs upward through the GST to the metal interconnect and downward through the heater resistor to the lower contact structure. Additional heat loss occurs by flow from the heater to surrounding material. The GST thermal conductivity is ~0.5 W/mK. A typical heater material like TiAlN has a thermal conductivity ~10 W/mK [189]. One would expect that the downward heat flow would dominate because of the large difference in thermal conductivity. The thermal efficiency can be improved by reducing the contact area, by increasing the length of the heater resistor, and by decreasing the thermal conductivity of the resistor material. According to Hudgens [189], reduction of contact area is by far the most effective measure that can be taken.

The magnitude of current required to switch the cell state is at present a major concern, and much effort is being given to finding ways to reduce the current. For experimental cells based on 180-nm technology the current was typically greater than 1 mA. For practical products to emerge the current needs to be on the order of 100 µA. Each phase change resistor requires that some type of switch, typically a transistor, be in series with the resistor in order to control current flow. For maximum cell density, it is desirable to implement that switch as a minimum feature device. This goal is incompatible with driving high RESET currents. Table 13.9 (derived from Lowrey et al. [190]) shows the magnitude of drive current that can be expected for various channel width transistors at different CMOS technology nodes.

Programming current scales downward as the contact area between the chalcogenide and the heater is reduced [200, 201]. Novel cell structures can reduce the contact area, either through the use of thin films, sublithographic techniques, or a combination of the two. Figure 13.117 presents several approaches for current reduction that are presently being evaluated by various research groups. In Figure 13.117, M1 and M0 refer to metal layers 1 and 0, and ILD is interlayer dielectric.

TABLE 13.9. MOS Select Transistor Width Requirements versus Drive Current

Technology node		180 nm	130 nm	100 nm	65 nm	45 nm	32 nm
I_{DS} per micron of channel width		900 µA	900 µA	900 µA	900 µA	1200 µA	1500 µA
MOS transistor width	F	162	117	90	58.5	54	48
	2F	324	234	180	117	108	96
	4F	648	468	360	234	216	192
	6F	972	702	540	351	324	288

Source: After [190].

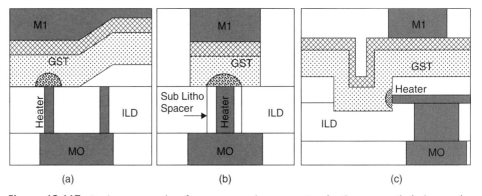

Figure 13.117. Design approaches for programming current reduction presently being evaluated [193].

Figure 13.117(a) uses thin-film technology to define a thin vertical heater. A thin trench (microtrench) is formed over the top surface of the heater and GST is deposited into the trench. The small intersection of the GST with the vertical heater establishes a contact cross section smaller than that which could be defined by lithography (i.e., sublithographic) [202]. Figure 13.117(b) employs a spacer-defined sublithographic heater cell. Here the bottom electrode contact (BEC) is first defined lithographically, and then spacers are formed that reduce the window to sublithographic dimensions. Chemical vapor deposition TiN is used to fill the BEC, and the heater structure is finished off by a chemical mechanical polishing (CMP) process [203]. Figure 13.117(c) shows a thin-film-defined heater and edge contact to the GST [204]. The amount of research to find practical solutions to the programming current problem is extensive, and no doubt these three example approaches will be rapidly followed by many other initiatives.

Reduction of programming current requirements can also be accomplished by adjustment of the GST material properties. Nitrogen doping of the GST during deposition has been found to suppress grain growth during crystallization. Smaller grain size improves the thermal stability of the material and increases the resistivity. Resistivity can be controlled over a wide range by varying the amount of nitrogen incorporated into the film. The resistivity of the GST is an important factor in achieving lower programming currents. The temperature of the active region is determined by I^2R heating, and the resistivity of the GST in comparison to the other resistances in the series circuit is a factor in establishing the desired localized impact of a given amount of current. Referring to Figure 13.111, R_H and R_V should be larger than R_C and R_0 in order to improve the efficiency of the writing process. Figure 13.118 shows the GST resistivity versus nitrogen concentration reported by Horii et al. [203].

Oxygen doping of the GST has also been used to reduce the programming current. Matsuzaki et al. [205] reported reducing the RESET current to 100 µA in PCM cells compatible with a standard 130-nm CMOS process. The design made use of 180-nm-diameter planar tungsten BEC. No effort was made to achieve sublithographic contact area. The oxygen was introduced by changing the GST sputtering

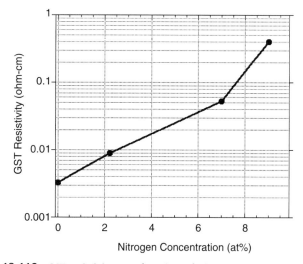

Figure 13.118. GST resistivity as a function of nitrogen concentration [203].

conditions, and material analyses indicated that the oxygen was distributed uniformly as germanium-oxide throughout the GST film. Grain size was smaller than that for undoped GST, and the rate that a-GST would convert to c-GST in elevated ambients was retarded (i.e., the thermal lifetime was extended).

13.9.5 Memory Array and Support Circuitry

The memory array and support circuitry for a phase change random-access memory is similar to that used in conventional random-access memories. The array consists of a tightly packed x-y arrangement of the cells where horizontal lines called wordlines (WLx) are brought out to allow selection of a word, and vertical lines called bitlines (BLx) are brought out associated with the individual bits in the word. Each cell has a selection device in series with the resistive chalcogenide stack. The most common choices for the selection device are an MOS transistor or a bipolar junction transistor (BJT). Figure 13.119 shows these two array arrangements. Very large arrays have been demonstrated using both approaches. The BJT approach offers the smaller cell size and has the potential of matching the cell size of Flash memory.

Once an array structure has been established, the circuitry required to form a complete integrated memory device is a matter of straightforward CMOS design. The obvious functions that must be implemented include address decoding down to the column and row select signals, operating mode control logic, input data path design, output data path design, SET and RESET current generation and control, and readout sense circuitry design. Subtleties creep into the design where measures are taken to control the distribution of SET and RESET resistances to maintain read margins in the face of process variation and layout-imposed factors.

Figure 13.120 points out the basic support circuit functions that interface to the array. The addressing function is common to READ, SET, and RESET. Logic decoding must be performed that ends up in selecting a particular WL and BL. If the operating mode is READ, a low-voltage source is selected and the sense amplifier is enabled to determine whether the BL voltage indicates that the cell is in a

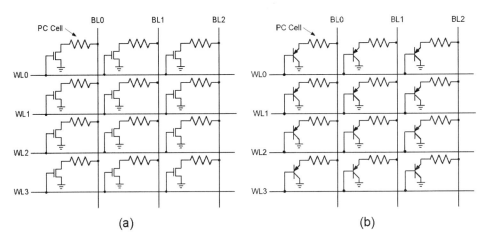

Figure 13.119. Phase change memory array configurations: (a) array with FET switch and (b) array with BJT switch.

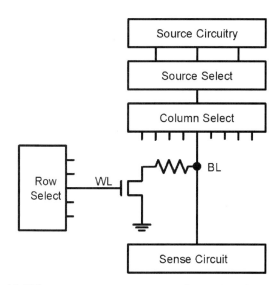

Figure 13.120. Basic support circuitry interfacing directly to the array.

high or low resistance state. If the operating mode is SET or RESET, the appropriate source must be selected and that waveform applied to the cell.

Because of the very large size of the arrays needed to compete in the nonvolatile memory marketplace, it is inadequate to simply design a chip to operate with a nominal design cell. The transitions from a-GST to c-GST and from c-GST to a-GST will differ from cell to cell for a variety of reasons, and the resulting distributions of low- and high-resistance states needs to be understood and managed.

Oh et al. [191] presented a strategy for handling the difference in SET and RESET current due to resistance in the bitline. Cells located close to the source side of the bitline would be exposed to higher current than cells near the bottom of the bitline, resulting in either underprogramming or overprogramming cells at the extreme ends. In this particular device there were 1024 cells tied to the bitline. To manage this problem the magnitude of the applied source was adjusted as a function of the wordline address. Four different source levels were used where the 256 cells closest to the source received the lowest level, the next 256 cells a higher level, and so forth.

More complex schemes for resistance distribution control similar to those used with Flash are feasible and may be applied to PCM. For example, doing SET and RESET operations using a series of narrow pulses with a read after each pulse may be a viable scenario depending on the target write time requirement. Such approaches will be implemented by individual companies for specific chips and are likely to be closely protected by patents.

13.9.6 Fabrication

Several major semiconductor vendors have established CMOS-compatible phase change memory processes. The universal development strategy was to prepare a process module that contains all the chalcogenide-oriented items. First, the CMOS active layers are fabricated, then the chalcogenide module is added, and finally the

TABLE 13.10. Process Flowchart

Blocks	Modules
Front end of line (FEOL)	Shallow trench isolation
	Wells implantation
	MOSFET definition
	BJT formation
	Salicide formation
PCM formation	Precontact
	Microtrench or lance
	Metal 0 (GST/cap)
Back end of line (BEOL)	Contact / via 0
	Line 1
	Via 1 / line 2
	Via 2 / line 3
	Passivation and Al cap

Source: From [206].

upper interconnect layers are produced. One example of this approach was reported by Intel and STMicroelectronics [206].

Here the target was the 90-nm technology node. Table 13.10 summarizes the overall process flow in terms of modules. The $12F^2$ cell made use of a vertical PNP-BJT switching element. The cell x pitch was 220 nm and the y pitch was 440 nm for a cell area of $0.0968\,\mu m^2$. The PNP emitter and base contacts were salicided using $CoSi_2$. The heater was landed on a tungsten plug and could be either a microtrench or a lance type. The chalcogenide was GST225 capped by a Ti/TiN barrier. For the microtrench heater the program currents could be as low as $400\,\mu A$ because of the small 400-nm^2 contact to the chalcogenide.

While this example is peculiar to Intel/STMicroelectronics, it illustrates the overall approach to PCM fabrication followed by the industry. Economics dictates that the PCM be prepared for fabrication as a compatible add-on to an existing established CMOS process.

13.9.7 Scaling

The PCM 1R-1T cell poses two somewhat separate scaling problems: scaling the chalcogenide resistor stack and scaling the series switch. Scaling the switch is, of course, a problem common to CMOS technology in general since the transistor is not a foreign element. The ability to scale the transistor is constrained by the requirement that it must be designed and sized to accommodate the voltage and current required to RESET the cell. That voltage may well exceed the supply voltage for the technology node. If a MOSFET select transistor is employed, the current requirement may force such a wide channel that the overall PCM cell size is seriously impacted.

Pirovano et al. [200] reported on a detailed analysis of PCM scaling using a physics-based electrothermal model of a cell verified by measurements conducted on sample devices. The logical foundation of the study was the observation that the RESET current can be scaled downward by scaling the contact area between the GST and the heater. This was confirmed by measurements performed on sample

PHASE CHANGE MEMORIES

devices (see Fig. 13.121). In this experiment the contact area was altered without changing the thickness of the GST and/or the heater material. This reduction in current was ascribed mainly to the increased thermal resistance at the contact. With a larger thermal resistance a smaller amount of power is required to raise the temperature, and thus the melting temperature will be reached at a lower current level.

Changes to the device geometry or materials can alter the electrical and thermal resistance, and thus can change the RESET current. Pirovano et al. [200] reasoned that changing all the material dimensions uniformly (isotropic scaling) by a factor k (larger than 1) would have the impact predicted in Table 13.11.

For scaling the switch [200] used guidelines from the International Technology Roadmap for Semiconductors (ITRS) 2002 update. For a BJT switch the scaling factors are $1/k^2$ for the emitter area, and $1/k$ for the base thickness and the emitter current. For a MOSFET switch the scaling factor is $1/k$ for the gate length, the gate width, and the saturation current.

The authors pointed out that this is a constant-voltage scaling scheme, and at small dimensions problems would be encountered with the gate oxide reliability of a MOSFET switch. A more detailed design of the switch devices is necessary beyond the 65-nm node.

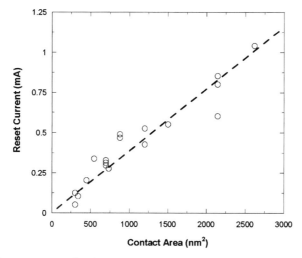

Figure 13.121. Experimentally observed RESET current as a function of contact area [200].

TABLE 13.11. Scaling Rules for GST and Heater Parameters

Parameter	Factor	Parameter	Factor
GST/heater contact area	$1/k^2$	Thermal resistance	k
GST thickness	$1/k$	ON-state resistance	k
Heater height/thickness	$1/k$	Threshold voltage	$\log(k)$
SET resistance	k	Programming voltage	1
RESET resistance	k	Programming current	$1/k$

Source: From [200].

While scaling by a factor k is an interesting concept that can be explored using a computer model, it needs to be understood that it is not simple to accomplish. The dimensions of the individual elements are determined by different processes. It is not just a matter of changing the lithographic minimum feature size. The important conclusion is that the PCM cell appears to function when the dimensions are reduced.

This leads to the next logical question. Are there specific phenomena that will limit the size of cells? Thermal cross talk (also called proximity disturb) is one such potential problem. The thermal cross-talk idea is concerned with the possibility that heating a cell to the GST melting temperature could raise the temperature of an adjacent cell in the amorphous state (RESET) to the extent that crystallization would be initiated. If the temperature did rise to that point, there is a possibility that repeated heating of this nature could have the cumulative effect of converting the cell to the crystalline state (SET), that is, loss of data.

Pirovano et al. [200] examined thermal cross talk using simulations and experiments and determined that at least down to the 65-nm node it was not a problem. Heating one cell caused negligible temperature rise in the adjacent cell. Later analyses showed that the decrease in temperature at radial distances from the active volume when normalized to the dimension of the heating area falls on a universal curve (see Fig. 13.122). When the contact area is scaled the same amount as the cell, thermal cross talk will not be an issue [207].

Another possible ultimate scaling limit is the minimum volume of phase change material that can remain stable in the low- and high-resistance states. Experiments using an atomic force microscope (AFM) probe indicated that spot sizes as small as 5 nm on a 15-nm pitch would be stable [193].

Table 13.12 provides an example of scaling assuming constant voltage across the chalcogenide stack and a diode selector device [207]. The vertical scaling of the diode parameters was 10% per generation. This resulted in an increase of over 50% per generation in the cell resistance and in the output signal level. It is believed that

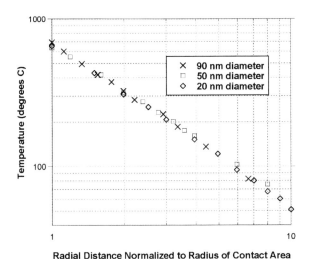

Figure 13.122. Temperature at RESET versus normalized radial distance from the contact area [207].

TABLE 13.12. PCM with Diode Selector Device Constant-Voltage Scaling Example

Technology node	nm	130	90	65	45	32	22
DRAM $\frac{1}{2}$ pitch	nm	130	90	65	45	32	22
Electrode contact dimension	nm	47	33	23	16	11	8
RESET current	µA	500	294	171	99	57	35
Chalcogenide thickness	nm	60	50	40	30	30	30
Chalcogenide resistance	kΩ	1.6	2.5	3.8	5.5	9.6	12.2
Heater resistance	kΩ	0.8	1.4	2.3	4.0	7.1	11.4
Total resistance	kΩ	2.4	3.9	6.1	9.5	16.7	23.6
Diode area	nm^2	18,590	8,910	4,648	2,228	1,091	535
Deep trench depth	nm	440	396	356	321	289	260
Buried WL depth	nm	390	351	316	284	256	230
Shallow trench depth	nm	200	180	162	146	131	118
P$^+$ junction depth	nm	120	108	97	87	79	71
Vertical diode depth	nm	270	243	219	197	177	159
WL depth under trench	nm	190	171	154	139	125	112
Diode junction vertical	A/cm^2	2.69	3.30	3.68	4.48	5.19	6.57
Diode vertical *IR* drop	V	0.070	0.077	0.078	0.085	0.089	0.101
Buried WL junction lateral	A/cm^2	20,200	19,100	17,100	15,900	14,400	14,200
Buried WL lateral *IR* drop	V	0.20	0.12	0.07	0.04	0.02	0.01
Diode forward voltage drop	V	1.37	1.30	1.25	1.22	1.21	1.21

Source: After [207].

the reduced signal level remains within the capabilities of modern-day sense circuitry and should not be a cause for concern.

13.9.8 Reliability

Reliability information on chalcogenide-based memory is available from cell level structures and experimental prototype devices of various bit capacities. Intrinsic endurance and retention has been studied in detail, and some test results are available in the literature. Figure 13.123 shows the resistance window for a PC memory element that has accumulated 10^{12} SET–RESET cycles. It has been observed that cycling endurance depends on the magnitude and duration of the RESET current pulse [207, 208].

Figure 13.124 [193] shows that overheating the cell by applying high current reduces the number of cycles before failure, yielding cells stuck in the low-resistance state. Failures also occur where the cell is stuck in the high-resistance state. This type of failure is possibly due to an open circuit in the contact region (chalcogenide delamination from the heater) after a large number of cycles [207, 208].

Intrinsic retention in a phase change memory is, of course, a matter of the stability of the two phase states. The c-GST state is the energetically favored state and is not a concern. Over time atomic diffusion occurs and tends toward further crystallization of the material. It is the a-GST state that is subject to a natural decay toward crystallization. It has been shown that the crystallization process has a high activation energy of 3.47 eV.

Figure 13.125 shows that high-temperature retention measurements can be extrapolated to estimate the retention time at lower temperatures. In this case the

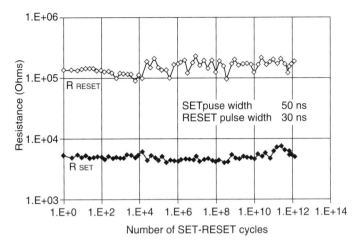

Figure 13.123. SET and RESET resistance vs. accumulated SET–RESET cycles for a PC memory cell [207].

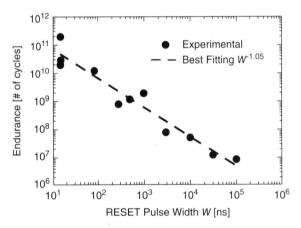

Figure 13.124. Endurance capability as a function of RESET pulse energy [193].

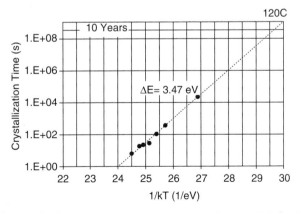

Figure 13.125. PCM cell intrinsic retention characteristic [193].

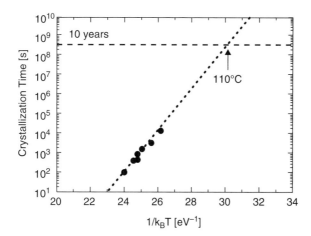

Figure 13.126. PCM array retention data [193].

product goal of more than 10 years retention at 85°C is exceeded by orders of magnitude. Figure 13.126 shows similar intrinsic retention information derived from array measurements [193]. It is expected that array retention in volume production will be dominated by extrinsic factors.

PCM arrays have been examined for possible disturb effects. Thermal proximity disturb effects were discussed above as an effect that could possibly limit scaling and were shown to be a manageable issue. Kim and Ahn [208] also investigated thermal proximity as a possible disturb effect in arrays by repeatedly writing a checkerboard pattern. Comparison of the RESET resistance distributions before and after 1×10^6 cycles confirmed that thermal proximity was not causing a significant resistance change.

Read disturb was explored by setting the read voltage at 0.4 V and performing 1×10^9 read operations without any failures [208]. To avoid a read disturb, the voltage applied across the memory cell must be kept lower than the threshold voltage. Depending on the resistance of the amorphous state for the manufactured cell, the threshold voltage can range from 0.5 to 1.2 V. If the threshold voltage is exceeded, a negative resistance will be encountered and the current will increase causing crystallization to begin.

For some users the radiation hardness of PCM is a matter of concern. Because the storage mechanism is not charge oriented, the chalcogenide cell portion of a PCM device is inherently hard to total ionizing dose and single event effects. This has been confirmed by radiation testing [209, 210]. Of course, those portions of the chip that consist of normal CMOS are subject to the same radiation concerns as any other CMOS device.

13.9.9 Products

Phase change memory products are expected to enter production in 2008. At the present time (2006) several vendors have produced prototype chips that are being characterized and used as learning vehicles. The largest chip to date [211] is a 512-Mb product intended for Flash NOR-type applications.

13.9.10 Summary

At the time this summary was written PCM was undergoing intense research and development by several major semiconductor device manufacturers. Studies continue to examine cell physics as it relates to functionality and to reliability. The materials issues are significant, and the electronic equilibrium and transport properties as well as the electrothermal and structural dynamics of the chalcogenide need to be well understood. Approaches to reduce the current required for writing are being explored. This is an important practical problem that impacts cell size and device architectural decisions. High-density array manufacturing is being defined. The integration of the chalcogenide stack within a CMOS process has been implemented in various ways, and volume manufacturability and reliability issues are being addressed.

At this stage of development it can be said that PCM has demonstrated the basic capabilities of fast read and direct write, high intrinsic endurance and retention, low voltage, and moderate energy operation. It is also believed that the cell can be produced at the 22-nm node. Now the attention of the industry is shifting to the challenges of proving reliability and high-volume manufacturability.

REFERENCES

1. K. Naruke, S. Taguchi, and M. Wada, "Stress Induced Leakage Current Limiting to Scale Down EEPROM Tunnel Oxide Thickness," *IEEE Electron Devices Meeting, Tech. Dig.*, pp. 424–427, 1988.
2. K. Kim, "Technology for Sub-50 nm DRAM and NAND Flash Manufacturing," *IEEE Electron Devices Meeting, Tech. Dig.*, pp. 323–326, 2005.
3. Y. Shin, J. Choi, C. Kang, C. Lee, K.-T. Park, J.-S. Lee, J. Sel, V. Kim, B. Choi, J. Sim, D. Kim, H.-J. Cho, and K. Kim, "A Novel NAND-type MONOS Memory using 63 nm Process Technology for Multi-Gigabit Flash EEPROMs," *IEEE Electron Devices Meeting, Tech. Dig.*, pp. 327–330, 2005.
4. B. Eitan, P. Pavan, I. Bloom, E. Aloni, A. Frommer, and D. Finzi, "NROM: A Novel Localized Trapping, 2-Bit Nonvolatile Memory Cell," *IEEE Electron Device Letters*, Vol. 21, No. 11, pp. 543–545, 2000.
5. B. Eitan, G. Cohen, A. Shappir, E. Lusky, A. Givant, M. Janai, I. Bloom, Y. Polansky, O. Dadashev, A. Lavan, R. Sahar, and E. Maayan, "4-bit per Cell NROM Reliability," *IEEE Electron Devices Meeting, Tech. Dig.*, pp. 547–550, 2005.
6. E. Maayan, D. Ran, J. Shor, Y. Polansky, Y. Sofer, I. Bloom, D. Avni, B. Eitan, Z. Cohen, M. Meyassed, Y. Alpern, H. Palm, E. S. Kamienski, P. Halbach, D. Caspary, S. Riedel, and R. Knofler, "A 512 Mb NROM Data Storage Memory with 8 MB/s data rate", *IEEE International Solid State Circuits Conference*, 2002, Vol. 2, pp. 76, 77, 407, 408, 3–7 Feb. 2002.
7. Y. Sofer, M. Edan, Y. Betser, M. Grossgold, E. Maayan, and B. Eitan, "A 55 mm2 256 Mb NROM Flash Memory with Embedded Micro-controller using an NROM-Based Program File ROM," *IEEE International Solid State Circuits Conference*, 2004, Vol. 1, pp. 48, 512, 15–19 Feb. 2004,
8. Y. Polansky, A. Lavan, R. Sahar, O. Dadshev, Y. Betser, G. Cohen, E. Maayan, B. Eitan, F.-L. Ni, Y.-H. J. Ku, C.-Y. Lu, T. C.-T. Chen, C.-Y. Liao, C.-H. Chang, C.-K. Chen, W.-C. Ho, Y. Shih, W. Ting, and W. Lu, "A 4 bits per cell NROM 1 Gb Data Storage Memory," *IEEE International Solid State Circuits Conference 2006*, pp. 448–458, 6–9 Feb. 2006.

9. I. Bloom, P. Pavan, and B. Eitan, "NROM—A New Non-Volatile Memory Technology: From Device to Products," *Microelectronic Engineering*, Vol. 59, pp. 213–223, Nov. 2001.

10. J.-R. Hwang, T.-L. Lee, H.-C. Ma, T.-C. Lee, T.-H. Chung, C.-Y. Chang, S.-D. Liu, B.-C. Perng, J.-W. Hsu, M.-Y. Lee, C.-Y. Ting, C.-C. Huang, J.-H. Wang, J.-H. Shieh, and F.-L. Yang, "20 nm Gate Bulk-finFET SONOS Flash," *2005 IEEE Electron Devices Meeting*, pp. 154–157, 5–7 Dec. 2005.

11. International Technology Roadmap for Semiconductors, 2005 Edition, available online at www.itrs.net/reports.html.

12. E. Lusky, Y. Shacham-Diamand, I. Bloom, and B. Eitan, "Characterization of Channel Hot Electron Injection by the Subthreshold Slope of NROM Device," *IEEE Electron Device Letters*, Vol. 22, No. 11, pp. 556–558, 2001.

13. H.-T. Lue, T.-H. Hsu, M.-T. Wu, K.-Y. Hsieh, R. Liu, and C.-Y. Lu, "Studies of the Reverse Read Method and Second-bit Effect of 2-bit/cell Nitride-Trapping Device by Quasi-Two-Dimensional Model," *IEEE Trans on Electron Devices*, Vol. 53, No. 1, pp. 119–125, Jan. 2006.

14. L. Larcher, P. Pavan, and B. Eitan, "On the Physical Mechanism of the NROM Memory Erase," *IEEE Trans. Electron Devices*, Vol. 51, No. 10, pp. 1593–1599, Oct. 2004.

15. G. Ingrosso, L. Selmi, and E. Sangiorgi, "Monte Carlo Simulation of Program and Erase Charge Distributions in NROM Devices," *IEEE European Solid State Research Conference*, pp. 187–190, 24–26 Sept. 2002.

16. E. Lusky, "The Spatial Distribution and Emission Mechanisms of Locally Trapped Charge in Oxide-Nitride-Oxide (ONO) Stacked Dielectrics for Non-Volatile Semiconductor Memories," Ph.D. dissertation, Tel-Aviv University 2002.

17. Y.-H. Shih, S. C. Lee, H. T. Lue, M. D. Wu, T. H. Hsu, E. K. Lai, J. Y. Hsieh, C. W. Wu, L. W. Yang, H. Y. Hsieh, K. C. Chen, R. Liu, and C.-Y. Lu, "Highly Reliable 2-bit/cell Nitride Trapping Flash Memory using a Novel Array-Nitride-Sealing (ANS) ONO Process," *IEEE Electron Devices Meeting 2005*, pp. 551–554, 5–7 Dec. 2005.

18. Y. Roizin, E. Pikhay, and M. Gutman, "Suppression of Erased State V_t Drift in Two-Bit Per Cell SONOS Memories," *IEEE Electron Device Letters* Vol. 26, No. 1, pp. 35–37, Jan. 2005.

19. A. Shappir, E. Lusky, G. Cohen, I. Bloom, M. Janai, and B. Eitan, "The Two-Bit NROM Reliability," *IEEE Trans. on Device and Materials Reliability*, Vol. 4, No. 3. pp. 397–403, Sept. 2004.

20. W. J. Tsai, N. K. Zous, C. J. Liu, C. C. Liu, C. H. Chen, T. Wang, S. Pan, C.-Y. Lu, and S. H. Gu, "Data Retention Behavior of a SONOS Type Two-bit Storage Flash Memory Cell," *IEEE Electron Devices Meeting*, pp. 32.6.1–32.6.4, 2–5 Dec. 2001.

21. T. Wang, W. J. Tsai, S. H. Gu, C. T. Chan, C. C. Yeh, N. K. Zous, T. C. Lu, S. Pan, and C. Y. Lu, "Reliability Models of Data Retention and Read-Disturb in 2-bit Nitride Storage Flash Memory Cells," *IEEE Electron Devices Meeting*, pp. 7.4.1–7.4.4, 8–10 Dec. 2003.

22. E. Lusky, Y. Shacham-Diamand, I. Bloom, and B. Eitan, "Electrons Retention Model for Localized Charge in Oxide–Nitride-Oxide (ONO) Dielectric," *IEEE Electron Device Letters*, Vol. 23, No. 9, pp. 556–558, Sept. 2002.

23. E. Lusky, Y. Shacham-Diamand, A. Shappir, I. Bloom, and B. Eitan, "Traps Spectroscopy of the Si3Ni4 Layer Using Localized Charge-Trapping Nonvolatile Memory Device" *Applied Physics Letters*, Vol. 85, No. 4, pp. 669–671, 26 July 2004.

24. A. Shappir, Y. Shacham-Diamanda, E. Lusky, I. Bloom, and B. Eitan, "Lateral Charge Transport in the Nitride Layer of the NROM Non-Volatile Memory Device," *Microelectronics Engineering*, Vol. 72, pp. 426–433, 2004.

25. L. Larcher, G. Verzellesi, P. Pavan, E. Lusky, I. Bloom, and B. Eitan, "Impact of Programming Charge Distribution on Threshold Voltage and Subthreshold Slope of NROM

Memory Cells," *IEEE Trans. on Electron Devices*, Vol. 49, No. 11, pp. 1939–1946, Nov. 2002.

26. A. Shappir, Y. Shacham-Diamand, E. Lusky, I. Bloom, and B. Eitan, "Subthreshold Slope Degradation Model for Localized-Charge-Trapping Based Non-Volatile Memory Devices," *Solid-State Electronics*, Vol. 47, No. 5, pp. 937–941, 2003.

27. A. Shappir, D. Levy, Y. Shacham-Diamand, E. Lusky, I. Bloom, and B. Eitan, "Spatial Characterization of Localized Charge Trapping and Charge Redistribution in the NROM Device," *Solid-State Electronics*, Vol. 48, No. 9, pp. 1489–1495, 2004.

28. E. Lusky, Y. Shacham-Diamand, I. Bloom, and B. Eitan, "Investigation of Channel Hot Electron Injection by Localized Charge Trapping Nonvolatile Memory Devices," *IEEE Trans. on Electron Devices*, Vol. 51, No. 3, pp. 444–451, Mar. 2004.

29. S. H. Gu, T. Wang, W.-P. Lu, W. Ting, Y.-H. J. Ku, and C.-Y. Lu, "Characterization of Programmed Charge Lateral Distribution in a Two-bit Storage Nitride Flash Memory Cell by using a Charge-Pumping Technique," *IEEE Trans. on Electron Devices*, Vol. 53, No. 1, pp. 103–108, Jan. 2006.

30. R. Daniel, Y. Shaham-Diamand, and Y. Roizin, "Interface States in Cycled Hot Electron Injection Program/Hot Hole Erase Silicon-Oxide-Nitride-Oxide-Silicon Memories," *Applied Physics Letters*, Vol. 85, No. 25, pp. 6266–6268, Dec. 2004.

31. E. Lusky, Y. Shacham-Diamand, A. Shappir, I. Bloom, G. Cohen, and B. Eitan, "Retention Loss Characteristics of Localized Charge-Trapping Devices," *IEEE International Reliability Physics Symposium*, pp. 527–530, 25–29 April 2004.

32. A. Shappir, E. Lusky, G. Cohen, I. Bloom, and B. Eitan, "Retention After Cycling in the NROM Non-Volatile Memory Device," *Proceedings Non-Volatile Semiconductor Memory Workshop*, Monterey, CA, pp. 77–78, Aug. 2004.

33. E. Kooi, J. G. van Lierop, and I. A. Appels, "Formation of Silicon Nitride at a Si-SiO$_2$ Interface during Local Oxidation of Silicon and during Heat-Treatment of Oxidized Silicon in NH$_3$ Gas," *J. Electrochemical Soc.*, Vol. 123, No. 7, 1976, pp. 1117–1120.

34. A. Shappir, E. Lusky, G. Cohen, and B. Eitan, "NROM Window Sensing for 2 and 4-bit per cell Products," *Proceedings Non-Volatile Semiconductor Memory Workshop*, Monterey, CA, pp. 68–69, Feb. 2006.

35. M. Janai, B. Eitan, A. Shappir, E. Lusky, I. Bloom, and G. Cohen, "Data Retention Reliability Model of NROM Nonvolatile Memory Products," *IEEE Transactions on Device and Material Reliability*, Vol. 4, No. 3, pp. 404–415, Sept. 2004.

36. M. Yoshimaru, N. Inoue, H. Tamura, and M. Ino, "Effects of Deposition Temperature on the Oxidation Resistance and Electrical Characteristics of Silicon Nitride," *IEEE Trans. on Electron Devices*, Vol. 41, No. 10, p. 1747, Oct. 1994.

37. W. J. Metz and J. R. Anderson, "Ferroelectric Storage Devices," *Bell Laboratories Record*, pp. 335–337 and 339–342, Sept. 1955.

38. J. R. Anderson, "Electrical Circuits Employing Ferroelectric Capacitors," U.S. Patent No. 2,876,436 (1959).

39. C. A. Paz de Araujo, J. D. Cuchiaro, L. D. McMillan, M. C. Scott, and J. F. Scott, "Fatigue-Free Ferroelectric Capacitors with Platinum Electrodes," *Nature*, Vol. 374, No. 6523, pp. 627–629, April 1995.

40. S. L. Miller and P. J. McWhorter, "Physics of the Ferroelectric Nonvolatile Memory Field Effect Transistor," *J. Appl. Phys.*, Vol. 72, No. 12, pp. 5999–6010, Dec. 1992.

41. T. A. Rabson, T. A. Rost, and H. Lin, "Ferroelectric Gate Transistors," *Integrated Ferroelectrics*, Vol. 6, No. 15, 1995.

42. H. Ishiwara, "Current Status and Prospects of FET-type Ferroelectric Memories," *FED Journal*, Vol. 11 Supplement, pp. 27–40, 2000.

43. H. T. Lue, C.-J. Wu, and T.-Y. Tseng, "Device Modeling of Ferroelectric Memory Field-Effect Transistor (FeMFET)," *IEEE Trans. Elec. Dev.* Vol. 49, No. 10, pp. 1790–1798, 2002.
44. J. H. Choi, J. Y. Lee, and Y. T. Kim, "Formation of Y2O3 Interface Layer in a YMnO3/Si Ferroelectric Gate Structure," *Appl. Phys. Letters*, Vol. 77, No. 24, 4028–4030, 2000.
45. W. S. Kim, S.-M. Ha, J.-K. Yang, and H.-H. Park, "Ferroelectric-Gate Field Effect Transistors using $Nd_2Ti_2)7/Y_2O_3$/Si Structures," *Thin Solid Films*, Vol. 398, pp. 663–667, Nov. 2001.
46. T. P. Ma and J.-P. Han, "Why is Nonvolatile Ferroelectric Memory Field-Effect Transistor Still Elusive," *Elec. Dev. Letters*, Vol. 23, No. 7, pp. 386–388, July 2002.
47. H. L. Stadler, "Ferroelectric Switching Time of BaTiO3 Crystals at High Voltages," *J. Appl. Phys.* Vol. 29, No. 10, pp. 1485–1487, Oct. 1958.
48. J. F. Scott, A. J. Hartmann, R. N. Lamb, F. M. Ross, A. DeVilbiss, C. A. Paz De Araujo, M. C. Scott, and G. Derbenwick, "Some New Results on Strontium Bismuth Tantalate Thin-Film Ferroelectric Memory Materials," *Materials Research Society Symposium Proceedings*, Vol. 433, pp. 77–84, 1996.
49. A. D. DeVilbiss et al, "High Frequency Testing Methodology and Investigation of The High Frequency Behavior of Layered Perovskite and PZT Ferroelectric Capacitors," paper presented at *Eighth International Symposium on Integrated Ferroelectrics*, Tempe, AZ, 18–20 Mar., 1996.
50. P. Larson, G. J. M. Dormans, D. J. Taylor, and P. J. van Veldhoven, "Ferroelectric Properties and Fatigue of PbZr0.51Ti0.49O3 thin Films of Varying Thickness: Blocking Layer Model," *J. Appl. Phys.*, Vol. 76, No. 4, pp. 2405–2413, 15 Aug. 1994.
51. T. Davenport, "FRAM: Advanced Feature Set Pulls the Technology into the High Density Domain," *Proceedings of the Non-Volatile Memory Technology Symposium 2002*, Jet Propulsion Laboratory Publication 02–21, p. 171, 2002.
52. T. Sumi, N. Moriwaki, G. Nakane, T. Nakakuma, Y. Judai, Y. Uemoto, Y. Nagano, S. Hayashi, M. Azuma, E. Fujii, S.-I. Katsu, T. Otsuki, L. McMillan, C. Paz de Araujo, and G. Kano, "A 256 kb Nonvolatile Ferroelectric Memory at 3 V and 100 ns," *IEEE International Solid State Circuits Conference*, pp. 268–269, 16–18 Feb. 1994.
53. R. Dat, D. J. Liechtenwalner, O. Auciello, and A. I. Kingon, "Imprint Testing of Ferroelectric Capacitors used for Non-Volatile Memories," *Integrated Ferroelectrics*, Vol. 5, pp. 275–286, 1994.
54. T. Otsuki and K. Arita, "Quantum Jumps in FeRAM Technology and Performance," *Integrated Ferroelectrics*, Vol. 17, pp. 31–43, 1997.
55. Y. Shimada, K. Arita, E. Fujii, Y. Nasu, Y. Nagano, A. Noma, Y. Izutsu, K. Nakao, K. Tanaka, T. Yamada, Y. Uemoto, K. Asari, G. Nakane, A. Inoue, T. Sumi, T. Nakakuma, S. Chaya, H. Hirano, Y. Judai, Y. Sasai, and T. Otsuki, "Advanced LSI Embedded with FeRAM for Contactless IC Cards and its Manufacturing Technology," *Integrated Ferroelectrics*, Vol. 27, pp. 1335–1358, 1999.
56. W. L. Gallagher and S. S. P. Parkin, eds. "Spintronics," *IBM J. of Res. and Devel.*, Vol. 50, No. 1, pp. 1–166, Jan. 2006.
57. M. Kryder, ed. "Magnetic Information Storage Technology," *Proc. of the IEEE*, Vol. 74, No. 11, Nov., 1986.
58. C. D. Mee and E. D. Daniel, eds. *Magnetic Recording, Volume I: Technology*, McGraw-Hill, New York, 1987.
59. T. Kagami, T. Kuwashima, S. Miura, T. Uesugi, K. Barada, N. Ohta, N. Kasahara, K. Sato, T. Kanaya, H. Kiyono, N. Hachisuka, S. Saruki, K. Inage, N. Takahashi, and K. Terunuma, "A Performance Study of Next Generation's TMR Heads Beyond 200 Gb/in^2," *IEEE Trans. Magn.*, Vol. 42, No. 2, Part I, pp. 93–96, Feb. 2006.

60. S. Mao, Y. Chen, F. Liu, X. Chen, B. Xu, P. Lu, M. Patwari, H. Xi, C. Chang, B. Miller, D. Menard, B. Pant, J. Loven, K. Duxstad, S. Li, Z. Zhang, A. Johnston, R. Lamberton, M. Gubbins, T. McLaughlin, J. Gadbois, J. Ding, B. Cross, S. Xue, and P. Ryan, "Commercial TMR Heads for Hard Disk Drives: Characterization and Extendibility at 300 Gb/in^2," *IEEE Trans. Magn.*, Vol. 42, No. 2, Part I, pp. 97–102, Feb. 2006.

61. R. R. Katti, "Current-in-plane Pseudo-Spin-Valve Device Performance for Giant Magnetoresistive Random Access Memory Applications," *J. Appl, Phys.*, Vol. 91, No. 10, pp. 7245–7250, May 15, 2002.

62. R. R. Katti, A. Arrott, J. Drewes, W. Larson, H. Liu, Y. Lu, T. Vogt, and T. Zhu, "Pseudo-Spin-Valve Device Performance for Giant Magnetoresistive Random Access Memory Applications," *IEEE Trans. Magn.*, Vol. 37, No. 4, pp. 1967–1969, July, 2001.

63. J. Gadbois, J.-G. Zhu, W. Vavra, and A. Hurst, "The Effect of End and Edge Shape on the Performance of Pseudo-Spin Valve Memories," *IEEE Trans. Magn.*, Vol. 34, No. 4, pp. 1066–1068, July, 1998.

64. J. Shi, T. Zhu, M. Durlam, E. Chen, S. Tehrani, Y. Zheng, and J. Zhu, "End Domain States and Magnetization Reversal in Submicron Magnetic Structures," *IEEE Trans. Magn.*, Vol. 34, No. 4, pp. 997–999, July, 1998.

65. E. Y. Chen, S. Tehrani, T. Zhu, M. Durlam, and H. Goronkin, "Submicron spin valve magnetoresistive random access memory cell," *J. Appl. Phys.*, 81 (8), pp. 3992–3997, 15 April 1997.

66. S. Tehrani, E. Chen, M. Durlam, M. DeHerrera, J. M. Slaughter, J. Shi, and G. Kerszykowski, "High-Density Submicron Magnetoresistive Random Access Memory," *J. Appl. Phys.*, 85 (8), pp. 5822–5827, 15 April 1999.

67. S. Tehrani, J. M. Slaughter, E. Chen, M. Durlam, J. Shi, and M. DeHerrera, "Progress and Outlook for MRAM Technology," *IEEE Trans. Magn.*, Vol. 35, No. 5, pp. 2814–2819, Sep., 1999.

68. H. Boeve, C. Bruynseraede, J. Das, K. Dessein, G. Borghs, J. De Boeck, R. Sousa, L. Melo, and P. Freitas, "Technology Assessment for the Implementation of Magnetoresistive Elements with Semiconductor Components in Magnetic Random Access Memory (Mram) Architectures," *IEEE Trans. Magn.*, Vol. 35, No. 5, pp. 2820–2825, Sep., 1999.

69. M. Johnson, "Magnetoelectronic Memories Last and Last," *IEEE Spectrum*, pp. 33–40, Feb., 2000.

70. G. Prinz, "Hybrid Ferromagnetic Semiconductor Devices," *Science*, Vol. 250, pp. 1092–1097, 1990.

71. E. C. Stoner and E. P. Wohlfarth, "A Mechanism of Magnetic Hysteresis in Heterogeneous Alloys," *Phil. Trans. Roy. Soc.*, Vol. 240, A. 826, pp. 599–642, 4 May 1948.

72. J. K. Watson, *Applications of Magnetism*, John Wiley and Sons, New York, 1980.

73. M. Scheinfein, "LLG Micromagnetics Simulator," Computer Simulation Program, Version 1.45, http://llgmicro.home.mindspring.com, 1998.

74. A. V. Pohm, B. A. Everett, R. S. Beech, and J. M. Daughton, "Bias Field and End Effects on the Switching Threshold of Pseudo Spin Valve Memory Cells," *IEEE Trans. Magn.*, Vol. 33, No. 5, pp. 3280–3282, Sep., 1997.

75. B. A. Everett and A. V. Pohm, "Single Domain Model for Pseudo Spin Valve MRAM Cells," *IEEE Trans. Magn.*, Vol. 33, No. 5, pp. 3289–3291, Sep., 1997.

76. D. Jiles, *Introduction to Magnetism and Magnetic Materials*, Chapman and Hall, London, 1991.

77. S. Chikazumi and S. Charap, "*Physics of Magnetism,*" Robert Kreiger Publishing Company, Malabar, Florida, 1984.

78. R. C. O'Handley, *Modern Magnetic Materials: Principles and Applications*, John Wiley and Sons, New York, 2000.

REFERENCES

79. S. Wolf and D. Treger, eds. Proceedings of the Spins/Spintronics Review, a DARPA-sponsored Program. Long Beach, CA, Sep. 3–7, 2001.
80. S. Wolf and D. Treger, eds. Proceedings of the Spins/Spintronics Review, a DARPA-sponsored Program. Delray Beach, FL, September 30–October 4, 2002.
81. D. Talbot, "Computing's New Spin," *Technology Review*, pp. 39, Jan./Feb., 2001.
82. S. A. Wolf, D. D. Awschalom, R. A. Buhrman, J. M. Daughton, S. von Molnar, M. L. Roukes, A. Y. Chtchelkanova, and D. M. Treger, "Spintronics: A Spin-Based Electronics Vision for the Future," *Science*, Vol. 294, pp. 1488–1495, 16 Nov. 2001.
83. S. Tehrani, B. Engel, J. M. Slaughter, E. Chen, M. Durlam, P. Naji, R. Whig, J. Janesky, and J. Calder, "Recent Developments in Magnetic Tunnel Junction MRAM," *IEEE Trans. Magn.*, Vol. 36, No. 5, pp. 2752–2757, Sep., 2000.
84. B. Heinrich and J. A. C. Bland, eds. *Ultrathin Magnetic Structures, Applications of Nanomagnetism*, Vols. III and IV, Springer, New York, 2005.
85. H.-J. Kim, S. C. Oh, J. S. Bae, K. T. Nam, J. E. Lee, S. O. Park, H. S. Kim, N. I. Lee, U.-I. Chung, J. T. Moon, and H. K. Kang, "Development of Magnetic Tunnel Junction for Toggle MRAM," *IEEE Trans. Magn.*, Vol. 41, No. 10, pp. 2661–2663, Oct. 2005,.
86. D. Lammers, "MRAM Debut Cues Memory Transition," *EETimes*, p. 1, 14, July 10, 2006.
87. R. R. Katti, "Attractive Magnetic Memories," *IEEE Circuits and Devices*, Vol. 17, No. 2, pp. 26–34, Mar., 2001.
88. L. Savtchenko, B. N. Engel, N. D. Rizzo, M. F. Deberrera, and J. A. Janesky, "Method of Writing to Scalable Magnetoresistance Random Access Memory Element," U. S. Patent 6,545,906, April 8, 2003.
89. B. N. Engel, J. A. Janesky, and N. D. Rizzo, "Magnetoresistive Random Access Memory with Reduced Switching Field," U.S. Patent 6,633,498, Oct. 14, 2003.
90. M. Durlam, D. Addie, J. Akerman, B. Butcher, P. Brown, J. Chan, M. DeHerrera, B. N. Engel, B. Feil, G. Gryukewich, J. Janesky, M. Johnson, K. Kyler, J. Molla, J. Martin, K. Nagel, J. Ren, N. D. Rizzo, T. Rodriguez, L. Savtchenko, J. Salter, J. M. Slaughter, K. Smith, J. J. Sun, M. Lien, K. Papworth, P. Shah, W. Qin, R. Williams, L. Wise, and S. Tehrani, "A 0.18um 4Mb Toggling MRAM," *IEEE Electron Devices Meeting Tech. Digest*, pp. 995–997, 2003.
91. T. andre, T. W. Andre, J. J. Nahas, C. K. Subramanian, B. J. Garni, H. S. Lin, A. Omair, and W. L. Martino, "A 4Mb 0.18um 1T1MTJ Toggle MRAM with Balanced Three Input Sensing Scheme and Locally Mirrored Unidirectional Write Drivers," *IEEE J. of Solid-State Circuits*, Vol. 40, No. 1, pp. 301–309, Jan. 2005.
92. H. Kubota, A. Fukuhima, Y. Ootani, S. Yuasa, K. ando, H. Machara, K. Tsunekawa, D. D. Djayaprawira, N. Watanabe, and Y. Suzuki, "Magnetization Switching by Spin-Polarized Current in Low-Resistance Magnetic Tunnel Junction with MgO (001) Barrier," *IEEE Trans. Magn.*, Vol. 41, No. 10, pp. 2633–2635, Oct. 2005.
93. T. Tsunoda and D. Mauri, "Magnetic Tunnel Junctions Using Reactively Sputtered Al2O3 Barriers," *IEEE Trans. Magn.*, Vol. 41, No. 10, pp. 2658–2660, Oct., 2005.
94. R. W. Dave, G. Steiner, J. M. Slaughter, J. J. Sun, B. Craigo, S. Pietambaram, K. Smith, G. Grynkewich, M. DeHerrera, J. Akerman, and S. Tehrani, "MgO-based Tunnel Junction Material for High-Speed Toggle Magnetic Random Access Memory," *IEEE Trans. Magn.*, Vol. 42, No. 8, pp. 1935–1939, Aug. 2006.
95. B. N. Engel, J. Akerman, B. Butcher, R. W. Dave, M. DeHerrera, M. Durlam, G. Grynkewixh, J. Janesky, S. V. Pietambaram, N. D. Rizzo, J. M. Slaughter, K. Smith, J. J. Sun, and S. Tehrani, "A 4-Mb Toggle MRAM Based on a Novel Bit and Switching Method," *IEEE Trans. Magn.*, Vol. 41, No. 1, pp. 132–136, Jan. 2005.
96. R. M. Bozorth, *Ferromagnetism*, D. Van Nostrand, New York, 1978.

97. A. B. Glaser and G. E. Subak-Sharpe, *"Integrated Circuit Engineering,"* Addison-Wesley, Reading, MA, 1977.
98. D. V. Averin and K. K. Likharev, "Single-electronics: Correlated Transfer of Single Electrons and Cooper Pairs in Small Tunnel Junctions," *Mesoscopic Phenomena in Solids*, (Eds.), Elsevier, Amsterdam, pp. 173–271, 1991.
99. H. Grabert and M. Devoret, (Eds.) *Single Charge Tunneling*, Plenum, New York, 1992.
100. L. P. Kouwenhoven, C. M. Markus, P. L. McEuen, S. Tarucha, R. M. Westervelt, and N. S. Wingreen, "Electron Transport in Quantum Dots," in L. Sohn (Ed.), *Mesoscopic Electron Transfer*, Kluwer, Dordrecht, pp. 105–215, 1997.
101. K. K. Likharev, "Single-Electron Devices and their Applications". *Proc. IEEE*, Vol. 87, pp. 606–632, Apr. 1999.
102. P. D. Dresselhaus, J. Li, S. Han, L. Ji, J. E. Lukens, and K. K. Likharev, "Measurement of Single Electron Lifetimes in a Multi-Junction Trap," *Phys. Rev. Letters*, Vol. 72, pp. 3226–3229, May 1994.
103. L. Ji, P. D. Dresselhaus, S. Han, K. Lin, W. Zheng, and J. Lukens, "Fabrication and Characterization of Single-Electron Transistors and Traps," *J. Vac. Sci. Technol. B*, Vol. 12, pp. 3619–3622, Nov.–Dec. 1994.
104. L. Guo, E. Leobandung, and Y. Chou, "A Single-Electron Transistor Memory Operating at Room Temperature," *Science*, Vol. 275, pp. 649–651, Jan. 1997.
105. A. Nakajima, T. Futatsugi, K. Kosemura, T. Fukano, and N. Yokoyama, "Room Temperature Operation of Si Single-Electron Memory with Self-Aligned Floating Dot Gate," *Appl. Phys. Letters*, Vol. 70, pp. 1742–1744, Mar. 1997.
106. J. J. Welser, S. Tiwari, S. Rishton, K. Y. Lee, and Y. Lee, "Room Temperature Operation of a Quantum-Dot Flash Memory," *IEEE Electron Device Letters*, Vol. 18, pp. 278–280, June 1997.
107. S. Tiwari, F. Rana, H. Hanafi, A. Hartstein, E. F. Crabbé, and K. Chan, "A Silicon Nanocrystals Based Memory," *Appl. Phys. Letters*, Vol. 68, pp. 1377–1379, Mar. 1996.
108. S. Tiwari, F. Rana, K. Chan, L. Shi, and H. Hanafi, "Single Charge and Confinement Effects in Nano-Crystal Memories," *Appl. Phys. Letters*, Vol. 68, pp. 1232–1234, Aug. 1996.
109. H. I. Hanafi, S. Tiwari, and I. Khan, "Fast and Long Retention-Time Nano-Crystal Memory," *IEEE Trans. on Electron. Dev.*, Vol. 43, pp. 1553–1558, Sep. 1996.
110. K. K. Likharev, "Electronics Below 10 nm," in *Nano and Giga Challenges in Microelectronics*, J. Greer, A. Korkin, and J. Labanowski, (Eds.), Elsevier, Amsterdam, pp. 27–68, 2003.
111. D. B. Strukov and K. K. Likharev, "Architectures for Defect-Tolerant Nanoelectronic Crossbar Memories," *J. of Nanoscience and Nanotechnology*, Vol. 7, pp. 151–167, Jan. 2006.
112. K. Yano, T. Ishii, T. Sano, T. Mine, F. Murai, T. Kure, and K. Seki, "A 128-Mb Early Prototype for Gigascale Single-Electron Memories," *IEEE International Solid State Circuits Conference Dig. of Tech. Paper*, pp. 344–345, 1998.
113. K. Yano, T. Ishii, T. Hashimoto, T. Kobayashi, F. Murai, and K. Seki, "Room-Temperature Single-Electron Memory," *IEEE Trans. on Electron. Dev.*, Vol. 41, pp. 1628–1638, Sep. 1994.
114. T. Ishii, K. Yano, T. Sano, T. Mine, F. Murai, T. Kure, and K. Seki, "Single-Electron Memory for Giga-to-Tera Bit Storage," *Proc. IEEE*, Vol. 87, pp. 633–651, Apr. 1999.
115. A. N. Korotkov and K. K. Likharev, "Ultradense Hybrid SET/FEY dynamic RAM: Feasibility of Background-Charge-Independent Room-Temperature Single-Electron Digital Circuits," in: *Proc. of 1995 Int. Semicond. Dev. Res. Symp.* University of Virginia, Charlottesville, pp. 355–358, 1995.

116. C. D. Chen, Y. Nakamura, and J. S. Tsai, "Aluminum Single-Electron Nonvolatile Floating Gate Memory Cell," *Appl. Phys. Letters*, Vol. 71, pp. 2038–2040, Oct. 1997.

117. K. K. Likharev, "Single-Electron Transistors: Electrostatic Analogs of the DC SQUIDs," *IEEE Trans. on Magn.*, Vol. 23, pp. 1142–1145, Mar. 1987.

118. A. N. Korotkov, "Analysis of Integrated Single-Electron Memory Operation", *J. Appl. Phys.* Vol. 92, pp. 7291–7295, Dec. 2002.

119. C. J. Amsinck, N. H. Di Spigna, D. P. Nackashi, and P. D. Franzon, "Scaling Constraints in Nanoelectronic Random-Access Memories," *Nanotechnology*, Vol. 16, pp. 2251–2260, Oct. 2005.

120. G. Dearnaley, A. M. Stoneham, and D. V. Morgan, "Electrical phenomena in amorphous oxide films," *Rep. Prog. Phys.*, Vol. 33, pp. 1129–1200, 1970.

121. Seo, M. J. Lee, D. H. Seo, E. J. Jeoung, D.-S. Suh, Y. S. Young, I. K. Yoo, I. R. Hwang, S. H. Kim, I. S. Byun, J.-S. Kim, J. S. Choi, and B. H. Park, "Reproducible Resistance Switching in Polycrystalline Nio Films," *Appl. Phys. Letters*, Vol. 85, pp. 5655–5657, Dec. 2004.

122. I. G. Baek, D. C. Kim, M. J. Lee, H.-J. Kim, E. K. Yim, M. S. Lee, J. E. Lee, S. E. Ahn, S. Seo, J. H. Lee, J. C. Park, Y. K. Cha, S. O. Park, H. S. Kim, I. K. Yoo, U-In Chung, J. T. Moon, and B. I. Ryu, "Multi-Layer Cross-Point Binary Oxide Resistive Memory (Oxrram) for Post-nand Storage Applications," *IEEE Electron Devices Meeting Tech. Digest*, Paper No. 31.4, 2005.

123. D. C. Kim, S. Seo, S. E. Ahn, D.-S. Suh, M. J. Lee, B.-H. Park, I. K. Yoo, I. G. Baek, H.-J. Kim, E. K. Yim, J. E. Lee, S. O. Park, H. S. Kim, U-I. Chung, J. T. Moon, and B. I. Ryu, "Electric Observations of Filamentary Conductions for the Resistive Memory Switching in Nio Films," *Appl. Phys. Letters*, Vol. 88, pp. 202–102, May 1–3, 2006.

124. B. J. Choi, D. S. Jeong, S. K. Kim, C. Rohde, S. Choi, J. H. Oh, H. J. Kim, C. S. Hwang, K. Szot, R. Wasser, B. Reichenberg, and S. Tiedke, "Resistive Switching Mechanism of Tio_2 Thin Films Grown by Atomic-Layer Deposition," *J. Appl. Phys.*, Vol. 98, pp. 03–37, Aug. 2005.

125. D. Lee, H. Choi, H. Sim, D. Choi, H. Hwang, M.-J. Lee, S.-A. Seo, and I. K. Yoo, "Resistance Switching of the Nonstoichiometric Zirconium Oxide for Nonvolatile Memory Applications," *IEEE Electron Device Letters*, Vol. 26, pp. 719–721, Oct. 2005.

126. H. Sim, D. Choi, D. Lee, M. Hasan, C. B. Samantray, and H. Hwang, "Reproducible Resistance Switching Characteristics of Pulsed Laser-Deposited Polycrystalline Nb2O5," *Microelectron. Eng.*, Vol. 80, pp. 260–263, Jun. 2005.

127. A. Chen, S. Haddad, Y.-C. Wu, T.-N. Fang, S. Lan, S. Avanzino, S. Pangrle, M. Buynoski, M. Rathor, W. Cai, N. Tripsas, C. Bill, M. VanBuskirk, and M. Taguchi, *IEEE Electron Devices Meeting Tech. Dig.*, Paper No. 31.3, 2005.

128. L. P. Ma, J. Liu, and Y. Yang, "Organic Electrical Bistable Devices and Rewritable Memory Cells," *Appl. Phys. Letters*, Vol. 80, pp. 2997–2999, Apr. 2002.

129. L. P. Ma, S. Pyo, J. Ouyang, Q. Xu, and Y. Yang, "Nonvolatile Electrical Bistability of Organic/Metal-Nanocluster/Organic System," *Appl. Phys. Letters*, Vol. 82, pp. 1419–1421, Mar. 2003.

130. J. Ouyang, C. W. C. W. Chu, C. R. Szmanda, L. Ma, and Y. Yang, "Programmable Polymer Thin Film and Non-Volatile Memory Device," *Nature Materials*, Vol. 3, pp. 918–922, Dec. 2004.

131. L. P. Ma, Q. Xu, and Y. Yang, "Organic Nonvolatile Memory by Controlling the Dynamic Copper-Ion Concentration within Organic Layer," *Appl. Phys. Letters*, Vol. 84, pp. 4908–4910, Jun. 2004.

132. L. D. Bozano, B. W. Kean, V. R. Deline, J. R. Salem, and J. C. Scott, "Mechanism for Bistability in Organic Memory Elements," *Appl. Phys. Letters*, Vol. 84, pp. 607–609, Jan. 2004.

133. R. Sezi, A. Walter, R. Engl, A. Maltenberger, J. Schumann, M. Kund, and C. Dehm, "Organic Materials for High-Density Non-Volatile Memory Applications," *IEEE Electron Devices Meeting Tech. Dig.*, Paper No. 10.2.1, 2003.
134. Y.-S. Lai, C.-H. Tu, D.-L. Kwong, and J. S. Chen, "Bistable Resistance Switching of Poly (N-vinylcarbazole) Films for Nonvolatile Memory Applications," *Appl. Phys. Letters*, Vol. 87, pp. 122101 1–3, Sep. 2005.
135. N. B. Zhitenev, A. Sidorenko, D. M. Tennant, and R. A. Cirelli, "Chemical Modification of the Conducting States in Polymer Nanodevices," *Nature Nanotechnology*, Vol. 2, pp. 237–242, Apr. 2007.
136. C. P. Collier, E. W. Wong, M. Belohradsky, F. M. Raymo, J. F. Stoddart, P. J. Kuekes, R. S. Williams, and J. R. Heath, "Electronically Configurable Molecular-Based Logic Gates," *Science*, Vol. 285, pp. 391–394, Jul. 1999.
137. C. P. Collier, G. Mattersteig, E. W. Wong, Y. Kuo, K. Beverly, J. Sampaio, F. M. Raymo, J. F. Stoddart, and J. R. Heath, "A [2]Catenane-Based Solid State Electronically Reconfigurable Switch," *Science*, Vol. 289, pp. 1172–1175, Aug. 2000.
138. Y. Chen, G.-Y. Jung, D. A. A. Ohlberg, X. Li, D. R. Stewart, J. O. Jeppesen, K. A. Nielsen, J. F. Stoddard, and R. S. Williams, "Nanoscale Molecular-Switch Crossbar Circuits," *Nanotechnology*, Vol. 14, pp. 462–468, Apr. 2003.
139. C. Li, D. Zhang, X. Li, S. Han, T. Tang, C. Zhou, W. Fan, J. Koehne, J. Han, M. Meyyappan, A. M. Rawlett, D. W. Price, and J. M. Tour, "Fabrication Approach for Molecular Memory Arrays," *Appl. Phys. Letters*, Vol. 82, pp. 645–647, Jan. 2003.
140. W. Wu, G.-Y. Jung, D. L. Olynick, J. Straznicky, Z. Li, X. Li, D. A. A. Ohlberg, Y. Chen, S.-Y. Wang, J. A. Liddle, W. M. Tong, and R. S. Williams, "One-Kilobit Cross-Bar Molecular Memory Circuits at 30-nm Half-Pitch Fabricated by Nanoimprint Lithography," *Appl. Phys. A*, Vol. 80, pp. 1173–1178, Mar. 2005.
141. Y-C. Chen, C. F. Chen, C. T. Chen, J. Y. Yu, S. Wu, S. L. Lung, R. Liu, and C-Y. Lu, "An Access-Transistor-Free (0T/1R) Non-Volatile Resistance Random Access Memory (RRAM) Using a Novel Threshold Switching, Self-Rectifying Chalcogenide Device," *IEEE Electron Devices Meeting Tech. Dig.*, Paper No. 37.4.1, 2003.
142. M. Kund, G. Beitel, C.-U. Pinnow, T. Röhr, J. Schumann, R. Symanczyk, K.-D. Ufert, and G. Müller, "Conductive Bridging Ram (Cbram): An Emerging Non-Volatile Memory Technology Scalable to Sub 20 nm," *IEEE Electron Devices Meeting Tech. Dig.*, Paper No. 31.3, 2005.
143. J. G. Simmons and R. R. Verderber, "New Conduction and Reversible Memory Phenomena in Thin Insulating Films," *Proc. Roy. Soc. A*, Vol. 301, pp. 77–102, 1967.
144. K. K. Likharev, A. Mayr, I. Muckra, and Ö. Türel, "CrossNets: High-Performance Neuromorphic Architectures for CMOL Circuits," *Ann. New York Acad. Sci.*, Vol. 1006, pp. 146–163, 2003.
145. H. Park, J. Park, A. K. L. Lim, E. H. anderson, A. P. Alivisatos, and P. L. McEuen, "Nanomechanical Oscillations in a Single-C60 Transistor," *Nature*, Vol. 407, pp. 57–60, Sep. 2000.
146. S. P. Gubin, Y. V. Gulyaev, G. B. Khomutov, V. V. Kislov, V. V. Kolesov, E. S. Soldatov, K. S. Sulaimankulov, and A. S. Trifonov, "Molecular Clusters as Building Blocks for Nanoelectronics: The First Demonstration of a Cluster Single-Electron Tunneling Transistor at Room Temperature," *Nanotechnology*, Vol. 13, pp. 185–194, Apr. 2002.
147. N. B. Zhitenev, H. Meng, and Z. Bao, "Conductance of Small Molecular Junctions," *Phys. Rev. Letters*, Vol. 88, pp. 226801 1–4, Jun. 2002.
148. J. Park, A. N. Pasupathy, J. I. Goldsmith, C. Chang, Y. Yaish, J. R. Petta, M. Rinkoski, J. P. Sethna, H. D. Abruna, P. L. McEuen, and D. C. Ralph, "Coulomb Blockade and the Kondo Effect in Single-Atom Transistors," *Nature*, Vol. 417, pp. 722–725, Jun. 2002.

149. S. Kubatkin, A. Danilov, M. Hjort, J. Cornil, J. L. Bredas, N. Stuhr-Hansen, P. Hedegart, and T. Bjornholm, "Single-Electron Transistor of a Single Organic Molecule with Access to Several Redox States," *Nature*, Vol. 425, pp. 698–701, Oct. 2003.

150. N. B. Zhitenev, W. R. Jiang, A. Erbe, Z. Bao, E. Garfunkel, D. M. Tennant, and R. A. Cirelli, "Control of Topography, Stress and Diffusion at Molecule-Metal Interfaces," *Nanotechnology*, Vol. 17, pp. 1272–1277, Mar. 2006.

151. H. B. Akkerman, P. W. M. Blom, D. M. de Leeuw, and B. De Boer, "Towards Molecular Electronics with Large-Area Molecular Junctions," *Nature*, Vol. 441, pp. 69–72, May 2006.

152. K. K. Likharev and D. B. Strukov, "CMOL: Devices, Circuits, and Architectures," in: G. Cuniberti, G. Fagas, and K. Richter, (eds.), *Introducing Molecular Electronics*, Springer, Berlin, 2005, pp. 447–477.

153. T. Dadosh, Y. Gordin, R. Krahne, I. Khirich, D. Mahalu, V. Frydman, J. Sperling, A. Yacoby, and I. Bar-Joseph, "Measurement of the Conductance of Single Conjugated Molecules," *Nature*, Vol. 436, pp. 677–680, Aug. 2005.

154. C. M. S. Torres, S. Zankovych, J. Seekamp, A. P. Kam, C. C. Cedeno, T. Hoffmann, J. Ahopelto, F. Reuther, K. Pfeiffer, G. Bleidiessel, G. Gruetzner, M. V. Maximov, and B. Heidari, "Nano-imprint Lithography: an Alternative Nanofabrication Approach," *Materials Science & Eng. C*, Vol. 23, pp. 23–31, Jan. 2003.

155. J. Choi, K. Nordquist, A. Cherala, L. Casoose, K. Gehoski, W. J. Dauksher, S. V. Sreenivasan, and D. J. Resnick, "Distortion and Overlay Performance of UV Step and Repeat Imprint Lithography," *Microelectron. Eng.*, Vol. 78–79, pp. 633–640, 2005.

156. S. R. J. Brueck, "There are No Limits to Optical Lithography," in: *International Trends in Applied Optics*, A. H. Guenther (ed.), SPIE Press, Bellingham, WA, pp. 85–89, September 2002.

157. H. H. Solak, C. David, J. Gobrecht, V. Golovkina, F. Cerrina, S. O. Kim, and P. F. Nealey, "Sub-50 nm Period Patterns with EUV Interference Lithography," *Microelectron. Eng.*, Vol. 67, pp. 56–62, Jun. 2003.

158. G.-Y. Jung, E. Johnson-Halperin, W. Wu, Z. Yu, S.-Y. Wang, W. M. Tong, Z. Li, J. E. Green, B. A. Sheriff, A. Boukai, Y. Bunimovich, J. R. Heath, and R. S. Williams, "Circuit Fabrication at 17 nm Half-Pitch by Nanoimprint Lithography," *Nano Letters*, Vol. 6, pp. 351–354, Mar. 2006.

159. M. R. Stan, P. D. Franson, S. C. Goldstein, J. C. Lach, and M. M. Ziegler, "Molecular electronics: From Devices and Interconnect to Circuits and Architecture," *Proc. IEEE*, Vol. 91, pp. 1940–1957, Nov. 2003.

160. S. Das, G. Rose, M. M. Ziegler, C. A. Picconatto, and J. E. Ellenbogen, "Architectures and Simulations for Nanoprocessor Systems Integrated on the Molecular Scale," in: *Introducing Molecular Electronics*, G. Cuniberti, G. Fagas, and K. Richter, (eds.), Springer, Berlin, pp. 479–512, 2005.

161. A. DeHon and K. K. Likharev, *Proc. ICCAD'05, IEEE*, Piscataway, NJ, pp. 375–381, 2005.

162. P. J. Kuekes, G. S. Snider, and R. S. Williams, "Crossbar Nanocomputers," *Sci. Amer.*, Vol. 293, pp. 72–75, Nov. 2005.

163. K. K. Likharev, "CMOL: A New Concept for Nanoelectronics," *Proc. of the 12th Int. Symp. on Nanostructures Physics and Technology*, St. Petersburg, Russia, 2004.

164. K. K. Likharev, Interface, "CMOL: A Silicon-Based Bottom-Up Approach to Nanoelectronics," Vol. 14, pp. 43–45, May 2005.

165. D. B. Strukov and K. K. Likharev, *Proc. FPGA'06*, Association for Computing Machinery, New York, pp. 131–137, 2006.

166. Ö. Türel, J. H. Lee, X. Ma, and K. K. Likharev, "Neuromorphic Architectures for Nano-Electronic Circuits," *Int. J. of Circuit Theory and Applications*, Vol. 32, pp. 277–302, Sep.–Oct. 2004.
167. J. H. Lee and K. K. Likharev, "Cross Nets as Pattern Classifiers," *Lecture Notes in Computer Science*, Vol. 3512, pp. 446–454, 2005.
168. D. B. Strukov and K. K. Likharev, "Defect-Tolerant Architectures for Nanoelectronic Crossbar Memories," *J. of Nanoscience and Nanotechnology*, Vol. 7, pp. 151–167, Jan. 2007.
169. K. Likharev, "Layered Tunnel Barriers for Nonvolatile Memory Devices." *Appl. Phys. Letters*, Vol. 73, pp. 2137–2139, Nov. 1998.
170. K. K. Likharev, "NOVORAM: A New Concept for Fast, Bit-Addressable Nonvolatile Memory Based on Crested Barriers," *IEEE Circuits and Devices*, Vol. 16, pp. 16–21, June 2000.
171. D. J. DiMaria, "Graded or Stepped Energy Band-Gap-Insulator MIS Structures (GI-MIS or SI-MIS)," *J. Appl. Phys.*, Vol. 50, pp. 5826–5829, Sep. 1979.
172. F. Capasso, F. Beltram, R. J. Malik, and J. F. Walker, "New Floating-Gate AlGaAs/GaAs Memory Devices with Graded Gap Electron Injector and Long Retention Times," *IEEE Electron. Dev. Letters*, Vol. 9, pp. 377–379, Aug. 1988.
173. A. N. Korotkov and K. K. Likharev, "Resonant Fowler-Nordheim Tunneling Through Layered Tunnel Barriers and Its Possible Applications," *IEEE Electron Devices Meeting Tech. Dig.*, pp. 223–226, Dec. 1999.
174. J. D. Casperson, L. D. Bell, and H. A. Atwater, "Material Issues for Layered Tunnel Barrier Structures," *J. Appl. Phys.*, Vol. 92, pp. 261–267, July 2002.
175. B. Govoreanu, P. Blomme, M. Rosmeulen, J. Van Houdt, and K. De Mayer, "VARIOT: A Novel Multilayer Tunnel Barrier Concept for Low-Voltage Nonvolatile Memory Devices," *IEEE Electron Dev. Letters*, Vol. 24, pp. 99–101, Feb. 2003.
176. B. H. Koh, W. K. Chim, T. H. Ng, J. X. Zheng, and W. K. Choi, "Quantum Mechanical Modeling of Gate Capacitance and Gate Current in Tunnel Dielectric Stack Structures for Nonvolatile Memory Application," *J. Appl. Phys.*, Vol. 95, pp. 5094–5103, May 2004.
177. S. J. Baik, S. Choi, U-I. Chung, and J. T. Moon, "Engineering on Tunnel Barrier and Dot Surface in Si Nanocrystal Memories," *Solid-State Electronics*, Vol. 48, pp. 1475–1481, Sep. 2004.
178. S. H. Hong, J. H. Jang, T. J. Park, D. S. Jeong, M. Kim, C. S. Hwang, and J. Y. Won, "Improvement of the Current-Voltage Characteristics of a Tunneling Dielectric by Adopting a $Si_3N_4/SiO_2/Si_3N_4$ Multilayer for Flash Memory Application," *Appl. Phys. Letters*, Vol. 87, pp. 152106 1–3, Oct. 2005.
179. B. Govoreanu, P. Blomme, J. Van Houdt, and K. De Meyer, "Enhanced Tunneling Current Effect for Nonvolatile Memory Applications," *Jpn. J. Appl. Phys.*, Part 1, Vol. 42, pp. 2020–2024, Apr. 2003.
180. Y. Liu, S. I. Shim, F. C. Yeh, X. W. Wang, and T. P. Ma, "Barrier Engineering for Non-Volatile Memory Applications," *SISC 2006*, San Diego, CA, paper 3.2, Dec. 2006.
181. J. Buckley, B. De Salvo, G. Ghibaudo, M. Gely, J. F. Damlencourt, F. Martin, G. Nicotra, and S. Deleonibus, "Investigation of SiO_2/HfO_2 Gate Stacks for Application in Non-Volatile Memory Devices," *Solid State Electronics*, Vol. 49, No. 11, pp. 1833–1840, Nov. 2005.
182. E. Cimpoiasu, S. K. Tolpygo, X. Liu, N. Simonian, J. E. Lukens, K. K. Likharev, R. F. Klie, and Y. Zhu, "Aluminum Oxide Layers as Possible Components for Layered Tunnel Barriers," *J. Appl. Phys.*, Vol. 96, pp. 1085–1093, July 2004.
183. X. Liu, V. Patel, Z. Tan, J. E. Lukens, and K. K. Likharev, "High-Quality Aluminum-Oxide Tunnel Barriers for Scalable, Floating-Gate Random-Access Memories (FGRAM),"

International Conference on Memory Technology and Design, 7–10 May, 2007, Giens, France, pp. 235–237, 2007.

184. R. G. Neale, D. L. Nelson, and G. E. Moore, "Nonvolatile and Reprogrammable, The Read-Mostly Memory is Here," *Electronics.* Vol. 56, 1970.

185. S. R. Ovshinsky, "A History of the Phase Change Technology," in *Memories Optiques et Systemes*, 1994.

186. T. Zhang, B. Liu, J.-L. Xia, Z.-T. Song, S.-L. Feng, and B. Chen, "Structure and electrical properties of Ge2Sb2Te5 Thin Film used for Ovonic Unified Memory," *Chinese Physical Letters*, Vol. 21, p. 3, 2004.

187. N. Yamada, E. Ohno, K. Nishiuchi, and N. Akahira, "Rapid-Phase Transitions of GeTe-Sb2Te3 Pseudobinary Amorphous Thin Films for an Optical Disk Memory," *Journal of Applied Physics*, Vol. 69, pp. 2849–2856, Mar. 1, 1991.

188. N. Yamada and T. Matsunaga, "Structure of Laser-Crystallized $Ge_2Sb_2+xTe_5$ Sputtered Thin Films for Use in Optical Memory," *J. of Applied Physics*, Vol. 88, pp. 7020–7028, Dec. 15, 2000.

189. S. J. Hudgens, "OUM Nonvolatile Semiconductor Memory Technology Overview," *Materials Research Society Symposium, Tech. Dig.*, Vol. 918, Paper 0918-HO5-01-G06-01, 2006.

190. T. A. Lowrey, S. J. Hudgens, W. Czubatyj, C. H. Dennison, S. A. Kostylev, and G. C. Wicker, "Characteristics of OUM Phase Change Materials and Devices for High Density Nonvolatile Commodity and Embedded Memory Applications," *Materials Research Society Symposium HH, Tech. Dig.*, pp. HH2.1.1 to HH2.1.12, 2003.

191. H.-R. Oh, B.-H. Cho, W. Y. Cho, S. Kang, B.-G. Choi, H.-J. Kim, K.-S. Kim, D.-E. Kim, C.-K. Kwak, H.-G. Byun, G.-T. Jeong, H.-S. Jeong, and K. Kim, "Enhanced Write Performance of a 64-mb Phase-Change Random Access Memory," *IEEE J. of Solid-State Circuits*, Vol. 41, pp. 122–126, 2006.

192. M. Gill, T. Lowrey, and J. Park, "Ovonic Unified Memory—A High-Performance Nonvolatile Memory Technology for Stand-Alone Memory and Embedded Applications," *IEEE International Solid-State Circuits Conference*, Paper 12.4, San Francisco, 2002.

193. G. Atwood and R. Bez, "Current Status of Chalcogenide Phase Change Memory," in *IEEE Device Research Conference Digest*, pp. 29–33, 2005.

194. A. Pirovano, A. L. Lacaita, D. Merlani, A. Benvenuti, F. Pellizzer, and R. Bez, "Electronic Switching Effect in Phase-Change Memory Cells," *IEEE International Electron Devices Meeting, Tech. Dig.*, pp. 923–926, 2005.

195. S. R. Ovshinsky and D. Adler, "Local Structure, Bonding and Electric Properties of Covalent Amorphous Semiconductors," *Contemporary Physics*, Vol. 19, pp. 109–125, 1978.

196. S. R. Ovshinsky, "Localized States in the Gap of Amorphous Semiconductors," *Physical Review Letters*, Vol. 36, pp. 1469–1472, 1976.

197. S. R. Ovshinsky, "Lone-Pair Relationship and the Origin of Excited States in Amorphous Chalcogenide," *International Topical Conference on Structure and Exciton of Amorphous Solids Tech. Dig.*, pp. 31–36, 1976.

198. M. Kasner, D. Adler, and H. Fritzsche, "Bonding Bonds, Lone-Pair Bonds and Impurity Sites in Chalcogenide Semiconductors," *Physical Review Letters*, Vol. 28, pp. 355–357, 1976.

199. M. Kasner, D. Adler, and H. Fritzsche, "Valence-Alternation Model for Localized Gap States in Lone-Pair Semiconductors," *Physical Review Letters*, Vol. 37, pp. 1504–1507, 1976.

200. A. Pirovano, A. L. Lacaita, A. Benvenuti, F. Pellizzer, S. Hudgens, and R. Bez, "Scaling Analysis of Phase-Change Memory Technology," *IEEE International Electron Devices Meeting Tech. Dig.*, pp. 29.6.1–29.6.4, 2003.

201. Y. N. Hwang, S. H. Lee, S. J. Ahn, S. Y. Lee, K. C. Ryoo, H. S. Hong, H. C. Koo, F. Yeung, J. H. Oh, H. J. Kim, W. C. Jeong, J. H. Park, H. Horii, Y. H. Ha, J. H. Yi, G. H. Koh, G. T. Jeong, H. S. Jeong, and K. Kim, "Writing Current Reduction for High-Density Phase-Change RAM," *IEEE International Electron Devices Meeting*, pp. 37.1.1–37.1.4, Dec. 2003.

202. F. Pellizzer, A. Pirovano, F. Ottogalli, M. Magistretti, M. Scaravaggi, P. Zuliani, M. Tosi, A. Benvenuti, P. Besana, S. Cadeo, T. Marangon, R. Morandi, R. Piva, A. Spandre, R. Zonca, A. Modelli, E. Varesi, T. Lowrey, A. Lacaita, G. Casagrande, P. Cappelletti, and R. Bez, "Novel Micro-Trench Phase-Change Memory Cell for Embedded and Stand-Alone Non-Volatile Memory Applications," *IEEE Symposium on VLSI Technology Tech. Dig.*, pp. 18–19, 2004.

203. H. Horii, J. H. Yi, J. H. Park, Y. H. Ha, I. G. Baek, S. O. Park, Y. N. Hwang, S. H. Lee, Y. T. Kim, K. H. Lee, U. I. Chung, and J. T. Moon, "A Novel Cell Technology Using N-Doped GeSbTe Films for Phase Change RAM," *IEEE VLSI Symposium on Technology Tech. Dig.*, pp. 177–178, 2003.

204. Y. H. Ha, J. H. Yi, H. Horii, J. H. Park, S. H. Joo, S. O. Park, U. I. Chung, and J. T. Moon, "An Edge Contact Type Cell for Phase Change RAM Featuring Very Low Power Consumption," *IEEE Symposium on VLSI Technology Tech. Dig.*, pp. 175–176, 2003.

205. N. Matsuzaki, K. Kurotsuchi, Y. Matsui, O. Tonomura, N. Yamamoto, Y. Fujisaki, N. Kitai, R. Takemura, K. Osada, S. Hanzawa, H. Moriya, T. Iwasaki, T. Kawahara, N. Takaura, M. Terao, M. Matsuoka, and M. Moniwa, "Oxygen-Doped GeSbTe Phase-Change Memory Cells Featuring 1.5 V/100 Microa Standard 0.13 Micron CMOS Operations," *IEEE International Electron Devices Meeting Tech. Dig.*, pp. 738–741, 2005.

206. F. Pellizzer, A. Benvenuti, B. Gleixner, Y. Kim, B. Johnson, M. Magistretti, T. Marangon, A. Pirovano, R. Bez, and G. Atwood, "A 90 nm Phase Change Memory Technology for Stand-Alone Non-Volatile Memory Applications," *IEEE Symposium on VLSI Technology Tech. Dig.*, pp. 122–123, 2006.

207. S. Lai, "Current Status of the Phase Change Memory and Its Future," *IEEE International Electron Devices Meeting Tech. Dig.*, pp. 10.1.1–10.1.4, 2003.

208. K. Kim and S. J. Ahn, "Reliability Investigations for Manufacturable High Density PRAM," *IEEE Reliability Physics Symposium Tech. Dig.*, pp. 157–162, 2005.

209. J. D. Maimon, K. K. Hunt, L. Burcin, and J. Rodgers, "Chalcogenide Memory Arrays: Characterization and Radiation Effects," *IEEE Trans. on Nuclear Science*, Vol. 50, pp. 1878–1884, 2003.

210. S. Bernacki, K. Hunt, S. Tyson, S. Hudgens, B. Pashmakov, and W. Czubatyj, "Total Dose Radiation Response and High Temperature Imprint Characteristics of Chalcogenide Based RAM Resistor Elements," *IEEE Trans. on Nuclear Science*, Vol. 47, pp. 2528–2533, 2000.

211. J. H. Oh, J. H. Park, Y. S. Lim, Y. T. Oh, J. S. Kim, J. M. Park, Y. J. Song, K. C. Ryoo, D. W. Lim, S. S. Park, J. I. Kim, J. H. Kim, J. Yu, F. Yeung, C. W. Jeong, J. H. Kong, D. H. Kang, G. H. Koh, G. T. Jeong, H. S. Jeong, and K. Kim, "Full Integration of Highly Manufacturable 512 Mb PRAM Based on 90 nm Technology," *IEEE International Electron Devices Meeting Tech. Dig.*, Paper 2.6, 2006.

INDEX

A

Acceleration factor (AF), 468–469, 557–559, 568, 569
Access time, 7, 23, 24, 26–27, 39, 41–43, 45, 387, 398–400
Activation energy, 135–136, 137, 467–469, 480, 489–490, 557
Active volume, 712, 714, 717, 724
Address bus, 43, 73
Address fuses, 78, 79, 81
Address path configuration, 91, 98–99
Address word, 6
Advanced technology attachment (ATA) interface, 47, 48, 51, 52, 53, 551
AF (acceleration factor), 468–469, 557–559, 568, 569
A-GST. *See* GST (germanium antimony tellurium)
AHI (anode hole injection) model, 160, 161, 167, 168, 571, 572
Algorithm skips, 91, 97, 98, 105
Algorithms. *See* Erase algorithm; Program algorithm
Alternative floating nodes, 195–196
Amorphous metal-oxide films, 698
Amorphous state, 709, 710–711, 715, 724, 727
AMR (anisotropic magnetoresistance) effect, 679
Analog blocks, 67, 71–73
AND-type Flash, 69, 71
Anisotropic magnetoresistance (AMR) effect, 679
Anode hole injection (AHI) model, 160, 161, 167, 168, 571, 572
Antenna ratios, 531, 532, 533
APC (augmented product code), 87
Array blocking, 25, 27–28, 55, 186
Array direct access, 91, 103–104
Array operation, 71, 110, 118, 119, 186–187
Arrays. *See* Cell arrays
Arrhenius model, 135, 136, 468, 472, 473, 557, 559, 567
Asperities, 148, 149, 150, 474, 512, 513
Asymmetrical blocking, 25, 29, 56
Asynchronous Flash, 39, 40, 41, 42, 43

ATA (advanced technology attachment) interface, 47, 48, 51, 52, 53, 551
Augmented product code (APC), 87
Autoclave test, 556

B

B4-Flash (back-bias-assisted band-to-band tunneling-induced hot-electron injection), 363
Back-bias-assisted band-to-band tunneling-induced hot-electron injection (B4-Flash), 363
Background operation (BGO):
 and DINOR circuits, 314, 326, 327–328
 emulating EEPROM, 327–328
 emulating static RAM, 327–328
 in Flash memory architecture, 67–68
Background pattern discrepancy (BPD), 283–284
Ball grid array (BGA) packages, 36, 37, 38
Band structure, 9, 10, 145
Bandgap reference, 118–119, 120, 121
Band-to-band hot-electron (BBHE) injection cell operation, 332–333
Band-to-band tunneling:
 back-bias-assisted band-to-band tunneling-induced hot-electron injection (B4-Flash), 363
 current-induced degradation and charge trapping, 460–462
 hot-electron injection
  Chung research studies, 353, 357
  and DINOR programmed cells, 332–333, 334
  Ohnakado research studies, 349–353, 354, 355
  and P-channel cells, 329–332, 334
  as P-channel Flash operation, 338, 339, 340–341, 342
  Shen research studies, 353, 355, 356
 physics overview, 158–159
 in tunnel-erase-induced oxide degradation, 459–460
Barrier engineering. *See* Tunnel barrier engineering
Bathtub curve, 539, 541, 553, 564

Nonvolatile Memory Technologies with Emphasis on Flash. Edited by J. E. Brewer and M. Gill
Copyright © 2008 the Institute of Electrical and Electronics Engineers, Inc.

741

Battery life, 25, 29, 36, 45, 46, 403
BBHE (band-to-band hot-electron) injection cell operation, 332–333
BBT. *See* Band-to-band tunneling
BGA (ball grid array) packages, 36, 37, 38
BIOS, 24, 29, 37, 39, 44, 466, 543
Bipolar Transistor Selected (BIST) P-channel cells, 359, 360, 361–362
BIST (Bipolar Transistor Selected) P-channel cells, 359, 360, 361–362
Bit-by-bit verify circuits:
    DINOR, 314, 315, 321, 333
    NAND EEPROM, 237–247, 248, 265, 266, 286
Bitline pitch, 16, 225–226, 255, 265, 270, 272
Bitlines. *See also* DINOR (divided bitline NOR)
    capacitive coupling, 265–266, 270, 288, 292
    defined, 5
    main bitline, 69, 70, 316, 321
    in NROM arrays, 15–16
    shielded, 265, 270–272, 273, 286, 288
    sub-bitlines, 69, 70, 316, 321, 339, 352
Block erase, 30, 34, 57, 226, 233–235
Block locking, 26, 31–32, 50, 57
Block reclamation, 35
Blocking:
    array, 25, 27–28, 55, 186
    block locking, 26, 31–32, 50, 57
    virtual, 34
Block-oriented memory, 7, 8, 224
Block-to-block disturb, 495–496
Blue-ray disks, 708
Boltzmann constant, 119, 135, 141, 468, 490, 557, 633
Boot block, 28, 29, 32, 66
Boot code, 28, 29, 36, 44, 59–60
Boot sector, 98–99
Bootstrapping techniques, 110, 111, 118
Bottom boot, 59, 99
Breakdown model, 164, 168–169
BTBT. *See* Band-to-band tunneling
Buffer circuits, 122–123
Bulk erase, 28, 30, 189, 392, 393
Bulk-limited conduction, 143–144
Burn-in:
    and process defects, 539–543
    and reliability monitoring, 564–565
Burst mode, 27, 41–43
Burst read, 40, 68, 272–273, 274, 285, 657
Bus cycles, 27, 57, 74
Bus widths, 25, 27, 40
By 8 mode, 100–101
By 16 mode, 89, 100–101
Byte programming, 30, 32, 297, 338

C

Cache, 8, 398, 399
Calling sequence, 33
CAM (content-addressable memory), 547–548
Capacitance model, 10, 140–143
CBC (channel boost capacitance) cell, 258–263
CD (critical dimension) control, 189, 698
Cell arrays. *See also* Flash memories; Multilevel cell (MLC) technology
    asymmetrical *vs.* symmetrical, 28–29
    differential sensing, 202–203
    DINOR, 316–320
    disturb concept, 186–187
    dividing into blocks, 28–29, 166
    P-channel, 343–345
    SuperFlash architecture and design, 212–214
    timing, 203–204
Cell density, 698, 718
Cell erase, 64, 209, 214, 456, 463, 493, 524, 602
Cell phones, 19–20, 37
C-GST. *See* GST (germanium antimony tellurium)
Chalcogenides, 12, 708, 709
Channel boost capacitance (CBC) cell, 258–263
Channel erase, 184–185, 188, 191, 206, 456, 457, 502, 561, 572, 602
Channel hot carrier (CHC) mechanism, 423–425
Channel hot-electron (CHE) injection:
    for compacting Flash cells, 65
    effect of silicon nitride as tunnel dielectric, 422–423
    vs. FN tunneling, 10
    as method of programming Flash memory, 447
    overview, 150–153
    for programming Flash cells, 64
Channel hot-electron (CHE) programming-induced oxide degradation:
    degradation mechanism of Flash EEPROM memory programming, 452–454
    gate current, 449–450
    overerase correction programming, 450–452
    trap reduction through nitridation, 454–455
Channel hot-hole-induced hot-electron (CHHIHE) injection:
    Chung research study, 353, 357
    Hsu research study, 345–349
    as P-channel Flash operation, 338, 339–340
Charge balance equation, 140, 141
Charge loss:
    and accelerated data retention bake tests, 467–473
    and activation energy, 467–469
    due to charge detrapping, 477
    due to contamination, 474, 475, 478
    due to cycling-induced oxide damage, 477
    due to oxide defect, 473, 474, 478
    intrinsic, 135–136, 467, 473, 478, 483, 607
    minimization as goal, 446
    new extrapolation law, 472–473
    overview, 135–138

through ONO, 474–477, 478
voltage-accelerated data retention test, 469–472
Charge loss/gain mechanisms, 473–477
Charge placement, 596, 597, 603–606, 608, 609–610, 614
Charge pumps:
   and DINOR circuits, 321–326
   heap, 117–118, 326, 327
Charge storage ability, 595–596
Charge-based memory devices, 2–3, 12
CHE. *See* Channel hot-electron (CHE) injection; Channel hot-electron (CHE) programming-induced oxide degradation
Chemical corrosion failure mechanisms, 556
Chemical mechanical polishing (CMP), 200, 253, 318
Chemical vapor deposition (CVD):
   impacts of intermetal dielectric and passivation films on Flash memory reliability, 533–536
   in silicon nitride tunnel dielectrics, 409, 410
CHHIHE. *See* Channel hot-hole-induced hot-electron (CHHIHE) injection
Chip controller, 72–73
Chip scale packages (CSPs), 37–38
CMOL (CMOS molecular circuits), 700–701
CMOS devices, 354, 356, 358, 530–533, 696, 700–701, 702, 703
CMOS molecular circuits (CMOL), 700–701
CMP (chemical mechanical polishing), 200, 253, 318
Code + data usage model, 23, 24, 49, 50, 54–62
Code and data, 20, 21, 22, 23–24, 26, 31, 54, 55–57, 62
Code and data storage, 13, 22, 23, 24.28
Code integrity, 39, 40, 45
Code security, 40
Code storage:
   execute-in-place (XIP) usage model, 38, 39–43, 45, 51
   socket Flash for, 543
   store and download (SnD) usage model, 38, 40, 43, 44–45
Code usage model, 23–24, 49, 50
Coercive voltage, 663, 670
Coffin-Manson equation, 559
Cold electrons, 10
Column decoder, 67, 397, 399
Column defects, 75
Column failures, 75
Column programming path, 67, 202
Column redundancy, 75, 80–81
Command user interface (CUI), 73, 91, 92
CompactFlash, 51, 52, 53, 543
Compaction, 63, 65, 74, 97–98, 105, 392–393
Complementary metal-oxide-semiconductor (CMOS), 354, 356, 358, 530–533, 696, 700–701, 702, 703

Compression, 48, 91, 106–108
Conduction band, 9, 144, 145, 146, 147, 148, 349, 437, 449, 450, 459, 460, 633, 703, 704
Conduction mechanisms, 10, 131, 414, 416, 436, 471, 499, 636
Contactless cell arrays, 316–317
Content-addressable memory (CAM), 547–548
Control algorithms, 33, 90
Control bus, 73
Control circuitry, 6
Control logic, 67, 73–75, 597
Controller block, 73
Counters block, 74
Coupling coefficients, 11
Coupling noise, 270–272
CPU (central processing unit), 8
Crested barrier, tunnel barrier engineering:
   and FN tunneling, 194
   NOVORAM/FGRAM memories, 703–706
   silicon nitride as dielectric, 437–439
Critical dimension (CD) control, 189, 698
Crystal growth, 196, 714
Crystalline state, 12, 709, 711, 724
Crystallization temperature, 710, 712, 714
CSPs (chip scale packages), 37–38
CUI (command user interface), 73, 91, 92
CVD (chemical vapor deposition):
   impacts of intermetal dielectric and passivation films on Flash memory reliability, 533–536
   in silicon nitride tunnel dielectrics, 409, 410
Cycling:
   erase overview, 29–30
   evaluations, 541
   life test, 539–543
   stress, 13
Cycling-induced data retention issues, 477–481
Cycling-induced degradations, 447–466

D

DAHC (drain avalanche hot-carrier), 450
Data bus, 22, 51, 73, 74, 100, 101, 399
Data compression, 91, 106–108
Data flow parameter, 7
Data logging, 23
Data organization, 7
Data output width, 7
Data path configuration, 91, 99–101
Data refresh operation, 3
Data retention:
   activation energy, 467–468
   bake tests, 467–473
   characteristics related to tunnel oxide and floating-gate poly texture, 481–484
   and charge-loss, 467–477
   defined, 466
   duration, 467
   effects, 571–572
   ferroelectric memories, 672–673

Flash EEPROM cycling-induced issues, 477–481
impact of soft errors, 484–486
magnetic memories, 387
multilevel cell (MLC) technology, 607
new extrapolation law, 472–473
nonvolatile memory (NVM) overview, 2, 9, 12–13
NROM (nitrided ROM) memories, 636–640, 641
overview, 135–138, 466
reliability issues, 466–486
SuperFlash, 217–218
temperature-accelerated test, 467–469
voltage-accelerated test, 469–472
Data retention storage life (DRSL) test, 554–555
Data storage, 46–53
Data usage model, 24, 49, 50
DC erase disturb, 187
DDR (double data rate), 43
Defective cells, 81, 88, 104, 548, 549
Defectivity model, 75
Defects:
  categories, 547
  density, 475, 484, 512, 525, 538, 542–543
  mapping, 547–548, 549
  reason for failure rate, 539, 541
Defects per million (DPM), 493
Degradation:
  cycling induced, 447
  induced by band-to-band tunneling, 460
  induced by charge trapping, 461
  induced by tunnel erase, 456
  during programming, 452
Density, Flash memory, 27
Depletion check, 74, 93, 94, 97
Depletion mode, 5, 6, 11, 158, 230, 272, 429, 516
Design for testability. *See* DFT (design for testability)
Destructive readout (DRO), 4, 666, 695, 696
Detrapping-related retention effects, 572
Device technology generations, 16–18
DFT (design for testability):
  address path configuration, 98–99
  array direct access and stresses, 103–104
  data compression, 106–108
  data path configuration and trimming, 99–101
  external control, 97
  fuse cell implementation, 92–93
  high voltages external forcing and monitor, 101–103
  high voltages trimming, 94–96
  internal pattern write and verify, 105–106
  internal state machine algorithm skips, 97–98
  overview, 89–91
  sense amplifier reference trimming, 93–94
  test entries, 91–92
  test organization, 91–92
  timings trimming, 96–97
DIBL (drain-induced barrier lowering), 626, 654
Dielectric requirements, 191–192
Dielectric scaling:
  alternative floating nodes, 195–197
  interpoly dielectric (IPD), 195
  overview, 191
  requirements, 191–192
  tunnel oxide, 192–195
Dielectrics for tunneling. *See* Tunnel dielectrics
Differential to single-ended converter, 112, 113
Digital cameras, 47, 48, 50, 53, 134, 224–225
Digital versatile disks:
  random access memory (DVD-RAM), 708
  rewritable (DVD-RW), 708
DINOR (divided bitline NOR):
  array architecture, 316–320
  and background operation (BGO), 326, 327–328
  band-to-band hot-electron (BBHE) injection cell operation, 332–333
  cell characteristics, 314–315
  charge pump circuit, 321–326
  circuit for low-voltage operation, 321–327
  defined, 313
  emulation of EEPROM and SRAM using BGO, 327–328
  endurance, 315, 333
  erasing, 313–314
  and gate-protected polysilicon diodes (GPPDs), 323–324, 325
  heap pump circuit, 326, 327
  high-voltage generation, 321–326
  low-voltage read, 320–321
  operation, 313–314
  and P-channel cells, 328–334, 350
  programming, 313–314
  read access speed, 321
  technology features, 320–321
  as type of Flash memory array, 64, 69
  virtual ground array (VGA) cell structure, 317–318
  voltage detect circuit, 324–326
  wordline boost scheme, 326–327
Diodes, programmable, 698–700
DIP (dual in-line) packages, 36
Disks, strengths and weaknesses, 21, 24, 46
Disturb effects, 138–139
Disturb immunity, 138
Disturbs:
  block-to-block, 495–496
  defined, 186, 487
  effect of cycling, 490, 494–495
  erase, 139, 187, 206, 210, 495

gate, 139, 338, 389, 390, 391, 393–394, 492–494, 630
interblock, 186
intrablock, 186–187
NAND array, 229–230
NROM array, 630–632
overview, 186
program, 211, 229–239, 280–281, 283, 339, 491–495, 575–576
read, 339, 487–490
SST cells, 219–220
SuperFlash, 210, 211–212
Divided bitline NOR. See DINOR (divided bitline NOR)
Double data rate (DDR), 43
Double partitioned memory arrays, 67. See also Background operation (BGO)
DPM (defects per million), 493
Drain avalanche hot-carrier (DAHC), 450
Drain disturb, 138, 187, 338, 389, 390, 391, 393, 491, 571, 572
Drain-induced barrier lowering (DIBL), 626, 654
DRAM (dynamic random-access memory), 3, 16, 21, 24, 42, 43, 44, 45, 66, 87, 142, 446, 485, 486, 539, 592, 594, 618, 619, 620, 623, 665, 667, 674, 677, 696, 706, 707
Driver information systems, 45–46
DRO (destructive readout), 4, 666, 695, 696
DRSL (data retention storage life) test, 554–555
Dual in-line (DIP) packages, 36
Dual partition Flash devices, 31
Dynamic random-access memory. See DRAM (dynamic random-access memory)

E
ECC (error correction coding), 87–89, 548–552
EEPROM (electrically erasable read-only memory). See also Flash memory
64-bit NAND example, 225–226
comparison of NAND and NOR types, 223, 224–225, 227
defined, 2
emulating, 327–328
vs. EPROM, 179
history, 14, 23–24, 223, 224
NAND technology overview, 223–226
relationship to Flash memory, 2, 54, 55, 594
strengths and weaknesses, 21
Effective electron temperature model, 151
Effective yield, 76, 78
Eitan, B., 15, 16
Electric charge quantization, 689–691
Electric field energy model, 166, 168
Electrode-limited conduction, 144
Electron discreteness impedance restriction, 689–690

Electron trap generation, 163–164, 166–168, 454, 477, 510
Electrostatic discharge (ESD), 67, 122, 555
Electrostatic potential, 140, 141
Embedded Flash:
advantages over stand-alone Flash memory, 375–376
applications, 377–383
by device type, 377–379
by end product, 380–382
by function, 379–380
by usage, 382–383
cells
1.5T cell, 374
2T cell, 374
2TS cell for low-voltage operation, 389–391
examples, 386
MoneT cell for high density and high speed, 391–394
SCSG cell for manufacturability, 386–389
special requirements and considerations, 383–385
defined, 49, 373, 543
design
array architecture, 399–400
cell selection, 400–401
design techniques for embedded Flash module, 398–403
Flash module design for embedded applications, 396–398
high-speed circuit techniques, 401–402
low-voltage design, 403
process consideration, 401
special requirements and considerations, 394–396
system architecture, 398–399
disadvantages compared with stand-alone Flash memory, 376–377
logic-based vs. memory-based processes, 385
overview, 49, 373–375
size and process complexity trade-off, 384–385
vs. stand-alone Flash memory, 47, 375–377
Endurance:
as aspect of nonvolatile memory (NVM), 13, 387
cycling, 454, 467, 544, 555
DINOR (divided bitline NOR), 315, 333
ferroelectric memories, 671–672
Flash memory limitations, 623
floating-gate devices, 133–135
magnetic memories, 387
NROM (nitrided ROM) memories, 639–640
overview, 13, 133–135
SST memory cells, 218–219
SuperFlash, 218–219
Energy barrier, 9–10, 144–146, 148, 150, 151, 183, 195, 447, 450, 486, 600, 601, 607, 709

Engine spark timing control, 380
Enhancement mode, 5, 11, 15, 213, 230–231, 272, 429, 516
EPROM (erasable programmable ROM):
    comparison with Flash memory, 601
    defined, 2
    vs. EEPROM, 179
    evolution of Flash memory from, 21–22
    history, 21, 23
    strengths and weaknesses, 21
    UV, 179–180
Erase algorithm, 27, 65, 67, 74, 97–98, 101, 105–106, 462, 544, 545–546, 547
Erase blocks, 35, 186–187, 233–235, 389–391, 393–394, 552, 598
ERASE command, 74
Erase counts, 35
Erase cycles, 29, 49–50, 57–58, 133, 139, 480, 515, 538, 544, 623, 642
Erase distribution, 104, 183, 187–190, 191, 462, 466, 481, 528, 530
Erase disturb, 139, 187, 206, 210, 495
Erase margin, 132, 491, 545, 641
Erase methods, 332, 353, 447, 456–459, 461, 479
Erase operation:
    alternative memory technologies, 623, 626, 627, 628, 629, 631, 636, 642, 643, 654, 671, 691, 703, 708, 716
    as basic Flash cell operation, 63, 64, 90
    basic NAND cells, 227–228
    block sizes, 28–29
    block-erase, 30, 34, 57, 226, 233–235
    bulk-erase Flash devices, 28
    and compaction, 65, 97–98
    conduction mechanisms, 10
    cycling, 29–30, 57–58
    and DINOR Flash memory technology, 314, 319, 321, 328
    distribution control, 187–190
    EEPROM history, 14, 27
    and electric fields, 216–217
    and embedded Flash memory, 387, 392, 544–547
    EPROM devices, 21–22
    erratic erase, 462–465, 570–571
    and error correction, 88
    and ETOX technology, 180, 183–185, 188, 189, 190, 191, 200, 205–206
    and Flash memory architecture, 66, 67, 68, 70, 71, 73, 74
    and Flash memory limitations, 623
    in floating-gate devices, 11–12
    and multilevel cell digital memories, 601, 605
    and NAND Flash memory technology, 233, 248
    and NOR Flash memory technology, 5, 179, 180, 183–185, 188, 189, 190, 191, 200, 205–206, 208, 215, 217

NROM devices, 15, 626, 627, 628, 629, 631, 636, 642, 643
overerase correction programming, 450–452
overview, 11–12, 131
and P-channel Flash, 338, 339, 343, 345, 353
physics of mechanism, 143–157, 159
poly-to-poly, 512–517
and reclamation, 35
reliability issues, 445, 447, 452, 456, 457, 458, 459, 461, 462, 478, 479, 482, 495, 523, 527, 537, 544, 547
and row redundancy, 79, 82–84, 86, 93, 96
self-boosted NAND inhibit scheme, 233–235
software issues, 32–33
SuperFlash EEPROM cells, 209–210
and SuperFlash technology, 208, 215, 217
and suspends, 30–31, 57
and tunnel dielectrics, 430, 431, 432, 433, 434, 437, 440
Erase pulse, 12, 74, 79, 96–97, 135, 447–448, 544, 545–546
Erase suspend, 30, 31, 57, 58, 74
Erase through oxide (ETOX). See ETOX (EPROM tunnel oxide)
Erase verify operation, 74, 93, 97, 98, 103, 228, 396, 448, 451, 453, 517, 544, 546
Erratic erase, 462–465, 570–571
Error correction, 87–89, 548–552
ESD (electrostatic discharge), 67, 122, 555
ETOX (EPROM tunnel oxide):
    array configuration, 603
    cycling-induced degradations in Flash memories, 447–466
    embedded erase and program algorithms, 544–547
    erase operation, 183–185, 601–602
    erased cell vs. programmed cell, 182–183
    Flash cell technology, 179, 180–206
    Flash memory background, 599–603
    history, 14–15, 612
    manufacturing, 612
    and multilevel cell (MLC), 592–594
    overview, 592–594
    program operation, 183, 184, 601
    read operation, 602–603
    relationship to UV-EPROM, 180–182
    sensing operation, 182–183
    stacked gate SAS etch process and erase distribution - reliability issues, 528–530
Execute-in-place (XIP) usage model, 38, 39–43, 45, 51

F

Face centered cubic (fcc) sublattice, 711, 712
FACE (Flash algorithmic control engine), 608–609
Failure in time (FIT), 553, 566, 675
Failure rate calculations, 565–570

INDEX **747**

Fast bits, 188, 387, 527
FAT (file allocation table) file sectors, 35
FCBR (full chip burst read) operation, 272–273, 274
Fcc (face centered cubic) sublattice, 711, 712
FDI (Flash Data Integrator), 58
Feature size, 190, 255, 262
Ferroelectric memories:
    capacitor cells, 665–667
    cells and arrays, 664–670
    die and test cost, 675–676
    endurance, 671–672
    febrication, 670–671
    field-effect transistor cells, 660, 664–665
    imprint behavior, 673
    nonvolatile characteristics, 671–673
    overview, 658–660
    products, 676–677
    programming speed, 669–670
    programming voltage, 667–669
    reliability, 674–675
    retention, 672–673
    scaling, 673–674
    stacked cells, 665–666
    storage mechanism, 660–664
FETs (field-effect transistors):
    ferroelectric, 660, 664–665
    N-channel, 416
    P-channel, 416
Few-electron phenomenon:
    as alternative memory technology, 689–696
    impact on scaling, 576–579
FGRAM (floating-gate RAM), 706–707
Field-accelerated testing, 137–138
Field-effect transistors (FETs):
    ferroelectric, 660, 664–665
    N-channel, 416
    P-channel, 416
Field-enhancing tunneling injector EEPROM cells. *See* SuperFlash
File allocation table (FAT) file sectors, 35
Filing systems, Flash, 35
Film thickness, 191–197
Find operation, 36
Fin-FET NROM cells, 655
FIT (failure in time), 553, 566, 675
Flash, origin of term, 14, 21
Flash algorithmic control engine (FACE), 608–609
Flash cards, 24, 46, 47, 50, 51, 52, 71
Flash Data Integrator (FDI), 58
Flash memories. *See also* Embedded Flash; NAND Flash technology; NOR Flash memories; SuperFlash
    alternative technologies overview, 617–619
    applications overview, 19–38
    architecture overview, 66–68
    arrray architectures, 69–71

    asymmetrical *vs.* symmetrical, 28–29
    basic cell operations, 63–65
    basic models, 140–143
    basic operating principles, 130–143
    capacitor model, 10, 140–143
    in cell phones, 19–20
    characteristics, 130–143
    circuit techniques, 108–123
    command set, 74
    control logic, 73–75
    cost considerations, 596–597
    cycling-induced degradations, 447–466
    data retention, 466–486
    design for testability, 89–108
    device attributes, 26–32
    disturbs, 487–496
    embedded *vs.* removable, 47
    error correction, 87–89, 548–552
    ETOX cell technology, 180–206
    evolution from EPROM, 2, 21–22
    floating-gate structure, 3, 9–12
    future applications, 45–46
    history, 14–18, 179
    how it works, 594–596
    key circuit building blocks, 200–206
    limitations, 619–624
    list of attributes, 25–26
    overview, 13–16
    physics, 129–171
    redundancy, 75–86
    reliability overview, 445–447
    reliability testing and screening, 552–570
    removable, 47, 50–53
    role of software, 20, 25–26, 29, 31, 32–36
    scaling issues, 190–200, 257, 258, 260, 262–263, 264, 265
    split-gate memory technology, 180, 206–216
    storage, 23–24, 38–62
    strengths and weaknesses, 21
    terminology, 131–140
    typical layout, 619, 620
    usage models, 23–24, 38–46, 49, 50, 54–62
Flash memory core (FMC), 375, 394, 395, 396–398
Flash translation layer (FTL), 35, 51
FlashFile architecture, 29
Floating-gate devices:
    capacitance model, 10, 140–143
    endurance, 133–135
    erasing, 183–185
    Flash cells as transistors, 63–65, 594
    history, 596
    isolation, 136, 197
    N-channel, with deposited filicon nitride tunnel dielectric, 425–429
    nitride, 195–196, 624–658
    noncrystal storage nodes, 196–197
    operational overview, 130–131

P-channel, with deposited silicon nitride
    tunnel dielectric, 429–432
  reliability issues, 526–528
  replacing, 196–197
  role of electrode, 3
  silicon nitride as tunnel dielectric, 409
  structural overview, 3, 9–12
Floating-gate RAM (FGRAM), 706–707
Floating-gate-to-control-gate isolation, 136
Floating-gate-to-floating-gate coupling,
    574–575
FMC (Flash memory core), 375, 394, 395,
    396–398
FN tunneling. *See* Fowler-Nordheim (FN)
    tunneling
4-bit NROM memory cells, 645–650, 655,
    657
4-level NAND cell array, 288–293, 306
Fourier transform infrared (FTIR) spectrum,
    411–412
Fowler-Nordheim (FN) tunneling:
  for comparing Flash cells, 65
  equation for erase distribution, 188
  for erasing Flash cells, 64
  as method of erasing Flash memory, 447
  as method of programming Flash memory,
    447
  in NAND EEPROMS, 223
  overview, 10, 145–148
  as P-channel Flash operation, 338, 339, 340,
    341, 342, 343
  for programming Flash cells, 64
FP (Frenkel-Poole) conduction, 414–415, 416,
    471
Frenkel-Poole (FP) conduction, 414–415, 416,
    471
FTIR (Fourier transform infrared) spectrum,
    411–412
FTL (Flash translation layer), 35, 51
Fujitsu, 409, 676, 677
Full chip burst read (FCBR) operation,
    272–273, 274
Full-width at half-maximum (FWHM), 633
Fuse bits, 73. *See also* Fuse cells
Fuse cells, 78–79, 92–93, 94, 96, 98, 99, 100
FWHM (full-width at half-maximum), 633

G
Garbage collection, 58, 276
Gate disturb, 139, 338, 389, 390, 391, 393–394,
    492–494, 630
Gate-induced drain leakage (GIDL), 341,
    575–576, 633, 635, 636, 637
Gate-protected polysilicon diodes (GPPDs),
    323–324, 325
Germanium antimony tellurium (GST) alloys,
    709–712, 714, 715–717, 718, 719–720, 721,
    722, 724, 725

Ge-Sb-Te. *See* Germanium antimony tellurium
    (GST) alloys
Giant magneto-resistance (GMR) effect,
    679–683, 687, 689
GIDL (gate-induced drain leakage), 341,
    575–576, 633, 635, 636, 637
GMR (giant magneto-resistance) effect,
    679–683, 687, 689
GPPDs (gate-protected polysilicon diodes),
    323–324, 325
Gray code, 82, 85, 89
GST (germanium antimony tellurium) alloys,
    709–712, 714, 715–717, 718, 719–720, 721,
    722, 724, 725
Guard rings, 123

H
Hamming codes, 550–551
Hard disk drives, 21, 40, 44, 46, 48, 62, 87, 596,
    678
Hardware redundancy, 549
HAST (highly accelerated stress test), 556
Hcp (hexagonal close-packed) structure, 712
Heap charge pump, 117–118, 326, 327
Heater resistor, 712, 717–718
Hexagonal close-packed (hcp) structure, 712
HfO_2, 194, 408, 434–435
High on-chip voltages, 109
High voltages external forcing, 91, 101–103
Higher injection MOS (HIMOS), 159, 386, 512,
    521–523, 524
Highly accelerated stress test (HAST), 556
High-temperature operating life (HTOL) test,
    554
High-voltage source with grounded gate erase
    (HSE), 456–459, 461, 462, 465
High-voltage transistors:
  MMOS and PMOS, 254, 265, 282
  overview, 536–537
  process defects, 539–543
  reliability in Flash memory products, 537–539
  role of cycling and burn-in, 539–543
HIMOS (higher injection MOS), 159, 386, 512,
    521–523, 524
Hole fluence, 160–162
Hole injection mechanism, 160–161, 167, 465,
    493, 654
Horizontal parity, 550
Host processor, 67, 73–74
Hot-carrier injection, 10, 162, 164, 191, 196, 363
Hot-electron injection:
  back-bias-assisted band-to-band tunneling-
    induced hot-electron injection (B4-
    Flash), 363
  band-to-band hot-electron (BBHE) injection
    cell operation, 332–333
  channel hot-electron (CHE) injection, 10, 64,
    65, 150–153, 422–423, 447

channel hot-hole-induced hot-electron (CHHIHE) injection, 338, 339–340, 345–349, 353, 357
Chung research studies, 353, 357
and DINOR programmed cells, 332–333, 334
Ohnakado research studies, 349–353, 354, 355
and P-channel cells, 329–332, 334
as P-channel Flash operation, 338, 339, 340–341, 342
secondary impact ionization initiated channel hot-electron injection, 156–157
Shen research studies, 353, 355, 356
substrate hot-electron injection (SHEI), 144, 153–154, 163–164
Hot-hole injection:
channel hot-hole-induced hot-electron (CHHIHE) injection, 338, 339–340, 345–349, 353, 357
reliability concerns, 432 (*See also* Substrate hot-hole injection (SHHI))
substrate hot-hole injection (SHHI), 154, 163
Hot-induced breakdown, 557
HSE (high-voltage source with grounded gate erase), 456–459, 461, 462, 465
HTOL (high-temperature operating life) test, 554
Hybrid CMOS/nanodevice resistive memories, 696, 700–701, 702, 703
Hydrogen release model, 166, 167–168
Hysteresis curve, 662–663, 684
Hysteretic charging curve, 691

I

IDE (integrated device electronics) interface, 51
Impedance restriction, 689
Incremental step pulse programming (ISPP), 247, 283
Infant mortality, 543, 564
Injection:
back-bias-assisted band-to-band tunneling-induced hot-electron injection (B4-Flash), 363
band-to-band hot-electron (BBHE) injection cell operation, 332–333
channel hot-electron (CHE) injection, 10, 64, 65, 150–153, 422–423, 447
channel hot-hole-induced hot-electron (CHHIHE) injection, 338, 339–340, 345–349, 353, 357
hot-carrier injection, 10, 162, 164, 191, 196, 363
secondary impact ionization initiated channel hot-electron injection, 156–157
source-side injection (SSI), 144, 155–156, 512, 517–525
substrate hot-electron injection (SHEI), 144, 153–154, 163–164
substrate hot-hole injection (SHHI), 154, 163

Injection mechanisms (overview), 135, 144, 145, 147, 156, 338, 654
In-line plasma charging damage, 530–533
Input/output buffers, 122–123, 228, 375–376
In-system reprogrammability, 19
Integrated device electronics (IDE) interface, 51
Intel:
Advanced+Boot Block architecture, 29, 32
Flash Data Integrator (FDI), 58
FlashFile architecture, 29
history of ETOX, 14–15, 17, 180, 612, 623
NOR technology, 22, 612
PCM fabrication, 721–722
role in Flash origins, 21
SSD design, 87
StrataFlash technology, 27, 41, 591, 594–596, 598, 599, 608, 612, 613
Interblock disturb, 186
Interface states, 192, 452–454, 498, 521, 522, 524
Interface traps, 160, 162–163, 167–168, 412, 416, 426, 448, 453–454, 502, 511, 521, 523, 524, 578
Interlaced vertical parity (IVP), 87
Interleaving, 40, 41, 57, 226
Intermetal dielectric layer, 533, 534–535
Internal pattern write and verify technique, 91, 98, 105–106, 107
Internal threshold voltage, 141
International Technology Roadmap for Semiconductors (ITRS), 16–18, 192, 702, 723
Interpoly dielectric (IPD), 191, 192, 195
Intrablock disturb, 186–187
Intrinsic charge loss, 12–13, 135–136, 467, 473, 478, 483, 607
Intrinsic failure mechanisms, 134, 554
Inverted-read operation, 276–278
I/O buffers, 122–123, 228, 375–376
I/O pads, 67, 376, 395
Ion contamination, 137, 142
IPD (interpoly dielectric), 191, 192, 195
Iridium oxide, 660
Islands, 408, 690, 693, 694, 696
Isolation of floating gates, 136, 197
ISPP (incremental step pulse programming), 247, 283
ITRS (International Technology Roadmap for Semiconductors), 16–18, 192, 702, 723
IVP (interlaced vertical parity), 87

J

JEDEC. *See* Joint Electron Device Engineering Council (JEDEC) standards
Jet-vapor deposition (JVD):
gate-quality film properties, 411–417
NOR overview, 192–193
for silicon nitride film, 409–411

Joint Electron Device Engineering Council (JEDEC) standards, 554, 574
JVD. *See* Jet-vapor deposition (JVD)

L

Lanthanum strontium cobalt oxide, 660
LATCH signal, 113–114
Latch-up testing, 543, 555
Lateral channel electric field, 10, 450, 453
Lateral redistribution, 633, 636
Lead integrity test, 556
Lead zirconate titanate (PZT), 659–660, 668, 670, 671, 673, 674
Level shifting circuits, 385
Life tests, 220, 467, 539, 541, 542, 554–555, 557, 561, 565–570, 573
Linear Flash card interface, 51
Linked lists, 34–35
Local oxidation (LOCOS), 198
Lockdown, 32, 60
LOCOS (local oxidation), 198
Low-temperature operation life (LTOL) test, 554
LTOL (low-temperature operation life) test, 554
Lucky electron model, 151, 601

M

Magnetic cores, 678, 687
Magnetic disks, 8
Magnetic memories:
 die and test cost, 688–689
 fabrication, 386–387
 magnetic GMR cell scaling, 687
 magnetic random-access memory (MRAM) with giant magnetoresistive devices, 679–684
 magnetic random-access memory (MRAM) with magnetic tunnel junction devices, 684–685
 magnetic tunnel junction scaling, 687–688
 nonvolatile characteristics
  endurance, 387
  retention, 387
 overview, 678–679
 programming speed, 686
 programming voltage, 685–686
 reliability, 688
 scaling, 687–688
Magnetic spin-valve devices, 679
Magnetic thin films, 679
Magnetic tunnel junction (MTJ) devices, 679, 684–685, 686, 687–688, 689
Magnetoresistance, 678, 679, 682, 684–685, 687, 688
Main bitline, 69, 70, 316, 321
Main memory, 8

Managers. *See* Media managers; Software data managers
Mass storage, 13, 46, 68, 283, 447, 543, 549, 551, 552
Media managers, 20, 30, 32, 34–36, 58, 59, 60–61
MEDICI device simulator, 330
Melting temperature, 709, 711, 712, 714, 723, 724
Memory. *See also* DRAM (dynamic random-access memory); Flash memories; Nonvolatile memory (NVM); SRAM (static random-access memory)
 common technologies, 21
 elementary concepts, 2–9
 hierarchy, 8
 history, 14–18, 592
 ideal, characteristics, 618–619
 low-cost goal, 592–594
 main, 8
 nonvolatile overview, 1–18
 storage overview, 9–12
Memory cards, 24, 46, 47, 50–53, 71
Metastable state, 711, 712
Micron, 87
MiniSD memory card, 53
Mirrorbit, 658
Mitsubishi, 69, 346
Mixed signal design implementation, 608–611
MLC. *See* Multilevel cell (MLC) technology
MMC (MultiMedia Cards), 51, 53
Mobile applications, 19, 29, 225, 313
Mobile ions, 137, 466, 467, 534, 535, 564
MoneT cell, 386, 391–394, 401
Monte Carlo method, 77
Moore, Gordon, 16
Moore's law, 1, 16, 614
MOS-C charge pump, 114–118
Motorola one-transistor cell. *See* MoneT cell
MRAM (magnetic random-access memory):
 with giant magnetoresistive devices, 679–684
 with magnetic tunnel junction devices, 684–685
MTJ (magnetic tunnel junction) devices, 679, 684–685, 686, 687–688, 689
Multilevel cell (MLC) technology:
 comparison of 1-bitcell and 2-bitcell product features, 598
 cost considerations, 592–594
 digital memories, 591–614
 evolution of technology, 592, 596–599
 Flash memory background, 599–603
 history, 592, 596–599
 key features, 599
 low-cost design implementation, 611–612
 low-cost process manufacturing, 612
 mixed signal design implementation, 608–611
 MLC concept, 596–599
 MLC operation, 603–607

INDEX   751

overview, 206, 591–592
power supplies, 613
precise charge placement, 603–606
precise charge retention, 607
precise charge sensing, 606–607
programming speed, 613
read speed, 613
reliability, 613
role of ETOX, 592–594
scaling, 614
SST cells in, 216
standard product feature set, 612–613
Multilevel NAND:
    array noise suppression technology, 286–293
    circuit technology, 283–286
    double-level V_{th} select gate array architecture, 290–293
    high-speed programming, 301–307
    three-level, 297–301
MultiMedia Cards (MMC), 51, 53
Multipartition Flash devices, 31

N
NAND Flash technology:
    array architecture, 5–6, 69, 70–71, 231–237, 286–293, 344–345
    basic cell structure and operation, 227–231
    bit-by-bit verify circuits
        simple, 237–242
        sophisticated, 242–247
    booster plate technology, 256–258
    channel boost capacitance cells, 258–263
    circuit/technology interactions, 270–283
    for data storage, 47–48
    floating-gate-to-floating-gate coupling, 574–575
    full chip burst read operation, 272–273
    key circuits, 270–283
    multilevel, 283–307
    negative V_{th} cells, 263–268
    overcoming energy barriers, 10
    overprogram elimination scheme, 247–252
    overview, 22, 23, 223–226
    P-channel Flash, 344–345
    process and scaling issues, 252–269
    process flowchart, 367, 368–370
    for removable Flash media, 50–51
    shallow trench isolation, 252–256
    shielded bitline sensing method, 270–272
    side-wall transfer transistor cell, 293–296
    similarities with NOR, 13, 15
    source line programming scheme, 278–283
    symmetric sense amplifier with page copy function, 276
    wordline spacing, 268–269
Nanocrystal memory, 12, 195, 196, 197, 363
Natural decay, 13, 725
N-channel devices, 416, 425–429

NDRO (nondestructive readout), 4, 696
Negative gate channel erase (NGCE), 456–459
Negative gate with positive source erase (NGSE), 456–459, 461, 462, 465
Negative level shifter, 110–111
Negative resistance, 715, 727
Negative threshold voltage, 65, 263, 451, 544
Negative voltage:
    charge pump circuits, 117
    switching, 110
Negative-gate erase, 184, 191, 206, 456, 491, 495, 537
Neobit, 364, 366
NeoFlash, 363, 366, 386
NGCE (negative gate channel erase), 456–459
NGSE (negative gate with positive source erase), 456–459, 461, 462, 465
Nitridation, 454–455
Nitride floating gates, 195–196
Nitride traps, 12, 363, 632–633
N-metal-oxide-semiconductor (NMOS) field-effect transistors (FETs), 625–626
NMOS (N-metal-oxide-semiconductor) field-effect transistors (FETs), 625–626
Noise:
    bitline capacitive coupling, 265–266, 270, 288, 292
    in cell array during program verify, 266
    cell array suppression, 286–293
Non-charge-based memory devices, 12
Nondestructive readout (NDRO), 4, 696
Nonremovable Flash memory, 46–47
Nonvolatile memory:
    NOVORAM, 694, 703–707
Nonvolatile memory (NVM):
    endurance aspect, 13
    EPROMs as, 21
    Flash overview, 13–16
    floating-gate structure, 3, 9–12
    funtional capability classifications, 2
    in ITRS, 16–17
    overview, 1
    retention aspect, 2, 9, 12–13
    storage aspect, 9–12
    unique aspects, 9–13
Nonvolatility, 1, 2, 9, 13, 32, 49, 138, 606, 607, 613, 678, 689
NOR Flash memories. *See also* ETOX (EPROM tunnel oxide); SuperFlash
    array architecture, 5, 6, 69, 343–344
    common erase bias schemes, 456–459
    for data storage, 47–48
    and erase operation, 11
    overcoming energy barriers, 10
    overview, 22
    P-channel Flash, 343–344
    process and scaling issues, 190–200
    process flowchart, 364, 367, 368

for removable Flash media, 50–51
similarities with NAND, 13, 15
NOVORAM (nonvolatile RAM), 694, 703–707
NROM (nitrided ROM) memories:
   4-bit cells, 645–650, 655, 657
   adjustable read, 642, 648
   array architecture, 16, 628–630
   array disturbs, 630–632
   cell structure and operation, 625–628
   data retention, 636–640, 641
   endurance, 639–640
   fabrication, 650–652
   manufacturing process, 650–652
   nitride trapping levels, 633–634
   overview, 4, 15, 16, 625
   process flow outline, 650
   products, 655–658
   quad technology, 645–650, 655, 657
   reliability, 638–645
   scaling, 652–655
   spatial distribution of localized charge, 635–636
   storage mechanism, 632–638
   threshold voltage shifts, 632
   window sensing, 641–643, 647
Nucleation, 410, 714
NVM. See Nonvolatile memory (NVM)

O

One/zero discrimination, 3
ONO structure, and charge loss, 474–477, 478
On-off regulation, 120
Optical bandgap, 716
Optical disks, 708, 710, 712
Output buffers, 42, 87, 101, 122, 123, 551
Overprogram elimination scheme, 247–252
Overstressing, 12, 109, 543, 569
Ovonyx, Inc., 709
Oxide breakdown, 168–171
Oxide charge trapping, 164–165
Oxide defects, 136–137, 217, 466, 473, 478, 512, 573, 588, 625, 639
Oxide degradation:
   CHE (channel hot-electron) programming-induced, 449–455
   electron trap generation, 163–164
   hole fluence, 160–162
   interface trap creation, 162–163
   overview, 159–160
   oxide charge trapping, 164–165
   stress-induced leakage current, 165–166
   trap generation mechanism, 166–168
   tunnel-erase-induced, 456–462
Oxides:
   integrity, 216–217
   plasma-induced damage, 530–533
   reliability issues, 216–217

P

Package size, embedded Flash device, 49
Packaging, Flash device, 36–38
Packaging qualification, 561
Page buffers, 27, 68, 224, 226, 235, 236, 243, 248, 249, 250, 284–286
Page copy function, 273–274, 276, 277, 278
Page mode, 27, 41, 42, 44, 398, 399
Page-program operation:
   in bit-by-bit verify circuit, 243–244, 245
   in high-speed multilevel programming, 301–307
   in multilevel NAND circuit, 285–286
Page-read operation, 229, 283, 285
Parameter blocks, 66
Parity check, 549–550
PASHEI programming technique, 427–428
PC cards. See PCMCIA cards
P-channel technology:
   DINOR archiecture
      band-to-band hot-electron (BBHE) injection cell operation, 329–333
      overview, 328
   field effect transistors, 416
   Flash memories
      array architecture, 343–345
      device structure, 338
      embedded memory, 354, 363–366
      erase mode, 339
      evolution, 345–366
      history chart, 346
      illustrated, 338
      multibit storage, 363
      operations, 338–343
      overview, 337–338
      program disturb mode, 339
      program mode, 339
      read disturb mode, 339
      read mode, 339
      scalability, 362–363
   floating-gate, with deposited silicon nitride tunnel dielectric, 429–432
PCM. See Phase change memories (PCM)
PCMCIA cards, 37, 51, 52–53, 543
PCs (personal computers), 8–9
PDA (postdeposition annealing), 411
Peak stress, 12
Percolation model for breakdown, 169, 170
Perovskite unit cell, 660–661
Personal Computer Memory Card International Association (PCMCIA) cards, 37, 51, 52–53, 543
Personal computers (PCs), 8–9
PF (Poole-Frenkel) conduction, 136, 144, 192, 475–476, 636, 637
Phase change memories (PCM):
   array and support circuitry, 720–721
   basic cell structure, 712

cell operation concepts, 712–714
electrical characteristics, 714–720
fabrication, 721–722
GST phase change material, 709–712
overview, 707–709
products, 727
programming waveforms, 714–720
reliability, 725–727
scaling, 722–725
storage mechanism, 709
Phase diagrams, 709–710
Plasma CVD using tetra-ethoxy-silane (P-TEOS), 533–534
Plasma-induced oxide damage, 530–533
Plastic leaded chip carriers (PLCCs), 36, 37
Plastic small outline package (PSOP), 37
Plate line, 665, 666, 667, 669
Plated wire, 678, 687, 689
PLCCs (plastic leaded chip carriers), 36, 37
Poisson distributon, 77, 566
Polyoxide conduction, 144, 148–150
Polyoxide tunneling, 133, 150
Polyoxides, 144, 148, 149, 150
Poly-to-poly erase:
    charge trap-up, 514–517
    formation of injectors, 512–514
    reliability issues, 512–517
Poole-Frenkel (PF) conduction, 136, 144, 192, 475–476, 636, 637
Positive source erase, 184, 456
Postdeposition annealing (PDA), 411
Posterase repair, 189, 465, 466
Power-loss recovery, 35, 58
PPOT (pressure pot operating test), 556
Precise charge placement:
    analog circuit blocks for, 609–610
    charge placement algorithm, 608–609
    mixed signal design implementation, 608–611
    overview, 603–606
Precise charge retention, 607
Precise charge sensing:
    analog circuit blocks for, 610
    parallel scheme, 610–611
Preconditioning, 189, 545, 556, 566
Preprogram operation, 79, 82, 83, 84, 85
Preprogramming, 74, 189, 207, 545
Pressure pot operating test (PPOT), 556
Primary memory, 8, 60
Product qualification flow, 561–564
Production test flow, 559–561
Program algorithm, 74, 81, 98, 237, 543, 544, 546–547, 627
PROGRAM command, 74
Program disturb:
    in basic NAND cells, 229–230
    effect of cycling, 494–495
    latest phenomenon in NAND Flash memory, 575–576

overview, 491
as reliability issue, 491–495
in source line programming scheme, 280–281, 283
Program margin, 453, 642
Program operation:
    as basic Flash cell operation, 63, 64, 90
    basic NAND cells, 228–229
    embedded Flash, 544–547
    and ETOX technology, 183, 200, 204–205
    high-speed multilevel programming, 301–307
    and intrablock disturbs, 186–187
    NROM (nitrided ROM) memories, 626, 627, 628, 629
    overview, 131
    page program operation, 236–237
    and P-channel Flash, 338–343
    physical mechanisms, 64
    self-boosted NAND inhibit scheme, 235–237
    soft, 189–190
    source line programming scheme, 278–283
    SuperFlash, 210–212
    symmetric sense amplifier with page copy function, 274–275
    three-level NAND architecture, 297, 299
Program path, 81, 204–205
Program suspend, 30
Program verify operation, 74, 93, 105, 237–247, 297, 299, 448
PROGRAM/ERASE RESUME command, 74
PROGRAM/ERASE SUSPEND command, 74
Programmable diode technologies, 696, 698–699
Programming voltage:
    ferroelectric memories, 667–669
    Flash memory limitations, 619–623
    magnetic memories, 685–686
Pseudo-binary tie line, 709, 710
Pseudo-spin valve (PSV), 680–683, 687
PSOP (plastic small outline package), 37
PSV (pseudo-spin valve), 680–683, 687
P-TEOS (plasma CVD using tetra-ethoxy-silane), 533–534
Pulse width generator, 96–97
PZT (lead zirconate titanate), 659–660, 668, 670, 671, 673, 674

Q

Quad NROM, 645–650, 655, 657
Quantization of gate charge, 691, 693
Quantum unit of resistance, 690

R

RAM (random-access memory), overview, 7, 21, 23. *See also* DRAM (dynamic random-access memory); MRAM (magnetic random-access memory); SRAM (static random-access memory)
Ramp of the erase pulse, 135

Random access memory digital versatile disks (DVD-RAM), 708
Random background charge, 693, 695, 696
Random telegraph signals (RTS), 576–579
Rapid thermal CVD (RTCVD), 409
READ ARRAY command, 74
Read disturb, 339, 487–490
Read operation:
  64-Mbit NAND EEPROM, 237
  as basic Flash cell operation, 3, 4, 5, 6
  basic NAND cells, 230–231
  and ETOX technology, 182–183, 200, 202–204
  full chip burst, 272–273
  NAND cell vs. SWATT cell, 294–295
  NROM (nitrided ROM) memories, 626
  symmetric sense amplifier with page copy function, 273–278
  three-level NAND architecture, 297, 298
READ STATUS REGISTER command, 74
Readout, destructive vs. nondestructive, 4, 695–696
Readout operation, 4, 5, 132, 133
Ready/busy signal, 7
Recall operation, 3, 4, 5, 243–244
Reclamation, 35
Recovery effects, 134
Redundancy:
  advanced design, 81–86
  in columns, 75, 80–81
  defectivity, 75
  Flash memory overview, 75
  fuse design, 78–79
  hardware, 549
  process variations, 75
  as reliability issue, 547–548, 549
  in rows, 75, 79–80, 82–86, 93, 96
  yield improvement, 75–76
  yield simulator, 77–78
Reference cells, 112, 113, 203, 602–603, 607, 610–611, 629, 647
Reference current generation, 112, 113
Reference voltage generator, 72, 118, 119
Reliability:
  acceleration models
    temperature acceleration model, 557
    temperature and humidity acceleration model, 559
    temperature cycling acceleration model, 559
    voltage acceleration model, 557–558
  ferroelectric memories, 674–675
  Flash memories
    burn-in, 539–543, 564–565
    design and system impacts, 543–552
    failure rate calculations, 565–570
    floating-gate devices, 526–528
    high-voltage periphery transistors, 536–543
    process impacts, 525–536
    product qualification flow, 561–564
  qualification methods, 573–574
  screening and qualification, 552–570
  hot-hole injection concerns, 432
  magnetic memories, 688
  multilevel cell (MLC) issues, 596–597, 613
  NROM (nitrided ROM) memories, 638–645
  oxide issues, 216–217
  phase change memories (PCM), 725–727
  SuperFlash
    contact integrity, 217
    data retention, 217–218
    disturbs, 219–220
    dynamic burn-in, 220
    endurance, 218–219
    life testing, 220
    overview, 216
    oxide integrity, 216–217
  testing
    acceleration models, 557–559
    data retention storage life (DRSL) test, 554–555
    electrostatic discharge, 555
    endurance cycling, 555
    failure rate calculations, 565–570
    highly accelerated stress test (HAST), 556
    high-temperature operating life (HTOL) test, 554
    latch-up, 555
    lead integrity, 556
    low-temperature operation life (LTOL) test, 554
    overview, 552–554
    pressure pot operating test (PPOT), 556
    production test flow, 559–561
    solderability testing, 556
    temperature cycle test, 556
    temperature humidity bias (THB) test, 555–556
    thermal shock, 556
    wafer sort and screen tests, 559–561
Removable Flash storage, 47, 50–53
Removable media, 47, 50–53
Removal of electrons, 11
RESET state, 709, 715
Resistive memory:
  expected performance, 701–702
  hybrid CMOS/nanodevice resistive memories, 700–701, 702, 703
  overview, 696–698
  programmable diode technologies, 698–700
Retention. See Charge loss; Data retention
Retention time, defined, 2, 12–13. See also Data retention
Reverse tunneling, 211–212
Rewritable compact disks, 708
Rewritable digital versatile disks (DVD-RW), 708

ROM (read-only memory):
    defined, 2
    vs. EPROM, 23
    strengths and weaknesses, 21
Row decoder:
    defined, 67
    level shifter, 201
    NAND arrays, 231–233
    NAND technology, 225–226
    redundancy design, 82–86
    staggered, 231–233
Row defects, 75
Row failures, 75
Row redundancy, 75, 79–80, 82–86, 93, 96
RTCVD (rapid thermal CVD), 409
RTS (random telegraph signals), 576–579

S

SAC (self-aligned contact), 197, 199
Saifun, 15, 625, 650
Samsung, 22, 225
SanDisk, 87, 496, 512
SAP (self-aligned poly), 191, 197, 199–200
SAS (self-aligned source):
    scaling technique, 197–198
    stacked gate reliability issues, 528–530
SA-STI (self-aligned shallow trench isolation), 252
Saturated drift velocity, 147
SBT (strontium bismuth tantalate): 660, 668, 670, 671, 673, 675, 662
Scaling:
    architectural aspect, 197–200
    dielectrics, 191–197
    feature size aspect, 190–191
    ferroelectric memories, 673–674
    film thickness aspect, 191–197
    Flash memory limitations, 623–624
    impact of few-electron phenomenon, 576–579
    impact of random telegraph signals, 576–579
    magnetic memories, 687–688
    multilevel cell (MLC) technology, 614
    NAND Flash technology, 257, 258, 260, 262–263, 264, 265
    NOR Flash technology, 190–200
    NROM (nitrided ROM) memories, 652–655
    P-channel technology, 362–363
    phase change memories (PCM), 722–725
    self-aligned contact (SAC) technique, 199
    self-aligned poly (SAP) technique, 199–200
    self-aligned source (SAS) technique, 197–198
    shallow trench isolation (STI) technique, 198–199
    SuperFlash, 214–216
    and tunnel dielectrics, 407–439
    tunnel oxide, 192, 193–195, 407, 511

Schottky emissions, 144, 471
SCSG (source-coupled split-gate) cell, 386–389, 401
SD (Secure Digital) memory cards, 51, 52–53
Secondary impact ionization initiated channel hot-electron injection, 156–157
Secondary impact ionization (SII), 141, 144, 156–157
Secondary memory, 8
Secondary store, 8
Secure Digital (SD) memory cards, 51, 52–53
SEDC (single-bit error detection and correction). *See* Single-bit errors
Self-aligned contact (SAC), 197, 199
Self-aligned poly (SAP), 191, 197, 199–200
Self-aligned shallow trench isolation (SA-STI), 252
Self-aligned source (SAS):
    scaling technique, 197–198
    stacked gate reliability issues, 528–530
Self-assembled monolayers, 698
Self-boosted-erase inhibit scheme, 233–235
Self-convergence-erasing scheme, 544
Self-convergent programming scheme, 353, 355, 356, 363
Self-limiting program scheme, 350
Semiconductor devices. *See also* Memory; Nonvolatile memory
    generations, 16–18
    history, 16–18
    nonvolatile drives, 8
Sense-and-latch (SL) unit, 284–285
Sensing schemes, 112–114, 614, 629
Sequential trap-assisted tunneling, 499
SET state, 709, 714, 715
Shadow array, 380
Shallow trench isolation (STI), 182, 197, 198–199, 252–256, 293, 385
Sharing of interconnection overhead, 15
SHEI (substrate hot-electron injection), 144, 153–154, 163–164
SHHI (substrate hot-hole injection), 154, 163
Shielded bitlines, 265, 270–272, 273, 286, 288
Shrink small outline package (SSOP), 37
Side-wall transfer transistor (SWATT), 293–296
SII (secondary impact ionization), 141, 144, 156–157
SILC. *See* Stress-induced leakage current (SILC)
Silicon dioxide, as tunnel dielectric, 408–409
Silicon nitride:
    advantage as tunnel dielectric, 422–423
    chemical vapor deposition, 409, 410
    as deposited tunnel dielectric, 417–425
    jet-vapor deposition, 409–411
    properties of gate-quality JVD films, 411–417
    as tunnel dielectric for N-channel floating-gate devices, 425–429

    as tunnel dielectric for P-channel floating-gate devices, 429–432
    as tunnel dielectric for scaled Flash memory cells, 407–439
Silicon oxide nitride oxide silicon (SONOS), 3, 12, 16, 195, 363, 366, 432–434, 617–618
Silicon Storage Technology, Inc., 206, 386, 496, 512. *See also* SuperFlash
Silicon-rich oxide (SRO), 347, 348, 408
Single-bit errors, 7, 550–551
Single-bit failures, 75, 87, 480, 625, 642, 645, 647, 675
Single-electron charging energy, 690
Single-electron memories, 618, 689–693
Single-electron transistor, 689, 691, 694, 695–696, 698
Single-electron trap, 691, 692, 700
Single-poly floating-gate cells, 136
SiO_2-based tunnel oxides, 408–409
SL (sense-and-latch) unit, 284–285
Small outline packages (SOPs), 36–37
SmartMedia (SSFDC), 51–52, 53
SMT (surface mount technology), 36
Snapback, 538, 543
SnD (store and download) usage model, 38, 40, 43, 44–45
SNNNS structure, 433–434
Socket Flash, 543, 551, 562
Sodium, 535
Soft errors, 484–486
Soft programming, 65, 188, 189–190
Software, as Flash attribute, 20, 25–26, 29, 31, 32–36
Software data managers, 24, 29, 50, 54, 56–57
Software stack, 58, 59–60
Soft-write effects, 138, 139
Solderability testing, 556
Solid-state disks (SSD), 87, 596–597, 598
SONOS (silicon oxide nitride oxide silicon) transistors, 3, 12, 16, 195, 363, 366, 432–434, 617–618
SOPs (small outline packages), 36–37
Source lines:
    array noise suppression suppression technology, 286–293
    programming scheme, 278–283
Source-coupled split-gate (SCSG), 386–389, 401
Source/drain hot-carrier injection disturbance, 576
Source-erase stacked gate Flash, 14
Source-side injection (SSI):
    charge trapping, 519–525
    gap region formation, 517–519
    overview, 144, 155–156
    reliability issues, 512, 517–525
Spansion, Inc., 22, 658, 699
Spatial charge distribution, 632

Spatially localized C_20 lone-pair valence band tail states, 716
Special test mode decoder, 91, 92
Spin polarization, 679, 685
Spin-valve-type devices, 679
Split-gate cells:
    applications, 220
    characteristics, 133, 138
    embedded Flash technology, 374, 385, 386–389, 401
    overview, 15
    reliability issues, 219, 517–519
    source-coupled (SCSG), 386–389, 401
    and source-side injection (SSI), 517–519
    SuperFlash, 179, 206, 209, 212–213, 214, 215, 216, 513
SRAM (static random-access memory), 20, 21, 38, 41, 42, 43, 44, 66, 87, 327–328, 485, 486, 539, 592, 594, 619, 623, 707
SRO (silicon-rich oxide), 347, 348, 408
SSD (solid-state disks), 87, 596–597, 598
SSFDC (SmartMedia), 53
SSI. *See* Source-side injection (SSI)
SSOP (shrink small outline package), 37
SST memory cell. *See* SuperFlash
SST (Silicon Storage Technology, Inc.), 206, 386, 496, 512. *See also* SuperFlash
ST Micro, 22, 722
Stacked gate structure, 13, 14, 206, 255, 268–269, 317, 528–530
Staircase programming pulses, 289
State machines, 8, 75, 91, 97–98, 384, 446, 543, 545, 547. *See also* Write state machines
Static RAM. *See* SRAM (static random-access memory)
Status output signal, 7
STI (shallow trench isolation), 182, 197, 198–199, 252–256, 293, 385
STMicroelectronics, 22, 722
Storage, 2, 9–12, 38–62
Store and download (SnD) usage model, 38, 40, 43, 44–45
StrataFlash technology, 27, 41, 591, 594–596, 598, 599, 608, 612, 613
Stress-induced leakage current (SILC):
    and FN tunneling, 447
    as limiting factor for tunnel oxide scaling, 511
    microscoopic characteristics, 508–510
    in oxynitride, 510–511
    related retention effects, 571–572
    in thin oxide after bipolarity stress, 502–508
    trap-assisted, 496
    tunnel oxide overview, 496–497
    uniform, in thin oxide, 407, 497–502
Strontium bismuth tantalate (SBT), 660, 662, 668, 670, 671, 673, 675
Sub-arrays, 2, 66, 399–400
Sub-bitlines, 69, 70, 316, 321, 339, 352

INDEX

Substrate hot-electron injection (SHEI), 144, 153–154, 163–164
Substrate hot-hole injection (SHHI), 154, 163
Subthreshold slope, 133–134, 162, 523
SuperFlash:
  applications, 220
  array architecture and operation, 212–214
  cell cross sections and layout, 207–208
  charge transfer mechanisms, 208–209
  embedded cell manufacturability, 401
  erase operation, 209–210
  erase threshold control and distribution, 214–216
  history, 179–180
  key circuit interactions, 215
  multilevel cell implementation, 216
  overview, 206–207
  process scaling issues, 214–216
  program operation
    overview, 210–211
    program-disturb, 211
    punch-through disturb, 212
    reverse tunnel disturb, 211–212
  reliability
    contact integrity, 217
    data retention, 217–218
    disturbs, 219–220
    dynamic burn-in, 220
    endurance, 218–219
    life testing, 220
    overview, 216
    oxide integrity, 216–217
Surface microdefects, 482, 484, 485, 526
Surface mount technology (SMT), 36
Suspends, 26, 30–31, 56, 57
SWATT (side-wall transfer transistor), 293–296
Switch-mode regulation, 119, 120
Symmetric sense amplifier, 273–278
Symmetrical blocking, 25, 29, 49
Synchronous burst mode, 43
Synchronous Flash, 41, 42–43
Syndrome generator, 87
System Flash, 29, 57, 543, 551

T
Tail bits, 159, 482–484, 485, 571, 577
TDDB (time-dependent dielectric breakdown), 216, 413, 476, 502, 547, 557, 564, 565
Technology node, defined, 16
Temperature acceleration model, 557
Temperature and humidity acceleration model, 559
Temperature cycle test, 556
Temperature cycling acceleration model, 559
Temperature humidity bias (THB) test, 555–556
Temperature-accelerated testing, 135–137, 138, 467–469, 506, 557, 559
Tessera Corporation, 38

Test entry, 91–92
Textured oxides, 408
Textured poly floating gate (TPFG), 154
Textured polyoxides, 148, 150, 515
Textured polysilicon gate, 512
THB (temperature humidity bias) test, 555–556
Thermal cross talk, 724
Thermal energy, 9, 710
Thermal shock test, 556
Thermochemical E model, 557
Thick organic films, 698
Thin films, 679
Thin small outline package (TSOP), 37
Threshold voltage:
  array noise suppression technology, 287–290
  NAND cell vs. SWATT cell, 295
  NROM (nitrided ROM) memories, 632
  overview, 132–133
  and readout operation, 132, 133
  transient characteristics, 133
Time-dependent dielectric breakdown (TDDB), 216, 413, 476, 502, 547, 557, 564, 565
TMR (tunneling magnetoresistance), 679, 685, 687–688
Top boot, 99
Toshiba, 14, 21, 87, 225
TPFG (textured poly floating gate), 154
Transconductance degradation, 421, 432, 447, 448, 459, 530, 531
Transient equation, 143
Transistors:
  field-effect (FET), 416, 625, 660, 664–665
  Flash cells as, 63–65, 594
  floating-gate concept, 3–4
  high-voltage
    MMOS and PMOS, 254, 265, 282
    overview, 536–537
    process defects, 539–543
    reliability in Flash memory products, 537–539
    role of cycling and burn-in, 539–543
  side-wall transfer (SWATT), 293–296
  single-electron, 689, 691, 694, 695–696, 698
Trap density, 162, 163, 165, 168, 169, 192, 413, 454, 500, 526
Trap generation mechanism, 166–168
Trap-assisted SILC, 496
Trapping of hot holes, 164
Triangular barrier, 145, 194, 471, 703–704
Trimming:
  in analog circuits, 366
  and data path configuration, 99–101
  DFT overview, 91, 92
  high voltages, 94–96
  reference, 93–94
  timings, 96–97
Triple-well process, 69, 70, 110, 367, 401, 457, 538, 539

TSOP (thin small outline package), 37
Tunnel barrier engineering:
  crested barrier
    and FN tunneling, 194
      NOVORAM/FGRAM memories, 703–706
      silicon nitride as dielectric, 437–439
  few-electron memories, 694
  magnetic memories, 684, 686
  multiple barriers, 437–439
  NOR memories, 192, 193–195
  NOVORAM/FGRAM memories, 703–706
  silicon nitride overview, 437
  single-electron memories, 691, 693
  tunnel oxide scaling, 192, 193–195
  U-shaped barrier, 439
Tunnel dielectrics:
  defined, 408
  deposited silicon nitride
    early work, 409–410
    for N-channel floating-gate devices, 425–429
    overview, 417–425
    for P-channel floating-gate devices, 429–432
    potential advantage, 422–423
  graded bandgap, 409
  high-K dielectric benefits, 434–436
  overview, 407–408
  for scaled Flash memory cells, 407–439
  SiO_2 as, 408–409
  for SONOS cells, 432–434
  thickness limitations, 619–620, 621
Tunnel oxide process. *See also* Tunnel barrier engineering
  dielectrics for scaled Flash memory cells, 407–439
  JVD nitride overview, 192–193
  overview, 192
  reliability issues, 526
  scaling limitations, 192, 407, 511
  SILC as limiting factor, 511
Tunnel-erase-induced oxide degradation:
  band-to-band current-induced degradation and charge trapping, 460–462
  band-to-band tunneling current, 459–460
  erase methods, 456–459
  erratic erase, 462–465
  overview, 456
Tunneling magnetoresistance (TMR), 679, 685, 687–688
2TS cells, 386, 389–391, 401, 403

U

Usage models, 23–24, 38–46, 49, 50, 54–62
U-shaped barrier, tunnel barrier engineering, 439
UV-EPROMs, 22, 69, 179–182, 466, 473

V

Valence alternation pair (VAP) states, 716
VAP (valence alternation pair) states, 716

Verify read operation, 229, 238, 240–241, 243, 244, 267, 274–275, 276, 303, 305
Vertical parity, 87, 550
Virtual blocking, 34
Virtual drain effect, 155–156
Virtual ground array (VGA), 16, 316–320, 625, 627, 629, 647, 656
Voice recording, 20, 134, 358
Volatile memory, 2, 3, 623
Voltage acceleration model, 557–558
Voltage detectors, 92, 324–326
Voltage level shifting, 109–111
Voltage multiplication, 114–118
Voltage regulation, 119–122
Voltage supply, as Flash memory attribute, 29
Voltage-accelerated test, 469–472
Von Neumann model, 8

W

Wafer sorts, 203, 548, 559–560
Weak cell detection, 104
Wear leveling, 35, 50, 58, 543, 547, 552
Wearout mechanisms, 35, 497, 541, 543, 553, 687
Weibull distribution, 170, 218
Wentzel-Kramers-Brillouin (WKB) approximation, 464
Window closure, 13, 134, 426, 447–448, 453, 454, 514, 521–522
Wiring resistance degradation, 547
Word compression, 107–108
Wordline pitch, 16, 225, 231, 257, 269
Wordline sweeping read (WSR) scheme, 285
Wordlines:
  boost scheme for DINOR circuits, 326–327
  defined, 5
  NAND spacing issue, 268–269
  in NROM arrays, 15–16
  and row decoder circuits, 201, 231
  in row redundancy, 79–80
Word-oriented memory, 7, 8
Write protection, 26, 31, 32, 40, 57
Write state machines (WSM), 25, 27, 30–31, 39, 56, 58, 73
Write suspends, 30, 56, 57
Write-read cycles, 13
WSM (write state machines), 25, 27, 30–31, 39, 56, 58, 73
WSR (wordline sweeping read) scheme, 285

X

XIP (execute-in-place) usage model, 38, 39–43, 45, 51

Y

Yield improvement, 75–76
Yield simulator, 77–78

ABOUT THE EDITORS

Joe E. Brewer holds a Master of Science degree in Electrical Engineering and has been engaged in nonvolatile memory development for the past four decades. Most of his career was spent with Westinghouse Electric (now Northrop Grumman Corporation) until retirement in May 1998. In 2000, he joined the faculty of the Electrical and Computer Engineering Department of the University of Florida. He was an early contributor to the IEEE Standards for MNOS Arrays and Floating Gate Arrays and led a series of Nonvolatile Memory Technology Workshops hosted at Westinghouse through the 1980s and 1990s. He and William Brown edited the 1998 IEEE Press book *Nonvolatile Semiconductor Memory Technology* and is co-editor with Stuart K. Tewksbury of the IEEE Press Book Series on Microelectronic Systems. He is the Electron Device Society liaison and the CPMT Society liaison to the IEEE Press. From 1996 to 2006, he was an editor for the *IEEE Circuits & Devices Magazine* and he has served as guest editor for several special topics in the *IEEE Transactions on Components, Packaging and Manufacturing Technology*. He is a professional engineer registered in Maryland and Florida and a Fellow of the IEEE. Mr. Brewer holds 7 U.S. patents and has authored more than 100 conference and journal papers.

Manzur Gill holds a Ph.D. in Electrical Engineering and has had more than 25 years of experience in high-tech industry and nonvolatile memory development. He was a senior member of technical staff at Texas Instruments, a nonvolatile memory design manager at National Semiconductor, and senior program manager in Flash memory development in NVM Exploratory Technologies and Ovonics Technology Development at Intel Corporation. He has managed international technology development teams and mentored industry-sponsored student development at universities. Dr. Gill has authored more than 30 technical publications for international journals and workshops and holds more than 75 issued patents. In 2005, he left the high-tech world to return to the land of his birth, Pakistan, to join the faculty of Forman Christian College in Lahore—now a chartered university—to help bring about a paradigm shift in college-level education in Pakistan. At Forman, he is chief advancement officer and a professor of Physics. He can be reached by email at manzurgill1@yahoo.com

---

Nonvolatile Memory Technologies with Emphasis on Flash. Edited by J. E. Brewer and M. Gill
Copyright © 2008 the Institute of Electrical and Electronics Engineers, Inc.